THE INTEGRATION OF GEOLOGY, GEOPHYSICS, PETROPHYSICS AND PETROLEUM ENGINEERING IN RESERVOIR DELINEATION, DESCRIPTION AND MANAGEMENT

Proceedings of the First Archie Conference,
held October 22-25, 1990, in Houston, Texas, U.S.A.,
Sponsored jointly by the:
American Association of Petroleum Geologists
Society of Exploration Geophysicists
Society of Petroleum Engineers
Society of Professional Well Log Analysts

Conveners: Robert Sneider, Wulf Massell, Rob Mathis, Dennis Loren, and Paul Wichmann

Published by
The American Association of Petroleum Geologists
Tulsa, Oklahoma, U.S.A.

Published 1991

ISBN: 0-89181-814-6

Association Editor: Susan Longacre
Science Director: Gary D. Howell
Publications Manager: Cathleen P. Williams
Special Projects Editor: Anne H. Thomas
Editorial Assistant: Marilyn B. Michaels

Cover design by Rusty Johnson.

The Archie Conference is sponsored by the American Association of Petroleum Geologists, the Society of Exploration Geophysicists, the Society of Petroleum Engineers, and the Society of Professional Well Log Analysts through the auspices of the Intersociety Coordinating Committee (ICC). This collective-society effort has two primary goals. First, it seeks to provide an integrated, multidisciplinary approach to examining and solving the scientific problems related to petroleum exploration and production. Second, it seeks to promote synergism within the sciences in general and the petroleum industry in particular.

American Association of Petroleum Geologists

Society of Exploration Geophysicists

Society of Petroleum Engineers

Society of Professional Well Log Analysts

Dedication to Gus E. Archie

Gus Archie

This conference is dedicated to the late Gus Archie—an innovator, researcher, manager and leader in the petroleum industry for over 45 years. His pioneering work on well logging and his research and management in the 1930s–1960s at Shell Development Company epitomizes the essence of the integration of geoscience and petroleum engineering.

Archie is best known by the industry for his studies of the resistivity of rocks, which have been the basis for quantitative log interpretation over the past 45 years. In the early 1940s he developed and published work that is the foundation for what is now called the "Archie Equation." [1,2] Also, he pioneered the development of the determination of porosity and hydrocarbon saturation from well logs. He added the geological (rock) contribution to well log interpretation by dovetailing rock type data and petrophysical lab data with well logs. His classification of complex carbonates and pore types is used by many companies and individuals. [3,4]

Those who knew Archie well would tell you "His modesty was nearly as legendary as his ingenuity. He shared his ideas and views openly and freely and let others receive credit for ideas that were his. For example, he initiated and guided Shell's pioneering work on hydrocarbon evaluation of low resistivity, shaly sands. He encouraged a systematic integrated study of the geological, petrophysical and engineering properties of reservoir rocks—The Rock Catalogue. He was very instrumental within Shell in fostering development/production geology and encouraging close cooperation among geoscientists and engineers in research and operations. He supported and encouraged the development and use of seismic for field-scale appraisal and integrated reservoir description/characterization for water-, steam- and chemical-flooding.

Archie would totally support and applaud the idea and theme of the conference. Having the conference named after him would have embarrassed this humble and modest man.

REFERENCES

1. Archie, G.E.: "Electrical Resistivity Log as an Aid in Determining Some Reservoir Characteristics," Trans. AIME (1942) 146, 54-67.
2. Archie, G.E.: "Electrical Resistivity as an Aid in Core Analysis Interpretation," Bull., AAPG (1947) 31, 350-66.
3. Archie, G.E.: "Introduction to Petrophysics of Reservoir Rock," Bull., AAPG (1950) 1-34.
4. Archie, G.E.: "Classification of Carbonate Reservoir Rocks and Petrophysical Considerations," Bull., AAPG (1952) 36, 278-98.

Contents

MULTIDISCIPLINARY STUDIES

RESERVOIR CHARACTERIZATION: PETROPHYSICAL FORMATION EVALUATION AND ROCK DESCRIPTION

RESERVOIR CHARACTERIZATION AND MODELING

RESERVOIR PROPERTIES PREDICTION

FRACTURED RESERVOIRS: ORIGIN, DESCRIPTION, EVALUATION

CREATING, MANAGING, AND EVALUATING MULTIDISCIPLINARY TEAMS

INTEGRATED FIELD STUDIES

Rejuvenation of a Mature Giant, North Ward Estes Field, Texas

VALUE OF 2-D AND 3-D SEISMIC IN RESERVOIR DELINEATION AND DEVELOPMENT

UTILIZATION OF GEOSCIENCE AND ENGINEERING TECHNOLOGIES TO INCREASE HYDROCARBON RECOVERY

INTRODUCTION TO THE CONFERENCE

On behalf of the Intersociety Coordinating Committee of the AAPG, SEG, SPE and SPWLA, I welcome you to the first Archie Conference. The Intersociety Coordinating Committee (ICC) is composed of the presidents of the four sponsoring societies. The 1989 ICC authorized this conference in the hope that it would demonstrate the effectiveness of multidisciplinary approaches to solving the world's energy problems. The 1989 AAPG Research Committee agreed to sponsor the Conference.

The ICC asked me to accept responsibility for organizing the conference and program. I picked Wulf Massell, Rob Mathis, Dennis Loren and Paul Wichmann as co-conveners and to represent the interests of the membership of the four sponsoring societies. With the help of many colleagues, the five of us selected a list of potential subjects, conference participants and authors. We five selected the final papers.

To make a conference smooth-running and successful, there is a lot of work behind the scenes. I especially want to acknowledge the help and advice of Mr. Ray Buchanan of Royal/Dutch Shell. The mammoth amount of requisite correspondence has been handled efficiently by Miss Debbi Boonstra of the AAPG Education Department, and Ramona Sneider. I thank them sincerely for their superb effort.

The conveners thank all of you for your participation and contributions. We believe strongly that the Archie Conference will promote multidisciplinary work and a communication amongst the disciplines. The subjects that we will discuss and share highlight the challenges, opportunities, and solutions in developing our hydrocarbon resources wisely and in optimizing hydrocarbon recovery.

Robert M. Sneider

MULTIDISCIPLINARY RESERVOIR STUDY OF THE BIMA FIELD, OFFSHORE N.W. JAVA, INDONESIA

Gary S. Woodling*
John G. Kaldi**
Randall C. Roe***
Kristina I. Oentarsih***

* ARCO Oil and Gas Company, Bakersfield, California
** ARCO Oil and Gas Company, Plano, Texas
*** Atlantic Richfield Indonesia, Inc., Jakarta

ABSTRACT

Bima Field, offshore northwest Java, is a sizeable reservoir containing reserves of approximately 700 MM bbls OOIP with a 50 BCF gas cap. At present only the northern 1/3 of the field is developed, with 7 platforms and 54 producing wells, of which 20 are horizontal. The field has multiple drive mechanisms and high viscosity oil (21 cp), resulting in rapid GOR and water-cut increase after 3 years of production. The high stakes (both reserves and facility investments) and the reservoir's complexities, make an effective reservoir management scheme critical. For this reason an integrated geological, geophysical and engineering description was carried out to provide a 3-D Reservoir Simulation Model to evaluate development options. Geologically, the Oligo-Miocene age Batu Raja Limestone was deposited on the Seribu Platform, a basement-controlled, fault-bounded structure. The Upper Batu Raja carbonate build-up is thickest on the structurally highest parts of the platform where the rock comprises a series of "cleaning upwards" cycles (muddy deposits overlain by progressively more grain-rich sediments). A Lower Miocene drop in sea-level caused subaerial exposure of much of the platform and leaching by meteoric fluids. This diagenetic event resulted in contrasts in the reservoir quality (porosity, permeability, fluid saturations) at various intervals of the Upper Batu Raja. Based on these dissimilarities, the reservoir was zoned into 6 model layers. Once zonation was established, well logs could be calibrated to whole and sidewall core. A dense grid of seismic data were used to map the Batu Raja structure. From these data, color seismic inversion sections were produced and calibrated to the well logs. The calibrated seismic data were then used to map the top of structure, the carbonate build-up's edges, the total thickness of the Upper Batu Raja (needed to control aquifer size in the model) and the thickness of the main pay zone (layers 1-3). Engineering reservoir description began with a detailed compilation of capillary pressure, relative permeability, production and DST data. The 3-D simulation model required special treatments, including varying the GOC depths to honor separate gas cap closures; making permeability pressure dependent in poorly-consolidated zones; and setting up horizontal well completion treatments. Results suggest that water injection into the oil rim and gas cap is an effective approach toward maximizing recoveries and minimizing gas cap resaturation. However, waterflood reserves are sensitive to injection timing. The synergistic approach of geological, engineering and geophysical input into the Bima reservoir study has had impact by delivering a reservoir management tool that can evaluate future development expansion and possible gas sales. The simulation model can also track fluid migration during the field's producing life. The geological/geophysical model led to an enhanced understanding of Batu Raja depositional and diagenetic processes that has potential in regional exploration strategies.

STUDY OBJECTIVES

The reservoir study was undertaken to maximize the economic worth of the Bima field and it's investments through the following:

1. Develop a geologic model for Bima Batu Raja deposition and diagenesis with regional exploration applications.
2. Construct a quantitative reservoir description and 3D fullfield simulation model.
3. Evaluate current and future development options using the fullfield model to improve ultimate recovery and reservoir management. These options include waterflood and gas reinjection as well as development expansion and possible gas sales.

THE INTEGRATION CONCEPT

Numerous disciplines (e.g. geology, geophysics, log analysis, reservoir engineering and production operations) are typically involved in any reservoir study. However, organizational structure and different scientific backgrounds usually hinder communication and interaction between groups. Normally, the description may be accurate enough to answer the engineering questions at hand. However, in fields with complex geology and/or sophisticated recovery plans the interaction must be enhanced to obtain the most accurate description possible with minimum revision.

Characterization of the Bima field used a "team" concept where all contributors were equally responsible for the final product. The process was coordinated by reservoir engineering to insure adaptability of the description to reservoir modeling.

FIELD PERFORMANCE

The Bima field is located 50 miles north of Jakarta (Fig. 1). The structure is a faulted anticline with an average sub-sea depth of -2550 ft ss. Formation dip is only 4-5° resulting in an areally extensive gas cap (Fig. 2). Initial reservoir pressure was 1135 psi at the gas-oil contact (GOC) of -2512 ft ss. The water-oil contact (WOC) is -2645 ft ss providing a 133 ft oil column. Reservoir temperature is 153°F.

The most prolific production comes from the semi-consolidated uppermost layer ("A" Zone) shown on the attached type log (Fig. 3). This zone has recorded drill stem test (DST) rates in excess of 2500 BOPD and contains 2/3rds of the OOIP. The lower zone ("C" Zone) contains the remaining oil and provides only 1/4th of the total field rate. Both zones are areally continuous over most of the field.

Production at startup in 1987 (Phase-1) was 25,000 BOPD which declined rapidly to 10,000 BOPD by early 1990 (Fig. 4). Water cut has increased to 50% with a corresponding rise in gas/oil ratio (GOR) to 2000 scf/STB over 3 years. The oil has a low solution gas/oil ratio (130 scf/STB) requiring artificial lift by electrical submersible pumps (ESP). Increasing water and gas production is a result of the adverse mobility ratio of the viscous (21 cp) crude.

The historical gas production rate has been 15-20 million scf/D, all of which is flared. Gas production has been responsible for 2/3rds of the reservoir voidage raising a concern for long-term gas cap resaturation and pressure depletion.

GEOLOGIC MODEL

Deposition

The upper Batu Raja formation at Bima is a carbonate buildup deposited upon a local fault-controlled basement high on the Seribu platform located between the Sunda and Ardjuna sub-basins. Strontium isotope dating (Miller and Kaldi, 1990) brackets the age of Batu Raja limestone deposition to 21-26 million years ago (Late Oligocene to Early Miocene).

Tectonic uplift of the Seribu basement block to shallow water depths (<200 ft) during the Late Oligocene initiated carbonate development. Carbonate production was less pronounced in the subsided blocks (e.g. Sunda and Ardjuna Basins) where excessive water depth and poor water clarity curtailed carbonate growth, resulting in the deposition of mainly clastics (Fig. 5A).

A general rise in sea level during Batu Raja time (Haq et al., 1987) caused carbonate production rates to accelerate to match the relative deepening of water. This resulted in relatively thick accumulations of Batu Raja carbonates on blocks that were subsiding. Considerable volumes of clastic input from adjacent horsts were added to these carbonates. These thick "dirty" carbonate packages can be found in areas such as the Ardjuna Basin south flank and the E-15 Graben (Ponto et al., 1987).

A mix of thick and thin carbonate deposits as well as Oligocene Talang Akar clastics were being deposited at the same time in different sub-environments (Atkinson and Kaldi, 1990) These sediments graded into each other laterally and vertically. The influx of clastic fines increased from erosion of locally exposed basement rocks uplifted by reactivated faults. In addition, any carbonates present on the exposed highs were subjected to erosion and diagenesis by meteoric (fresh) waters.

Influx of detrital material into the grabens eventually overwhelmed the carbonate secreting organisms, thereby halting carbonate production. However, local basement highs remained as active sites of carbonate sedimentation until a eustatic rise in sea level eventually caused the cessation of most carbonate production leading to the subsequent deposition of deep-water shales.

On the Seribu Platform, variations in Batu Raja thickness conform to variations in the paleo-structural depth of the basement. Wells drilled atop local basement highs (e.g. ZUC-5 and ZUF-5) have a thicker carbonate development (>150 ft) than wells situated above a relatively lower basement (e.g. ZUB-5, ZUE-5 and ZUG-5 with thicknesses <100 ft). In areas of low basement the thickest Talang Akar sands are present below Batu Raja thins and the thickest ZU-24 shales overlie Batu Raja thins.

On logs there is a clear break between the upper and lower Batu Raja formations. Geologically, the break marks the cessation of mixed carbonate and clay-rich sediments of the lower Batu Raja and the clean limestones of the upper Batu Raja. The clay-free upper Batu Raja began to develop initially in the southern part of the field and proceeded through time towards the north, making the boundary a lithologic marker rather than a time marker (Fig. 6).

Batu Raja deposition occurred on a broad carbonate platform interior, in relatively quiet waters. The sediments are comprised of 3 main lithofacies:

- Argillaceous Mudstone-Wackestone
- Skeletal Wackestone-Packstone
- Coral-Rich Grainstone-Packstone

These lithofacies were deposited in "cleaning upwards" vertically repetitive cycles (each cycle having progressively more carbonate and less clay content). Primary (depositional) porosity was fairly high, ranging from 25 to 35%.

No indication of framework reef development has been recognized in any core or logs examined. However, localized small patch reefs can be inferred from the presence of coral grainstones. By analogy to recent sedimentation in the Thousand Islands area of offshore N.W. Java, these grainstones probably formed as storm deposits on the margins of exposed parts of the buildups. The modern grainstones have very limited lateral or vertical extent, occurring only on island fringes or near patch reefs.

Carbonate deposition of the upper Batu Raja limestone was followed by a transgressive phase during which the ZU-24 and G-2 (Gumai) shales were deposited throughout offshore N.W. Java. The interpretation of these shales as transgressive is based on their widespread distribution, uniform thickness and the consistent presence of glauconite.

Diagenesis

Diagenesis due to subaereal exposure during changing sea levels is responsible for much of the reservoir-quality rock found in the upper Batu Raja. A fall in sea level (possibly equivalent to the third order fall noted by Haq, et al. 1987) caused emergence and subaerial exposure of the paleo-topographically high carbonate buildup (Fig. 5B). Exposure provided a thin (5-20 ft) subaerial weathering zone (with local soil development) and a transitional zone of meteoric diagenesis (10-30 ft).

Meteoric diagenesis consisted of dissolution of mineralogically unstable aragonitic grains to form moldic and vuggy pores and micro-porosity. These diagenetic textures were superimposed upon all original depositional facies resulting in homogenization of prior textures. A faint lamination, representing original bedding, is still visible in most cores. In addition, many platy forams lie in what was most likely their original depositional position. Moldic porosity, produced by leaching of aragonitic grains is common in this zone. There has also been considerable dissolution of the fine grained carbonate matrix (former aragonitic components). This micro-scale dissolution enhanced the overall porosity and permeability of the sediment.

Organic detritus has been observed in all core through the "A" Zone. In addition, a plant rootlet was found in the ZUB-5 core further suggesting local soil development. Local chemical and biological processes within the active soil horizon along with leaching by meteoric fluids caused dissolution and brecciation of the carbonate rock. Porosity was enhanced slightly (33-35%). However, permeabilities were raised from originally 3-5 md to 1000+ md. At the base of the "A" Zone porosity and permeability were enhanced preferentially within burrows, whereas the matrix surrounding the burrows is low reservoir quality mudstone or wackestone.

A mixing zone between the base of the fresh water phreatic lens and top of sea water brines was the site for additional dissolution and dolomitization due to fluctuating salinity and pH. This may explain the origin of the lower pay zone ("C" Zone). Seasonal fluctuation in the depth of the fresh water table also provided some locally-enhanced permeability intervals between the "A" and "C" Zones in some areas. These zones form discontinuos pay intervals in the reservoir today.

Comparison with Other Batu Raja Fields

The faunal assemblage found throughout the Batu Raja section at Bima reflects deposition in a relatively low-energy environment dominated by carbonate mud (micrite). The Bima buildup contains benthonic foraminifera which are well adapted to life in low energy, quiet water environments. They may have lived in association with non-calcareous algae and/or sea grasses which aided in trapping the carbonate mud in the buildup. This restricted environment limited development of typical shallow-water "reef" assemblages found elsewhere offshore Java.

In contrast, the Rama Field (Ardila and Kuswinda, 1982) and Krisna Field, (Wight and Hardian, 1982) both located 30 miles to the west of Bima in the Sunda Basin, as well as the Arun Field of Northern Sumatra (Jordan and Abdullah, 1988) produce hydrocarbons from the Batu Raja formation (or its equivalent). However, in these fields the buildups are composed mainly of corals and contain a more diverse fauna. This faunal difference probably reflects the more "open" marine position of the Rama, Krisna and Arun paleo-highs during Batu Raja deposition and relatively less influx of clastics at these locations. Corals at Bima thrived only locally as isolated patch reefs on the edges of the buildup. In summary, reservoir quality was better at time of deposition for these fields than at Bima. Reservoir quality at Bima is due to diagenesis rather than depositional facies.

Geologic Flow Units

The upper Batu Raja was divided into 6 geologic flow units for field-wide correlation between wells (Fig. 3). Average reservoir properties for each unit are summarized in Table 1. The "A" Zone was divided into the "Dark Brown" and the "Mottled" units to differentiate the subaerially exposed horizon from the zone of transitional diagenesis. These flow units honor lithologic boundaries as well as providing the vertical communication necessary for coning in the simulation model.

MODEL LAYERING

The geologic flow units formed the basis for developing the model layers (Fig. 3). The "Dark Brown" unit was split into 2 layers to allow a thin (4 ft.) layer at the top to help simulate gas tonguing observed in the field. The discontinuous pay observed in the "B" Zone (typically correlated between 2-3 wells) was combined with the "Mottled" unit (Layer-3). This honored field-wide hydrocarbon volumes while minimizing the total number of layers. The "Upper Tight" unit governed vertical communication between the "Mottled" and "C" Zone, and the "Aquifer" layer provided bottom water for coning. Table 1 compares the model layering with geologic flow units.

RESERVOIR PROPERTY MAPPING

Reservoir properties (gross thickness, porosity, permeability and net/gross ratio) were averaged for each layer from well logs. These layer properties were mapped for model initialization using a combination of well control, geologic interpretation and seismic interwell data (when available). Map control of reservoir limits was determined from seismic cross sections.

Permeability was estimated from log porosity using porosity-permeability crossplots developed for each flow unit. Net/gross ratio was determined using a bulk density (RHOB) cutoff of 2.25 g/cm^3 to remove calcite-cemented streaks. Average reservoir properties for all layers are summarized in Table 1.

Water saturation was not mapped but instead initialized using capillary pressure (P_c) functions developed for each flow unit (Fig. 7). Foot-by-foot water saturations from well logs were plotted against height above WOC then converted to the model P_c curves. Water saturations from log-based P_c curves agreed with core-based Leverett J-Function predictions.

INCORPORATION OF SEISMIC

Seismic cross sections are typically used for mapping structure and identifying faults. Seismic resolution is usually limited to bed thicknesses greater than 30 ft for Bima depths (-2550 ft ss) allowing only gross intervals to be mapped rather than model layers. However, on rare occasions, where velocity contrsts are significant between adjacent zones, model layers can be mapped. This was the case for the porous"A" Zone (layers 1&2) which occurs between low porosity shale (above) and tight limestone (below).

For the Bima field, extensive 2D seismic was available for mapping upper Batu Raja gross thickness (Fig. 8). In addition, seismic inversion techniques (Poggiagliolmi, et al., 1988) provided cross sections of acoustic impedance contrasts (Fig. 9) which were used to determine the following:

- "A" Zone Thickness
- Buildup Edge / Reservoir Limits
- Interwell Pore Volumes (Porosity-Thickness)
- Basement Topography
- Lithology Distribution (Sand, Shale & Carbonate)

Accuracies of seismic-derived thickness and pore volumes for the upper Batu Raja were surprisingly good when compared to wells drilled before and after seismic interpretation (Figs. 10 and 11). However, considerable differences in scale still existed between seismic and model layer resolution. Therefore, the following allocation scheme was devised to assign seismic values to model layers:

"A" Zone Thickness

The high permeability "A" Zone could be resolved from seismic cross sections. This provided a thickness map of the main pay which was important in fringe areas of the field where the layer dipped into the WOC resulting in edge water influx.

There was still a scale problem since the "A" Zone represents 3 model layers (2 within the "Dark Brown" and 1 in the "Mottled"). In addition, the "Mottled" unit is transitional between the "Dark Brown" pay and the "Upper Tight" non-pay. Seismic resolution could only distinguish a "fuzzy", irregular "A Zone" to "Upper Tight" transition. A comparison of the seismic "A" Zone thickness with well data thicknesses showed that inclusion of 100% of the "Dark Brown" and 50% of the "Mottled" resulted in the best agreement (Fig. 12). Therefore, thickness maps for the "Dark Brown" and "Mottled" units were first mapped using well control then normalized to agree with the seismic "A" Zone map using the following formula:

$$H_A = H_{db} + \frac{1}{2}\left(H_m\right) \qquad \text{..................(1)}$$

where: H_A = "A" Zone thickness (ft)
H_{db} = "Dark Brown" thickness (ft)
H_m = "Mottled" thickness (ft)

The resulting "Dark Brown" map is shown in Fig. 13. Note that in the northern part of the field water influx can occur due to communication with the WOC. These features are outside the limits of well control and therefore would have been missed during mapping had seismic data not been incorporated.

Aquifer Thickness

It is important to be able to predict "Aquifer" thickness (Layer-6) since it is responsible for bottom water drive. Field-wide cross sections revealed much of the thickness variation for the upper Batu Raja was due to changes in "Aquifer" thickness. Therefore, the upper layers ("Dark Brown" through "C" Zone) were first mapped as individual layers then summed together to represent the upper Batu Raja *without* the "Aquifer" layer. This sum was subtracted from the seismic Batu Raja thickness as follows:

$$H_{aq} = H_{UBR} - (H_{db} + H_m + H_{ut} + H_c) \quad ...(2)$$

where: H_{aq} = "Aquifer" thickness (ft)
H_{UBR} = Upper Batu Raja thickness (ft)
H_{db} = "Dark Brown" thickness (ft)
H_m = "Mottled" thickness (ft)
H_{ut} = "Upper Tight" thickness (ft)
H_c = "C" Zone thickness (ft)

The resulting "Aquifer" map (Fig. 14) contains much more detail than obtainable using only well control. In addition, the map reveals a thicker aquifer in the undeveloped southern part of the field (Phase-2) suggesting a greater potential for water coning in that part. This risk is heightened by the adverse mobility ratio of the viscous crude.

Pore Volume

Allocation of upper Batu Raja pore volume (porosity-thickness) to individual model layers was not successful given the differences in scale. Pore volume variation is due mainly to changing thickness since porosities typically range from 27 to 33%. Therefore, areal distribution of porosity is not an important characterization issue. However, accuracy of the seismic-derived pore volume is encouraging and the inversion technique may be valuable in other fields.

FULLFIELD MODEL CONSTRUCTION

Grid

The fullfield model was based on a 58 by 35 grid with 6 layers resulting in 12,180 total cells (Fig. 15). Phase-1 cells are 5.7 acres (500 ft by 500 ft) providing 3-4 cell interfaces between the 60-acre wells. The undeveloped Phase-2 cells have an aspect ratio of 500 ft by 2000 ft oriented along the structural crest. This provided grid refinement in the east-west direction parallel to gas and water contacts. Cell pore volumes were large enough to allow an IMPES solution (implicit pressures and explicit saturations) in the 3-phase black oil model.

Relative Permeability

As in the case for capillary pressure, different relative permeability functions were used for each layer. Water-oil relative permeability was initialized in the model using a primary drainage curve developed from core measurements.

Water-oil imbibition data were then normalized according to initial water saturation (S_{wi}) and input as a primary imbibition curve (Fig. 16). Scanning routines in the model allow imbibition to start from the primary drainage curve according to initial water saturation. This approach minimized the potential for flowable water in the model at field startup given the large transition provided by the capillary pressure curves (Fig. 7).

Aquifer Size

The geologic model suggested that diagenesis was due to subaereal exposure of the buildup. However, this exposure and resulting permeability enhancement, probably did not extend beyond the buildup boundaries. Therefore, the pore volume of model boundary cells below the WOC was increased by only a factor of 4 (Fig. 15) to account for a limited aquifer size in the "A" Zone. This provided an edge aquifer volume only slightly larger than the oil rim volume, yet large enough to maintain adequate water influx rates to match well production and pressure histories.

SPECIAL SIMULATOR TREATMENTS

Horizontal Wells

There are 20 producing wells at Bima with deviation angles greater that 80° from vertical. The horizontal section (average length of 1500 ft) is typically completed in the "A" Zone. The "C" Zone is used for the horizontal target if the "A" Zone contains gas. The justification for horizontal wells was to increase productivity index (PI), thereby minimizing gas and water coning with the viscous crude by limiting the drawdown pressure. In addition, the effective drainage of each well could be maximized given the limited number of well slots (9) on each offshore platform.

Horizontal wells at Bima were modeled with productivity terms assigned to grid cells along the horizontal section. Some wells crossed as many as 5 cells. Well productivities were calculated using the wellbore length and permeability averaged from the vertical and horizontal planes:

$$K_{well} = \sqrt{K_h * K_z} \qquad \text{..................(3)}$$

where: K_{well} = wellbore permeability (md)
K_h = cell horiz. permeability (md)
K_z = cell vertical permeability (md)

Vertical permeability was determined using an initial vertical/horizontal permeability (K_v/K_h) ratio of 0.05 (later history matched). Absolute permeabilities in the model were adjusted until the model-calculated PI matched observed field PI for each well.

Electrical Submersible Pumps (ESP)

Field operations require switching of wells from artificial lift (ESP) to natural flow when high gas/liquid ratios (GLR) are reached. This is to prevent mechanical damage in the pump during 2-phase flow. Simulator logic is required to switch bottom hole pressures (BHP) according to lift method.

Tubing hydraulic curves were used to vary the BHP for either pumping (300 psi) or natural flow (400-600 psi). The historical GLR threshold, before switching to natural flow, is 1000 scf/STB. The resulting hydraulics are depicted in Fig. 17 which includes the ESP curve (GLR<1000 scf/STB) and naturally-flowing curves (GLR>1000 scf/STB).

Pressure Dependent Permeability

The semi-consolidated "A" Zone exhibited core permeability reductions with increasing net overburden stress. Permeability reduction at maximum drawdown (300 psi BHP) is 20% below original values at initial reservoir pressure (1135 psi). Conversely, an injection pressure of 1700 psi will result in a permeability enhancement of 15% above initial values. These trends were applied in the model to both the inter-block transmissibilities and wellbore productivity terms under varying reservoir pressures. This treatment more closely honors overburden effects of the semi-consolidated formation.

Variable Gas-Oil Contact Depth

An isolated gas cap to the south (Phase-2) has a higher GOC (-2480 ft ss) than the main part of the buildup (-2512 ft ss) due to differences in structural closure. Pseudo gas-oil capillary pressure curves were implemented to provide variable contacts while still honoring a single GOC datum. A single GOC datum was required in the model for matching initial gas cap pressures with the PVT bubble point pressure thereby providing equilibrium for both thermodynamic and capillary effects.

Layer Pinchouts

The layers were laterally continuous over most of the field. The exception was the absence of the Upper Tight and "C" Zone in the north-south trending "low" between the ZUB and ZUE platforms (Fig. 2). The pore volume for these layers is zero in this region causing vertical transmissibilities to also be zero. Water coning in this region is known to exist between the "Aquifer" layer and "A" Zone. Therefore, "non-local" transmissibility connectors were used in the model for communication across the missing layers (Fig. 18). Non-local connectors are typically used for fault logic to account for layer juxtaposition across faults (Fig. 19).

HISTORY MATCH

Model permeabilities were adjusted until all well productivities were matched. This ensured realistic cell permeabilities while also achieving a productivity match. Representative wellbore skins (ranging from post-acid cleanup of -3 to 5 for damage) were assigned where available. Permeability multipliers of 1-5 times were typically required to match field PI. It is suspected that the porosity-permeability crossplots from the "A" Zone core were pessimistically biased towards lower permeabilities due to excessive sampling of the relatively easy-to-plug cemented intervals in an otherwise semi-consolidated core.

Oil rates were specified during the history match (January 1987 to January 1990). BHP was controlled during prediction using tubing hydraulics curves with a constant surface pressure. Acceptable matches of water cut and GOR were obtained for all 54 wells in the model. Most of the wells experienced gas and water coning or tonguing as typified in the match for Well ZUF-1 (Fig. 20).

The total field match is shown in Fig. 4. Field-wide reservoir pressures were limited given the short production history, but matched where available. The simulated Phase-1 oil-rim pressure of 800 psi agreed with observed field pressures (700 to 900 psi) after 2-1/2 years of production.

Lab-derived relative permeability curves provided an adequate match with the exception of gas relative permeability (K_{rg}). Lab K_{rg} measurements provided an unrealistically low exponent (n) of 1 for the following K_{rg} equation:

$$K_{rg} = \left[\frac{1 - S_l - S_{gc}}{1 - S_{wi} - S_{gc}}\right]^n \qquad \text{.................(4)}$$

where: n = exponent

S_l = liquid saturation

S_{gc} = critical gas saturation

S_{wi} = initial water saturation

Multiple exponents of 2 and 3 (Fig. 21) in different field areas improved the gas production match compared to the lab exponent of 1. It is believed that premature gas breakthrough during the lab experiment lowered the K_{rg} exponent. A critical gas saturation (S_{gc}) of 2% provided adequate gas movement in the model for matching field performance.

Pseudo relative permeability curves to account for gridding effects were determined unnecessary based on agreement between fine-grid and coarse-grid models under primary and secondary recovery. The good agreement was a result of the relatively small grid size (5.7 acres) and thin (4 ft) layer added to the model top to help gas tonguing.

An average K_v/K_h ratio of 0.05 was required for matching water and gas coning. Individual well K_v/K_h values were further adjusted to match gas and water rates. Little or no fault isolation was required during the history match or observed in field performance to date.

DEVELOPMENT OPTIONS

Projected ultimate primary recovery at Bima is 10% OOIP for the developed Phase-1 area. This marginal recovery is a direct result of the adverse mobility ratio of the viscous oil. The gas cap and aquifer volumes are not significant enough to provide long-term pressure support and their presence only adds to water and gas production problems.

Gas voidage rates of 15-20 million scf/D over the 3-year production life have raised concern over long-term gas cap resaturation and pressure depletion. Therefore, immediate reservoir management decisions rest on secondary recovery options (gas reinjection versus waterflood).

Gas Reinjection

Reinjection of produced gas was evaluated since significant reservoir energy is currently being flared under primary production. Gas production accounts for 2/3rds of the reservoir voidage. To date over 22 billion scf have been produced compared to the original Phase-1 gas cap (15 billion scf). Solution gas and influx of Phase-2 gas account for much of the gas production since the current size of the Phase-1 gas cap approximates the original size.

Gas reinjection would require only 1 injector on the crest of the structure. Injection rates in excess of 40 million scf/D were projected for the prolific "A" Zone. Gas reinjection would require fewer facility investments compared to waterflood and could be implemented in a faster time frame. However, an incremental recovery of only 1% OOIP (Phase-1) over primary was projected using a 1992 startup.

Limited recovery improvement from gas reinjection is due to a combination of unfavorable oil viscosity and low structural dip resulting in severe gas cycling. Incremental recovery could have been improved if implemented earlier in the field life. However, the process was judged uneconomic during earlier evaluations.

Waterflood

Phase-1 waterflood is determined to be the most economically viable option for improved recovery at Bima. Waterflood was simulated on 3 platforms in the oil rim (ZUA, ZUC and ZUF) using 3 injectors (converted producers) per platform. An additional platform in the gas cap (ZUE) was also converted to sea water treatment and injection. Gas cap injection accelerated the "blowdown" of gas in Phase-1 and helped minimize gas cap resaturation.

A Phase-1 injection rate of 60,000 BWPD (12 wells) was predicted using an injection BHP of 1700 psi (fracture gradient of 0.67 psi/ft). Economic shut-in limits of 2000 BOPD (field) and 95% water cut (by well) were imposed to control the project life.

Projected incremental recovery from Phase-1 waterflood varied from 3.4% to 4.5% OOIP (Phase-1) depending on injection startup (1992-1994). This difference was inversely related to the ultimate gas production at field abandonment (Fig. 22).

Future Evaluations

The model has the ability to evaluate future development opportunities as they arise including additional Phase-1 infill wells, Phase-2 expansion (primary and secondary) and gas cap blowdown under gas sales. Regional movement of reservoir fluids across the field can be tracked during the producing life to evaluate fullfield effects upon these projects. It should be noted that fullfield simulation can evaluate these scenarios individually or in combination with each other.

CONCLUSIONS

1. The synergistic integration of multiple disciplines during Bima reservoir characterization provided a more useful description:

- A "qualitative" geologic model was forced to become more "quantitative".
- The exhaustive description incorporated data from all scales (thin section - core - seismic).
- The description was more accurate and adaptable to reservoir simulation needs.
- History matching was simplified and accelerated.
- Resulting rate and recovery forecasts are more reliable.

2. Modeling of dominant Bima water drives (bottom and edge water influx) was improved by using seismic formation thickness allocated to model layers.

3. Deriving interwell pore volumes using seismic inversion techniques was attempted. However, only a qualitative allocation to Bima model layers could be included.

4. The geologic model for Bima deposition and diagenesis can be applied during evaluation of other Batu Raja limestone buildups in the Seribu platform region.

5. Phase-1 waterflood opportunities look attractive with an incremental recovery of 3.4 to 4.5% OOIP (Phase-1) above primary recovery (10% OOIP). However, these reserves are sensitive to injection timing due to gas cap resaturation during deferral.

ACKNOWLEDGEMENTS

We appreciate the technical contributions of Bob Crumb (Atlantic Richfield Indonesia, Inc.) and Rich Seader (ARCO International Oil and Gas Co.) in many areas of Bima characterization.

We also wish to thank ARCO Oil and Gas Company, Atlantic Richfield Indonesia, Inc. and Pertamina for permission to publish this paper.

REFERENCES

Ardila, L. E., and I. Kuswinda, 1982, The Rama Field: an oil accumulation in Miocene Carbonates, West Java Sea. Proc. ASCOPE Conference, Surabaya, 32pp.

Atkinson, C.D., and Kaldi, J.G., 1990, Tertiary Reservoirs of Offshore N.W. Java, Indonesia: examples of hydrocarbon accumulations in a mixed carbonate-clastic environment. (Abst.) AAPG Bull. v. 74, p. 601-602.

Fletcher, G.L., and Bay, K.W., 1975, Geochemical eEvaluation - N.W. Java Basin. Proceedings of the Indonesian Petroleum Association, 4th Ann. Conv., Jakarta, p. 211-241.

Haq, B.U., Hardenbol, J., and Vail, P.R., 1987, Chronology of fluctuating sea levels since the Triassic. Science, v. 235, p. 1156-1167.

Miller, N.R., and Kaldi, J.G., 1990, Strontium isotope chronostratigraphy and diagenesis of the Bima Field Batu Raja Limestone, offshore N.W. Java, Indonesia. (Abst.) AAPG Bull. 74, p. 728-729.

Jordan, C. F., Jr. and M. Abdullah, 1988, Lithofacies analysis of the Arun reservoir, North Sumatra, Indonesia. Giant Oil and Gas Fields, SEPM Core Workshop No. 12, SEPM, Tulsa, p. 89-118.

Poggiagliolmi, E., Checka, V.R., Roe, R.C., and Purantoro, R. 1988, Reservoir petrophysics of Bima Field, N.W. Java Sea. Proceedings of the Indonesian Petroleum Association, 17th Ann. Conv., Jakarta, p. 359-365.

Ponto, C.V., Wu, C.H., Pramoto, A., and Stinson, W.H., 1988, Improved interpretation of the Talang Akar depositional environment as an aid to hydrocarbon exploration in the ARII Offshore NW Java Contract Area. Proceedings of the Indonesian Petroleum Association, 17th Ann. Conv., Jakarta., pp. 397-422.

Wight, A. and Hardian, D., 1982, Importance of diagenesis in carbonate exploration and production, Lower Batu Raja carbonates, Krisna Field, Java Sea. Proceedings of the Indonesian Petroleum Association, 11th Ann. Conv., Jakarta, pp. 211-235.

Table 1. Summary of layer properties.

GEOLOGIC ZONE	LITHOFACIES FLOW UNIT	MODEL LAYER	SIMULATION PURPOSE	H THICK (ft)	Ø POR (%)	K PERM (md)	Swi WATER (%)
"A"	DARK BROWN	1	GAS TONGUE	4	31 to 36	100 to 1000	35 to 45
		2	MAIN PAY	5 to 25			
	MOTTLED	3	PAY TRANSITION	10 to 30	30 to 32	60 to 100	45 to 55
"B"	discontinuous pay UPPER TIGHT	4	GAS & WATER CONING	30 to 40	26 to 29	2 to 5	85 to 100
"C"	"C" ZONE	5	LOWER PAY	10 to 20	29 to 32	20 to 100	45 to 55
"D"	AQUIFER	6	BOTTOM WATER DRIVE	40 to 60	24 to 27	2 to 5	85 to 100

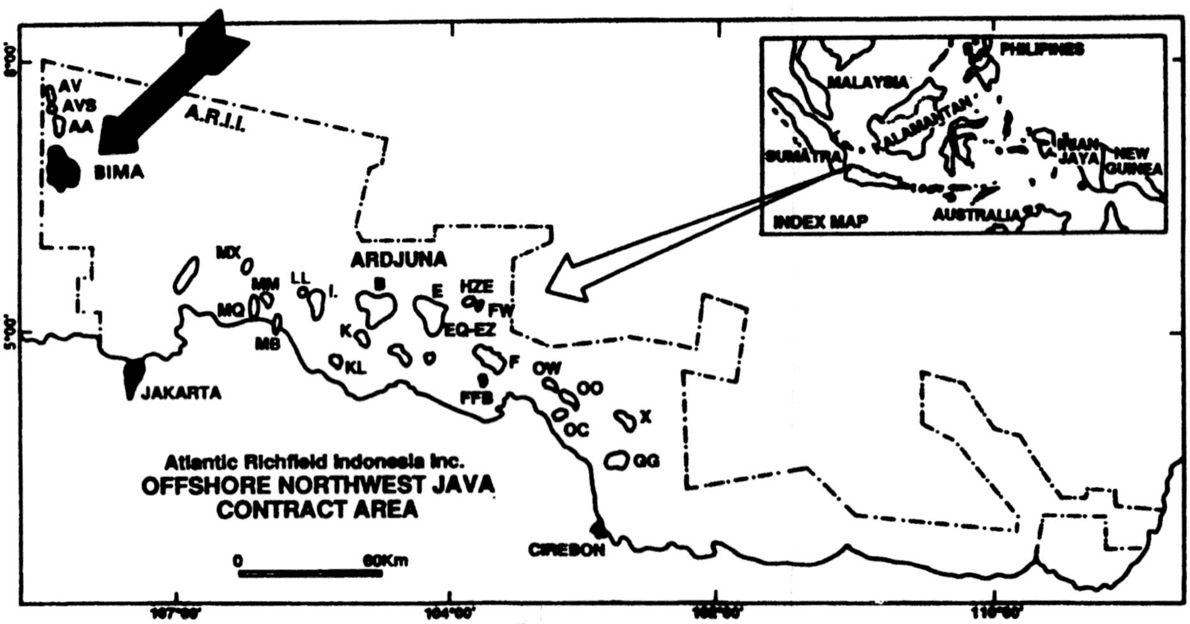

Figure 1. Bima Field location map.

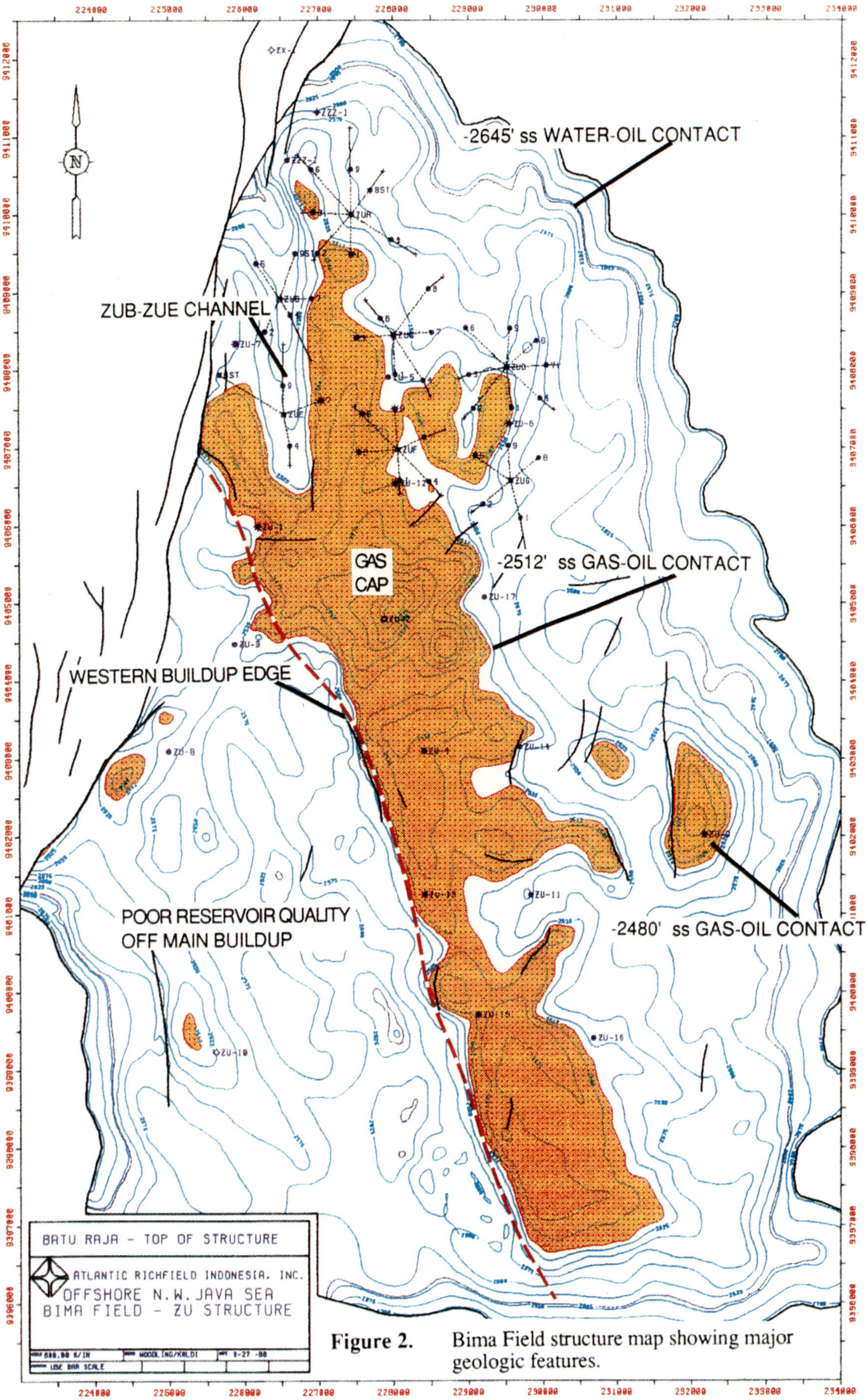

Figure 2. Bima Field structure map showing major geologic features.

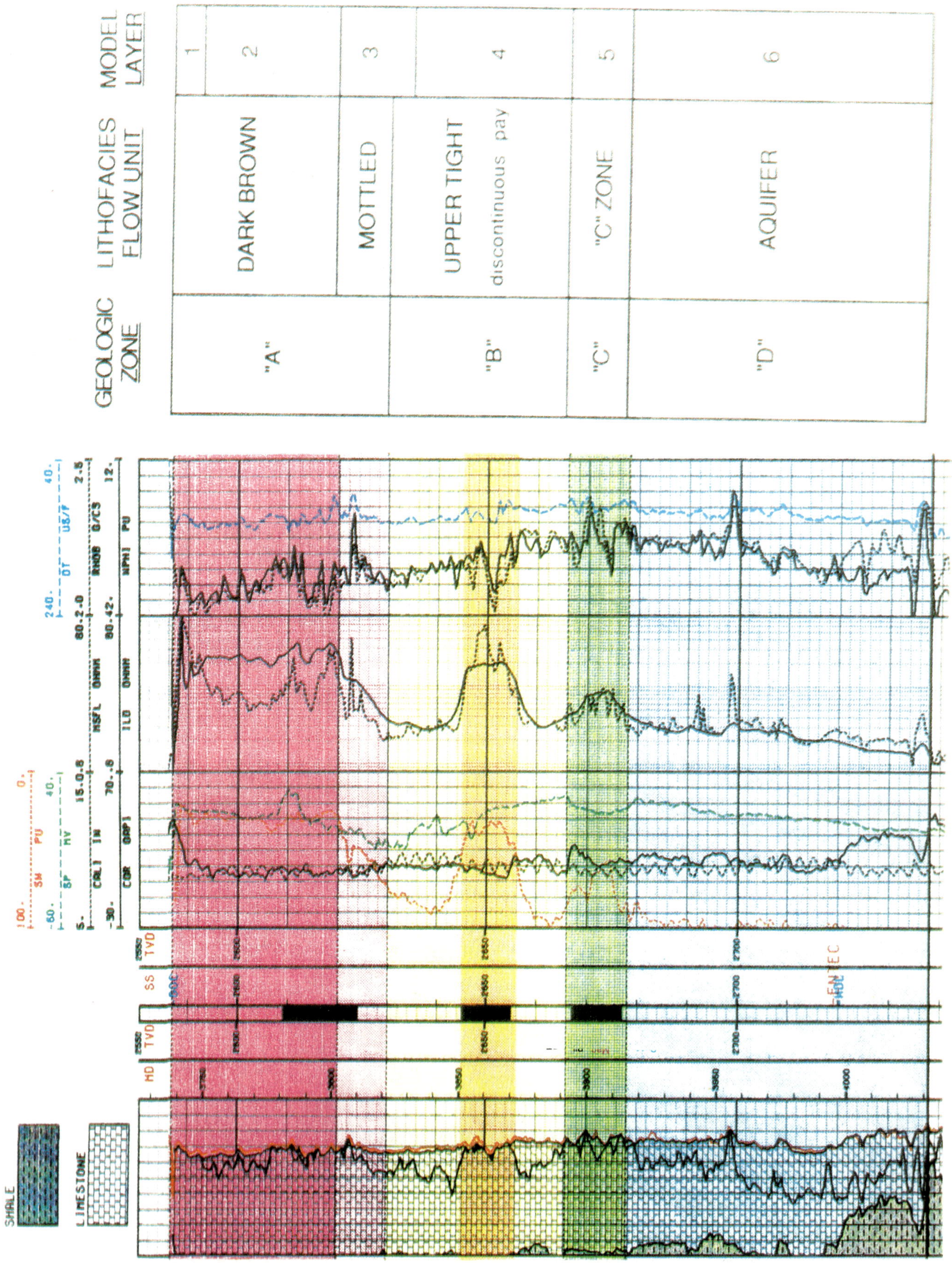

Figure 3. Upper Batu Raja type log with model layers.

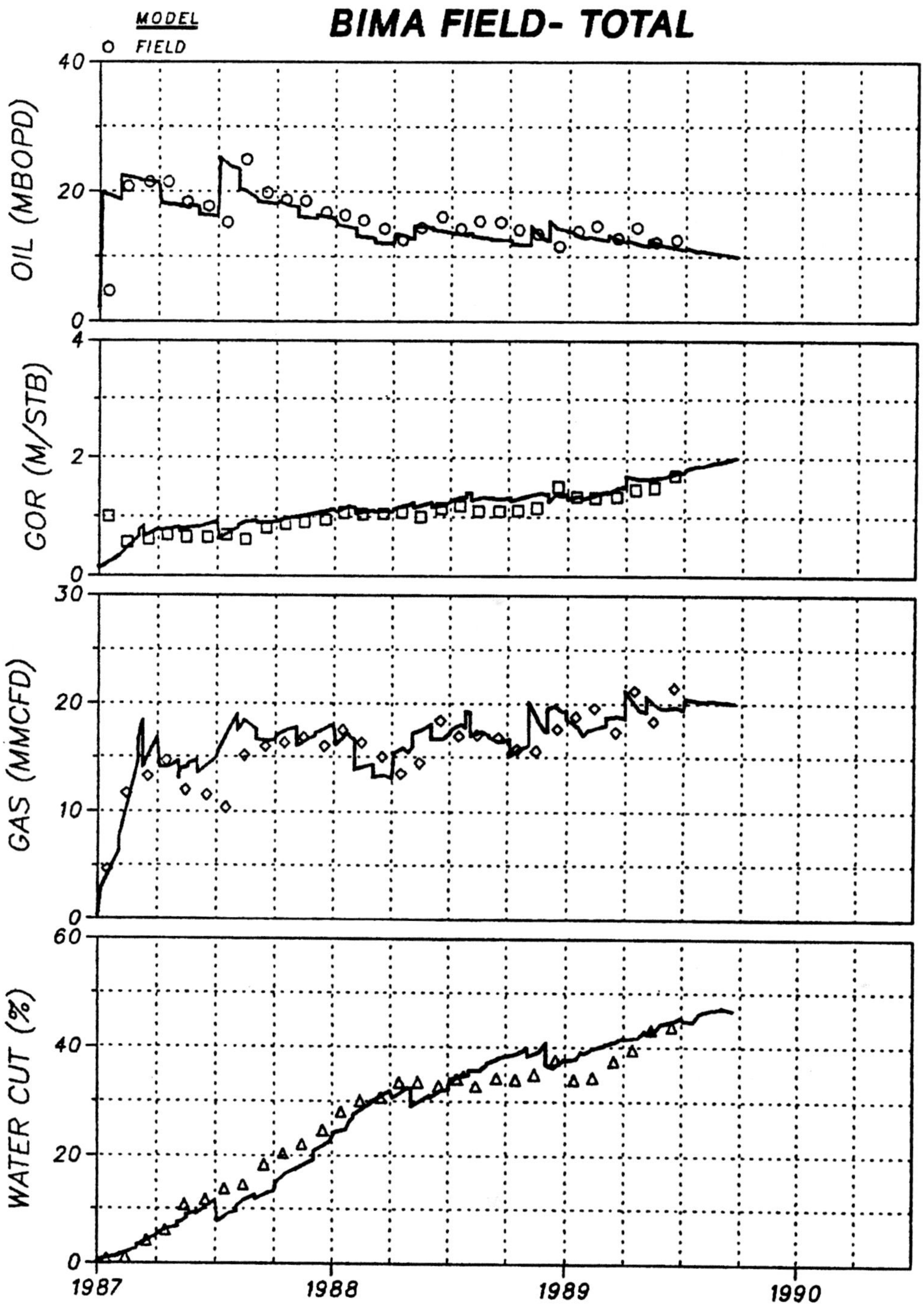

Figure 4. Bima Field performance compared to model history match.

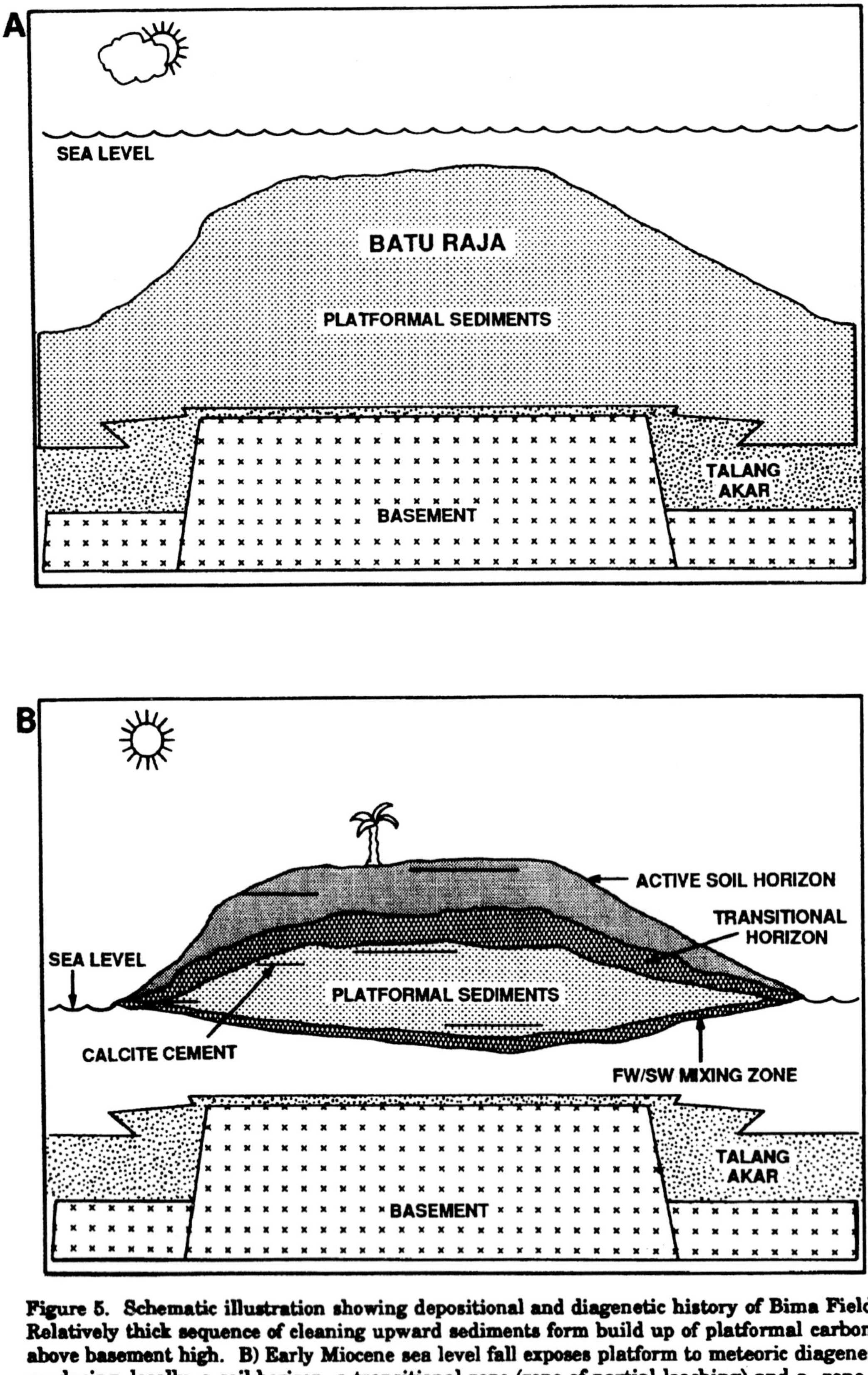

Figure 5. Schematic illustration showing depositional and diagenetic history of Bima Field. A) Relatively thick sequence of cleaning upward sediments form build up of platformal carbonates above basement high. B) Early Miocene sea level fall exposes platform to meteoric diagenesis, producing, locally, a soil horizon, a transitional zone (zone of partial leaching) and a zone where mixing of fresh (meteoric) and sea water resulted in porosity development

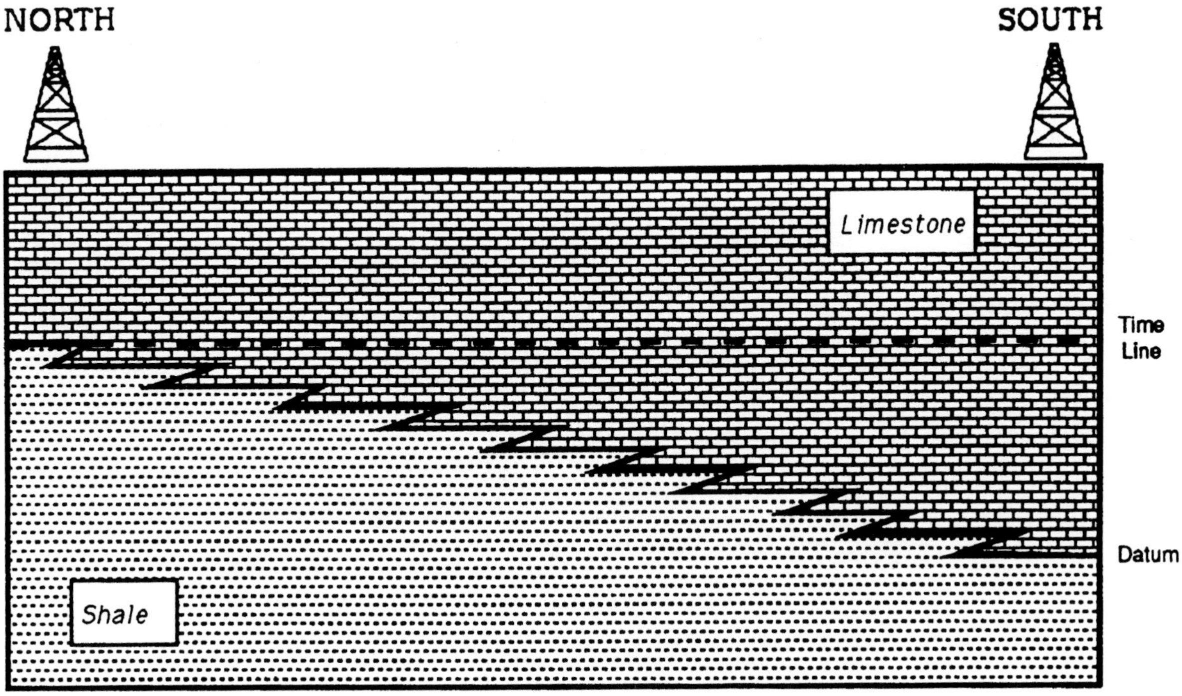

Figure 6. Bima Field carbonate-shale facies transition.

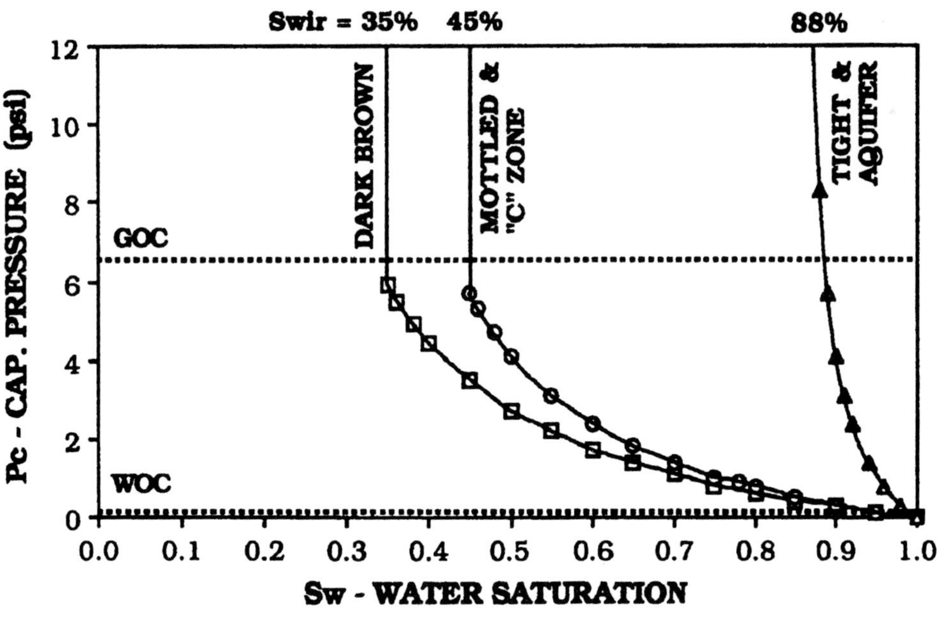

Figure 7. Model capillary pressure functions by rock type.

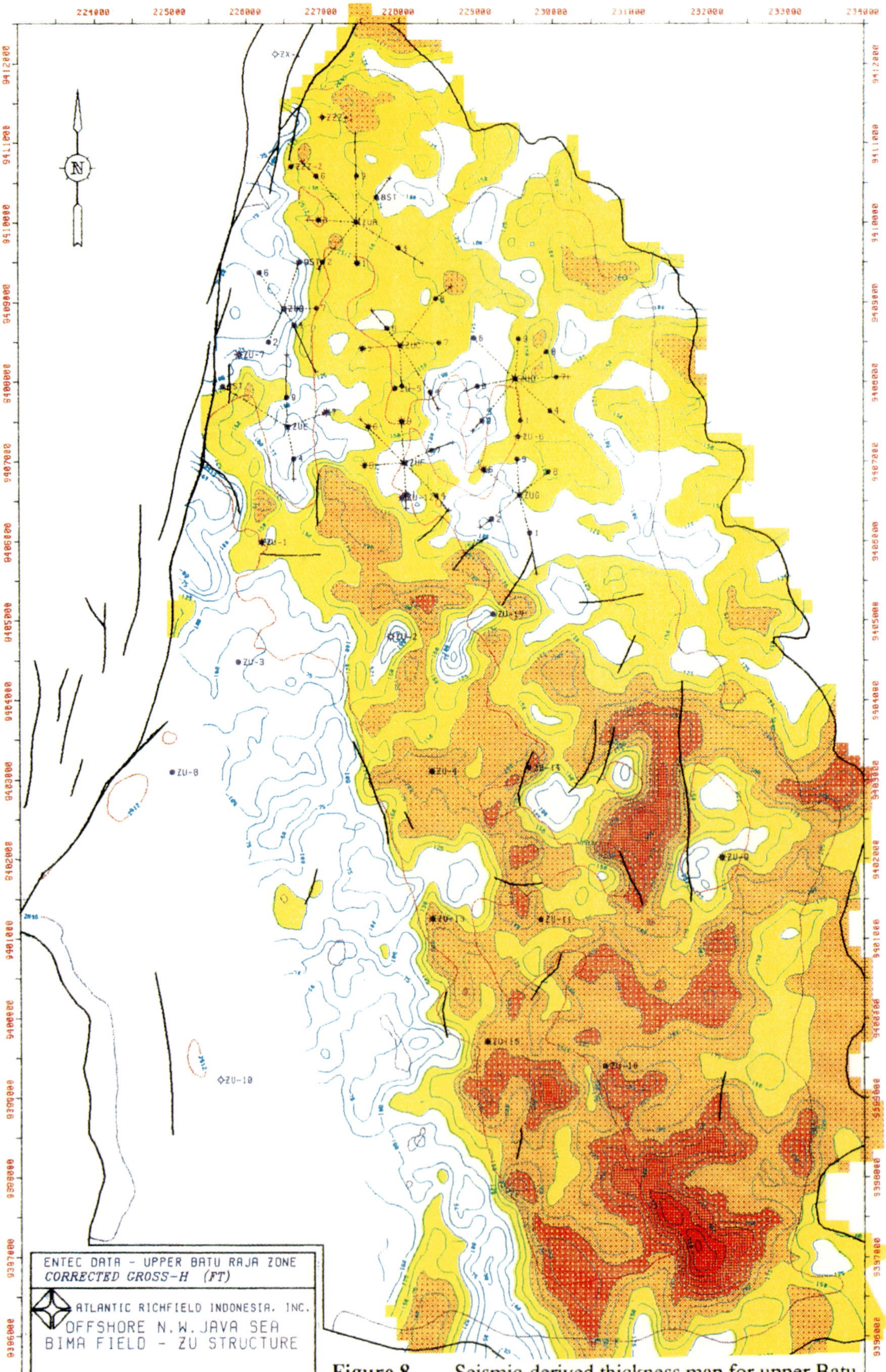

Figure 8. Seismic-derived thickness map for upper Batu Raja formation.

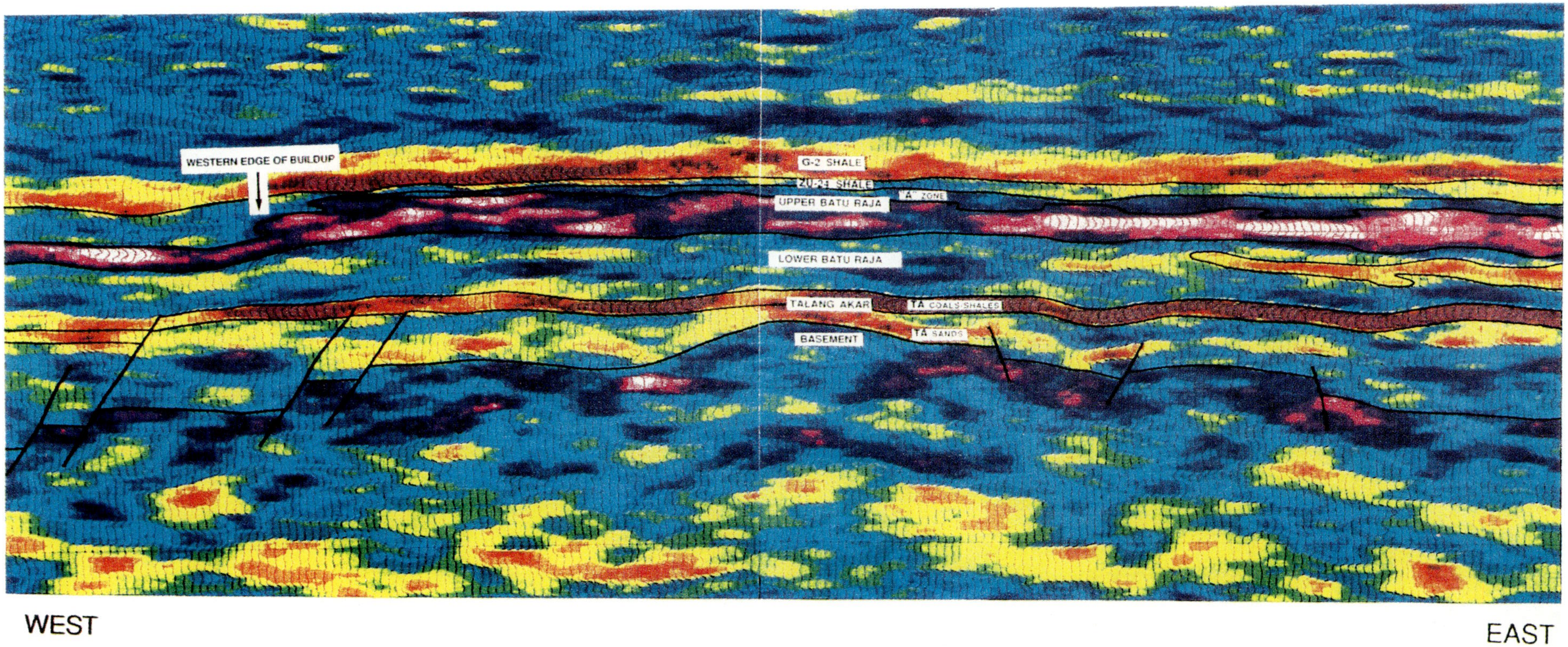

Figure 9. Example seismic cross section of acoustic impedance contrast showing lithology variation.

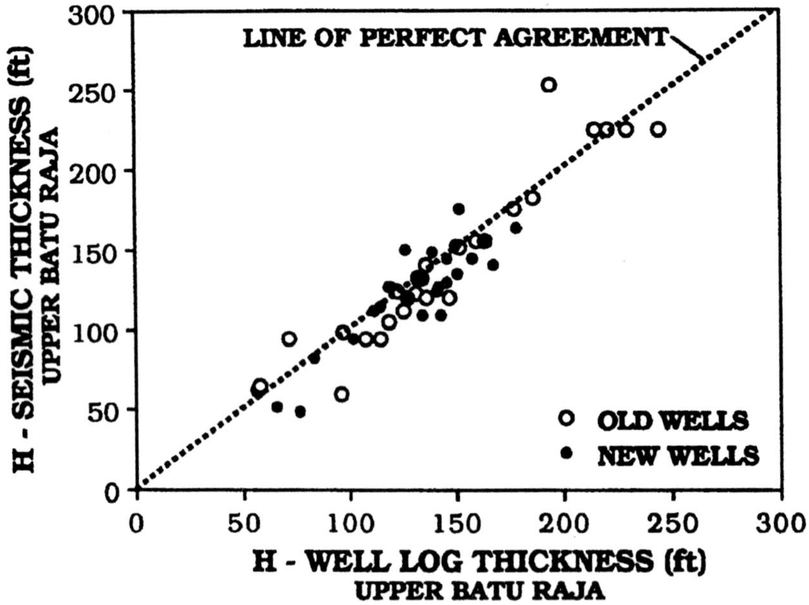

Figure 10. Seismic-derived thickness for upper Batu Raja compared to well values.

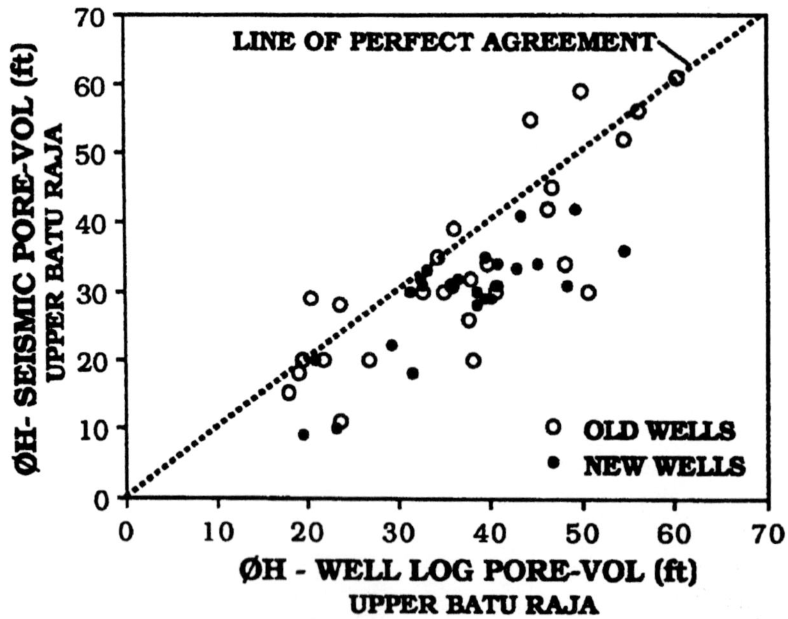

Figure 11. Seismic-derived pore volume for upper Batu Raja compared to well values.

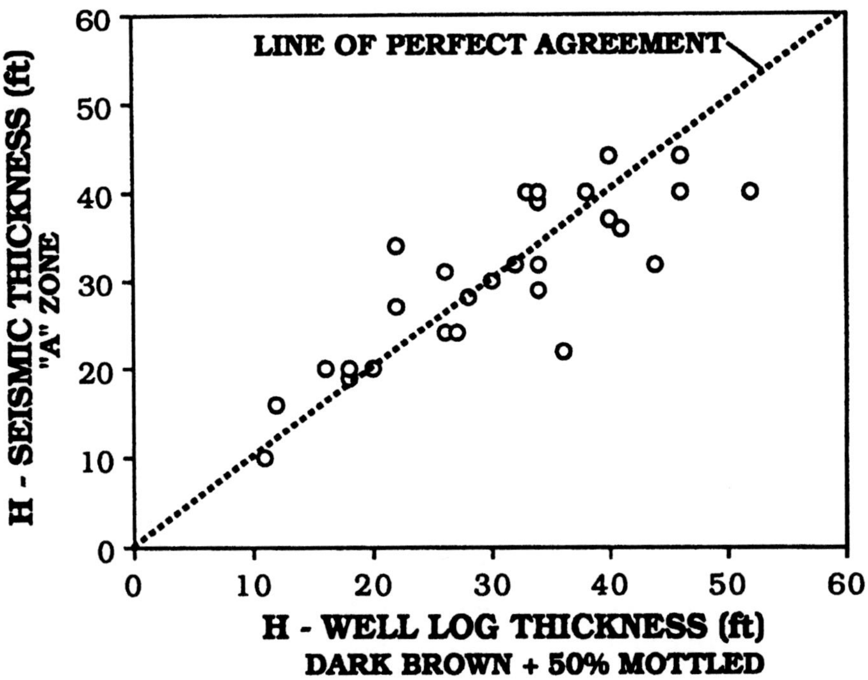

Figure 12. Seismic-derived thickness for "A" Zone compared to well values for Dark Brown and 50% of Mottled units.

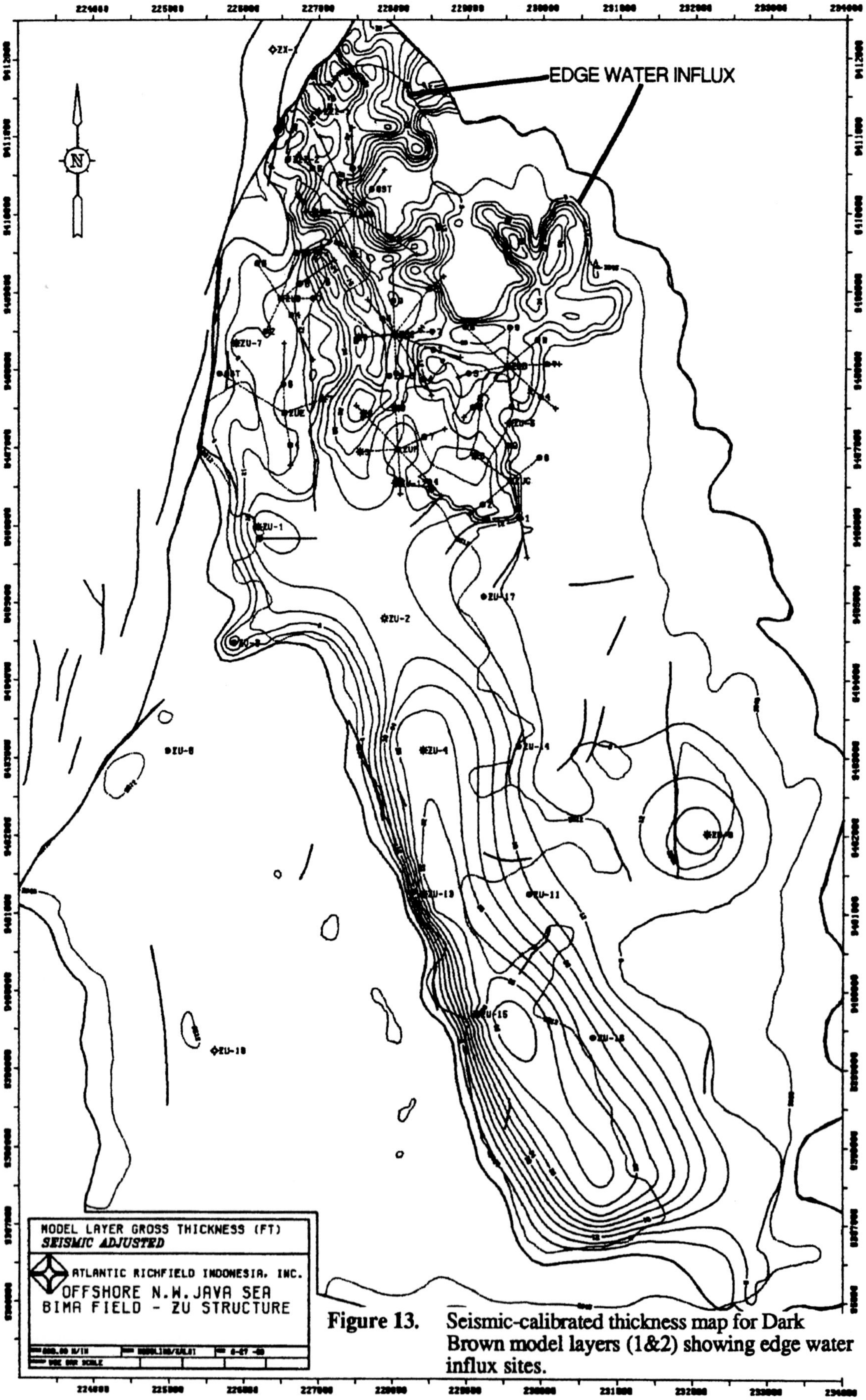

Figure 13. Seismic-calibrated thickness map for Dark Brown model layers (1&2) showing edge water influx sites.

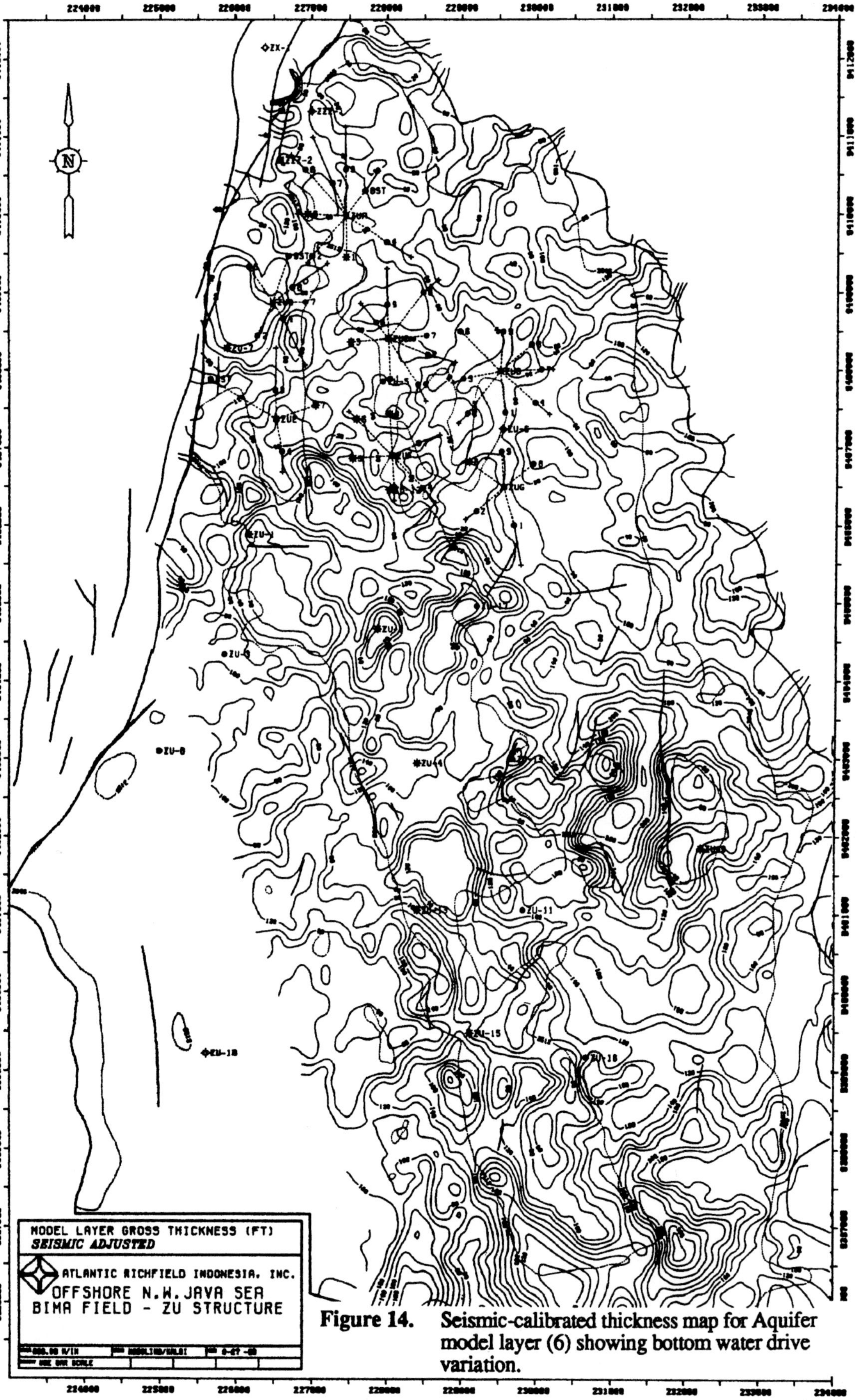

Figure 14. Seismic-calibrated thickness map for Aquifer model layer (6) showing bottom water drive variation.

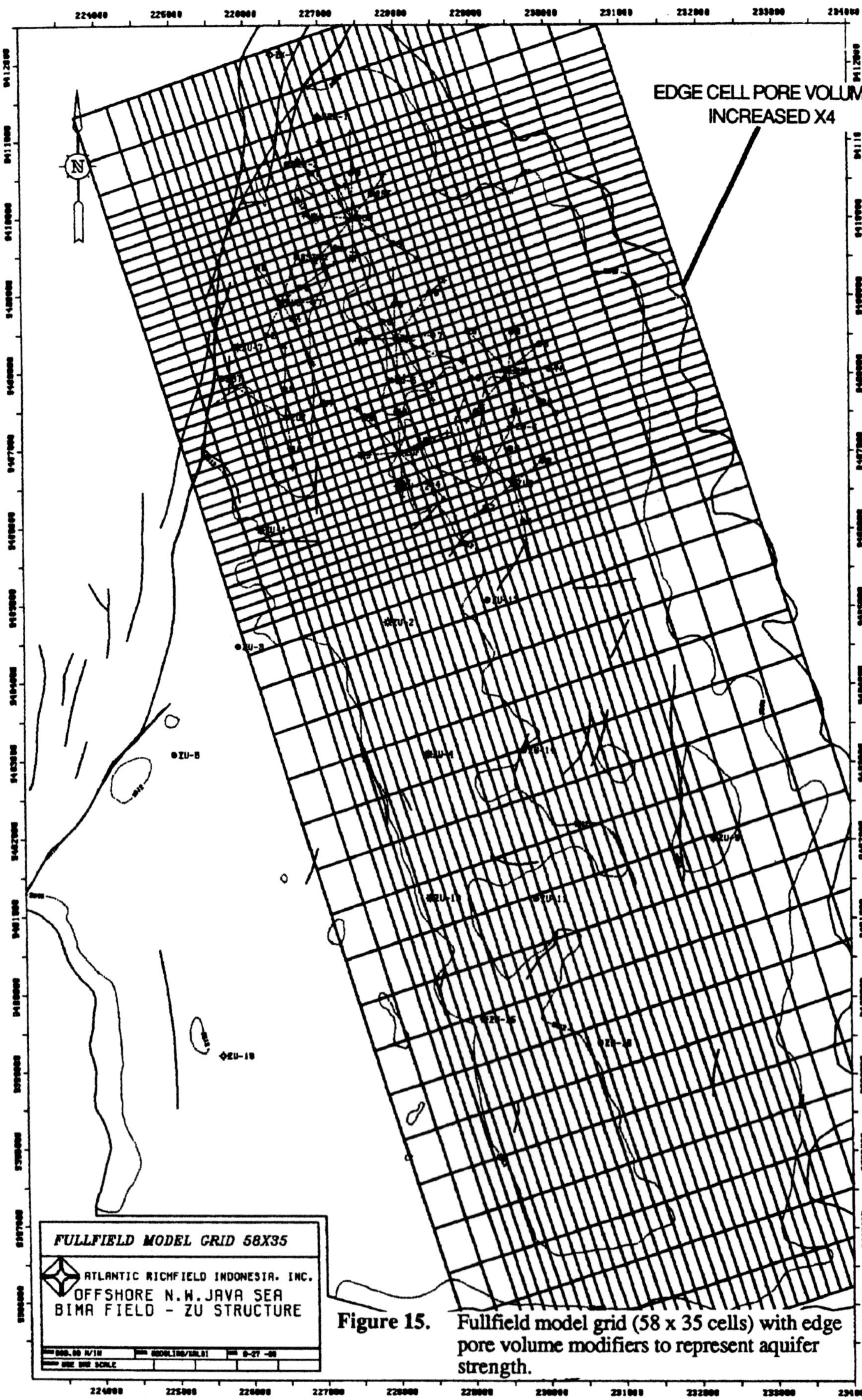

Figure 15. Fullfield model grid (58 x 35 cells) with edge pore volume modifiers to represent aquifer strength.

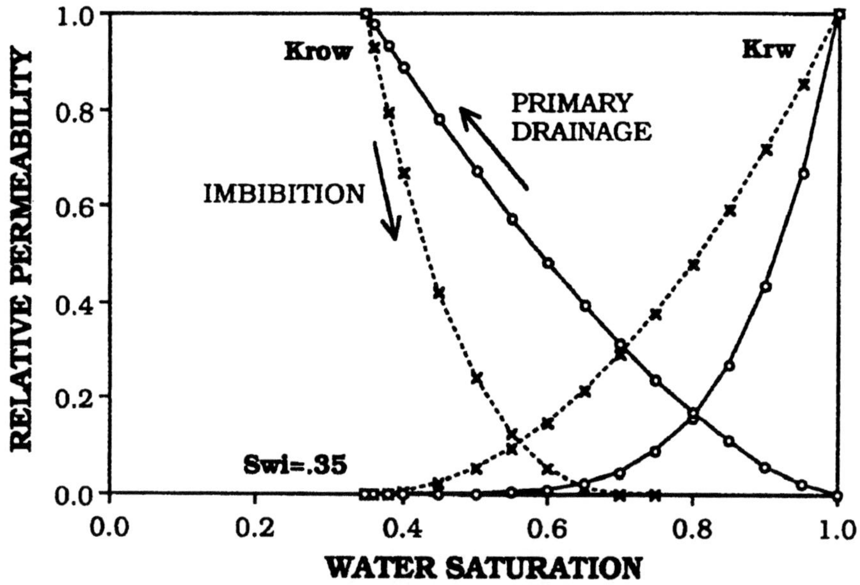

Figure 16. Water-oil primary drainage and imbibition relative permeability for Dark Brown unit.

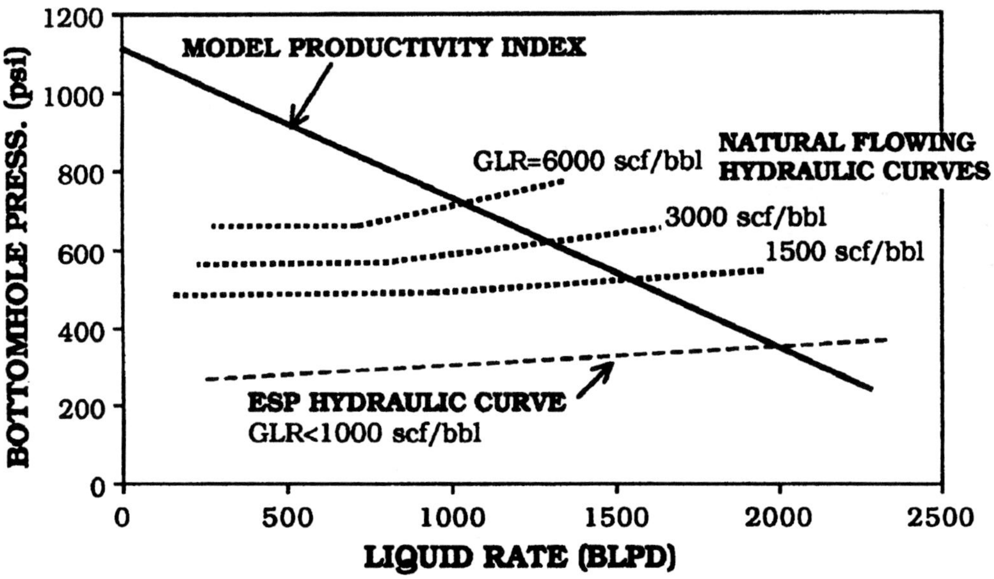

Figure 17. Tubing hydraulic curves for natural flowing or electrical submersible artificial lift.

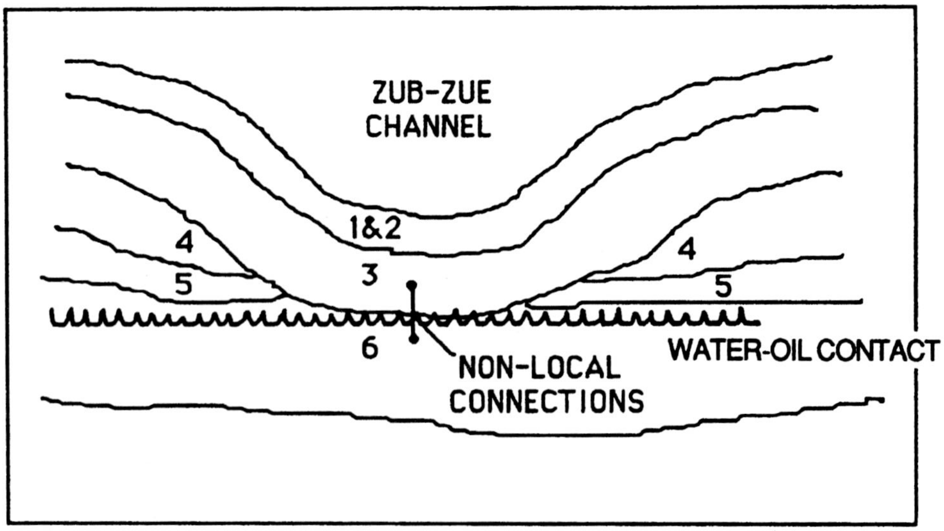

Figure 18. Non-local transmissibility connections for communication across pinchout.

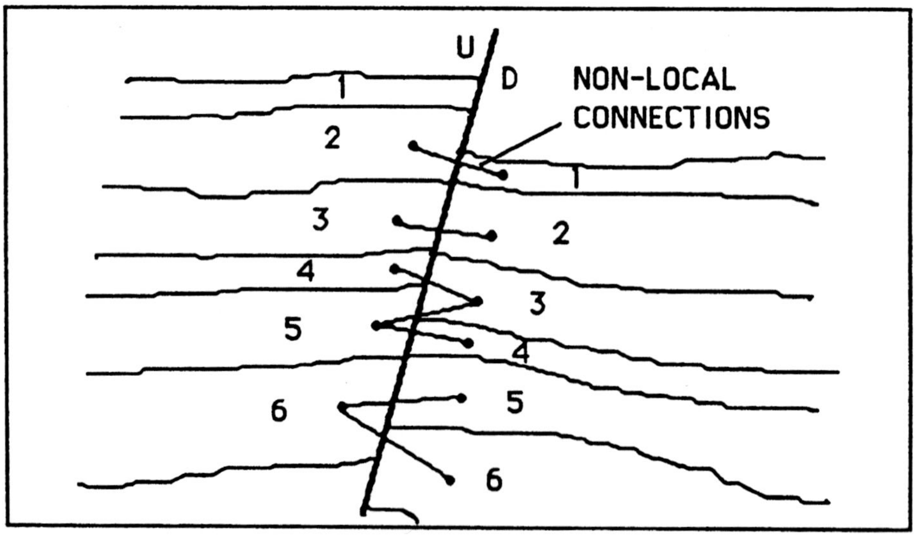

Figure 19. Typical non-local transmissibility connectors for communication across fault.

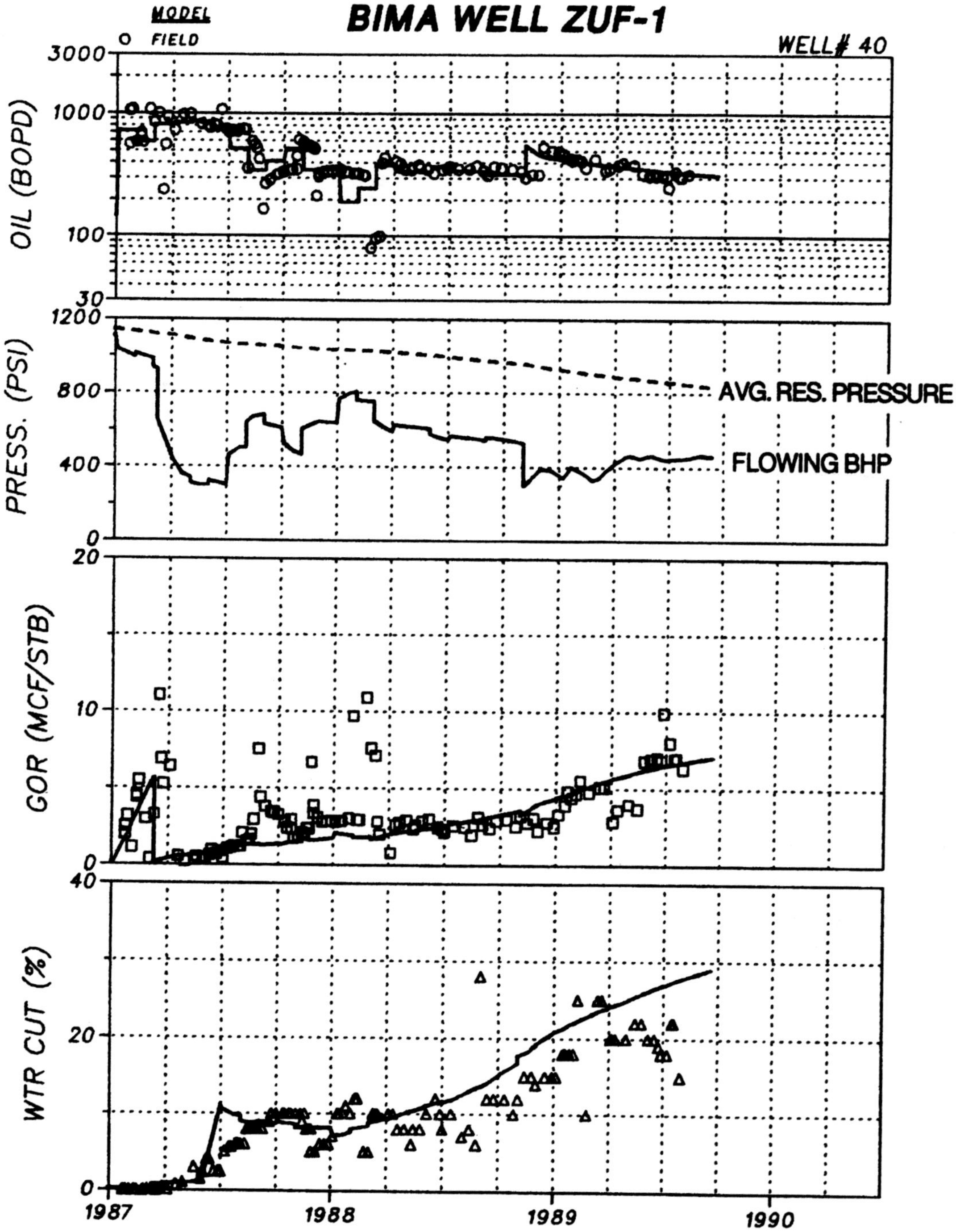

Figure 20. History match of well ZUF-1 showing typical match of both gas and water production.

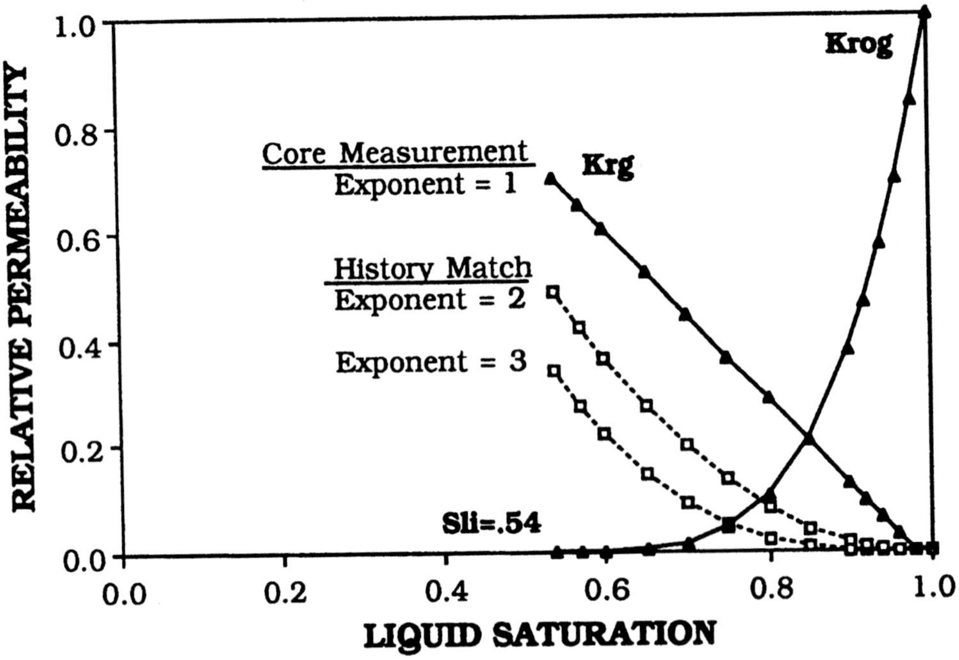

Figure 21. Comparison of history matched and lab-derived gas relative permeability for Dark Brown unit.

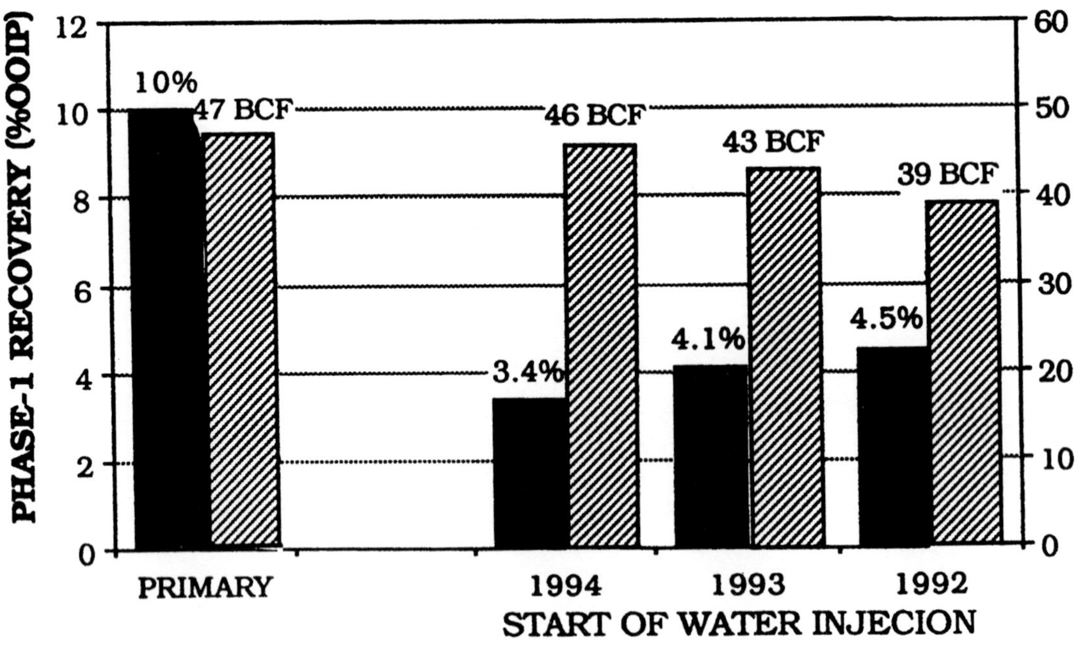

Figure 22. Primary and waterflood incremental oil recovery (Phase-1 OOIP) as a function of waterflood start year and ultimate gas production.

Intan Field Development—A Synergistic Approach

Mike Farhadi

In 1987 and 1988 MAXUS Southeast Sumatra Inc. discovered two fields in the Asri Basin offshore Indonesia with initial reserve estimates of 300 MMBO. Early development of the Intan Field along with the creation of a reservoir management team resulted in a final development plan for Intan that was successful in saving capital and time as well as minimizing undrained areas. The experience and knowledge gained from the modeling and development of the Intan Field is being applied effectively to the development of the Widuri Field with reserves of 250 MMBO.

The Intan Field is a NE-SW trending, faulted, anticlinal feature located on the northwest flank of the Asri Basin. A study of the environment of deposition reveals the reservoir sands to have been deposited by meandering and low sinuosity rivers flowing from the northwest and from the northeast into the Asri Basin. The discovery well Intan-1 was drilled in October 1987 and encountered 74 feet of net oil pay in the early Miocene Talang Akar formation. Intan-1 flowed at a combined rate of 5845 BOPD from two flow tests. Five additional delineation wells were drilled to confirm the size of the structure depicted by the seismic interpretation and to establish reservoir boundaries. Development drilling of the Intan A platform started in March 1989. Production commenced from 9 wells on the Intan A platform ahead of schedule on June 20, 1989 into a temporary production, storage, and export facilities while permanent facilities were being constructed.

Shortly after the first barrel of oil was produced, a reservoir management team comprised of geophysicist, geologist, petrophysicist and petroleum engineers was formed to reevaluate reservoir parameters and to construct a comprehensive reservoir model. By the end of 1989, all of the reservoir characterization work was complete and construction of a 3-dimensional black oil reservoir simulation model was in progress. This model was completed in February 1990, prior to further development drilling from the Intan B platform. The model was used to predict the future performance of the existing wells on the Intan A platform and predicted performance for different development patterns for the B platform. As a result of being able to look at several development scenarios, the number of wells and the placement of those wells were optimized so as to minimize undrained areas.

AN INTEGRATED APPROACH TO RESERVOIR CHARACTERIZATION IN THE ILLINOIS BASIN

D. F. Oltz, H. E. Leetaru, B. Seyler and S. T. Whitaker

INTRODUCTION

The Illinois State Geological Survey (ISGS) possesses reservoir characterization capabilities that derive from the use of geological and engineering data and techniques including: porosity and permeability measurements, core-flow experimentation, petrography, scanning electron microscopy and energy dispersive x-ray, x-ray diffraction, workstations and computer modeling of geological, geochemical and engineering data. Additionally, a large core and sample library is maintained by the Survey. The ISGS research program to improve and enhance oil recovery in the Illinois Basin utilizes these capabilities and facilities in a multidisciplinary geological and engineering approach to statewide reservoir characterization.

Hydrocarbon production in Illinois has followed a pattern typical of mature basins (fig. 1). Early discoveries were based on surficial geology and resulted in a peak in hydrocarbon production in 1908-1910. The introduction of seismic exploration methods in the late 1930s resulted in significant new discoveries and a production peak in 1940. The application of water-flood technology and hydraulic fracturing during the 1950s led to a lesser production peak in 1956. Since that time, however, production has continued to decline, with minor fluctuations due to variations in the price of oil (fig. 1). The application of improved and enhanced recovery methods could reverse this trend.

The Department of Energy (DOE) has shown in their nationwide analysis that two classes of oil, mobile and immobile, remain in reservoirs after normal production. In Illinois, DOE estimates some 1.5 billion barrels of mobile and 4.5 billion barrels of immobile oil exist in known reservoirs. Illinois ranks ninth in the nation in unrecovered mobile oil. Illinois' total hydrocarbon resource is shown in Figure 2. The present ISGS research program is directed toward determining the distribution and the methods to recover some portion of the mobile oil fraction. Extraction of immobile oil will require application of enhanced oil recovery (EOR) technologies, but much of the new information about the distribution and controls on the mobile oil fraction will be transferred to the immobile.

Our approach has been to define several areas of scientific inquiry, select two target reservoirs, and plan a short- to long-term research agenda. The multidisciplinary tasks necessary to implement such a program are illustrated in the chart (fig. 3). An end result of any oil and gas applied research effort is field testing. The mature Illinois Basin is a likely candidate for testing improved and enhanced recovery research hypotheses since there are 143,000 wells in Illinois with spacing on 10, 20 and 40 acres. Varying amounts of data exist on these wells and yield abundant control for testing models.

This project addresses:

1) improved understanding of controls on reservoir heterogeneities
2) increased technology transfer
3) improving efficient recovery of oil and gas from Illinois reservoirs

One mission of the Oil and Gas Section is to advise the state in promoting wise and efficient development of its resources. Demand for liquid hydrocarbons is projected to increase, therefore attention to increasing production in this and other mature basins is needed. Significant incremental recovery could result through modification of existing (conventional) techniques. Conventional well spacing and conventional operating techniques will not drain most Illinois reservoirs. The interrelationships between well density (spacing), recovery efficiency and ultimate recovery are controlled by the heterogeneity of the reservoirs.

The portion of the basin fill under study is within the Mississippian Valmeyeran and Chesterian Series. The partial stratigraphic column (fig. 4) shows the two subject formations, the Cypress and the Aux Vases. Reservoirs from these formations are among the most productive in the basin, and are the subject of a dozen detailed field studies in progress. The Beech Creek (or Barlow) and the Ste. Genevieve Formations that are used for interval mapping are also shown.

This paper briefly illustrates three facets of the current project: first, a field where there are severe limitations on the available data; second, an improved methodology for local use of old electric logs that make up much of our data base; and third, a field with a relatively large amount of data.

BARTELSO FIELD

Bartelso Field (fig. 5) in southwestern Illinois is one of many fields in the Illinois Basin that contains a limited amount of geologic and engineering data. The field was discovered in 1936 when a structural high indicated by coal borings was tested and oil was recovered from sandstone within the Mississippian Cypress Formation at a depth of slightly over 1000 feet. Three years later oil was also encountered in Silurian carbonates at a depth of about 2400 feet. Subsequent development revealed that the Silurian reservoir was a pinnacle reef, and that differential compaction of sediment around the reef resulted in structural draping of younger strata, including the Cypress (fig. 6). Cumulative production from both Silurian and Cypress reservoirs at Bartelso is approximately 4.5 million barrels of oil. This discussion is limited to the Cypress Formation.

The limitations on available data are due to several factors: 1) although the Cypress was developed on a 10-acre spacing, which normally would provide excellent control, many of the wells were not logged; 2) the entire Cypress section was not drilled in many wells; 3) production figures from Cypress and Silurian reservoirs are combined, making it impossible to determine the exact amount of production attributed to either individual horizon; 4) no pressure tests are available; and 5) only two partial cores from the Cypress are available from the field area. Logs that were run are predominantly spontaneous potential (SP) resistivity logs.

Detailed stratigraphic correlations of wireline logs within the field indicate that the Cypress Formation may be separated into four individual sandstone intervals each separated by thin shale layers which apparently correlate to regional shale beds. The four sandstone intervals are arbitrarily labeled in ascending order, pink, purple, brown and red (fig. 7). The upper three sandstone intervals produce hydrocarbons at Bartelso; the youngest (red) sand is the major reservoir in the field.

The lowest interval (pink) displays a characteristic blocky signature on SP-resistivity logs and comprises a relatively clean sandstone package. This interval is approximately 40 to 50 feet thick in the field area and contains numerous discontinuous shale beds. The lack of penetrations into this sandstone, the absence of cores, and the paucity of wireline logs in particular, have made it impossible to accurately determine the distribution of the sandstone(s) within the pink interval at Bartelso. Regional mapping indicates the sandstones from this interval underlie the entire field area and probably represent deltaic deposits.

The overlying sand interval (purple) is generally 10 to 15 feet thick and is relatively continuous throughout the field (fig. 8). Interpretations from wireline logs indicate that shales are rare in this interval, and that unique compartments are difficult to define due to the lack of adequate control. Relative homogeneity of this sandstone interval is indicated by the fact that there appears to have been one oil-water contact originally at -550 feet.

The distribution pattern of the sandstones within the purple interval suggests an upper shoreface deposit, possibly coalesced beach sands, that have been reworked by waves and/or currents. It is probable that these sandstones were derived from the underlying sandstones of the pink interval.

The interval overlying the purple sand (brown) is dominated by siltstones and shales, although a thin sandstone is present locally in the field. This sandstone displays a more suppressed SP deflection than the other Cypress sandstones, partly due to thin-bed resolution problems on the SP-resistivity logs.

Distribution of sandstone within the brown interval (fig. 9) suggests a small influx of clastics into quiet water. The sandstone apparently is derived from a northwesterly source and may represent a very small delta formed by ebb tidal currents, or by a small distributary. The abundance of shales and silts in this interval suggests the main environment of deposition was either lagoonal or estuarine. The original oil-water contact for this interval was approximately -540 feet and allows for an isolated trap along the western, downdip flank of the field.

The uppermost Cypress sandstone (red) interval at Bartelso comprises well-formed, discontinuous sand lenses that are locally stacked. A relatively thick shale separates this interval from the underlying brown interval and the overlying Beech Creek (Barlow) Limestone.

Sandstone distribution within the red interval is more easily defined than in other intervals due to more data. Detailed stratigraphic correlations based on continuous, but thin, shale layers enable several individual sandstones to be differentiated within this interval (figs. 10A and 10B). The distinct orientation of these sandstones in a northeast-southwest direction, together with their isolated nature, suggests that they are marine bars.

The northern bar complex comprises at least two individual stacked bars that are separated by a thin shale layer. The effectiveness of this thin shale as a seal is exhibited by drillstem tests that indicated gas in the lower red sandstone whereas the upper red sandstone in the same wells tested only oil. The gas-oil relationships can best be explained by having the updip, gas-saturated edge of a lower bar underlying an oil-saturated portion of an upper bar. In fact, the sandstone thicknesses shown on the isopach maps in Figures 10A and 10B suggest that the bar trends could be further subdivided into several smaller shingled sand lenses. Unfortunately, the poor resolution that is characteristic of the logs run in the field, and the lack of cores through the interval, make further subdivision speculative.

Core analyses reveal that clean Cypress sandstones have relatively consistent porosities ranging from 16% to 22%. Permeabilities vary from 50 millidarcies to over 600 millidarcies. These variations are largely due to facies changes within the sandstones.

Examinations of two partial cores that encountered sandstones from the red and purple intervals and of well cuttings from the entire Cypress indicate that silica is the primary Cypress cementing agent. Quartz overgrowths have modified the shape of original grains by increasing their angularity. The lack of clay rims makes it difficult to determine the original grain boundaries.

Feldspar grains, visually estimated at 3% to 6% of the rock, show various stages of dissolution. This dissolution has increased porosity and microporosity to a minor degree.

Calcite cement is rare and is most commonly observed as syntaxial overgrowths on echinoderm fragments. The lack of calcite suggests that it is not a major factor in porosity reduction.

Clays constitute less than 2% of the rock in clean Cypress sandstones and, in order of decreasing abundance, consist of kaolinite, chlorite and illite. Kaolinite appears to be concentrated along bedding planes and near partially dissolved feldspars. Chlorite and illite tend to be disseminated throughout the samples. Under current operating practice, clays do not appear to be detrimental to production.

Results of this study indicate that the Cypress is composed of four separate reservoir types that become progressively more marine upward through the sequence. These reservoir types are, in ascending order, deltaic, upper shoreface, lagoonal, and offshore marine. Each reservoir type should have its own development program to optimize recovery. Secondary recovery methods should likewise be designed for the type of reservoir that is being exploited. Sweep efficiencies will obviously be influenced by the geometries of the individual reservoir. As a result, uniformly spaced wells on a secondary recovery project would not efficiently drain the multiple reservoir types that exist at Bartelso. Separate waterflood projects would be required for each reservoir.

Ten acre spacing is apparently adequate to drain sandstones from the purple interval of the Cypress Formation at Bartelso Field. The western limit of the lobate reservoir within the brown interval at Bartelso has an irregular geometry that may not lend itself to efficient drainage by a simple 10-acre well spacing. Additionally, it is conceivable that compartmentalization may, at a local scale, preclude efficient drainage from the brown interval with this spacing. The bar complex of the red interval at Bartelso shows a tendency for reservoir shingling and was probably not efficiently drained by a 10-acre well pattern. It is likely that off-pattern producers that were drilled due to the waterflood project significantly increased the effectiveness of hydrocarbon recovery within these sandstones.

As illustrated by the Bartelso Field study, wireline logs are commonly the only source of data in many fields in the Illinois Basin. It is therefore necessary to develop methods that can maximize their utilization. One such method was developed at the ISGS for using electric logs from the Aux Vases Formation, which has historically presented problems for log interpretation.

OLD ELECTRIC LOGS

Logging tools for measuring porosity were rarely used in Illinois before 1960, so it is necessary to utilize old electric logs for detailed reservoir analysis. For example, in a typical field such as King Field, Jefferson County, Illinois, less than five neutron or micrologs were run out of the 163 wells drilled. One suite of modern logs was run, but this well did not penetrate the reservoir facies.

Old electric logs combine the SP and the normal and lateral resistivity curves. By 1956, the induction log began replacing the electric log as the primary resistivity measurement tool (Hilchie 1979), although in the Illinois Basin electric logs continued to be run during the early 1960s. Old electric logs can be used effectively to estimate porosity and water saturation from the Upper Valmeyeran (Mississippian) Aux Vases sandstone. A more detailed explanation of these empirical methods using old electric logs can be found in Leetaru (1990).

The distribution of the 70 wells from which both Aux Vases core data and electric log data were collected is shown in Figure 5; these wells are from Franklin, Hamilton, Jefferson, and Wayne Counties in south-central Illinois. The log response of the Aux Vases sandstone within each well was subdivided into zones of similar electrical properties that were calibrated with the core analyses. Most wells in this study had core from only one zone. In total there were 73 zones (or data points) used in this study.

Because the actual value of SP on the log is not an absolute number, SP values cannot be directly compared among different wells. Normalizing SP values against an internal standard can compensate for changes in the scale of the log, the mud resistivity, and the size of the borehole, thus allowing direct comparisons of SP values between different drill holes.

Clean sandstones from the Cypress Formation, which occur about 200 feet above the Aux Vases, were used in this investigation to normalize (standardize) the SP curves. Cypress sandstones are typically light gray to white, fine to medium grained, angular, and friable. Analyses of numerous Cypress cores reveal that porosity values are relatively consistent; therefore, the Cypress is especially suitable as a standard for normalizing SP.

The first step in normalizing the SP measurements is to establish a shale baseline (SP_{min}) through the average SP curve in a thick shale. A clean sand baseline (Sp_{max}) is also established which, for this study, was defined from the Cypress sand with the largest negative millivolt value. The millivolt values for the shale baseline and the clean sand baseline established for each log record were input with the SP for the zone of interest in the normalizing equation (1). The Normalized Spontaneous Potential (NSP) values are unitless and range from 0 to 100.

$$NSP = \frac{|SP_{log}| - |SP_{min}|}{|SP_{max}| - |SP_{min}|} \times 100 \qquad [1]$$

Where

SP_{max} = average maximum SP reading (MV) for a clean Cypress sandstone
SP_{min} = average SP (MV) at the shale baseline
SP_{log} = SP value (MV) for the zone of interest

The relation between NSP and measured core porosity in all 73 zones from the study area has a Pearson correlation coefficient (r) of 0.83 (fig. 11). The data exhibit considerable scatter or deviation from the least-squares regression line. More than a 5% porosity difference can occur between the porosity measured by core analysis and the predicted porosity calculated using the best fit regression line. However, most wells show a significantly smaller amount of error. The calculated standard error of estimate is 2.6% porosity. As shown in Figure 11, the two lines drawn parallel to the regression line at a vertical distance equal to the standard error of estimate define a region that includes two-thirds of the points from a given sample.

A logarithmic relationship is observed between average core porosity and core permeability (fig. 12). This relationship is linearized by using the log of the permeability value. A direct correlation also exists between the NSP and log of the permeability but the relationship is not linear (fig. 13). Kolodzie (1980) observed a similar relationship between permeability and NSP, and estimated permeability by using the NSP. The wide distribution of measured permeability values for the Aux Vases when plotted against NSP suggests this method cannot be used in regional reservoir studies to predict permeability. For example, in Figure 13 for an NSP value in the mid-60s, measured permeability ranges from <10 md to >100 md. Aux Vases sandstone reservoirs with permeability <10 md are not commercial wells, whereas those at 100 md can be prolific producers.

This study suggests that the method of contrasting NSP to permeability is generally a poor predictive tool when applied over broad regions, however, for local areas such as portions of Jefferson County, the NSP-permeability method may be useful (fig. 14). The NSP cross plots may work here because the Aux Vases was formed in a similar diagenetic and depositional environment. Subsequent work may document a regionally distributed relationship between NSP and permeability which would allow semiquantitative predictions of some reservoir characteristics.

Once porosities are calculated, Pickett Plot analyses can be used to determine the water saturation of Aux Vases sandstones. This method was applied in an analysis of King Field, which has produced more than 4 million barrels of oil. All of the logged wells that penetrated the Aux Vases interval were plotted on the Pickett Plot (fig. 15). The porosity was calculated using the NSP method; resistivity was measured from the long normal.

When cementation exponent (m) is constant, porosity and resistivity from those wells that either tested water or were drilled below the oil-water contact should ideally plot along a straight line on log-log graph paper (Lang 1973). At King Field, the oil-water contact is not well defined, and some of the wells that have been interpreted as wet may contain oil. All of the wells drilled into the postulated water zone plot below s_w >50 percent. The values obtained in this Pickett Plot analysis are more scattered than an analysis using modern logs, probably due to inherent error in using estimated porosity from the SP.

In summary, this study suggests that comparison of NSP porosity-permeability relationships in combination with Pickett plot analysis, offers a valuable method for estimating key reservoir properties in areas of limited data. In other areas, such as at Zeigler Field, data on actual reservoir properties are abundant.

ZEIGLER FIELD

Zeigler Field (fig. 16) exhibits many of the features that have made the evaluation of Aux Vases reservoirs problematic. Among these phenomena is the very low resistivity commonly found in highly oil productive zones. This has led to erroneous water and oil saturation calculations resulting in questionable reserve estimates. Detailed reservoir characterization integrating geologic and engineering parameters was used in this study to address some of these problems.

Zeigler Field was discovered in 1963 when oil-bearing sandstones within the Aux Vases Formation were encountered at a depth of approximately 2250 feet. The field lies approximately 3½ miles north of the Cottage Grove Fault System. By 1990, 1.6 million barrels of oil had been produced from the field with individual wells producing an average of 60 thousand barrels of oil. Peripheral wells that were marginal producers were converted to injection wells and the field was immediately placed on waterflood. As waterflooding progressed, permeability barriers were found in several locations in the main field, leading the operator, V. R. Gallagher, to deduce that the main body of the field comprised four to five poorly connected lenses.

The lack of structural closure and the geometry of the reservoir suggests that the trap at Zeigler Field is stratigraphic. Figure 16 shows both a structure map on top of the Beech Creek "Barlow" Limestone, a significant marker bed located above the Aux Vases Formation, and a structure map on top of the Ste. Genevieve Limestone, directly underlying the Aux Vases. The Aux Vases sandstone isopach map (fig. 17) illustrates that the clean sandstone coincides with limits of production in the field.

The depositional environment of the Aux Vases sandstone reservoir at Zeigler Field is interpreted to be a tidal sandbar deposited on very fine-grained, green, ripple-bedded, shallow marine sandstone which, in turn, was deposited on oolitic Ste. Genevieve limestone.

The reservoir tidal sandbar facies is overlain and grades laterally into non-reservoir tidal flat, sand flat and interbar deposits in a shallow marine, mixed siliclastic-carbonate environment. These interpretations are based on observations in cores, cross-sections, isopachs and outcrop study. The tidal sandbar facies is composed of fine to medium grained crossbedded sandstone with bioturbated zones. Permeability ranges from 50 to 750 millidarcies. The recent development of the Mack lease in the northeast quadrant of 7S-2E-Section 19 provided an opportunity for an in-depth study of the geologic and engineering characteristics of one of the isolated sand lenses. The remainder of this paper will focus on work completed using data from the Mack lease.

The structural fence diagram (fig. 18) illustrates the geologic aspects of the reservoir, including the convex upward geometry of the tidal sandbar facies in the northeast quadrant of T7S-R2E sec.19. The orientation of the fence diagram is shown on the isopach map (fig. 17). The geologic cross sections (fig. 19) include a north-south structure cross section (B-B') controlled by the V. R. Gallagher Mack #1, #3 and #6 wells, and an east-west structure cross section (A-A') controlled by the Mack #1, #2 and #5. The oil-saturated and oil-stained zone is associated with the sandbar facies. The zone separating the sandbar facies from the underlying Ste. Genevieve Limestone has less than 0.1 millidarcies of permeability and porosity ranging from 1% to 18%. It is an offshore marine, very fine grained, ripple-bedded sandstone, with green chlorite shale laminae. The zone above and lateral to the sandbar facies may act as a seal and consists of tidal flat, sand flat or interbar deposits containing poor reservoir qualities.

The V. R. Gallagher Mack #1 well contains the largest variation in reservoir qualities and facies within the Mack lease. Figure 20 is a composite representation of the Mack #1 well showing the relationship of flow units to permeability and facies variations. The producing zone in this well is confined to the sandbar facies and contains three flow units. Flow unit 1 contains the highest permeability in the reservoir, from 517 to 750 millidarcies. While porosity is not as variable as permeability, the highest porosities (up to 28%) are also found in flow unit 1. Flow units 2U and 2L are also in the oil-productive portion of the sandbar facies, but are less permeable than flow unit 1. Permeability in 2U ranges from 131 md to 385 md. Permeability in flow unit 2L ranges from 62 md to 178 md.

Flow unit 3L corresponds with the offshore marine, ripple-bedded, very fine-grained sandstone that separates the sandbar facies from the underlying Ste. Genevieve Limestone (fig. 20). This unit possesses the lowest permeability, usually 0.1 md or less.

Flow unit 3U, located at the top of the reservoir above the sandbar, corresponds with sandflat and tidal-flat facies. This flow unit also possesses low permeability, ranging from 24 md to 15 md and is not oil productive.

Cluster analysis was used to determine flow units with similar properties. There is a high degree of agreement among data points in flow units 1 and 3L and less similarity between flow units 2U and 2L.

The flow unit structure cross-sections in Figure 21 show vertical and lateral relationships of flow units in the Mack lease. The north-south flow unit cross-section (B-B') includes Mack 6, 1, and 3 and is comparable to the geologic cross-section in Figure 19. The east-west flow unit cross-section (A-A') of the Mack 5, 2, 1 wells is comparable to the geologic cross section in Figure 19. The third flow unit cross-section (C-C') of the S. Plumfield 1, Mack 2 and Mack 3 shows good agreement of flow units 2U and 2L with the oil saturated portion of the sandbar facies. Flow unit 3L is below the oil producing sandbar zone and forms a permeability barrier. Unit 3U overlies the sandbar flow units, has poor permeability and porosity, and forms a permeability barrier above the oil productive zone. Flow units 1, 2U, and 2L correspond with the oil producing portion of the sandbar facies. Flow unit 1, containing the best reservoir qualities, is confined to the Mack #1. Flow units 2U and 2L have a greater areal distribution and comprise the bulk of the oil productive zone.

There is good agreement of flow unit boundaries with facies; however, agreement is incomplete. An example of good, but incomplete, agreement between geologic facies and flow unit boundaries is observed when comparing the W-E (A-A') geologic cross section in Figure 19 with the flow unit cross section of the same wells in Figure 21. The geologic cross section shows a three-foot thick section of the sandbar facies in the V. R. Gallagher Mack #5 well while the flow unit cross-section of the same well shows that flow units 2U, 2L and 1 associated with the sandbar facies are not present in the Mack #5 well. This apparent discrepancy occurs because the sandstone in Mack #5 contains sedimentary structures, such as cross-bedding, that categorize it as a tidal-sandbar. However, it lacks the high permeability associated with flow units 1, 2U or 2L.

Permeability variations in flow units within the sandbar facies are attributed in part to diagenetic events. Early to late diagenetic events associated with shallow and intermediate burial depths respectively are largely responsible for the development of porosity and permeability within the oil saturated zone of the sandbar facies. The paragenetic sequence for the sandbar facies is shown in Figure 22.

Early shallow burial processes include pervasive calcite cement infilling most pores, minor silica cementation primarily at grain to grain contacts, and formation of some precursor clay coatings. Later diagenetic processes include dissolution of early calcite cement, dissolution of large amounts of feldspar, and transformation and formation of clay rims. Development of good porosity and permeability is due to these later-stage events which occurred at intermediate (<5000 ft) burial depths. Dissolution of calcite cement and feldspar may have been enhanced by fluids associated with petroleum migration.

X-ray diffraction analyses show that authigenic illite, chlorite and mixed-layered illite-smectite are present in amounts ranging from 2% to 10% in the oil-productive portion of the reservoir. While clays usually comprise less than 10% bulk volume, SEM/EDX analyses show that the clays coat most grains. These clay coatings are significant because water bound in the clays is responsible for low resistivities observed on logs and the resulting erroneously high water saturation calculations. Resistivity logs for the Mack #1 illustrate this phenomenon (fig. 23). The resistivity of the oil producing zone is approximately 2 ohm/meters, indicating a high water saturation and low oil saturation. This well has been producing approximately 100 barrels of oil per day since completion in January, 1987 with a negligible amount of water. Studies of samples from this zone by SEM/EDX show that most sand grains are coated with authigenic chlorite and/or illite.

The best reservoir qualities in Zeigler Field are found in the sandbar facies. Diagenesis and sedimentary processes are responsible for variations found within the sandbar facies. There are significant vertical and lateral heterogeneities, manifested as distinctive flow units within the oil productive portion of the sandbar facies, which can result in compartmentalization of Aux Vases sandstone reservoirs.

RESULTS

As the study progresses, we anticipate being able to accomplish the following:

1. Better assess the state's remaining oil resource, validate the number, and use it in planning documents, including the state's energy policy.

2. Document characteristics of brines, oils, and clays for use in geochemical and reservoir models.

3. Enhance the State's data base by increasing its size, accessibility, and user friendliness.

4. Expand the utility of old E-logs.

5. Test some of the drilling and completion fluids commonly used in the basin in core-flow experiments. Since a single formation varies considerably across the basin, variations in response to certain fluids are also expected.

6. Characterize reservoirs in sufficient geologic and engineering detail to be able to recommend appropriate IOR and EOR technologies. Map facies variations; construct a classification system to encompass reservoirs with like properties; provide a manual of reservoir-specific drilling, completion and stimulation techniques to accommodate heterogeneity across the basin.

In the Illinois Basin, most of the exploration, development and production is accomplished by independents. These companies do not maintain research staffs and could benefit from having access to new data, recent technological advances and new ideas about the habitat of oil.

REFERENCES

Hilchie, D. W., 1979, Old electric log interpretation: Institute for Energy Development, Tulsa, Oklahoma, 161 p.

Kolodzie, S., 1980, Analysis of pore throat size and use of the Waxman-Smits equation to determine OOIP in Spindle Field, Colorado: Presented at Society of Petroleum Engineers meeting, Dallas, Texas, September 1980, SPE paper 9382.

Lang W. H., 1973, Porosity-resistivity cross-plotting: The Log Analyst, January-February, v. 14, n. 1., p. 16-20.

Leetaru, H. E., 1990, Application of old electric logs in analysis of Aux Vases Sandstone (Mississippian) Reservoirs in Illinois, Illinois Petroleum 134, Illinois State Geological Survey, Champaign, Illinois.

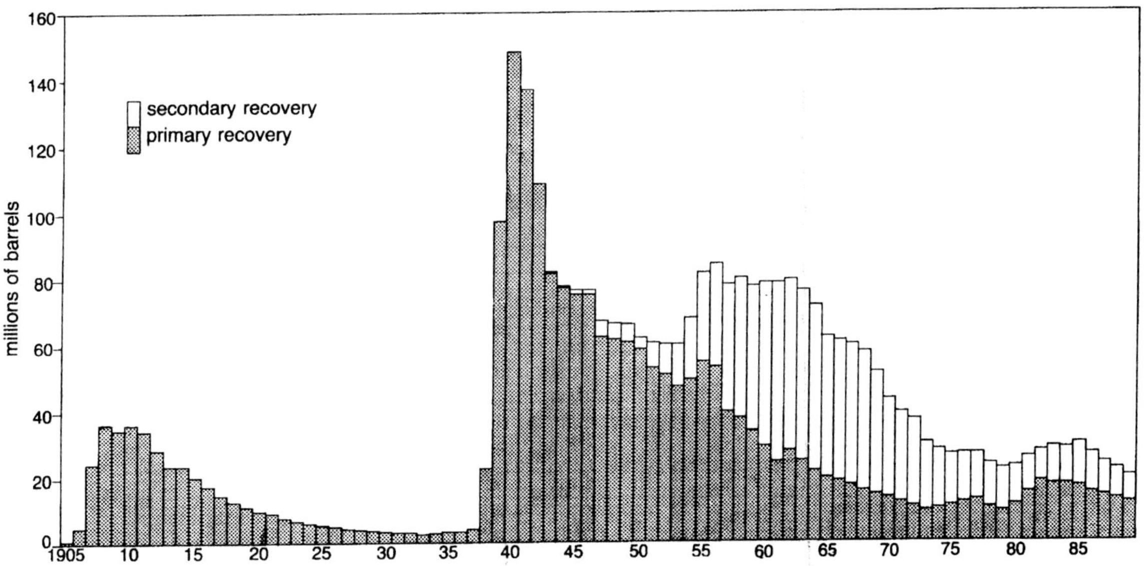

Figure 1 Yearly oil production in Illinois.

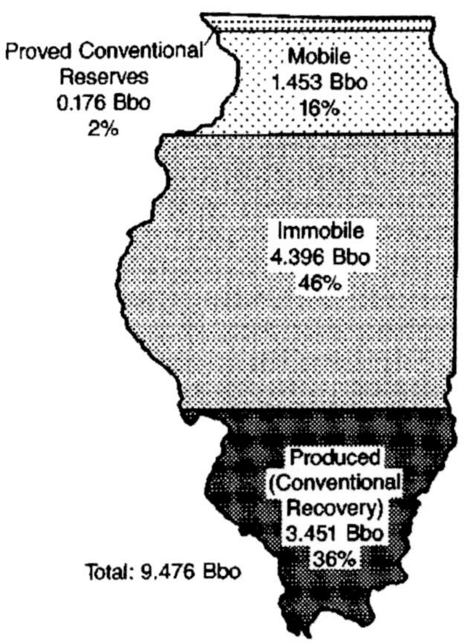

Figure 2 Illinois oil resources.

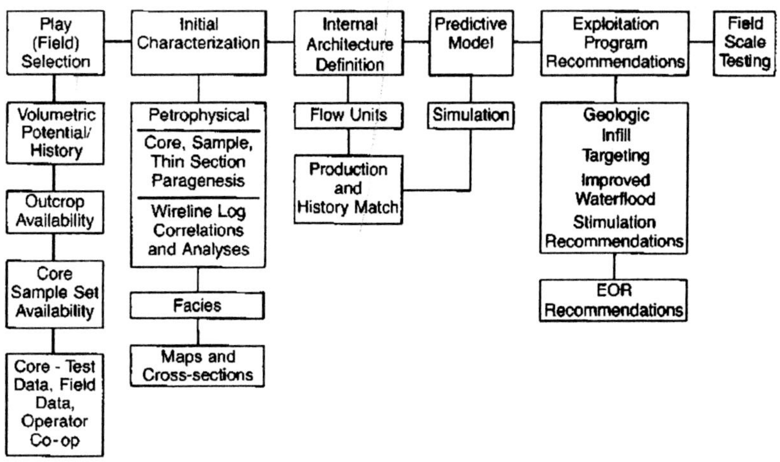

Figure 3 Illinois State Geological Survey integrated reservoir analysis.

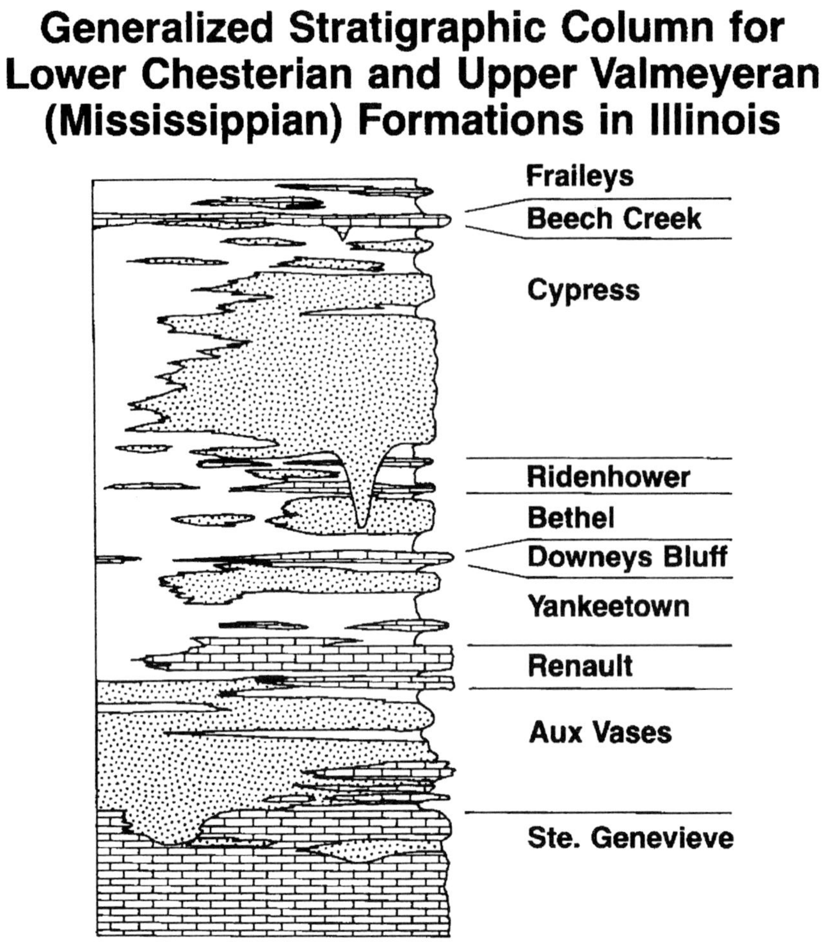

Figure 4 Abbreviated stratigraphic column of formations studied.

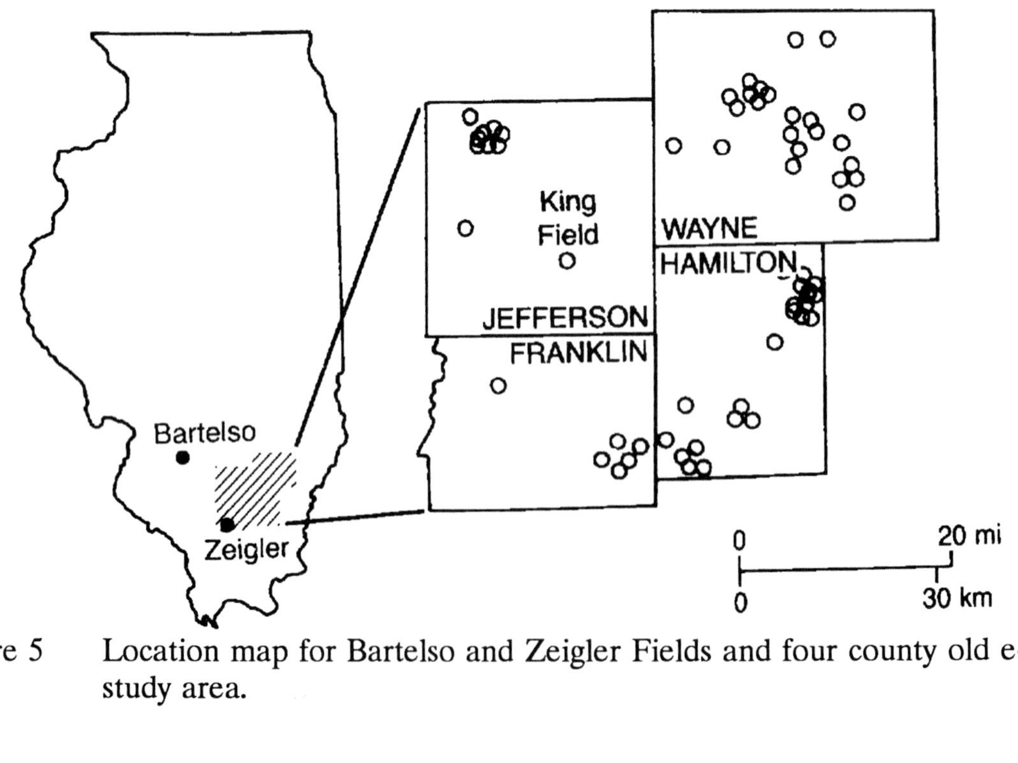

Figure 5 Location map for Bartelso and Zeigler Fields and four county old e-log study area.

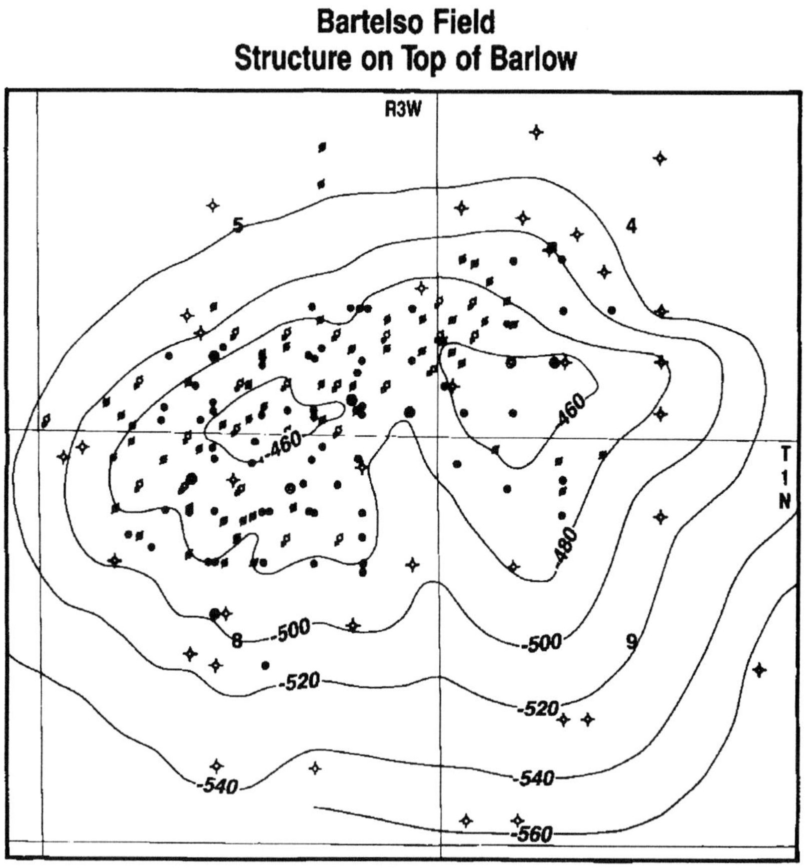

Figure 6 Barlow structure at Bartelso Field.

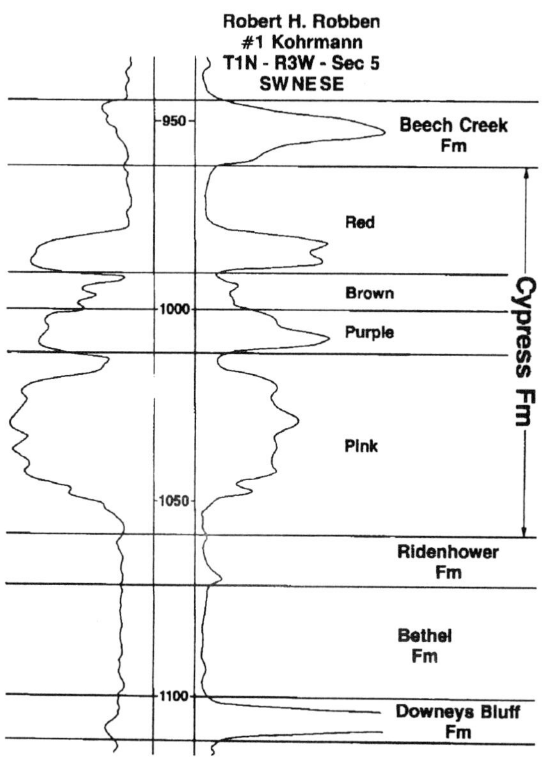

Figure 7 Type log for Bartelso Field showing the four intervals within the Cypress Formation.

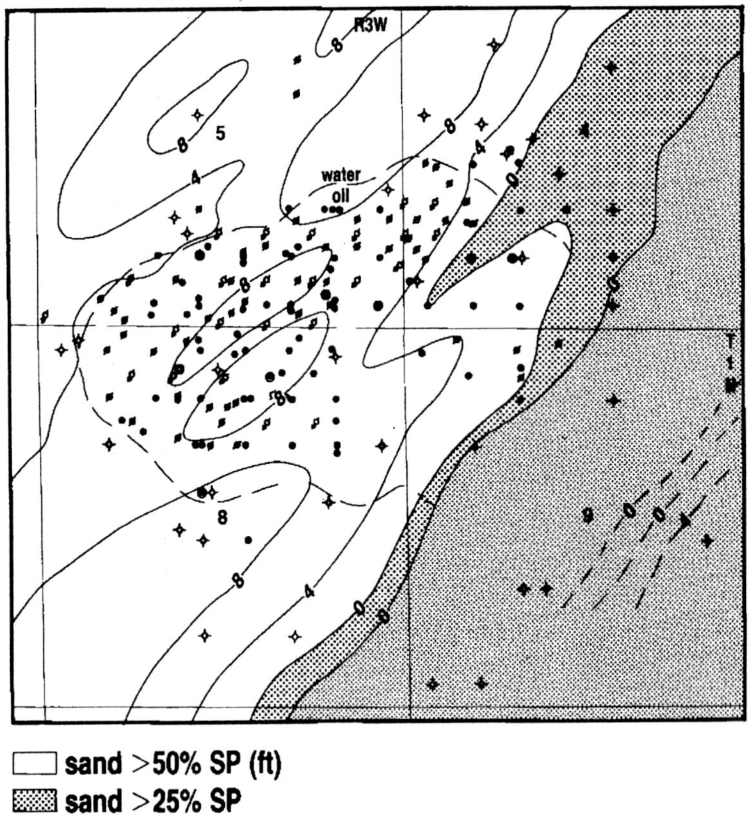

Figure 8 Net sand isopach within Cypress purple intervals at Bartelso Field.

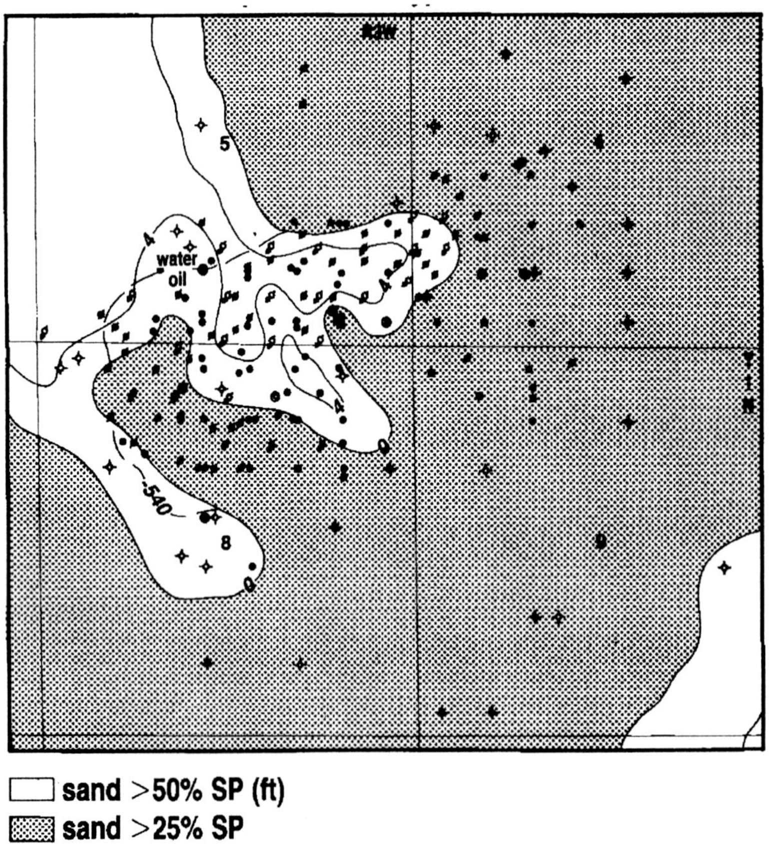

Figure 9 Net sand isopach within Cypress brown interval at Bartelso Field .

A

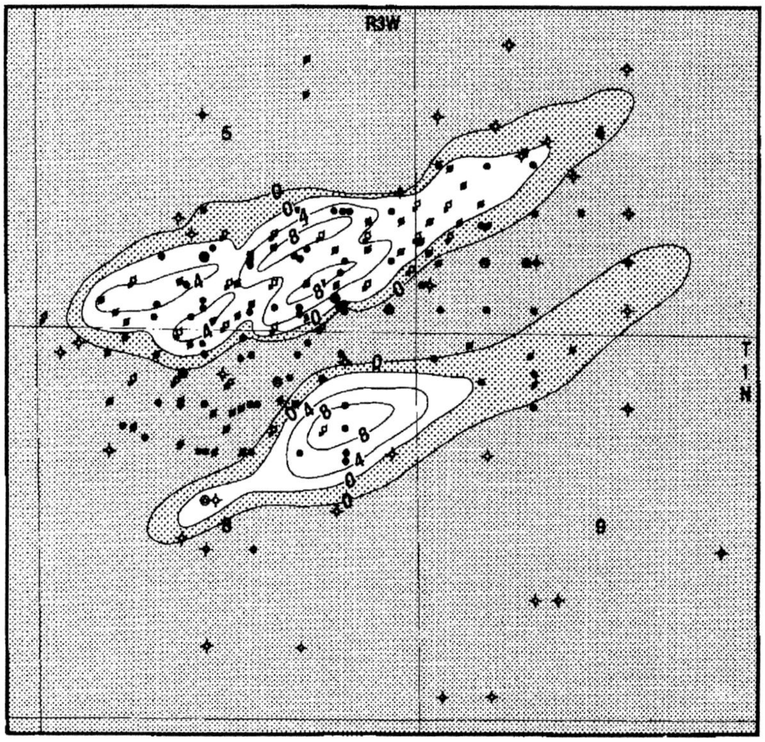

sand >50% SP (ft)
sand >25% SP

B

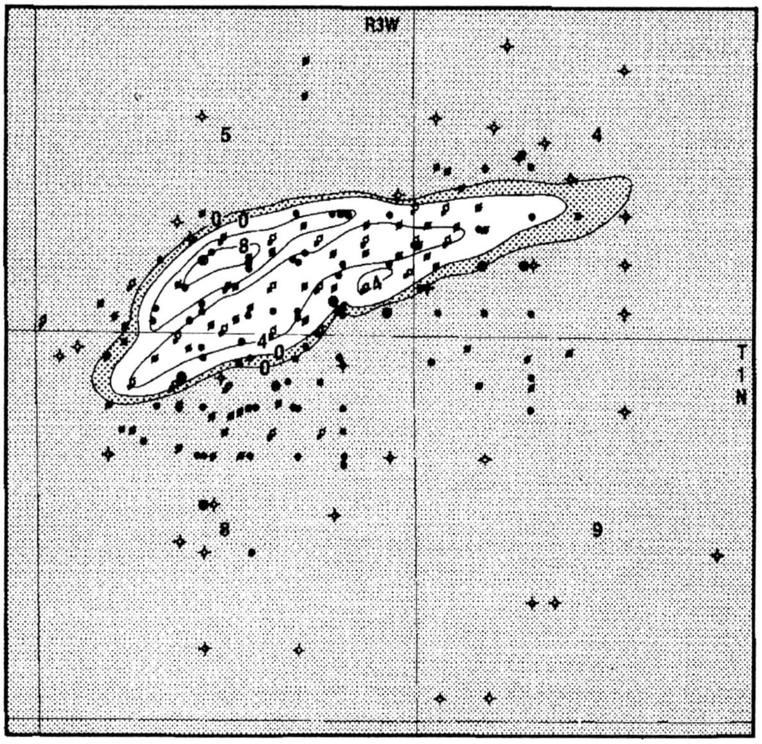

sand >50% SP (ft)
sand >25% SP

Figure 10A, 10B (10A) Cypress lower red sand isopach; (10B) Cypress upper red sand isopach at Bartelso Field.

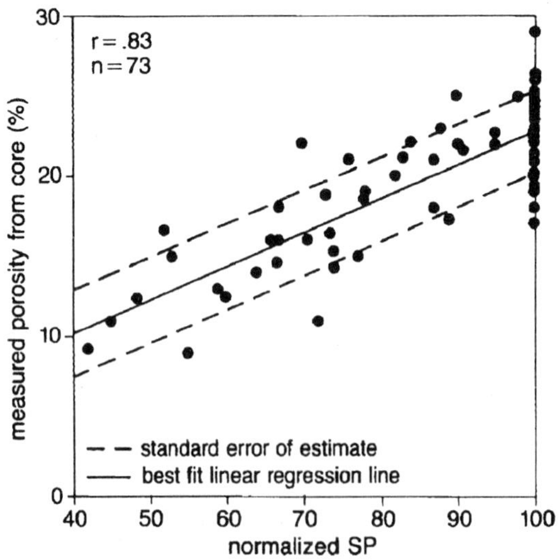

Figure 11 Average measured core porosity relative to NSP for all counties in study area. Two-thirds of data values will fall within the bounds of the standard error of estimate (from Leetaru, 1990).

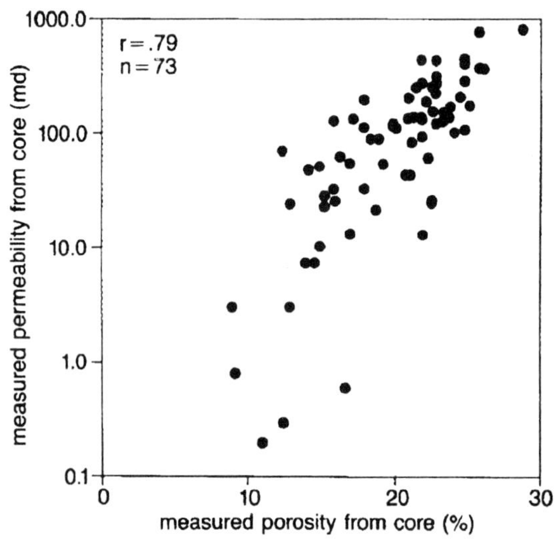

Figure 12 Average measured core permeability relative to average measured core porosity (from Leetaru, 1990).

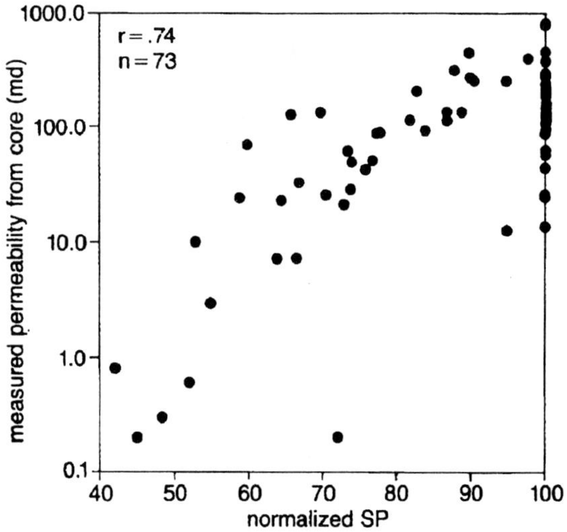

Figure 13 Average measured core permeability relative to NSP for all four counties (from Leetaru, 1990).

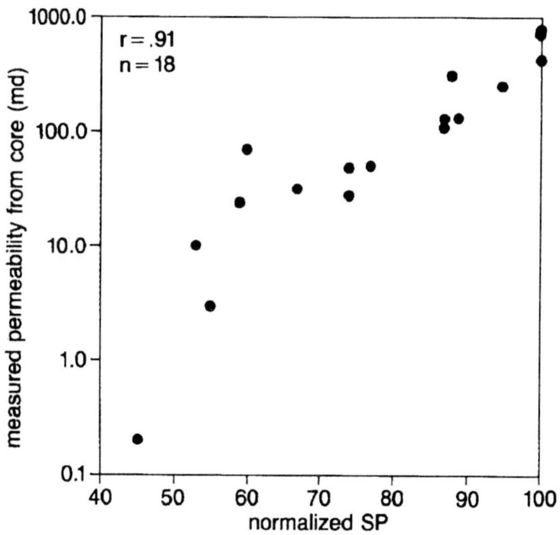

Figure 14 Average measured core permeability relative to NSP for Jefferson County (from Leetaru, 1990).

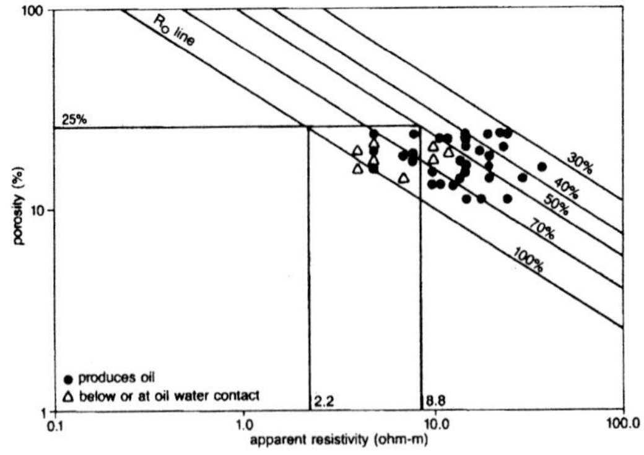

Figure 15 Pickett Plot of estimated porosity relative to apparent R+ from the short normal for King Field, Jefferson County. The slope of the R_0 line is equal to a cementation exponent of approximately 1.7 (from Leetaru, 1990).

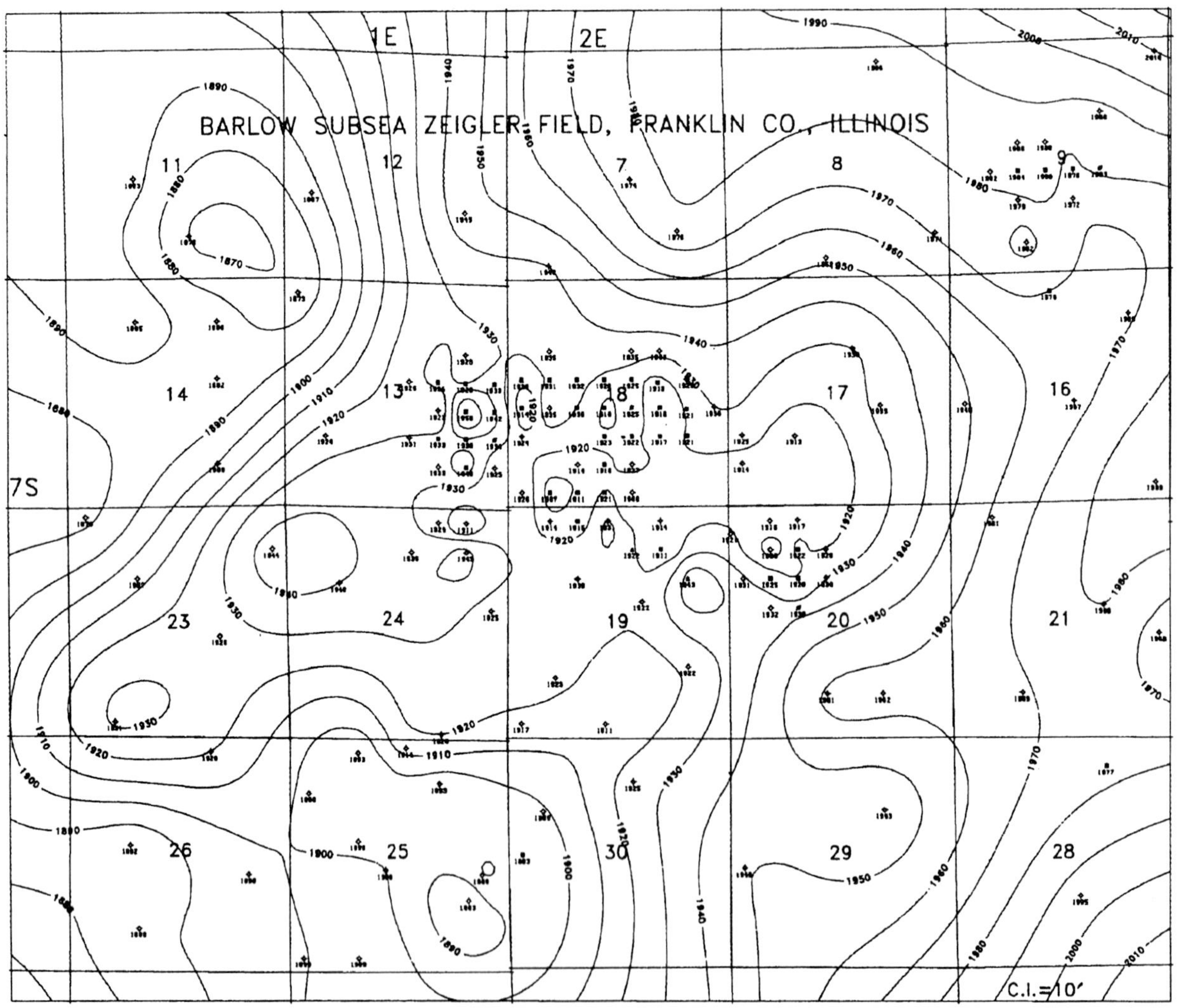

Figure 16 Structure on Barlow (Beech Creek) Limestone at Zeigler Field in Franklin Co., Illinois. Contour interval is 10 feet.

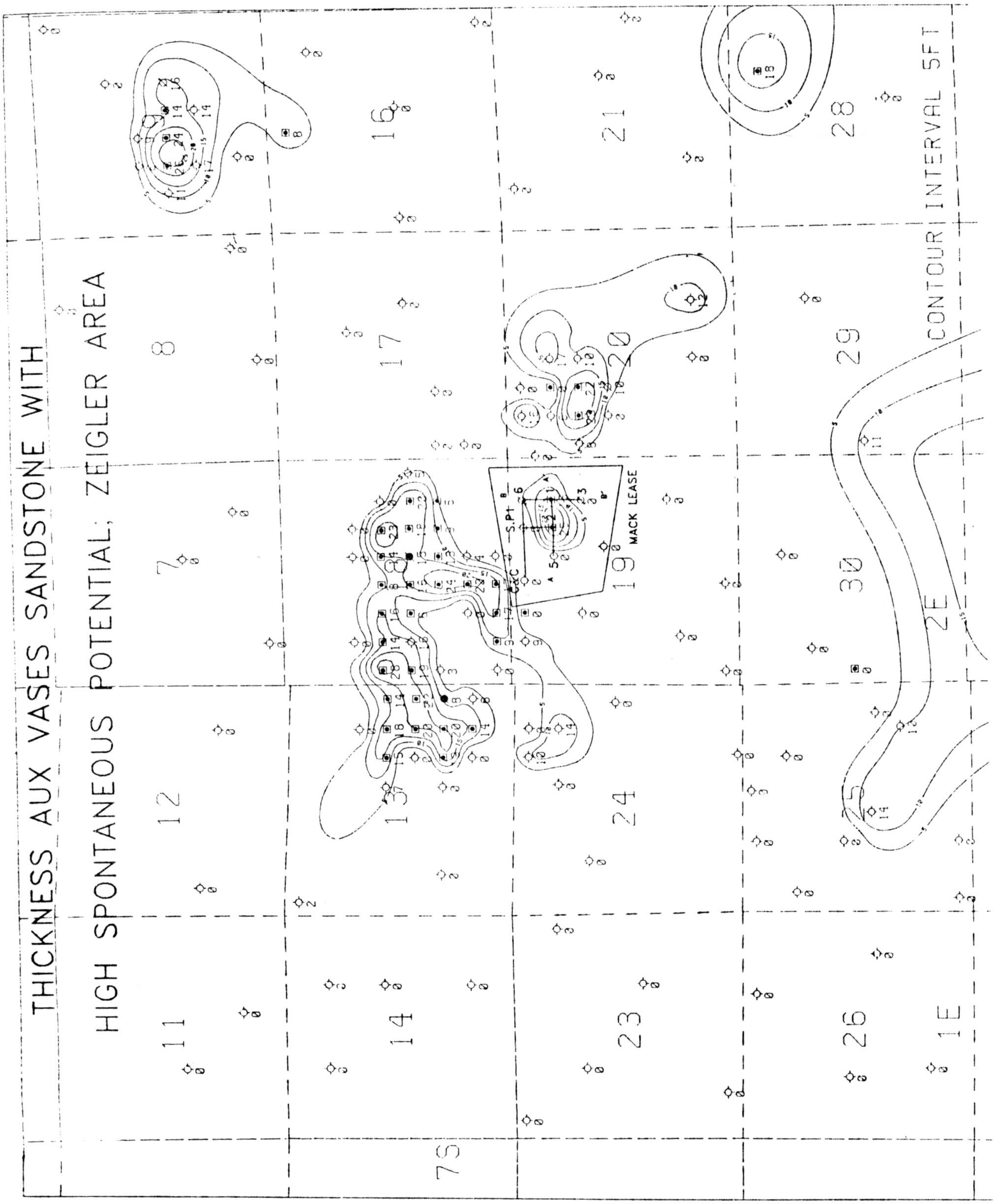

Figure 17 Isopach thickness map of Aux Vases sandstone with high spontaneous potential deflection. Note good correlation of producing wells with clean sandstone. Map shows location of wells in cross sections and fence diagram on Mack lease in northeast quadrant T7S-R2E-19.

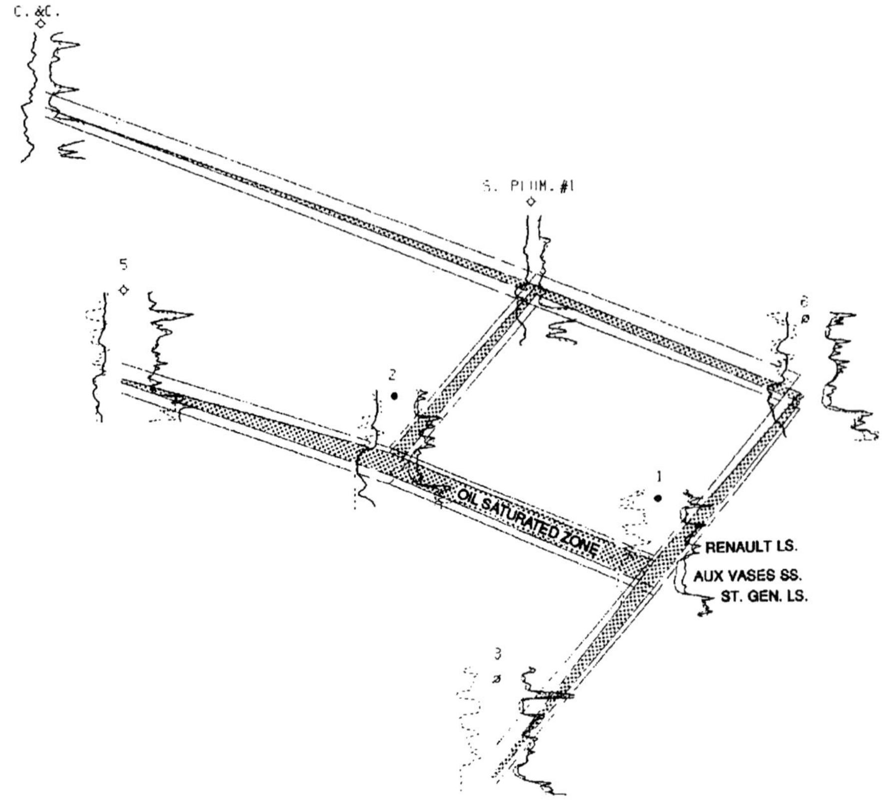

Figure 18 Structure fence diagram of wells in the Mack lease. These wells correspond with those shown in the Mack lease in the previous figure. The oil saturated portion of the convex upward sandbar facies is highlighted.

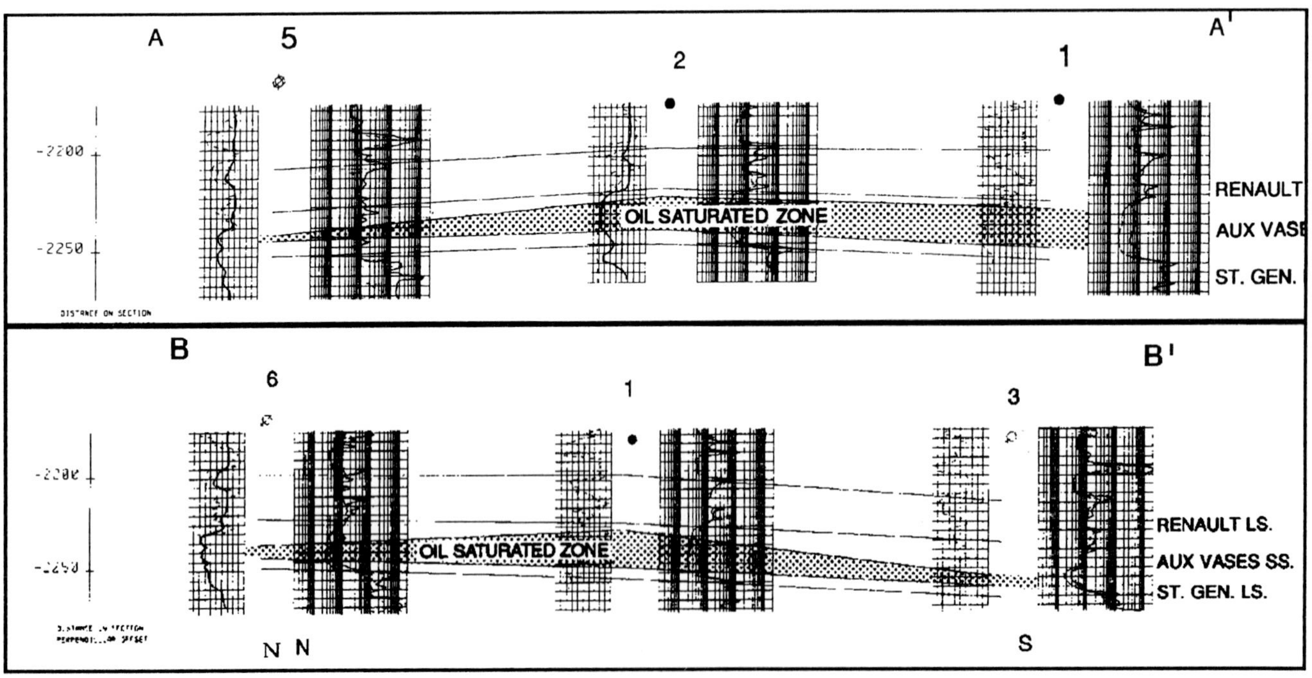

Figure 19 West-East (A-A') structure cross section and north-south (B-B') structure cross section of Mack lease well locations shown in Figure 17. The upward convex geometry of the sandbar facies is shown.

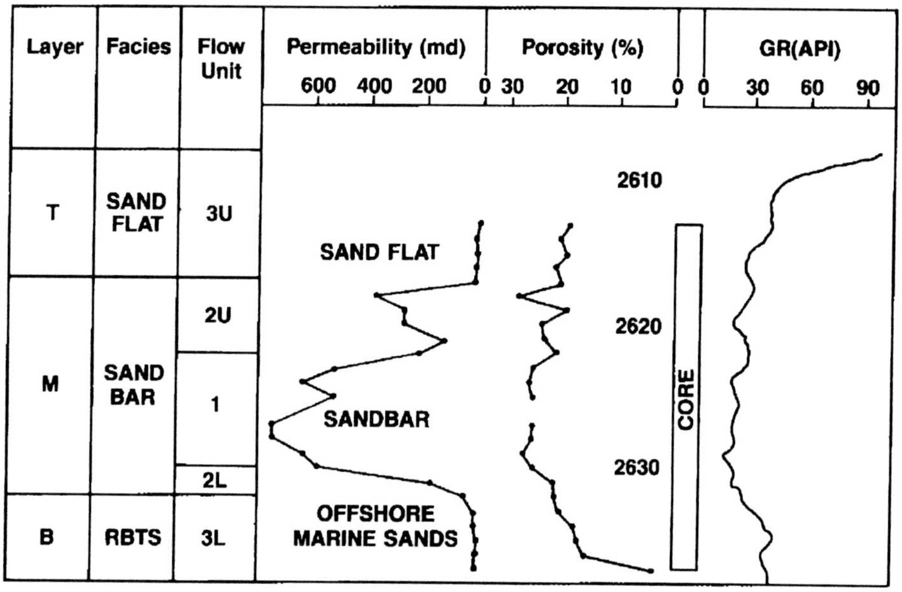

Figure 20 Composite representation of flow units and their relationship with facies variations Mack #1. Flow units are based on permeability variations.

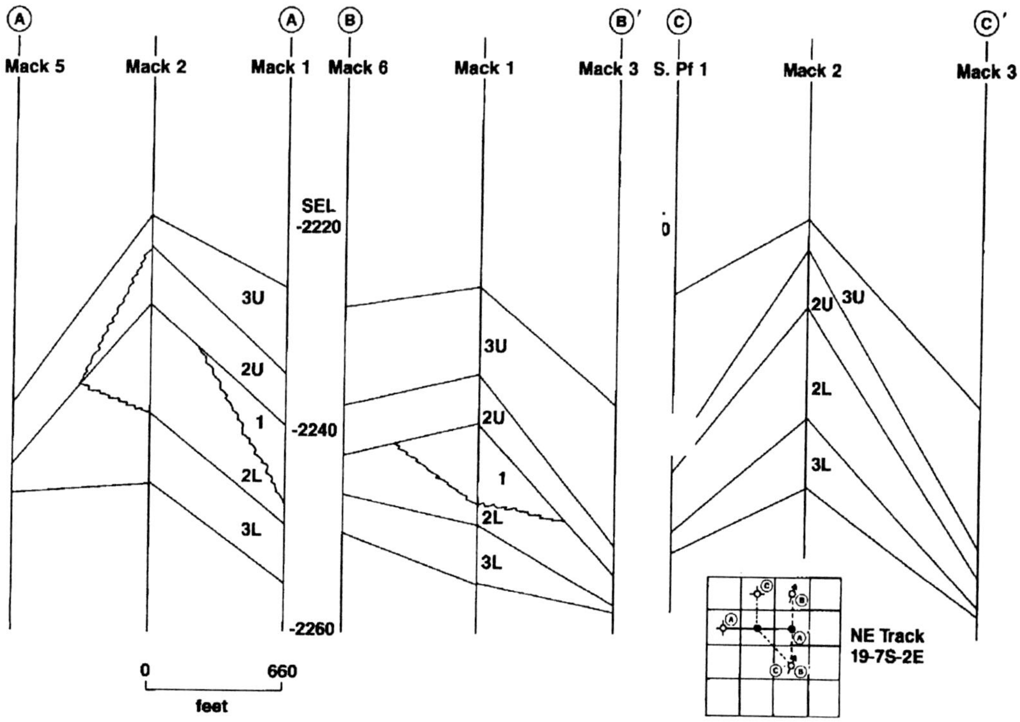

Figure 21 Flow unit cross section. Flow unit cross sections of the Mack lease can be compared to geologic cross sections (fig. 19) of the same wells in cross sections A-A' and B-B'. Illustrates the potential for compartmentalization due to lateral and vertical heterogeneities in these wells.

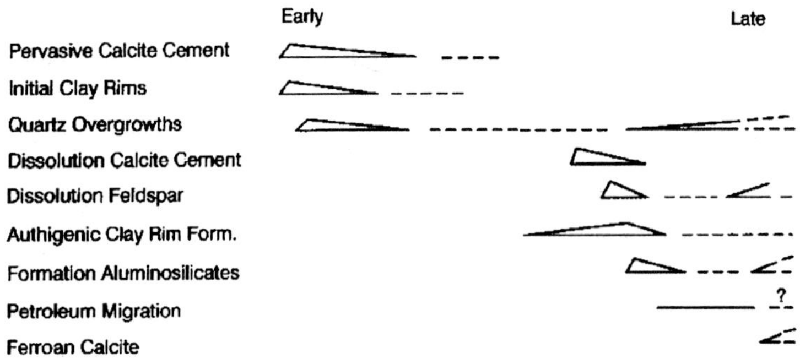

Figure 22 Paragenetic sequence showing early to late stage diagenetic events that affected the development of permeability and porosity in the reservoir sandbar facies at Zeigler Field.

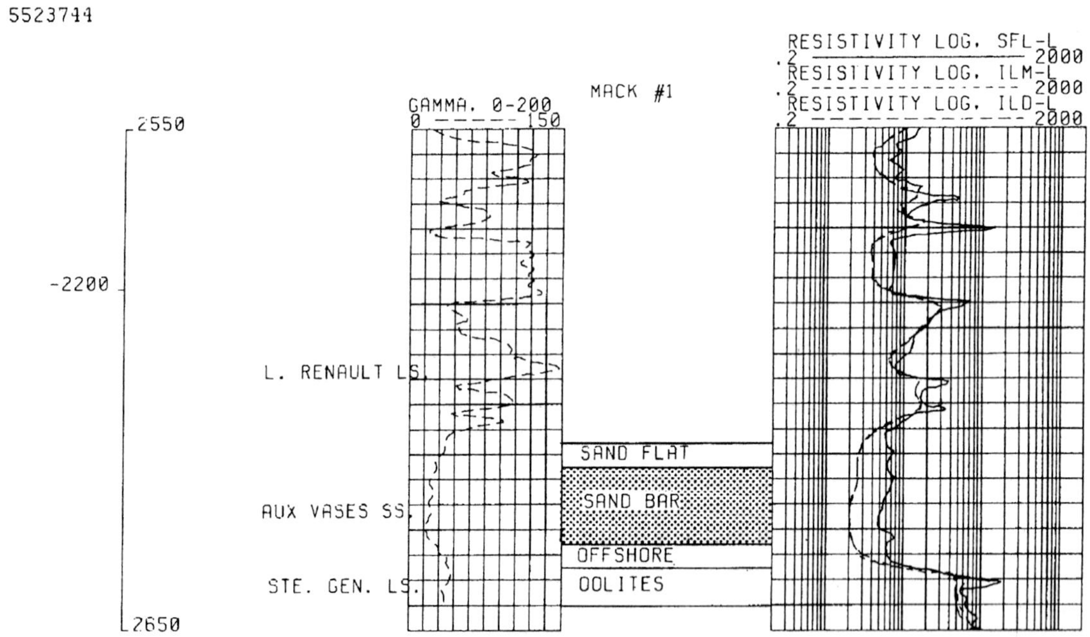

Figure 23 Resistivity logs, Mack #1 well demonstrating a typically low resistivity (~2 ohm/meters) of a highly oil productive zone. This response is due to water bound in authigenic clays coating and cementing sand grains in the reservoir.

Interdisciplinary Effort Optimises Field Developments in the U.K. North Sea Brae Area

T M ALQASSAR* - SEG - LL&E, London
W GALLACHER* - SPE - LL&E, London
S DE MOSS+ - SPE - LL&E, New Orleans

Prepared for the Archie Research Conference, Houston, TEXAS October 22-25 1990

ABSTRACT

The Brae fields, operated by Marathon Oil (UK) Limited, are situated in Block 16/7 in the UK, North Sea. The first of the Brae fields was discovered in 1975 with the testing of a rich gas condensate accumulation, later named North Brae. Further drilling led to the discovery of three oil fields, West, Central and South Brae, and appraisal drilling concentrated on South Brae.

After four wells had been drilled to appraise South Brae a fan delta sedimentary model was developed to explain the geology. The South Brae plan of development was based on this model. Early development drilling in 1983/4 came up with major surprises and indicated that the reservoir system was more complex than originally assumed. As the planned production profile appeared difficult to achieve a combined study involving geophysical, geological and reservoir engineering was instigated to understand the geological model and achieve optimal well locations. The available 2D seismic although lacking in quality and resolution was an essential part of this study. After a new geological model had been developed further development wells were more successful. This enabled the field to achieve predicted production rates and enabled an effective waterflood to be put in place to improve recovery.

By 1982 North Brae had been appraised and the development plan for this reservoir was assessed using a compositional reservoir simulator. Gas cycling with partial pressure maintenance was selected as the best development option. Prior to the platform being installed in 1987 a 3D seismic programme was completed. With experience gained from South Brae and integrating geological and geophysical interpretations a refined geological model was rapidly and efficiently developed. Subsequent development drilling was highly successful and the field produced at plateau rates earlier than expected. Reservoir simulation incorporating the complex geology confirmed that gas cycling was indeed the best option and that excellent recoveries could be expected.

The multi-disciplinary study on South Brae successfully achieved its objectives and using the experience gained from this study enabled the development of North Brae to be rapidly optimised with excellent results.

* LL&E (UK), Inc., 40A Dover Street, LONDON, W1X 3RB, England
+ LL&E, 909 Poydras Street, NEW ORLEANS, LA 70160, USA

INTRODUCTION

Block 16/7 is situated in the UK North Sea approximately 160 miles from the North East coast of Scotland, (Figure 1). A consortium of companies acquired this block in the 3rd round of licensing in 1970.

The first exploratory well 16/7–1 was drilled in 1974 and encountered a significant hydrocarbon column in the Upper Jurassic. The well tested gas condensate in 1975 at an aggregate liquid rate of 21,500 barrels of condensate per day. This discovery was later named North Brae. The next two wells both encountered black oil, the 16/7–2 well discovery tested oil in the younger Paleocene formation and this structure was later named West Brae. The 16/7–3 discovered oil in sandstones and conglomerates and this accumulation was later named Central Brae. By 1976 additional seismic had been acquired and parts of the blocks were relinquished under the terms of the licence with the remaining part becoming 16/7a. Well 16/7a–8, drilled in the South of the block, encountered oil in the Upper Jurassic. This well was the South Brae discovery well and tested at an aggregate rate of 33,000 bopd. Appraisal drilling concentrated on South Brae and a further four wells were drilled to delineate the field. It was not until 1980 that the second well 16/7a–14 was drilled on North Brae. All these wells, except 16/7–2, lie at the Western Margin of the Viking Graben and encountered hydrocarbons within the Upper Jurassic coarse clastics which are sourced and sealed by the Kimmeridge Clay. The exploration, appraisal wells and hydrocarbon accumulations are shown in figure 2.

Marathon Oil (UK) Limited is the operator of the Brae Field Complex and has a working interest of 38%. The other partners of the Brae Group include Britoil Plc (20%), Bow Valley Petroleum Limited (14%), Kerr McGee Oil (UK) Plc (8%), Gas Council (Exploration) Limited (7.7%), LL&E (UK) Inc (6.3%), Sovereign Oil and Gas Plc (4%) and Norsk Hydro Oil and Gas Limited (2%).

The interpretation and opinions of both the operator and the joint venture partners are expressed at technical level at frequent partner meetings. These forums enable all the partners to express their views and the application of a multi–disciplinary approach from all the participants has led to the efficient solving of problems and to the successful development of the Brae area.

SOUTH BRAE

The field was discovered in 1977 with the drilling and testing of well 16/7a–8. This well encountered 880 feet of total net pay and tested at an aggregate rate of 33,000 barrels of oil per day. A well schematic is shown in figure 3 which indicates the sand/conglomerate nature of the Brae formation. A further four appraisal wells, 16/7a–10,11,12 and 13 were drilled between 1977 and 1978. The position of these wells are shown in figure 4. All these wells tested oil from the Upper Jurassic sands and conglomerates at depths of 12–13,500 feet subsea. The API gravity of the oil is 37 degrees.

With data from these wells the sedimentary model adopted to explain the reservoir distribution was a partly emergent fan delta. This model envisaged coarse clastic sediments at the entry point close to the major fault zone to the West. These

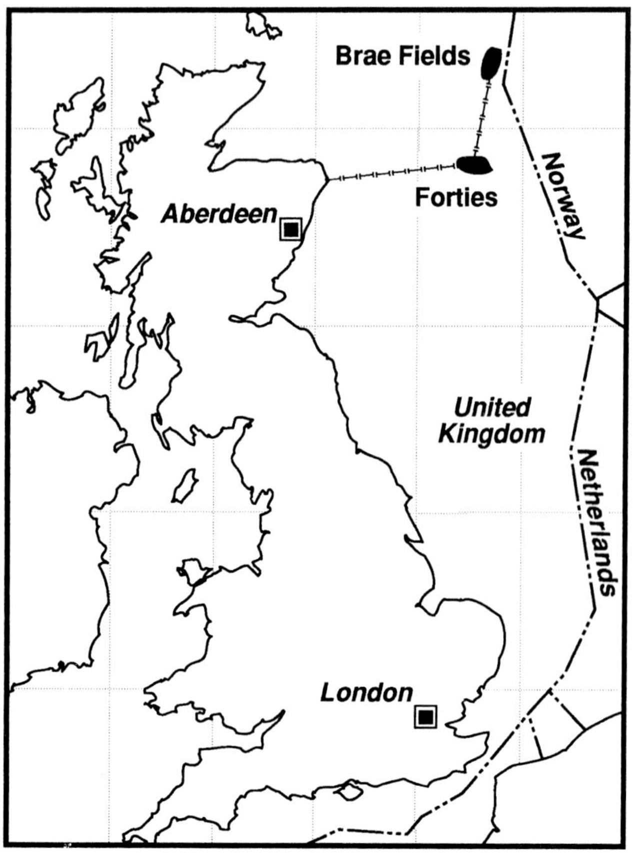

Figure 1: Brae fields location map

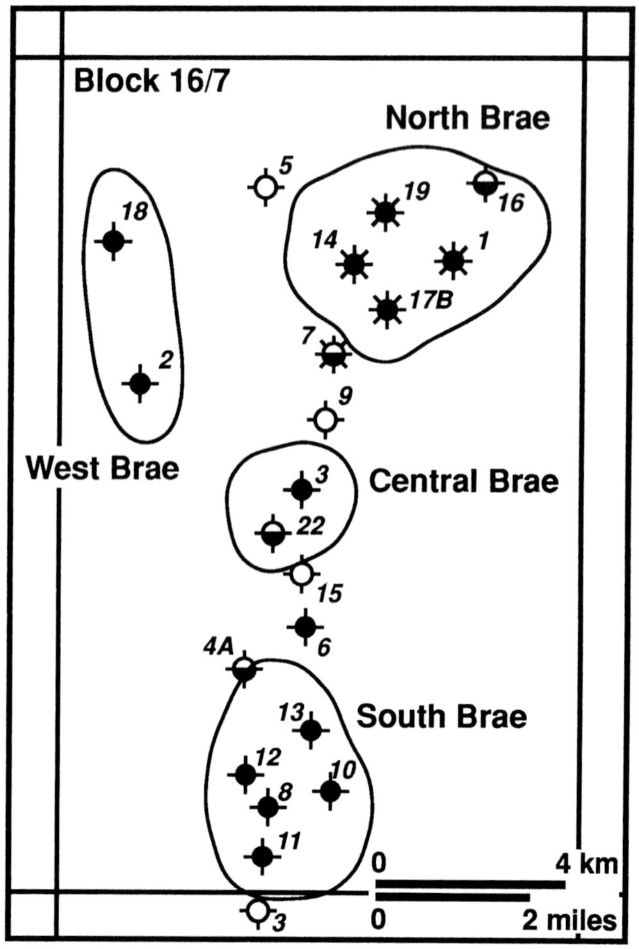

Figure 2: Brae development area

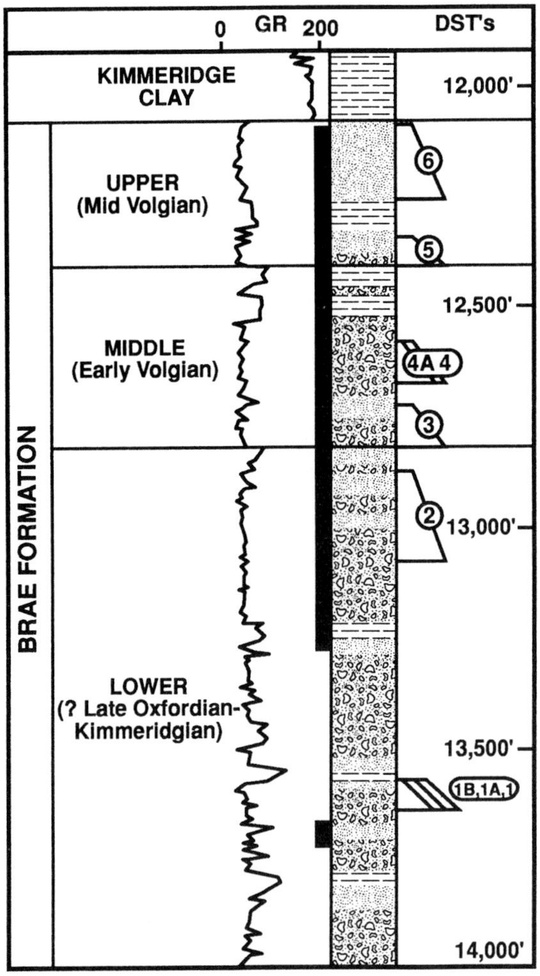

Figure 3: South Brae discovery well 16/7a-8

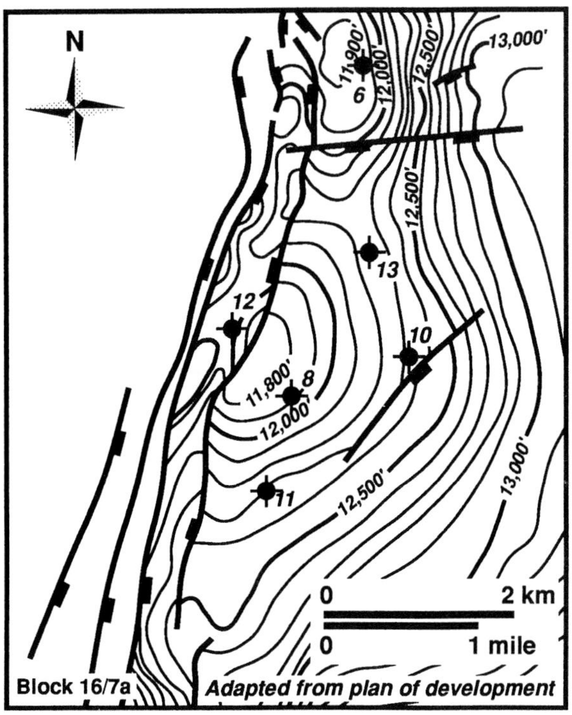

Figure 4: South Brae Base Cretaceous unconformity depth structure map

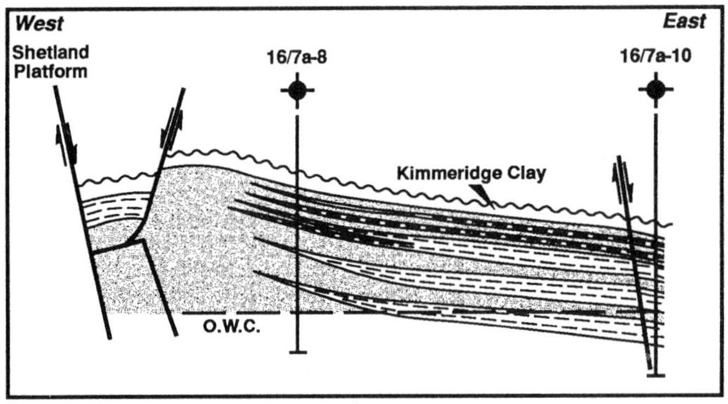

Figure 5: South Brae diagramatic cross section

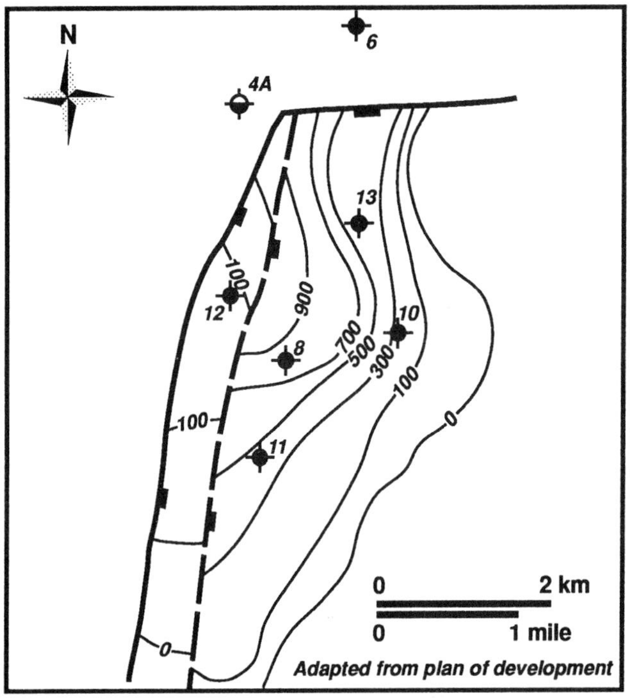

Figure 6: South Brae pre-development net pay map

coarse clastic conglomerates and sandstones were thought to thin outwards finally being replaced by laminated siltstones and shale. The geological cross section through wells 16/7a-8 and 16/7a-10 shown in figure 5 demonstrates the change in lithology from West to East. The geological model developed to explain these changes in lithology resulted in a simple arcuate fan shape reservoir distribution with maximum accumulation at the centre of the fan. Figure 6 shows a net pay map and the locations of the exploration/appraisal wells. The productivity of the wells and the potential reserves indicated that a commercial development of the field was possible.

A plan of development based on this geological model was submitted to the Department of Energy for their approval. The development plan allowed for nineteen production wells with crestal gas injection and peripheral water injection for pressure maintenance and sweep. Reservoir simulation indicated that reserves in the order of 300 million barrels of liquids (including oil and NGL's) could be produced over the life of the field. In January 1980 the development plan was approved by the Department of Energy. A single steel jacket was installed over the field in June 1982. The design production rate of the platform facilities was 120,000 bopd and it was anticipated that the

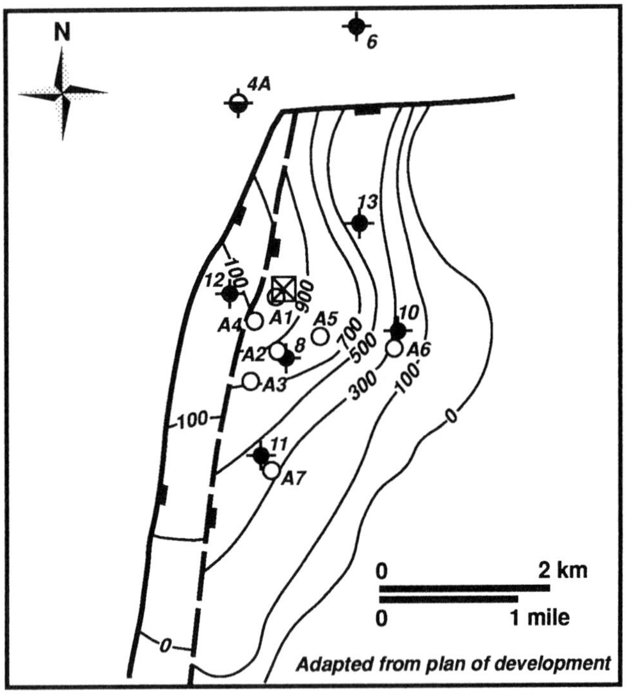

Figure 7: South Brae first seven development wells

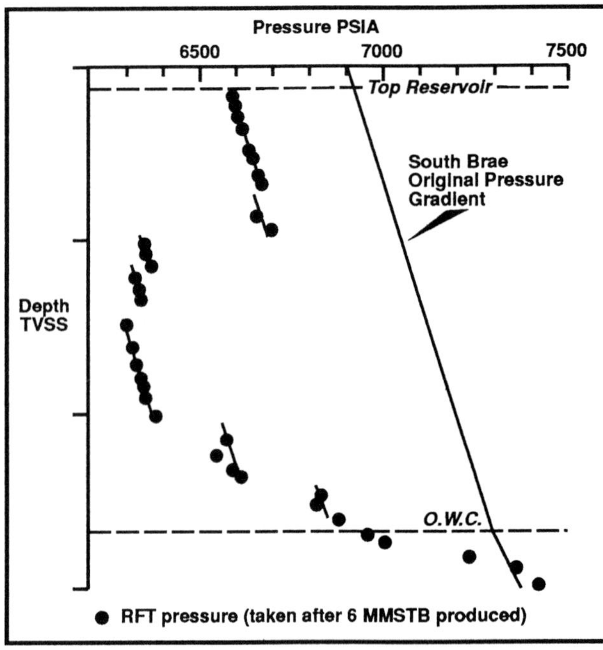

Figure 8: 16/7a-A3 RFT pressures

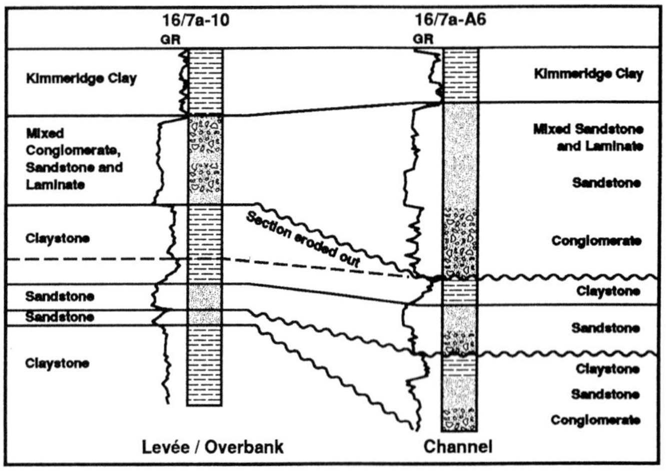

Figure 9: Wells 16/7a-10 & A6 geological correlation

first four development wells would achieve the predicted plateau rate of 100,000 bopd within four months from the start of production. Development drilling commenced in March 1983 with first oil in July 1983.

The first development well was drilled vertically below the platform. The location was considered to be at the centre of the fan, and was expected to encounter maximum pay. The well results were disappointing as the well encountered less coarse clastics than expected. The results of this well implied that either the geology of the field was more complex than originally assumed or the fan was much smaller than previously mapped. The position of this well and the next six development wells is shown in figure 7.

The third well, A3 came up with a further surprise, RFT pressures taken in this well indicated that the field was displaying multi–reservoir behaviour,illustrated in figure 8. This figure indicates differential depletion after only 6 MMSTB of production. The original geological model and development plan had assumed that there were no vertical restrictions to flow and that the reservoir would act as one homogeneous unit. The evidence of layered reservoir behaviour had immediate effects on both the production and water injection philosophy. The next major surprise came with the drilling of well A6. This well was drilled as a twin to the delineation well 16/7a–10 and was only 650 feet away from it. The well encountered nearly double the net pay of 16/7a–10 with coarse clastics replacing the finely laminated sediments. A geological correlation between these two wells is shown in figure 9. This was the first indication of the reservoir being a more complex submarine fan channel system with considerable vertical and lateral variations in lithology.

At this time it was clearly necessary to understand the results of the wells if producers and water injectors were to be optimally located in order to effectively develop the field. A multi–discipline study was instigated to re–evaluate the geological and reservoir aspects of the field. Using the data from the early wells the study confirmed the presence of channel systems. These coarse clastic channel systems have erosional bases and complex channel fill deposits with fine grained units tending to overlie coarse grained units. This new interpretation was supported by seismic where the coarse clastic channel shows less compaction than those out of the channel. The existing 2D seismic was lacking in quality and resolution to map within the reservoir, however it was adequate to define areas of channel development. A typical 2D seismic line is shown in figure 10. Re–interpretation of the existing 2D seismic and correlation of the layers using open hole logs and RFT's together with cased hole logs led to an improved understanding of the channels and the reservoir system. Wells drilled in the channels were highly productive and those drilled to the edge were poor producers. Identification of the position of the channels was necessary to both improve the production rate and to locate water injection wells for pressure support. Initial production from the development was disappointing and not as high as planned. Production rates did not exceed 70,000 barrels of oil per day within the first four months of start up of production. Well locations which were subsequently drilled to test the new model were highly

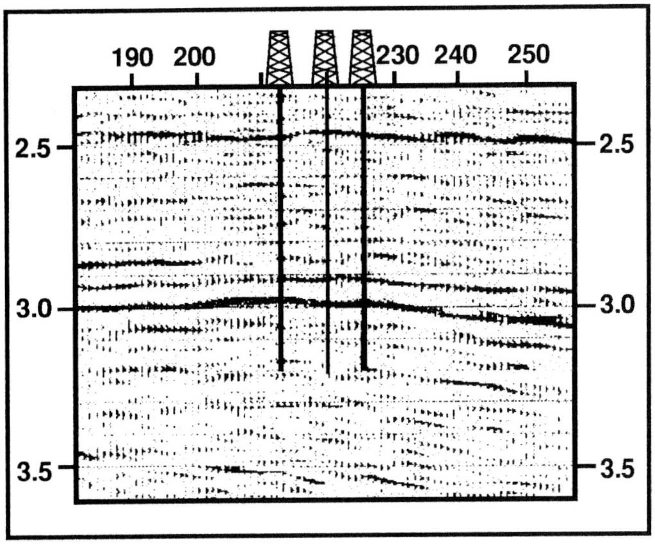

Figure 10: South Brae 2D Seismic section

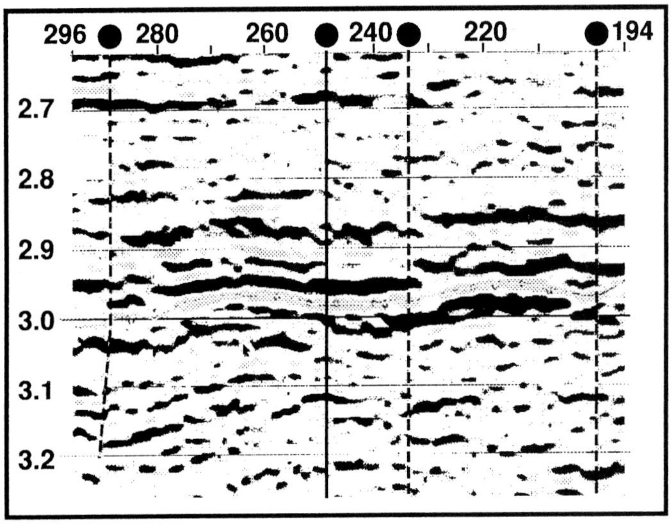

Figure 11: South Brae reconstructed seismic line from the 3D survey

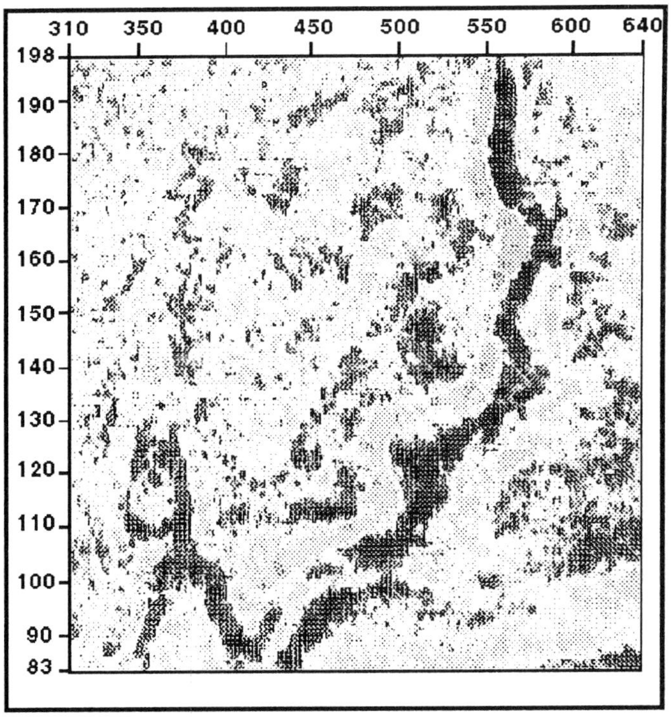

Figure 12: South Brae time slice

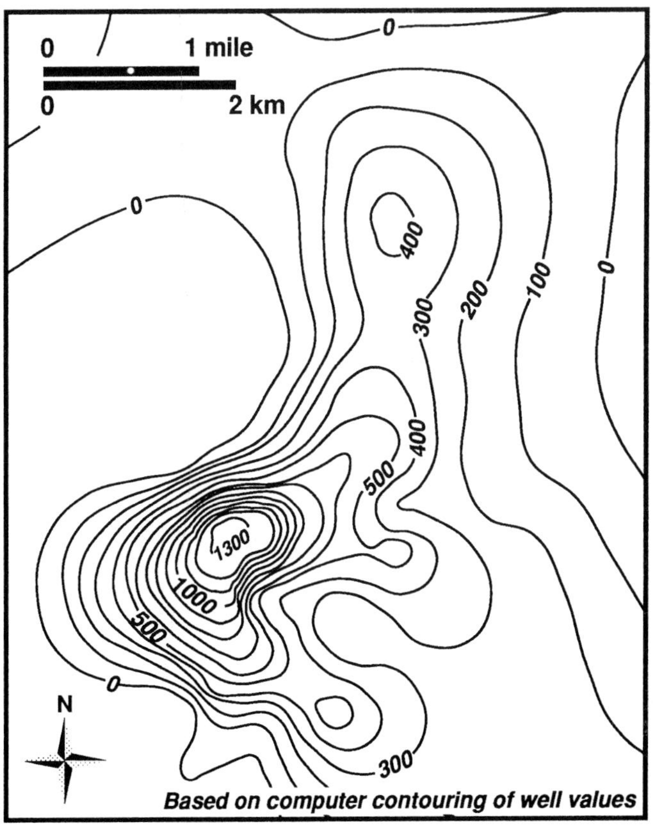

Figure 13: South Brae current Net Pay map

successful. These wells confirmed the model and were excellent producing wells. By late 1986 after optimally locating the wells the field was consistently producing above the plateau rate.

In 1985 3D seismic was shot over the area and this became available for use in 1986. A reconstructed line from the 3D survey is shown in figure 11. The use of this 3D seismic proved invaluable in identifying the channels and confirmed the revised interpretation. Wells post 3D seismic continued to both confirm the geological model and be highly successful. The success of these wells enabled high production rates to be maintained to late 1989. A time slice from the 3D is shown in figure 12. This figure indicates the seismic response to the channels. A revised net pay map shown in figure 13 indicates the existence of three main channels and shows considerable differences to that in the original plan of development.

NORTH BRAE

The first well drilled in 1974 in the Brae area was well 16/7–1. This well discovered North Brae and encountered over 400 feet of net pay. The well schematic is shown in figure 14. The well tested gas condensate at an aggregate rate of 21,500 barrels of condensate per day. In view of the condensate nature of the hydrocarbons the field was not appraised until 1980 when well 16/7a–14 was drilled to the West of the discovery well. This well was followed in 1981 by wells 16/7a–17b and 19.

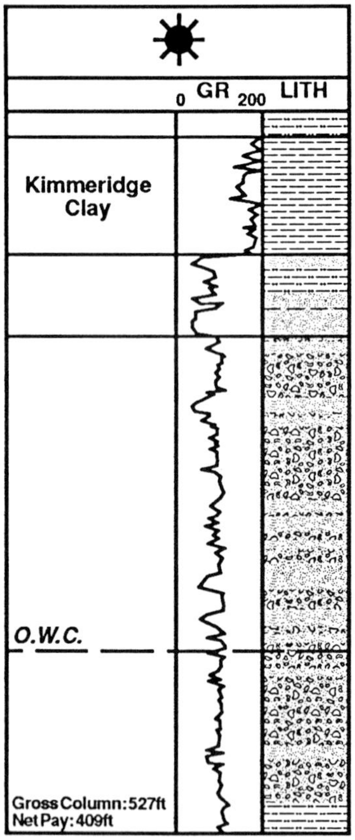

Figure 14: Well 16/7-1

These wells indicated variable sand, shale and conglomerate content, figure 15 shows a geologic schematic of these wells. 2D seismic was available during the appraisal/pre–development phase and three geological interpretations were postulated, all of which could explain the results from the initial wells. These interpretations are shown in figure 16, and positions of these wells are indicated on a net pay map, figure 17. The first possibility was an erosional base submarine channel where the crestal wells 16/7–1 and 16/7a–14 are in the channel and the other two wells 16/7a–17b and 19 are outside the channel. The second possibility was a lateral facies change where there is a conglomerate core with lateral sand halos. These sand halos shale out at the edge of the fan. The third possibility to explain the presence of the conglomerate was that the centre of the field had been controlled by tectonic faults. Without additional drilling the model could not be confirmed but

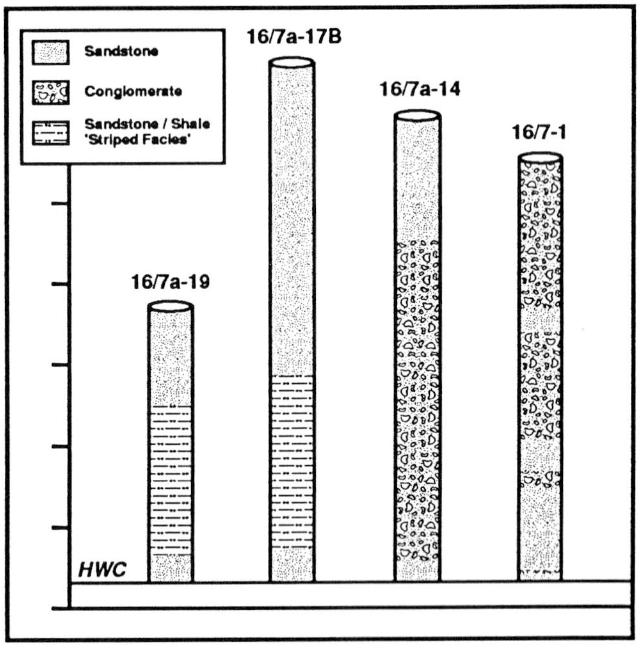

Figure 15: North Brae appraisal wells

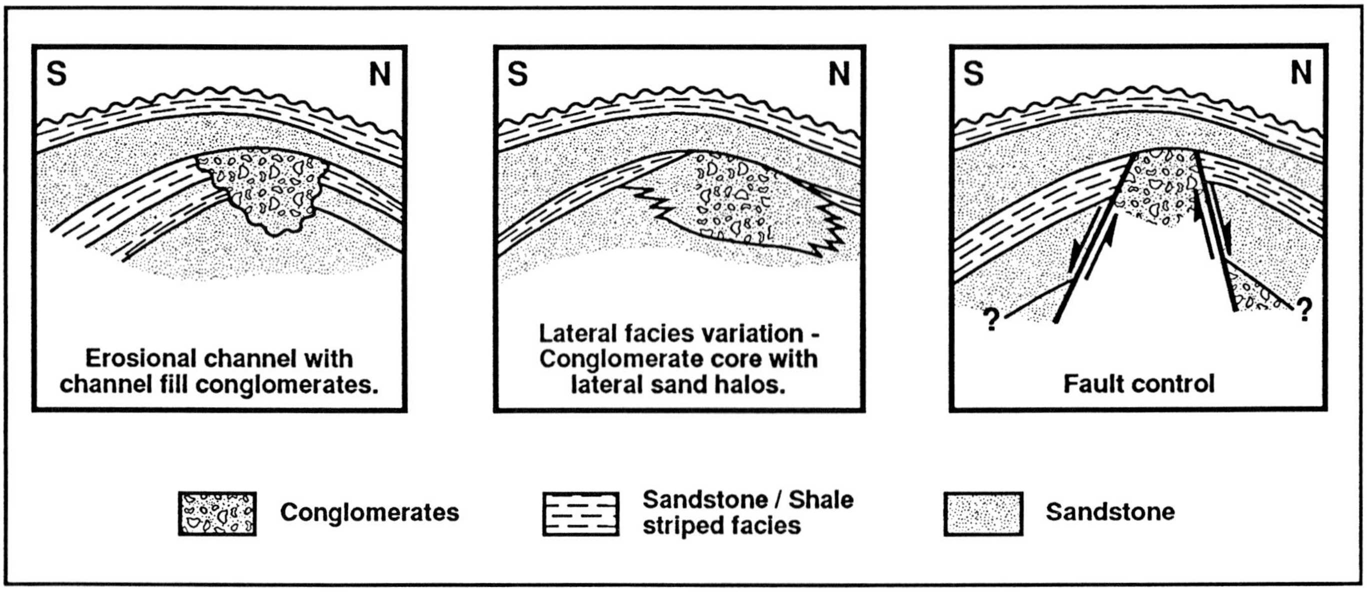

Figure 16: North Brae alternative geological model interpretations

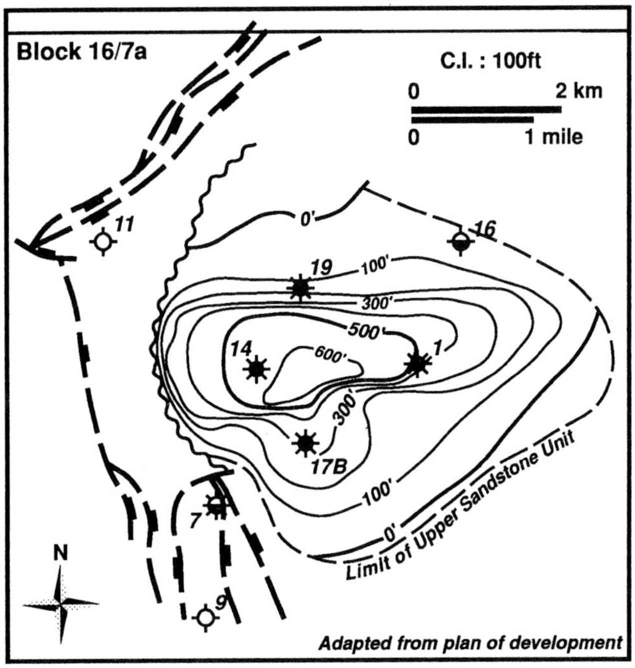

Figure 17: North Brae pre-development Net Pay map

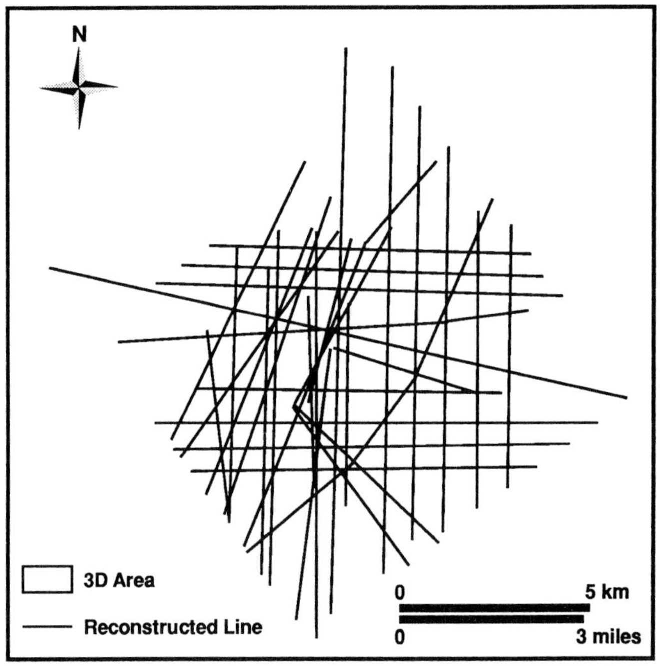

Figure 18: North Brae 3D seismic grid

similar wet gas in place and reserves were estimated from all three models. The appraisal drilling had been completed before South Brae started production, consequently there was no benefit at this stage from the South Brae development experience.

The liquid reserves on North Brae are determined with the use of a fully compositional reservoir simulator. The liquid content of the reservoir fluid increases with depth and crestal gas injection with downdip peripheral production was envisaged as the optimal means of developing the reservoir. The simulation model indicated sufficient reserves and well productivity to commence a development and indicated that gas cycling with partial pressure maintenance optimised liquid recovery. This development scheme was submitted to the Department of Energy in January 1983. The Department of Energy approved the plan of development in October 1983 and the platform was installed some seven miles North of South Brae in 1987. The North Brae field is the first gas condensate field to be developed in the North Sea and its single platform represents the largest gas processing facility ever installed offshore. The design production rate is 440 million cubic feet of wet gas per day with a maximum liquid rate of 90,000 barrels per day. The rich gas is processed to recover both condensate and natural gas liquids. The residual dry gas is compressed and injected back into the reservoir. North Brae also receives gas from South Brae to aid in the recycling effort. The development plan envisaged twelve producers to achieve the plateau rate of 75,000 barrels per day in early 1990. It was initially planned to recycle the gas for nine years to maximize liquid recovery.

During the construction phase for the North Brae development experience was being gained from the South Brae development which demonstrated the importance of seismic data in successfully locating channels and optimizing development well locations. Prior to development drilling a full 3D seismic survey was shot and in 1986 the processed seismic was available for interpretation. Figure 18 shows the 3D seismic grid which was shot as a series of SE–NW lines, this figure also shows constructed seismic lines from the workstation which are used to demonstrate certain aspects of the interpretation or to tie in certain wells. The ability to construct these lines is one of the most significant contributions of the 3D seismic to fully understand and map the geology of complex reservoirs such as North Brae. The use of amplitude strength as an indication of unit thickness is also a useful aid in the interpretation and was made even more practical with the advent of the workstation. Figure 19 shows the Base Cretaceous time structure map based on the 3D seismic interpretation and figure 20 shows the relative amplitude superimposed on the structure map. The 3D seismic can be used to identify channels and the figures 21,22 and 23 are all examples indicating channels. Before development drilling had commenced a well developed

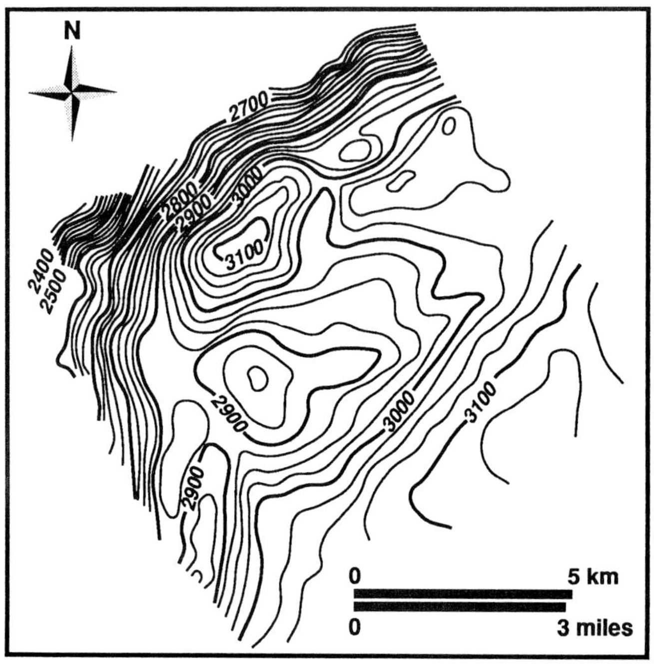

Figure 19: North Brae Base Cretaceous time structure map

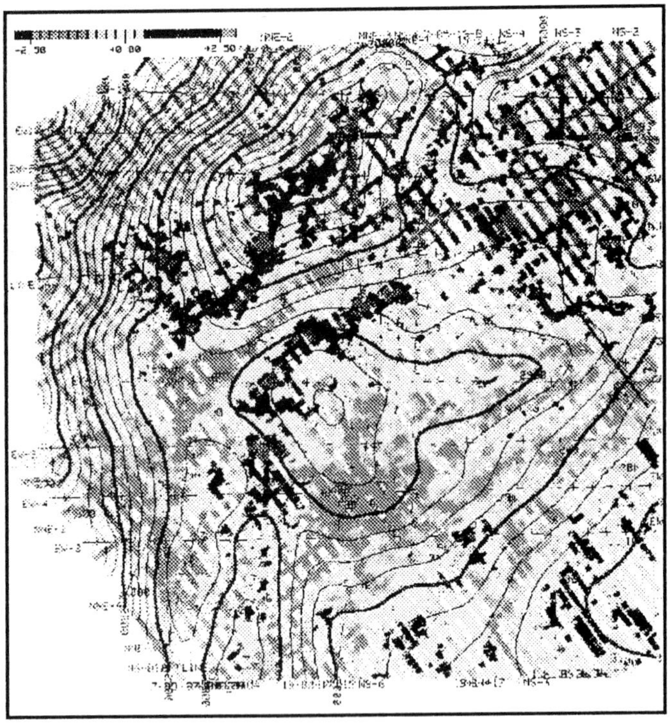

Figure 20: North Brae time structure map with relative amplitude

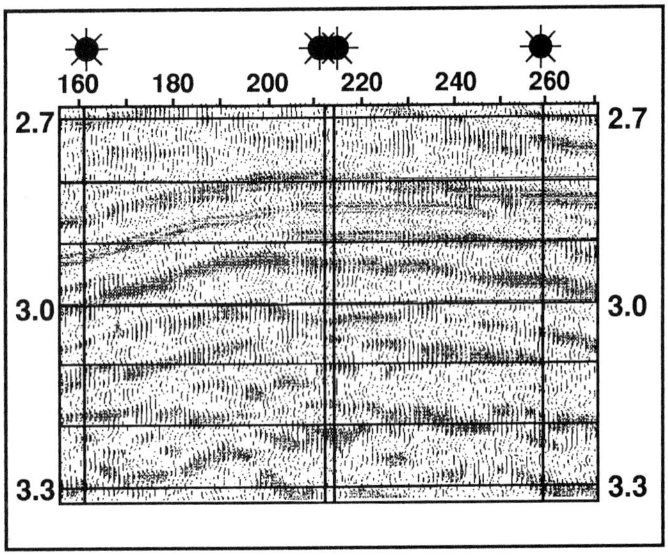

Figure 21: North Brae seismic line to show channel and erosional base

geological model for the North Brae field was in place. The quality of the 3D seismic was such that the top reservoir and details within the reservoir could be mapped.

The success of the seismic and geological interpretation in identifying the position of the channels enabled the development wells to be drilled in optimal locations. The field came on stream in April 1988 some five months earlier than planned and also achieved plateau production much earlier than expected. Plateau rates have been maintained since December 1988 at around 75,000 barrels of condensate per day. The productivity of the wells has been so high that there has been considerable flexibility for further delineation of the field. The results from the development wells have refined the model and confirmed that the erosional base submarine channel is indeed an

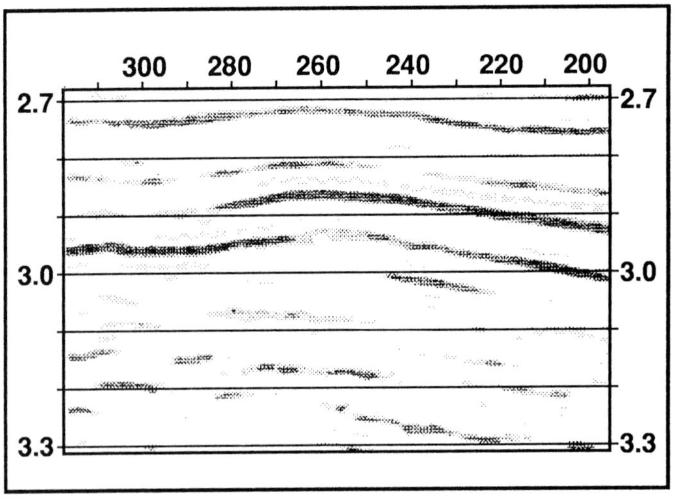

Figure 22: North Brae seismic line to show channel

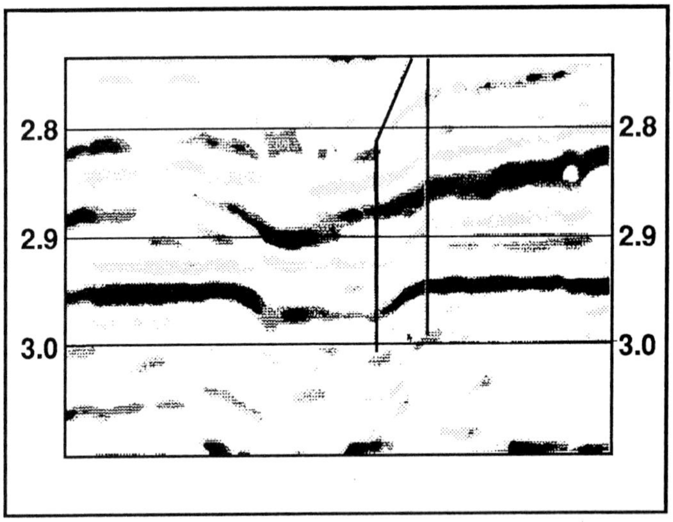

Figure 23: North Brae flattened seismic line to show channel

accurate description of the reservoir system, figure 24.

The use of 3D seismic before the start of development drilling together with the rapid integration of data from early development wells made it possible to refine the geological model for North Brae early in the development. This enabled the field to be developed with considerable success. Currently production is limited by topside facilities and the surplus capacity of the wells has enabled the field to be further delineated. A vertical well drilled solely for conglomerate production will have poor productivity even though the conglomerate is deep in the reservoir and is therefore in the richest part of the reservoir. A horizontal well has recently been drilled in the conglomerate core,this well has a horizontal section of some 1600 feet. At the time of writing this paper the well has not been tested however the well should be a long term rich condensate producer. This well will have distinct benefits in maintaining the liquid profile on North Brae and improving the liquid recovery from the field.

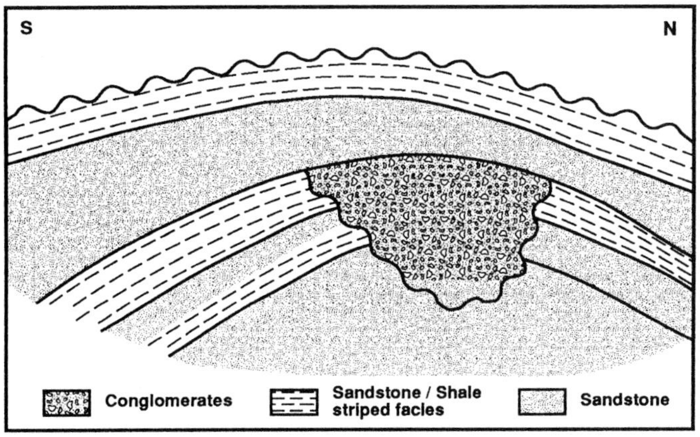

Figure 24: North Brae geological cross section

In view of the revised geological model it was considered necessary to verify the reserves and confirm the development philosophy. In order to determine these reserves it was necessary to update the fully compositional reservoir simulation of the field. As the geology of North Brae is complex the construction of the model did not lend itself to the traditional 'layer cake' approach. At the time of the simulation model there was no option for 'non–neighbour connections' where one grid cell can connect to another which is not immediately adjacent to that cell. Therefore, careful attention had to be paid to the layering in the model in order to accurately honour the geology of the field.

To illustrate the complexity of the geology of the field a geological cross section perpendicular to the axis of the conglomerate body's, shown in figure 24. It is evident from this figure that vertical flow does not take place in a 'layer cake' manner. If it was not for the conglomerate core, the upper and lower sands would be separated by a shaley layer and would not be in vertical communication with each other. As the conglomerate is permeable, albeit, much less permeable than the sands, vertical communication in the field is afforded via the conglomerate core.

The first step in building the simulation model was the construction of maps of the lithological layers. These maps were necessary to translate the geological model into simulation input. Thickness maps of the sand and shale layers before they were eroded by the conglomerate were constructed using well information and the geological depositional model. The isopach of the sand body immediately below the main shale is shown in figure 25. Similar maps were drawn for each of the lithological units including the conglomerate core.

The post erosional isopach of each sand and shale layer was generated by the computer in the following manner and is illustrated in figure 26 A matrix of grid depth values was generated by adding the thickness of the layers overlying the conglomerate to a structure map of the top of the reservoir. The surface identified as Grid 1 is shown in the figure as 'Sand Top' where the conglomerate is absent and 'Uneroded Sand Top' where the conglomerate is present. Grid 2 was determined by adding to Grid 1 the isopach of the uneroded sandstone as shown previously in figure 25. The surface identified as Grid 3 is the bottom of the Conglomerate and is calculated by adding the conglomerate thickness to the uppermost surface. The isopach of the sand layer was found by

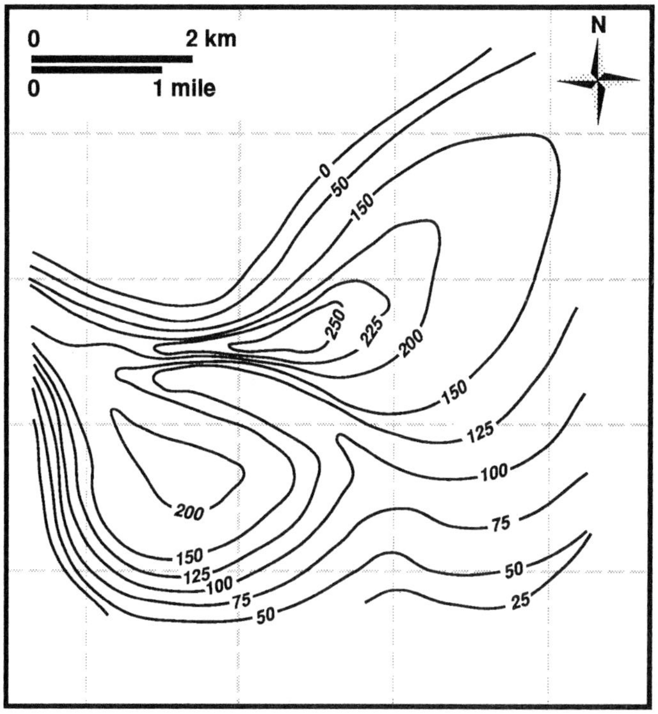

Figure 25: North Brae restored gross thickness map

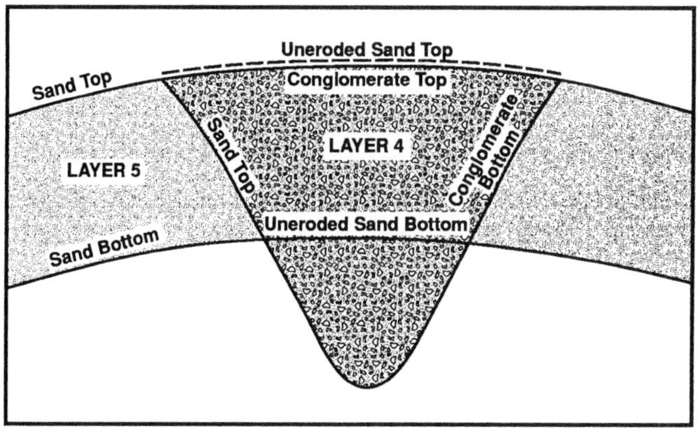

Figure 26: North Brae layer relationships

subtracting Grid 3 from Grid 2. Where the sand is missing the grid subtraction results in a negative value so these numbers were removed with a filter. The remaining grid is the eroded sandstone isopach and is illustrated in figure 27.

The digital map data was converted mathematically to the 'layer cake' format required by the simulator. A total of nine layers and three rock types were used in the model. This conversion results in simulation grid values which are variable to reflect the changes in lithology within each model slice. A computer generated cross section perpendicular to the conglomerate core is shown in figure 28. This figure illustrates that the representation of the reservoir units in the model does indeed honour the geological model.

The liquid content of North Brae increases with depth,illustrated in figure 29. Representative samples of the North Brae reservoir fluid were used to calculate an equation of state for use in the model. A satisfactory match using the Peng–Robinson correlation was achieved. A total of eight pseudo components were used to describe the fluid composition in the reservoir. The simulator uses the reservoir PVT properties along with the reservoir description to predict the production of wet gas from each well. A separate set of mathematical constants to relate the plant separation of the fluids is used to convert the wet gas to dry gas and liquids. The simulation model is 'flashed' through the 'topside process equipment' to give condensate and dry gas in place.

The reserves in the development plan for North Brae were achieved with gas cycling with partial pressure maintenance. On North Brae there are four crestal gas injection wells with the producers drilled downdip from these wells. The closer the producing interval is to the hydrocarbon water contact, the richer the fluid produced. All available gas after processing on the platform is injected back into the reservoir. Each development well after being drilled had RFT pressures taken in both the hydrocarbon leg and the aquifer. Early wells indicated no vertical restrictions to flow but later wells indicated that there were indeed vertical transmissibility barriers. The pressure depletions from original pressure provided a means of confirming the geological and engineering assumptions in the simulation. The calculated pressures for a well at each grid centre were compared to the RFT pressures. In order to achieve a better match of pressures it was necessary to adjust the vertical permeability between layers and also increase the size of the aquifer. These changes are justified as the size of the aquifer is unknown and shale breaks have been identified from open hole logs. The pressure match obtained with these adjustments was excellent and is demonstrated in figure 30.

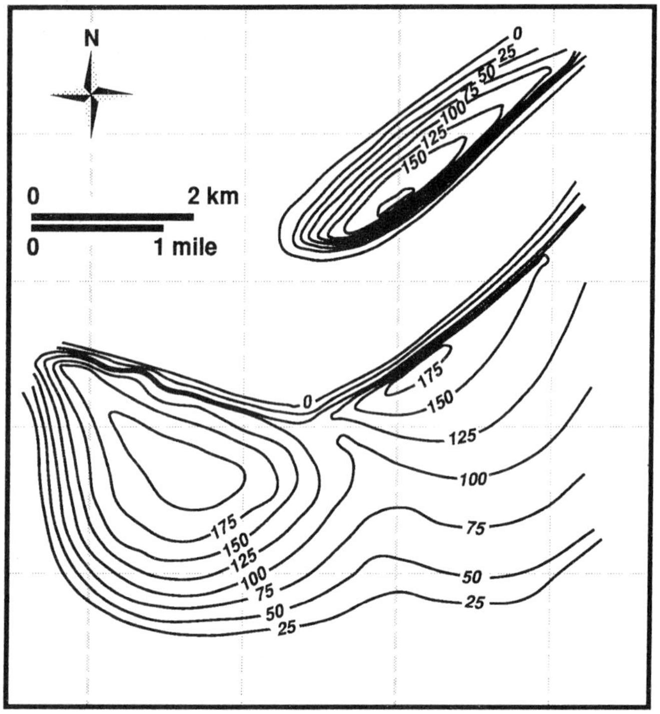

Figure 27: North Brae gross thickness map

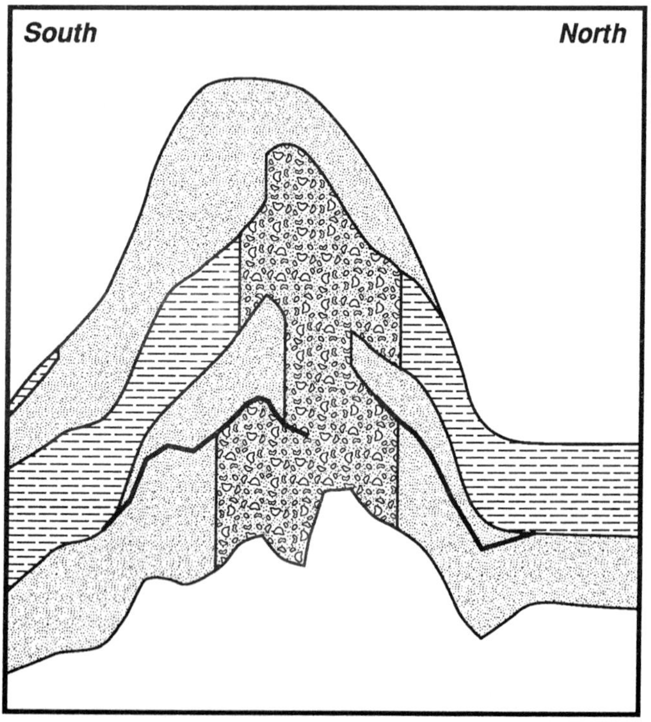

Figure 28: North Brae simulation model geological cross section

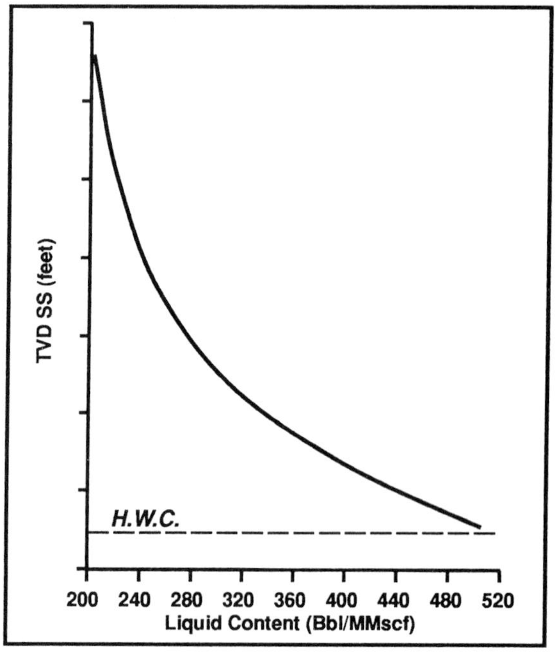

Figure 29: North Brae liquid content with depth

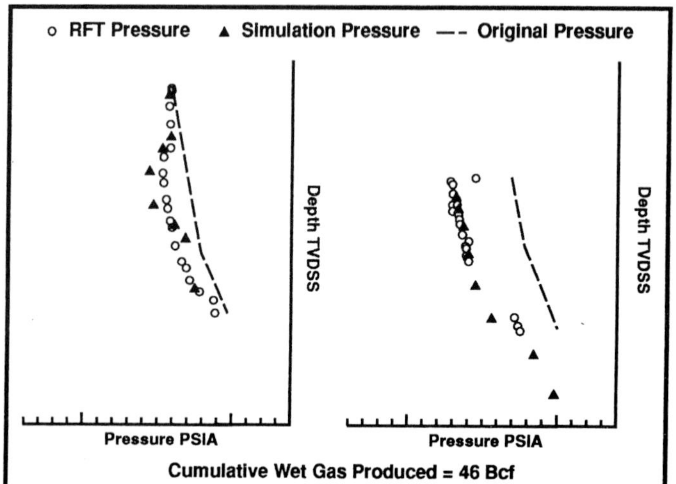

Figure 30: North Brae RFT pressure plots

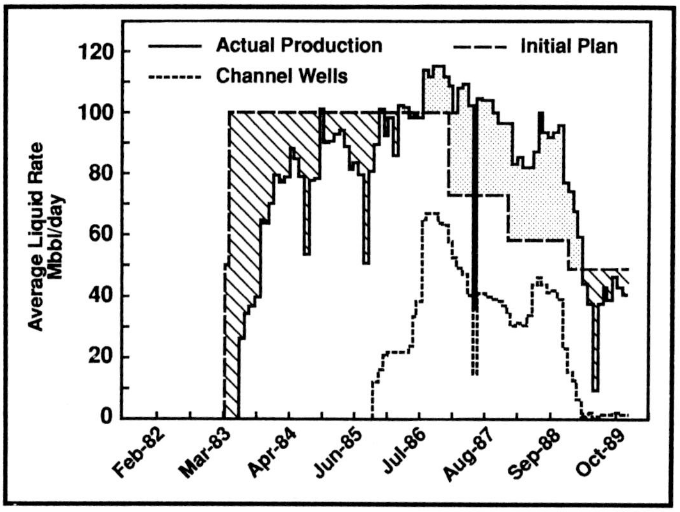

Figure 31: South Brae production profile

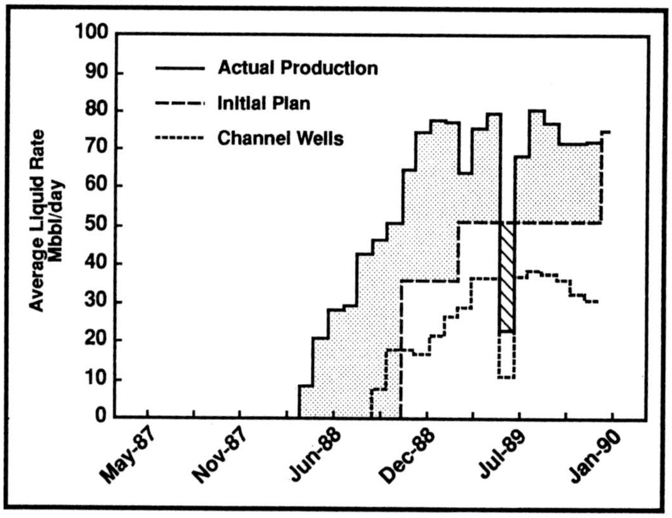

Figure 32: North Brae production profile

This figure shows pressures taken from wells drilled approximately 18 months after production start up after some 46 BCF of gas had been produced. This match of pressures provided sufficient confidence in the model to use it as a predictive tool.

The model predicts that with gas cycling as a development method, the pressure in the reservoir declines slowly during the recycling period and that overall liquid recovery is excellent. Even though the geology of North Brae was complex, a reservoir simulation model was successfully constructed to accurately honour the geology. The original plan of gas cycling as a development strategy was confirmed to be the best option.

CONCLUSIONS

The Brae area is a geologically complex area with each field showing different characteristics. Although the South Brae field was thought to be fully appraised, the development plan for the field contained a very simple view of the geology of the field. The development wells came up with major surprises and the predicted plateau rate appeared difficult to achieve. The available 2D seismic enabled considerable advances to be made in refining the geological model but the process of understanding the geology, maximizing production rates and maintaining pressure support was efficiently brought about by all disciplines working together. The success of the study enabled the development wells to be optimally located and the field achieved plateau production albeit with some delay. Figure 31 illustrates the production from South Brae and highlights the delay in reaching plateau production. This figure also indicates the significant contribution of the channel wells to the total production.

The value of 3D seismic was recognised during the South Brae study and as part of the development plan for North Brae extensive 3D seismic was shot, processed and interpreted before the start of development drilling. The success in identifying the correct geological model and the position and direction of the channel enabled development drilling to proceed with confidence. Although the geology of the field is complex, reservoir simulation confirmed the reserves and indicated that recycling gas maximised liquid recovery. The success of defining the geological model before development drilling commenced enabled the development well locations to be optimised. The profile from North Brae, figure 32, indicates that the production from the field reached plateau production earlier than expected. Again this figure also indicates the contribution of two channel wells to overall production rates. Currently production from North Brae is constrained by the topside facilities. This surplus capacity has enabled the field to be further delineated. A horizontal well has been drilled in the conglomerate core in the rich condensate part of the reservoir. This well should be a long term rich condensate producer and should improve the liquid recovery from the field. The experience gained from drilling and completing this well will have distinct benefits for other fields in the Brae area.

The development of South and North Brae fields indicates that with multi–discipline teams working together even geologically complex fields can be developed with success. If returns on investment are to be maximised on capital intensive projects such as these it is imperative that synergy exists to provide solutions to problems that require a multi–disciplinary approach.

ACKNOWLEDGMENTS

We thank Marathon and the Brae partners for their permission to publish this paper. The opinions in this paper are those of the authors and do not necessarily reflect the views of all the partners. The success of the development of the Brae Area is due to all disciplines within all the companies working towards a common goal.

REFERENCES

1. Harms, J.C., Tackenburg, P., Pickles, E. and Pollock, R.E. 1981. The Brae oilfield area, In Illing, L.V. and Hobson, G.D. (Eds.) *Petroleum Geology of the Continental Shelf of North-west Europe*, Heyden, London, pp. 352-357.
2. Stow, D.A.V., Bishop, C.D., Mills, S.J. 1982. Sedimentology of the Brae oilfield, North Sea: Fan models and controls, *Journal of Petroleum Geology* **5**, 129-148.
3. Turner, C.C., Cohen, J.M., Conell, E.R. and Cooper, D.H.,1987. A depositional model for the South Brae oilfield, In Brooks, J. and Glennie, K. (Eds.) *Petroleum Geology of North-west Europe*, Graham & Trotman, pp. 853-864.

THE PROCESS AND VALUE OF PETROPHYSICAL INTEGRATION

R. W. Hietala

Petroleum exploration and development programs can benefit significantly through the integration of petrophysical analysis. Quantitative assessments of hydrocarbon volumes and anticipated production characteristics can improve the exploration and development decision process and reduce risk. Canadian Hunter's success at applying this philosophy reflects three key ingredients: people, technology and a work environment which fosters innovation.

Petrography is fundamental to all aspects of successful formation evaluation. Rock framework components and secondary constituents are characterized for their impact on log response. The descriptions of pore types and volumes guide the selection of critical parameters affecting fluid flow and electrical properties.

Log analysis techniques begin with a model which incorporates the petrographic data. Improvements in data resolution yield better quantitative models to describe the pore space available for hydrocarbon saturation. Permeability measurements calibrated to in- situ reservoir conditions provide the basis for productivity estimates.

Drill stem tests and production tests are used to calibrate the well log calculations of porosity, water saturation and producibility. Comparisons to known results provide a reference for the well or formation being considered.

Technology transfer is enhanced through specialists working on exploration problems through to the development phase and back into exploration. The pragmatic application of "state of the art" technology must be practised routinely for this approach to be successful. A receptive environment reaching through to the top levels of management is an underestimated requirement in achieving results through petrophysical integration.

It is an honor to be part of this tribute to recognize the contributions of Gus Archie, the first true petrophysicist. This series of papers will show the power of integrating the geosciences. Representation and support is strong from the geological, geophysical, engineering and petrophysical communities. This paper will not leave the reader with any new equations for reservoir evaluations. These pursuits are being covered elsewhere in the series. Canadian Hunter is a company committed to a process of petrophysical integration and it is the purpose of this paper to explain the process to you.

Canadian Hunter Exploration's founders decided a new approach was needed to generate exploration ideas in the mid 1970's. At that time the traditional tools were seismic and subsurface geology incorporating limited analysis of rock and well log data. High quality work efforts were being performed by many companies but usually at the mature levels of field development, or in "post-mortem" evaluations. The company's goal was to routinely provide geologists with thorough formation evaluations. In other words, there was a demand for petrophysical interpretations to be complete, accurate and timely to influence decision making. A system was required that would impact acreage acquisition, casing point and well completion decisions.

PETROPHYSICAL INTEGRATION

Definition

The definition of petrophysical integration involves synergy. (Figure 1) The goal is to combine interpretation of rock, log and test information within a suitable geological model.

The value of this approach is to significantly reduce exploration and development risk in a business always noted for high levels of risk.

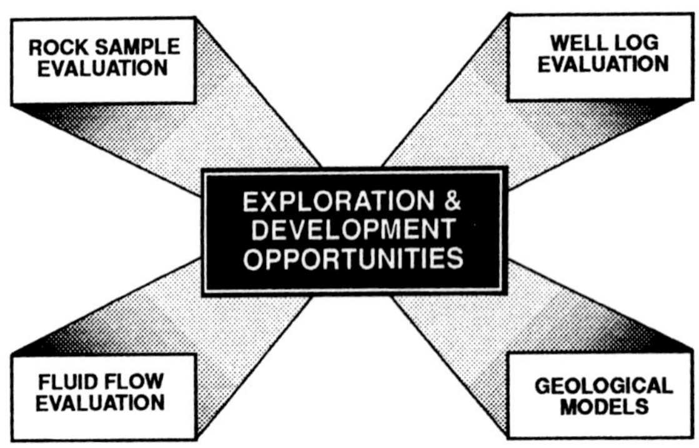

Team Selection

Selecting the petrophysical team is one of a company's most important tasks. The petrophysical team at Canadian Hunter was started with very seasoned log and rock analysts. These expert analysts had been involved in many exploration and development programs. During the drilling of numerous wells they acquired the analogs the company would use in its programs. Too much emphasis cannot be placed on the value of practical hands on experience in a fast moving exploration effort.

A blend of geologists and engineers form the nucleus of the petrophysical team. The team has been successful in deliberately integrating service company ideas with major oil company techniques. When forced with the decision to expand, the company has preferred to hire the most experienced individuals that could be identified. "Forced with the decision to expand," is emphasized because staff levels must be kept at a minimum. Teams need to be small in size to be effective.

Consultants have contributed during high work load projects, by injecting new ideas and providing skills lacked internally. Companies often find it difficult to hire consultants out of financial concern or internal pride. A belief is held in identifying the best person for the job whoever he or she may be, and genuinely seeking their advice. It is appropriate to suggest that there is room for top consultants in all organizations.

Rocks are Basic

The industry's technological advances continually provide new measurements and sensors to indirectly image the reservoir. One can lose sight of the simple fact that it is a real rock that is being described. (Figure 2)

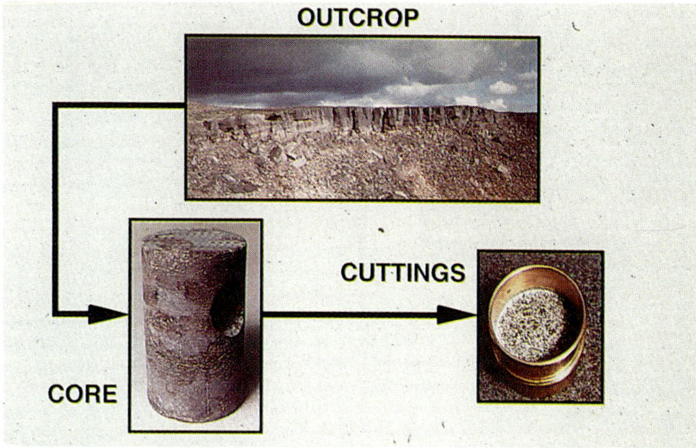

A great deal of effort is put into direct visual descriptions of the rock that the sensors are indirectly measuring. This basic belief in rock and log integration was reinforced through the guidance of Dr. Robert Sneider. Bob pioneered the description skills and communication medium to visually characterize clastic sequences for rock typing purposes.

Rock Typing

The rock typing system is a method to rank drill cuttings by permeability ranges. The general classifications are identified as Type I, II and III. (Figure 3) Type I is further subdivided into Ia, Ib and Ic and Id. Canadian Hunter routinely applies this rock typing system to all examinations of drill cuttings. It is the Type I rocks that really interest an exploration company.

TYPE	CHARACTERISTICS	
I	• VISIBLE Ø (common - abundant) • PINPOINT Ø (common - abundant) • PORE INTERCONNECTION • GRAINS WILL DISLODGE EASILY	A 100 + md B 10 - 100 md C 1 - 10 md D .5 - 1 md
II	• VISIBLE Ø (scattered) • PINPOINT Ø (common - abundant) • GRAINS DIFFICULT TO DISLODGE	
III	• VISIBLE Ø (none - isolated) • PINPOINT Ø (none - scattered) • VERY WELL CONSOLIDATED • ABUNDANT PORE FILLING MATERIAL	

The rock typing system is based on a reference to core. Plugs of known porosity, permeability and lithology are used to make comparators for drill cuttings analysis. (Figure 4)

Petrographers use these comparators in their training and in daily evaluation work. The comparators are critical to acquiring good, visual interpretations.

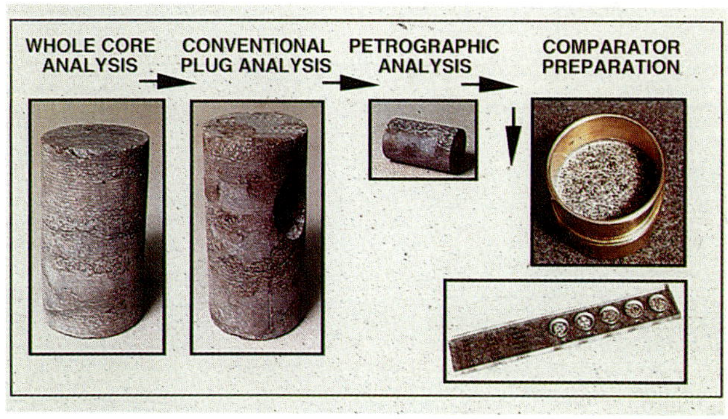

Cuttings analysis sheets are used by the petrographers to record their analysis. (Figure 5) Complementing the lithology and permeability determination is a complete

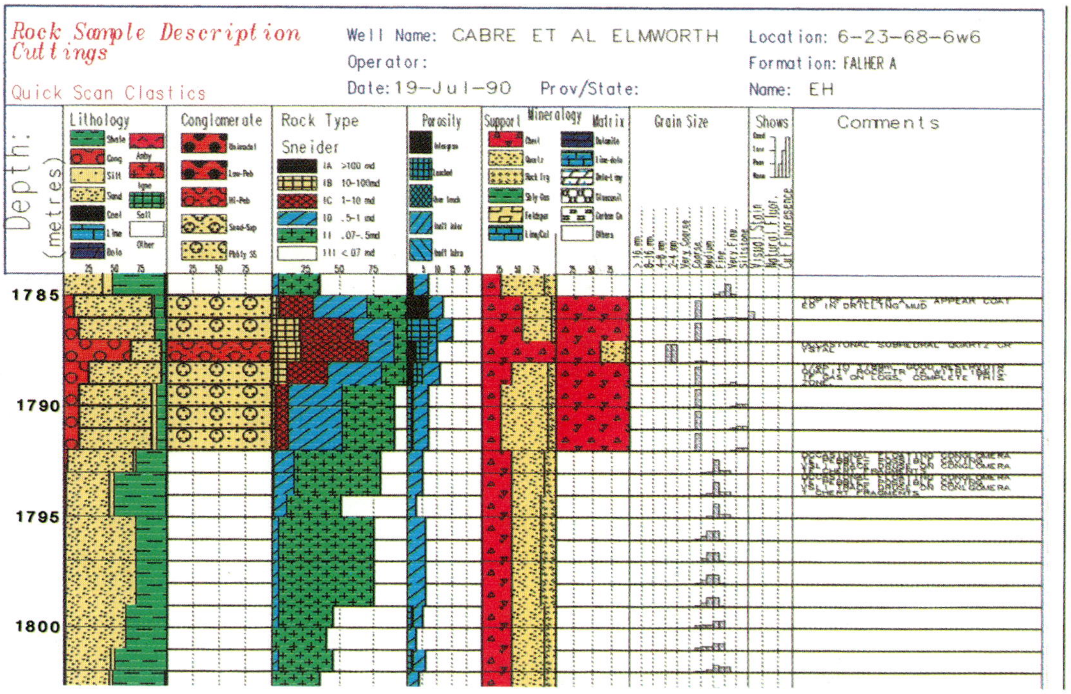

characterization of porosity, mineralogy, cements, clays and grain size. This information is currently captured through entry into lap-top computers.

Petrographic data are treated as borehole curves to be integrated with well log information. Rock typing enhances the volumetric interpretation obtained from well logs. The spatial distribution of matrix, cements, clays and porosity significantly affects estimates of hydrocarbon volume and producibility.

Petrographic Training

The routine examination of drill cuttings on a large scale is hard work. A committed staff of nine experts meet the petrographic analysis requirements. (Figure 6) All are internally trained, with industry consultants supplementing the training program. On average, it takes three years of intense microscope work to train a specialist who has the

initial aptitude and interest.

Petrographers receive intense instruction in the areas of core analysis, mineralogy, thin section description and SEM interpretation. In addition, significant effort is focused on field analog comparisons. Petrographers involved in day to day projects are required to make visual interpretations of cores and cuttings in order to recommend an action the company should take on a given project. The petrographers' recommendations directly influence acreage acquisition and well decisions.

Rock Catalogs

Few geologists have been exposed to various forms of rock catalogs. (Figure 7) They are invaluable. The

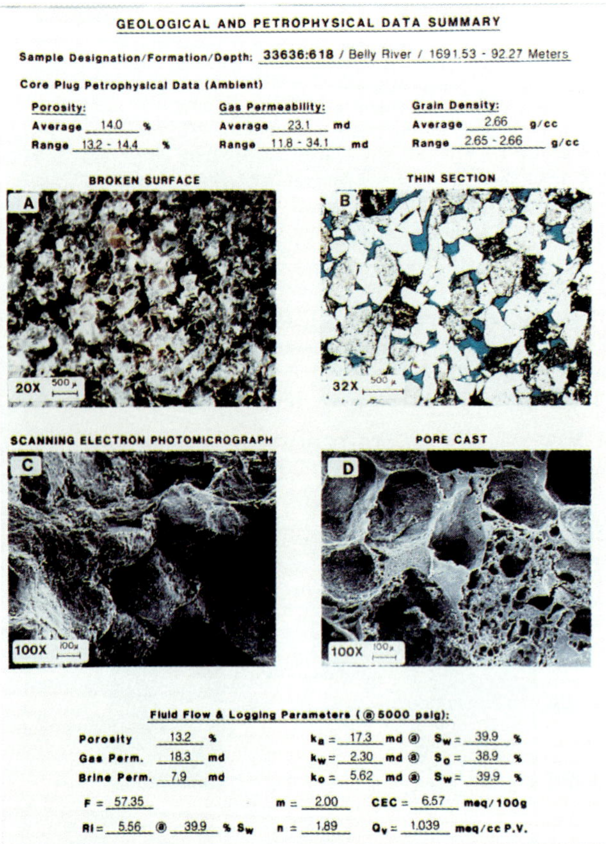

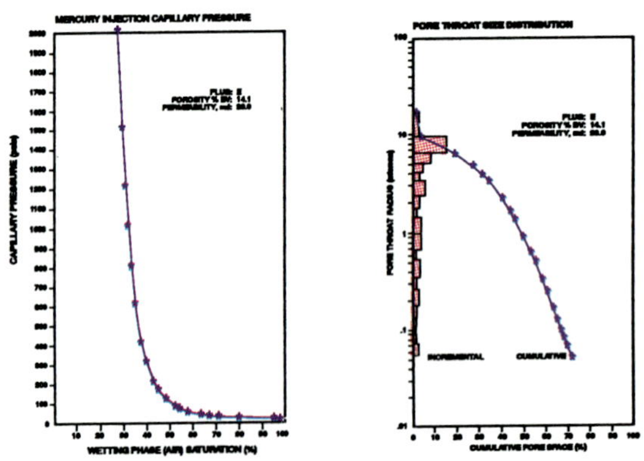

petrographers visually identify a similar rock in a catalog to match the rock from an evaluation problem under study.

The company has developed a smaller rock catalog to focus on porosity, permeability, capillary pressure and stress relationships. This has provided benefits in cost and time. Specifically, petrographers use the rock catalog comparisons to provide interpretations in four key areas. The areas are:

1. Cementation factor - "m"
2. Porosity and permeability changes from bench to in-situ conditions
3. Estimate of an equivalent capillary pressure curve
4. Assessment of mineralogy and impact of completion fluids

SEM Analysis

The incorporation of routine SEM technology provides a

further quantification of rocks. (Figure 8) The system Canadian Hunter uses, incorporates SEM analysis, elemental analysis and image processing. Having the SEM located on the same floor as the petrophysical team greatly enhances problem solving capabilities. Many companies use SEM systems, but few are able to have the equipment used routinely by their own petrographers and petrophysicists. This is essential to the integration process.

The ability to estimate a cementation factor from drill cuttings is often questioned. However, on numerous occasions visual estimates of "m" have closely agreed with core analysis values. The company has confidence in this petrographic interpretation technique and routinely applies it.

This SEM photo (Figure 9) illustrates two samples of different cementation factors. There are marked differences in the surface topography and porosity distribution seen in the two samples. The higher cementation factor sample has greater surface relief and contains larger pores within the uniform pore network. The acquired SEM data is matched with the previously mentioned rock catalog.The value of routine rock catalog usage is to provide the log analysts with quantitative parameters which are otherwise approximated or their variability ignored.

m=1.79

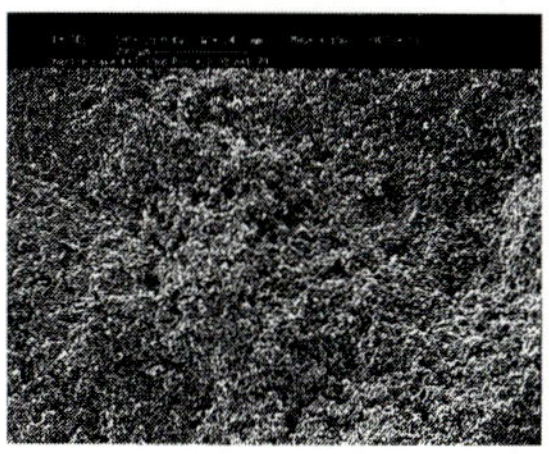

m=2.12

Exploration Scans

The use of a scan mode to evaluate large areas of exploration data is of considerable benefit. Log analysts will examine and determine manually or by computer all porosity in a well. Calculations of lithology, porosity, water saturation, reserves and estimates of flow rate are routinely made and compared.

The resultant values of reserves and flow rates are mapped by geological formation. Geologists are then able to see the big picture and define trends and anomalies for closer scrutiny. Many companies focus the formation evaluation efforts on a small number of zones or key wells. A problem is created when a set of geological filters limits the data set available to the petrophysical analyst. He must then merely react, limiting the potential of his expertise. A preference is for the consideration of all available petrophysical and petrographic data in an area to generate a total formation evaluation. This meets with the previously stated goal to routinely provide the geologists with thorough formation evaluations.

During the exploration log scan, petrographers seek out all available core and drill cuttings data. A well by well comparison is made between the rock descriptions and the log scan analysis. The methods usually reinforce and complement each other, but when there are differences, an opportunity to generate new exploration models often results.

Interpretation Systems

Detailed log analysis is performed on a variety of interpretation systems. Canadian Hunter incorporates, service company interpretation systems to complement the internal evaluation system known as CHELOG.

The philosophy for interpretation systems, is that they are tools, to be used by the analyst; the choice of the appropriate one is left to his discretion. An internal software focus is placed on what is unique to the way Canadian Hunter looks at rock-log analysis. A continued effort is maintained to minimize re-inventing the wheel. A close working relationship with key service companies is of considerable value and benefits can be realized by openly consulting their advice.

Log Analysis Models

Log interpretation models centre on resolving the interrelationships between volumes, measurements and parameters. Log analysis capabilities have advanced significantly from the procedures of one or two "chart-book" calculations, to more in-depth petrophysical modelling. The recording of high resolution log data greatly improves rock-log comparisons. Thin bed processing techniques more accurately represent porosity and resistivity data, and enhance volumetric calculations. Lithology, pore volume, fluid distribution and permeability can be routinely modelled. The following illustration (Figure 10) typifies a solution obtained from the CHELOG system.

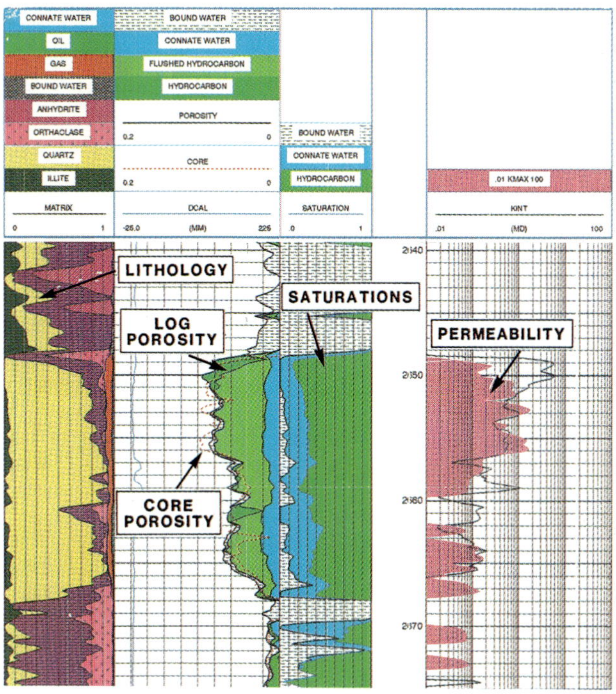

These detailed analyses are passed on to the geologists for incorporation into the prospect evaluation, or assessment of development opportunities.

Water Chemistry

Time and money is often spent acquiring log and core information in good faith with little attention paid to another set of critical parameters. The parameters are those of formation water chemistry. Archie noted the importance of studies on hydrocarbon reservoirs, impermeable layers and zones containing water. Knowledge of all three zones is required to evaluate hydrocarbon potential.

A procedure and interpretive system has been developed to screen water and filtrate analyses of drill stem and production tests. The interpretation system characterizes pure formation water and drilling fluid filtrates, together with their mixtures. (Figure 11)

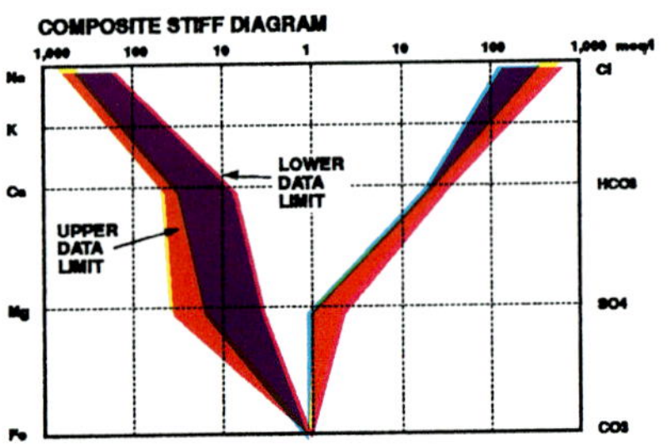

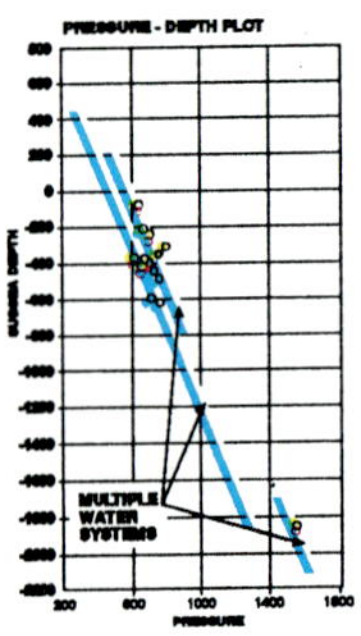

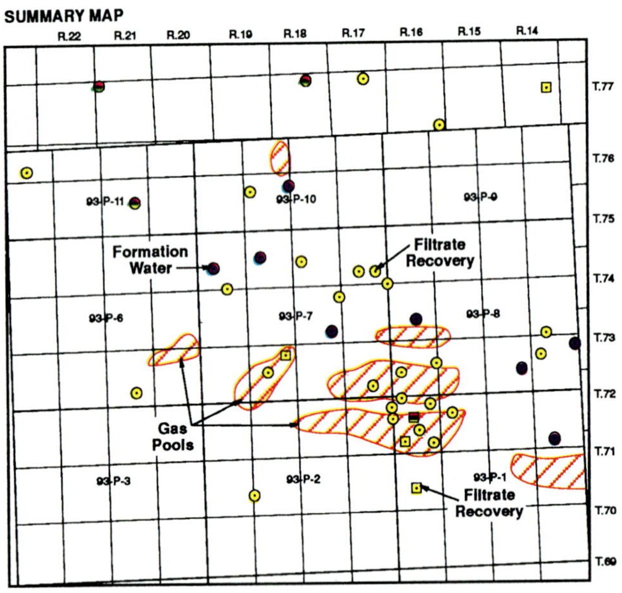

Consideration of ion concentrations, ion ratios and pH levels are used to separate formation waters from recoveries containing varying levels of filtrate contamination. A template for representative waters is created in the form of STIFF diagrams using the screened analysis. Pressure depth plots are used to guide water typing interpretations. A formation map is made to easily communicate the resulting interpretations. Obviously the end point values include an estimate of Rw, but the recognition of filtrates and mixtures can lead to opportunities when combined with DST and hydrodynamic evaluation.

Permeability Analysis

Routine comparisons of core porosity to core permeability are generated. This often yields the shotgun patterns that everyone is so familiar with. However, unique relationships are often revealed by re-evaluating the crossplots using petrographic descriptions and looking at the data as subsets of lithology, pore type or cement distributions. Regressions are then performed to derive a suitable permeability relationship. Log calculated volumes are used to estimate the permeability relationship and produce a core equivalent bench permeability curve. The relationship of bench permeabilities to stressed conditions is estimated from the rock catalogs or measured on plugs.

Canadian Hunter has worked very closely with a small independent lab to acquire stressed measurements.The result of this effort was the development of an instrumentation system to measure permeability on small plugs. The plugs are measured for single phase gas permeability at elevated gas and overburden pressures. This system is now a routine service provided by a major core analysis company.

For a set of Cretaceous sand plugs (Figure 12) there are significant reductions in gas permeability from bench to reservoir conditions. A rock catalog plug of the same reservoir is also shown. Calibrations to in-situ conditions are necessary to estimate flow rates from petrophysical data.

IN-SITU PERMEABILITY REDUCTION EFFECTS

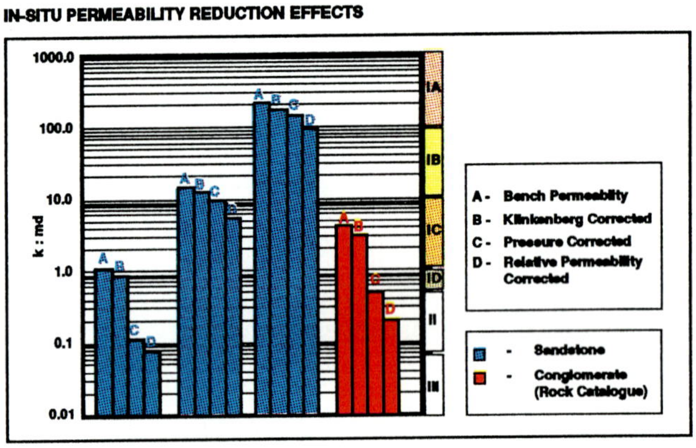

Image Analysis

A routine evaluation system to extend the use of thin sections has been recently developed. A large part of this technology is proprietary at present, but will be published later this year by Robert Gies, of Canadian Hunter. His work is briefly introduced here.

The viewing of thin sections in the back-scatter electron mode of the SEM allows a further quantification of the pore space. The pore space in back scatter mode is black. Matrix and pore fillers grade from gray to white depending on their density. A basic image analysis system detects the porosity of the thin section and generates a pore size distribution curve. Mr. Gies has extended the published works of Ben Swanson and Bert Thomeer. The result estimates permeability with an acceptable degree of accuracy for exploration and development purposes at Canadian Hunter.

The interpretation system has been calibrated against the thin section data base of an industry rock catalog. The following (Figure 13) illustrates the resultant relationship

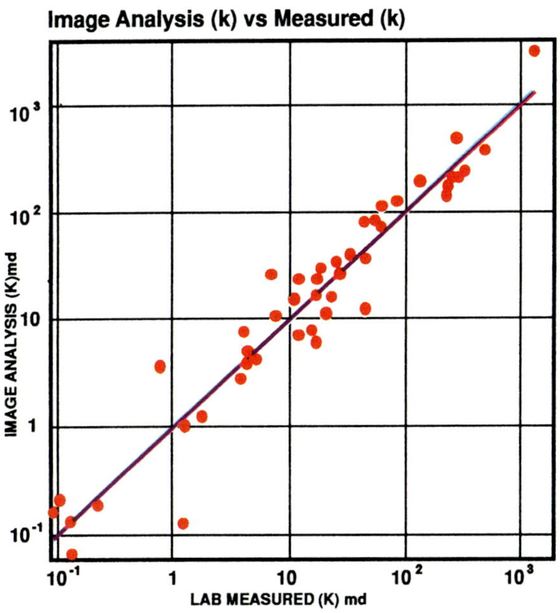

between image analysis and measured permeability using forty-six world wide samples. This methodology is being used to make real time operational decisions.

The use of rotary sidewall core samples and a quick turn around in thin section preparation will support this evaluation technique. Many workers are focusing on image analysis of rocks and pore systems. The technique described here was developed with practical goals in mind, to assist in porosity characterization, permeability estimation and to approximate a capillary pressure curve. This technology improves our reservoir evaluations in a timely and cost effective manner. The estimation of hydrocarbon producibility is enhanced through the integration process. (Figure 14)

Estimating Flow Rates

The interpretation system Canadian Hunter uses to relate petrophysical results to flow rates is called DELIVER. The program performs a rigorous numerical treatment of wellbore and reservoir conditions, but presents the results in a simple graphical form. (Figure 15)

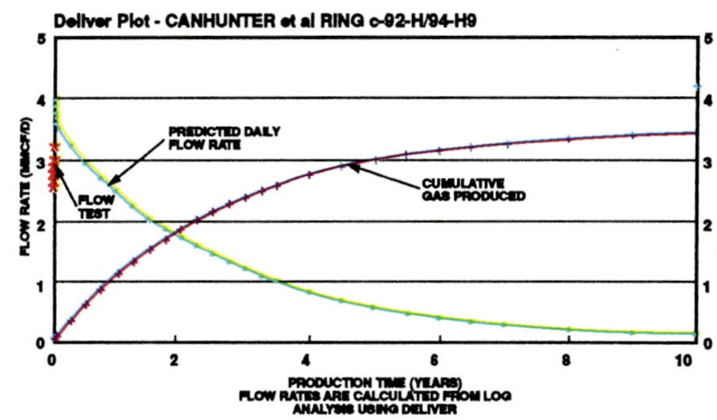

The two curves shown are flow rate and cumulative hydrocarbon production with time. The estimate of recoverable reserves is directly related to reservoir quality through this analysis system. There are sometimes ways to arrive at a more quantifiable answer through reservoir simulators but these take hours. However, DELIVER estimates are performed on an exploration evaluation basis on many zones and many wells and the work is usually completed in minutes. In evaluating old wells, an alternative is to use the initial flow rates from scout cards. Often, the scout card rates have been found to be inaccurate.

The quantification of producibility results from an attempt to bring basic reservoir engineering techniques into a front line exploration and development program.

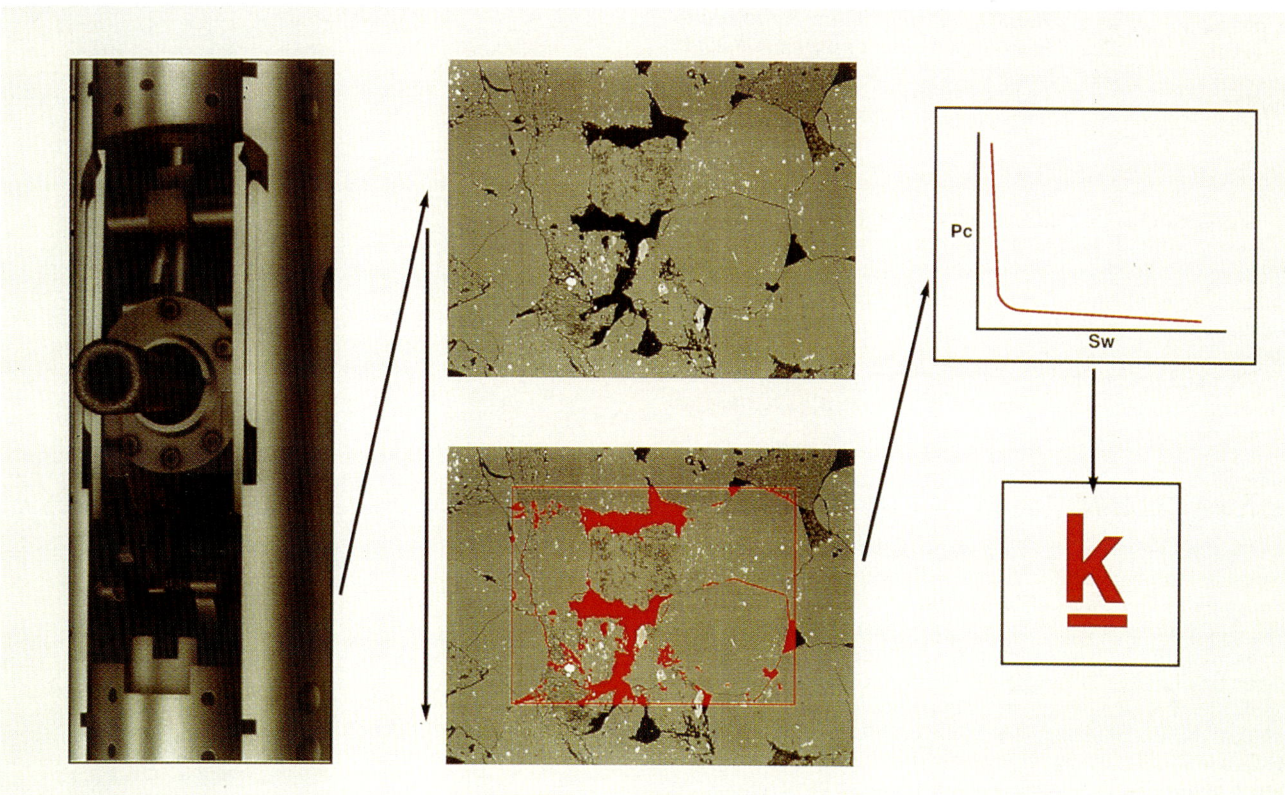

COMPUTER STRATEGIES

The exploration and development opportunities in the 1990's will be challenges. The vast amounts of new and old data will require the pragmatic use of computers. The goal of using computers to turn data into information should be emphasized.

Canadian Hunter takes advantage of significant computer resources, but maintains a basic attitude to their use. An interpreter must be able to do something better or faster through computerization, or a computer solution should not be chosen over a manual one. Computers are excellent for visual representation. The company has invested in developing a multi-well petrophysical evaluation system. A multi-well analysis system provides benefits in quality control of basic data, more consistent analysis and the ability for geologists to recognize trends. The prudent use of computer evaluation systems will allow for more "thinking time" which is vital in the integration process.

TEAM ORGANIZATION

It has been beneficial to have a focused petrophysical group within the company. Technology development is more effective when a team of related professionals are located close to each other. The ability to maximize effort on priority projects is also enhanced. The interaction and flow of ideas between group members has developed the present evaluation system. The petrophysical team works on development and exploration programs facilitating the flow of technical advances to diverse project areas. Such cross-departmental flow is sometimes difficult with specialists working in isolated environments.

As efforts expand to new development areas, the company has, and will continue to incorporate petrophysical team members within project groups. This will maintain a continued closeness to the daily problems of a project group as a company grows. The ability to rotate specialists between teams will also provide a career broadening experience and allow for increased job fulfilment.

FUTURE POTENTIAL

Significant opportunities exist in further petrophysical integration through the calibration to seismic responses. Reservoir evaluation is a process of analyzing measurements and observations at various scales. The extension from the SEM to cuttings to cores to logs and into seismic characterization is a natural one. This understanding has the potential to add previously hidden reserves to all companies pursuing exploration and development opportunities.

MANAGEMENT PHILOSOPHY

The material presented here touches on petrophysical methods and techniques. Many will recognize very familiar approaches to formation evaluation problems. The key to the integration process is to take a company's respective capabilities, and apply them routinely in a front line operating mode. A frequently mentioned comment by the senior staff of companies, is that they have the technological pieces, but do not routinely use them. To use the available technology effectively, the top management of a company must believe in it, routinely ask for it, and have consistent faith even when results are not positive. In the long run the process of petrophysical integration will pay off in reducing risk and identifying large hydrocarbon reserves yielding quality production.

Acknowledgements

The petrophysical team members are truly the ones who make our integration process possible. The information presented here is a snapshot of the work efforts of many people. Log analysis is performed by Ted Connolly, Dave Curwen, Jim Earley, Monty Fryt, Greg Reid, Steve Stretch, Don Westacott and Dennis Woon. Petrographic analysis contributions come from Shirley Dawson, Lorraine Druyff, Ray Gueder, Doug Hayden, Howie King, Grant Loss, Bev Creed and Diana Bogle. SEM and image analysis development is made possible through Bob Gies and Jon McGovern. Systems and technology development are the responsibilities of Steve Cheshire, Sid Ott and Phil Reed. Technical support contributions come from Hazel Boyd, Judy Bunyan and Larry Sopko. Core analysis is co-ordinated by Krystyna Hoffman-Fedenchuk and Joe Sobochan. Our consultants have always added to our capabilities.

The author would like to thank Canadian Hunter for allowing this information to be published. Special recognition is warranted for John Masters and Jim Gray for their support and direction of the petrophysical effort at Canadian Hunter.

The Archie Approach to Characterizing Carbonate Reservoirs

F. Jerry Lucia, D. G. Bebout and C. Kerans

Modern methods of characterizing carbonate reservoirs include geologic and engineering studies of various types and at a range of scales, all focused on describing the volume and distribution of oil and gas saturation and the flow characteristics of the reservoir. Such activities include (1) constructing the geologic reservoir framework at the sequence scale, (2) making detailed facies maps at the parasequence scale, (3) analyzing petrophysical data through integration of rock-fabric, facies, porosity, permeability, and water saturation, (4) mapping petrophysical parameters, (5) interpolating deterministic data between wells using geostatistical methods, (6) analyzing production data, and (7) inputting data into computer flow simulators for performance evaluation.

In 1952, G. E. Archie published a landmark paper showing that petrophysical properties can be related to rock textures. Although this relationship forms a fundamental basis for all reservoir characterization studies, it is often overlooked, resulting in beautiful geologic reservoir descriptions that cannot be converted into engineering parameters or in exquisite engineering studies that defy geologic understanding. This is particularly true in carbonate reservoirs, where correlations are uncertain, pore geometries complex, and log analysis perplexing.

Research on several carbonate fields in the Permian Basin and published data from fields in other basins and from other geologic ages, have shown the significant rock-fabric parameters for carbonate reservoir characterization to be (1) size and sorting of depositional and diagenetic particles, (2) volume of interparticle pore space, (3) volume of separate-vug pore space, and (4) presence or absence of touching-vug porosity. The three-dimensional spatial distribution of these four parameters together with reservoir height control the volume and flow characteristics of the reservoir. Combinations of gamma ray, neutron, density, acoustic, and resistivity wireline logs can be used to estimate these parameters in uncored intervals once transforms have been developed from cored wells.

The Dune and Farmer (Grayburg) fields in West Texas are examples of carbonate reservoirs in which particle size and sorting together with interparticle porosity are dominant influences on reservoir volume and flow. The distribution of particle size and sorting is controlled by upward-shoaling depositional and dolomitization processes, whereas the distribution of interparticle pore space results from cementation/compaction processes. In Dune field, the grainstone intervals are the most permeable; in Farmer field, however, the coarse-grainstone intervals tend to be cemented with anhydrite and gypsum, leaving the wackestone intervals the most permeable.

Vuggy porosity has a dominant influence on reservoir volume and flow when dissolution processes are significant. Separate-vug pore space is particularly important where subaerial exposure of grainstone shoals has inverted the pore space from intergranular to moldic. In Taylor-Link (San Andres) field of West Texas, both touching-vug and intergranular pore space are present. The distribution of particle size and sorting is controlled by upward-shoaling depositional processes, whereas interparticle and touching-vug pore space was produced by karst as well as compaction/cementation processes. In contrast, touching-vug porosity formed by prolonged karst dissolution is the predominate process in the Ordovician Ellenburger fields of West Texas. All these examples show how the relationship between petrophysical parameters and rock fabrics determines the important geologic parameters in carbonate reservoirs and can be used to convert geologic observations into numerical engineering data.

UTILITY OF MECHANICAL FACIES FOR ROCK CLASSIFICATION, CHARACTERIZATION, AND CORRELATION

K. A. Alhilali and G. Shanmugam
Mobil Research and Development Corporation
Dallas Research Laboratory
P. O. Box 819047
Dallas, Texas 75381-9047

ABSTRACT

Mechanical facies, which represent the rocks' dynamic and mechanical properties, can be used to characterize and subdivide reservoir formations at the borehole and in areas between wells. For example, the mechanical facies of the Travis Peak-Cotton Valley sequence of east Texas and western Louisiana can be used to distinguish and accurately map poor reservoir calcite-cemented zones. Mechanical facies are derived from acoustic logs where both compressional and shear wave velocities are obtained.

The advantages of using mechanical facies for reservoir characterization, classification, and subdivision are:

1. Mechanical facies are a quantification of a rock's in-situ physical properties and can reflect the dynamic characteristics of the reservoir.
2. Mechanical facies may provide useful information away from the borehole by combining acoustic logging and tomographic survey data to classify and characterize reservoir zones laterally as well as vertically.
3. Mechanical facies can be used for well-to-well correlation.
4. Mechanical facies provide a continuous record and can be obtained in uncased or, under proper conditions, cased wells.

The versatility of mechanical facies for reservoir description and characterization is illustrated by several examples from diverse areas and for varying lithology.

I. INTRODUCTION

The last few years have seen increased recognition of the importance of reservoir characterization in all phases of hydrocarbon production and management. As more oil and gas fields become candidates for secondary and enhanced oil recovery (EOR) projects, it is critical for the reservoir engineer to fully understand each specific rock heterogeneity and its relation to the dynamic characteristics of the reservoir. Reservoir characterization essentially dictates the strategy for EOR projects. EOR processes are complex, require large front-end capital inputs, and thus are less "forgiving" of mistakes. To reduce this technical and economic risk, reservoir description and characterization are critical elements in planning an EOR project. In offshore fields with large well spacing and the likelihood of significant reservoir heterogeneity between wells, detailed characterization is needed at the wellbore and away from the well.

Present techniques for reservoir characterization rely heavily on core description and logs, such as natural gamma, resistivity, and density. It is proposed that mechanical facies, which are significant in-situ rock properties, can be used effectively to subdivide and characterize reservoirs. Mechanical facies relate to the mechanical aspects of a rock body that may be recognized on mechanical properties logs. The expectation is that fundamental rock units with the same mechanical properties are likely to be part of a pattern and have similarity in a hydrocarbon reservoir. Furthermore, mechanical facies provide a continuous record and may be obtainable in old cased wells where core data are scant. Therefore, mechanical facies may be utilized by reservoir geologists and engineers, in conjunction with other data, to estimate reservoir parameters and build a model to predict the geometry of the reservoir and its dynamic behavior.

The primary object of mechanical facies is not to identify lithology, determine mineralogy, isolate lithofacies, or locate altered zones. Although mechanical facies are partly a manifestation of these formation properties in varying degrees, they are not very useful for deriving such properties. For example, the mechanical facies of a formation very often do not correspond with its lithofacies. The main objective of mechanical facies is to subdivide a rock formation based on its mechanical properties for the purpose of improved reservoir description and characterization.

Mechanical facies represent a bulk physical property of the rock and often reflect the dynamic characteristics of the reservoir, such as grain size, porosity, permeability, cementation, and clay content (i.e, the rock components and how they are put together). Mechanical facies are derived from the compressional and shear acoustic velocities of the formation. These velocities are obtained by running suitable acoustic logging tools. Although no examples are given in this paper, the velocities may be obtainable from a vertical seismic profile in a well or an acoustic tomography survey between wells. The wells can be uncased or, under favorable conditions, cased.

In this paper, several examples are shown to illustrate the significance of the mechanical facies as a reservoir property and its use in subdividing and characterizing the reservoir. The mechanical facies can be used to:

(a) improve porosity estimates from acoustic logs,
(b) obtain better permeability estimates from porosity, and
(c) subdivide a reservoir formation for well-to-well correlation and description.

The examples in this paper represent an initial endeavor. More work is needed to determine the true utility of mechanical facies as a tool in reservoir description and characterization, and as input data for reservoir simulation of a field.

II. MECHANICAL FACIES

Elastic moduli, such as Young's modulus, bulk modulus, and shear modulus, are measures of a rock's mechanical properties and hence can be used to characterize the rock. Zones characterized through the use of mechanical properties are called the mechanical facies of the reservoir.

We have found the ratio of a rock's incompressibility to its shear modulus, which is proportional to the sigma function (F), to be a robust mechanical properties parameter for dividing a reservoir into mechanical facies. We have observed that the sigma function is strongly influenced by grain size, mineral composition, and how the grains are put together as a rock unit. In other words, parameters which affect reservoir quality, such as grain size, porosity, permeability, cementation, and clay content, influence the magnitude of the sigma function. Furthermore, the sigma function is a dimensionless quantity, with a large range of sensitivity (from 1 to 40 for rock formations). Also, it is a bulk physical property of the rock, which can be quantified and used for well-to-well correlation.

The sigma function is computed from the ratio of the compressional and shear velocities using the following formula:

$$F = \frac{1+\nu}{1-2\nu}\,, \qquad (1)$$

where

$$\nu = \frac{0.5r^2 - 1}{r^2 - 1}$$

is Poisson's ratio, and

$$r = \frac{\text{compressional velocity } (V_p)}{\text{shear velocity } (V_s)}\,.$$

Equation (1) is derived from the ratio of the incompressibility (k) to the shear modulus (μ); these are defined as

$$k = \frac{\nu E}{(1+\nu)(1-2\nu) + 2\mu/3}\,,$$

$$\mu = \frac{E}{2(1+\nu)}\,,$$

where E is Young's modulus. By simple algebra the ratio k/μ can be reduced to the following:

$$\frac{k}{\mu} = \frac{2}{3}\left(\frac{1+\nu}{1-2\nu}\right)$$

$$= \frac{2}{3}F$$

The compressional and shear wave velocities are obtained from running suitable acoustic logging tools, or possibly from a vertical seismic profile (VSP) in a well or a cross-hole tomography survey. Thus, mapping mechanical facies may be accomplished in areas between wells to achieve quantitative reservoir characterization in two dimensions. With this technique, physical description of a reservoir formation away from the borehole is possible.

III. EXAMPLES

Selected examples are presented here to show the versatility of using mechanical facies to subdivide a reservoir for description and characterization. However, many of these examples are from individual wells. Mechanical facies have not been used, as yet, in a field-wide reservoir study where it is anticipated they will provide the most benefits.

A. Improved Porosity Estimates From Acoustic Logs

Porosity estimates derived from acoustic logs (compressional or shear) in complex or mixed lithology are often highly inaccurate. We have found that the precision of these estimates can be significantly improved if the formations are subdivided into classes based on the value of the sigma function before the estimates are made. A separate transform relating slowness (the reciprocal of velocity) to porosity is then used for each class. In other words, formations with a sigma function of, say, 2.6 will have a different transform than formations with a sigma function of 3.6. To illustrate this concept, we will examine porosity estimation in carbonate lithology using compressional slowness (DTp) and shear slowness (DTs).

Figure 1 is a crossplot of (core-measured) porosity and DTp (measured by acoustic logging) for a large number of carbonates from various fields around the world. These carbonates include limestone, dolomite, chalk, dolomitic limestone, and cherty limestones. The best least-squares line fit and the zone where porosity can be estimated within two porosity units (indicated by two dashed lines) are included in the figure. The large scatter in the data and the magnitude of the error in some of the estimates signify the problem with the acoustic log as a tool for reliable porosity estimation even in carbonate lithology. The unreliability of the estimates is due to the variability of the composition of the matrix and, hence, the assumption that a single matrix velocity can be used for all carbonates cannot be valid. The estimates can be improved considerably if the matrix composition and, hence, its velocity is known. When matrix composition is unknown, the value of the sigma function can be used to divide the formations into groups, providing shear slowness data are available. Each group will then have different mechanical properties. If the mechanical properties reflect the different composition and matrix structure of each group, which they often do, then estimates of porosity, based on a least-squares line fit for each group, will be more reliable. Each of these groups will have a different matrix velocity estimate and will lead to better porosity estimates.

Based on the value of the sigma function, the carbonate data were grouped into five classes (mechanical facies), and a best straight line fit for DTp and porosity was then determined for each class. Figure 2 includes all data for which the sigma function is less than 2.66; notice the least-squares line fit for this group of data is significantly different from the line determined for all the data in Figure 1. Figure 3 shows all data for which the sigma function is greater than 3.61; notice the least-squares line fit is different again. Furthermore, Figure 3 indicates that the sensitivity of DTp to porosity for carbonates with sigma function greater than 3.61 is very low; it is not advisable to use DTp to estimate porosity for this class of carbonates. Nonetheless, when porosities are estimated separately for each mechanical facies and compared with core-measured porosity, significant improvement is achieved as can be seen from Figure 4. This figure shows that by having an independent slowness-porosity correlation for each of the five mechanical facies, instead of one for all carbonates, the reliability of porosity estimates from acoustic logs is markedly enhanced. Furthermore, the value of the sigma function can be an indication of how reliable a porosity estimate obtained from the acoustic logs can be. Note that in analyzing this set of data no assumption was made as to the actual lithology or mineralogy of the carbonates.

B. Estimation of Permeability From Porosity

Of all the formation parameters that are used for reservoir management and development, permeability is one of the most critical. Permeability is required to determine the optimal drainage points and production rate, to specify optimal completion and perforation design, and to devise EOR patterns and injection conditions. Commonly, permeability estimates are derived from porosity estimates via an empirical transform. Since permeability-porosity correlation is often very poor, the permeability estimates from such a technique are unreliable.

An intrinsic permeability-porosity relationship for a given formation is highly dependent on grain size distribution. A rock's mechanical properties, to a large extent, are also controlled by the size of the grains and how they are "put together." Therefore, mechanical facies can be used to divide a reservoir formation prior to deriving a permeability-porosity transform. By devising a separate transform for each mechanical facies of the reservoir, estimation of permeability from porosity can be improved significantly.

To illustrate this technique, Figure 5 is a cross-plot of core-measured porosity and core-measured permeability for 52 carbonate cores from a single lithofacies (shoal/shelf carbonates). The figure shows that the excellent porosity of these carbonates is not always associated with good permeability, and any attempt to estimate permeability from porosity using a single transform for this reservoir will not be successful.

Examination of the sigma value of these cores revealed that two distinct mechanical facies are present (see Figure 6). One mechanical facies, which we call W, has a sigma function value around 2.80, while the other mechanical facies, which we call Q, has a sigma function value around 2.96. When these two mechanical facies are separated, it can be seen that mechanical facies W has good porosity and permeability, while mechanical facies Q has good porosity, but relatively poor permeability (see Figure 7). Therefore, to obtain good permeability estimates from porosity, we must use two different transforms, one for mechanical facies W and another one for mechanical facies Q.

The results of the estimation are summarized in Figure 8 which shows the benefits of utilizing mechanical facies when porosity is being used to estimate permeability. The circles in the figure indicate the predicted values of core permeabilities when all data are used to derive a transform (fourth-order polynomial), while the triangles represent the predicted values of core permeabilities when two separate transforms, one for facies W and another for facies Q, are derived (fourth order polynomial in each case). Notice how useless the permeability estimates (circles) are when mechanical facies are not considered (present existing method). Furthermore, Figure 8 shows that most low permeability streaks in this reservoir can be identified from the estimates when mechanical facies are used.

C. Reservoir Description and Well-to-Well Correlation

Example 1: Edwards Formation, Darst Creek Field, Central Texas

Figure 9 shows the sigma function plotted together with gamma, neutron porosity and bulk density, oil saturation, and tube (Stonely) wave logs for the Edwards and Georgetown formations in a Darst Creek well. The tube wave log is an acoustic log and represents the slowness of the tube (Stoneley) wave which can be used as an indication of porosity and fluid mobility in the formation. The character and the value of the sigma function were used to divide the two formations into several mechanical facies as indicated in the middle track.

Two distinct mechanical facies (A and B) can be recognized in the Georgetown limestone. The high value of the sigma function for facies B indicates the argillaceous nature of this zone, which was confirmed by core examination.

The Edwards formation is divided into three mechanical facies (A, B, and C). Facies A and B constitute the upper Edwards, while facies C is the lower Edwards. Core-measured grain density suggests that the upper part of facies A in the Edwards limestone has been dolomitized, yet it has had almost no effect on the quality of the reservoir as indicated by the values of the sigma function and porosity logs. The significant drop in the sigma function in the lower Edwards (facies C) indicates the siliceous nature of the carbonates in this zone. The extremely low values between 2,650 and 2,658 ft. may represent ancient hard grounds. These hard grounds are normally burrowed and contain chert nodules and siliceous mud. This conclusion was confirmed by core examination (see Figure 10). None of the other logs in Figure 9 contain any indication of the presence of these chert nodules which can be a significant factor in the hydrodynamics of the reservoir. Also notice that the natural gamma log (left-hand track) is almost featureless and is not very useful for subdividing, characterizing, or correlating these formations from well to well.

Figure 11 shows the results of the conventional logs and core analysis. Notice the correlation between the lithology derived from logs and cores (right-hand column on Figure 11) and the mechanical facies.

Example 2: Carbonate-Cemented Sandstone of the Travis Peak and the Cotton Valley Formations - Waskom Field, East Texas

Figures 12 and 13 are plots of various logs for the Travis Peak and Cotton Valley sequence. As shown by the lithology log (extreme left side of each figure), these formations consist of sand/shale sequences with the sand becoming calcareous (mostly calcareous cement) in certain intervals.

In the left track of each figure the sigma function log, natural gamma log, and SP log are plotted. The natural gamma log and sigma function log are very similar wherever sands are clean and contain only a minor calcareous cement (see Figure 13 for example). However, when a sand becomes calcareous, the sigma function increases significantly, while the natural gamma log remains unaffected. For example, the natural gamma log for the sand in the interval 8,060-8,148 ft. of Figure 12 is nearly constant, while the sigma function in the interval 8,132-8,148 ft. where the sand is clean and free of calcite cement is significantly lower than it is in the zones where calcite cement is present.

Therefore, the sigma function in this case can be used very effectively to map clean sand zones and distinguish poor-reservoir calcite cemented intervals in the Travis Peak and the Cotton Valley sequence of east Texas and western Louisiana. This is significant not only from a logging point of view, but also shows the possibility that such zones can be detected and mapped using vertical seismic profiles (VSP) or even surface seismic data (compressional and shear).

Similar experience has been reported in the upper Cretaceous Tuscaloosa sand[8]. Two distinct types of sandstones are recognized in the Tuscaloosa sand: (1) silica-cemented, well-indurated, and fairly porous sandstone (porosity = 7-22) and (2) chlorite-cemented, semi-friable, and very porous sandstone (porosity = 22-32). The first type of Tuscaloosa sand has a sigma function value around 2.53, while the second type has a sigma function around 10. Therefore, the sigma function can be used to distinguish the two types of reservoir rock which is critical for the management and the exploitation of this very important gas reservoir. The Tuscaloosa sand is an important reservoir target in Louisiana, Mississippi, and Alabama.

Example 3: Gulf Coast Sand/Shale Sequence Callaghan Ranch Field

Figure 14 is a plot of various logs for a moderately consolidated Tertiary sand/shale sequence from a well in the Callaghan Ranch Field, south Texas. The similarity between the natural gamma log and the sigma function log (lower part of figure) is obvious. The advantages of the sigma function over the natural gamma log to subdivide and characterize *these sand/shale sequences* are:

1. Since the sigma function represents a physical bulk mechanical property of the formation, it can always be used on a uniform basis as a quantitative measure to determine the quality of the reservoir. For example, based on the values of the sigma function for sand A in Figure 14 (8,040-8,050 ft.) and sand B (8,250-8,270 ft.), it can be confidently stated that sand A has a better reservoir quality than sand B. Further, using the difference between the values of the sigma function in these two sands, it is possible to determine how much "better" sand A is. This is not always possible when the natural gamma log is used.

2. In a similar manner, the value of the sigma function can be used to classify shales according to their mechanical strength and stability. This can be extremely useful in drilling and enhanced recovery projects.

Depending on the degree of consolidation of sand/shale sequences, the value of the sigma function is lowered by the presence of gas. Therefore, gas correction may be required in some cases. Such a correction is made using the well-known Gassman equation.

Example 4: Gulf of Mexico Unconsolidated Sand/Shale Sediments

In loosely consolidated sediments, there exist two factors which can affect the magnitude of the sigma function significantly; namely, gas content and pressure variations. For shallow formations (less than 6,000 ft.) where the overburden pressure gradient is high, the sigma function gradient must be computed and removed before the sigma function log can be used to characterize the formations. In geopressured zones, not only does the shear velocity drop sharply and significantly, but also it becomes low in energy. Compressional velocity also drops, but the decrease is smaller and seems to be transitional. Therefore, in geopressured zones, shear velocity and, hence, the sigma function cannot always be measured reliably. The magnitude of the "gas" correction can be estimated from the Gassman equation as previously mentioned. This has not been done for the examples given here.

Figure 15 is a comparison of the natural gamma and the sigma function logs from a Grand Banks well, offshore Louisiana, which shows that the sigma function can be used to subdivide lithology just as well as the natural gamma log. The top of the geopressure in this well is at about 5,990 ft. Below this depth, the value of the sigma function is lower for both sands and shales compared to similar normally-pressured sands and shales. The sigma function log indicates that the zone from 5,680-5,880 ft. is more shaley than is indicated by the gamma log. No cores are present in this interval to determine the source of this difference. Minor gas is present in the sand from 5,535-5,600 ft. The sigma function is not a good indicator of the presence of gas; however, a decrease in the sigma function accompanied by a significant increase in the compressional slowness (DTp) can be an indicator of the presence of gas and light hydrocarbons.

Figure 16 is a plot of the sigma function from a Green Canyon well, offshore Louisiana, together with sidewall measured permeability. The sigma function log shows a thin interbedded sand/shale sequence at 10,140-10,240 ft. Lower values of the sigma function are an indication of a good quality reservoir (good permeability) and possibly gas (e.g., 10,330-10,350 ft.).

Example 5: Austin Chalk, Darst Creek Field, Central Texas

The Austin Chalk formation is a relatively thick sequence of Upper Cretaceous limestone. Its textural spectrum varies between lime mudstone, mud-supported wackestone, and wackestone. Figure 17 shows the sigma function, together with the core-measured grain density for the Austin Chalk in the Darst Creek Field. Notice the significant contrast in the sigma function between the overlying Pecan Gap formation, which is an argillaceous organic-rich dark lime mudstone, and the light colored, almost pure limestone (calcium carbonates) of the Austin Chalk. Based on the shape and value of the sigma function, the Austin Chalk can be subdivided into seven mechanical facies (see Figure 17). The high value of the sigma function of facies C indicates that this unit contains clay. This is confirmed by the SEM examination of the core taken at a depth of 2,237 ft. (see Figure 18). Units A, C, D, and G represent old hard grounds which are burrowed and may contain glauconitic intraclasts. However, only C facies contain Kaolinitic clay. Figure 19 shows micrographs of the various mechanical facies found in the Austin chalk.

Example 6: Sand-rich Submarine Fan Complex of the Potter Formation, San Joaquin Valley, California

The lithofacies of the Potter sand consist of deep-marine braided channel, meandering channel, levee, and crevasse splay deposits. These lithofacies are all rich in sand but vary in grain size, thickness, and permeabilities across the reservoir. Some of these facies are known to be lenticular and wedge out laterally. The sand varies from gravel and coarse sand to fine sand with little or no clay. Therefore, a gamma log is not useful for reservoir characterization or well-to-well correlation in this area (see Figure 20). Grain size and permeability variations appear to govern the mechanical facies here to a large extent, as can be seen in Figure 21, which is a north/south profile relating the sigma function log in Well S to the core description in two adjacent wells. No cores were taken in Well S.

Example 7: Fractured Granitic Rock, Cajon Pass, California

Figure 22 shows the variation of the sigma function in fractured granitic basement rock from the Cajon Pass No. Fed. 2-26 Well in southern California. Lithology and fracture density distribution are shown in the top of the figure. The compressional velocity in this interval is also shown in the lower part of the figure for comparison. Notice the significant decrease in the value of the sigma function for the Mylonite granite (MG) as we go from the 1,900-1,950 m zone to the 2,000-2,040 m zone. This decrease is caused by an increase in the density of fracture in the deeper zone.

Figure 22 clearly shows that rock type, as well as the density and the pattern of fracturing, determines the value of the sigma function for these formations. Therefore, the sigma function log can be used to distinguish highly fractured zones in this type of formation and to correlate, between wells, the various types of basement rocks which exist in this area.

IV. SUMMARY AND CONCLUSIONS

The potential application of mechanical facies as a rock property for the purpose of reservoir description and characterization is significant. For the first time, reservoirs can be subdivided based on the bulk mechanical property of the rock which often reflects the dynamic characteristics of the reservoir, the key in reservoir development, simulation, and modeling. Furthermore, using mechanical facies, it may be possible for the first time to characterize reservoirs in-between wells by using the techniques of seismic tomography and vertical seismic profiling (VSP). Thus, it is now possible to characterize reservoirs in two dimensions.

It has been shown, by way of examples, how mechanical facies can be used to obtain better estimates of porosity and permeability from acoustic logs, delineate calcite cemented zones in the Travis Peak-Cotton Valley reservoirs of east Texas, determine the degree of sand shaliness in Gulf Coast and Gulf of Mexico reservoirs, delineate fracture zones in highly consolidated and crystalline rocks, and distinguish paleo hard grounds in the Austin Chalk, etc. These properties are the key to exploration for, and the development of, hydrocarbon reservoirs in the mentioned areas.

However, mechanical facies are not equivalent to, or a substitute for, lithofacies. Although in some cases there may be some correspondence between the two types of facies, such as the one described for the Potter sand in the San Joaquin valley, they are often not related.

Furthermore, mechanical facies can be used for simple well-to-well correlation and to classify shale formations for the purpose of identifying "problem shales" during drilling, completion, and hydraulic fracturing.

Mechanical facies are derived from the compressional and shear velocities of the rocks. They are simple to obtain at the well site and are a byproduct of running a suitable commercially available acoustic tool, or possibly from a vertical seismic profile (VSP) in a well or a tomographic survey between wells.

V. ACKNOWLEDGMENT

Several discussions with and comments from D. M. Williams and J. R. Bulau were very helpful in developing this technique. This paper includes several important modifications suggested by C. C. Purdy and J. A. Quirein.

REFERENCES

1. Baltensperger, P.A., 1990, "Case Study: Defining the Morrow (Pennsylvanian) Using Multicomponnent Seismic Data" presented at the 60th. Annual SEG Meeting.
2. Benzig, W. M., 1978, "V_S/V_P Relationship in Carbonates and Sandstones Laboratory Data," presented at 48th Annual SEG Meeting.
3. Castagna, J. P., Batzle, M. L., and Eastwood, R. L., 1985, "Relationship Between Compressional Wave and Shear Wave Velocities in Clastic Silicate Rocks," *Geophysics*, 50, No. 4, pp. 571-581.
4. Domenico, S. N., 1984, "Rock Lithology and Porosity Determination from Shear and Compressional Wave Velocity," *Geophysics*, 49, No. 8, pp. 1188- 1195.
5. Dowla, N. et al., 1990, "Laboratory Investigation of Saturation Effect on Mechanical Properties of Rocks," presented at 31st Annual SPWLA symposium, pp. 1-22, paper EE.
6. Eastwood, R. L., and Castagna, J. P., 1983, "Basis for Interpretation of V_P/V_S Ratios in Complex Lithologies," presented at 24th Annual SPWLA symposium, pp. 1- 17.
7. Gregory, A. R., 1976, "Fluid Saturation Effects on Dynamic Elastic Properties of Sedimentary Rocks," *Geophysics*, 41, No. 5, pp. 895-921.

8. Hosseini, M. ., and Hayatdavoudi, A., 1986, "Reservoir Characterization of Tuscaloosa Sand by Mineralogical and Petrophysical Data," SPE Formation Evaluation, December, 1986, pp. 584-594.

9. Inderweisen, P.L., and Lo, T., 1990, "Cross-Hole Seismic Tomographic Imaging of Reservoir Inhomogeneities in the Midway Sunset Field, California" Presented at the 60th. Annual SEG meeting.

10. Iverson, P. W., Fahmy, B. A., and Smithson, S. B., 1989, "V_p/V_s from Mode Converted P-SV Reflections," *Geophysics*, 54, No. 7, pp. 843-852.

11. Johnston, D. H., 1987, "Physical Properties of Shale at Temperature and Pressure," *Geophysics*, 52, No. 10, pp.1391-1401.

12. Pardus, Y.C. & et. el., 1990 "Vp/Vs and Lithology in Carbonate Rocks: A case study in the Scipio Trend in Southern Michigan" Presented at the 60th. Annual SEG Meeting.

13. Robertson J. D., 1987, "Carbonate Porosity from S/P Travel Time Ratio," *Geophysics*, 52, No. 10, pp. 1346-1354.

14. Prakla-Seismos, 1983, "The Vp/Vs Ratio, Lithological and Facial Statements," Prakla-Seismos Report No. 2, pp. 18-24.

15. Savre, W. C., and Burke, J. A., 1963, "Determination of True Porosity and Mineral Composition in Complex Lithologies with the Use of Sonic, Neutron, and Density Surveys," *The Log Analyst*, 4, No. 1, pp. 439-474.

16. Schmoker, J. W., and Schenk, C. J., 1988, "Facies Composition Calculated from the Sonic, Neutron, and Density Log Suite, Upper Part of the Minnelusa Formation, Powder River Basin, Wyoming," *The Mountain Geologist*, 25 No. 3, pp. 103-112.

17. Sonneland, L., Barkved, O., Olsen, M., and Snyder, G., 1989, "Application of Seismic Wavefield Attributes in Reservoir Characterization," GECO Report.

18. Suyama, K., Imai, T., Ohtomo, H., Ohta, K., and Tahashi, T., 1984, "Delineation of Structure in Alluvium and Dilluvium Using SH-Wave Reflection and VSP Methods," Presented at the 54th. Ann. Inter. Mtg. of the SEG, Atlanta.

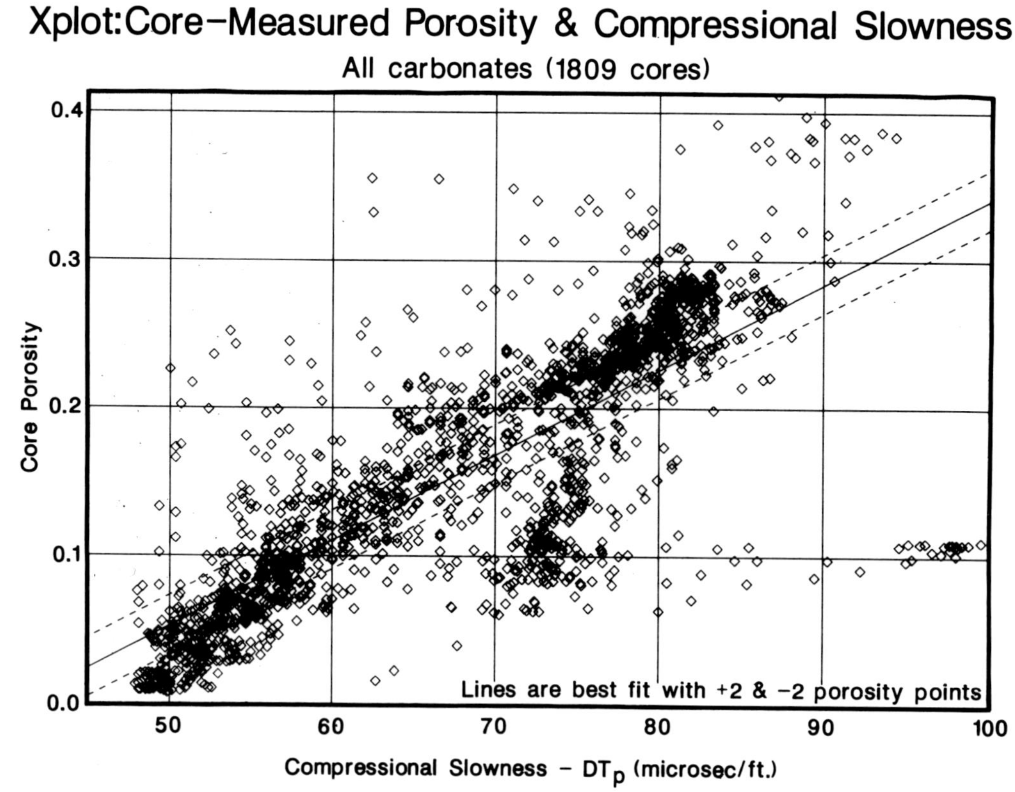

Figure 1

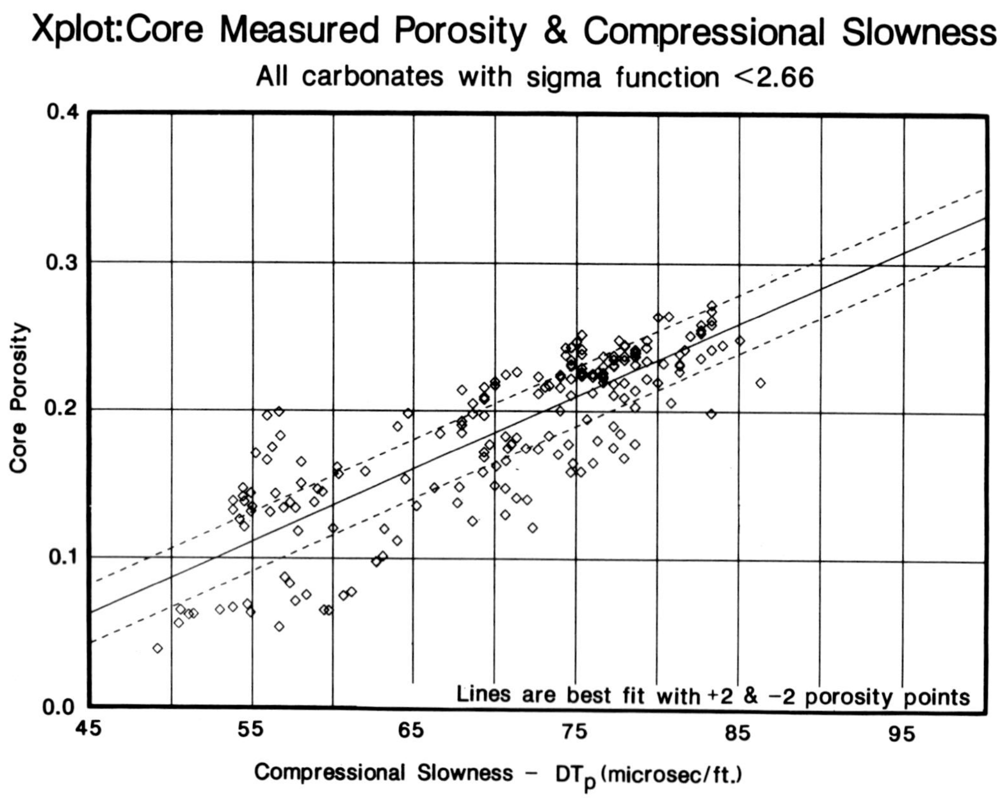

Figure 2

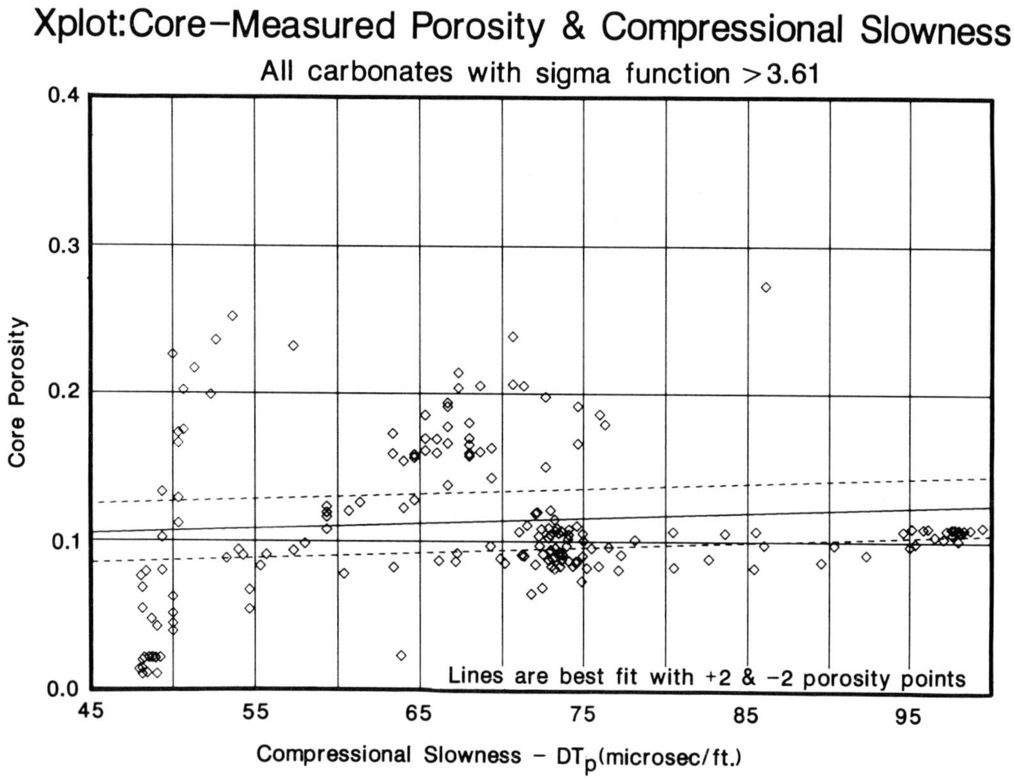

Figure 3

Porosity Estimation Error Comparison: All Carbonates

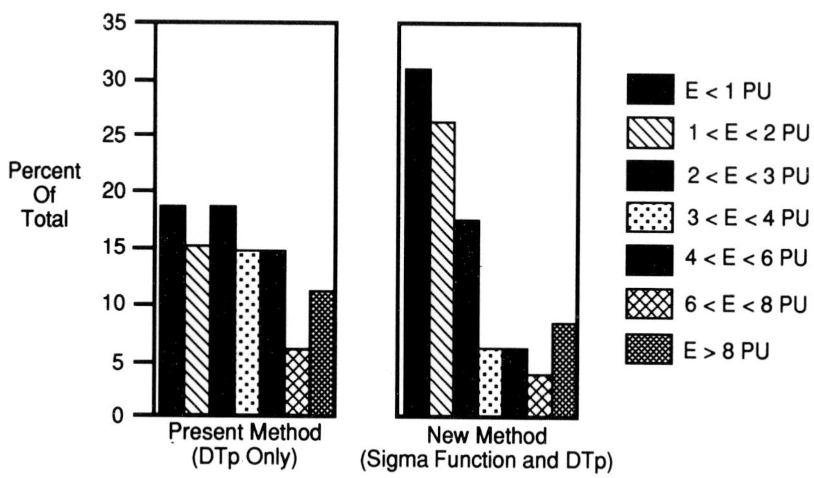

Estimation Error E (In Porosity Units); Total = 1809 Points In Each Case

Figure 4

Xplot of Core-Measured Permeability & Porosity

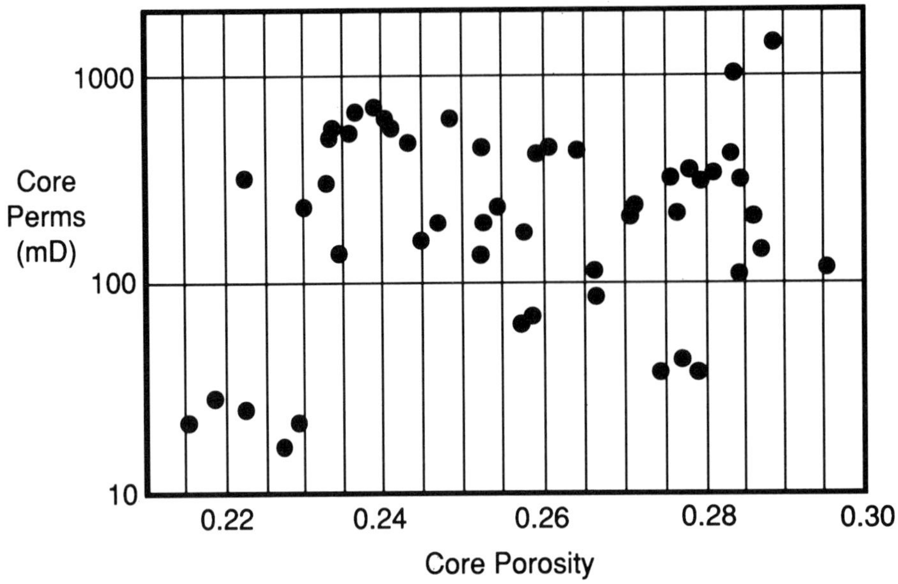

Figure 5

Xplot of Porosity & Sigma Function

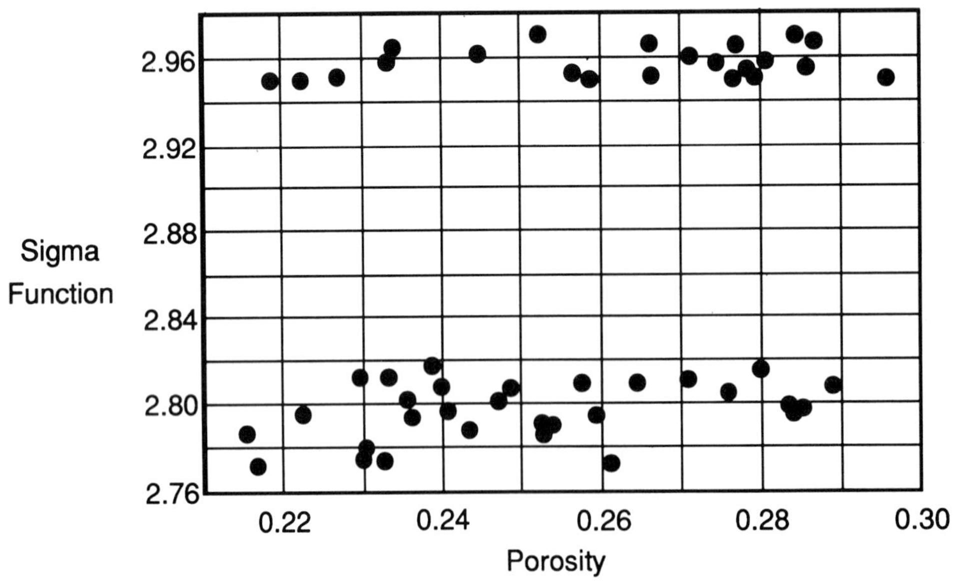

Figure 6

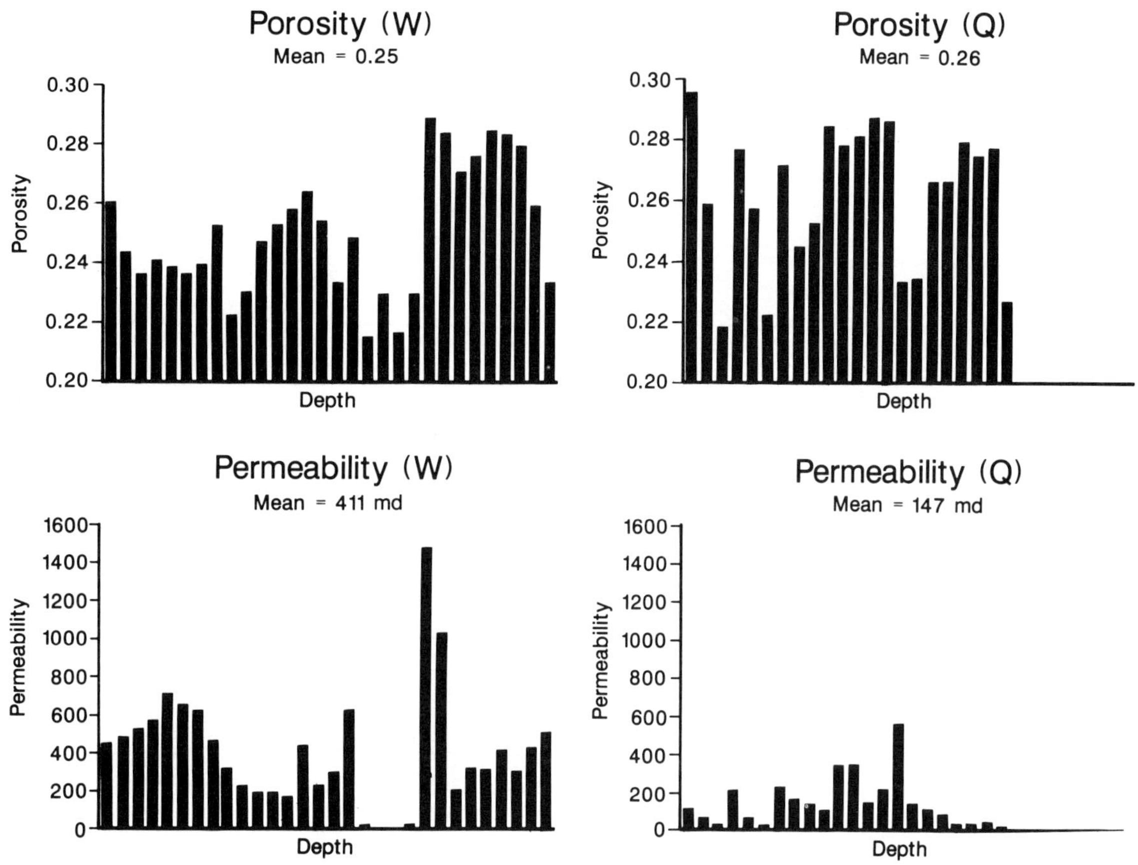

Figure 7

Permeability Estimation From Porosity (Carbonates)

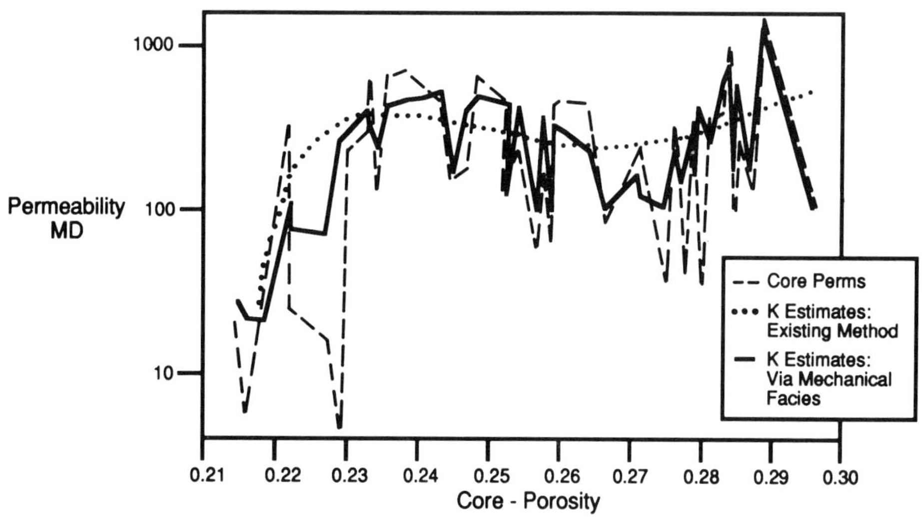

Figure 8

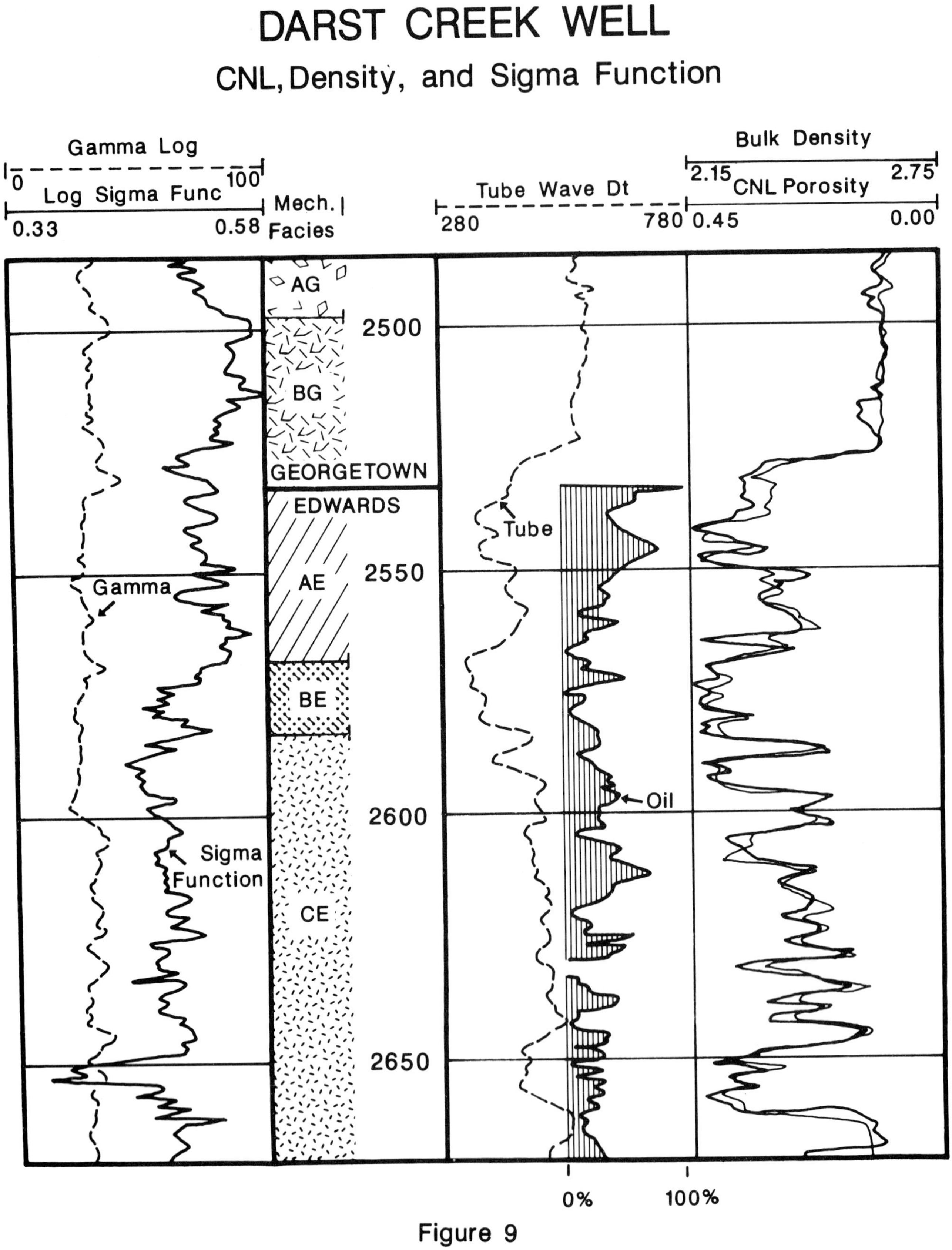
DARST CREEK WELL
CNL, Density, and Sigma Function
Gamma Log
0
100
Log Sigma Func
0.33
0.58
Mech. Facies
Tube Wave Dt
280
780
Bulk Density
2.15
2.75
CNL Porosity
0.45
0.00
AG
2500
BG
GEORGETOWN
EDWARDS
AE
2550
BE
CE
2600
2650
Gamma
Sigma Function
Tube
Oil
0%
100%

Figure 9

DARST CREEK FIELD
GUADALUPE CO., TEXAS
2647'– 62'

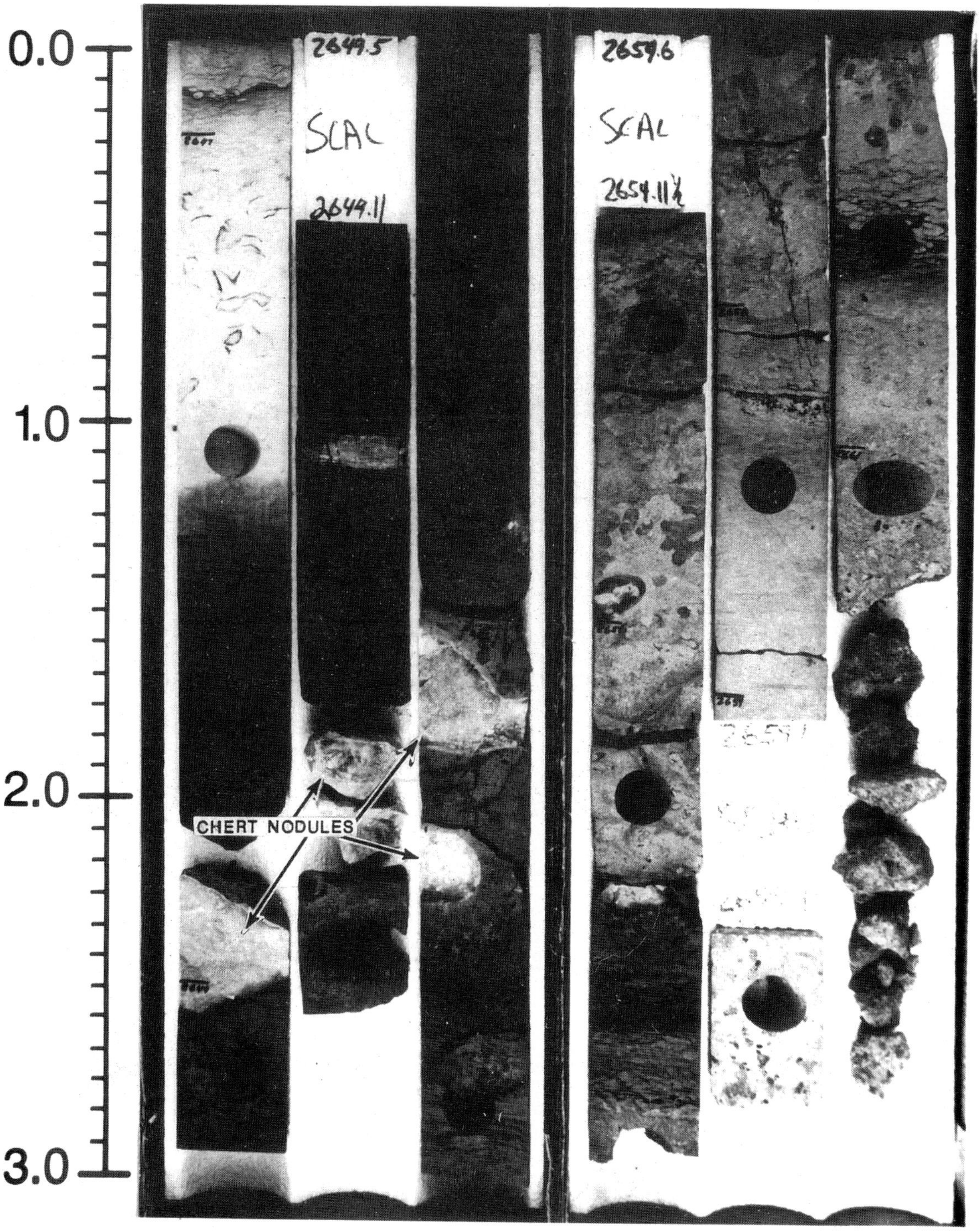

Figure 10

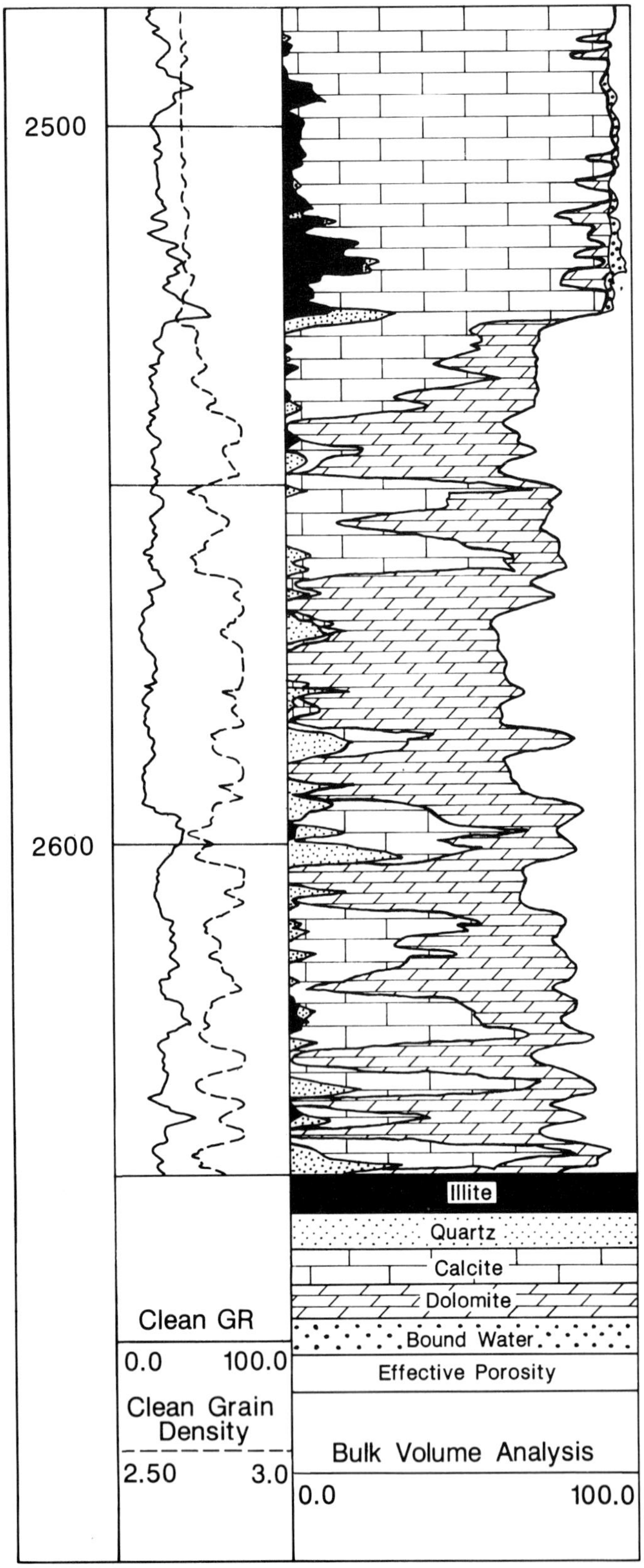

Figure 11

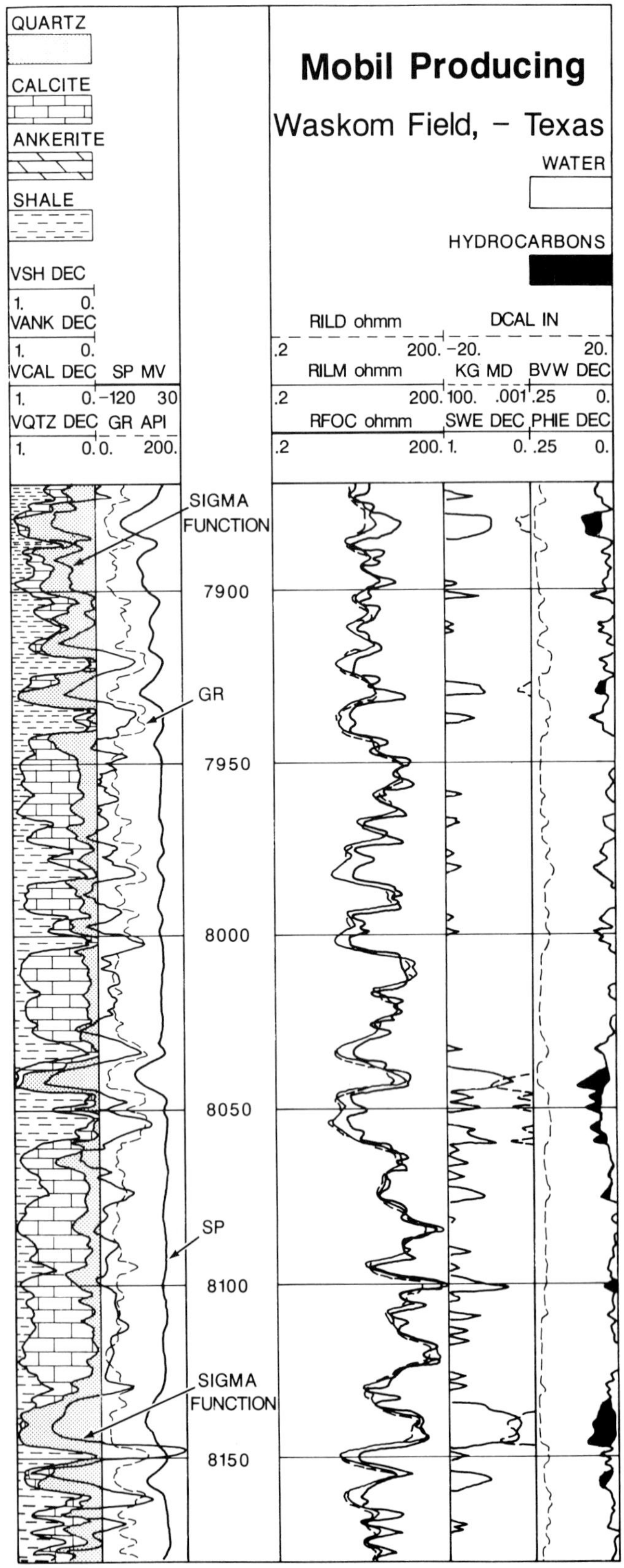

Figure 12

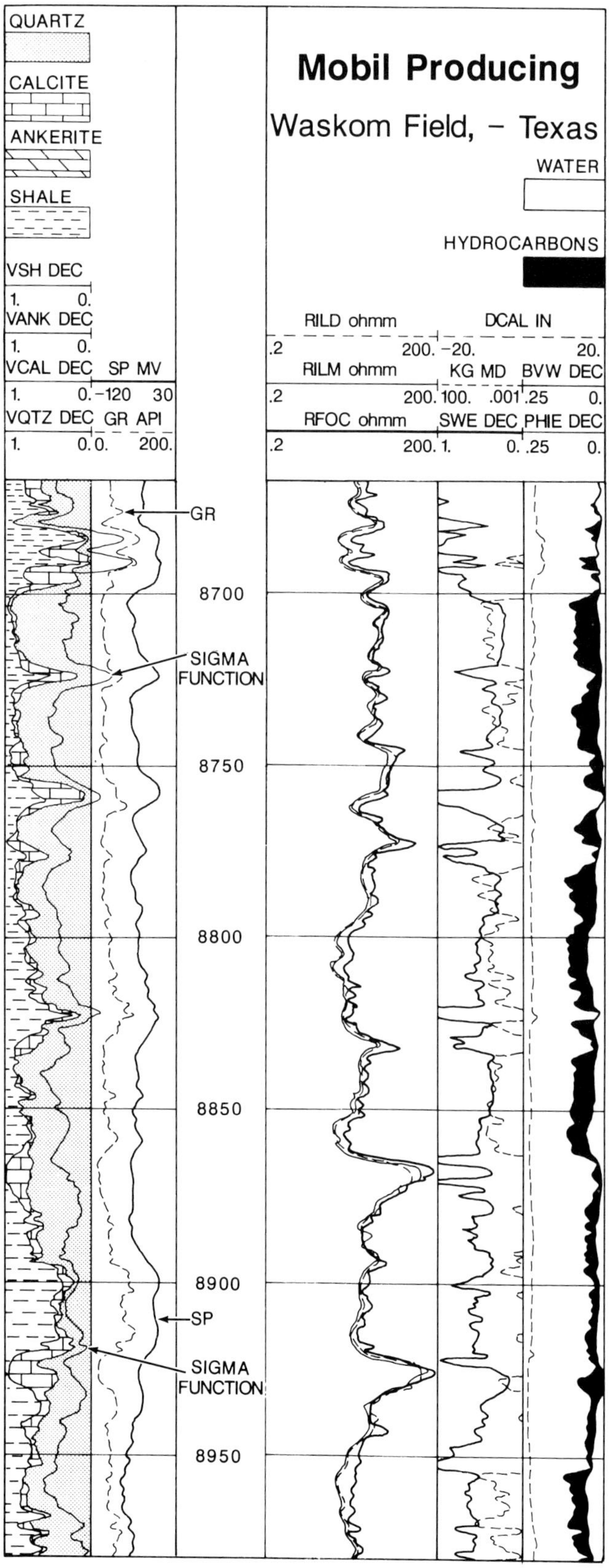
QUARTZ
CALCITE
ANKERITE
SHALE
VSH DEC
1. 0.
VANK DEC
1. 0.
VCAL DEC
SP MV
1. 0. -120 30
VQTZ DEC
GR API
1. 0. 0. 200.
Mobil Producing
Waskom Field, - Texas
WATER
HYDROCARBONS
RILD ohmm
DCAL IN
.2 200. -20. 20.
RILM ohmm
KG MD
BVW DEC
.2 200. 100. .001 .25 0.
RFOC ohmm
SWE DEC
PHIE DEC
.2 200. 1. 0. .25 0.
GR
8700
SIGMA FUNCTION
8750
8800
8850
8900
SP
SIGMA FUNCTION
8950

Figure 13

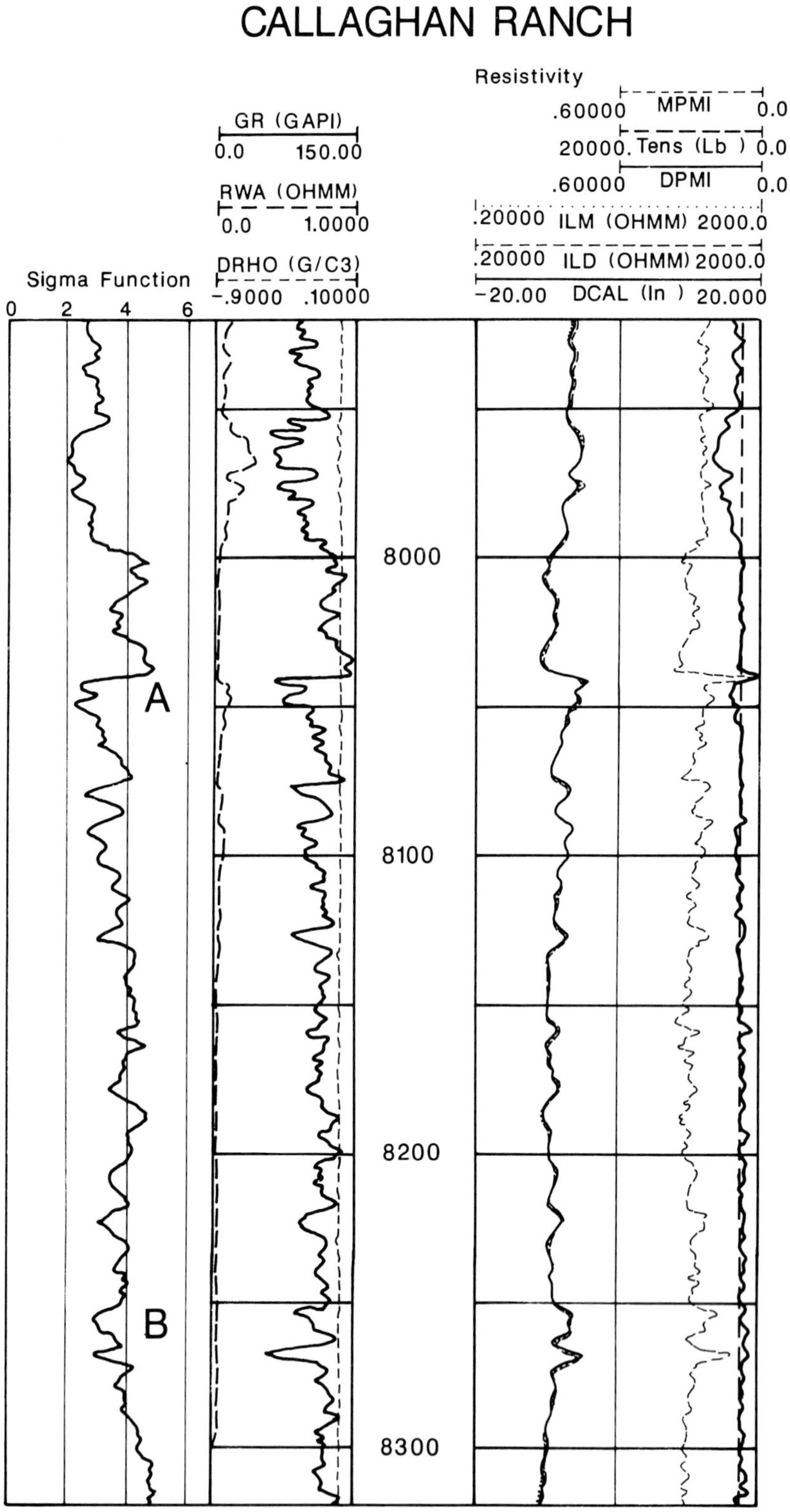
CALLAGHAN RANCH
Resistivity
.60000 MPMI 0.0
20000. Tens (Lb) 0.0
.60000 DPMI 0.0
.20000 ILM (OHMM) 2000.0
.20000 ILD (OHMM) 2000.0
-20.00 DCAL (In) 20.000
GR (GAPI)
0.0 150.00
RWA (OHMM)
0.0 1.0000
DRHO (G/C3)
-.9000 .10000
Sigma Function
0 2 4 6
8000
8100
8200
8300
A
B

Figure 14

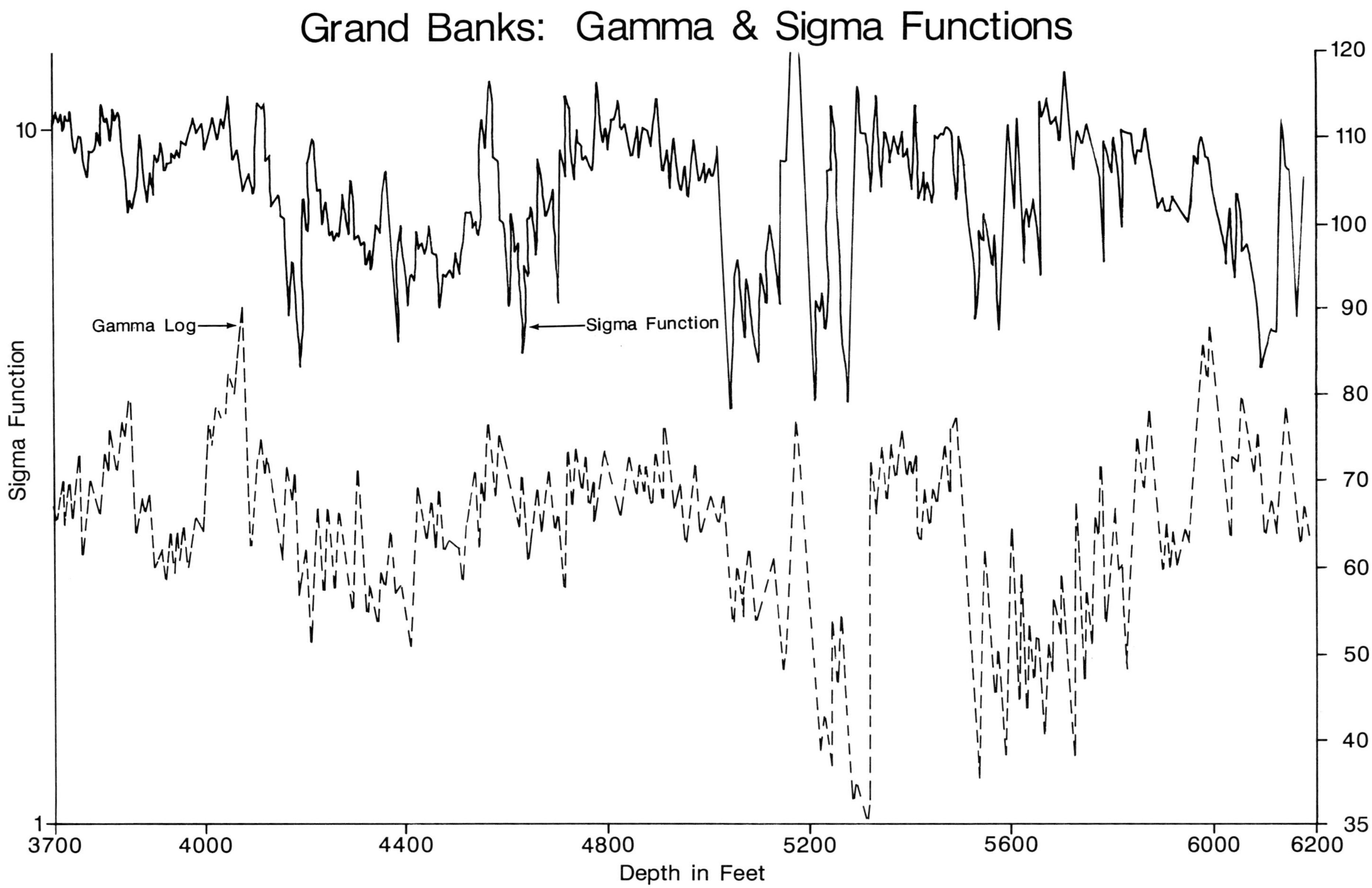
Grand Banks: Gamma & Sigma Functions
Gamma Log
Sigma Function
Sigma Function
Gamma
10
1
120
110
100
90
80
70
60
50
40
35
3700
4000
4400
4800
5200
5600
6000
6200
Depth in Feet

Figure 15

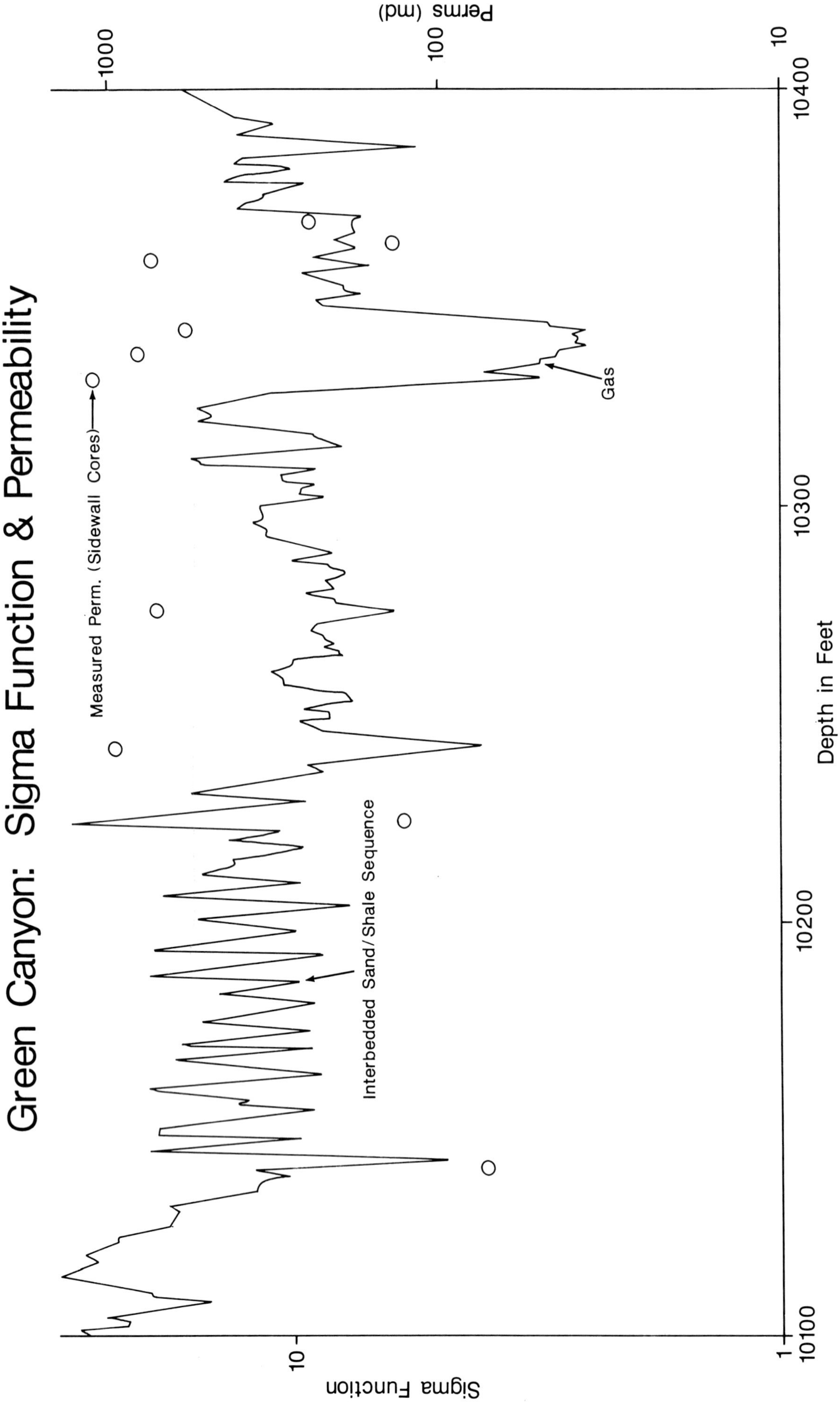
Green Canyon: Sigma Function & Permeability
Measured Perm. (Sidewall Cores)
Interbedded Sand/Shale Sequence
Gas
Perms (md)
1000
100
10
Sigma Function
10
1
10100
10200
10300
10400
Depth in Feet

Figure 16

Darst Creek: Core Grain Density & Sigma Function

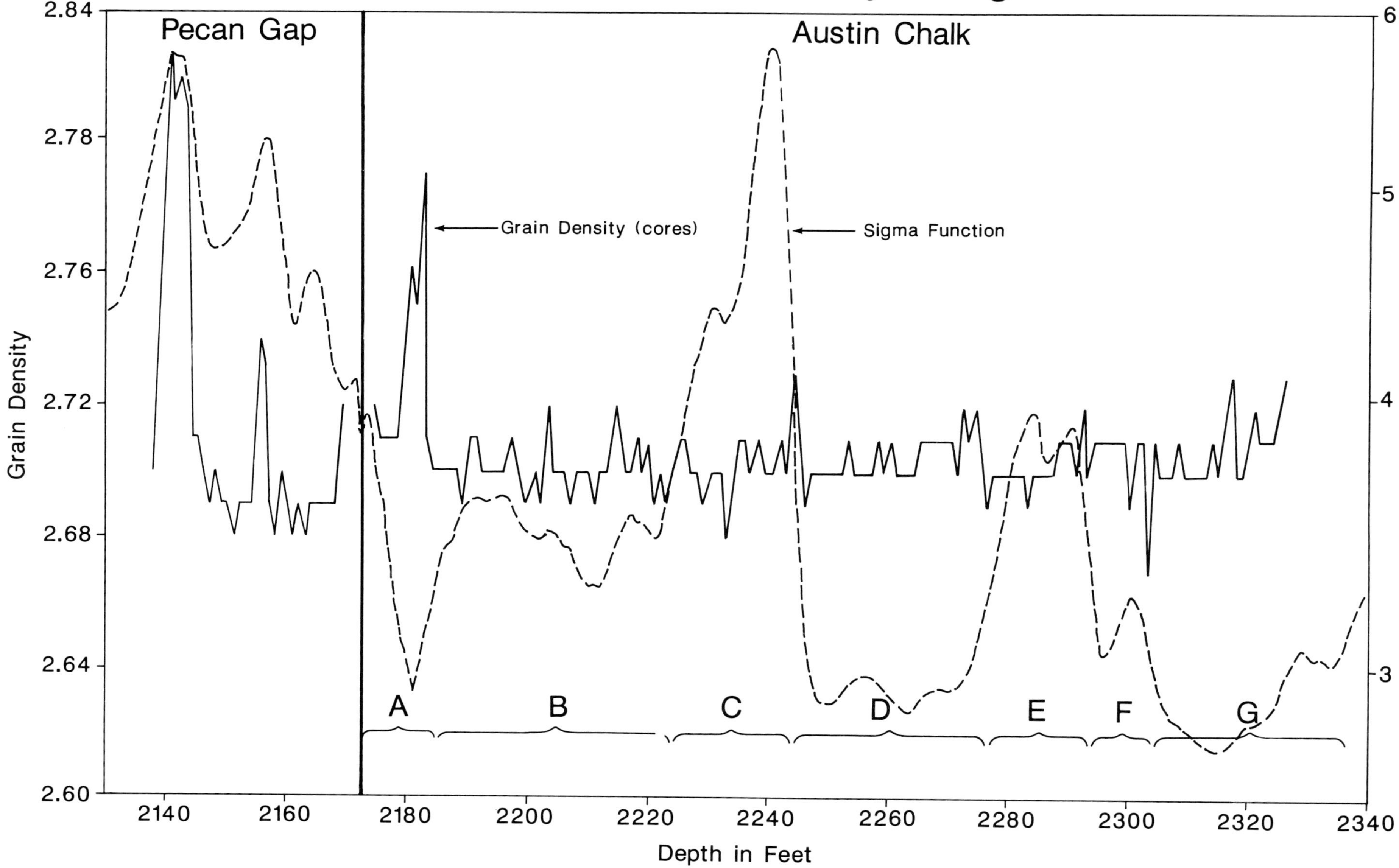

Figure 17

Darst Creek Well

2237 Feet (Austin Chalk)

Kaolinite booklets (E - J2 - 6) are numerous on a fracture plane.

Scale bar (in lower right) equals 10 microns

Figure 18

Darst Creek Well

Contrasting Three Mechanical Facies In The Austin Chalk Facies A, C, and D

2183 Feet

Scale bar (in lower right of each micrograph) equals 10 microns

2230 Feet

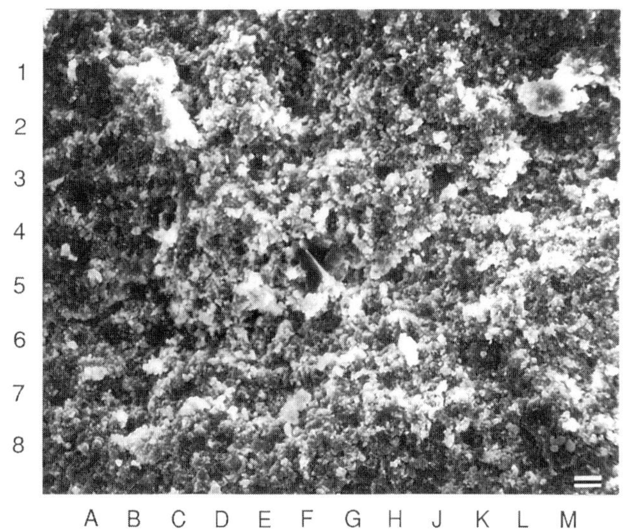

2248 Feet

Figure 19

WELL A

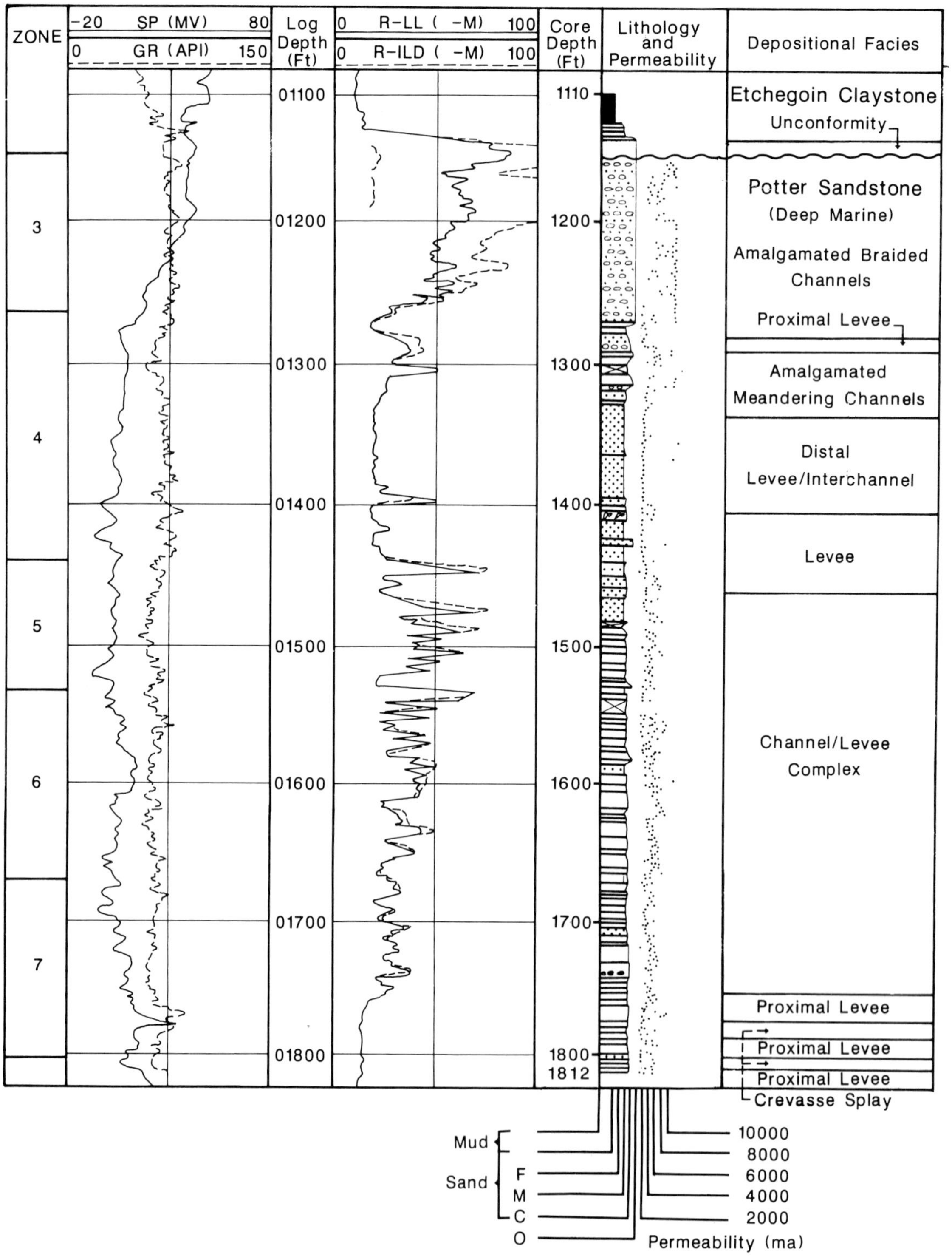

Figure 20

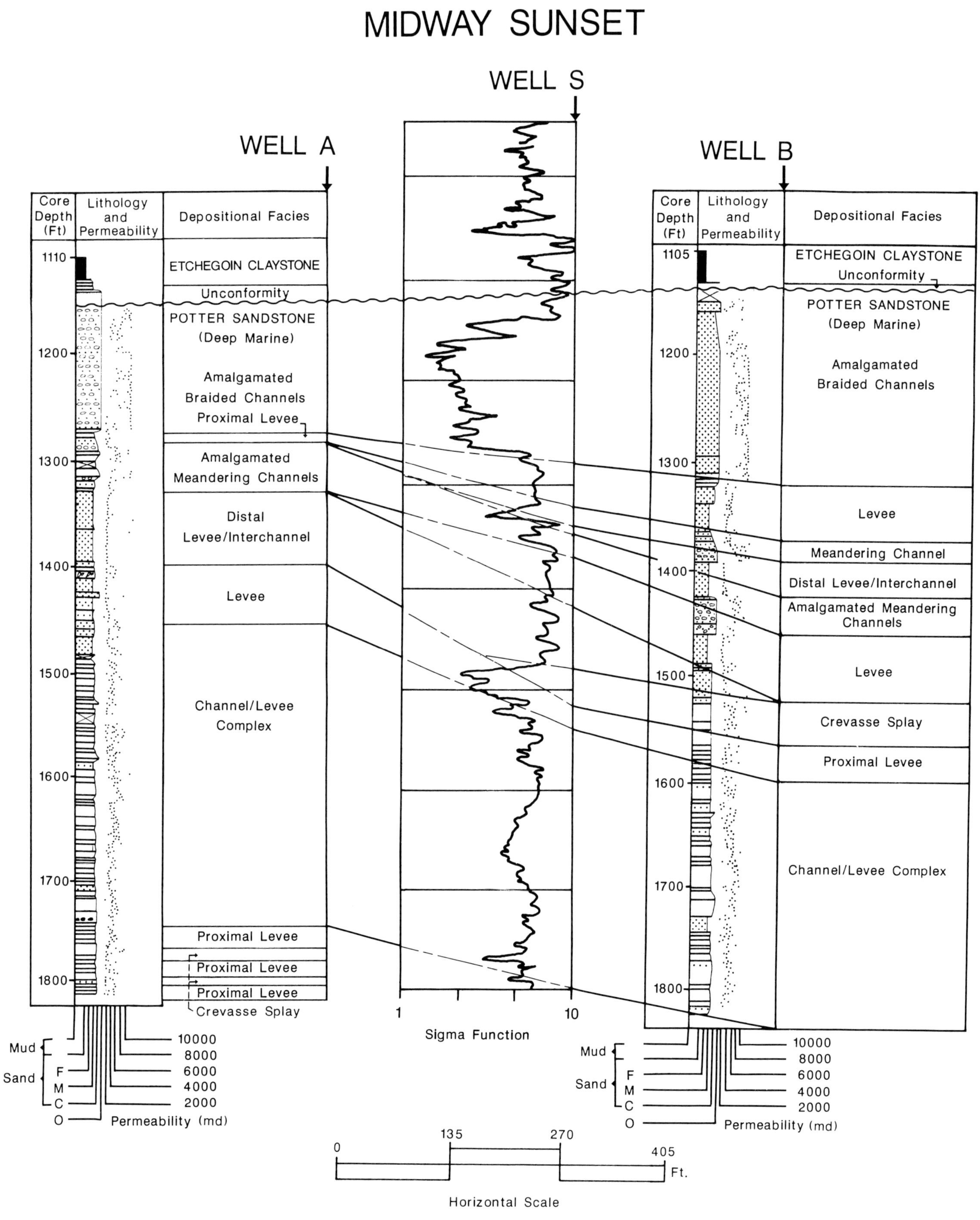

MIDWAY SUNSET
WELL S
WELL A
WELL B
Core Depth (Ft)
Lithology and Permeability
Depositional Facies
ETCHEGOIN CLAYSTONE
Unconformity
POTTER SANDSTONE (Deep Marine)
Amalgamated Braided Channels
Proximal Levee
Amalgamated Meandering Channels
Distal Levee/Interchannel
Levee
Channel/Levee Complex
Proximal Levee
Proximal Levee
Proximal Levee
Crevasse Splay
Meandering Channel
Distal Levee/Interchannel
Amalgamated Meandering Channels
Levee
Crevasse Splay
Proximal Levee
Channel/Levee Complex
Sigma Function
1
10
Mud
Sand
F
M
C
O
10000
8000
6000
4000
2000
Permeability (md)
0
135
270
405
Ft.
Horizontal Scale

Figure 21

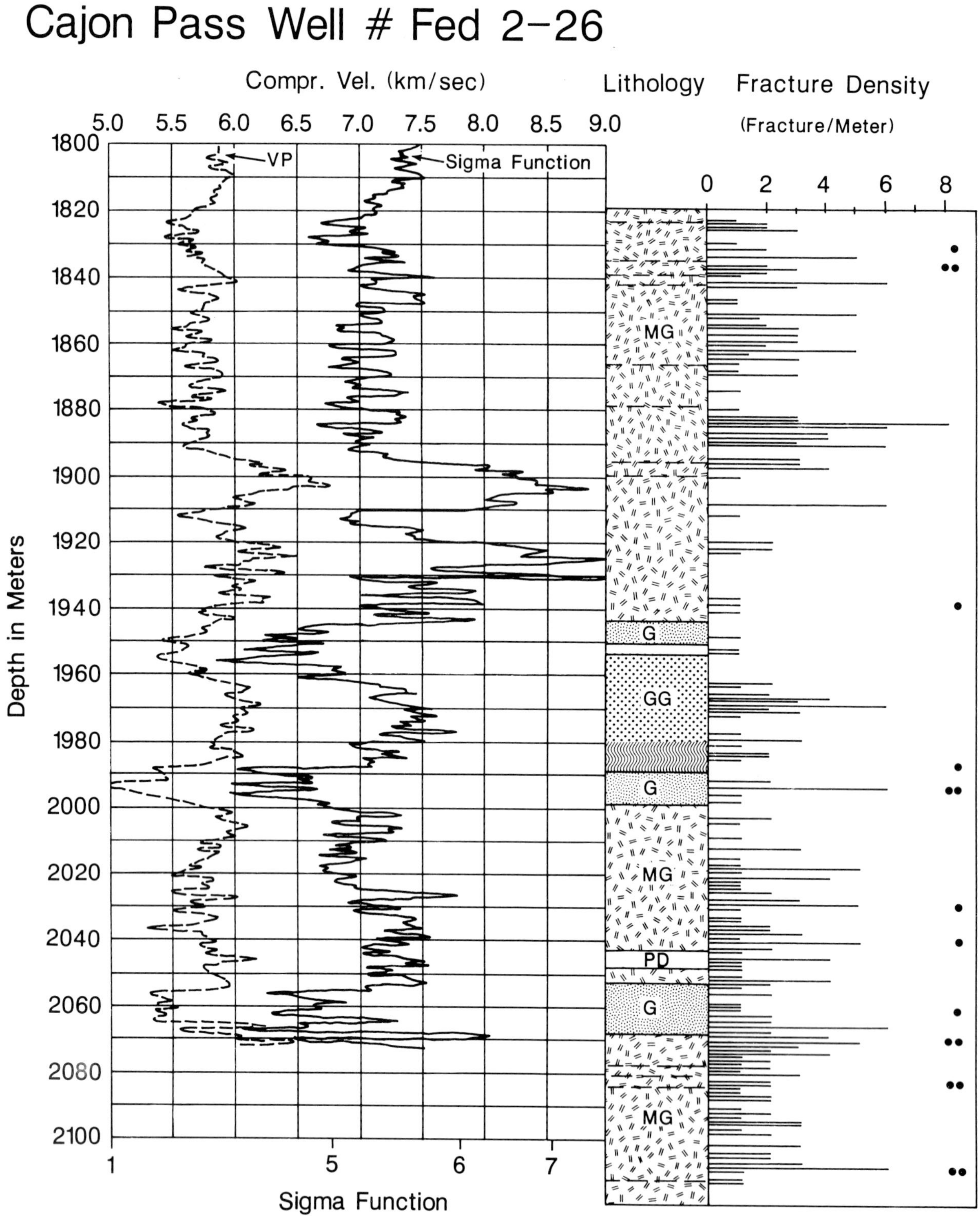

Cajon Pass Well # Fed 2-26
Compr. Vel. (km/sec)
5.0
5.5
6.0
6.5
7.0
7.5
8.0
8.5
9.0
VP
Sigma Function
Lithology
Fracture Density
(Fracture/Meter)
0
2
4
6
8
Depth in Meters
1800
1820
1840
1860
1880
1900
1920
1940
1960
1980
2000
2020
2040
2060
2080
2100
MG
G
GG
PD
1
5
6
7
Sigma Function

Figure 22

Integration of Borehole Data for Improved Reservoir Characterization

Roy D. Nurmi

Most reservoirs are undersampled, thus making integration of all available data necessary for reservoir characterization. Although there is a wealth of data in individual wells, it's effective integration generally requires a good understanding of measurements strengths and limitations, including vertical resolution and volume measured. The importance of understanding the details of measurements has been demonstrated by recent reservoir studies from Egypt to India. The heterogeneous variation of porosity in carbonate reservoirs (at every scale of examination from microns to kilometers) requires careful integration to adequately characterize such reservoirs. The routine evaluation of the standard measurements (core, well log, and testing) in such systems (Cretaceous and Tertiary) are often found to be erroneously interpreted because there are many types of heterogeneities. Moreover, fractured carbonate reservoirs require even more careful integration of core, well log data, and well testing data, especially when the analysis includes thin permeable streaks as are present in some giant Khuff gas reservoirs of The Gulf. These and other giant reservoirs need detail geological reservoir studies and high resolution measurements. Core petrophysical data and well log data are often both needed and should be routinely integrated rather than only being compared for quality control.

The geological and geometrical analysis of borehole data sometimes allows the projection of structural and/or sedimentological information away from the borehole. 'Similar fold' modelling can be used to construct structural cross-sections around individual wells using dip data on a geological workstation. The value of this approach has been demonstrated in the analysis of faulting, including growth faults in The Gulf, Nile, and Godavari delta systems. Integration and analysis on a geological workstation of fractures, lithology, and structure provides a better efficiency in reservoir characterization. Likewise, the sedimentological analysis of bedding revealed by borehole imagery can suggest the depositional origin, orientation, and geometry of some sandstone reservoir deposits. First, however, the electrical or acoustic imagery must be understood in both petrophysical and geological terms.

APPLICATIONS OF OUTCROP GAMMA-RAY LOGGING TO FIELD DEVELOPMENT AND EXPLORATION

Douglas W. Jordan
Roger M. Slatt
ARCO International Oil and Gas Company 2300 West Plano Parkway Plano, Texas 75075

Robert H. Gillespie
Anthony E. D'Agostino
Mark H. Scheihing
ARCO Oil and Gas Company 2300 West Plano Parkway Plano, Texas 75075

ABSTRACT

Gamma-ray logs of outcrops have been generated using two techniques. These techniques demonstrate the applicability of outcrop logging to better understand reservoir facies architecture and exploration type problems.

The first logging technique employs the use of a standard logging truck and gamma-ray sonde. The truck is positioned near the top of the cliff face and the sonde is lowered to the bottom of the cliff. Gamma-ray counts are recorded as the sonde is raised at a constant rate.

The second logging technique employs the use of a commercially available, hand-held, gamma-ray scintillometer. The tool measures total radiation at the outcrop. Equally-spaced measurements are made along the section and are displayed as a function of depth below a reference point.

Examples of gamma-ray logging experiments conducted on turbidites of the Jackfork Group (Pennsylvanian) in central and southern Arkansas demonstrate: (1) the reliability, potential pitfalls, and degree of uncertainty associated with correlating subsurface gamma-ray logs, (2) the nature of vertical stratification and bed cyclicity in these types of sequences as interpreted from logs, (3) the relation of gamma-ray response to lithology and reservoir quality, and (4) the seismic expression of these types of strata.

Perhaps the most powerful application of outcrop gamma-ray logging is the ability to improve interpretations and correlations of subsurface well logs by comparing them with gamma-ray logs obtained from analogous outcrops. Examples from the Long Beach Unit of the Wilmington Oil Field, California, and Point Mugu (Santa Barbara Channel), California demonstrate this application.

INTRODUCTION

Geologists in the past have sometimes measured gamma radiation of outcrops with a hand-held scintillometer in order to evaluate lithology and organic content of beds, as well as to correlate strata between outcrop sections and subsurface wells (Provo et al., 1977; Howe, 1989). The purpose of this paper is to expand upon these efforts by utilizing two techniques to generate gamma-ray logs of outcrops. These techniques measure the characteristics of the strata responsible for a particular gamma-ray log response and demonstrate applications and potential pitfalls of interpreting vertical sequences, depositional environments, lateral continuity, and reservoir quality from gamma-ray logs. Analysis of outcrop gamma-ray logs from a locality considered to be analogous to an oil field or exploration prospect can help reservoir geologists/engineers and explorationists obtain a much better understanding of their field or prospect.

Most subsurface logs represent data at a single point in space, but these data may change rapidly in a very short distance away from a borehole. It is therefore important to have a knowledge of the depositional system in order to reduce risk in interpreting log patterns for resource prediction and evaluation. Logging of strata exposed in outcrops where the geometry and stratification character of the deposit can be observed is the subject of this paper. The logging techniques are presented first, followed by some examples of their general application, then by specific field development and exploration applications.

GAMMA-RAY LOGGING TECHNIQUES

The two different techniques utilized for outcrop gamma-ray logging are used under different conditions. The first technique employs a standard scintillometer tool (sonde) run from a typical logging truck. This technique is used to log cliff or quarry faces that are inaccessible by foot. The second technique uses a commercial, hand-held gamma-ray scintillometer to log readily-accessible outcrops.

Truck-Mounted Gamma-Ray Sonde

The gamma-ray sonde that is used in this study is about 7 feet long and weighs approximately 100 pounds. This sonde is sufficiently sturdy so that the source crystal can withstand the impacts it receives as the tool is lowered and raised along the rock face.

The logging truck is driven to the top of the quarry or cliff to a pre-selected logging site, which is typically a smooth, vertical cliff with few obstructions. A small tripod with a sheave-wheel on top is positioned at the cliff edge to guide the logging cable over the edge. The sonde is lowered down the vertical face by cable to the base of the cliff (Fig. 1). The sonde is then raised at a constant rate (10 ft/min) and gamma-ray measurements are continuously recorded. The logging rate is slower than conventional (borehole) rates since, on a cliff face, the gamma-ray tool is exposed to less than half of the rock mass normally encountered in a borehole.

A rope is attached to the base of the sonde and is used to guide the sonde past protuberances as it is drawn upward (Fig. 1). This technique is also necessary to ensure that the tool is kept in close proximity to the cliff face.

Gamma-ray measurements can be affected by minor borehole washouts (especially when used with a heavy mud); however, extraordinarily large cave-ins or solution caves can significantly decrease the count rate. Tests at the outcrop were performed to determine the maximum distance away from a cliff face that the gamma-ray tool could be positioned before the count rate is adversely effected. Repeated measurements of a single sample site show that the sonde may be up to 2 feet away from the face and have little effect on the count rate. However, the variance in count rate significantly increases beyond this distance.

Other steps useful to ensure quality outcrop logs include: (l) having photographs of the cliff available during logging, (2) hanging a scale marked at one-foot increments over the cliff during photographing and logging (Fig. 1), (3) providing walkie-talkie communication between the logging truck and the ground observer, and (4) marking on the sonde the exact gamma-ray recording point (mid-way down the length of the sonde) so that the observer on the ground can mark on the photographs, and/or relay to the logging truck via walkie-talkie, the exact location on the cliff of any key measurement site.

Upon completion of logging, the data can be displayed in a standard depth format (Fig. 2).

Hand-Held Gamma-Ray Scintillometer

For this type of logging, a lightweight, portable ScintrexTM GIS-S Integrating Gamma Spectrometer is used. This instrument has the capability of measuring elemental or total radiation. The instrument is allowed to warm up and, after calibration, the "Total Count" mode is selected. Background radiation is considered to be relatively constant. Further details on the mechanics and specifications of the instrument are found in the ScintrexTM manual.

Typically, measurements are made at two foot intervals. Caution is exercised to ensure that true vertical thickness is measured to define each two foot increment. Selected stratigraphic sections are also logged at half foot intervals, which provides more detail. This aspect is discussed in a later section of this paper.

Tests in the field were conducted to determine the optimal sampling rate and number. Three, 5, and 10 measurements with sampling rates of 1, 3, 10, and 100 seconds (the only rates available on the ScintrexTM model that was used) at each station were compared. For example, a gamma-ray reading at a 10 second sampling rate was found to be almost exactly the same as a 3 second reading multiplied by 3.33. It was concluded that the most efficient method of obtaining data is to take five total count readings for three seconds each. The highest and lowest values are discarded, and the other three readings are averaged to give the gamma-ray value at that station. The data can then be displayed in standard well log curves as a function of stratigraphic depth, utilizing a scale calibrated in "counts per second". But since the original data are collected at 3 second sampling rates, it is recognized that the scale is actually displaying "counts per 3 seconds".

GENERAL APPLICATIONS

The outcrop examples presented in this section of the paper are from the Pennsylvanian Jackfork Group in Arkansas (Fig. 2). Superb exposures in quarries and dam sites contain sheet-like, laterally continuous (on the scale of the outcrop) to lenticular, laterally discontinuous turbidite strata. For a review of the geologic framework of the Jackfork Group refer to Jordan et al. (in press).

Big Rock Quarry in North Little Rock, Arkansas (Fig. 2) exposes a 3,000 foot long by 20~ foot high quarry face comprising predominantly laterally discontinuous strata that were deposited by turbidity currents in channels and depressions on a continental slope or base-of-slope environment (Jordan et al., in press). Three gamma-ray logs, spaced 500 and 600 feet apart (Fig. 2), were measured along the northeast quarry wall. These data were obtained with a logging truck. A close-up view of the stratification and gamma-ray log is shown in Figure 3.

Description of a detailed lithologic section was not possible because of the steepness of the quarry wall. However, Figure 4 compares log 1 with a subsurface core of the same interval taken approximately 50 feet behind the quarry wall several years ago by Shell Oil Company and described by Link and Stone (1986). The correlation between the log response and gross lithology is good. The exception to this occurs in the thinly-bedded interval between 20 and 40 feet. The log response suggests this interval to be a thick sandstone sequence.

The three logs (see Fig. 2) taken at this site illustrate potential problems in well log correlation of laterally discontinuous strata. On several field trips participants have correlated these logs prior to examining the outcrop. Typically four or five distinct stratigraphic intervals are subdivided on the logs (Fig. 5A). However, when the quarry is examined, most beds are not laterally continuous at the 500 and 600 foot spacing of the logs, in fact, the correlations shown in any of the three cases in Figure 5 are erroneous. A better correlation, made after studying the quarry face, is shown in Figure 6. In comparing this correlation to that shown in Figure SA, it is particularly noteworthy that the thick sandstone from 88-168 feet on Log ~2 is a discontinuous sandstone slump upon which younger sandstone beds onlap, and (2) the coarsening/cleaning upward sandstone unit at 030 feet on Log #3 is a laterally discontinuous sandstone bed near the top of the quarry face. In the subsurface, these beds, though thick, would not extend between typically-spaced wells (40 acre spacing).

Log Correlation of Laterally Continuous Strata

The DeGray Lake Dam Spillway, near Arkadelphia, Arkansas (see Fig. 2 for location) exposes a 1,000 foot vertical succession of steeply-dipping turbidite strata. These strata record repeated cycles of turbidite sedimentation in a deep marine basin (Jordan et al., in press). The succession is superbly exposed and easily accessible along two spillway walls (east and west) spaced approximately 300 feet apart. The entire vertical section has been measured in detail by Morris (1977) (Fig. 7) and Brecken (1988); individual measured segments have been published by Moiola and Shanmugam (1984).

The overall vertical sequence grades upwards from thin-bedded sandstones and shales at the base, through thicker-bedded sandstones, into massive pebbly sandstones/conglomerates at the top. Several major cycles of sandstone deposition are separated by shales (10-30 feet thick) of hemipelagic or debris flow origin.

Figure 7 shows gamma-ray logs obtained with the hand-held scintillometer at two foot sample spacings for the east and west walls of the spillway. Two approaches were used to determine that these strata are laterally continuous over relatively long distances (hundreds of feet to miles). First, because the pebbly sandstone beds at the top of the sequence form resistant ridges, they can be traced continuously along strike in the field and on air photos over a distance of several miles (Brecken, 1988). Secondly, the thickness of individual sandstone and shale beds was measured within two stratigraphic intervals along both sides of the spillway: (l) on the eastern wall (DeGray East) and western wall (DeGray West) of the spillway at 320-350 and 330-360 feet (log depth), respectively (Figs. 7 and 8), and (2) on the eastern wall and western wall of the spillway at 365-460 and 375-470 feet, respectively (Figs. 7 and 9). Correlation of individual beds across the 300 feet separating the two spillway walls was accomplished by noting distinctive sedimentary features and by comparing the position and character of beds on opposite sides of the spillway within the context of the entire vertical sequence.

Because the distance between the spillway walls was known, measurements of the thickness of individual beds on both walls provided a means of estimating their rate of change of thickness, and consequently their lateral continuity (this technique is explained in more detail by Jordan et al., in press). Figures 10 and 11 show equations and sample smoothed curves derived from measurements of 80 beds which relate the changes in thickness of sandstone and shale beds as a function of distance from a specified reference point. These data show that even relatively thin beds are laterally continuous for distances of several hundreds to thousands of feet. This long-distance lateral continuity contrasts with the limited lateral continuity (many beds <500 feet) observed at Big Rock Quarry. In the subsurface, most of these beds would probably be laterally continuous between typically-spaced wells (40 acre spacing).

Thick stratal packages are easily correlated on gamma-ray logs over the 300 feet separating the east and west walls of the spillway (Fig. 7). In order to compare the level of bed resolution at two foot sample spacing to logs with a finer level of resolution, gamma-ray logs were obtained at the two previously discussed stratigraphic intervals with the hand-held scintillometer at half-foot sampling intervals.

For the upper sequence (Fig. 8), the log obtained at a two foot sample spacing only resolves two sandstone intervals (labelled A and B) which can be identified and correlated, but at half-foot spacing the beds can be further subdivided into individual correlative packages (Al, A2, Bl, B2, and B3) and beds (a, b, c, and d), even though there is considerable change in thickness of some beds (e.g. A2, B2, B3) over 300 feet (Fig. 8).

The lower stratigraphic interval can be interpreted on the two-foot log (Fig. 7) as a typical thickening/cleaning-upward lobe (Shanmugam and Moiola, 1988; 1991). However, the greater detail of the half-foot log shows that this sequence is really a composite of several small thickening/cleaning-upward or thinning/dirtying-upward sequences (Fig. 9). The gamma-ray responses of most individual beds are correlative across both sides of the spillway. Individual sandstone beds in the lower part of the sequence tend to alternate vertically between being thickest on the west or east wall of the spillway (Fig. 9). These observations suggest an alternative to the traditional interpretation that thickening/cleaning-upward and thinning/dirtying-upward vertical sequences of turbidites comprise lobes and channels, respectively (Selley, 1979; Shanmugam and Moiola, 1988). As illustrated in Figure 12, the same type of vertical sequence and gamma-ray log character (and vertical rock sequence) can be obtained when sandy turbidite flows are deposited on the seafloor such that as one bed is deposited, the next younger bed is laid down in the adjacent topographic depression created by the first bed. Switching back and forth of individual bed axes (resulting in compensation cycles of Mutti and Normark, 1987) may provide a complex arrangement on gamma-ray logs (and outcrops~ of thickening/cleaning-upward and thinning/fining-upward sequences at a variety of scales. Such sequences can be misinterpreted as laterally continuous lobes and laterally discontinuous channels, respectively.

At Hollywood Quarry (see Fig. 3), located about 7 miles west of the DeGray Lake Dam Spillway, fifteen gamma-ray logs were obtained with the logging truck around the periphery of the quarry (Fig. 13). This quarry exposes both lenticular and sheet-like turbidite strata; in addition, some faults cut through the quarry. This logging exercise was designed to develop a scaled-down representation of well logs and spacing patterns in the subsurface. With such a representation, it is possible to demonstrate potential correlation pitfalls which might be encountered in areas with these types of geological complexities.

Figure 14 shows the log correlations at Hollywood Quarry. The beds strike about North 35 degrees East and dip 7 degrees to the Northwest so that the beds on the east wall of the quarry dip beneath those on the west wall. The upper part of the southern half of the west wall comprises laterally discontinuous strata. The northern half of the west wall contains more laterally continuous, but highly faulted beds. The east wall contains thick, lenticular sandstones and pebbly sandstones which are probably correlative with the pebbly sandstones at the top of DeGray Lake Dam Spillway. In this example, if only logs were available, the correlations shown in Figure 14 most likely would not have been made. Considering that these logs are typically spaced only a few hundred feet apart, this exercise demonstrates some of the uncertainty in well log correlations that might be anticipated in oil and gas fields of complex facies architecture and structure. Coring of development wells for detailed sedimentologic and stratigraphic analysis, and utilizing 3D seismic surveys for structural and stratigraphic interpretation, in addition to well testing, would reduce this uncertainty in such situations and aid in the understanding of interwell heterogeneity.

Reservoir Quality Characterization

Gamma-ray logs may provide a measure of clay content which may affect reservoir quality (porosity and permeability). Higher gamma-ray counts are usually interpreted as indicating relatively finer grain-size and/or relatively higher clay content of strata. However, not all clay minerals contain measurable radioactive elements and some sand-size minerals do contain radioactive elements, so that gamma-ray measurements are not a universally reliable indicator of reservoir quality (Rider, 1990). The Jackfork Group is very quartzose, lacks significant feldspar which might contain radioactive potassium, and contains non-radioactive illite clay (Jordan et al, in press). Therefore, these strata provide an opportunity to visually investigate the relationship between gamma-ray counts and clay content (i.e., the higher the gamma count, the poorer the reservoir quality).

To demonstrate this relationship, the gamma-ray logs at DeGray Lake Dam Spillway were first calibrated to relative reservoir quality on the basis of visual estimates of matrix (including clay) content (i. e., the lighter-colored the sandstone, the less the matrix content). Petrographic and gamma-ray analysis of seven Jackfork sandstone samples from DeGray Lake and another spillway at Dierks Lake (Table 1) demonstrate that the lower count rates of gamma rays are associated (in 6 out of 7 samples) with a lower amount of matrix (including clay) material and higher porosity.

The scale of relative reservoir quality for the DeGray Lake Dam Spillway is illustrated in Figure 7. The term relative reservoir quality is stressed because, in fact, absolute reservoir quality of these rocks is poor based upon porosity and permeability measurements and visual observations; porosities are normally less than 5~ and permeabilities are less than 0.01 md (Jordan et al., in press). Most porosity is occluded by cement (primarily siliceous) and matrix.

Gamma-ray log data collected at DeGray Lake Dam Spillway also suggest that thinner sandstones contain more matrix (are of finer average grain-size) than thicker sandstones, and thus have poorer reservoir quality. The relationship between bed thickness and grain-size of turbidites previously has been demonstrated by reservoir quality. The relationship between bed thickness and grain-size of turbidites previously has been demonstrated by Potter and Scheidegger (1966), Sadler (1982), Middleton and Neal (1989), and Slatt et al. (in press). A decrease in grain-size associated with a decrease in bed thickness is a result of both more and coarser-grained sediment being carried in suspension in a flow of greater turbulence than in a less turbulent flow. Lower energy flows produce thinner and finer-grained beds. A plot of bed thickness vs. gamma-ray counts for beds at the spillway shows that the thicker the bed, the lower the count and the cleaner the sandstone (Fig. 15). The high degree of scatter of data points (R squared equals 0.34) is due to the fact that the area of investigation of the gamma-ray scintillometer (1 foot in diameter; data from ScintrexTM manual) is greater than the thickness of some of the beds measured. Consequently, radiation from adjacent shale beds is detected.

Seismic Modelling Using Outcrop Gamma-Ray Logs

One dimensional synthetic seismograms of some Jackfork outcrops were constructed in order to give geologists and geophysicists a better appreciation of the seismic resolution of these types of geologic sequences. The vertical stratification characteristics of the sandstones and shales were interpreted from the gamma-ray logs. Sonic logs from exploration wells in the area provided average sandstone and shale velocities of 16,500 and 12,500 feet/sec, respectively. By combining these data, vertical velocity both lithologies. The logs were convolved with a zero phase wavelet using a four-point bandpass filter (4, 12, 35, and 55 Rz) to approximate the frequency bandwidth expected on land at 10,000 feet depth.

An example of a one-dimensional synthetic seismogram

Table 1. Gamma-ray readings and petrographic characteristics of seven samples from the Jackfork in Arkansas.

Sample No. (and Log Depth)	CPS	Q	QCM	FEL	MRF	CHT	MAT	OTH	POR	GS	SRT
DIERKSW 1 (590 ′)	440	33	13	6	22	2	16	7	1	VF	P
DIERKSW 2 (506′)	370	45	26	5	11	1	7	4	1	VF—F	M
DIERKSW 3 (430 ′)	295	45	22	2	12	1	2	11	5	F-M	M-W
DIERKSW 4 (230 ′)	270	36	33	3	7	10	8	2	1	F	M
DIERKSW 5 (34 ′)	230	42	40	4	6	1	1	2	4	F-M	W
DEGRAYE 1 (254 ′)	261 (860) **	44	35	2	9	3	3	1	3	F-M	M-W
DEGRAYE 2 (190 ′)	203 (670) **	40	40	3	7	2	2	1	5	F-M	W

KEY:
CPS: COUNTS PER 3 SECONDS
Q: QUARTZ
QCM: QUARTZ
FEL: FELDSPAR
MRF: METAMORPHIC ROCK FRAGMENT
CHT: CHERT
MAT: MATRIX
OTH: OTHER
CEMENT POR: POROSITY
GS: GRAIN SIZE
SRT: SORTING

**: TEN-SECOND READING; THREE—SECOND READING, RECORDED ABOVE, WAS NORMALIZED for the DeGray Lake Dam spillway section is shown in Figure 16. Positive reflections are associated with major sand-shale interfaces, and lower amplitude events are associated with more thinly-bedded intervals.

Through this modelling approach, the vertical resolution issue is addressed and not the lateral or spatial resolution of the seismic data. Although the seismic modelling is confined to one-dimensional synthetic seismograms, the data are amenable to generating two-dimensional seismic models where more than one gamma-ray log of an outcrop is available (e. g., Big Rock Quarry, Hollywood Quarry, etc.). Two dimensional changes observed in the seismic data would certainly assist in understanding lateral changes in seismic character as they relate to lithologic changes as long as it is understood that most of the stratal units encountered in these outcrops are too thin to be uniquely resolved on conventional seismic data.

COMPARING SUBSURFACE GAMMA-RAY LOGS OF STRATA WITH LOGS OF OUTCROP ANALOGS

The most important application of outcrop gamma-ray logging is the ability to improve subsurface well log interpretations and correlations by comparing logs with those of analogous outcrops. Two examples are presented which demonstrate this application.

Field Development Example

The Long Beach Unit of the giant Wilmington Oil Field, Los Angeles Basin, California produces oil from a 3000+ foot thick stratigraphic interval comprising Mio-Pliocene turbidite sands. The most prolific interval is the 1000 foot thick Ranger Zone, which has been described in detail by Slatt et al. (in press). Core lithologies, vertical sequences and facies, and gamma-ray log characteristics of the Ranger Zone bear striking resemblance to the 1,000 foot thick Jackfork Group section at DeGray Lake Dam Spillway (Fig. 17). Issues of concern in the Long Beach Unit are the nature of stratification and vertical bed cyclicity, the lateral continuity of thin sandstone and shale beds, and their resolution on gamma-ray and other logs. A better understanding of these issues has resulted from comparison of log and stratigraphic characteristics of the Ranger Zone with those of the Jackfork Group at the Spillway. This is particularly true when the comparisons are made at the outcrop.

The Miocene Rincon and Topanga formations exposed at Point Mugu, California offer an opportunity, using outcrop logging techniques, to delineate possible subsurface reservoir units in the adjacent Santa Barbara Channel. These units comprise over 1200 feet of interbedded sandstones, siltstones, claystones, and clay shales (Jordan and Marquard, 1990) that were deposited in a tectonically active basin. The lower part of the section originated as shelf deposits during sea-level rise (Fig. 18). An erosional surface flooring a deep-water canyon cuts these highstand deposits and marks a sequence boundary. The overlying strata consist of a deformed (by slumping) and undeformed section comprising thin- to thick-bedded sandstones derived from high-density turbidity currents and slumps.

The hand-held scintillometer was used to measure most of the flatlying section exposed along the coastline and shown in Figures 18 and 19A. The stratigraphic sequence of the units along the coastline has a log character similar to subsurface logs ~Fig. 19B), allowing them to be interpreted. Once the possible hydrocarbon-bearing zone (i. e., lowstand turbidite facies) is identified on other well logs in the offshore (and tied to seismic sections), the interval can be mapped and prospects generated.

Similar log-to-outcrop correlations (especially in areas where conventional core material is absent) and extrapolation of the interpretation to subsurface logs can provide the explorationist with a very useful tool in the search for hydrocarbon accumulations.

CONCLUSIONS

1. Two techniques have been utilized which produce reliable gamma-ray logs of outcrops. One technique uses a standard gamma-ray sonde run off of a logging truck to log vertical cliff or quarry faces. The second technique employs a hand-held gamma ray scintillometer.

2. These logging techniques are useful for demonstrating: (1) the reliability, potential pitfalls, and degree of uncertainty associated with correlating subsurface gamma-ray logs, (2) the nature of stratification and vertical bed cyclicity depicted on logs, ~3) the gamma-ray response of variable lithology and reservoir quality, and (4) the synthetic seismic response to these types of strata.

3. Outcrop gamma-ray logging is particularly applicable to improving interpretations and correlations of subsurface well logs by comparing them with outcrop gamma-ray logs of analogous strata.

ACKNOWLEDGMENTS

We wish to express our appreciation to James Ebanks, Jr., and Andrew Slatt for field assistance in obtaining some of the logs, Charles Stone for field guidance and for freely discussing his knowledge of the Jackfork Group, Douglas Lawson for his advice on the mathematics of estimating laterally continuity of strata, Galen Treadgold for his assistance in generating the synthetic seismogram, and Donald Lowe for sharing his sedimentological insights and providing certain measured stratigraphic sections. This manuscript benefitted from reviews by Jim Lorsong, Eric Pasternack, and Phil Lowry; however, the content of this paper is the sole responsibility of the authors. ARCO Oil and Gas Company and ARCO International Oil and Gas Company allowed us to present this information, and the drafting departments performed their usual high-quality work.

REFERENCES

1. Brecken, C. E., 1988, Sedimentology and facies of the Pennsylvanian Jackfork Group in the Caddo Valley and DeGray Quadrangles, Clark County, Arkansas: Ph. D. dissertation, The University of Tulsa, 134 p.

2. Howe, D., 1989, Surface gamma-ray profiling technique applied to Cretaceous Ferron Sandstone, east-central Utah (abs.): Amer. Assoc. Petrol. Geol., v. 73, p. 365.

3. Jordan, D. W., Lowe, D. R., Slatt, R. M., Stone, C. G., D'Agostino, A. E., Gillespie, R. H., end M. H Scheihing, in press, Scales of geological heterogeneity of Pennsylvanian Jackfork Group, Ouachita Mountains, Arkansas: Applications to field development and exploration for deepwater sandstones: Dallas Geol. Soc. Field Guidebook, 1991 Amer. Assoc. Petrol. Geol. Ann. Conv., Dallas.

4. Jordan, D. W., and R. S. Marquard, 1990, Integration of sequence stratigraphy and process sedimentology:

Miocene Rincon and Topanga Formations, Santa Barbara Basin near Point Mugu, California (abs.): Amer. Assoc. Petrol. Geol., v. 74, p. 689.

5. Link, M. H., and C. G. Stone, 1986, Jackfork Sandstone at the abandoned Big Rock Quarry, North Little Rock, Arkansas, in Stone, C. G. and Haley, B. R., 1986, Sedimentary and igneous rocks of the Ouachita Mountains of Arkansas: Guidebook Geol. Soc. Amer. Ann. Mtg., San Antonio, p. 1-8.

6. Middleton, G. W., and W. J. Neal, 1989, Experiments on the thickness of beds deposited by turbidity currents: Jour. Sed. Petrol., v. 59, p. 297-307.

7. Moiola, R. J., and G. Shanmugam, 1984, Submarine fan sedimentation, Ouachita Mountains, Arkansas and Oklahoma: Gulf Coast Assoc. Geol. Soc. Trans., v. 34, p. 175-182.

8. Morris, R. G., 1977, Flysch facies of the Ouachita Trough—with examples from the spillway at DeGray Dam, Arkansas, in Stone, C. G. (ed.), Symposium on the Geology of the Ouachita Mountains: Arkansas Geological Commission, v. 1, p. 158-168.

9. Mutti, E., 1985, Turbidite systems and their relations to depositional sequences, in Zuffa, G. G. (ed.), Provenance of Arenites: D. Reidel Publ. Co., Dordrecht, p. 65-93.

10. Mutti, E., and N. R. Normark, 1987, Comparing examples of modern and ancient turbidite systems: Problems and concepts, fn Leggett, J. K. and G. G. Zuffa (eds.), Marine Clastic Sedimentology: Concepts and Case Studies, p. 1-38.

11. Potter, P. E. and A. E. Scheidegger, 1966, Bed thickness and grain size: graded beds: Sediment., v. 7, p. 233-240.

12. Provo, L. J., Kepferle, R. C., and P. E. Potter, 1977, Three Lick Bed: Useful stratigraphic marker in the Upper Devonian Shale in Eastern Kentucky: Energy Research and Development Administration, Morgantown Energy Research Center, CR-77-2, 56 p.

13. Rider, M. H., 1990, Gamma-ray log shape used as a facies indicator: critical analysis of an oversimplified methodology, in Hurst, A., Lovell, M. A., and A. C. Morton, (eds.), Geological Applications of Wireline Logs: Geol. Soc. Spec. Publ. 48, p. 27-37.

14. Sadler, P. M., 1982, Bed-thickness and grain size of turbidites: Sediment., v. 29, p. 37-51.

15. Selley, R. C., 1979, Dipmeter and log motifs in North Sea submarine-fan sands: Amer. Assoc. Petrol. Geol. Bull., v. 63, p. 905-917.

16. Shanmugam, G., and R. J. Moiola, 1988, Submarine fans: Characteristics, models, classification, and reservoir potential: Earth-Sci. Rev., v. 24, p. 383-428.

17. Shanmugam, G., and R. J. Moiola, 1991, Types of submarine fan lobes: Models and implications: Submarine fans: Amer. Assoc. Petrol. Geol. Bull., v. 75, p. 156-179.

18. Slatt, R. M., Philllps, 5., Boak, J. M., and M. B. Lagoe, ln press, Scales of geological heterogeneity of a deep-water sand giant oil field, Long Beach Unit, Wilmington Field, California fn Rhodes, E. G., and T. F. Moslow, (eds.), Marine Clastic Reservoirs: Examples and Analogs: Springer-Verlag.

Figure 1 - Gamma-ray sonde (large arrow) being raised from the base of the west wall of Hollywood Quarry by a cable (small arrow) attached to the logging truck (not in view). The person at the base of the quarry wall is using a rope attached to the sonde to guide it past rock protuberances as it is drawn upward. This logging run produced the gamma-ray log that is superimposed upon the quarry wall in the photograph. Note the scale marked in one foot increments.

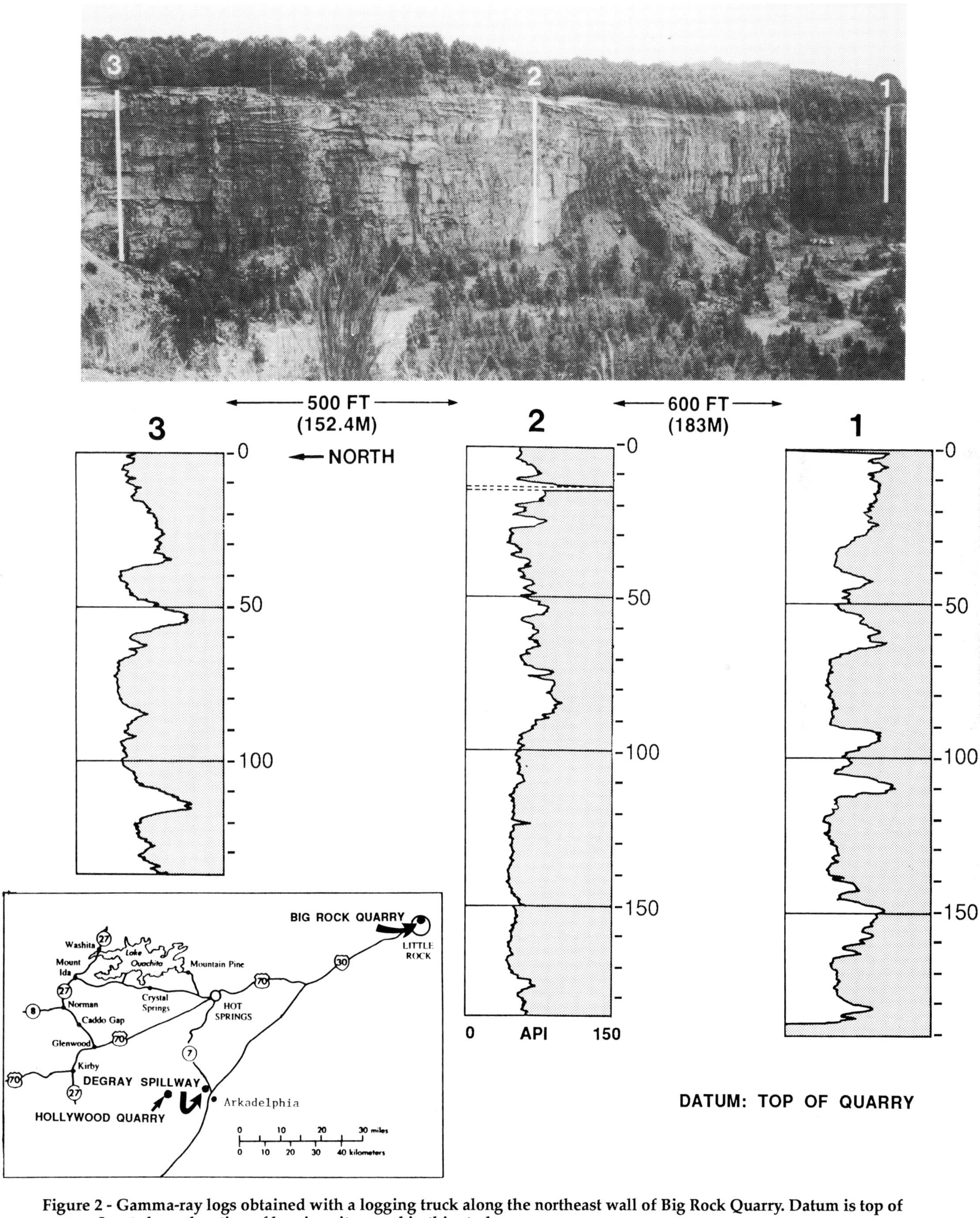

Figure 2 - Gamma-ray logs obtained with a logging truck along the northeast wall of Big Rock Quarry. Datum is top of quarry. Inset shows location of logging sites used in this study.

Figure 3 - Comparison of gamma-ray log #l at Big Rock Quarry with the quarry wall at the logging site. Arrows point to individual beds depicted on the log.

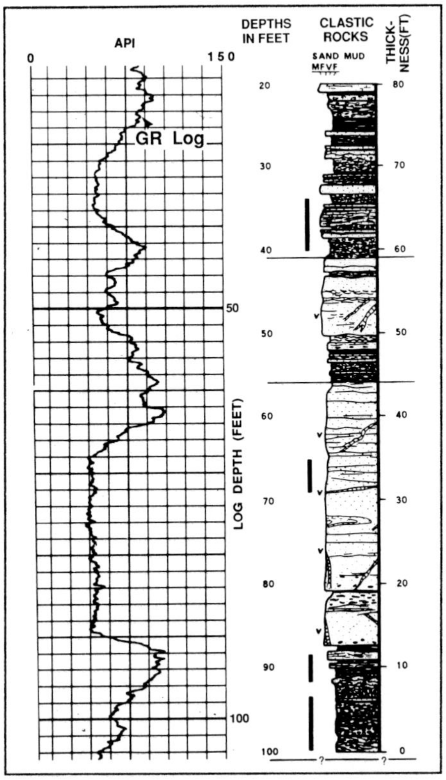

Figure 4 - Comparison of gamma-ray log #1 at Big Rock Quarry with a Shell Oil Company core taken 50 feet behind the quarry face at the logging site. Core description is after Link and Stone (1986).

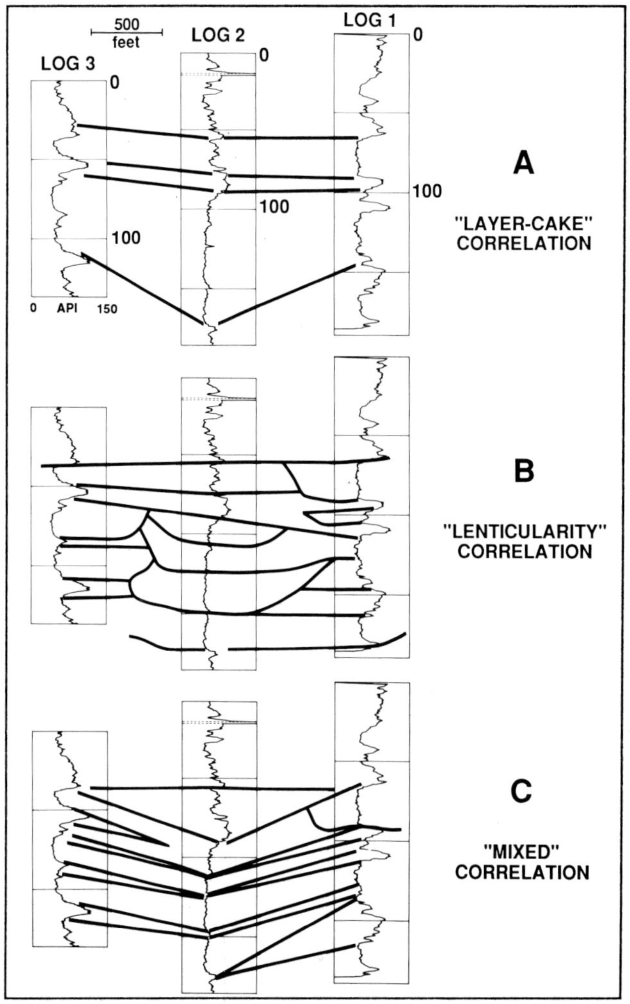

Figure 5 - Three possible, but incorrect, correlations of gamma-ray logs from Big Rock Quarry, illustrating (A) layer-cake (B) lenticular, and (C) mixed correlations.

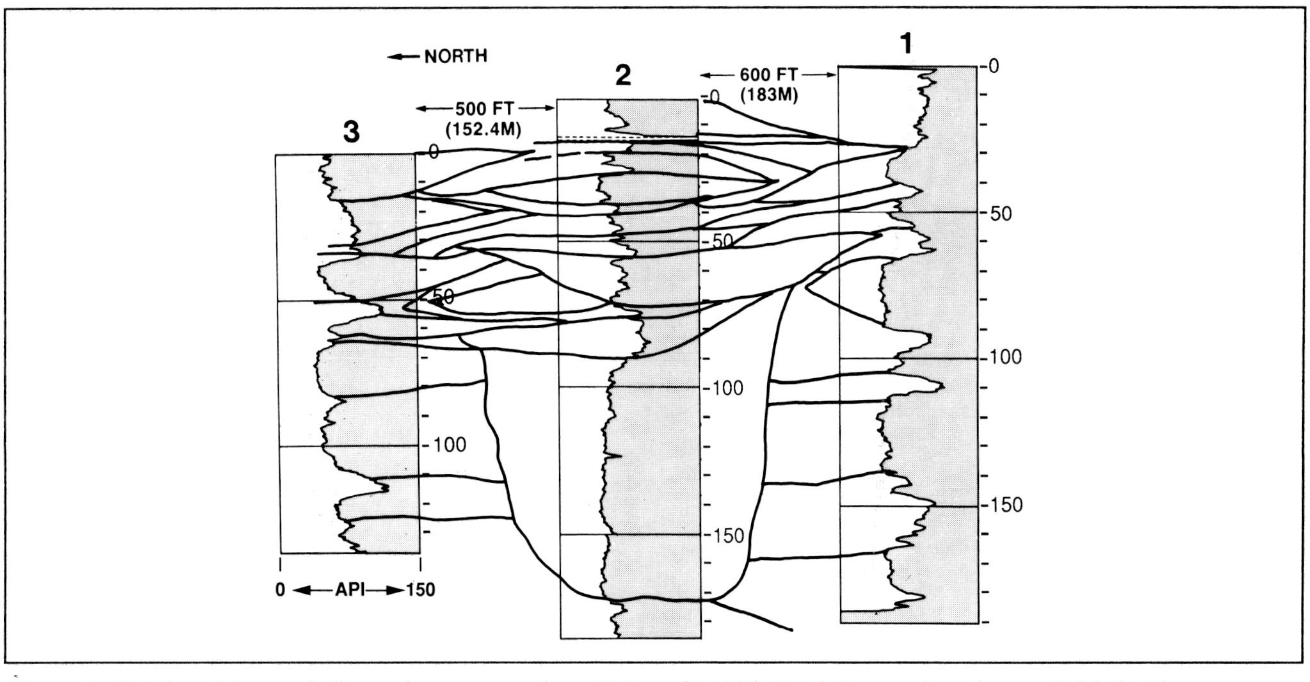

Figure 6 - Stratigraphic correlations of gamma-ray logs #1, 2, and 3 at Big Rock Quarry, based upon field sketches.

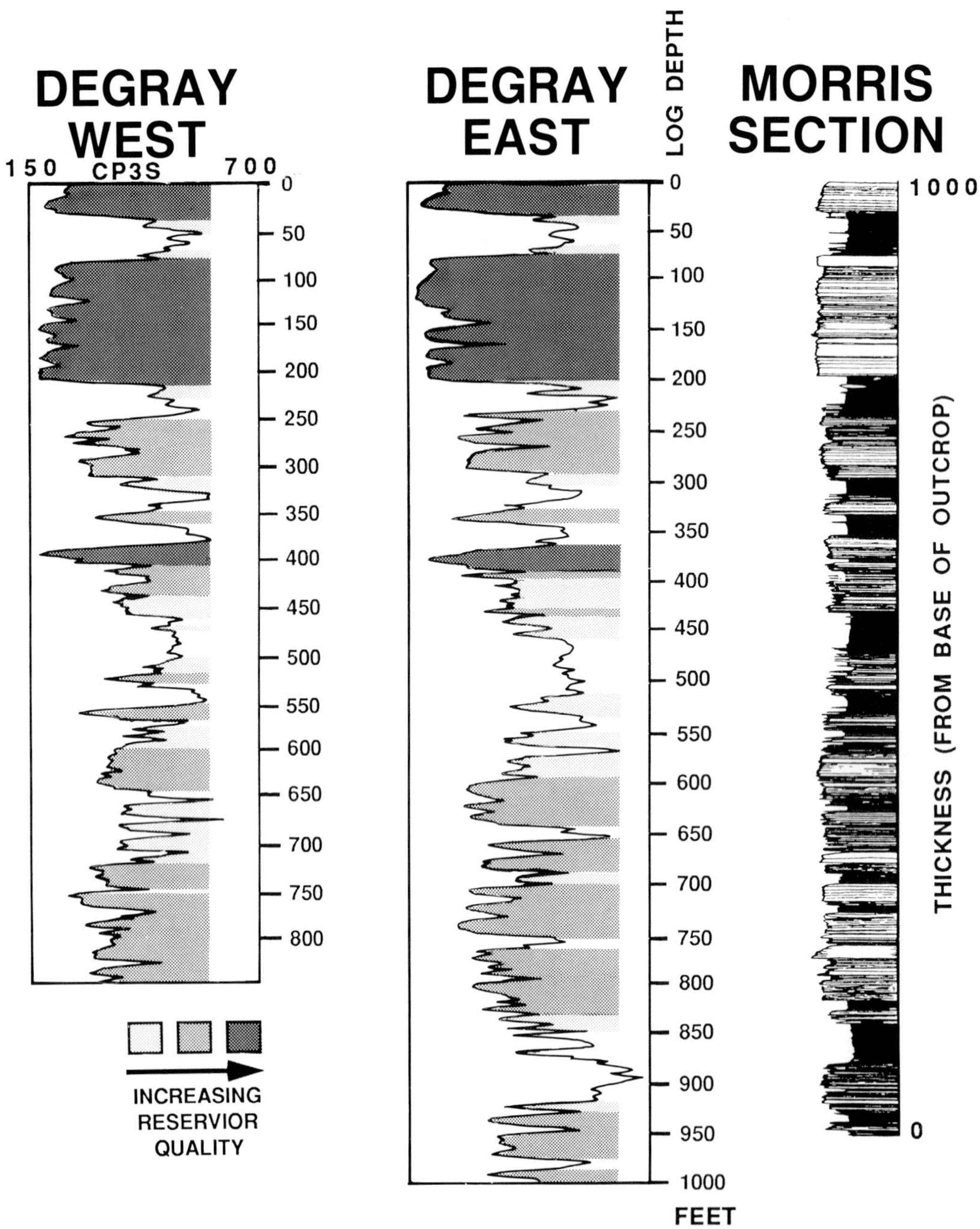

Figure 7 - Gamma-ray logs obtained with a hand-held scintillometer (two foot sample spacing) along the east and west walls of DeGray Lake Dam Spillway. The two walls are spaced about 300 feet apart. Measured section by Morris (1977) is shown for comparison ~there is a 5% error in measurement). Estimated relative reservoir quality is shown on the logs.

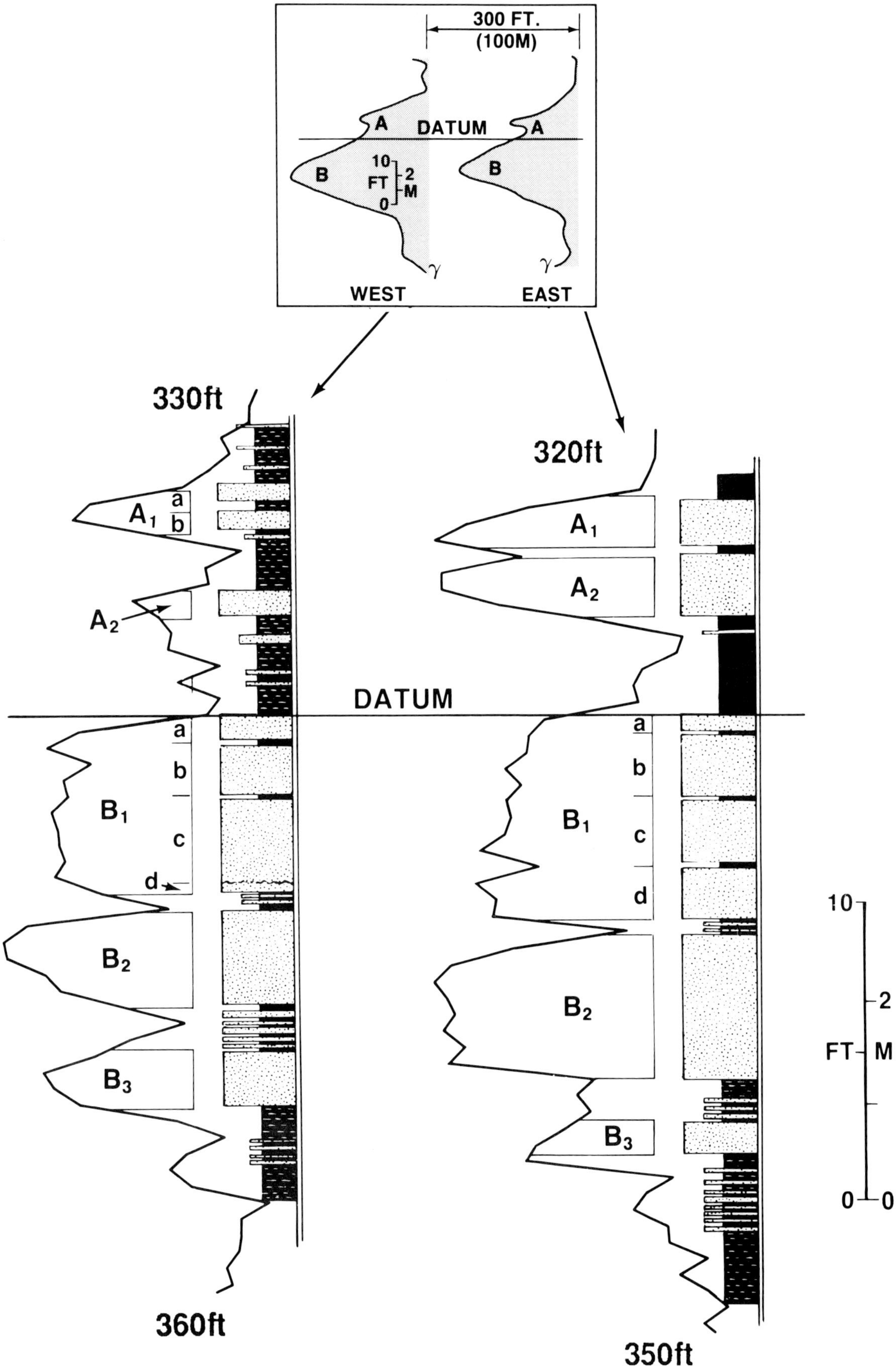

Figure 8 - Detailed stratigraphy and gamma-ray logs (half foot sample spacing) for the interval 320-350 and 330-360 feet on the eastern and western walls, respectively, of DeGray Lake Dam Spillway. See Figure 7 for the position of the interval within the entire spillway section. The inset shows the log interval obtained at two foot sample spacings. Detailed stratigraphy of the two major intervals, subdivided into smaller packages ~A1, A2, B1, B2, and B3) and beds (a,b,c, and d), was interpreted by D. R. Lowe (in Jordan et al., in press)

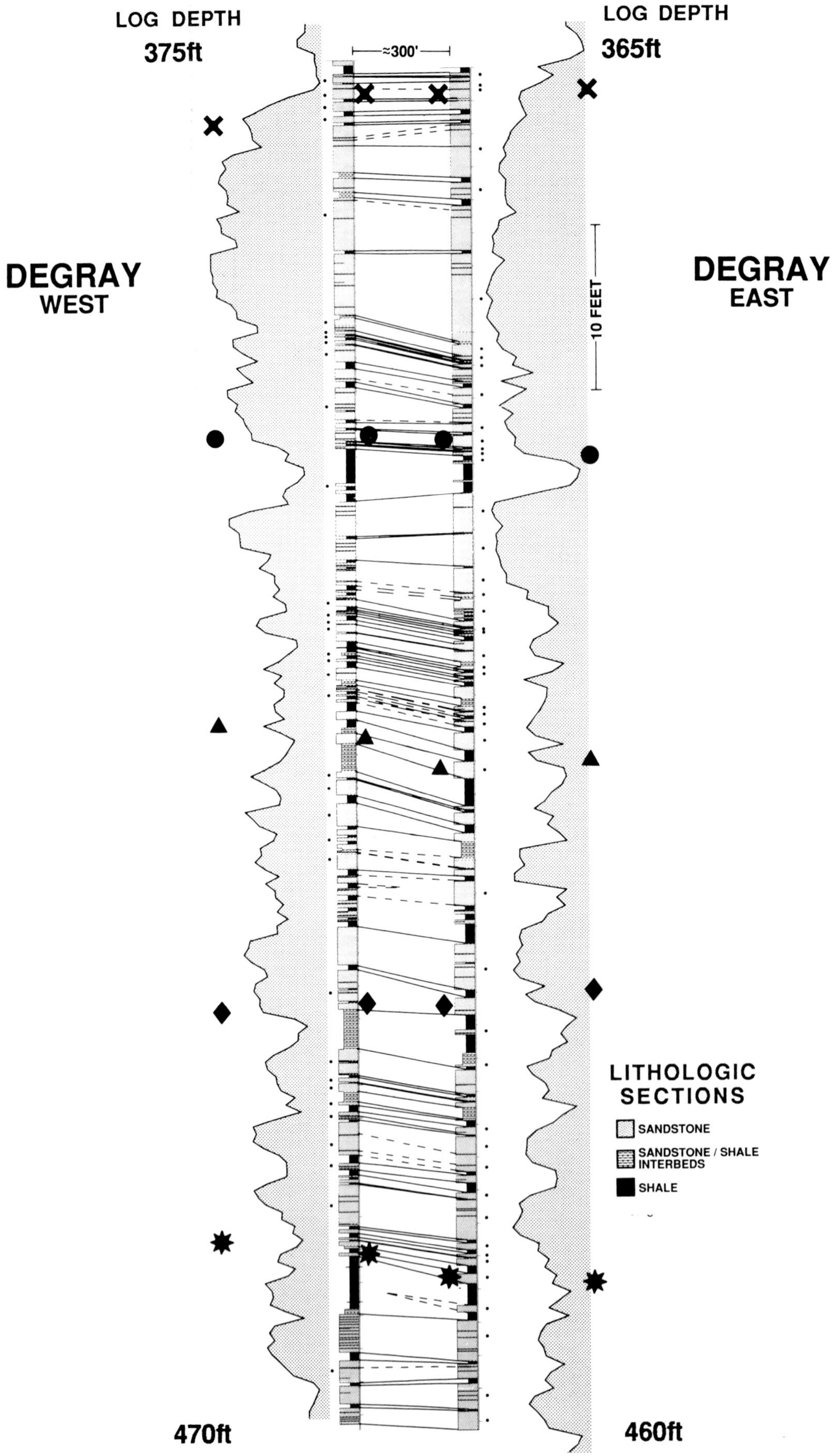

Figure 9 - Detailed stratigraphy and half foot gamma-ray logs for the interval 365-460 and 375-470 feet (log depth) along the eastern and western walls, respectively, of DeGray Lake Dam Spillway. See Figure 7 for the position of the interval within the entire spillway section. The large symbols refer to some individual beds correlated between the outcrops and logs. Small dots are located adjacent to the bed that is thickest (relative to the other wall of the spillway).

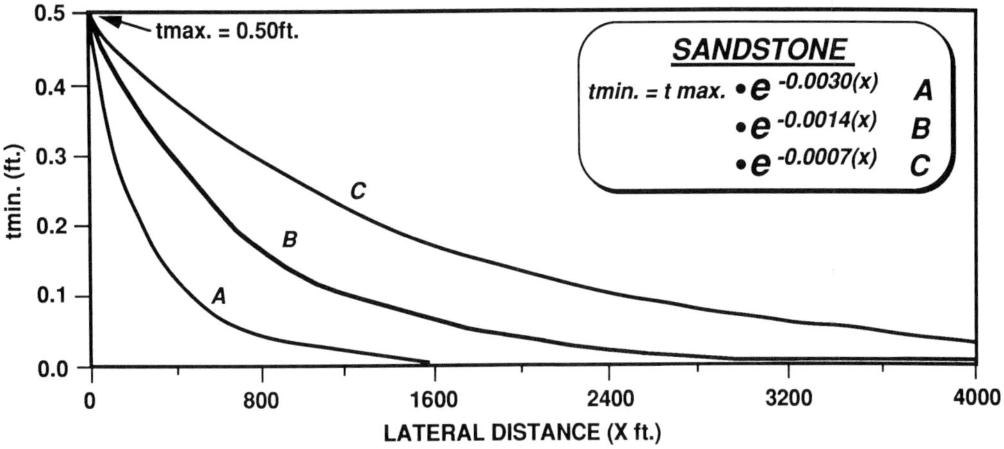

Figure 10 - Three empirical equations (A, B, and C) for sandstones at DeGray Lake Dam Spillway relating bed thickness on both sides of the spillway (t max and t min) to lateral distance or continuity of the bed (x). Curves A, B, and C show the relationship for a bed that has a maximum thickness (t max) of O . 5 feet. For example, using curve A, a bed O . 5 feet thick at one point will be 0.03 feet thick over a lateral distance along the bed of 800 feet. After Jordan et al. (in press).

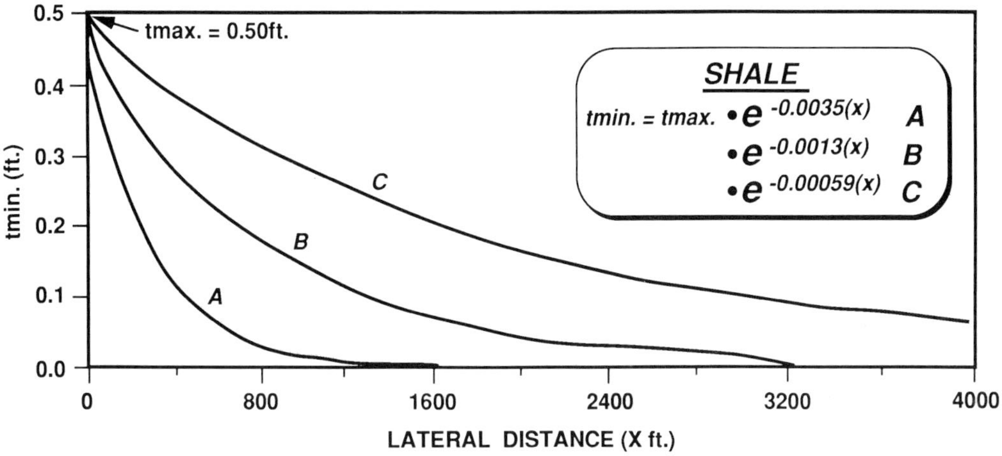

Figure 11 - Three empirical equations (A, B, and C) for shales at DeGray Lake Dam Spillway relating bed thickness on both sides of the spillway (t max and t min) to lateral distance or continuity of the bed (x). Curves A, B, and C show the relationship for a bed that has a maximum thickness (t max) of 0.5 feet. For example, using curve Ar a bed 0.5 feet thick at one point will be 0.02 feet thick over a lateral distance along the bed of 800 feet. After Jordan et al. (in press).

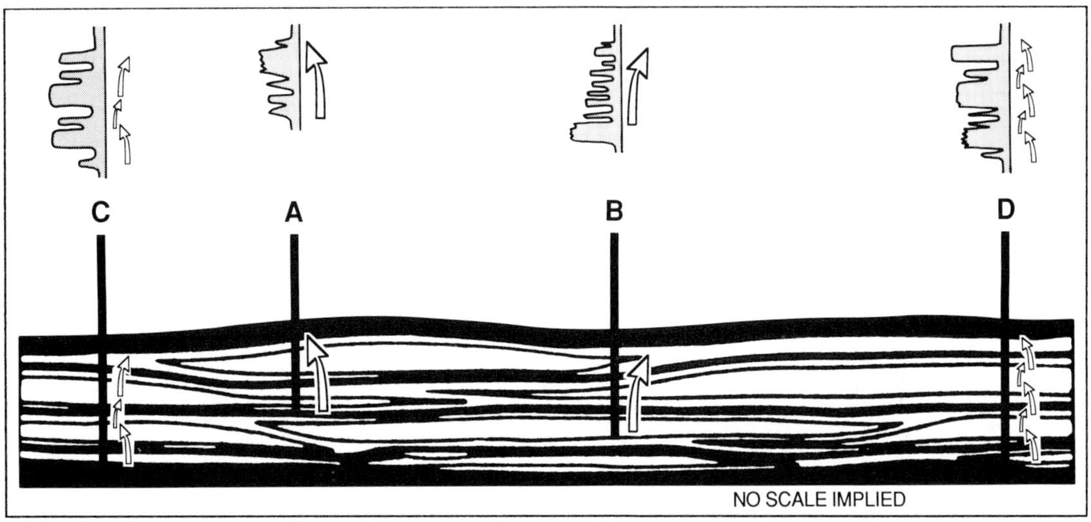

Figure 12 - Two dimensional view of sheet-like or lobate turbidite geometries which illustrates the development of compensation cycles (Mutti and Normark, 1987) as an individual bed is deposited in the adjacent topographic depression created by deposition of the immediately preceding bed. Some sample pseudo gamma-ray logs are shown to illustrate how thinning/dirtying- (small arrows) and thickening/cleaning(large arrows) upward vertical sequences can be formed by these compensation cycles. Lower diagram is modified from Mutti (1985).

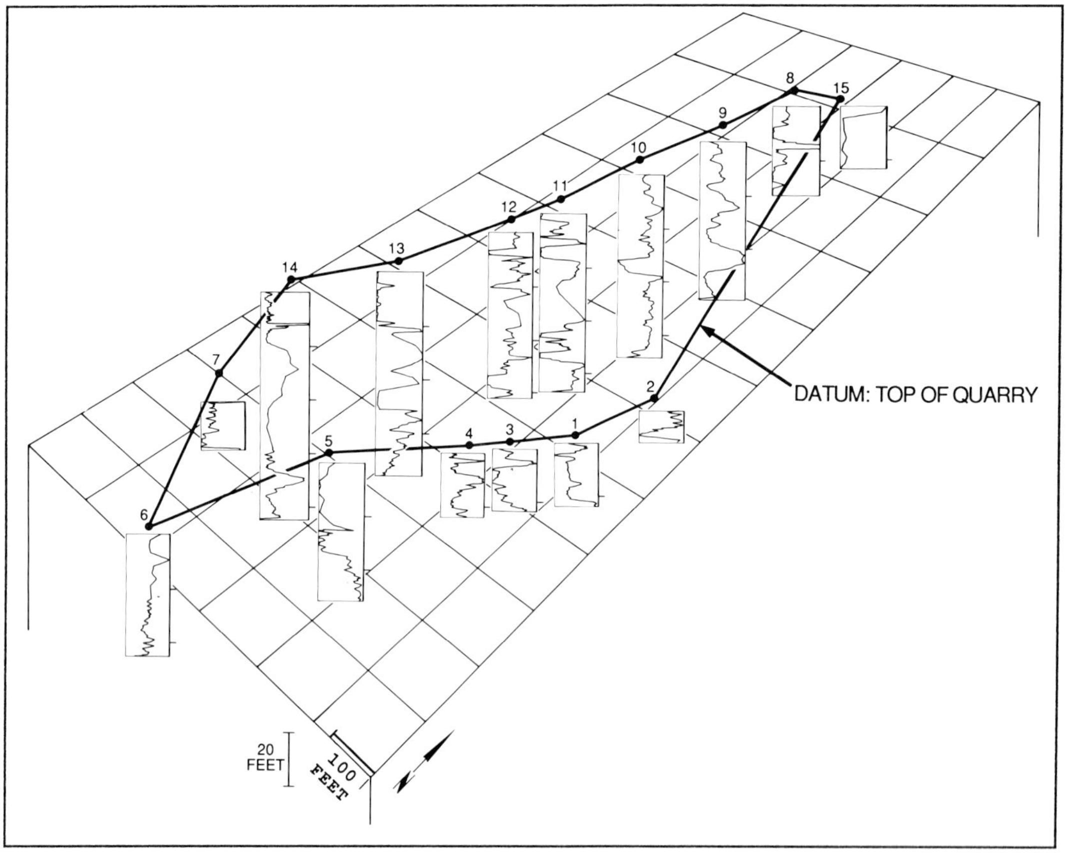

Figure 13 - Location of the fifteen gamma-ray logs obtained with the logging truck at Hollywood Quarry. Heavy line connects the top of the quarry wall.

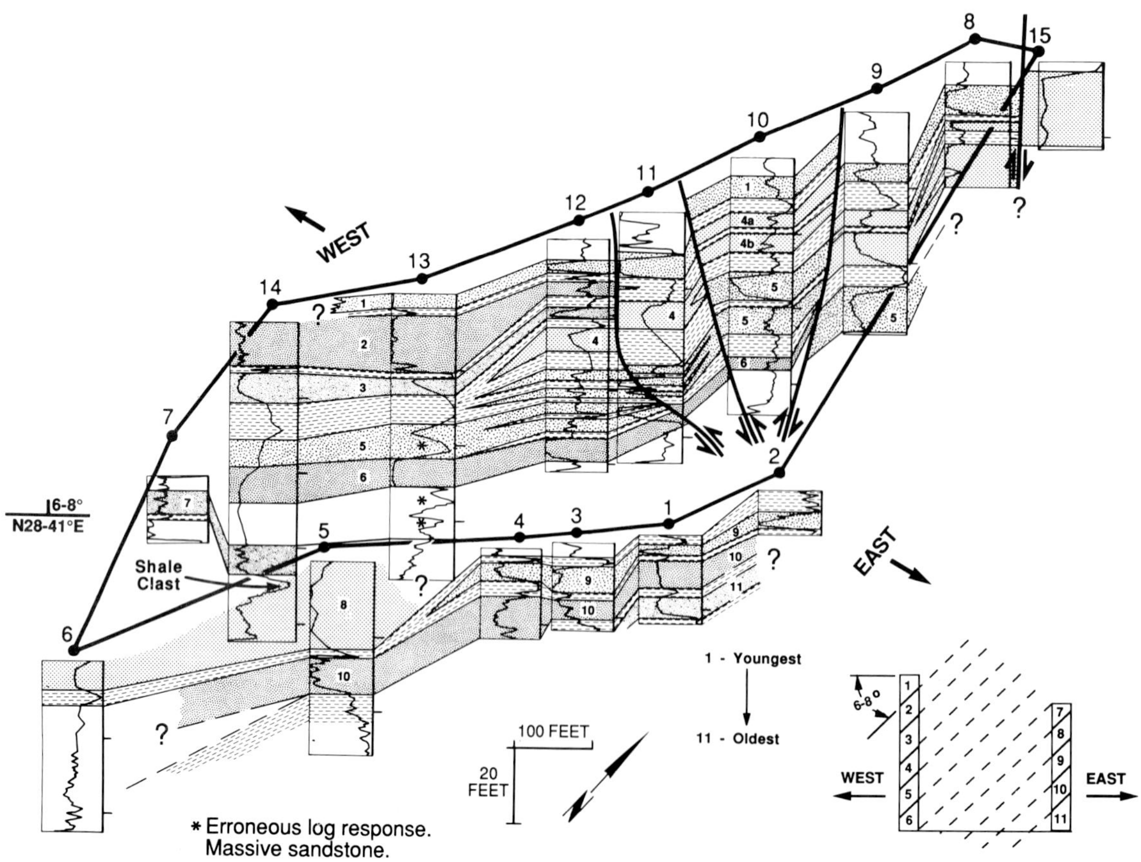

Figure 14 - Gamma-ray log correlations at Hollywood Quarry.

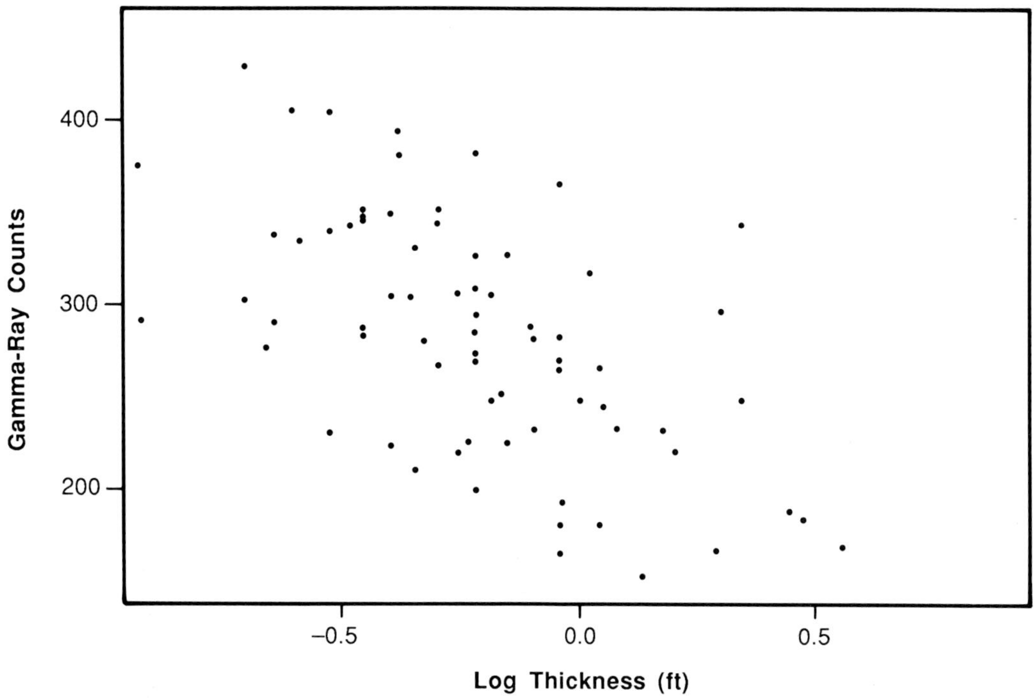

Figure 15 - Plot of gamma-ray counts vs. log of sandstone bed thickness for eighty sandstone beds measured along the east and west walls of the DeGray Lake Dam Spillway at 365-460 and 375-470 feet (log depth), respectively (Fig. 9). The thickness distribution of the sandstone beds was determined to be log normal, so the logarithm of the bed thickness was used to convert to a normal distribution.

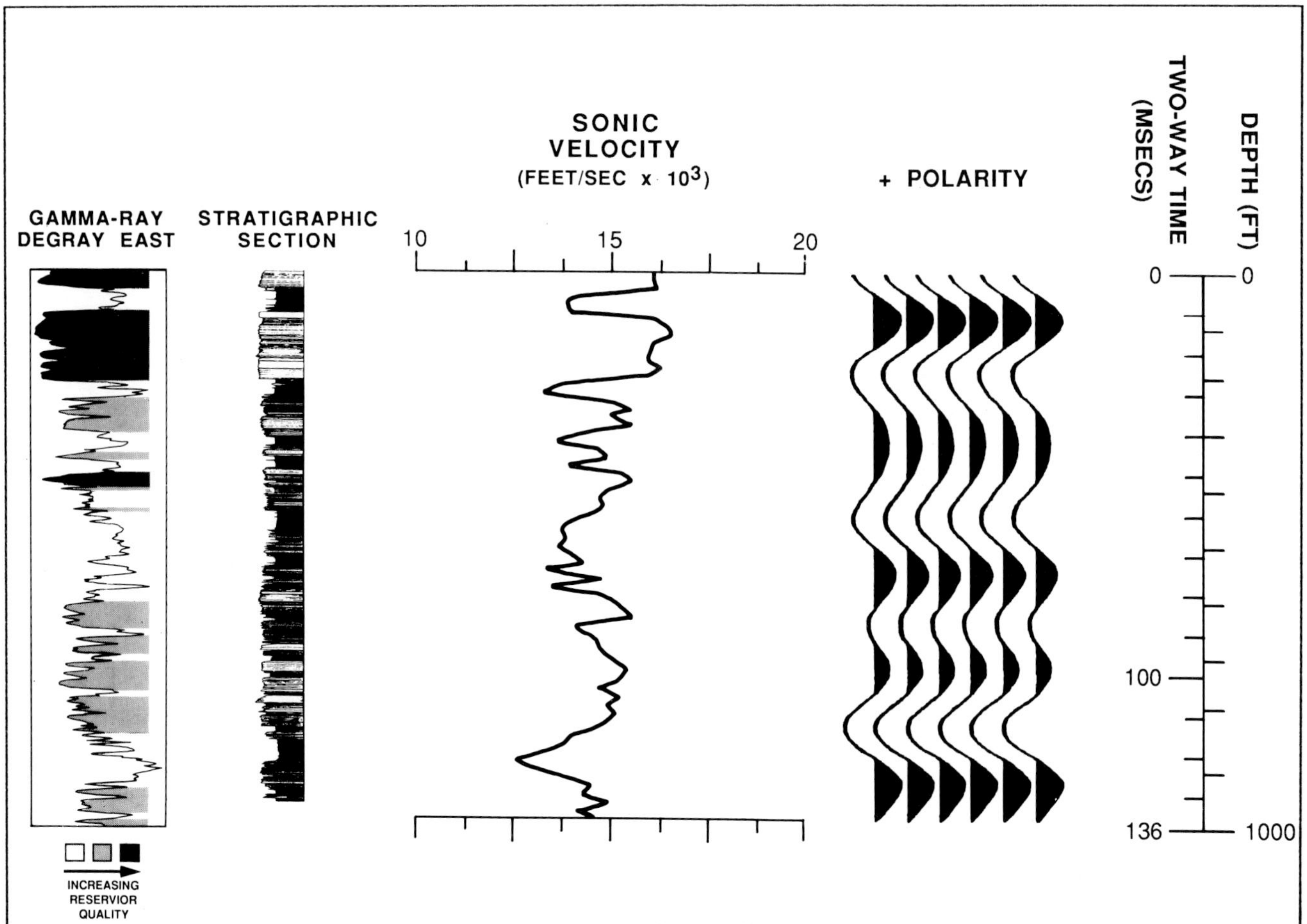

Figure 16 - Outcrop gamma-ray log, measured stratigraphic section, pseudo-sonic velocity log, and one-dimensional synthetic seismogram for the DeGray Lake Dam Spillway section.

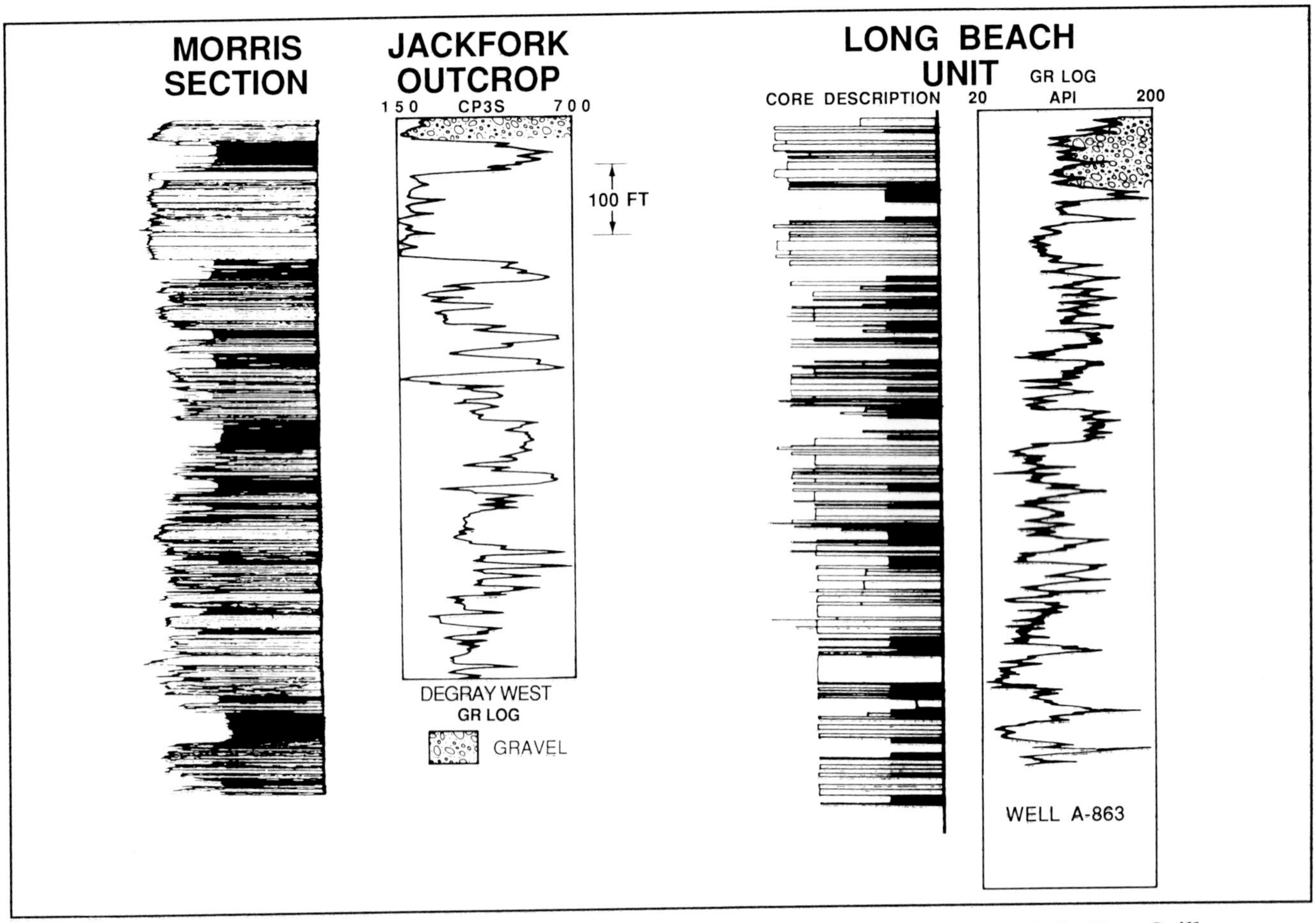

Figure 17 - Comparison of measured stratigraphic section and outcrop gamma-ray log at DeGray Lake Dam Spillway ~see Fig. 7) with a measured core section and well log gamma-ray profile of the Long Beach Unit, Wilmington Oil Field, Los Angeles Basin, California (after Slatt et al., in press).

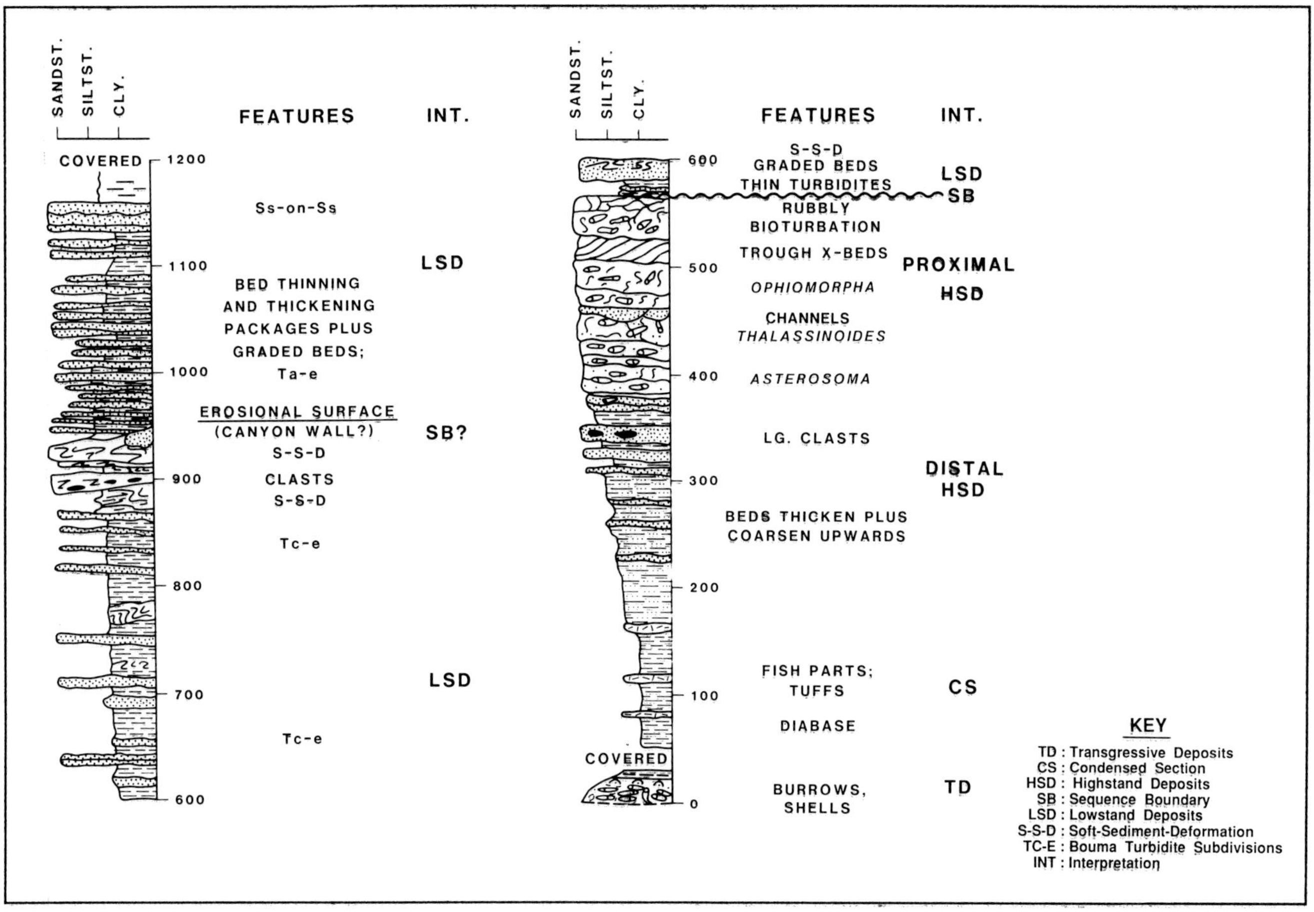

Figure 18 - Interpretation of continuous Miocene section exposed at Point Mugu, California.

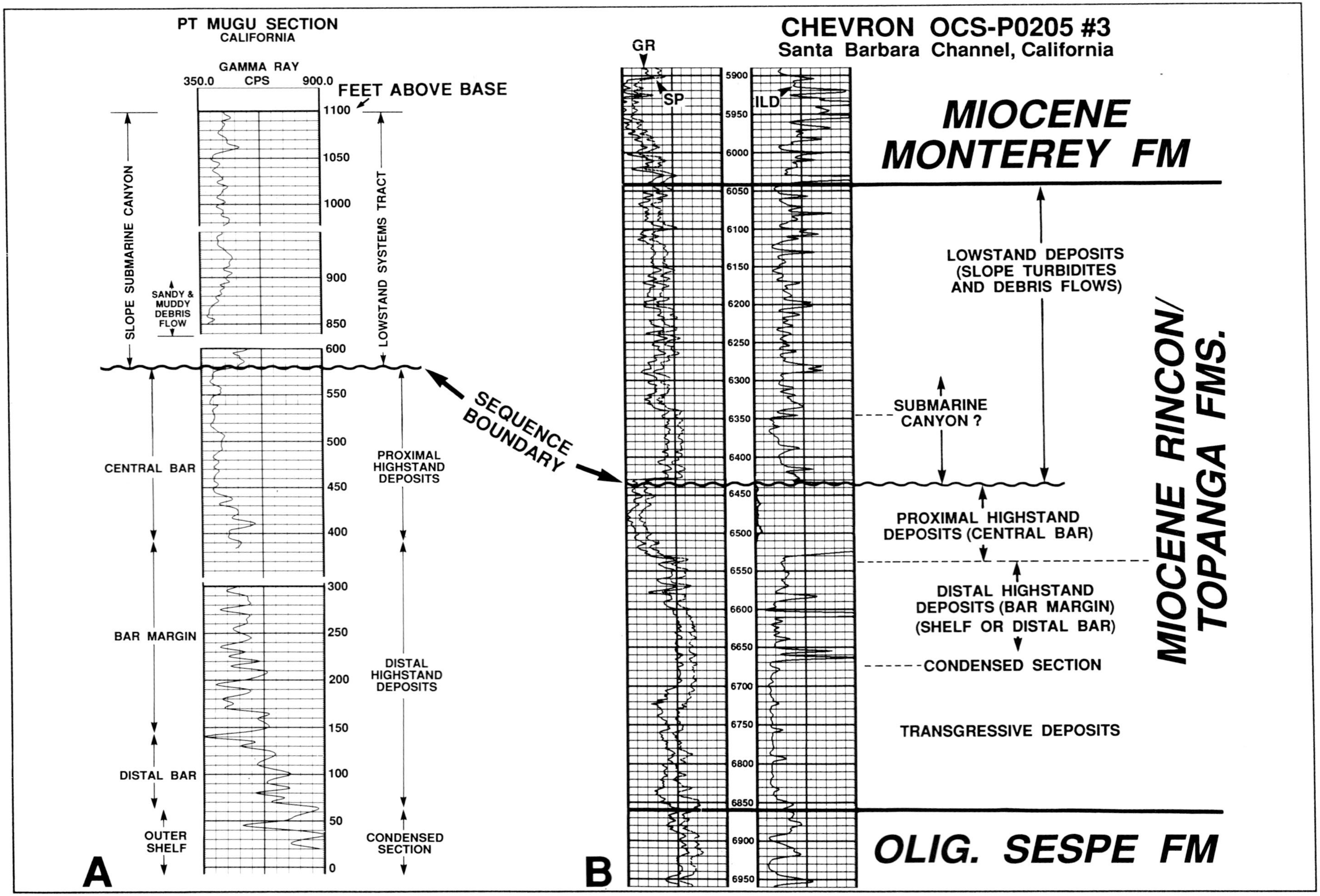

Figure 19 - A) Measured gamma-ray log from Pt. Mugu, California section. Log depths refer to measured outcrop section shown in Figure 18. Gaps denote covered or inaccessible portions of the section. Note interpretations of sequence stratigraphic units. B) Wireline logs from an offshore well in the Santa Barbara Channel, California. Stratigraphic units were inferred from log pattern, but only after the log patterns from the coastal exposures were understood and interpreted. Mapping of the turbidite sandstones in other wells may lead to definition of prospects.

Synergism in Formation Evaluation Enhances Detailed Reservoir Description

Bob A. Hardage
Walter H. Fertl
Atlas Wireline Services

Synergism in formation evaluation enhances reservoir characterization by combining data that describes reservoir properties at microscopic, macroscopic, megascopic, and gigascopic scales (Table 1). This paper describes new technologies that make valuable contributions to a reservoir data base at each of these four descriptive scale levels.

MICROSCALE RESERVOIR EVALUATION

As indicated in Table 1, cores and drill cuttings are the primary information sources used to describe reservoir properties on a microscale. Newly developed high frequency (250 kHz) sonic scanners can now produce continuous 360° acoustical reflection images of a borehole wall that provide much of the same information that can be obtained from actual cores. For example, the image in Figure 1 shows thin, detailed bedding and fracture systems from which many of the same sedimentary and tectonic interpretations can be made as is done with spatially oriented cores.

MACROSCALE RESERVOIR EVALUATION

At the macroscale level of reservoir evaluation, this paper will focus attention on only two developments in well logging, even though other macroscopic techniques involved in MWD, wireline testing and logging are equally important. The selected topics are thin bed analysis and clay typing.

Thin-Bed Analysis

Thinly bedded sand/shale sequences are encountered throughout the world, particularly as low-resistivity intervals in the U.S. Gulf Coast area. Despite apparent water saturation values ranging from 55 to 85%, as calculated by conventional log interpretation techniques, such intervals can be important hydrocarbon producers.

Recent technological advances in high-resolution logging instrumentation, increased data sampling rates, improvements In high-resolution signal processing techniques (e.g. deconvolution), and newly developed thin-bed interpretation methods now allow a more accurate reservoir description of such sequences. Improvements include a more reliable net pay count, enhanced well-to-well correlations, and better thin-bed agreement between well log, core, and tool data. Consequently, petrophysicists can now produce a more accurate identification and quantitative evaluation of oil and gas-bearing stringers, and seismic stratigraphers can now better identify the condensed sections that are so critical to sequence analysis. For example, the plots in Figures 2 and 3 show results achieved with one newly developed thin-bed resistivity tool which can resolve bed thicknesses as small as 2 inches.

Clay Typing

Shale formations and shaly clastic reservoir rocks contain varying amounts of different clay minerals that can be distributed in the host rock in laminated, dispersed, or granular modes. The respective volumes of specific clay types in a reservoir need to be known in order to decide what type of well fluid should be used to minimize clay swelling and iron precipitation, and to set production flow limits that do not result in pore plugging due to the migration of fines. Based on density, neutron, and natural spectral gamma ray logging data, reliable clay typing can be achieved. On a cation exchange capacity (CEC) versus hydrogen index (HI) crossplot, smectite, illite, and kaolinite/chlorite clay minerals group at distinct, clearly defined positions as shown in Figure 4. Incorporating Pe (photoelectric absorption index) data further differentiates between chlorite and kaolinite and allows quantitative estimates of the four major clay minerals (smectic, illite, chlorite, and kaolinite) to be made. Accurate typing of reservoir clays by detailed log data analyses of this nature will lead to the implementation of proper well completion practices, to improved reservoir descriptions, and ultimately to enhanced production.

MEGASCALE RESERVOIR EVALUATION

Referring to the list of megascopic disciplines listed in Table 1, we will restrict our discussion to those technologies just emerging in the areas of crosswell tomography and seismic-while drilling.

Crosswell Tomography

One of the current emphases in interwell imaging of reservoirs is the technique of crosswell tomography, where the seismic source and receiver are positioned in separate wells. Crosswell tomography is attractive to groups responsible for reservoir description because higher frequency seismic wavefields are recorded and signal attenuation is reduced when the source and receiver are deep in the earth. The frequency bandwidth required for a particular interwell imaging problem is basically a function of the source-to-receiver spacing. Short range propagation distances of a few hundred feet require a broadband, high frequency source/receiver system, and larger propagation distances require a lower frequency source/receiver combination.

An example of crosswell data recorded with one of the industry's long-range borehole sources, called a Resonant

Cavity Oscillator, is shown in Figure 5. The local stress produced by the device in the source well is less than 800 psi, which comforts engineers that are concerned about source-induced well damage. Even with this low stress level, robust compressional and shear wavefields are observed at this site at propagation distances as large as 4,000 ft.

Seismic-While-Drilling

In some reservoir development efforts where it is difficult to make accurate seismic depth maps due to structural complexity or to anomalous velocity behavior, it can be advantageous to use seismic-while-drilling measurements to guide the bit to its proper seismic target. In this technique, the seismic disturbances created by the drill bit are recorded by surface receivers and processed so as to identify the travel times from the surface to the bit. In field applications to date, this approach to real-drill-time-bit positioning has been used to ensure that the bottom hole location is on the proper side of reservoir fault, to indicate the onset of overpressuring ahead of the drill, and to alert drillers when to set core barrels to sample a seismic anomaly.

An example of seismic-while-drilling data is shown spliced into a seismic line crossing a drilling well ln Figure 6. At the time the drill bit reflection image (the center column) was produced, the bit had penetrated only to depth A in the stratigraphic section. Thus all events below A in the drill bit image occur below the bit, illustrating how the seismic-while-drilling technique can predict stratigraphic conditions ahead of the drill and assist in guiding the well to an optimum bottom hole location for efficient reservoir development and characterization. In this instance, the drill-bit-generated prediction is reasonably reliable down to about level B, which is a distance of approximately 1000 ft ahead of the bit.

GIGASCALE RESERVOIR EVALUATION

Surface seismic recordings comprise the bulk of the gigascopic scale data used in reservoir evaluation. The one gigascale reservoir elevation technique that will be emphasized here is a process that combines the advantages of 3-D seismic images with the concepts of an amplitude-vs-offset analysis in order to achieve an optimum interpretation of reservoir behavior.

3-D AVO Analyses

Seismic amplitude-vs-offset (AVO) analyses can be useful for estimating some reservoir properties between wells and for determining optimal locations for infill drilling. AVO analyses performed on 3-D seismic grids are particularly useful because the increased subsurface sampling of 3-D data allows detailed multi-azimuth AVO studies to be made which is invaluable if reservoir facies exhibit rapid lateral variations. It is also ln the area of AVO technology where there is one of the stronger bonds between macroscale data (specifically the Vp/Vs ratio provided by digital array acoustic logs) and gigascale measurements (i.e., surface seismic AVO behavior).

Shown in Figure 7 is a comparison of two horizontal time slices that follow along a potentially productive gas zone imaged by a 3-D seismic grid. The slice at the left is the classical bright spot analysis that is commonly used to identify the presence of gas. The slice at the right is an AVO analysis of the same gas reservoir reflection, and it shows that the productive gas zone is considerably smaller than what would be implied by the bright spot image. Ongoing drilling is confirming that the 3-D AVO analysis is a more reliable indicator of gas and a better tool for siting development wells in this particular field than is the bright spot technique alone.

REFERENCES

Fertl, W.H. and Chilingarian, G.V., 1990, Type and distribution modes of clay minerals from well logging data: Jour. Petrol. Sci. and Eng., Vol 3, p. 321-332.

Khokhar, R.W. and Johnson, W.M., 1989, A deep laterolog for ultrathin formation evaluation: SPWLA Annual Mtg paper, Brochure available from Atlas Wireline Services.

Van Sickle, K. and Valusek, J.E., 1990, AVO analysis of 3-D seismic data identifies untested reservoirs in old gas field: The Leading Edge, v. 9, No. 7, p. 18-22.

TABLE 1—INTEGRATION OF VARIOUS DISCIPLINES FOR IMPROVED HYDROCARBON RESOURCE VALUATION

SCALE OF INVESTIGATION	DISCIPLINES/PARAMETERS
MICRO	DRILL CUTTINGS; CORES (conventional, sidewall, percussion, rotary)
MACRO	MWD; WELL LOGS; WIRELINE TESTING
MEGA	WELL TESTING, BOREHOLE SEISMIC; TOMEX (SEISMIC- WHILE-DRILLING); CROSSWELL TOMOGRAPHY
GIGA	2-D/3-D SEISMIC; RESERVOIR SIMULATION; RESERVOIR ENGINEERING; REGIONAL FORMATION TEMPERATURE, PRESSURE, SAND/SHALE RATIO, AND SHALE RESISTIVITY RATIO (SRR)

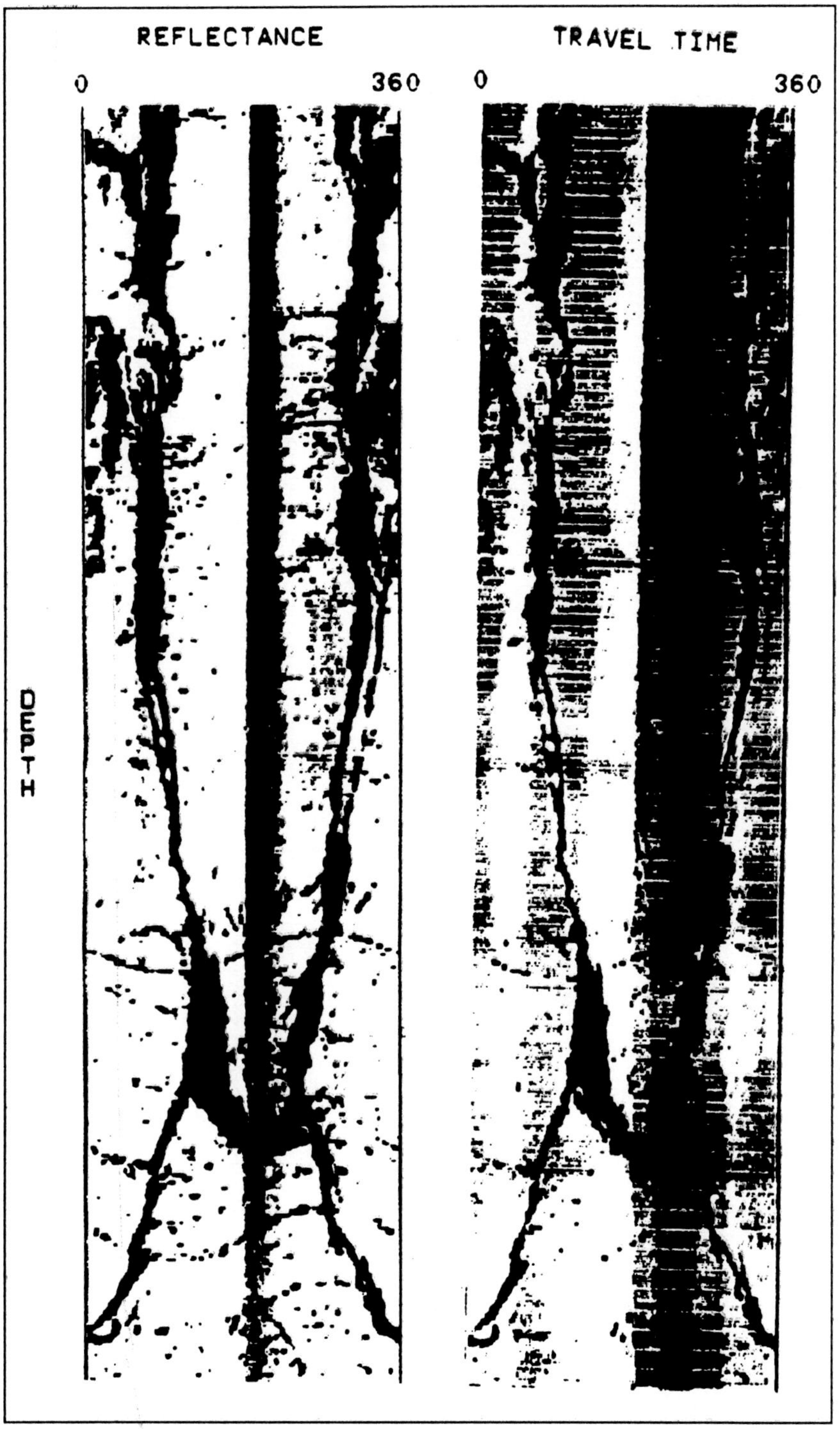

Figure 1: A high frequency acoustic image of a borehole wall. The vertical dimension is depth, and the image spans approximately 10 vertical ft. The horizontal dimension is angular position around the borehole (0° to 360°). Darker shades imply lower reflectance (left) and larger borehole diameter (right). The vertical, elongated, V-shaped feature extending almost the entire length of the display is a fracture.

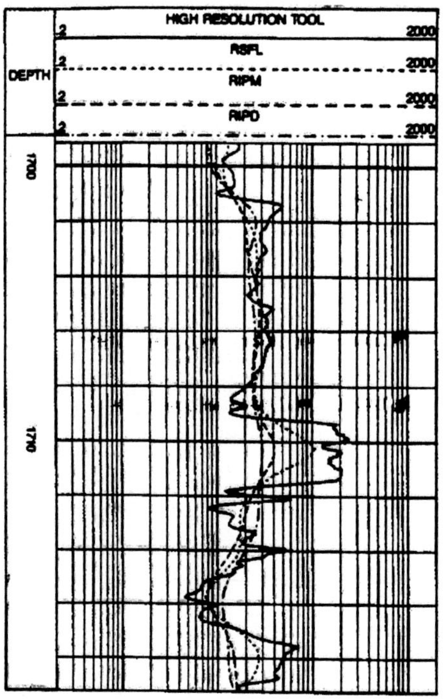

Figure 2: A comparison between the response of a high resolution resistivity tool (solid curve) and conventional resistivity tools. (Khokhar and Johnson, 1989).

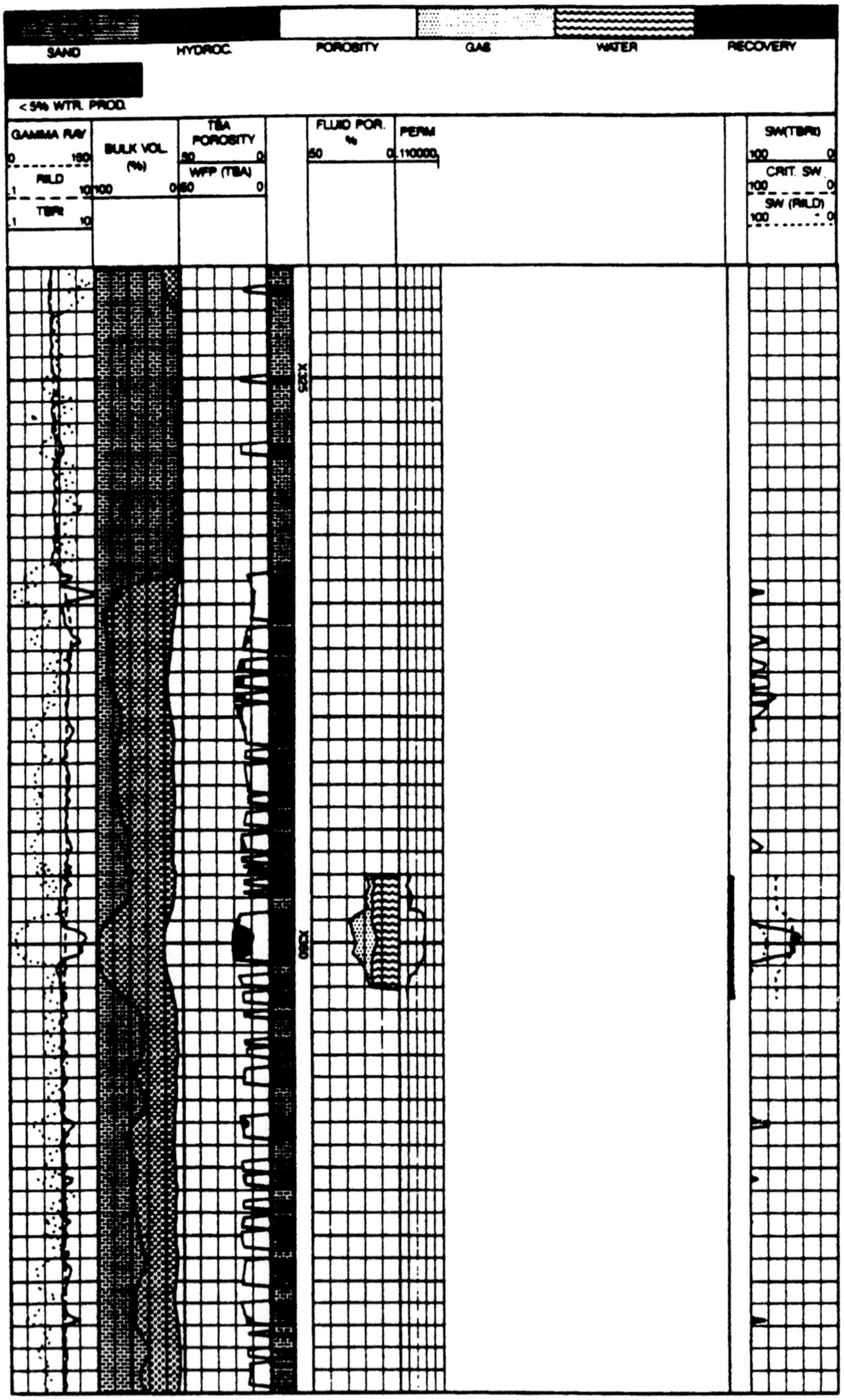

Figure 3: Conventional induction log data recorded in the 2-ft bed at X350 ft imply that S_w=80% (extreme right panel); whereas, the high resolution TBR_t curve shows that S_w=50%, which is slightly below the critical S_w (Khokhar and Johnson, 1989).

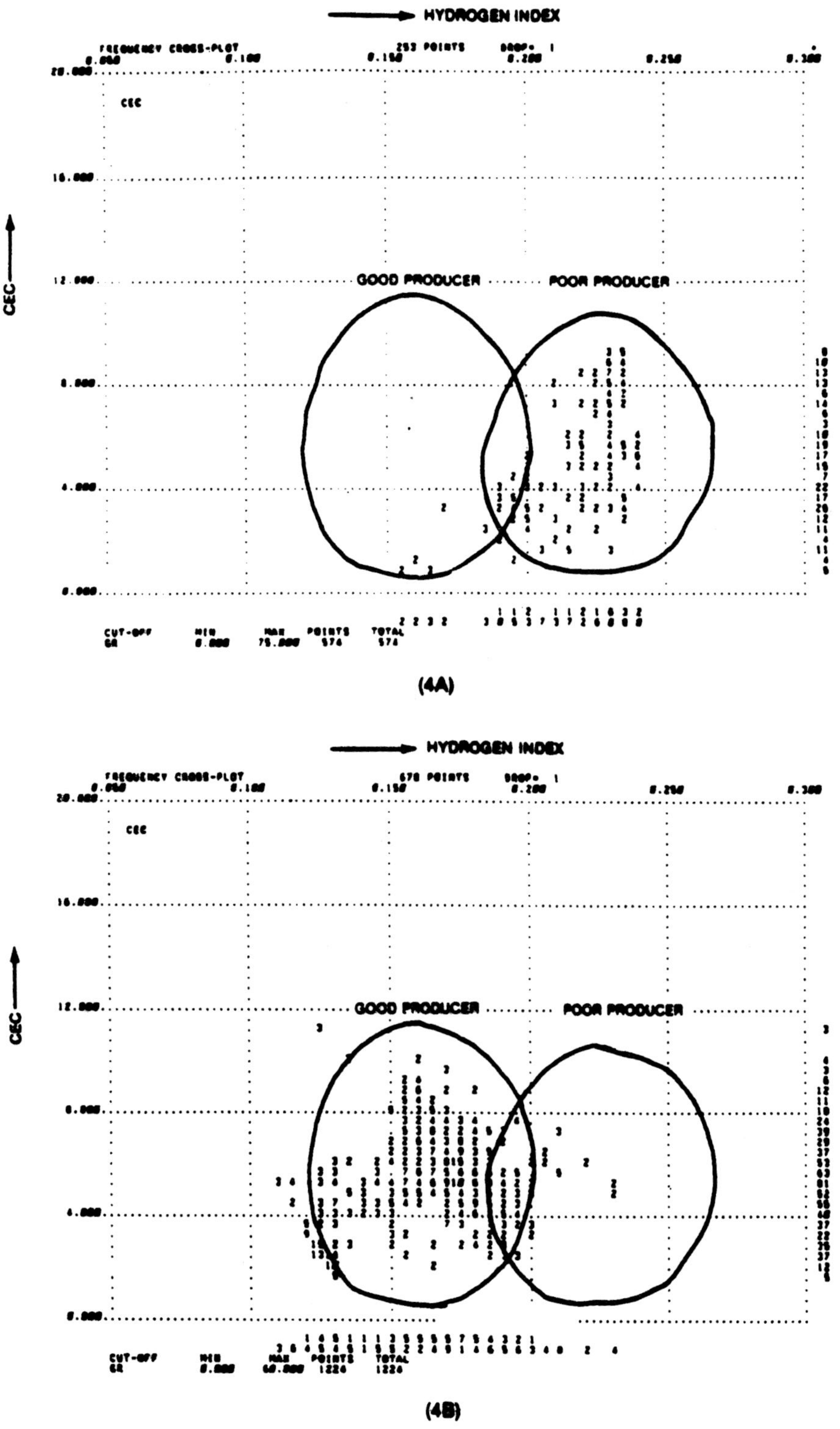

Figure 4: A crossplot of cation exchange capacity (CEC) versus hydrogen index which distinguishes illite clays (good producer well) from chlorite/kaolinite clays (poor producer well) (Fertl and Chilingarian, 1990).

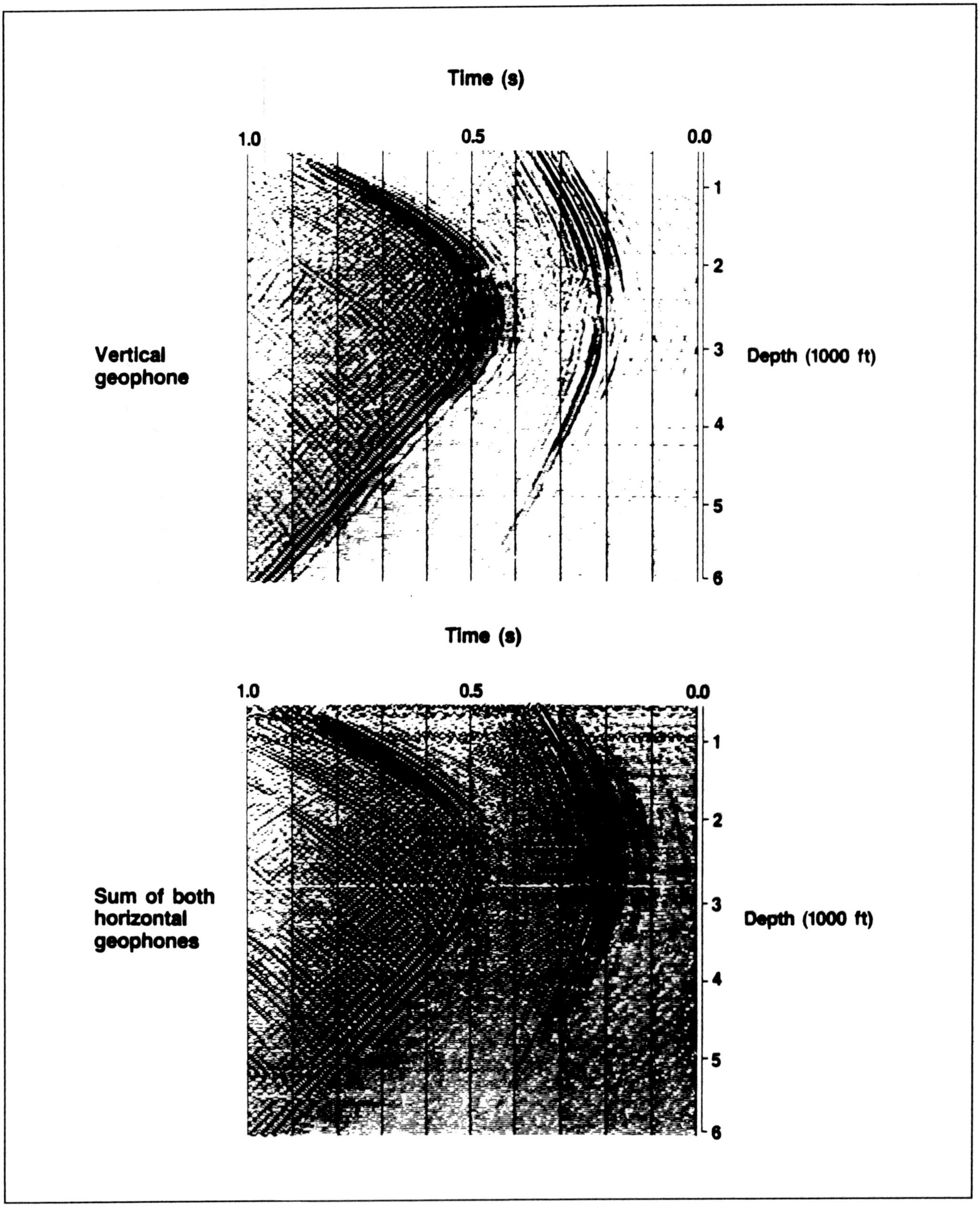

Figure 5: Crosswell wavefields recorded when a Resonant Cavity Oscillator borehole source was kept at a depth of 2500 ft in one well and a 3-component geophone was locked at vertical intervals of 25 ft, between the depths of 6000 ft and 500 ft, in a second well. The well-to-well spacing is 1800 ft, which means that the straight line distance between the source and the deepest receiver level is 4000 ft. The curved wavefront with an apex at 0.2s is the compressional arrival, and the wavefront whose apex is at 0.5s is the shear arrival.

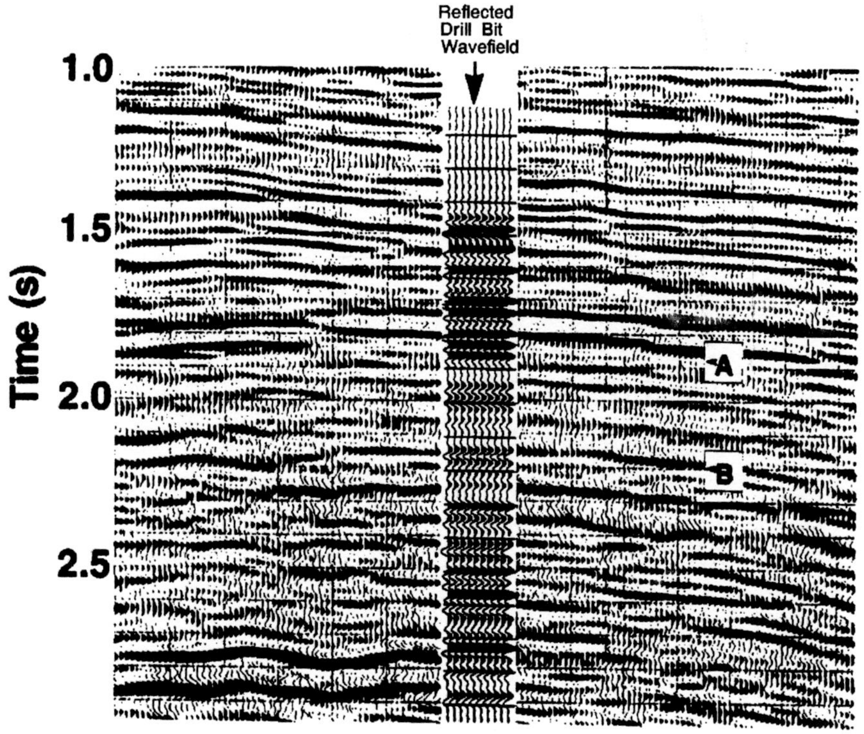

Figure 6: A comparison between conventional 2-D seismic data and reflections retrieved from a drill-bit-generated wavefield. The bit had progressed only to depth A when the drill bit reflection images were produced. Thus all events below A in the center strip are a prediction of seismic conditions ahead of the bit.

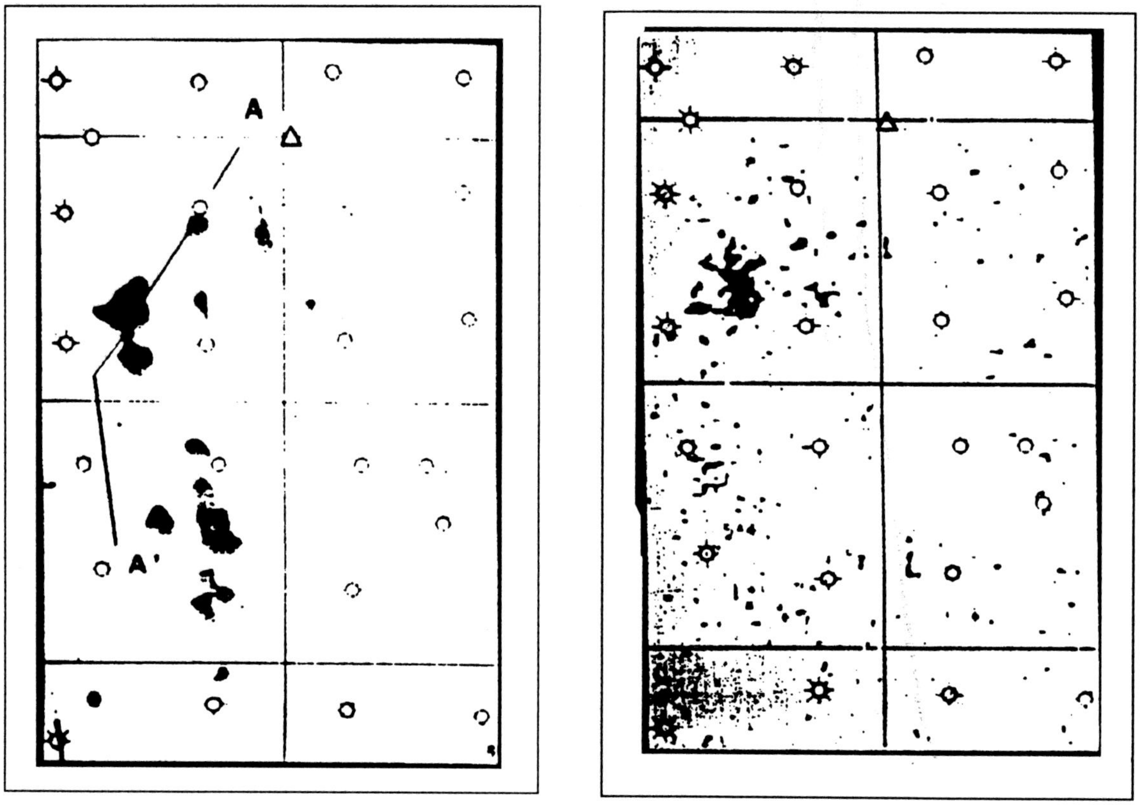

Figure 7: A comparison between a horizontal time slice of a 3-D seismic bright spot analysis and a 3-D AVO calculation for a gas-prone reflection interval. The dark areas indicate either (left) bright spots, or (right) positive gas-prone AVO responses (Van Sickle and Valusek, 1990).

RESERVOIR GEOLOGY OF THE TAYLOR SANDSTONE IN THE OAK HILL FIELD, RUSK COUNTY, TEXAS: INTEGRATION OF PETROLOGY, SEDIMENTOLOGY, AND LOG ANALYSIS FOR DELINEATION OF RESERVOIR QUALITY IN A TIGHT GAS SAND

C. L. VAVRA
M. H. SCHEIHING
J. D. KLEIN

ARCO OIL AND GAS COMPANY
RESERVOIR ENGINEERING ANALYSIS
PLANO, TEXAS

ABSTRACT

This case history deals with the delineation of reservoir quality in a tight gas sand, the Taylor Sandstone (Cotton Valley Group—Upper Jurassic) in Rusk County, Texas. The purpose of the study was to determine the geologic controls on reservoir performance and to provide a wireline log model of net pay for reserve calculations. This study was based on petrologic and sedimentologic analysis of five cored wells, and log analysis of 118 wells.

The Taylor interval consists of tightly cemented, very fine- to fine-grained quartzose sandstones interbedded with mudstones, siltstones and carbonates. Matrix porosity and permeability are quite low, even for tight gas sands: helium permeability rarely exceeds 0.1 md, and porosity is typically less than 10 percent. Cores examined to date appear virtually devoid of open natural fractures.

Six major rock types or petrofacies can be distinguished based on petrographic criteria. Each petrofacies is characterized by a unique combination of dominant pore geometry (pore size, shape, sorting and interconnectivity) and pore-filling mineralogy. Of the six petrofacies identified, three have reservoir potential: (1) Primary Macroporous Quartz-cemented, (2) Moldic Macroporous Quartz-cemented and (3) Microporous Clay-cemented. The Primary Macroporous Quartz-cemented Petrofacies is characterized by quartz-cemented sandstone with primary macropores interconnected by slot-like pore throats. This petrofacies has the highest reservoir quality. The Moldic Macroporous Quartz-cemented Petrofacies is characterized by virtually isolated secondary pores (molds) in quartz-cemented sandstone. Reservoir quality is intermediate. The Microporous Clay-cemented Petrofacies is characterized by sandstones with abundant clay cement and a pore geometry dominated by microporosity. This pore geometry results in the lowest reservoir quality of the reservoir petrofacies.

A wireline log model to calculate porosity was based on a shale-corrected neutron-density cross-plot solution, with corrections applied for matrix and calibration errors. The porosity model included identification of macroporous, microporous, and nonreservoir petrofacies based on the calculated porosity and V_{shale}. Different grain densities were used for each petrofacies, thus requiring an iterative solution. Water saturation calculations were carried out using Archie's equation with laboratory-determined coefficients *a*, *m*, and *n*.

Petrofacies types show a strong association with depositional environment, even though the primary pore network has been strongly altered by diagenesis. The depositional environment interpreted for the Taylor sandstone bodies is a barrier island with back-barrier, foreshore, shoreface and inner shelf sub-environments. Two such barrier island complexes are present in the Oak Hill Field. The Primary Macroporous Petrofacies is associated with clean, well-winnowed sandstones comprising foreshore and tidal channel/delta environments. The Moldic Macroporous Petrofacies occurs in upper shoreface and some back-barrier sandstones. The Microporous Petrofacies is associated with clay-rich, more poorly sorted, generally bioturbated sandstones that comprise lower shoreface, inner shelf and some back-barrier/lagoonal sandstones.

The Taylor interval was subdivided into correlative zones and sub-zones based on unconformities and sedimentological sequences observed in core that identified major sandstone bodies within the stratigraphic interval. This, combined with the wireline petrofacies model, permitted mapping and tracing petrofacies around the field. This information permitted the recognition of stratigraphic and geographic zones of good and intermediate reservoir quality (primary and moldic macroporosity) associated with foreshore, upper shoreface and tidal channel facies and its relationship to zones of poorer quality (microporous) reservoir rocks associated with lower shoreface, inner shelf and various back-barrier facies.

This study demonstrates the utility of an integrated petrologic, sedimentologic and wireline log analysis study in identifying the controls on reservoir quality at the pore level and extending this understanding to the interwell and fieldwide scale via a depositional model and log analysis.

INTRODUCTION

This paper summarizes a comprehensive geologic study of the Taylor sandstone (Upper Jurassic) of the Cotton Valley Group in the Oak Hill Field, Rusk County Texas (Figure 1). This combined petrologic, petrophysical and sedimentologic study was undertaken to determine geologic controls on reservoir performance and to provide input into reserves calculations in the Taylor interval within the Oak Hill Field. Specific goals of this study were to: (l) establish pore-level geologic controls on reservoir quality, (2) develop log models for porosity and water saturation, and analyze logs for 118 wells, and (3) determine reservoir geometry and zonation.

This study is based on detailed petrographic analysis of samples from four conventional cores (Figure 1), detailed log analysis of five cored wells, sedimentological descriptions of five cored wells, five correlation sections that tie the cored wells, and gross sand thickness maps of the area.

Stratigraphy and Paleogeographic Setting

Productive intervals of the Oak Hill Field in the East Texas Basin form part of the Cotton Valley Group of Late Jurassic to Early Cretaceous age. Lithostratigraphic formations in the Cotton Valley Group consist of (from the base up) the Bossier Shale, the Cotton Valley Sandstone and the Knowles Limestone (Figure 2). In the Oak Hill Field, an informal lithostratigraphic nomenclature is used to further subdivide the Cotton Valley Sandstone (Figure 2). The "Taylor Sandstone", which is the subject of this study, extends from the top of the Bossier Shale to the base of the "C" limestone. The Taylor interval is subdivided into five informal zones, which correspond to major mudstone-bounded sandstone packages. The "Upper Cotton Valley Sandstone", which overlies the "C" limestone, was not included in this study.

Regional analysis of paleogeography indicates that rocks comprising the Cotton Valley Sandstone in the Oak Hill Field area were deposited in an inner shelf through coastal plain setting characterized by strandplains and barrier islands. Sediment was supplied to these strandplains and barrier islands from fluvial-deltaic systems to the north and northeast of the study area (Figure 3).

General Characteristics of the Taylor Sandstone

The Taylor interval consists of tightly cemented, very fine- to fine-grained sandstones interbedded with mudstones, siltstones, and carbonates. Dominant sandstone cements are quartz, carbonate and clay. Matrix porosity and permeability are quite low, even for tight gas sands: helium permeability rarely exceeds 0. 1 md, and porosity is rarely greater than 10 percent.

Cores examined have few natural fractures and appear to be virtually devoid of open natural fractures. A few incipient, possibly open fractures have been identified, but the vast majority of natural fractures examined are completely cemented. Coring-induced fractures are common in all Cotton Valley cores. These fractures commonly interconnect pores that were isolated prior to fracturing. Because of this, coring-induced fractures can result in overestimation of porosity and permeability during conventional core analysis.

Measurement of helium permeability at simulated reservoir stress (3000 psi), which should close most coring-induced fractures, results in approximately an order of magnitude reduction of permeability compared to unstressed measurements. Least-squares regression of the data produces a well constrained (R^2= 0.89) equation of the form:

$$\log(k_S) = 0.88(\log(k_A)) - 0.90 \quad (1)$$

where k_S = helium permeability at simulated reservoir stress (3000 psi),
k_A= helium permeability at ambient pressure (≈400 psi).

Thus, for a sample with a permeability of 0.01 md measured at ambient stress, the expected permeability at simulated reservoir stress would be 0.002 md.

Helium porosity values measured during conventional core analysis do not require adjustment to reflect porosity at reservoir conditions. Porosity reductions that result from closing coring-induced fractures are counteracted by porosity increases from removal of salts that precipitated from pore water after core recovery.

PETROFACIES

Six major rock types or petrofacies in the Taylor sandstone can be distinguished based on petrographic criteria. A petrofacies is a body of rock distinguished primarily by its petrographic attributes such as composition, diagenesis and pore type. The concept of petrofacies is similar to that of lithofacies except that petrofacies are typically based on the microscopic character of the rock, whereas lithofacies are based on macroscopic features such as bedding and grain size. Each petrofacies is characterized by a unique combination of dominant pore geometry and pore-filling mineralogy; however, within any one petrofacies several pore types or pore-filling minerals are typically present. Because petrofacies are defined by pore geometry (which is an important control of reservoir properties) and mineralogy (which can control well log response) petrofacies can be used to subdivide the formation into units with different expected production behavior and log character.

The six petrofacies are named for the dominant pore geometry and pore-filling mineralogy: (1) Nonporous Carbonate-rich, (2) Nonporous Matrix-rich, (3) Nonporous Quartz-cemented, (4) Primary Macroporous Quartz-cemented, (5) Moldic Macroporous Quartz-cemented and (6) Microporous Clay-cemented (Table 1). Of these five petrofacies, only the Primary Macroporous Quartz-cemented, Moldic Macroporous Quartz-cemented and Microporous Clay-cemented Petrofacies have reservoir potential. Only the reservoir-quality petrofacies will be described here. For purposes of log analysis, the two macroporous petrofacies were considered together and the three nonreservoir petrofacies were combined into one petrofacies.

Primary Macroporous Quartz-cemented Petrofacies

The Primary Macroporous Quartz-cemented Petrofacies is defined by samples containing quartz as the major cement and primary, interparticle macroporosity as the principal pore type (Table 1). The pore geometry of this petrofacies is dominated by interparticle macropores interconnected by narrow slot-like pore throats that are bounded by faces of quartz overgrowths (Figure 4). The size, number and distribution of pores are controlled by grain size, sorting and extent of quartz cementation. Rocks assigned to the Primary Macroporous Quartz-cemented Petrofacies have the highest reservoir quality observed to date from the Oak Hill Field.

The Primary Macroporous Quartz-cemented Petrofacies is further characterized by relatively good helium porosity and permeability, low grain density, and low clay and pyrite contents (Table 1). Mercury injection capillary pressure (MICP) measurements indicate that pore throats are typically relatively well sorted, with 50 percent of the throats having radii greater than 0.45 μm (Figure 5). Water saturations, as determined by conventional core analysis, are substantially lower than those from other reservoir petrofacies (Table 1; see also Water Saturation section).

Samples assigned to this petrofacies occur most commonly in parallel-laminated sandstone. This petrofacies tends to be texturally and mineralogically more mature (better sorted and more quartzose) than other petrofacies. The paucity of secondary moldic pores is primarily due to the absence of soluble grains, which were apparently removed by mechanical abrasion and winnowing in the high energy depositional environment.

Moldic Macroporous Quartz-cemented Petrofacies

The Moldic Macroporous Quartz-cemented Petrofacies is defined by samples containing abundant quartz cement and a significant amount of moldic macroporosity (secondary voids formed by dissolution of feldspar, chert and organic fragments). The pore geometry of this petrofacies is dominated by scattered, virtually isolated secondary pores, which are typically lined by chloritic clay (Figure 6). The size, number and distribution of these pores are controlled, in part, by the size, number and distribution of soluble grains and by the diagenetic history of the rock. Mercury injection capillary pressure measurements indicate that pore throats are moderately well sorted and have a median radius of approximately 0.15 μm (Figure 5). Permeability is a complex function of the amount of porosity, the proportion of primary porosity and the amount of pore-bridging clay cement. The optimum combination is high, dominantly primary porosity and low clay content. This complex relationship makes it difficult to establish a model that can predict permeability values from porosity, as described below.

This petrofacies occurs most commonly in parallel-laminated and cross-stratified sandstones. Sandstones assigned to this petrofacies contain a higher proportion of chemically unstable feldspar, chert and organic fragments than those assigned to the Primary Macroporous Quartz-cemented Petrofacies. The dissolution of these unstable framework grains resulted in the moldic porosity that characterizes the Moldic Macroporous Quartz-cemented Petrofacies. Although these pores are typically lined by chloritic clay, the clay does not form a continuous network.

Microporous Clay-cemented Petrofacies

The Microporous Clay-cemented Petrofacies is defined by samples having pore geometries dominated by microporosity (pores smaller than about 10 μm) and containing significant amounts of clay cement (Figure 7). Clay cements, which contain much of the porosity in the form of interparticle microporosity, are dominantly illite and chlorite. mercury injection capillary pressure measurements indicate poorly sorted pore throat sizes with a median radius of approximately 0.04 μm (Figure 5).

This petrofacies occurs most commonly in bioturbated sandstones. Clay, introduced by burrowing organisms, has been recrystallized into clay cement, which prevented extensive cementation by quartz. Microporous clay cement occurs in virtually all the pores, forming a continuous network.

CONTROLS ON RESERVOIR QUALITY

Porosity and Permeability

In the Taylor sandstone, porosity and permeability are controlled by a complex interaction of depositional environment, initial sediment composition and diagenesis. Unlike many reservoirs where particular lithofacies are invariably reservoir and others nonreservoir, in the Taylor sandstone virtually any lithofacies can be reservoir or nonreservoir depending on diagenesis.

In manyclastic reservoirs, relationships between helium porosity and permeability are relatively easy to establish because the pore structure is strongly controlled by depositional grain size and sorting. However, in the Taylor sandstone, relationships between porosity and permeability are difficult to establish because of the strong diagenetic overprint, which has significantly altered the pore structure. Diagenetic processes that have modified pore structure in the Taylor reservoir petrofacies include: (l) compaction, (2) cementation by quartz, (3) dissolution of isolated detrital grains, (4) formation of clay pore linings and (5) cementation by microporous pore-bridging and pore-filling clay. Samples containing coring-induced fractures have higher reported unstressed permeabilities, but these values were excluded because they do not represent matrix permeability. Each of the above processes affects pore geometry and the relationship between porosity and permeability differently. This, combined with variations in the intensity of each diagenetic process from sample to sample and the potential presence of coring-induced fractures, results in very diffuse relationships between porosity and permeability (Figure 8).

Although the overlapping ranges of conventional porosity and permeability data suggest that the three reservoir petrofacies have similar reservoir quality, these data do reflect the relative or absolute producibility of the petrofacies *in situ* where they contain two phases (water and gas). Effective gas permeabilites of stressed samples at reservoir water saturations range from one to over four orders of magnitude lower than stressed gas permeabilites at S_W=0 (Figure 9). The Primary Macroporous Quartz-cemented Petrofacies has the most favorable relative permeability relationships because pore walls and throats are smooth and relatively large. Effective gas permeability of this petrofacies at reservoir conditions ranges from 0.01 to 0.002 md. The Moldic Macroporous Quartz-cemented Petrofacies has less favorable relative permeability due to the finer, more tortuous pore system and clay linings of the moldic pores. Effective reservoir gas permeability of this petrofacies ranges from 10^{-3} to 10^{-5} md. The Microporous Clay-cemented Petrofacies has the least favorable relative permeability, with effective reservoir gas permeabilites of less than 10^{-5} md.

These data suggest that although the Microporous Clay-cemented Petrofacies makes up approximately 75 percent of the reservoir volume, productivity is controlled by the distribution of the macroporous petrofacies.

Water Saturation

The Taylor sandstone differs from many clastic reservoirs in that water saturations are independent of total porosity (Figure 10). In many clastic reservoirs, porosity and water saturation are inversely related because as porosity increases, pore throat radius tends to increase, resulting in decreases in capillarity and water saturation. However, because of the strong diagenetic overprint in the Taylor, there is no relationship between pore throat size and total porosity.

Although water saturation is independent the total amount of porosity, water saturation is dependent on the type of porosity present. In particular, there is a strong inverse relationship between the proportion of primary interparticle porosity and water saturation (Figure 10), because pore throats associated with primary pores are substantially larger than those associated with moldic or micropores (Figure 5). These larger pore throats result in lower capillarity and, therefore, lower water saturation in rocks with a high proportion of primary interparticle porosity.

Because water saturation is partly controlled by pore type, which is used to define petrofacies, a systematic relationship should exist between water saturation and petrofacies. Analysis of conventional core plug data from a tightly constrained interval in one well (Figure 11) indicates that water saturations are consistently the lowest in Primary Macroporous Quartz-cemented Petrofacies (mean S_w = 35.5

percent), intermediate in the Moldic Macroporous Quartz-cemented Petrofacies (mean S_w = 54.4 percent), and highest in the Microporous Clay-cemented Petrofacies (mean S_w = 81.6 percent). Calculations correcting for the effects of confining pressure, unextracted salt, temperature, pressure and solution gas suggest that core-derived water saturations should be increased by approximately 12 percent of their value in order to reflect *in situ* values. If this is correct, the observed differences in water saturation among petrofacies should be even greater in the reservoir. These variations correlate well with variations in capillary pressure observed among the reservoir petrofacies (Figure 5).

WIRELINE LOG ANALYSIS

Wireline log models were developed using conventional core data from the five cored wells. Preliminary steps included environmental corrections to log data, depth shifting, and corrections to core data to account for changes from reservoir conditions. As previously noted, the porosity corrections (overburden and precipitated salt) are equal and opposite in sign, thereby cancelling each other. Core water saturations were multiplied by a factor of 1.12 to obtain reservoir conditions.

The first steps in the log analysis consisted of calculation of V_{shale}, bad hole flag, and a shale-corrected neutron porosity (Figure 12). The V_{shale} at each depth was the minimum of the V_{shale} calculated from the gamma-ray and the SP logs. Linear relationships were assumed between logs and V_{shale}. The bad hole flag was based on the caliper log, and was triggered whenever the hole diameter exceeded 12 inches or in rugose hole. The neutron porosity was corrected for shale effects using a correction that was linear with the V_{shale}.

Recognition of Petrofacies from Well Logs

Because petrofacies are defined by pore geometry, which is an important control of reservoir properties, and mineralogy, which can control well log response, petrofacies definition provides a critical link between log response and reservoir properties. Reservoir petrofacies are distinguished from nonreservoir petrofacies on wireline logs using a porosity log cutoff of 4 percent (Figure 13). This cutoff value is not meant to imply a formal definition of net pay, but rather was chosen as an optimistic reservoir indicator.

Macroporous Quartz-cemented (clay-poor) and Microporous Clay-cemented (clay-rich) intervals were distinguished using a cutoff of 10 percent on the volume of shale (V_{shale}) log, which is based on a corrected Gamma-Ray log. An upper V_{Shale} cutoff of 50 percent was chosen for the Microporous Petrofacies to eliminate shale washouts. For the purposes of this study, the Intercrystalline Macroporous Quartz-cemented Petrofacies and Moldic Macroporous Quartz-cemented Petrofacies were grouped together for log analysis. Analysis of data from the five cored wells indicates that log-assigned petrofacies are in good agreement with petrofacies determined by detailed petrographic analysis. The resulting petrofacies picks were used to map the reservoir, and to improve calculation of porosity and water saturation from wireline logs.

Porosity Model

Porosity was computed as the average of the density porosity and shale-corrected neutron porosity (Figure 12). The density porosity was initially computed with a grain density of of 2.68 g/cc and fluid density of 1.08 g/cc. The porosity and V_{shale} results were then used to determine petrofacies on a foot-by-foot basis. The porosity was redetermined using different values of grain density for each petrofacies: 2.66, 2.68, and 2.694 g/cc for macroporous, microporous, and nonreservoir petrofacies respectively. The petrofacies picks were then redetermined, and the process was repeated several times until stable results were obtained. A backup porosity was required for the depths where the bad hole flag was set, which was a common occurrence. Two different approaches were used for the backup porosity. The first was based simply on the V_{shale}, scaled to a maximum value in "clean sand" and to zero at V_{shale} of 100%. The second backup porosity was based on the sonic log when available, with different empirical relations depending on V_{shale}.

The final output porosity was the neutron density porosity as described above with the following modifications. If the bad hole flag was set, then the backup porosity was used. In addition, whenever the neutron-density porosity was greater than the backup porosity, then the average of the two were used. Finally, the output porosity was limited to a maximum of 14 porosity units.

The porosity calculations are very sensitive to tool calibration errors and effects of isolated carbonate streaks, which are not accounted for in the model. Procedures were developed to correct the neutron and density porosities for these problems. The approach for determining the corrections was based on the neutron-density crossplot, where valid data points would fall within a particular region. Depending on the location of outliers, correction terms were added to either the neutron or density porosity, and the porosity calculations were then repeated.

Comparison of log to core data indicated that the average log porosity was within 1 porosity unit of the average core porosity. The standard error of estimate of the log porosity values was 1.5 porosity units. The mean porosity for reservoir quality petrofacies was 6.0 porosity units. The log model was 85% to 90% successful in determining the petrofacies, based on comparison to core (Figure 13).

Water Saturation Model

Water saturation calculations were carried out using Archie's equation. Clay corrections were not required, due to the high conductivity of the formation brine, 0.015 ohm-m. The coefficients for Archie's equation were determined in a laboratory study with core-plugs of the microporous and macroporous petrofacies. Values of a=1.0, m=2.0, and n=1.3 were obtained. Because the deep induction log was used to compute water saturation, the porosity log was filtered so that the two logs had similar spatial frequencies.

Comparison of log to core saturation data indicated that log water saturations were consistently less than core. This systematic bias error was finally concluded to be due to invasion of mud filtrate into the core, thus increasing the core water saturation. The mean error in the log water saturations was approximately 10 saturation units, when compared to the core data. Because the actual saturations at reservoir conditions are not known, it was not possible to ascertain the true accuracy of the model. A study of the error in water saturation due to the error in porosity and formation resistivity indicates that the uncertainty in the saturation may be as large as 50% in the reservoir petrofacies.

DEPOSITIONAL ENVIRONMENT

On the basis of detailed core descriptions and wireline log correlations, the depositional model interpreted for the Taylor interval is that of a progradational barrier island with associated shoreface and inner shelf environments (Figure 14). Each environment has its own characteristic gamma-ray signature that can be used to identify and trace the major facies packages around the field (Figure 15).

At least two barrier island complexes with reservoir potential exist in the Oak Hill Field. The lowermost complex (complex "A") overlies a sequence of inner shelf to lower shoreface sandstones at the base of the Taylor interval (Figure 16). The barrier core (upper shoreface and foreshore) of barrier complex "A" is located at the northern part of the study area and appears to trend northeast--southwest. To the south, barrier core sandstones grade laterally into upper/lower shoreface sandstones and transitional/inner shelf sandstones and mudstones. To the north, lagoonal and back-barrier limestones, mudstones and sandstones are interbedded with and overlie barrier core sandstones. Progradation of barrier island complex "A" was terminated by a marine transgression that reworked the upper portions of the complex. Within the study area, the dominant facies of barrier complex "A" is the shoreface.

The uppermost complex (complex "B") is separated from the underlying complex by a transgressive horizon (Figure 16). Barrier core "B" is located in the southern part of the study area and appears to also trend northeast--southwest. To the north, barrier core "B" sandstones grade laterally into lagoonal and back-barrier facies consisting of lagoonal limestone and mudstone, tidal channel, flood tidal delta, and washover fan sandstones. Shoreface sandstones are not well-developed within the study area due to the southerly location of barrier core "B". Presumably, shoreface and shelf sandstones exist further to the south beyond the southerly extent of well control. Transgression drowned barrier island complex "B" and led to deposition of the "C" limestone in a restricted marine setting, probably produced within a large embayment.

Relationships Between Petrofacies and Depositional Environment

Reservoir petrofacies of the Taylor sandstone are closely linked to depositional environment in the Oak Hill Field. Rocks of the Primary Macroporous Quartz-cemented Petrofacies occur in the highest energy portions of the foreshore deposits of the barrier core and in channel deposits of tidal inlets and tidal deltas (Figure 15). This petrofacies tends to be texturally and mineralogically more mature (better sorted and more quartzose) than other petrofacies. Soluble grains were apparently removed by mechanical abrasion and winnowing in the high energy depositional environment.

Rocks of the Moldic Macroporous Quartz-cemented Petrofacies occur most commonly in somewhat lower energy but still well-winnowed (clay-poor) portions of the the upper shoreface, foreshore and beach in the barrier core, tidal channels and washover fans. Sandstones assigned to this petrofacies contain a higher proportion of chemically unstable feldspar, chert and organic fragments than those assigned to the Primary Macroporous Quartz-cemented Petrofacies.

Environments characterized by higher clay content, finer grain size, poorer sorting and common bioturbation, such as the inner shelf/lower shoreface and back-barrier sandstones, are typically dominated by the Microporous Clay-cemented Petrofacies (Figure 15). Where well-winnowed environments grade laterally into clay-rich environments, such as barrier core "A" grading into shoreface environments to the southeast, there is a gradual transition from the macroporous petrofacies to the Microporous Clay-cemented Petrofacies (Figures 15 through 18). Regional mapping of petrofacies and depositional environments suggests that the ratio of macroporous:microporous petrofacies is approximately 1:3 (Figures 17 and 18).

The relationship between petrofacies and depositional environment is complicated locally by diagenesis. For example, where well-winnowed back-barrier sandstones are overlain by lagoonal limestone, the sandstone is typically non-reservoir due to cementation by carbonate.

Reservoir Continuity and Zonation

Correlation of sandstone packages on regional cross sections indicates that individual sandstone units may extend well over one mile in both strike and dip directions except for tidal channel sandstones, which probably extend tens to hundreds of feet along strike and hundreds to thousands of feet in the dip direction. However, facies changes and local heterogeneities due to variations in diagenesis may occur within these sandstones that may limit reservoir continuity. For example, the presence of shell debris is strongly associated with calcite cementation in shoreface sandstones. Identical lithologies except for shell content may have very different reservoir properties due to calcite cementation (see Porosity and Permeability section).

SUMMARY

The Taylor interval in the Oak Hill Field consists of tightly cemented very fine- to fine-grained sandstone interbedded with mudstone, siltstone and carbonate. Reservoir sandstones are interpreted as deposits of two stacked barrier island systems. These systems trend east-northeast to west-southwest. Barrier island cores (upper shoreface and foreshore deposits) extend over a mile in strike and dip directions. Tidal channel and flood tidal delta deposits have less lateral continuity. Barrier island, tidal channel and tidal delta facies are characterized by quartz-cemented sandstone with isolated secondary pores and slot-like remnant primary pores (Primary and Secondary Macroporous Quartz-cemented Petrofacies). These deposits probably represent the highest quality reservoir rock.

Barrier island facies grade into lower shoreface and inner shelf facies to the southeast, and into burrowed back-barrier environments to the northwest. Sandstones in these facies have lateral continuity similar to the barrier island deposits, however, these facies are characterized by clay-cemented sandstones dominated by microporosity (Microporous Clay-cemented Petrofacies). Further to the southeast, lower shoreface and inner shelf sandstones interfinger with and grade into inner shelf mudstones.

Reservoir quality is controlled by diagenesis and depositional environment. The strong diagenetic overprint complicates porosity-permeability relationships and adds to reservoir heterogeneity. Regional mapping indicates that the ratio of microporous:macroporous petrofacies is approximately 3:1. However, the microporous rocks have substantially higher water saturation, which results in significantly lower effective gas permeability. Because of this, productivity is largely controlled by the distribution of the macroporous petrofacies.

Wireline logs can be used to accurately determine macroporous, microporous, and nonreservoir petrofacies. In addition, logs can be used to interpret porosity (to within 1.5 porosity units, or 25% accuracy), and water saturation (to within approximately 50% accuracy). The large error in water saturation is due to the large relative error in porosity. The problems associated with log interpretation in this environment demonstrate the need for well-calibrated porosity tools together with procedures to calibrate the porosity logs on a well-by-well basis using internally consistent criteria.

Reservoir petrofacies, defined by mineralogy and pore geometry, can be recognized on wireline logs. Depositional environments can be distinguished on logs based on characteristic gamma ray signature. Recognition of petrofacies and depositional environments on wireline logs permits the mapping of these features on a field-wide scale. When completed, maps of petrofacies thickness and average porosity by zone will help in: (1) determining the relative contribution of the two petrofacies to production, (2) identifying the geographic areas and stratigraphic intervals that should be most productive, and (3) refining reservoir engineering simulation models.

At the conclusion of this study, the following recommendations were made:

1) Infill drilling should concentrate on the main fairway of production, which coincides reasonably well with macroporous quartz-cemented sandstones of the upper shoreface to foreshore parts in the lower barrier island sequence. Intersection of tidal channel sandstones in the southern portion of the field may also provide additional production opportunities.

2) The depositional and petrographic models, discussed here, imply the existence, along shoreline trend, of other sand-rich depositional

environments such as strandplains and deltas, which may have similar porosity development. Given the appropriate structural closure, these are extension exploration opportunities. This study should provide a basis for extension exploration in the area.

ACKNOWLEDGMENTS

Numerous colleagues at ARCO contributed to the work described here. Thin-sections were prepared by S. Sanders. J. Rathmell and F. Smith shared their thoughts on aspects of reservoir engineering and special core analysis. J. Toney and L. Liang assisted with scanning electron microscopy. J. Bishop assisted with core and core plug cataloging. S. Zucker established and maintained the project computer database. X-ray diffraction analyses were made by C. Stuart, C. Mitchell and J. Pessagno. J. Twyman provided information on well locations.

REFERENCES

Coleman, J.L., Jr. and Coleman, C.J., 1981, Stratigraphic, sedimentologic and diagenetic framework for the Jurassic Cotton Valley Terryville massive sandstone complex, northern Louisiana: Transactions of the Gulf Coast Association of Geologic Societies, v. 31, p. 71-79.

Eversull, L.G., 1985, Depositional systems and distribution of the Cotton Valley blanket sandstones in northern Louisiana: Transactions of the Gulf Coast association of Geological Societies, v. 35, p.49-57.

McGowen M.K., and Harris, D.W., 1984, Cotton Valley (Upper Jurassic) and Hosston (Lower Cretaceous) depositional systems and their influence on salt tectonics in the East Texas Basin *in W.P.S.* Ventress et al (eds.) The Jurassic of the Gulf Rim: Proceedings of the third annual research conference Gulf Coast Section, Soc. Econ. Paleontologists and Mineralogists Foundation, p. 213-253.

Morgan, J.T., and Gordon, D.T., 1980, Influence of pore geometry on water-oil relative permeability: Jour. Pet. Tech. p. 1199-1208.

Newkirk, T.F., 1971, Pŏssible future petroleum potential of Jurassic western Gulf basin: American Assoc. Petrol. Geologists Memoir 15, p. 927-953.

Pittman, E.D., 1979, Porosity, diagenesis and reservoir capability of sandstone reservoirs, *in* Scholle, P.A. and Schluger, P.R. (eds.), Aspects of Diagenesis: Soc. Econ. Paleontologists Mineralogists Special Pub. 26, p. 159-173.

Randolph, P.L., and Soeder, D.J., 1986, Rock Matrix Analysis of Eastern Gas Shale and Western Tight Gas Sands: Institute of Gas Technology, DOE/MC/20342-2135,112 pp.

Walls, J.D., 1981, Tight gas sands: permeability, pore structure and clay: SPE/DOE paper 9871, 9 pp.

Wescott, W., 1984, Diagenesis of a tight gas formation: the Jurassic Cotton Valley Sandstone, East Texas Basin in: Presley, M.W. (ed.), The Jurassic of East Texas, East texas Geologic Society, p. 118-126.

Wood, D.H., 1981, Structural effects of salt movement in the East Texas Basin, *in:* Kreitler, C.W., (ed.), Geology and Hydrogeology of the East Texas Basin -- A Report on the Progress of Nuclear Waste Isolation Feasibility Studies: The University of Texas at Austin, Bureau of Economic Geology, Geological Circular 81-7, p. 21- 27.

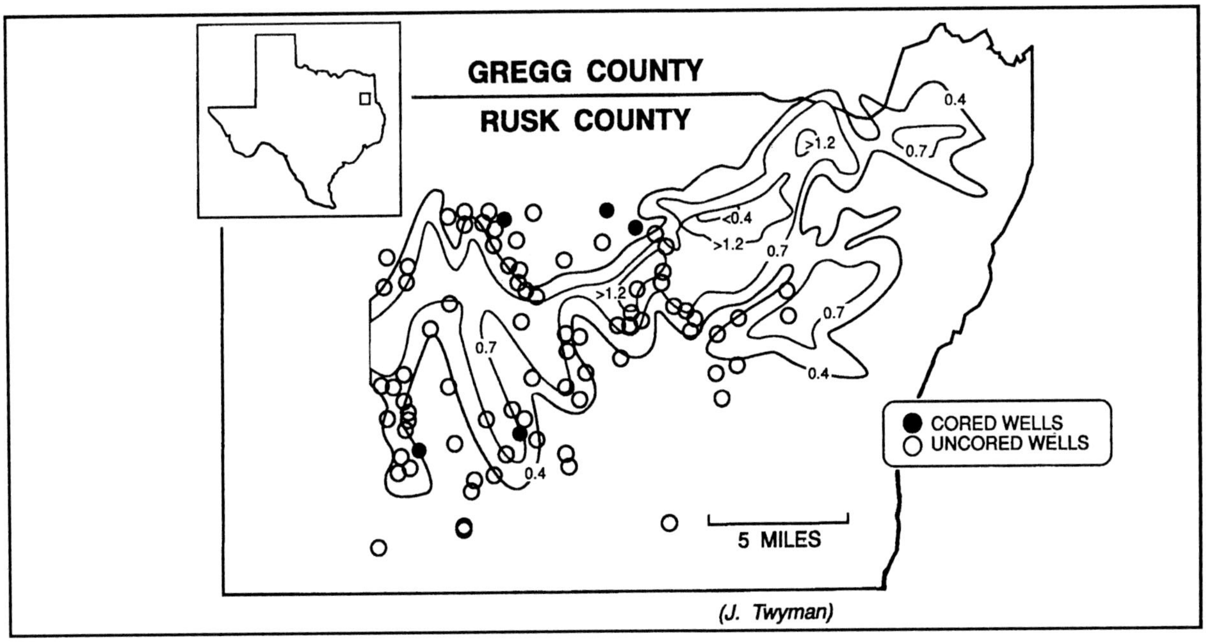

Figure 1. Locality map of the Oak Hill Field, showing log and core control. Contour lines indicate 36 month cumulative production in BCF. Inset map shows the position of the Oak Hill Field in East Texas.

<table>
<tr><th rowspan="2">SERIES</th><th rowspan="2">GROUP</th><th colspan="2">FORMATION</th></tr>
<tr><th>E.TEXAS BASIN</th><th>OAK HILL FIELD</th></tr>
<tr><td rowspan="2">LOWER CRETACEOUS</td><td rowspan="5">COTTON VALLEY</td><td>? KNOWLES LIMESTONE</td><td>"B" LIMESTONE</td></tr>
<tr><td rowspan="3">COTTON VALLEY SANDSTONE</td><td>UPPER COTTON VALLEY SANDSTONE</td></tr>
<tr><td rowspan="3">UPPER JURASSIC</td><td>"C" LIMESTONE</td></tr>
<tr><td>TAYLOR SANDSTONE</td></tr>
<tr><td>BOSSIER SHALE</td><td>BOSSIER SHALE</td></tr>
</table>

Figure 2. Comparison of the lithostratigraphic framework in the Oak Hill Field with that in the East Texas Basin. Modified from Wood (1981) and Wescott (1984).

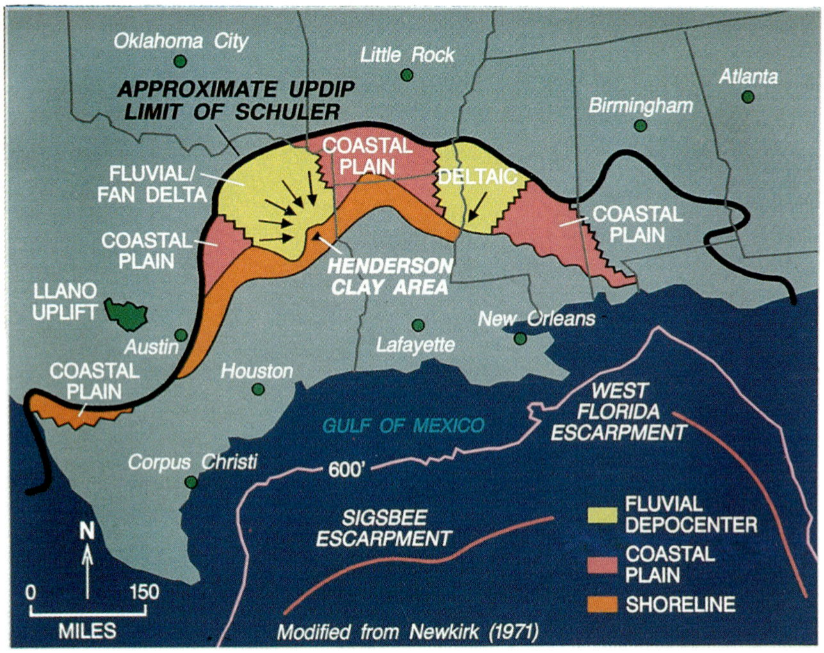

Figure 3. Regional paleogeography of the Gulf Coast area during Cotton Valley time. Transport direction of sediment in East Texas Basin indicated by arrows. Modified from Newkirk (1971) with data from Coleman and Coleman (1981), Eversull (1985) and McGowen and Harris (1984).

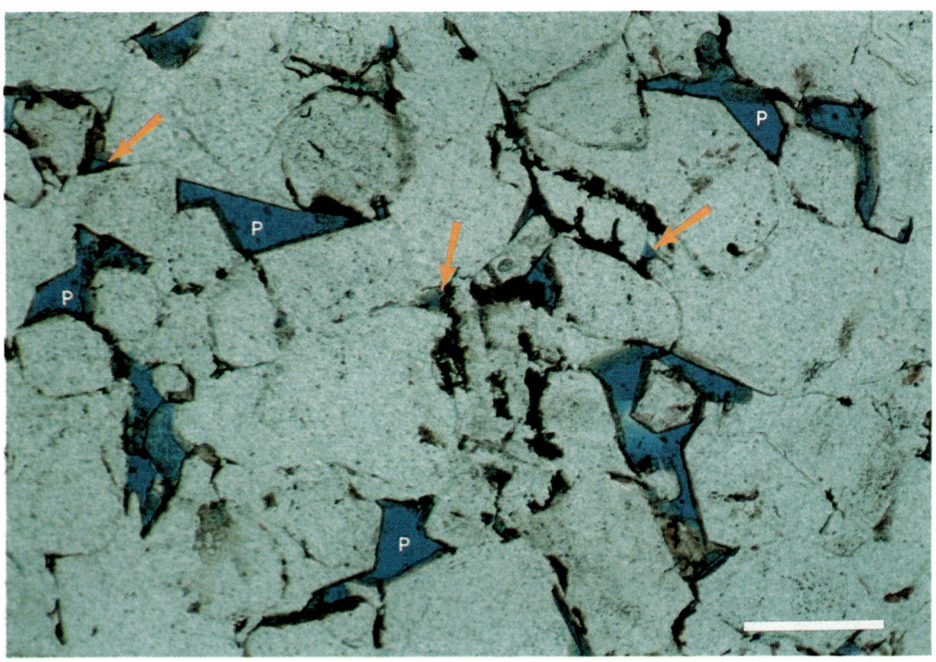

Figure 4. Photomicrograph of sample assigned to the Primary Macroporous Quartz-cemented Petrofacies. Primary interparticle pores (P) are bounded by smooth walls of quartz overgrowths. Pore throats (arrows) are typically larger than those in other petrofacies. Plane polarized light. Scale bar is 125 mm.

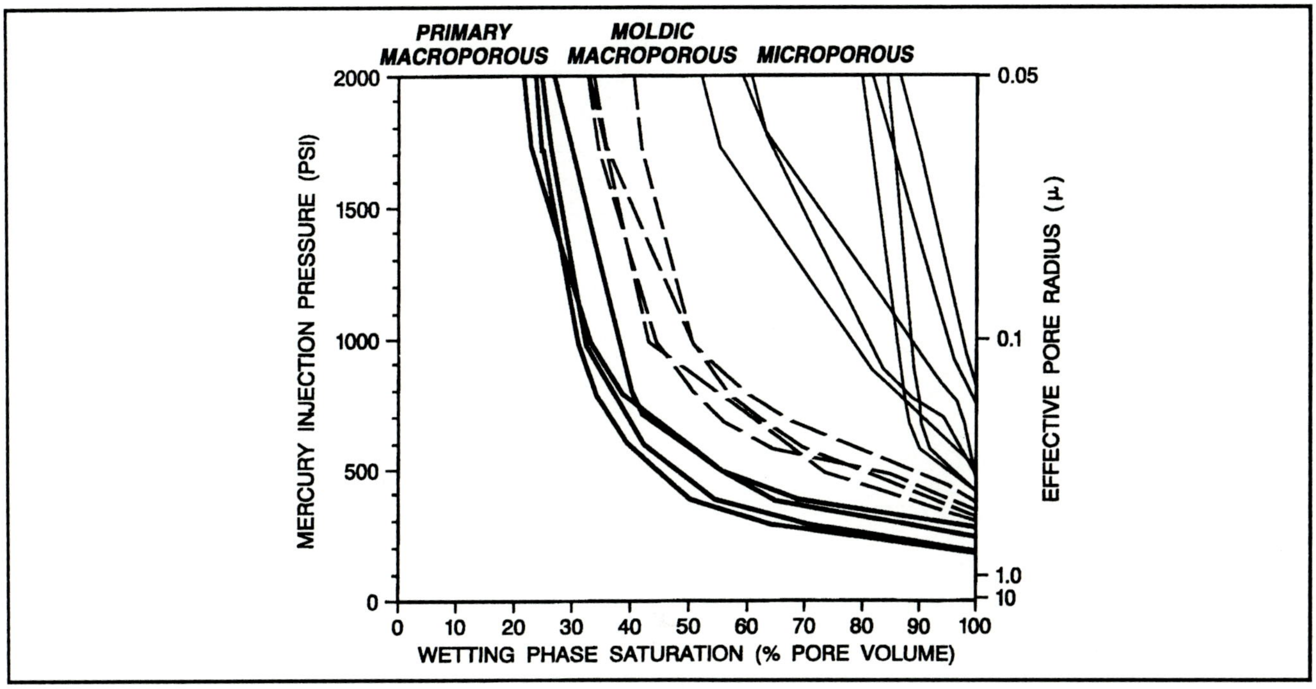

Figure 5. Mercury injection capillary pressure curves for reservoir petrofacies. Although these petrofacies are indistinguishable based on helium porosity (Table 1), they have significantly different effective pore throat size distributions as indicated by mercury injection capillary pressure. The median effective throat radius for the Primary Macroporous Quartz-cemented Petrofacies is 0.45 mm, compared to 0.15 mm for the Moldic Macroporous Quartz-cemented Petrofacies and 0.04 mm for the Microporous Clay-cemented Petrofacies.

Figure 6. Photomicrograph of a sample assigned to the Moldic Macroporous Quartz-cemented Petrofacies. Note moldic secondary pores (M), which are lined by chloritic cement. Pore throats (arrows) are significantly smaller than those in the Primary Macroporous Quartz-cemented Petrofacies. Plane polarized light. Scale bar is 125 mm.

Figure 7. Photomicrograph of sample of the Microporous Clay-cemented Petrofacies. Note abundant microporosity (blue). Plane polarized light. Scale bar is 125 mm.

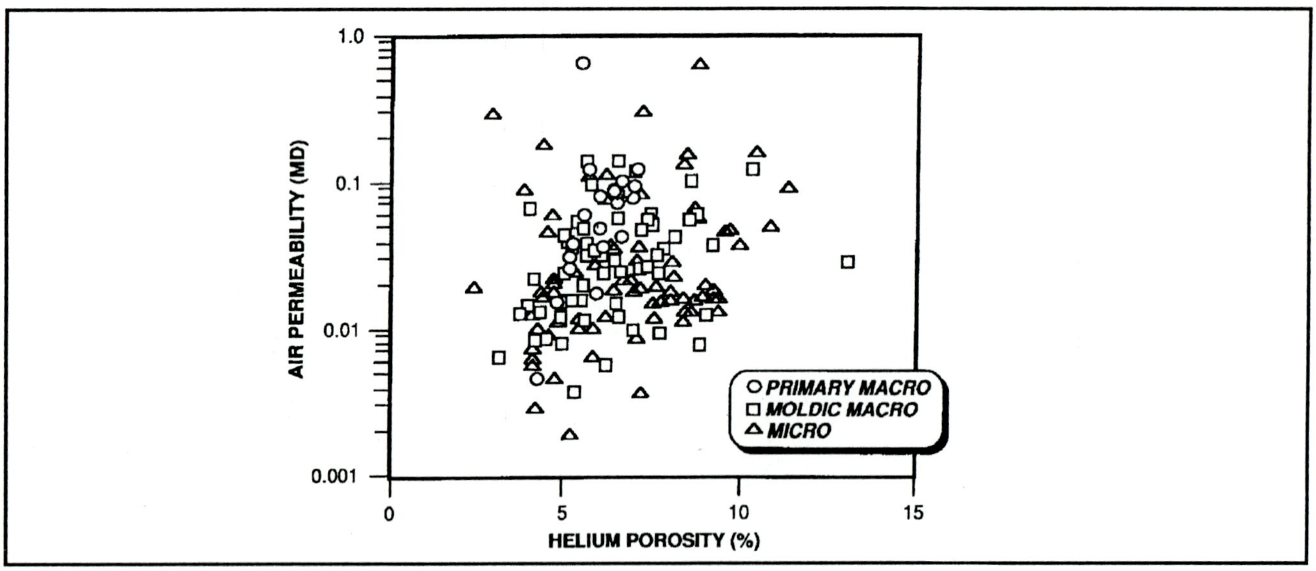

Figure 8. Porosity-permeability crossplot by petrofacies. Relationships are poor due to the strong overprint of several diagenetic processes, each with differing effects on porosity and permeability.

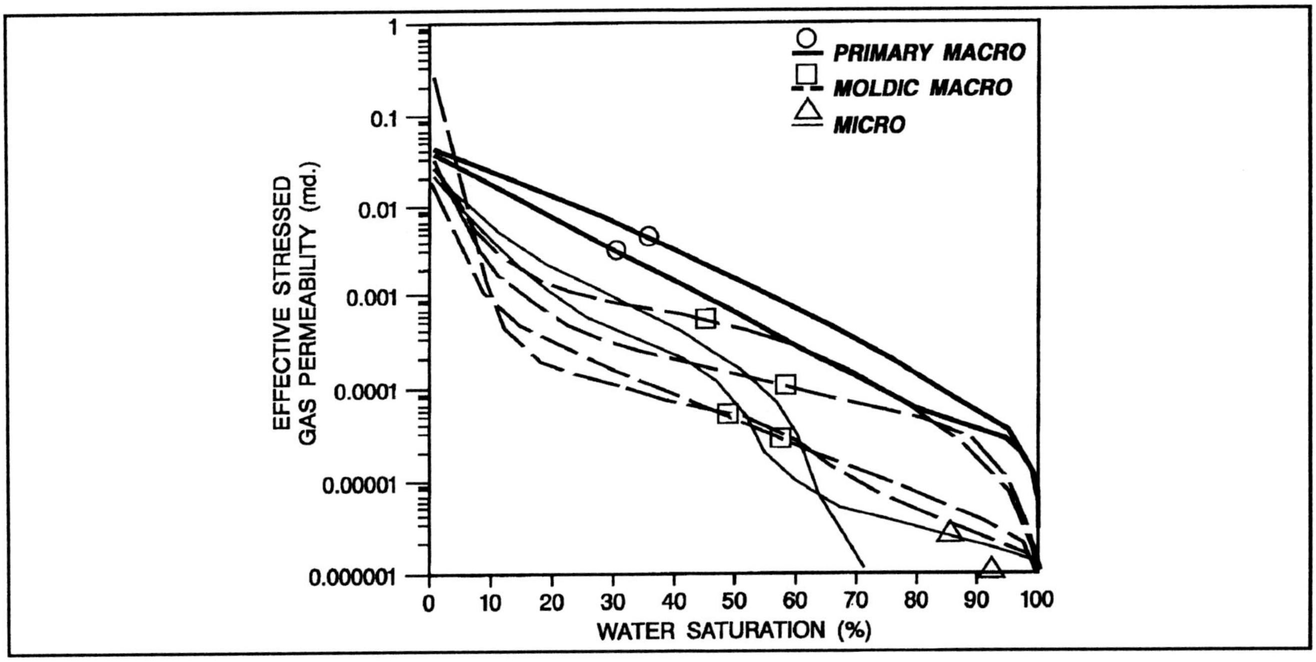

Figure 9. Effect of water saturation on gas permeability at reservoir stress. Points indicate estimated reservoir fluid saturation of individual samples based on capillary pressure and core saturation data. Effective gas permeabilities of stressed samples at reservoir water saturations range from one to over four orders of magnitude lower than stressed gas permeabilities at S_w=O.

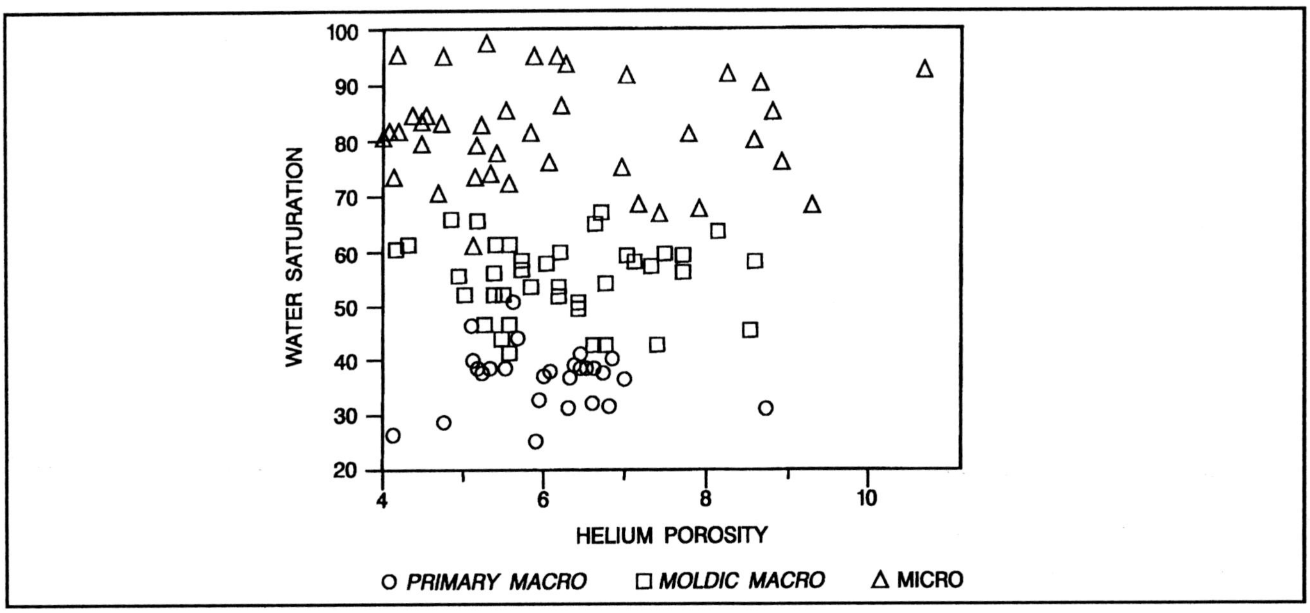

Figure 10. Relationships between core-derived water saturation and helium porosity. Water saturation is independent of helium porosity, but is strongly controlled by petrofacies, due to variations in effective pore throat size distributions. Thus, although the three reservoir petrofacies have statistically indistinguishable porosities, significant differences in productivity result from variations in water saturation.

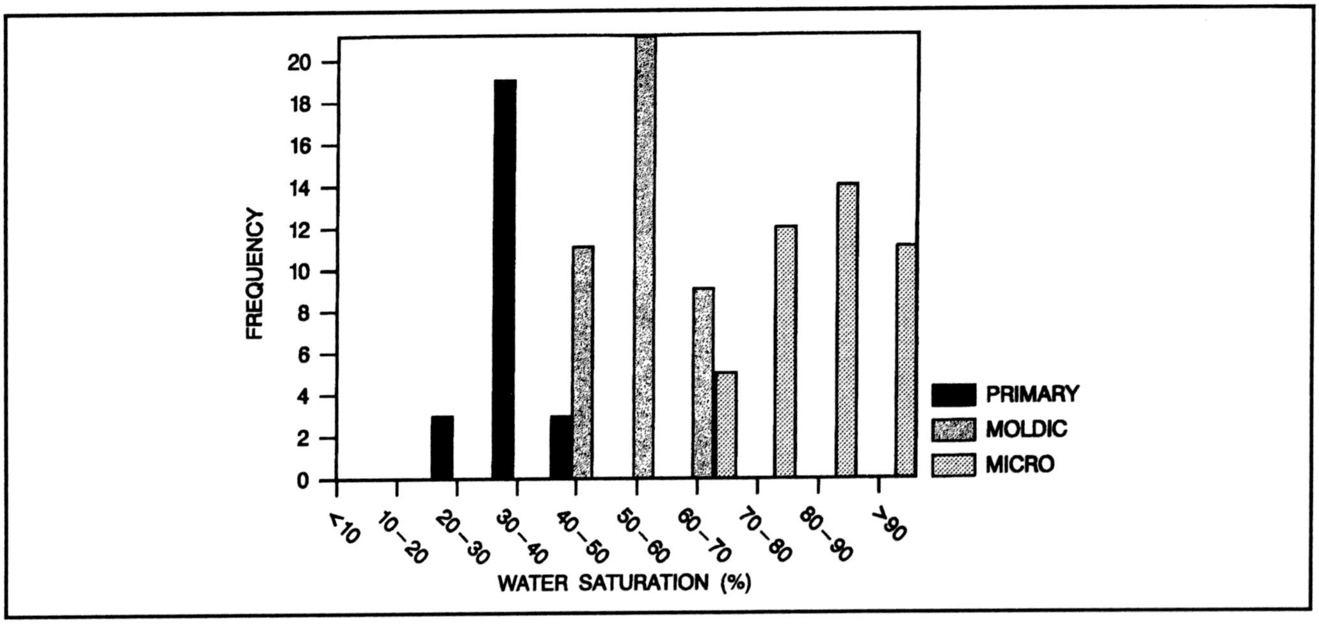

Figure 11. Histogram of core-derived water saturation by petrofacies from one well. The strong dependence of saturation on petrofacies results from significant differences in effective pore throat size distributions and, therefore, capillarity.

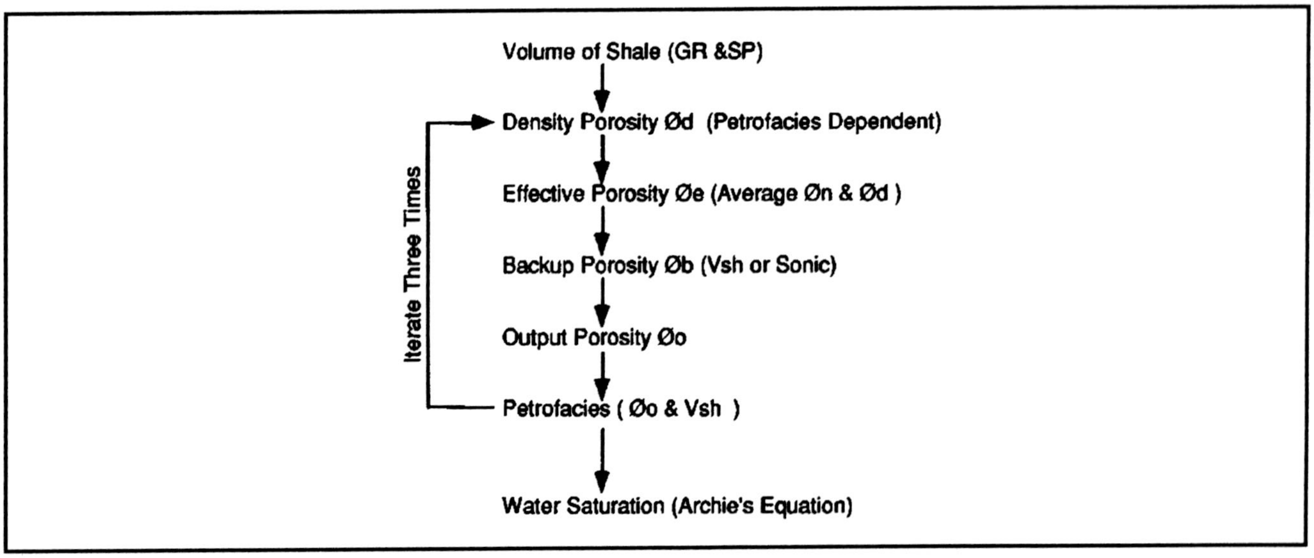

Figure 12. Flow chart of wireline log analysis procedure.

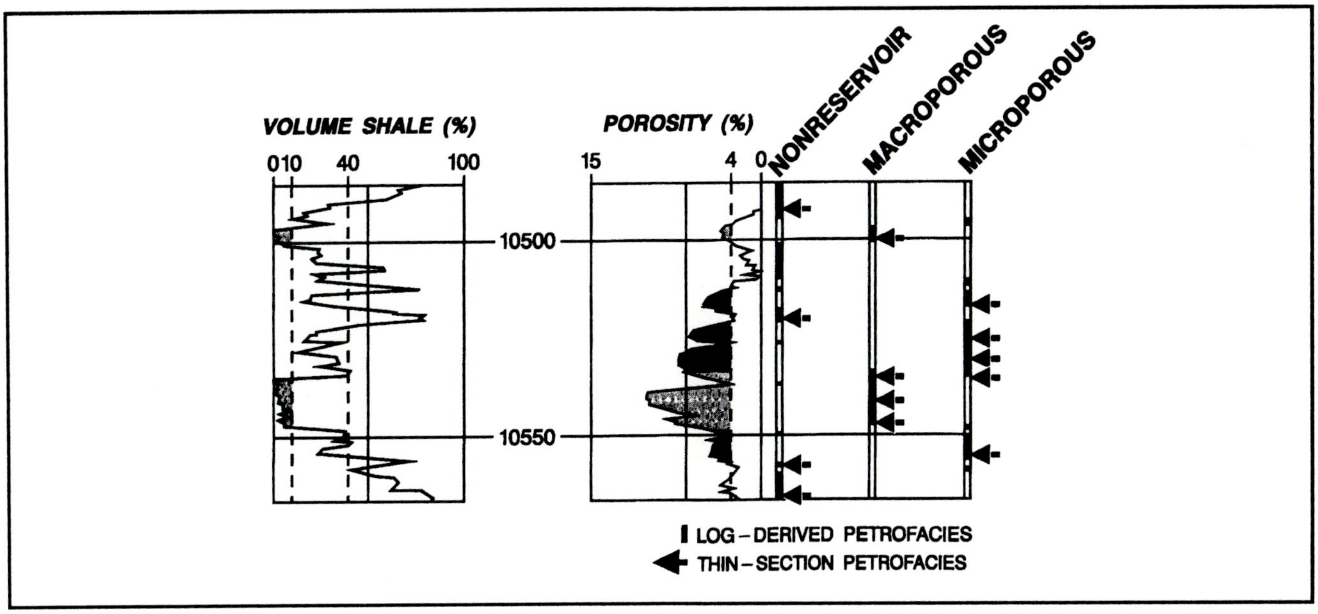

Figure 13. Comparison of petrofacies assigned by thin-section control with those assigned by the wireline log model. Logs were processed applying porosity and VShale cutoffs. The log model was 85-95 percent successful in determining the petrofacies.

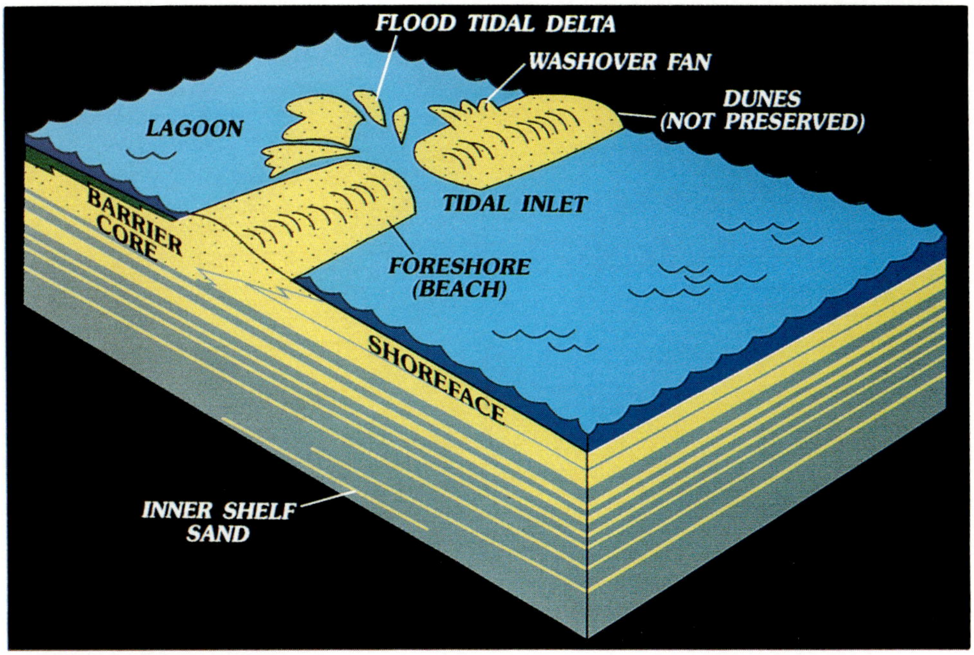

Figure 14. Schematic block diagram of the depositional model of sandstone bodies in the Taylor interval.

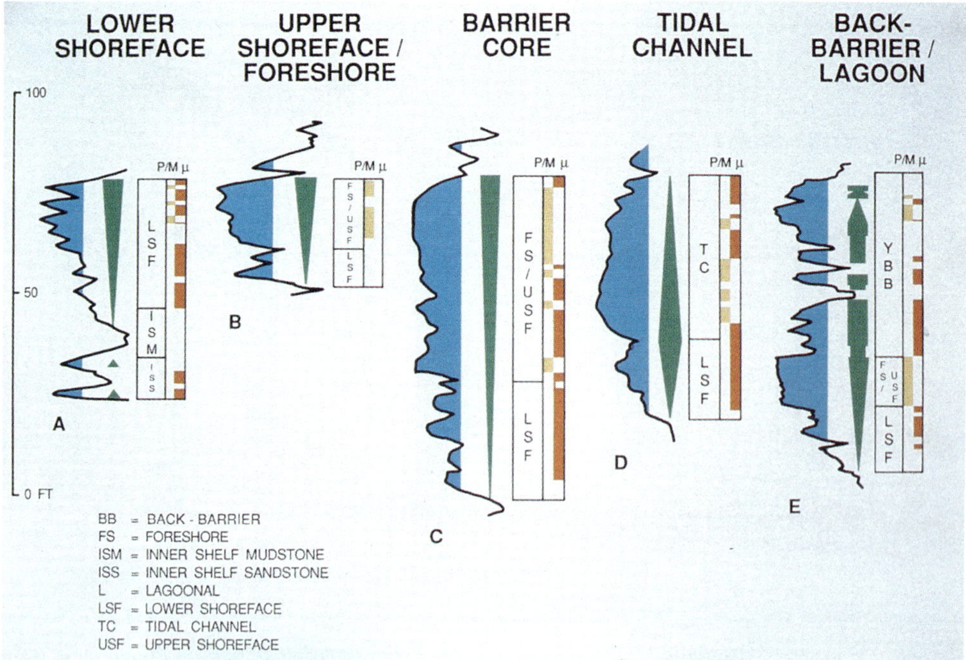

Figure 15. Gamma-Ray characteristics and petrofacies distribution of major subenvironments of the barrier island/shoreface/shelf facies tract represented by Taylor sandstone bodies in the Oak Hill Field. Green triangles represent coarsening and fining-upward sequences. "P/M" denotes distribution of the macroporous petrofacies; "ml" denotes the distribution of the Microporous Claycemented Petrofacies. The shaded regions on the Gamma-Ray logs indicates less than 40 percent V_{Shale}.

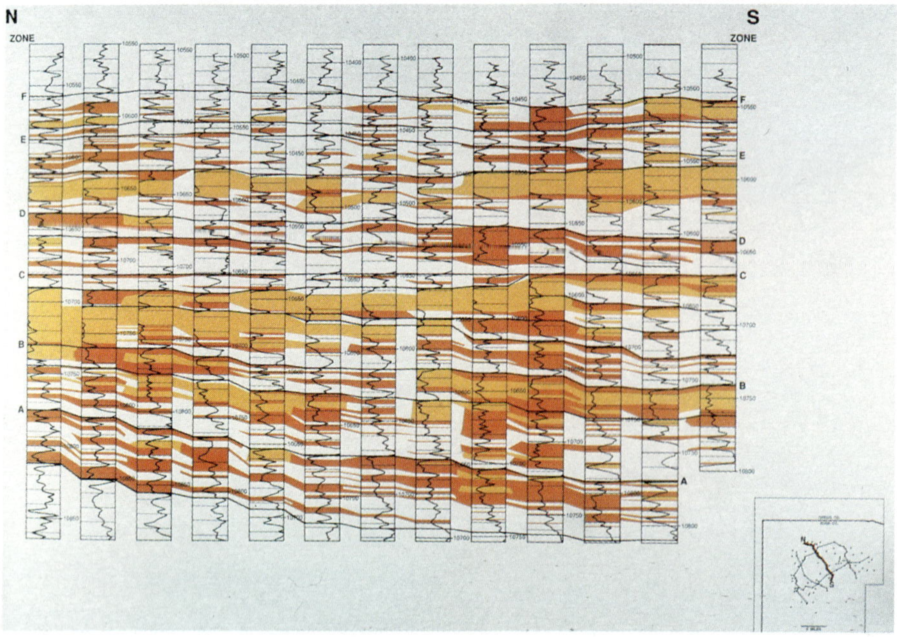

Figure 16. Diagrammatic, dip-oriented cross-section through the Oak Hill Field area showing the stratigraphic distribution of shelf, shoreface and barrier island complexes "A" and "B", location of the transgressive surface separating barrier island complex "A" from "B", and the reservoir zonation of the Taylor interval. Ye]low regions denote distribution of macroporous petrofacies, orange indicates location of microporous petrofacies, whereas nonreservoir petrofacies are white.

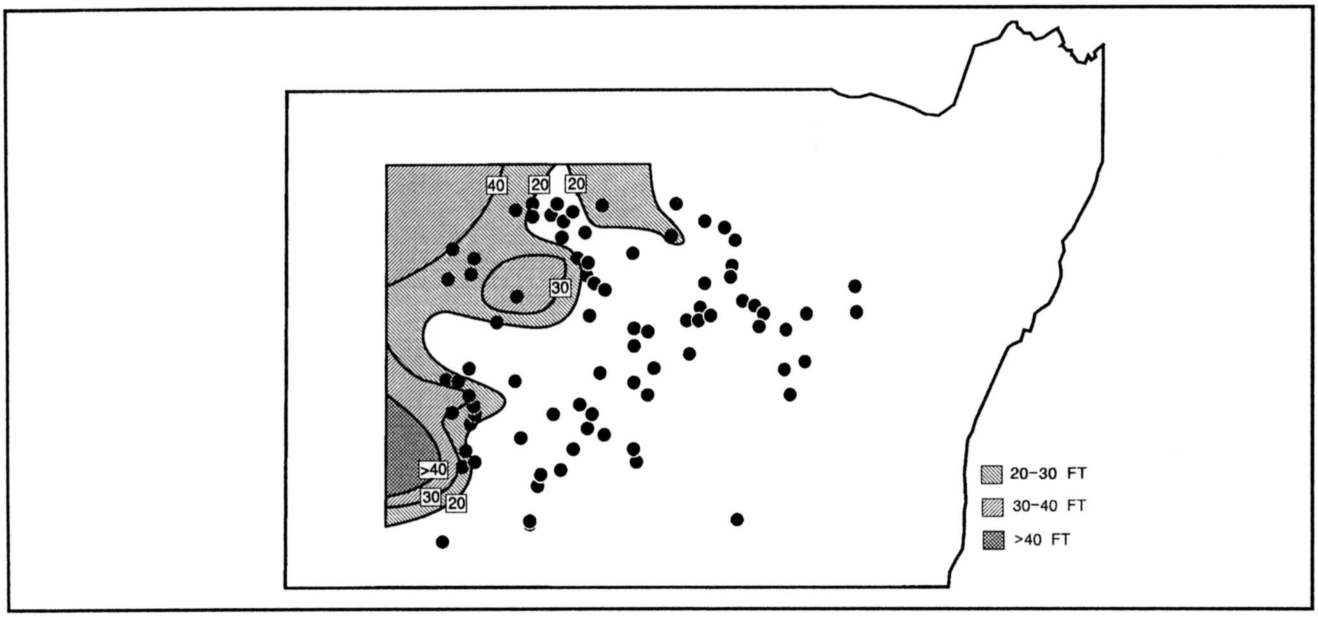

Figure 17. Map showing the distribution and thickness (in feet) of the Primary and Moldic Macroporous Quartz-cemented Petrofacies for one reservoir zone. The northeast- to southwest-trending thick of macroporous petrofacies corresponds to a barrier core sequence. Dots indicate well control, with petrofacies determined using the wireline log model.

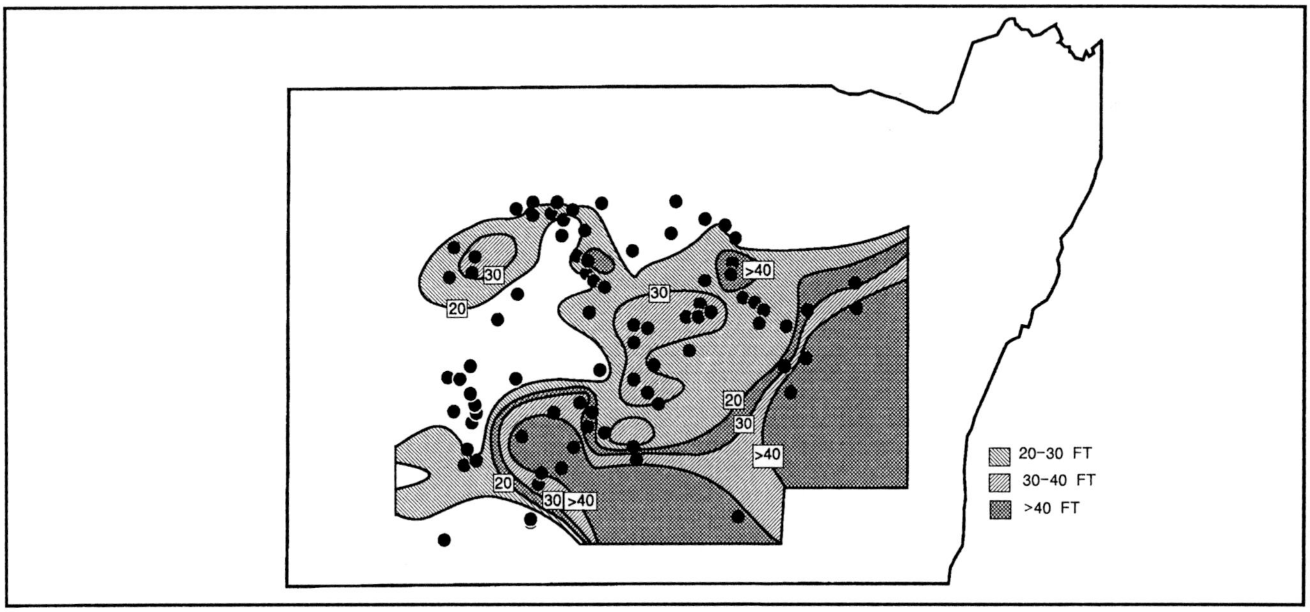

Figure 18. Map showing the distribution and thickness (in feet) of the Microporous Clay-cemented Petrofacies for same reservoir zone as shown in figure 17. Thickening microporous rocks to the southeast represent the transition from the clean, well-winnowed barrier core in the northwest to more clay-rich rocks of the lower shoreface and inner shelf.

TABLE 1. RESERVOIR PETROFACIES, CORE DATA AND COMPOSITION

	Primary Macroporous Quartz-cemented	Moldic Macroporous Quartz-cemented	Microporous Clay-cemented
Number of Samples	25	97	78
Helium Porosity (%)	6.1	6.2	6.1
Air Permeability (ambient, md.)	0.060	0.032	0.035
Air Permeability (3000 psi, md.)	0.011	0.006	0.007
Water Saturation	35	54	81
Grain Density	2.65	2.66	2.68
Effective Pore Throat Radius (μm)	0.45	0.15	0.04
Clay Matrix and Cement	1.0	2.5	13.9
Quartz Cement	10.0	13.8	4.4
Carbonate Cement	2.6	3.4	2.6
Primary Macroporosity	3.4	1.2	0.1
Moldic Macroporosity	1.1	2.0	0.0
Microporosity	1.8	2.7	7.3

NOTES: Helium porosity, permeability, water saturation and grain density are means from conventional core analysis data. Median effective pore throat radius is computed from mercury injection capillary pressure data. Compositional data are from petrographic analysis of thin-sections.

ACOUSTIC TOMOGRAPHY for RESERVOIR SURVEILLANCE

J.H. Justice, M.E. Mathisen,
A.A. Vassiliou, S. Singh*,
P.S. Cunningham and
J.R. Bulau

MOBIL R&D CORPORATION
*MOBIL E&P SERVICES, INC

ABSTRACT

As significant new hydrocarbon reserves become increasingly difficult to find, it may be true that our greatest potential resource for future development lies in our existing reservoirs. In order to fully exploit these remaining resources, attention is now being directed to the reservoir as a domain worthy of serious study and planning. Reservoir characterization in the future will incorporate the contributions of many disciplines in an effort to build accurate reservoir models and to update these models as new data sets are acquired and interpreted. These models, in turn, serve as a basis for planning optimal field development and management strategies.

An important role in this effort will lie with imaging technologies which can sense beyond the borehole and into the interwell volume. Cross borehole seismic tomography is a technology which shows promise of providing useful information on the interwell volume at reservoir scale. Future applications may lie in EOR monitoring as well as in reservoir characterization and description. When properly used, it may represent an important adjunct to other technologies for reservoir management.

INTRODUCTION

For most companies today, replacement of reserves is an issue of growing importance. This is particularly true as major discoveries (1 billion barrels of oil or 1 trillion cubic feet of gas) become increasingly rare. As exploration moves more and more into frontier areas of the world, often involving harsh or remote environments, or deep water, the cost of developing and delivering new reserves to market can be substantial. As a result, attention is now being directed to the enormous reserves currently remaining in our existing oil reservoirs. Many of these reservoirs are reaching or have reached primary depletion. In many cases, only thirty to forty percent of reserves have actually been produced. Using relatively inexpensive means of secondary production, this percentage can be increased, but substantial reserves may still remain.

Tertiary means of production can be expensive. When fluids are injected into the reservoir to stimulate further production, there is an associated cost. If these fluids migrate into non-reservoir zones or channel to production wells, bypassing large quantities of reserves, the program may become economically unviable and damage to the reserve may result. If zones which were thought to be in communication are actually separated by flow barriers then valuable reserves may remain unproduced until a better understanding of the reservoir can be gained. In all cases, the need exists for improved reservoir characterization, including detailed understanding of reservoir structure and dynamics.

Today we are beginning to recognize the reservoir as a subject worthy of serious study. Efforts are being made in most companies to bring geological, geophysical, petrophysical and engineering expertise to bear an understanding the reservoir environment, and to apply this understanding to the development of a comprehensive management plan for the reservoir. Reservoir characterization is emerging as a new and important area of inquiry, combining the skills and expertise of many disciplines to achieve improved understanding and management of our hydrocarbon resources. This emerging discipline will combine core studies with well log analysis, geological cross sections, production history, reservoir simulation and geophysical imaging technology to formulate plans for field development and to continually update those plans as new data sets are acquired. As a result of this effort, new technologies are being stimulated specifically for applications in reservoir management. In particular, new geophysical tools are being developed to better enable us to image and to monitor reservoir structure and dynamics.

Among the new tools being developed for this purpose is tomographic imaging technology. This new technology, often referred to as cross borehole tomography, has the potential to play a very important role in helping us to understand the interwell structure and dynamics of the reservoir on a scale which is intermediate between well logging and surface seismic imaging. The potential applications of this new technology are manifold and the possibility exists that it may become a cost effective tool for better reservoir characterization and management.

We are beginning to understand that the reservoir is in fact a complex and heterogeneous environment. The distribution of porosities and permeabilities may be quite complex and their effect on flow of multi phase fluids through the reservoir may be quite difficult to predict. The amount of data which can be acquired from a reservoir is never adequate to fully predict its performance. Most of the measurements which we are able to make are local measurements and tell us something about the reservoir only in the vicinity of the borehole. The interwell volume is often the subject for speculation and guess work. Of all measurements which can be made, only a few can tell us very much about the interwell environment. Included among these are surface seismic imaging and pressure transient testing. Each of these technologies still leaves room for speculation even when properly used. Cross borehole tomography offers an additional set of measurements which span the interwell volume and which can help to provide additional constraints to our speculation. While it cannot answer all questions pertinent to the development and management of the reservoir, it may represent an important adjunct to the technologies which we can bring to bear on this important issue. With better and more complete information about the reservoir comes the possibility of better reservoir management, resulting in improved recovery and enhanced profitability.

What is Cross Borehole Tomography?

Cross borehole tomography may be compared to the familiar CAT scan (Computer Assisted Tomography) used in medical diagnosis. The mathematical basis for computer assisted tomography derives from the work of Johann Radon, a German mathematician who published his results in the early part of the 20th century. It was not until the 1970's that it was understood how to combine physical measurements with the mathematics derived by Radon to produce tomographic images of a remotely sensed object.

In geophysical tomographic imaging there are two measurements which can be made and from which tomographic images can be reconstructed. These measurements are travel time of a seismic wave through the medium to be imaged, and attenuation of a seismic wave propagating through the medium to be imaged. In medical CAT scanning, the attenuation of X-rays propagating through the body to be imaged is the physical measurement. Since bones and tissues in the body absorb X-rays at different rates, a reconstructed image of the rate of X-ray attenuation corresponds to a medically useful image of the interior of the body. In geophysical tomography, the attenuation measurement is very difficult to obtain with good reliability. As a result, most geophysical tomographic imaging today relies on the measurement of travel time of a seismic wave propagating through the medium of interest, which may be a hydrocarbon reservoir. From this data, the velocity of propagation of the seismic wave at each point in the reservoir can be reconstructed.

Since there are two modes of propagation of seismic waves which are possible, we are often able to reconstruct not one but two images of the reservoir and these images may carry slightly different information about the reservoir and its contents. The two types of waves are known as compressional waves and shear waves. The velocity of propagation of each type of wave is determined by certain parameters which are characteristic of the reservoir environment. In a purely elastic medium, these parameters are related to the usual elastic moduli, including bulk modulus and shear modulus. In a more complex medium, these velocities may depend on many more parameters. As a result, one is not going to unravel all the mysteries of the reservoir with seismic measurements alone. These measurements must be complemented with many other types of data to constrain and to calibrate the interpretation of tomographic images. The problem of interpretation of seismic tomographic images is a new area of study and requires that we learn much more than is currently known about the relationship between the properties of rock/fluid systems and seismic wave propagation. It seems certain that important future advances will be made in this area of inquiry and already some progress can be reported. We will look at this issue in more detail later.

Let us now briefly consider how seismic tomographic data is obtained and processed to form an image of the seismic velocity fields. Cross borehole seismic tomography offers the advantage of being able to take seismic measurements in the reservoir environment itself rather than from a remote location such as the surface of the earth. To acquire tomographic data we lower a source of seismic energy into a wellbore and simultaneously lower seismic receivers in another wellbore nearby. The source and receivers will occupy stations in each wellbore which span the zone to be imaged. The resulting image which will be reconstructed represents a map of the seismic velocity fields in the interwell volume between the two boreholes.

During data acquisition, the source and receivers will occupy all possible combinations of source and receiver stations. Typically these stations will be located at regularly spaced intervals through the zone to be imaged. At each pair of source/receiver stations, the source is fired and the travel time of the seismic wave to the receivers is measured. If there are forty source stations and forty receiver stations, then sixteen hundred independent travel time measurements will be acquired and used to reconstruct the image.

To reconstruct the tomographic image of the seismic velocity fields between the boreholes, an initial model for the compressional and shear wave velocity fields is assumed. In the absence of no other information, these fields are usually uniform. Synthetic data is computed for the assumed velocity fields and compared with the observed data set. A mathematical procedure then updates the assumed model of the velocity fields until the synthetic and observed data are in satisfactory agreement. The resulting models of the compressional and shear wave velocity fields are then plotted and constitute the tomographic image. Resolution in these images is determined by the geometry of the survey and by the bandwidth of the seismic waves employed. New tools are being developed in an effort to provide the kind of resolution which is needed in many reservoir engineering applications. Today the resolution of most seismic tomographic surveys lies between the resolution provided by surface seismic surveys and the fine detail available from well logs.

Seismic Tomography and Reservoir Characterization

Seismic traveltime tomography reconstructs the seismic velocity field between two boreholes. To the extent that these velocity fields can be associated with reservoir parameters of interest, seismic tomography can be a tool which has some value in reservoir characterization and management. We are familiar with sonic well logs which measure the seismic velocity field in the neighborhood of the borehole. These logs are often used to infer information about lithology and even porosity of the sediments. Indeed, seismic velocities are often closely associated with lithology in the broad sense, including mineralogical content, grain size and sorting, cementation, porosity and fluid content, and overburden pressure to name a few. It is reasonable to expect that tomographic reconstructions of the velocity field between boreholes are influenced by these same factors. To the extent that we are able to calibrate seismic tomographic images, we may be able to infer variations in one or more of these parameters from the tomographic images of the velocity fields. In addition, changes induced in the reservoir by various processes may also induce changes in the observed seismic velocity fields.

In 1984, it was first observed that heating of sands saturated with heavy oils could cause a significant decrease in the observed velocity of seismic wave propagation in the sands. This observation suggested the use of repeated seismic measurements to monitor changes in the seismic velocity field caused by heating due to thermal enhanced oil recovery programs. Examples have been reported of the use of both surface seismic monitoring as well as cross borehole seismic monitoring for this purpose.

Seismic velocities are particularly sensitive to the presence of free gas in formations, and often the injection of gas, or the liberation of gas due to a decrease of the reservoir pressure field, can induce a change in the seismic velocity fields which can be detected. Using a sequence of cross borehole tomographic images, taken at different times, to detect changes in the seismic velocity field in a reservoir, as a result of physical changes taking place in the reservoir, is referred to as time-lapse tomography. In time-lapse tomography, we are primarily interested in the differences in the seismic velocity fields in successive tomographic images. Assuming that all else remains constant, these differences are associated with those reservoir parameters which are changing. These changes are often associated with enhanced oil recovery programs. If only a few reservoir parameters are changing, then interpretation of the successive tomographic images is relatively simple. On the other hand, if many reservoir parameters are changing, then additional information may be required to correctly interpret the changes seen on successive tomograms.

In order to better understand the relationship between seismic measurements and reservoir properties, several other techniques may be used. These techniques may be employed to plan a seismic surveillance program or to aid in the interpretation of seismic reservoir data. We may, for example, make seismic measurements on core samples being used to simulate an enhanced oil recovery program. Another technology, which we have been developing recently, extracts appropriate reservoir parameters from numerical reservoir simulations to predict changes in the seismic velocity field which would occur as a result of an enhanced oil recovery program. This technique often allows us to establish a cause and effect relationship between seismic measurements and changes occurring in the reservoir as the result of an enhanced oil recovery program. In addition, recent logs and production data can play an important role in understanding reservoir changes which are observed in the seismic velocity fields.

While enhanced oil recovery monitoring using seismic imaging technology may prove to be important in the future, an equally important application may also be found in the use of seismic imaging technology to delineate reservoir structure, including lithologic variations which may prove to be important in reservoir characterization. For example, identifying terminations of reservoir units, porosity development, structural features and faulting may play an important role in formulating field development plans and production strategies. As a result, seismic imaging should play an important role in reservoir characterization and surveillance in the future.

Examples of Tomographic Imaging

We have acquired seismic tomography data both for the purposes of enhanced oil recovery monitoring as well as for reservoir characterization. We shall briefly consider results from each of these applications in order to illustrate the kind of information which may be extracted from tomographic images. For the sake of discussion, we shall confine our attention here to compressional wave tomograms even though we have acquired and interpreted shear wave tomograms as well. We shall draw our examples from steam injection enhanced oil recovery programs in heavy oil sands as well as from the reservoir characterization of a mid-continent U.S. field containing of well-consolidated carbonate reservoirs.

Our first example illustrates results of time-lapse tomographic images taken in the early stages of development of this technology in a heavy oil reservoir undergoing steam foam injection. Figure 1 illustrates the location of the wells used in the survey. The injection well is located off the survey line near the center as shown in the well diagram. The first survey was acquired in October 1987, in the early stages of development of this technology, and before the injection program had begun. The resulting image is shown in Figure 2. The reservoir cross section is seen to be relatively homogeneous, with no large velocity variations apparent. Resolution in these early tomograms is quite low, so that no fine detail can be seen in the reservoir. The second tomogram was acquired after a six month interval, in March of 1988, and is shown in Figure 3. At this time changes in the velocity field were becoming apparent in the vicinity of well B as shown in the figure. Velocities can be seen to have decreased in the six month interval from the beginning of the project to the date of acquisition of the second tomogram. This decrease in velocity is associated with the injection program. The third survey was acquired one year after the initial survey and six months after the second survey and is shown in Figure 4. In this tomogram, it is quite apparent that velocities have decreased sharply in the vicinity of Well B. Although the tomograms are of very low resolution, a clear directionality is present in the tomograms, indicating that the zones in which the greatest change of velocity has occurred are oriented toward Well B more than toward Well A. This might indicate an azimuthally directional flow pattern away from the injector.

The second example is a more recent tomogram taken in a heavy oil project undergoing cyclic steam injection. A detailed geological interpretation has been made of this reservoir cross section and is shown imposed on the tomogram. Not all aspects of the geology interpretation can be discerned by changes in the velocity field in the tomogram. This is because some of the geologically discernable units in the reservoir actually exhibit similar velocities of seismic wave propagation. As shown in Figure 5, well logs define variations in stratigraphy, facies, structure and fluid properties at the survey wells. An unconformity (wavy line) which is defined by both electrical resistivity and sonic logs occurs at the top of the reservoir. Beneath the unconformity, sand stratigraphy and sedimentary facies are documented at each well by the resistivity logs. Correlations between wells are illustrated by dashed lines. The uppermost "Amalgamated Braided Channel Sand Facies" is, according to core description and core analysis data, the best reservoir unit. Sonic logs define a low velocity "Desaturated Zone" at the top of the sand. The zone occurs within the amalgamated sand braided channel facies, except at the top of the reservoir, where it extends into the underlying meandering sand channels. Velocities in this zone are lowest at the structurally high end of the reservoir.

To facilitate tomogram interpretation, the tomography velocity data is divided into five velocity fields which correlate with the log defined contacts and zones discussed above. Confidence in the accuracy of the velocity fields is increased by the strong correlation between the tomographic velocities, plotted as dots, and the sonic log. The desaturated zone at the top of the reservoir is imaged by the lower velocity fields with the lowest velocities occurring at the structurally high side of the tomogram. These sands have been significantly influenced by the steam process and have been referred to as “air sands”.

Most sands have velocities that are between 6,200 and 8,000 ft/sec. The heavy oil sands that have been partially affected by the steam process have velocities between 6,200 and 7,000 ft/sec. This is the broadest velocity field shown and cuts across several sand facies contacts. Heavy oil sands which have not been significantly affected by the steam process are imaged by a high velocity field of 7,000 to 8,000 ft/sec at the base of the formation. These velocity fields correspond in a qualitative way to hot saturated and cold saturated zones.

The highest velocity field 8,000 to 9,000 ft/sec. seen near the base of the formation is controlled by lithology rather than fluid content. Log and core data indicate that the higher velocities are caused by a sequence of thin carbonate cemented sands.

This example illustrates that single surveys across EOR projects, when calibrated with available well data, can also be used to document important variations in crosshole fluid distribution and reservoir properties. These single-survey interpretations are important because they provide a timely response to operating units in the field and help to clarify problems and identify objectives for subsequent surveys.

We now consider examples of tomographic imaging in several typical mid-continent carbonate reservoirs. A tomogram from both reservoirs is shown in Figure 6. The tomogram illustrates that changes in the reconstructed seismic velocity fields correlate with formation boundaries and Paleozoic carbonate reservoir zones defined by neutron logs. The formation and reservoir contacts are clearly defined at the wellbore and extend for the most part directly to the other well. The irregular surface at the top of the upper reservoir may represent a Paleokarst surface which is known to be a water flood thief zone. Reservoir heterogeneity is documented by variations in the lower reservoir neutron log signature and the discontinuous nature of the corresponding velocity fields in the tomogram. The uniform velocity field corresponding to the upper reservoir, in contrast, connects wells with similar log signature and documents a more homogeneous reservoir.

Applications of Seismic Tomography in Reservoir Characterization

Research in rock properties, and the development of new technologies, such as those related to reservoir simulation which were discussed earlier, are underway. With the success of these efforts, seismic cross borehole tomography promises to become an important tool for use in reservoir management for improved oil recovery.

Specific applications of this technology may be found in planning infill drilling, especially when porosity development is associated with a discernable variation in the seismic velocity field. When the movement of flood fronts can be monitored in the reservoir then permeability variations and barriers to fluid flow may be inferred. In addition, it is often possible to image faults and to map reservoir units using seismic tomographic imaging technology. When pressure fields are varying in the reservoir, it may be possible to monitor gas coming out of solution or going into solution, as well as the movement of gas caps in the reservoir. When flood fronts can be tracked, then it may be possible to identify bypassed oil or banked oil which could be recovered with appropriate changes in injection strategy or development drilling.

By tracking the movement of injected fluids in enhanced oil recovery programs, we can determine the extent to which these programs are meeting economic objectives and can better determine the economic viability of the programs. In particular, it may be possible to determine actions, which, if taken, would result in significantly improved economic performance of the program. In summary, studies have indicated that one of the greatest needs in reservoir engineering is the requirement to be able to infer reservoir parameters in the interwell volume, which is poorly sensed by most existing technologies today. Cross borehole tomography offers perhaps the best hope at present of accessing this volume and inferring information about it at a scale and resolution which can be used for planning improved oil recovery programs.

Economics of Cross Borehole Tomography

As practiced today, cross borehole tomography is a relatively expensive operation. This is due in part to the fact that tools being used for cross borehole tomography surveys were not originally designed for the purpose. In particular, the use of a single receiver with a single source can result in relatively long survey times to acquire data for tomographic image reconstruction. New multi-element receiver tools are being designed and are coming into use, which significantly cut the time required to acquire the tomographic survey. The goal of these developments is to reach a tomographic survey capability which will have a minimal impact on field operations and which can be carried out at the lowest possible cost. In addition, tools are being developed for high resolution imaging so that thin beds on the order of several meters thick, as well as other phenomena of a similar scale, can be resolved when required.

The ultimate value of any technology is determined by the value of the information which it provides in comparison to the actual cost of acquiring that information. Improving the economic efficiency of an enhanced oil recovery program by only a few percent, or increasing the recoverable reserves in a field, may well justify the cost of a cross borehole seismic surveillance program. We can surely improve our abilities to recover oil by making improvements in enhanced oil recovery technologies, but without adequate surveillance capability, many of these technologies may not realize their full potential. In particular, it seems unlikely that we could ever design a truly optimal reservoir management strategy without the means of inferring something about reservoir properties in the interwell volume. We have stated before and will state again that proper field management starts from the day the first well is drilled and should always include a reservoir surveillance monitoring program. Cross borehole tomography is a developing technology which promises to play an important role in reservoir characterization in the future.

EOR INJECTION PATTERN

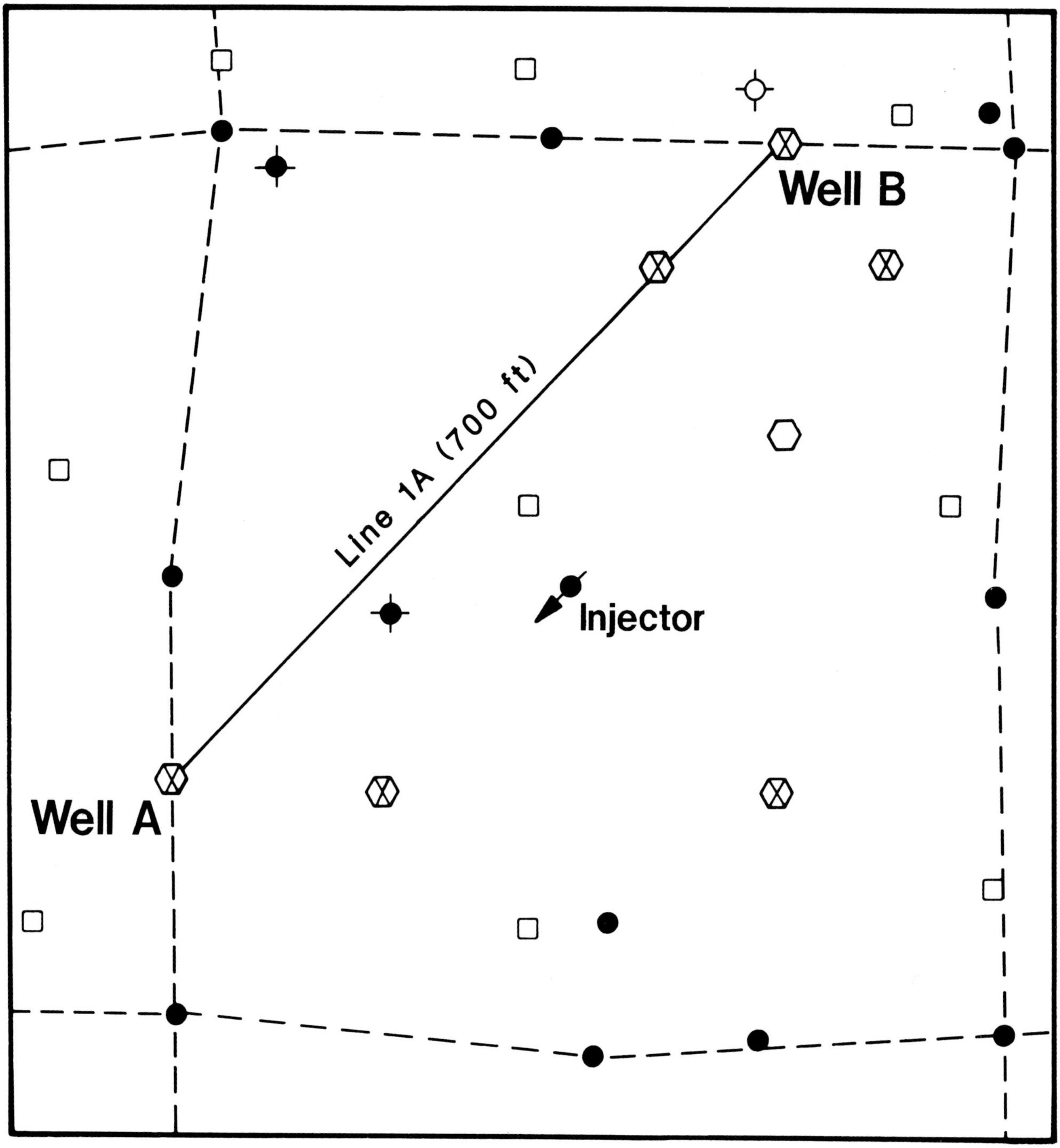

Figure 1

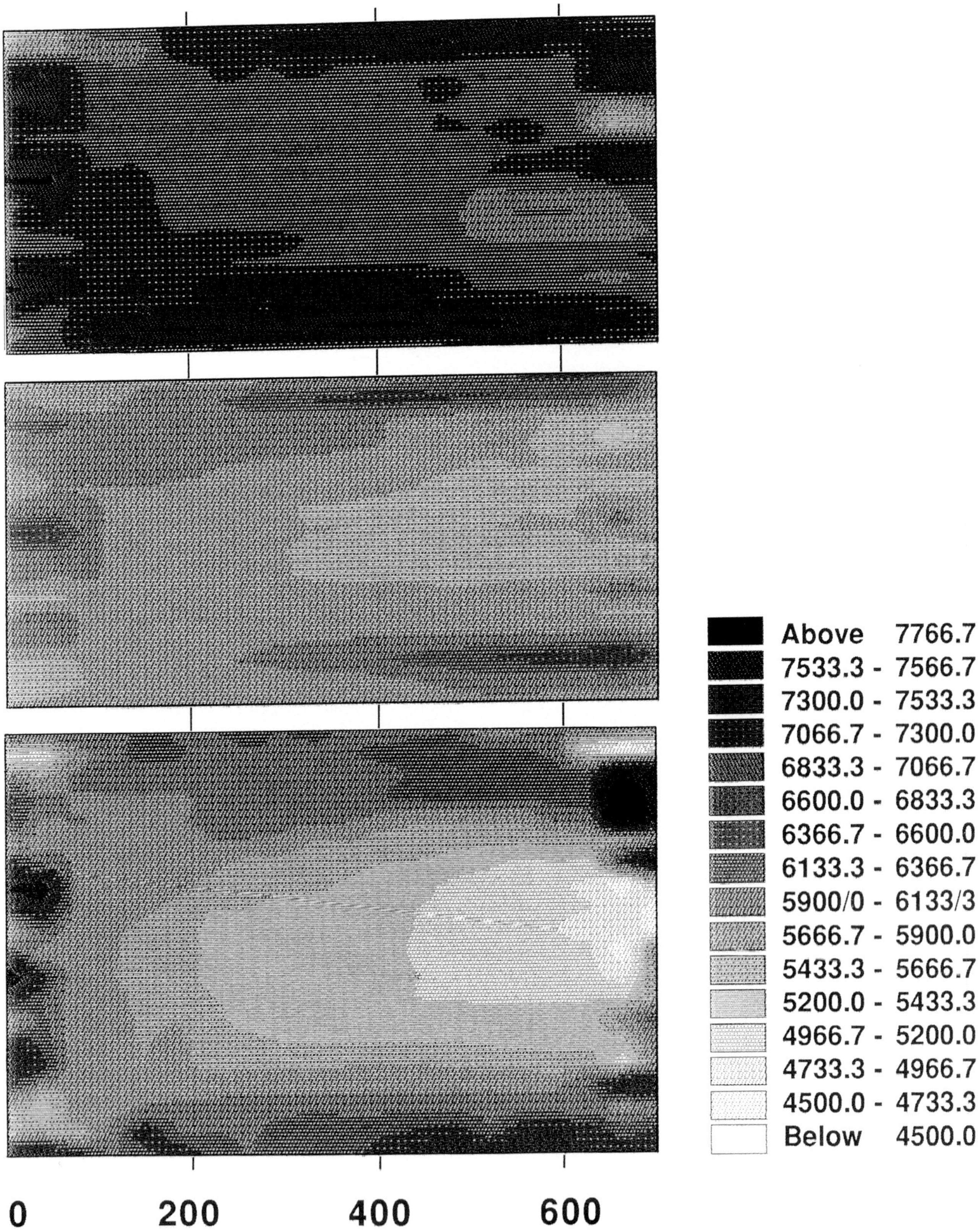

Figures 2(top), 3(middle), 4(bottom)

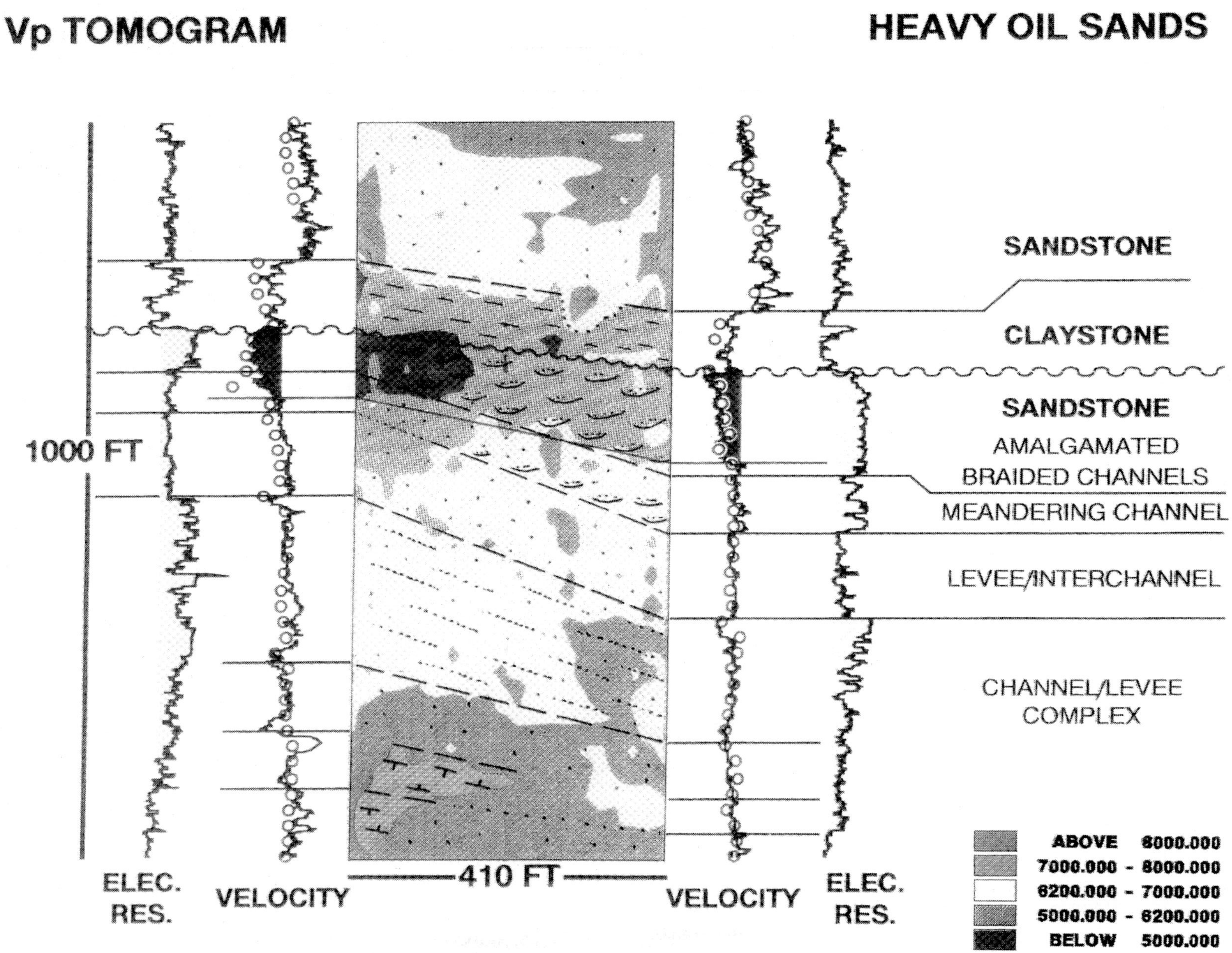

Vp TOMOGRAM
HEAVY OIL SANDS
1000 FT
SANDSTONE
CLAYSTONE
SANDSTONE
AMALGAMATED
BRAIDED CHANNELS
MEANDERING CHANNEL
LEVEE/INTERCHANNEL
CHANNEL/LEVEE
COMPLEX
ELEC.
RES.
VELOCITY
410 FT
VELOCITY
ELEC.
RES.
ABOVE 8000.000
7000.000 - 8000.000
6200.000 - 7000.000
5000.000 - 6200.000
BELOW 5000.000

Figure 5

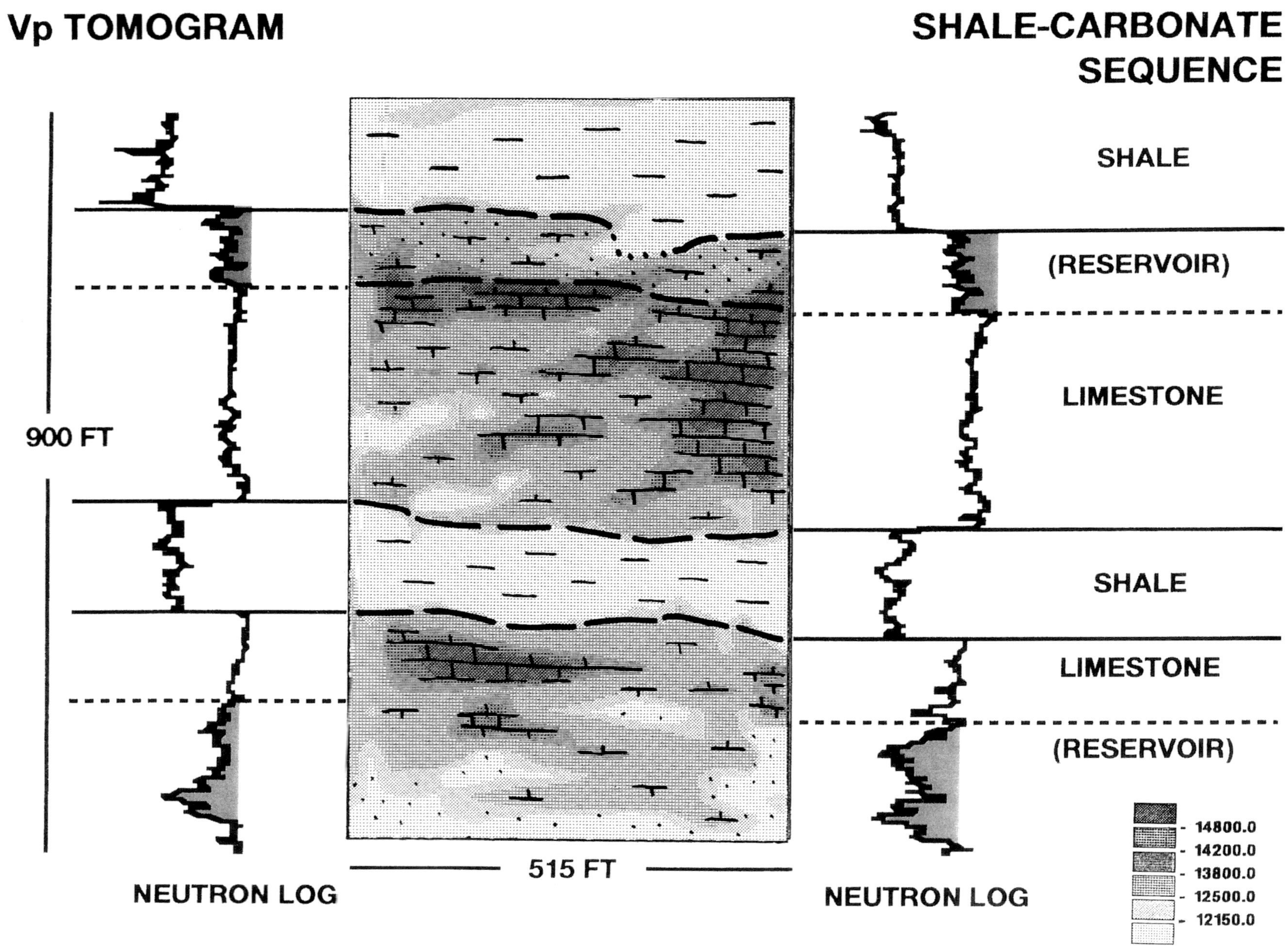
Vp TOMOGRAM
SHALE-CARBONATE SEQUENCE
900 FT
515 FT
NEUTRON LOG
NEUTRON LOG
SHALE
(RESERVOIR)
LIMESTONE
SHALE
LIMESTONE
(RESERVOIR)
14800.0
14200.0
13800.0
12500.0
12150.0

Figure 6

Parameterizing and Block-Averaging Electrical Characteristics of a Reservoir: an Essential part of Electrical/Electromagnetic Evaluation of Production Process

A.J. Mansure
Sandia National Laboratories, Albuquerque NM 87185

ABSTRACT

The geology of an oil reservoir is not homogeneous at the scale of resolution of an electrical/electromagnetic (E/EM) measurement or at the grid scale of an oil reservoir simulator. This paper discusses issues governing how the petrophysical laws governing resistivity of homogeneous rocks average to give bulk or grid scale resistivity. The bulk-averaged resistivity of a heterogeneous material measured by a given E/EM technique is a function of the direction of current flow relative to any structure of the material. The difference in bulk resistivity between current flowing across and parallel to geologic structure is frequently significant resulting in a material that is anisotropic. Failure to recognize the importance of the direction of current flow can lead to significant errors in interpretating E/EM measurements. The averaging of core-scale properties (porosity, saturation, etc.) to give representative block resistivity has been analyzed using field data to ensure that the statistics (average, variance, correlation, etc.) are geologically realistic. Merely plugging average saturation and porosity into Archie's law, while ignoring how the saturation and porosity are spatially distributed, can give block resistivities that are as much as 50% off. The analysis of field data was generalized from a bedded unit to an amorphous block by considering a bimodal distribution of pore sizes. For this example the resistivity calculated using average properties can be off by a factor of ten. This demonstrates that the question of how to average properties to determine bulk resistivity is a fundamental petrophysical question, not just an artifact of the field data used.

INTRODUCTION

The ultimate recovery of oil depends on process control. The fundamental information on which the process control should be based is the spatial and temporal changes in oil saturation that accompany the process. It is difficult to acquire adequate saturation data.[1] Thus, process control is typically based on presumed knowledge of the producibility of the reservoir. This presumed knowledge is frequently based on numerical simulators, the utility of which is severely limited by the lack of detailed geological data because the fluid zones of a reservoir are heavily dependent upon the local geologic variability, anisotropy, and heterogeneity.

Electrical resistivity logging is the primary method used for determining where to complete a well and estimate oil saturation; however, electrical logs can only see into the reservoir about 20 feet. Current practice is to use data from production and monitoring wells to provide information about the fluid zones of the process in between the wells.[1,2] However, geologic factors typically vary significantly, even on the closest of well spacings, resulting in highly irregular shapes for the process zones. Thus, in reservoir characterization, knowledge of lateral continuity and heterogeneity is critical for extending the reservoir description provided by electrical resistivity logging away from the wellbore.

An E/EM measurement technique that would monitor the growth of the production process in the space between wells and provide information about saturations and high permeability zones would be a valuable asset. Field trials of E/EM measurement techniques have been conducted;[3-6] however, there is uncertainty in the interpretation of the results. For such techniques to be useful, methods must exist for interpreting the measurements.

Previous work[7,8] has shown how Archie's law and shaly sand models (the relationships used to interpret well logs) can be used to understand steamflood resistivity changes of a homogeneous rock. The interpretation of E/EM monitoring of an oil recovery processes will be done on a scale large enough that the rock cannot be considered homogeneous. Thus, it is important to understand how core-scale properties are related to block or average resistivity. This understanding must consider both the nature of the heterogeneous block of rock (does it have structure, are the

*This work was performed by Sandia National Laboratories operated for the United States Department of Energy under contract DE-AC04-76DP00789.

heterogeneities dispersed, etc.) and how the measurement process samples the block or rock (how the current flows relative to any structure, etc.).

Topics discussed in this paper include: a) the importance in determining bulk resistivity of the direction of current flow relative to geologic structure, b) how the distribution of saturation and porosity affect bulk resistivity, c) anisotropy and its effect on interpretation of bulk resistivity, d) bulk resistivity and the nonuniqueness of inversions of E/EM measurements, and e) travel-time vs. amplitude measurements of bulk resistivity.

EFFECT OF DIRECTION OF CURRENT FLOW, OR CONDUCTIVITY VS. RESISTIVITY AVERAGES

It is important to know how an E/EM measurement depends on resistivity in order to bulk-average properly. For example, in the far-field a Controlled Source Audiofrequency Magnetotellurics (CSAMT) signal moves downward across beds probing the beds in series, which would seem to imply the resistivity of the beds should be combined in series. However, because the currents flow parallel to the beds, the proper bulk average is resistors in parallel. Failure to recognize this can lead to significant errors in calculating bulk resistivity.

The importance of the direction of current flow in proper averaging was demonstrated for field conditions by digitizing an induction log from Kern River field on ten-foot intervals to give a fine structure resistivity profile (see Figure 1). Block averaging of the profile was done considering the measurement to respond to the resistance of the individual beds both as resistors in series (labeled resistivity average on figure) and resistors in parallel (labeled conductivity average on figure). For regions of low variability in the original log (lower part of figure), the two averages are about the same. For regions of high variability (middle part of figure), the two averages are quite different. The division of the section into blocks for the two averages was varied in order to produce the best overall representation of the profile.

Frequency-domain plane-wave forward modeling was performed for the fine structure and the block average profiles (see Figure 2). The conductivity average gives very accurate surface impedances (<0.1% error) resistivity average gives surface impedances that are as much as 50% off.

From these calculations it is concluded that 1) it is very important that the block averaging be done considering the direction of current flow, and 2) "plane wave" measurements average out the fine structure so that only block or bulk properties are needed to describe the resistivity profile. This means that the overburden and underburden can be adequately described by a few "average" layers. Unfortunately, it also implies that it will not be possible to determine the fine structure of resistivity variations of the process.

An additional point that came out of this study is the importance of the underburden. This is shown in Figure 3. The curves labeled preflood and postflood show the apparent surface resistivity that results from forward modeling the actual resistivity fine structure of Figure 1 before and after a simulated resistivity change due to a steamflood at 240 m. For the curve labeled "incorrect underburden," the underburden resistivity structure was correct down to the flood, but below the flood it was determined by taking a resistivity average rather than a conductivity average of the actual resistivity structure. The point to note is that incorrectly representing the underburden creates an error bigger than the difference between the preflood and postflood curves.

AVERAGING CORE SCALE PROPERTIES TO DETERMINE BULK RESISTIVITY

Three different scales are important to the interpretation of electrical and electromagnetic diagnostic techniques: (1) the core scale or the scale at which Archie's law applies, (2) the block scale (tens of meters) or the scale at which reservoir simulators and electromagnetic forward and inverse programs grid the reservoir, and (3) the reservoir scale or the scale at which E/EM measurements are made (see Figure 4). Because petrophysical properties that determine resistivity (porosity, saturation, temperature, salinity, etc.) are not homogeneous over a block of rock, it is necessary to understand how the distribution of the properties affects resistivity. The importance of variations in these properties is best investigated by an evaluation of field data to ensure that the statistics (average, variance, correlation, etc.) are geologically realistic.

Figure 5 shows a resistivity profile from the Light Oil Steamflood at Elk Hills Naval Petroleum Reserve.[7] The "+" symbols are the resistivity values calculated at discrete depths using Archie's law. There is good agreement between these calculations and the logged resistivity (solid line on the figure). The average resistivity of the interval can be determined by averaging the logged value at each depth to get 6.9 ohms/m (vertical dashed line on Figure 5). However, if the saturation and porosity are averaged and then these values plugged into Archie's law, a resistivity of 3.4 ohm/m is predicted, a value that is about 50% off (vertical dotted line on Figure 5). The reason for the considerable difference in these two numbers has to do with the difference between the mean of a parameter and the expected value of a function for a parameter. For example Archie's law says that the functional dependence of resistivity on saturation (S_W) is S_W^{-2} for which

$$\text{Expected Value }\{S_W^{-2}\} \neq \{1/\text{Avg.}S_W\}^2 = \int S_W^{-2}P(S_W)dS_W, \cong \{1/\text{Avg.}S_W\}^2\{1+3\sigma^2+\dots\}$$

where $P(S_W)$ is the frequency of occurrence of water saturation S_W. That is, the expected value of the function S_W^{-2} is not in general equal to the square of $1/\text{Avg.}S_W$. Instead, it is equal to the integral of S_W^{-2} times $P(S_W)$. It is approximately given by the last of the above equations that multiplies $\{1/\text{Avg.}S_W\}^2$ by a factor that includes the standard deviation (σ) of the water saturation. For the example in Figure 5, if the expected value of the resistivity is determined this way, using both the average and standard deviations of the saturation and porosity, the resistivity determined agrees with the average of the logged resistivity.

In the above analysis, a subtle assumption was that the resistivity profile of Figure 5 was being probed by current that flows vertically so that the resistivity of the individual beds is to be averaged as resistors in series (simple average of logged interval). In the previous section, it was pointed out that sometimes we need a block average appropriate for current flowing parallel to the beds. To do this one must express Archie's law in terms of conductivity so that the functional dependence on water saturation is S_W^2. When this is done, the appropriate correction term is $\{1 + \sigma^2\}$. Thus, measurements probing parallel to the beds are less influenced by variability in the distribution of saturation.

The dependence of the bulk resistivity on the distribution of properties, rather than just the average properties, is a fundamental petrophysical issue, not just an artifact of the bedded nature or geological structure of the Elk Hills data. This can be demonstrated by examining a bimodal pore-size distribution example taken from the work of Morrow[9] (see Figure 6). The water drains from the coarse grain material leaving it partially saturated (S_w=7%) whereas the fine grain material is 100% saturated. If the saturation is averaged over the whole block and that value is plugged into Archie's law, a conductivity of 0.0085 mho/m is calculated. Assuming that the fine grain material is amorphously dispersed in the coarse grain material (i.e., there is no structure such as bedding), the effective conductivity of the block can be calculated taking into account the actual distribution of saturation using the Hanai-Bruggeman equation.[10-12] This gives a conductivity of 0.00098 mho/m. This demonstrates that merely plugging the average saturation into Archie's law can give a conductivity that is off by as much as a factor of ten. The problem is that Archie's law applies to a homogeneous sample or to the fine and coarse grain materials individually, but does not apply to the bulk material containing both materials.

INTERPRETATION OF BLOCK RESISTIVITIES

Because of the dependence of bulk resistivity on the distribution of rock properties and the direction of current flow, there is an uncertainty in the interpretation of block resistivity. In a given E/EM experiment the direction of current flow is known, but the distribution of rock properties (saturation, etc.) is not necessarily known. The heterogeneity or distribution of rock properties is what the measurement is trying to determine. Consider an experiment where horizontally flowing current interrogates a waterflood. Within the swept zone, water saturation is assumed to increase from 40% to 57% which changes the local resistivity from 10 Ω-m to 5 Ω-m in the process. Figure 7 shows three different displacement processes: a) where the water sweeps the upper half of the volume, b) where the displacement is "piston- like," and c) where the unswept volume is dispersed. The appropriate averaging for these three different displacements is a) parallel or conductivity average b) series or resistivity average, and c) Hanai-Bruggeman[10-12] or dispersed structural average. Figure 8 shows the block resistivity measured as a function of swept volume for these three different averaging methods. The figure shows that if the measurement determined that the block average resistivity is 7 Ω-m that it could be due to 43% sweep of the vertical section, 59% complete "piston like" displacement, or 53% efficient displacement with unswept volume dispersed.

This nonuniqueness in the relationship between block resistivity and the volume swept can be either a drawback in the utilization of E/EM methods to monitor oil recovery processes, or can be exploited as a means of distinguishing between the displacement processes.

ANISOTROPY

The effect of anisotropy on electrical measurements has long been recognized.[13-17] However, the bulk resistivity of rock has usually been considered isotropic unless specific evidence has shown it to be anisotropic. The justification for assuming an isotropic medium is the desire to simplify overwhelmingly complex E/EM measurements. To the degree that the log in Figure 1 is typical of regions where E/EM techniques will be used to monitor oil recovery processes, it would perhaps be better to assume bulk rock is anisotropic unless evidence that it can be considered isotropic. Assuming bulk rock is isotropic, when it is really anisotropic, for the goal of simplifying the measurement interpretation process, can be self-defeating if it leads to inversions of E/EM data that cannot be interpreted in terms of known geology. Schlumberger et al.[13] showed that "solutions that are unique under the assumption of homogeneous and isotropic beds are no longer unique when the beds are anisotropic."[18]

The block averaging of the log in Figure 1 showed a resistivity anisotropy ratio (square root of resistivity average divided by conductivity average) of 1.2 for current flow parallel and transverse to the beds. Figure 9 shows that this degree of anisotropy is sufficient to have a significant effect on the interpretation of a measurement made to monitor an oil recovery process. This figure compares the effects of an anisotropic overburden and a buried resistivity contrast (eg., oil recovery process being mapped) on the apparent surface resistivity measured between a buried electrode and an electrode moved across the surface of the earth. The anisotropic overburden mimics the effect of the buried process. Thus, failure to recognize that the bedded nature of an overburden makes the overburden behave anisotropically could cause the false identification of a buried target. The reason anisotropy is important to this experiment is that as the electrode moves across the surface, the current changes the direction in which it crosses the overburden beds.

Many proposed E/EM measurements to monitor oil recovery process, such as cross borehole tomography, involve interrogating the same block of rock in different directions and so invite the question of anisotropy. A significant point to be noted is that the process being monitored must, in general, be described by anisotropic permeability and so E/EM monitoring of that process would be expected to respond to the process anisotropically.

AVERAGE RESISTIVITY AND THE NONUNIQUENESS OF INVERSIONS OF E/EM MEASUREMENTS

The inversion of E/EM measurement data results in a coarse, or block- average, description of resistivity structure. Understanding how coarse this description is and the uniqueness of the inversion is an important part of understanding the scale to which core properties must be averaged and what the "inversion averaged resistivity" represents.

The non-uniqueness of inversions of frequency-domain surface-base electromagnetic soundings was investigated by forward-modeling downward- moving plane waves in a layered, laterally homogeneous earth. The first question addressed was, is the nonuniqueness of resistivity inversions commonly discussed due to 1) poor inversion algorithms, 2) noisy or incomplete data, or 3) more than one mathematical solution to the inversion problem? This question was addressed by comparing forward models of a three-layered earth. The calculations address the inversion of frequency dependent data to get the resistivity structure of a laterally homogeneous earth. It was found that two distinctly different subsurface resistivity structures could give the same complex surface impedance to better than

0.05%. Thus, regardless of which of the three possibilities above results in the nonuniqueness of the inversion, very different resistivity structures can be obtained starting from measured data sets that are within expected noise levels of each other. That is, the nonuniqueness is of practical significance and cannot be eliminated by just changing inversion algorithms.

Having established the practical significance of the nonuniqueness of inversion of a single data set, the question was addressed of whether pre- and poststeamflooding data, when used together, can be uniquely inverted to determine the resistivity change of a steamflood. To address this question a synthetic resistivity profile was used that had the general character of the resistivity structure at Kern River; however, the profile was kept simple so that the number of layers did not exceed the number of frequency points. The primary tools used were a USGS non-linear least-squares plane wave inversion program (NLSPW by Anderson[19]), and a simple forward-modeling program (PWRES written by Caffey[20]) that evaluates the algebraic expressions for the surface impedance of a layered media as given by Wait.[21]

The function of these two programs was compared for known data sets and shown to give accurate results. The inversions were done requiring that the resistivity above and below the steamflood remain unchanged, but the depth of the steamflood was left as a variable. The analyses showed that pre- and poststeamflooding data, even when used together, do not remove the depth degeneracy. That is, one cannot tell the difference between a thick, low- resistivity change and a thin, high-resistivity change. Independent knowledge of the depth of resistivity change is needed. The need for independent depth knowledge means that surface based frequency-domain measurements alone cannot be used to look for depth-dependent phenomena such as vertical sweep and steam override. Rather, frequency-domain measurements will give information about bulk quantity of steam under a given measuring location, but not its vertical distribution.

The next question addressed was whether pre- and poststeamflooding data together with independent depth knowledge will remove the degeneracy from the data. The inversions were done requiring that the resistivity above and below the steamflood remain unchanged and the depth of the top and bottom of the flood were input into the inversion. This inversion study is summarized on Figure 10. First, preflood and postflood resistivity structures were established based on Kern River Data (upper left quadrant of the figure). These resistivity structures were forward modeled to determine the apparent resistivity data (upper right quadrant of the figure). A joint inversion of these two data sets was done subject to the constraint that all the resistivity change is within the known depth interval of the flood (lower left quadrant on the figure). The inversion resistivity structure was forward modeled to determine how well it corresponds to the input apparent surface data (lower right quadrant on the figure). Note the input data to the inversion (points) are indistinguishable from the results of forward modeling the inversion resistivity structure (lines).

From this inversion study, it was found that independent knowledge of the depth of resistivity change removes the degeneracy of the inversion for a three-layer earth. Thus, if the depth of the layers is known from another source, the resistivity change of a steamflood imbedded in uniform over- and underburdens can be determined by inversion. However, for multilayered over- and underburdens (as shown on Figure 10), independent depth knowledge is not enough to make the inversion unique, but the results were somewhat encouraging. For this approach to the inversion process, the inversion resulted in an effective overburden resistivity profile that distorted the pre- and postflood resistivities of the steamflood in a similar manner. Thus, even though the pre- and postflood resistivities are not accurately estimated by the inversion; their ratio is a qualitative indicator of the steamflood resistivity change.

After the above work was completed with the synthetic data set, the study was repeated with the fine-structure resistivity profile from Kern River shown on Figure 1. It was found that the added complexity made it considerably more difficult to do the analysis (something that would be dependent upon the inversion used), but the same conclusions still apply.

TRAVEL-TIME VS. AMPLITUDE MEASUREMENTS OF RESISTIVITY

The dependence of bulk resistivity upon direction of current flow (how the signal propagates through the media) has already been discussed. In that discussion, simple concepts of series and parallel resistors were used to empirically investigate how a CSAMT measurement samples layered media to give a bulk resistivity. When the measurement uses transient or time-varying fields, the potential for dependency on frequency (or on how the measurement technique samples the fields) must be considered.

To investigate these effects, conceptual experiments were investigated for the one dimensional medium shown on Figure 11 that has a change in resistivity at z=b ($r(z)=r$, $z<b$; $r(z)=r'$, $z \geq b$). At the origin (z=0), there is a vertical current sheet (I). Two cases are considered for I, one where the magnitude of the current is sinusoidal (plane wave or PW case) and one where there is a step function (θ case) change in I ($I=\theta(t)$). At distance z=D, there is an electric field meter (M). For the PW case, M records the magnitude of the electric field, and for the θ case, M records either the amplitude of the maximum electric field detected or the time of arrival of the maximum electric field detected. (For the PW case, phase measurements could be made but would be difficult to interpret in this conceptual experiment.)

The analytic solutions for electric fields have been derived for both the PW and θ cases. An effective resistivity (r_{eff}) can be defined by equating the solutions for $r=r'=r_{eff}$ and $r \neq r'$. The meaning of this effective resistivity is that if the material on the left of z=D were replaced with a material with resistivity r_{eff} and the material on the right of z=D remains the same (resistivity r'), M would report the same measurement. Thus, in some sense r_{eff} is the average or bulk resistivity of the material between z=0 and z=D. Figure 12 shows this r_{eff} assuming r=1, r'=0.5 and b = 5. Interpretations of Figure 12 regarding how the different measurements sample the resistivity-heterogeneity differently and thus determine different bulk resistivities, of course, apply strictly only to the measurements reported here. However, the issues raised by these interpretations should give insight into general question of "is the average or bulk resistivity determined by an experiment a function of how the measurement technique samples the electric field?"

Figure 12 shows four different curves. Two are for the θ experiment and are labeled Amplitude (for the measurement of the amplitude of the maximum electric field detected) and Time (for the measurement of the time of arrival of the maximum electric field). Two are for the PW experiment and are labeled w=0.5 (for a sinusoidal angular frequency of 0.5) and w=25 (for a sinusoidal angular frequency of 25).

As a general philosophical point, r_{eff} is only an average property of the media if it is a function of the media and independent of the measurement technique. In surface electrical measurements (Wenner array, Schlumberger array, etc.), it is recognized that the "apparent surface resistance" reported is a function of the array being used (each array configuration has its own type curves) and thus is not strictly a property of the media. Inversions of the "apparent surface resistance" are done to try to determine resistivities that are properties of the media only. Unfortunately, the nonuniqueness of the inversions obscures the question of whether the inversions are independent of the array used. For the conceptual experiments described here, inversion is not required. The fact that the curves on Figure 12 are different indicates that r_{eff} is not truly a bulk average property of the media.

The two PW curves show that measurements at different frequencies sample the media differently giving different r_{eff}'s even though the resistivity of the material itself is not a function of frequency. The two θ curves show that different measurements of the same electrical excitation can give different r_{eff}'s. The deviations of r_{eff} from r=1 when M is on the left of the material change (z=b=5) show that the effective resistivity is influenced by material outside the region of measurement, and that the wrong bulk resistivity can be reported (even for a homogeneous material) because of heterogeneities outside the block of interest.

The discussion above, and the existence of multiple curves on Figure 12, indicate that r_{eff} form these experiments cannot be plugged into Archie's law to determine average saturation disregarding the method of measurement. Somehow the functional dependence of r_{eff} on the measurement technique must be incorporated into the analysis before Archie's law can be used.

A question that must be asked is whether the issues raised by this discussion are really significant, or do they exist because the analysis was too elementary. Perhaps an analysis of the PW experiment must be made including a spectrum of frequencies, or the θ-measurement must be analyzed using the full time-dependent signal. Certainly, such analysis would help, but may not necessarily eliminate the issues raised here. For the θ experiment it can be shown rigorously that no r_{eff} exists that can replace the material between I and M such that the field transmitted past M will have the same time dependence and amplitude as the original signal. Measurement of the complete signal would allow an "exact" determination of the position of b and the resistivity on either side assuming there was a single interface. If the material consisted of a number of layers, the structure could only be approximated. If this approximate structure is independent of the measurement, then Archie's law could be applied readily; if it is not, then the issue still exists.

SUMMARY AND CONCLUSIONS

Interpretation of E/EM measurements monitoring oil recovery processes has historically focused on inversion to produce a map or image of the process. While inversion of E/EM field data is critical, the success of E/EM techniques for monitoring oil recovery processes will depend upon synergism with geology and petrophysics. For the inversions and interpretations to be accepted, they need to be consistent with well logs and other geological knowledge. A map of an oil recovery process is of value if there is a relationship between resistivity and the petrophysical properties (saturation, porosity, etc.) that control the process. Thus, the resistivity structure from E/EM measurements needs to be consistent with Archie's law and petrophysics of the process.

The geology of an oil reservoir is not homogeneous at the scale of resolution of an E/EM measurement or at the grid scale of an oil reservoir simulator. It is, therefore, necessary to understand how the resistivity of rocks average to give bulk or grid-scale resistivity. Based on a knowledge of the direction of current flow, the resistivity of thin beds can be averaged to establish a bulk resistivity that is the same average determined by E/EM measurements. The fact that bulk rock is frequently bedded with material of varying resistivity results in a bulk material that is anisotropic. This means that the response of the bulk rock will depend upon the direction the E/EM measurement probes the rock.

The resistivity of bulk rock depends upon the distribution of petrophysical properties, rather than just the average values of the properties. Thus, merely plugging average saturation into Archie's law can give a conductivity that is off by as much as a factor of ten. The functional dependence of the petrophysical laws governing resistivity provides the mechanism for including this dependence on the distribution of the properties.

Neither the determination of petrophysical properties from resistivity nor the inversion of E/EM measurements are unique. However, by the application of constraints, such as knowledge of the distribution of fluids in the reservoir or knowledge of the depth of the reservoir, reasonable interpretations of petrophysical properties from E/EM measurements can be made. The nonuniqueness is not because the interpretation tools are inadequate, it is fundamental to the physical processes being analyzed. Thus, inversion techniques designed to remove the nonuniqueness, such as "least-structure inversions," may obscure the process rather than aid the interpretation.

Conceptual examples have been developed to compare amplitude and time-of- flight measurements of resistivity. The "average resistivity" measured by these two techniques differs, indicating that the resistivity measured by these two techniques is not truly a bulk property of the media, since it is also a function of the method of measurement. Thus, the "average resistivity" from these conceptual experiments cannot be plugged into Archie's law to determine average saturation with disregard for the method of E/EM measurement. Somehow the functional dependence of the resistivity on the measurement must be incorporated into the analysis before Archie's law can be used.

The successful application of E/EM measurements to monitoring oil recovery will require that bulk or block-scale resistivities, measured on a volume of rock that is heterogeneous and anisotropic, be interpreted in terms of water saturation and porosity. This paper provides an introduction to how core- scale rock properties of a heterogeneous material combine to establish bulk resistivity. Also discussed are issues regarding whether field E/EM measurements give uniquely interpretable measurements of bulk resistivity that are actually properties of the bulk rock.

NOMENCLATURE

$f(S_w)$	=	Frequency of occurrence of water saturation,
S_w	=	Water saturation (fraction),
σ	=	Standard deviation,
r	=	Resistivity,
θ(t)	=	Step function θ(t)=0, t<0; Θ(t)=1, t≥0.

REFERENCES

1. Thomas, E.C., J.E. Richardson, M.T. Shannon, and M.R. Williams, The Scope and Perspective of ROS Measurement and Flood Monitoring, JPT, pp 1398- 1406, November 1987.
2. Widmyer, R.H., Use of Monitor/Observation Wells in the Monitoring and Evaluation of Oil Recovery Projects, JPT, pp 967-975, August 1987.
3. Bartel, L.C., and Ranganayaki, H.F., Acquisition of controlled-source audiofrequency magnetotelluric (CSAMT) data at an active steam-drive site, Expanded Abstracts Society of Exploration Geophysicists Annual Meeting, Dallas, Texas, pp 213-215, 1989.
4. Witterholt, E.J., J.L. Kretzschmar, and K.J. Jay, The Application of Crosshole Electromagnetic Wave Measurements to Mapping of a Steam Flood, CIM 82-33-81, 33rd Annual Technical Meeting of the Petroleum Society of CIM, Calgary Alberta, June 6-9, 1982.
5. Pihlman, M. and P.E. Harben, Low Frequency Electromagnetic Cross-Borehole Imaging, UCRL—99769, International Symposium on Enhanced Oil Recovery, Maracaibo, Venezuela, February 19-22, 1989.
6. Bartel, L.C., Modeling and Analysis of the CSAMT Geophysical Technique Results to Map Oil Recovery Processes, SPE 11192, Proceedings 57th Annual SPE Meeting, New Orleans, LA, September 26-29, 1982.
7. Mansure, A.J., R.F. Meldau, and H.V. Weyland, Field Examples of Electrical Resistivity Changes During Steamflooding, SPE 20539, 65th Annual SPE Meeting, New Orleans, September 23-26, 1990.
8. Mansure, A.J., and R.F. Meldau, "Steam-Zone Electrical Characteristics for Geodiagnostic Evaluation of Steamflood Performance," SPE Formation Evaluation, September 1990, pp 241-247.
9. Morrow, N.R., Irreducible wetting-phase saturations in porous media, Chemical Engineering Science, v 25, pp 1799-1815, 1970.
10. Bruggeman, D.A.G., Berechnung verschiedener physikalischer konstanten von heterogenen Substanzen: Ann. Physik, v 24, pp 636-664, 1935.
11. Hanai, T, Theory of the dielectric dispersion due to the interfacial polarization and its application to emulsions: Kolloid-Zeitschrift, v 171, pp 23-31, 1960.
12. Hanai, T, Dielectric theory on the interfacial polarization for two phase mixtures: Bull Inst. Chem. Res., Kyoto Univ., v 39, pp 341-367, 1961.
13. Schlumberger, C., M. Schlumberger, and Leonardon, Some Observations Concerning Electrical Measurements in Anisotropic Media, and Their Interpretation, Am. Inst. Mining Metall. Engineers Tech. Pub. 505, v 110, pp 159-182, 1934.
14. Reddy, I.K., and D. Rankin, Magnetotelluric Response of Laterally Inhomogeneous and Anisotropic Media, Geophysics, v 40 N6, pp 1035-1045, Dec. 1975.
15. Moran, J.H., and S. Gianzero, Effects of formation anisotropy on resistivity-logging measurements, Geophysics, v 44 N7, pp 1266-1286, July 1979.
16. Campbell, D.L., Model for Estimating Electric Macroanisotropy Coefficient of Aquifers With Horizontal and Vertical Fractures, Geophysics, v 42 N1, pp 114-117, Feb. 1977.
17. Zonghou Xiong, Electromagnetic fields of electrical dipoles embedded in a stratified anisotropic earth, Geophysics, v 54 N12, pp 1643-1646, Dec. 1989.
18. Van Nostrand, R.G., and Cook, K.L., Interpretation of Resistivity Data, Geological Survey Professional Paper 499, US Gov. Printing Office, Washington, pp 23, 1966.
19. Anderson, W.L., Adaptive nonlinear least-squares solution for constrained or unconstrained minimization problems (Subprogram NLSOL), USGS Open-File Report 82-68, 1982.
20. Caffey, T.W.H., Sandia National Laboratories, Div. 6258, Albuquerque, NM.
21. Wait, J.R., Electromagnetic Waves in Stratified Media, Pergamon Press, Oxford, pp 1-11, 1970.

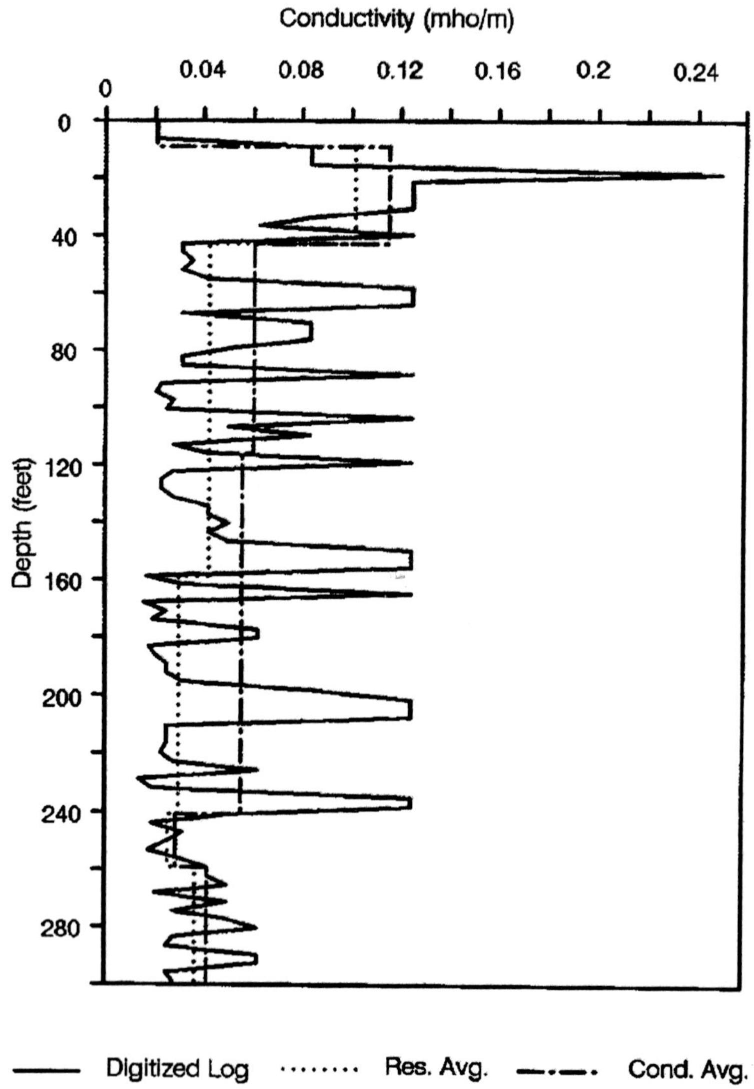

Figure 1: Effect of Conductivity Vs Resistivity Averaging

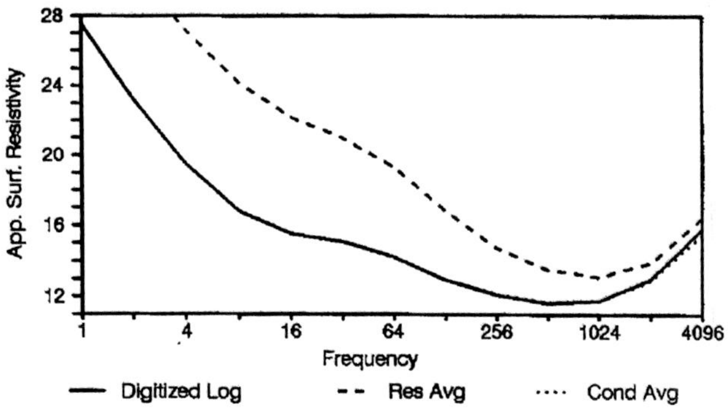

Figure 2: Effect of Averaging on Forward Modeling

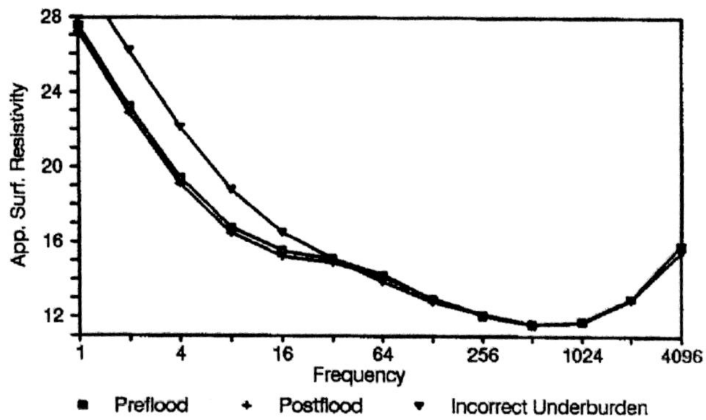

Figure 3: Effect of Underburden

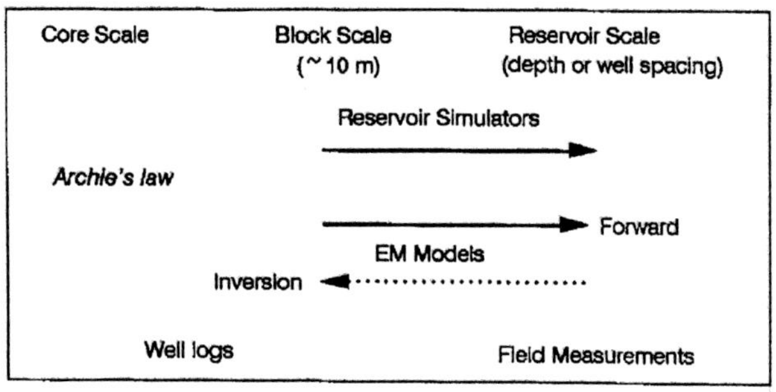

Figure 4: Important Scales for Interpreting E/EM Measurements

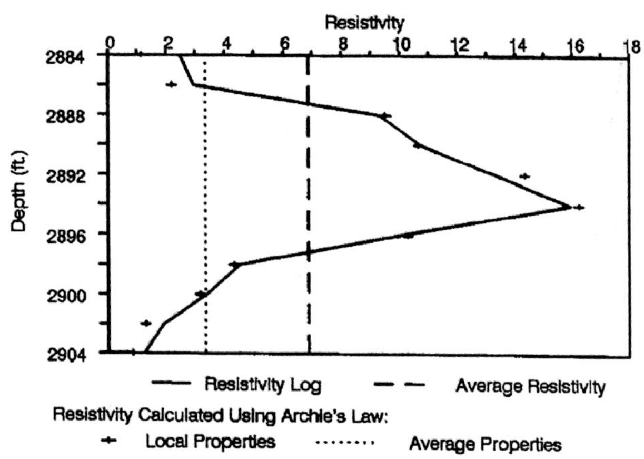

Figure 5: Average Properties Vs Block Average Resistivity

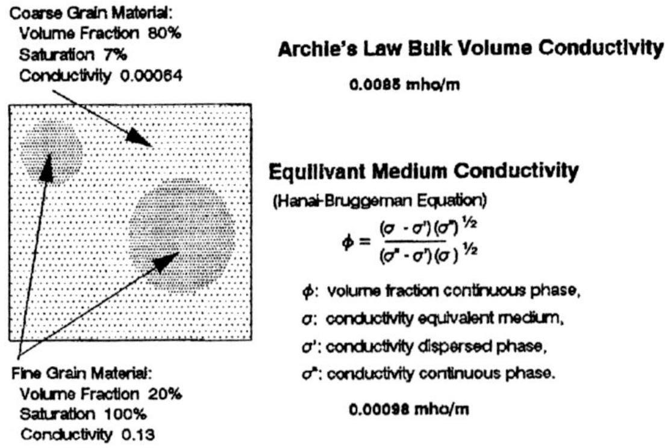

Figure 6: Effect of Bimodal Pore Size On Conductivity

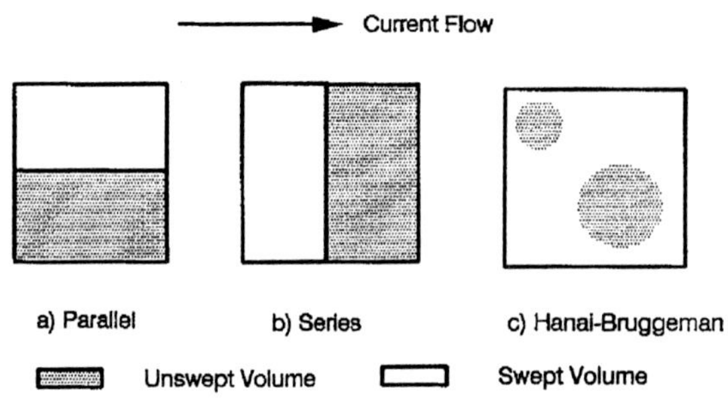

Figure 7: Various Swept Volumes

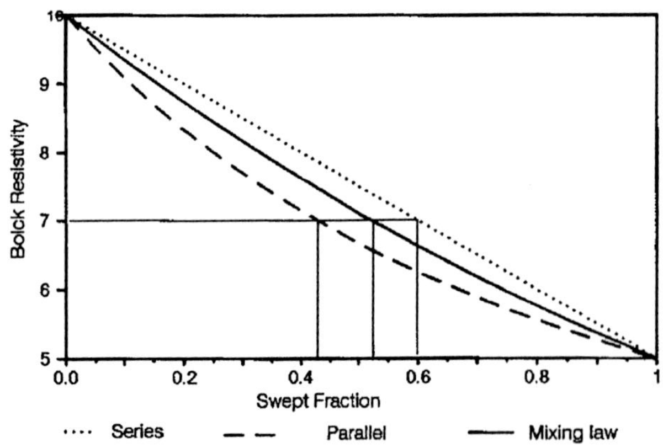

Figure 8: Resistivity of a Partially Swept Block

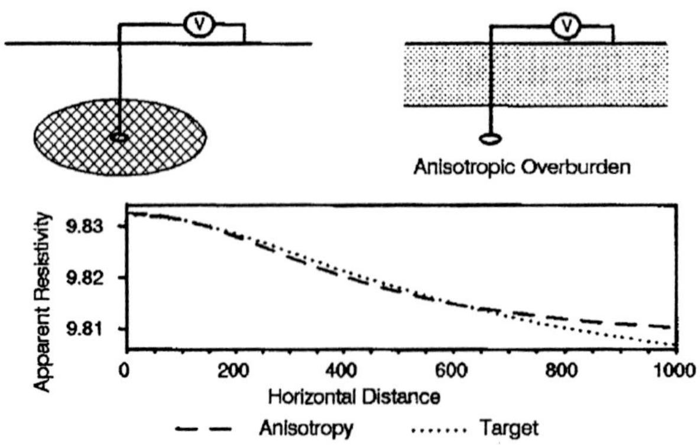

Figure 9: Anisotropy Can Mimic The Effects Of A Target

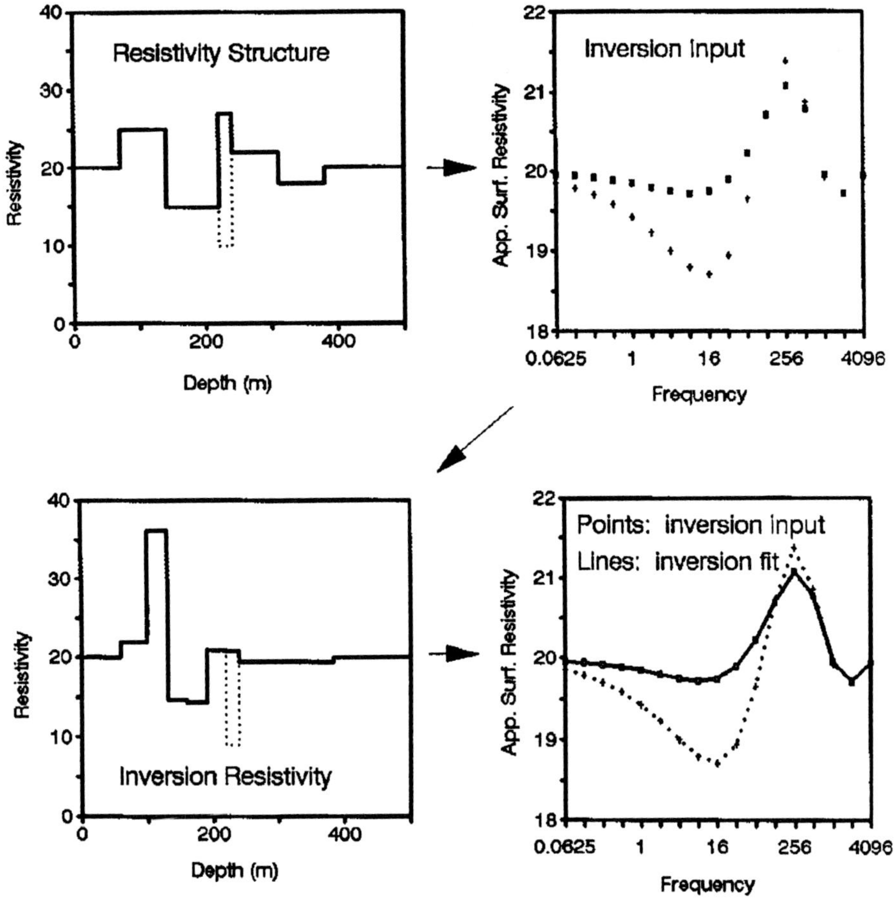

— ▪ Preflood ··· + Postflood

Figure 10: Resistivity Change Inversion With Depth Knowledge

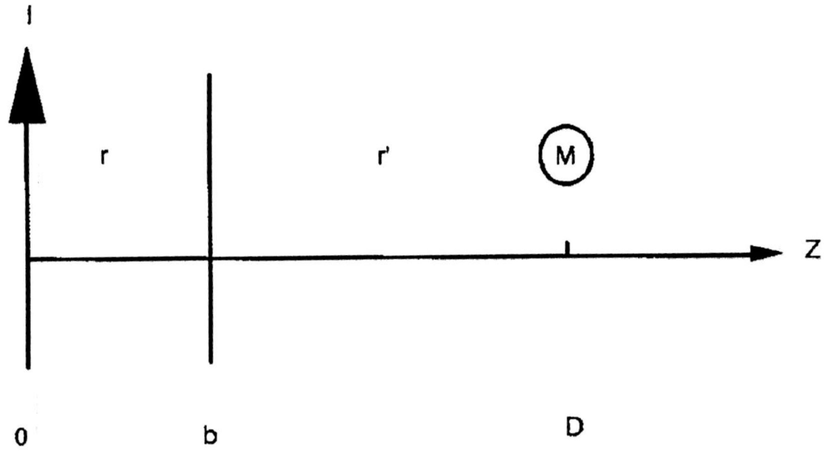

Figure 11: Configuration of PW and Θ Measurement Cases

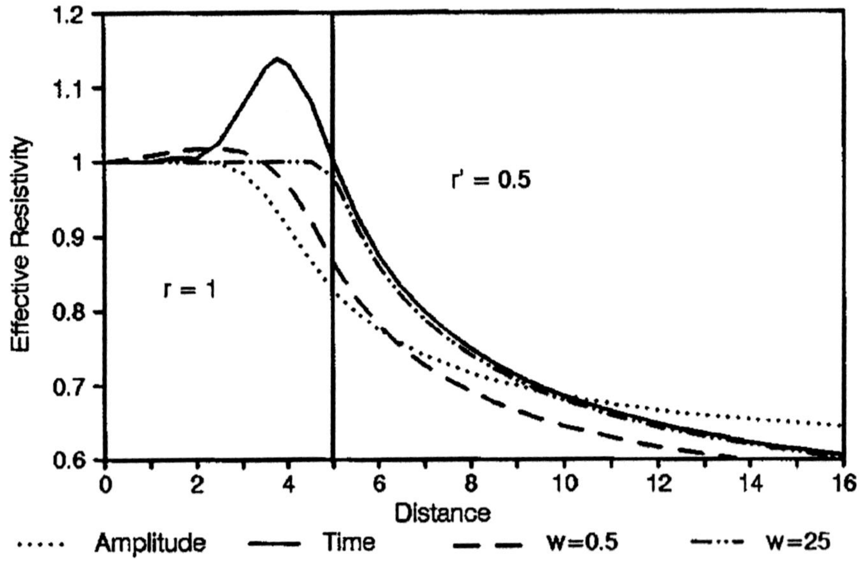

Figure 12: Effective Resistivity for PW and Θ Measurement Cases

Simulating Fluid Flow Through a Geologically Realistic Permeable Medium

Larry W. Lake

For the past five years, we have been studying the Page Sandstone outcrop in Northern Arizona as a prototype for eolian oil reservoirs and aquifers. The work has encompassed detailed geological studies and highly intensive descriptions of the distribution of hydraulic conductivity or permeability. We are now at the point of using this information to evaluate predictions about fluid flow through naturally-occurring sands. Work on this topic is the subject of this presentation.

The first step in the study was to establish a "truth case," a standard against which to measure the success of alternative procedures for flow prediction. Our truth case consisted of a highly detailed (more than 11,000 finite element nodes) numerical simulation of a miscible displacement through a portion of the outcrop wherein each nodal property was assigned according to the actual value existing at that point. This intensely deterministic simulation also allows us to evaluate the extent of geologic detail necessary for a good prediction; mobility ratio and the distribution of average permeability are the most important quantities governing fluid flow. Permeability anisotropy, fourth order-bounding surfaces and dispersion are much less important.

With the truth case in hand, we redo the simulations based on the amount of data that would normally be available for subsurface conditions. In this case, of course, it is the generation of interwell properties which is being tested. The work investigates two basic procedures: pseudofunctions and conditional simulation. The pseudofunction approach generates effective relative permeabilities (a concept borrowed from immiscible flow) to model the effects of heterogeneities on a gross or flow unit scale. The effective relative permeabilities now depend on the size of the flow units, the mobility ratio and the nature of the distribution of the permeability within the flow unit.

The conditional simulation approach generates interwell properties statistically and evaluates each procedure according to whether the distribution of the simulation results agrees with the results of the truth case. Conditional simulation also has the advantage that it allows risk assessment, but it is much more computationally intense than pseudofunctions. We have evaluated both fractal and traditional vaiogram approaches to random field generation, the key ingredient of conditional simulation. The fractal approach appears to generate the most realistic fields (compared to what is actually observed), an observation which leads to a general conceptual model for geologic media based on the concepts of geometric fractals.

A 3-D INTERACTIVE MODELING SYSTEM FOR INTEGRATED RESERVOIR INTERPRETATION

THOMAS J. LASSETER

Tech•Logic, Inc.
15325 189th Avenue NE
Woodinville, Wa. 98072 U.S.A.

ABSTRACT

Integrated interpretation requires the representation of the reservoir as a single consistent model which can be displayed and interactively edited in three dimensions. To be practical, high-performance 3-D graphics systems and the requisite support software are essential. Integrated data display and editing tools allow interpreters from the various disciplines to view and edit the model from their unique perspectives.

The model representation must support more complex topologies than the conventional "layer-cake" model if reservoir models and their associated performance predictions are to be realistic. Surfaces represented by detailed grids derived from maps and seismic interpretation are used to form the "framework" for the geological interpretation. The use of parametric curves and surfaces provides a compact and editable representation of the geometry allowing the interpreter to quickly modify the model when new data becomes available or alternative "realizations" are required.

With these requirements, a prototype 3-D interactive integrated reservoir interpretation system has been developed. Geophysical and petrophysical data can be displayed and edited in the time domain on ribbon sections between deviated wells. Complex geological models can be quickly developed using parametric curves and surfaces. Property distributions over geological units derived from zone-averaged values can be displayed and interactively edited. Geometry and property distributions can be displayed and output as mapped surfaces. Volumetrics capabilities and more direct interfaces for reservoir simulation are currently under development.

INTRODUCTION

Integration and interactivity are the keys to improved reservoir interpretation. Only through integration of the associated interpretation activities can the data be completely and consistently used. Only through interactivity can the interpretation activities be efficient enough to be practical.

The development of high-performance three-dimensional computer graphics systems coupled with high-performance computational systems has made integrated, interactive 3-D reservoir interpretation systems possible. This new technology has facilitated the development of innovative approaches to reservoir interpretation.

APPROACHES FOR INTEGRATED INTERPRETATION

It is often stated that integrated interpretation can be achieved by simply having applications communicate with each other. As an example, if seismic interpretation and petrophysical interpretation applications both generate maps in a standard format which they can exchange, then under this definition they are integrated. The problem is that both applications build separate and generally inconsistent models. It remains for the interpreters from the separate (and contending) disciplines to negotiate a consistent model. This "negotiation process" is time-consuming and inefficient. Neither interpretation domain (geophysical or petrophysical) provides the framework for representing and interpreting data from the other domain, so the two interpretation "factions" are left with paper logs, paper sections, and the "blackboard" to work out a consistent model.

The essence of integrated interpretation is a single, consistent model. It is not sufficient for applications to simply communicate with each other or share a common database. Each discipline has a unique perspective on the model based upon the associated data with unique contributions to make to the model definition. These perspectives overlap significantly, however, so model definition and editing must be done in a way which allows each discipline to evaluate whether proposed changes remain consistent from their perspective.

In order to provide a well-integrated consistent interpretation system, the following guidelines have been used:

1) all interpretation is done on a single self-consistent model;

2) graphical displays are designed to integrate as many data sources as is practical;

3) the use and juxtaposition of multiple windows provides additional integration where a single window would be impractical.

In subsequent sections, the details of this interpretation approach will be described.

INTEGRATED GEOPHYSICAL INTERPRETATION

Three-dimensional seismic interpretation has been conventionally done using vertical or time-slice sections. A major problem in seismic interpretation is making effective use of the well data. Wells are often highly-deviated making it difficult to tie the well data to planar seismic sections, particularly in complex geology. A better approach facilitated by high-performance 3-D graphics hardware is to construct ribbon sections between deviated wells and "cut" the 3-D seismic cube with these ribbon sections. This provides accurate ties between the well data and the seismic data as shown in Figure 1. Seismic data can also be displayed in the well-window allowing the interpreter to view the data both globally and in detail as shown in the figure. Synthetic seismograms could also be displayed to aid in the interpretation.

In order to generate these integrated 3-D seismic and well data displays, the well deviation survey must be converted from depth to time. Strictly speaking, a complete model is required to do this correctly. In practice, it seems reasonable to use sonic data, vertical seismic profiles, etc to achieve an approximate depth-to-time conversion. Once the well data is displayed in time in the well window together with the 3-D seismic data, the well data can be interactively adjusted so that geological markers match seismic markers. This effectively edits the velocity distribution which can be evaluated for consistency. Having accurately tied the 3-D seismic data and well data together at the wells, we can start from seismic markers at the wells and interpret events on the ribbon sections. This interpretation involves identifying both faults and horizons. When faults have been identified in wells, this can be used as important information for locating faults accurately on the connected ribbon sections.

Having defined and edited the time-to-depth relationship for each of the wells, we can approximately convert each ribbon section from time to depth in several ways. The time-to-depth relationship implies a velocity which can be used to convert the complete section. Where the section is defined in terms of correlatable layers (as we will discuss), the velocity distribution can be extracted from the interpolated distribution for each layer. Each of the sections used in the seismic interpretation could then be mapped to depth with significantly less effort than inverting the entire seismic cube. Interactive editing of the velocity distribution would allow the interpreter to modify and presumably improve the time-to-depth conversion.

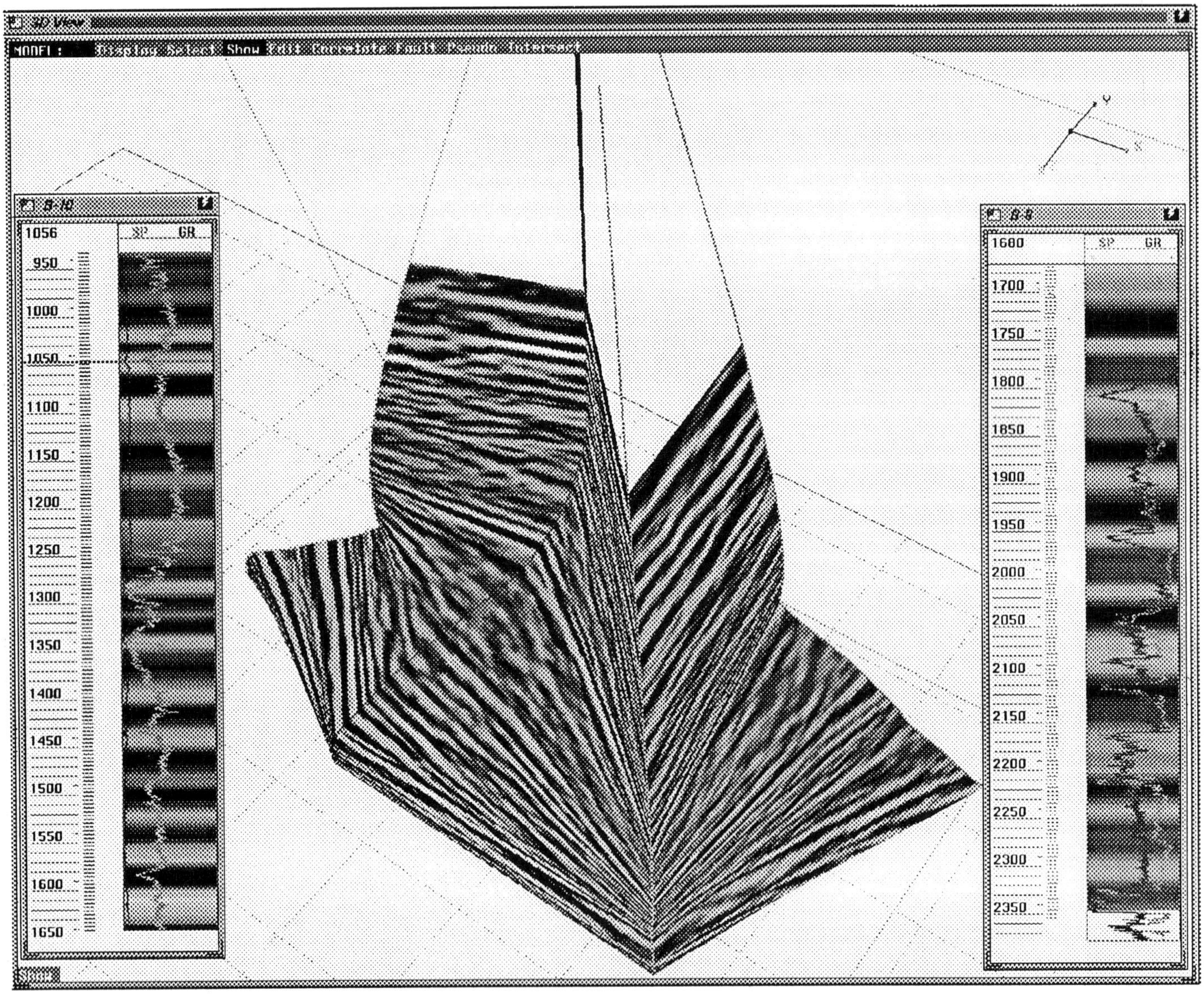

Fig. 1. 3-D seismic data display on ribbon sections between wells.

INTEGRATED PETROPHYSICAL INTERPRETATION

The prototype interpretation system allows the display of log and zone information in the well windows as was shown in Figure 1. Additional displays which would significantly improve integration and interpretation include lithological zones, core data, dipmeter data, production test data, biostratigraphic information, and general annotation. With these displays, the petrophysicist, petrologist, and geologist can develop consistent and detailed interpretation of rock structure, lithology, stratigraphy, and properties.

Computer-processed interpretation of petrophysical data is typically done on a well-by-well basis with the same procedure being applied to the entire well. From the processed logs a manual or automatic interpretation of lithology is typically performed. Stratigraphic zones are then identified and correlations made. At this point, the computer-processed petrophysical interpretation should be repeated, not on a well-by-well basis, but on a stratigraphic unit-by-unit basis. This approach should provide a more accurate "model-based" petrophysical interpretation. The improved lithological interpretation should in principle improve the geophysical and geological interpretation of the model.

INTEGRATED GEOLOGICAL INTERPRETATION

The geophysicist and the petrophysicist define the "framework" for the reservoir model. The geophysicist interprets the major structural and stratigraphic interfaces which can be identified from the seismic data. The petrophysicist can determine lithology and stratigraphy at the wells. The geologist is left to fill in the details. Regardless of the amount of seismic and well data available, the geologist must make heavy use of his knowledge of depositional processes to render any kind of detailed geological model. It is often the case that many different "realizations" are possible given the same data. As new data comes in, the geological model is often drastically affected. These observations emphasize the need for easy-to-use 3-D modeling tools for the geologist which allow him to quickly construct and edit a model which remains consistent with the data he has available.

Figure 2 shows a salt dome overlain by layers which have been faulted by the rising dome. This model uses 3-D parametric curves and surfaces which can be easily manipulated and modified. When a surface is edited, it can be recomputed and redisplayed in real time. This model was built with very little effort and again emphasizes the performance and interactivity which can be achieved with current 3-D graphics hardware.

Having defined the geometry of the model, the geologist must assist in the determination of property distributions within each unit which has been identified. Here again the data density is very low relative to the density desired. Furthermore, property measurements from core, logs, and seismic data are often suspect for a number of reasons. Permeability in particular is very difficult to derive from conventional data sources.

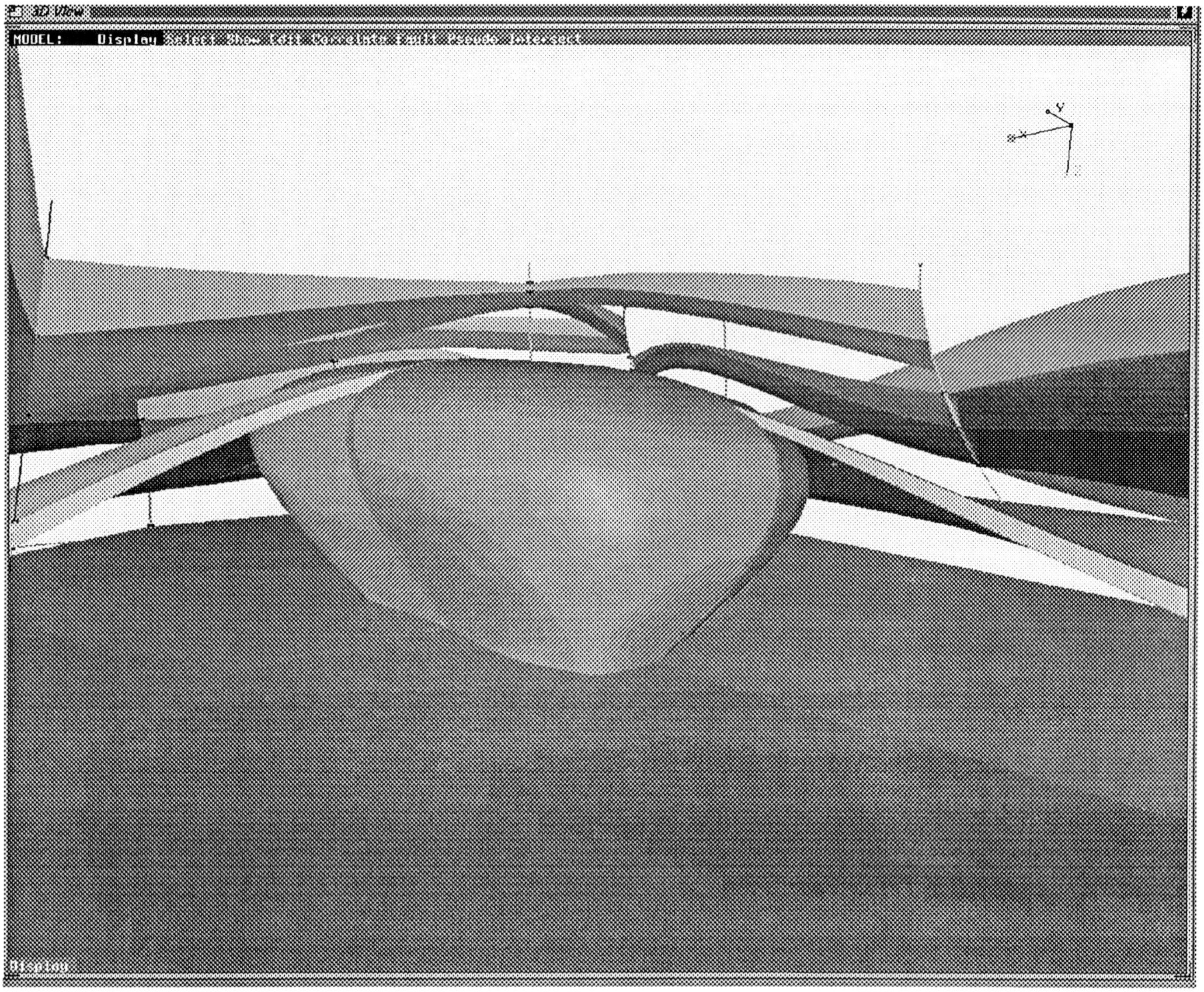

Fig. 2. Salt dome overlain by layers which have been faulted by the rising dome.

The property determination procedure begins with a display of the property interpolated over the units defined as is shown in Figure 3. The geologist often has a model for porosity and permeability distributions within each depositional type: fining upward with decreasing porosity and permeability for channels, for example. The available data and resulting interpolation should be evaluated in light of the appropriate model and interactively adjusted as necessary. Here again a number of "realizations" are possible and the tools should make the rendering of a number of realizations efficient.

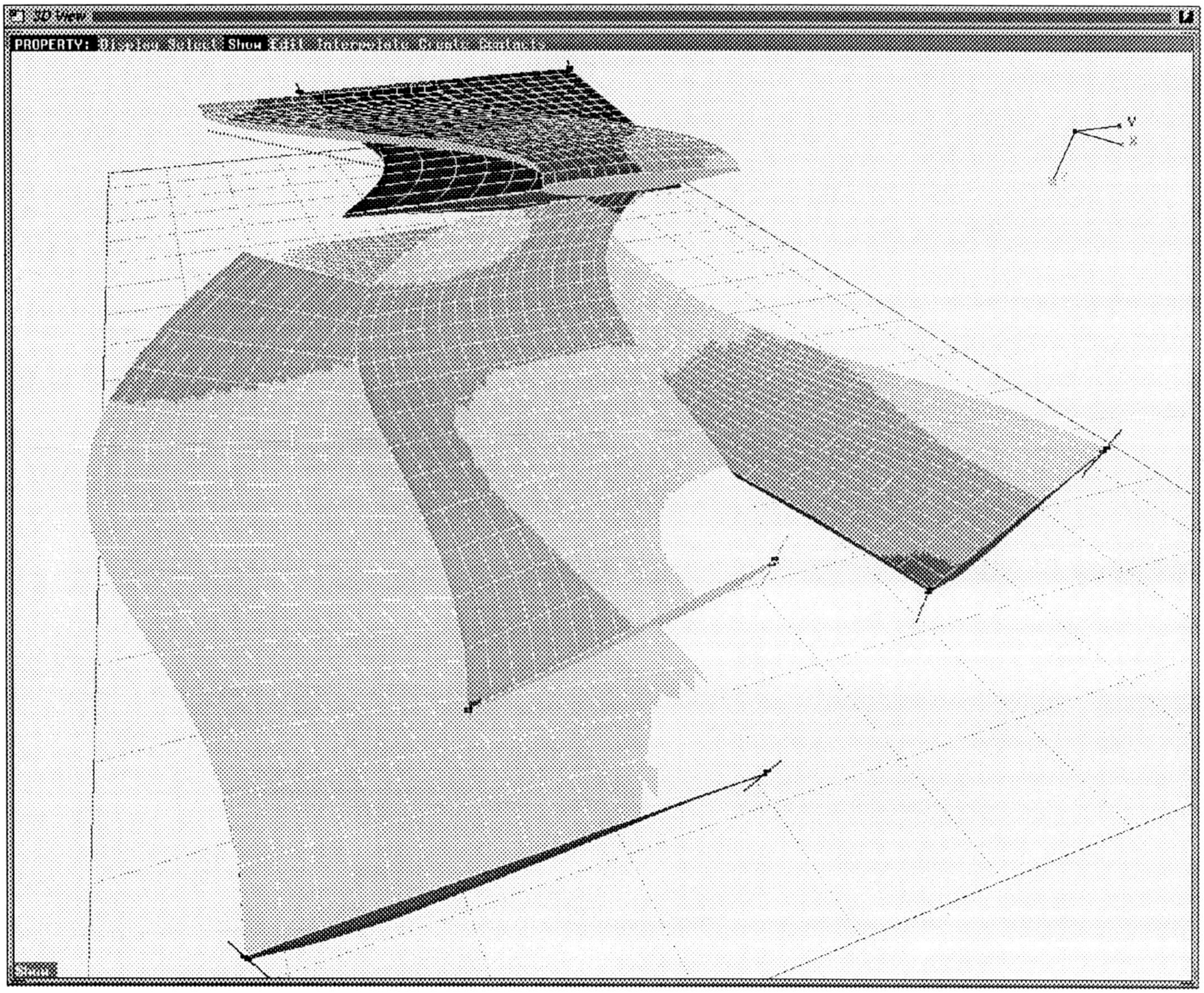

Fig. 3. Braided-channel sequence with interpolated permeability displayed.

INTEGRATED ENGINEERING INTERPRETATION

The model initially defined by the geophysicist and petrophysicist, and "filled-in" by the geologist arrives at the reservoir engineer. His responsibility is to identify the major "flow units" controlling the "hydraulics" of the reservoir and simplify the geometry of the model as much as possible while retaining its essential topology. Given this model his task is to determine the fluids-in-place, the reserves (how much can be recovered), and the producibility (how fast can it be recovered).

Conventionally, reservoir engineering models are almost always represented as layer-cakes. Producing horizons are typically modeled as single continuous layers when in fact they are a complex, partially disconnected network of flow units. This oversimplification of the geology often results in significant overestimates of reserves and producibility. It also is the major reason for lack of integration between the geologist and engineer. Once the engineer has converted the geological model to layers of grid blocks and subsequently modified grid-block properties in the reservoir history-matching process, the relationship between the engineering model and the geological model is often completely lost.

As was shown in Figure 3, a series of braided channels can be gridded for subsequent input into a reservoir simulation model. This model preserves the topology of the reservoir and yet is practical for reservoir simulation. As the geometry of the model is edited, the grid is automatically recomputed and redisplayed in real time.

The results of simulations run by the reservoir engineer can be easily shown as part of the unit display. The analysis of the results often leads to significant modifications of the geometry and/or property distributions. By preserving a single consistent model of the reservoir, the engineer can effectively communicate his model modification recommendations to the other interpretation team members and work in collaboration with them to define a new, consistent and presumably more accurate model.

An important task of the engineer is to determine the volume of hydrocarbons in place. For the modeling approach we have been describing, given the geometry of each unit, its associated porosity distribution, the geometry of the fluid contacts and the capillary pressure curves, the fluids-in-place for each unit can be computed. This is an improvement over the conventional approach in which fluids-in-place for the entire reservoir as a whole are computed. Computing on a unit basis allows the engineer to associate recovery factors with each unit depending on the production facilities associated with the unit, etc.

MODEL CONSTRUCTION AND EDITING TOOLS

The system we have been describing (IREX) is being developed by an industry consortium. It runs on the Silicon Graphics series of computers and makes efficient use of their high-performance three-dimensional graphics capabilities. Curve and surface definitions utilize parametric representation allowing for multivalued surfaces such as folds and reverse-thrust faults.

In the IREX prototype development, focus has been on evaluating interactive 3-D modeling functions. In addition, a high-performance flexible data management system has been developed which provides methods for allocating, storing, retrieving, and deleting data objects.

The graphics interface consists of five basic window types:

3-D. A 3-D view of the grid which shows an outline of a 3-D seismic data volume, wells (if highlighted), sections, maps, and parametric patches; the 3-D window is where all editing other than normal well definition and initial zoning takes place.

Map View. A map view of the model; well x-y values are defined in this view; the top window is also used for highlighting/unhighlighting wells and sections.

Well. A scalable and scrollable display of the well TVD, zoning, log curves, and amplitude extracted from the 3-D seismic.

Color Definition A window for selecting and editing color spectra associated with seismic, wells, zones, stratigraphy, and lithology.

Message. A window for data input and diagnostic message output.

The current system has the following basic model construction and editing tools:

Data loading. 3-D seismic data, well deviation surveys, well logs, and maps can be loaded from ASCII files.

Data management. The model and associated data can be saved at any time in a compact binary representation which can be subsequently reloaded.

Seismic display. Seismic amplitude data from a 3-D data set can be displayed on any arbitrary section as a series of Gouraud-shaded triangles offering smooth color interpolation without display artifacts.

Pseudo well definition. The 3-D path of a pseudo well is defined using the well window to pick time or depth and the map view window to pick x and y.

Zone definition. Individual well zones are interactively defined and edited in the well window. Once defined, zones can be edited in the 3-D window also.

Fault definition. Fault lines (both normal and reverse) can be defined. When added, up- and down-thrown block zones are automatically generated. The user can edit the displacement, slope, and curvature at the fault. Correlations can then be made along the fault surface on both the up- and down-thrown blocks. Faults crossing faults can be easily modeled with the system.

Correlation. In the 3-D window or in two well windows, the interpreter picks two zones in different wells belonging to the same stratigraphic or structural unit. Edges consisting of one or more parametric cubic curves are drawn between the respective tops and bottoms of the two zones.

Edge editing. Each parametric curve on an edge is defined by two end-points and two "handles" which control slope and curvature; the user can pick and move the points and/or handles.

Pseudo wells. Additional detail is added to a section between wells by specifying the path of a pseudo well on the section. Correlations crossed by the pseudo wells are automatically added to the pseudo well description as zones. Additional zoning can then be added to the pseudo wells.

Path editing. The paths of pseudo wells and fault lines defined on sections can be edited; a fault can be easily moved anywhere on a section for example.

Subunit definition. The geometry of uncorrelatable subunits can be specified by defining pseudo wells on existing sections, adding additional zones on the pseudo wells, and then correlating between these zones.

Fault surface definition. A series of fault lines can be connected by correlating zones on both sides of the fault surface. The cross-sections of units crossing the fault surface can be described on both sides of the fault. Faults crossing faults can be described by inserting fault lines on both sides of the fault surfaces with the appropriate vertical and lateral displacement.

Vertex definitions. A vertex can be inserted anywhere on an existing parametric curve, effectively breaking it into two curves each of which can be further edited.

Surface patch definitions. Three or four vertices can be selected on the same surface and a parametric patch or series of patches will be constructed. The patch is then computed along with normals so that a lighted display of the surface can be made.

Correlation display. Correlations on sections between wells can be shown as both wireframe or solid color fill; the latter uses a lighting model which gives the eye important visual cues to surface shape.

Surface display. The parametric patch surfaces can be shown as lighted and smooth-shaded surfaces.

As the model is defined and edited, an associated model topology is constructed and modified. Correlations and patches are formed into layers. Layers are collected into stratigraphic units as well as being organized into structural blocks. The interpreter can then display and edit the model by editing layer geometries and properties on a unit basis or on a block basis.

AN INTEGRATED INTERPRETATION CASE STUDY

Figure 4 shows an actual data set consisting of 3-D seismic data, three interpreted maps, and eleven wells with associated log curves. The maps are time maps interpreted on a seismic workstation. Velocity data was used to convert the well deviation surveys to time.

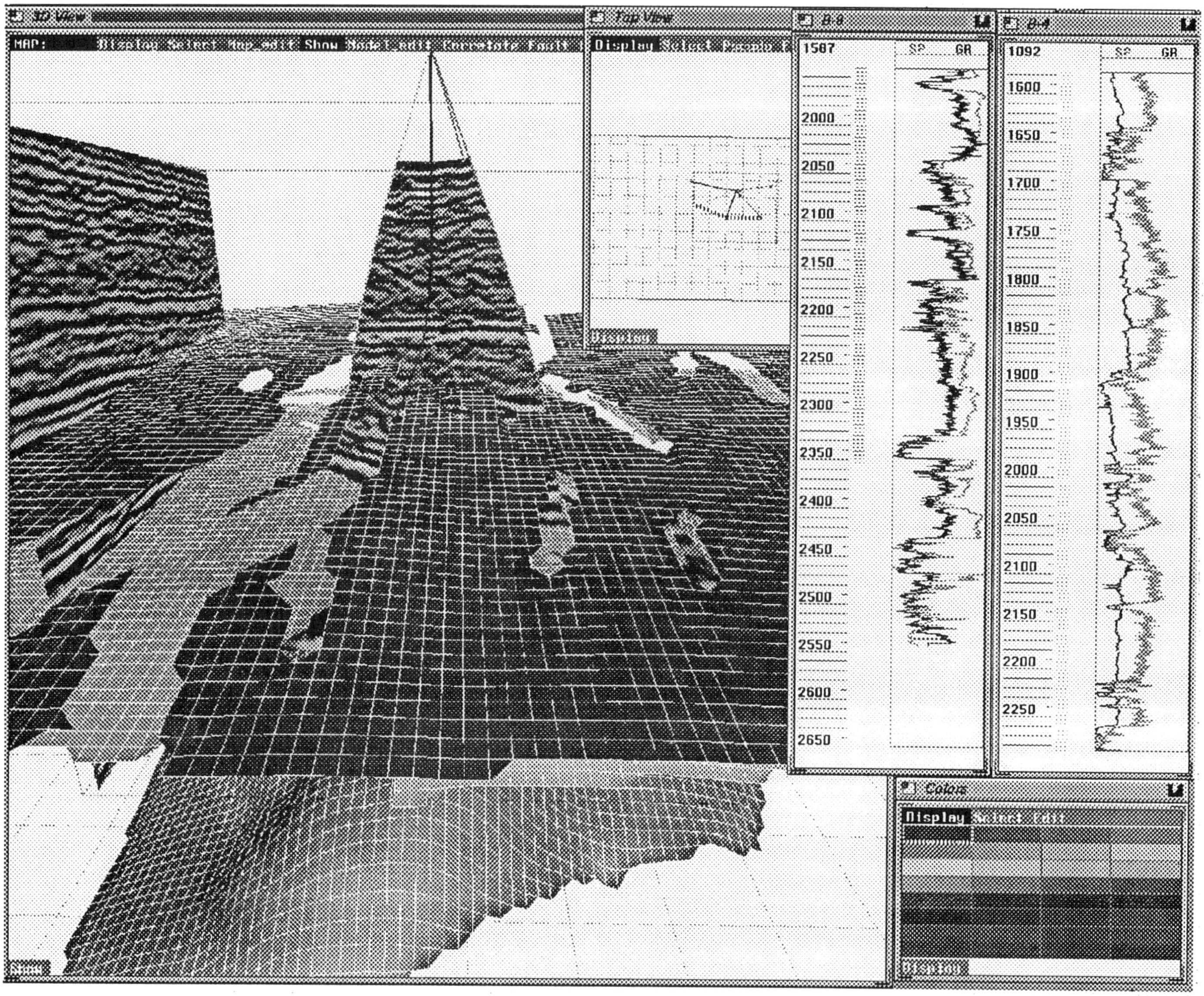

Fig. 4. Actual interpretation data set: 3-D seismic displayed as vertical, time-slice, and ribbon sections; well surveys and log data; interpolated maps.

Picking three or four wells, three- or four-sided patches are automatically cut from the intersected maps. Zones and correlations between mapped surfaces are automatically generated. Parametric patches approximating the map surface patches are also automatically computed. Ribbon sections are automatically built allowing the map patch edges to be displayed against the seismic data to validate the map interpretation. Where faults have been defined on the maps, the interpreter is required to build faults on the ribbon sections. The up- and down-thrown sides of the faults are then correlated thus defining individual fault blocks.

The zones that have been automatically built can be compared with the well log data again to validate the interpretation and can be edited as necessary. The log curves associated with each zone can be examined for subzones which can then be interactively defined and correlated. This would allow the geologist to construct a series of braided channel-sands within a bounding channel-sand unit as was shown in Figure 3.

In this case study, a log-derived permeability curve has been computed arbitrarily from the gamma-ray curve. In practice, the interpreter should be able to use a suite of property generation functions as well as generate customized functions. In Figure 5, the permeability curves for two of the wells are shown in the associated well windows along with the zone averages. The interpolated permeability map for the top layer is shown in the 3-D window. The user can interactively edit the permeability curves or the zone-averaged values. The interpolation is then automatically recomputed and redisplayed. Since we are currently using a simple interpolation scheme, the update and redisplay occurs virtually instantaneously.

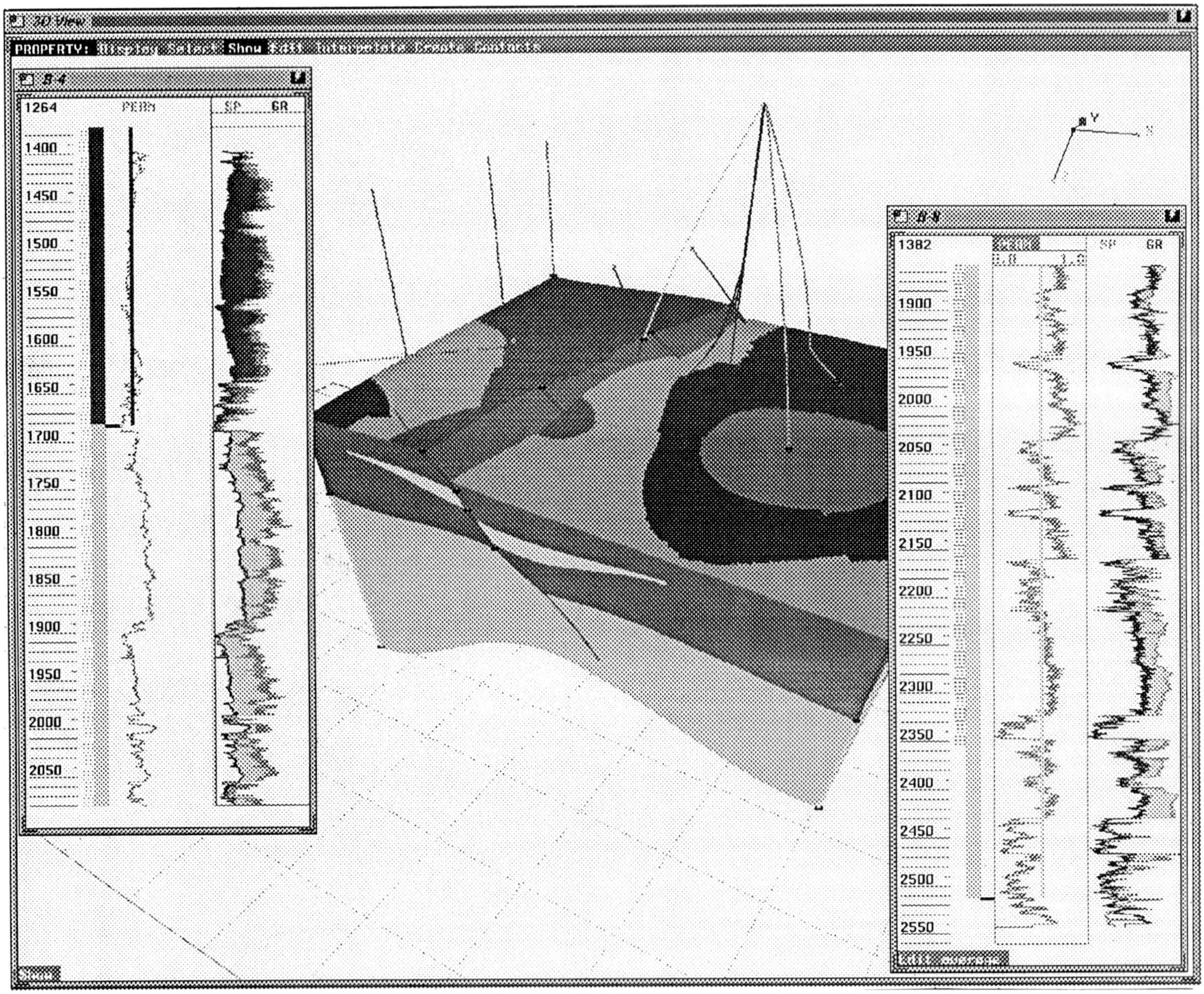

Fig. 5. Permeability display: curves and zone-averages in well window; interpolated distribution in 3-D window.

SUMMARY AND CONCLUSIONS

Integrated interpretation requires the representation of the reservoir as a single consistent model which can be displayed and interactively edited in three dimensions. To be practical, high-performance 3-D graphics systems and the requisite support software are essential. Integrated data display and editing tools allow interpreters from the various disciplines to view and edit the model from their unique perspectives. Facilitating the display of geophysical and petrophysical data in conjunction with reservoir simulation output as an example would allow the interpretation team to more effectively evaluate and ameliorate model inconsistencies.

Surfaces represented by detailed grids derived from maps and seismic interpretation are used to form the "framework" for the geological interpretation. Parametric curves and surfaces provide a compact and editable representation of the geometry allowing the interpreter to quickly modify the model when new data becomes available or alternative "realizations" are required. The model topology in terms of layers, units, and blocks is automatically generated and maintained.

The oil industry is facing more difficult interpretation problems as reservoir development focuses on more complex geological environments. The development of interactive 3-D interpretation systems used by multidisciplinary interpretation teams should help offset the increased uncertainties and risks.

Basin History and Diagenetic Evolution of the Missisauga Formation, South Sable Basin—Offshore Nova Scotia—A Petrographic and Petrophysical Approach

Mark B. Thomas

The Lower Cretaceous Missisauga sandstones, which form part of a major southward-prograding delta system, are the principal reservoir units in the South Sable Basin (Figure 1). In Missisauga sandstones of similar lithology and burial depth, porosity values are highly variable ranging from 12% in the Glenelg wells (3500 m) to over 20% in North Triumph (3850 m). Variations in porosity and the occurrence of porous and permeable rocks at depth can result from geopressuring, early hydrocarbon migration into the reservoir, and/or-the dissolution of a specific component or components in the rock.

To investigate porosity occurrence as a function of geopressuring, pressure-depth plots, as well as plots of shale interval transit time (ITT) versus depth were constructed. From these plots it is evident that hydropressuring exists from surface to the base of the Missisauga Formation. This study excludes deeper geopressured regions such as the Venture gas field. Petrographic examination of the Missisauga sandstones revealed that in the poorer quality reservoirs much of the primary porosity was destroyed by the combined effects of compaction and silica cementation, while some secondary porosity resulted from the dissolution of-feldspars and rock fragments. In the more porous and permeable sandstones the pore system resulted from the dissolution of a pervasive, texturally early ferroan calcite cement. The creation of porous reservoir units through a similar dissolution process has also been observed in sandstones from other areas such as the North Sea, offshore Newfoundland and Mesozoic reservoirs in Alberta. The porosity in these reservoirs depends not only on depth but also on the timing and amount of cementation as well as the degree of dissolution.

While the ITT plots indicated that the stratigraphic section to the base of Missisauga sandstones is hydropressured, the plots show two distinct trends (Figure 2). In those wells along the present day shelf edge, the slope of the plots is much steeper (i.e. little change in ITT with depth) than in those wells further back on the shelf. When comparing the slopes on the ITT plots to regional heat flow trends, it is observed that those wells with the steeper gradients along the shelf edge also have the highest heat flow. In addition, the most significant gas finds in the hydropressured portion of the basin have been along the shelf edge suggesting that gas generation is possibly a localized event related to heat flow.

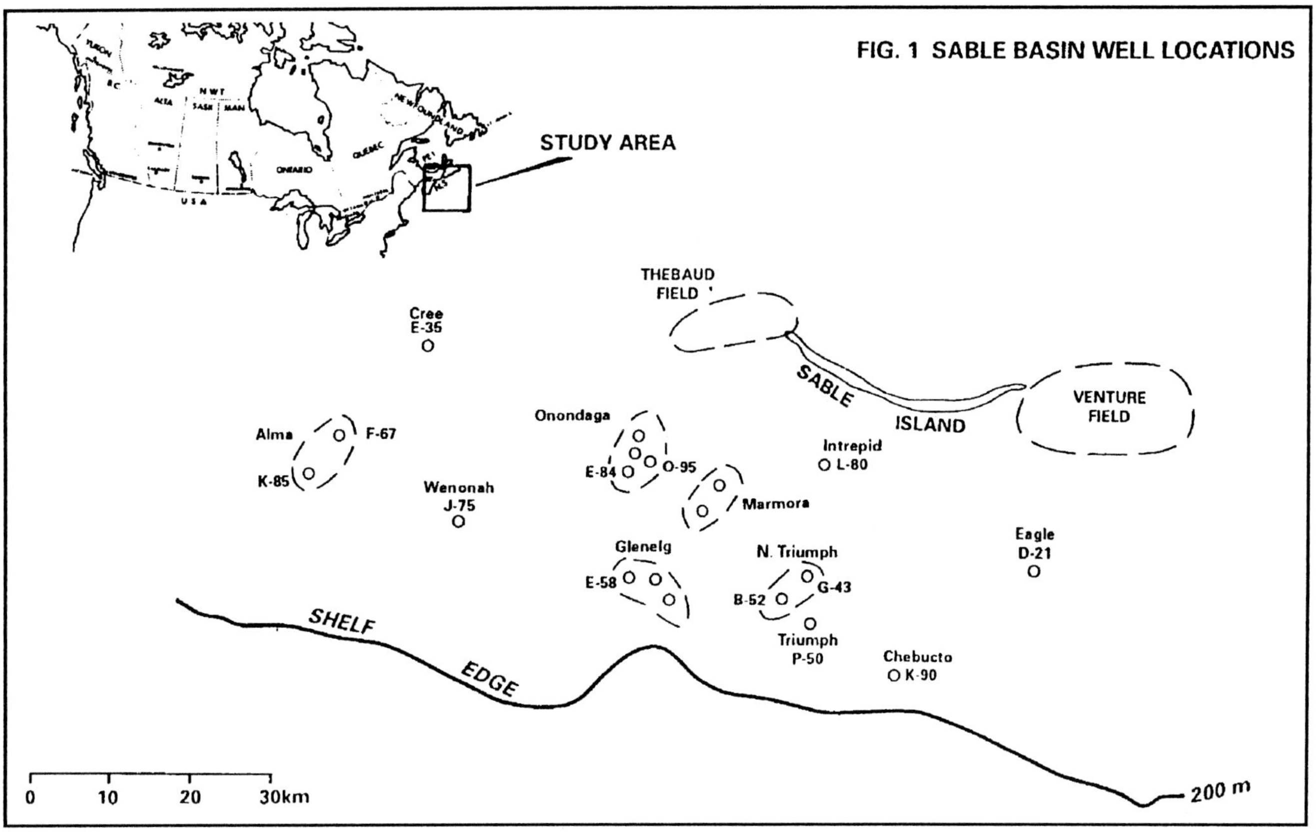

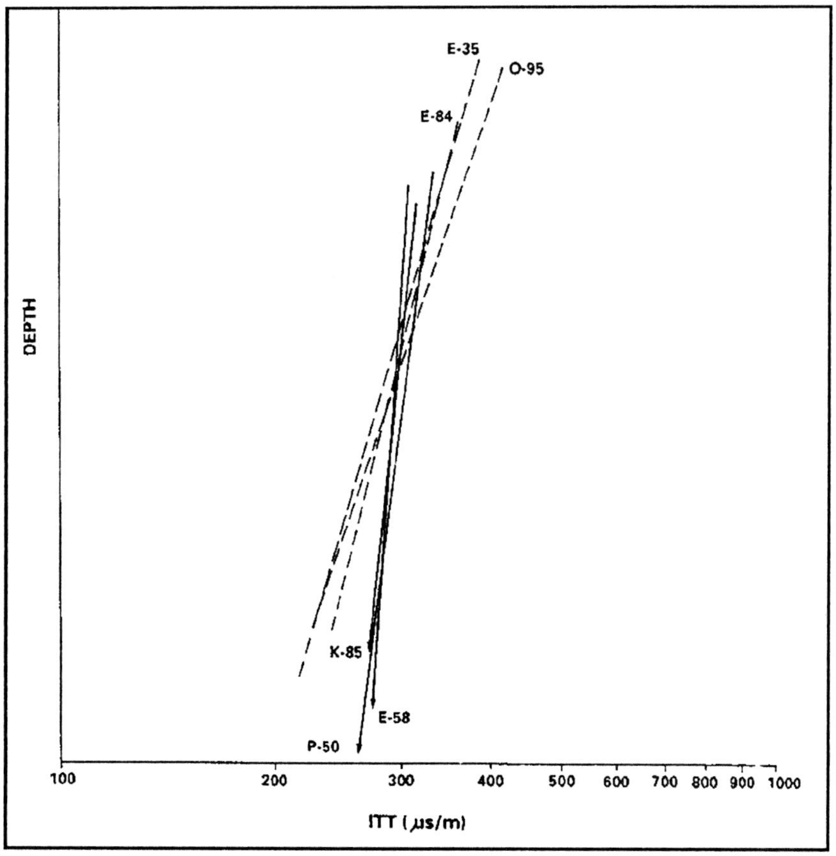

Figure 2: Trends of interval transit time vs depth

Resolution of Carbonate Reservoirs at Depth—An Integrated Approach to Lower Development Risk, Smackover Trend, Alabama

Timothy J. Petta and J. Gregory Bryant

Porous Smackover algal mounds associated with granite knobs produce oil throughout southwestern Alabama. A detailed gravity survey over Vocation Field demonstrates lateral density contrasts between reservoir and non-reservoir rocks can be measured even when the reservoir depth exceeds 14,000 feet (4270 m). A similar gravity survey combined with standard seismic and subsurface interpretation methods could have drastically increased the success rate and lowered development costs.

Vocation Field will produce nearly 5 million barrels of oil from high energy algal mound and ooid facies deposited on the flanks of the basement knob. Porosity is dolomite intercrystalline, moldic and interparticle. Most of the porosity was inherited from the original high energy rock fabric. Porosity values as high as 18 percent occur within the pay section. Effective porosity exceeds 250 feet on the flanks of the structure. The trap was not filled to the spill point. Water provides the reservoir drive and recovery is expected to exceed 40 percent of the oil in place. Eight of the sixteen wells have produced.

The basement high is a slight anomaly on regional gravity and aeromagnetic surveys. Seismic lines over the feature indicate structural closure but near the crest the Smackover and basement events are indistinguishable. The first wells were drilled near the crest and were successful discovery and confirmation wells. Only six of the remaining fourteen wells drilled were successful, a disappointing rate of 42 percent. Some of these wells were structurally low and wet, others structurally high and tight. Seismic data provided no clue as to whether porosity was present or the distribution of porosity.

A gravity survey with 1/8 mile (200 m) station spacing was collected during 1988. Second derivative maps of a filtered residual data set clearly indicate lateral density contrasts that are apparently caused by porosity variations in the Smackover carbonates. Using the detailed survey and the other data sets, operators could have eliminated four to six dry holes causing the development success rate to increase from forty two to seventy five percent.

THE FRACTURED RESERVOIR— A NEW DEFINITION (extended abstract)

Peter E. Gretener
Department of Geology & Geophysics
University of Calgary
Calgary, AB, T2N 1N4

Fractured reservoirs have been recognized for over 40 years. Generally they have been, and still are regarded as a special type of reservoir with only a limited contribution to the total oil/gas reserves. DRUMMOND (1964) estimated that only about 13% of all reserves are contained in such reservoirs. It is the purpose of this paper to challenge this view. In order to do this it is necessary to take a closer look at the different types of reservoirs that might be called fractured traps.

The fractures that are of particular interest are the joints observed in almost all sediments. SCHEIDEGGER (1979) states: "Joints are ubiquitous phenomena". Thus their actual presence, with highly variable spacing, is not a problem as any field geologist can confirm. Generally the orientation of these joints is vertical, or high angle, to bedding.

In order to be effective fluid conduits such joints must be OPEN. For reservoir depletion this means that they must be open NOW. One may question this additional requirement and argue that a joint once formed will never again close perfectly. Just how do we justify the necessity for gaping joints? Surely there are laboratory experiments that point in that direction. However, the simplest proof can be found in the process of artificial fracing, a daily operation in the oil patch for over 40 years, and a most successful one at that. The artificial fracture is produced in a positive (compressive) virgin stress environment by an excess pressure (injection pressure) within the fracture. Upon release of this excess pressure such a fracture will close in the presence of compressive natural rock stresses. The closure is prevented by the propping agent and the fact remains that the fracing process (a misnomer) is really a mini-dike forming operation. No better proof exists that demonstrates that only open fractures are efficient fluid conduits.

Natural open fractures cannot exist in the presence of a positive virgin stress situation. Total stress relief of at least one principal stress is a fundamental requirement for open fractures to occur. As will be shown later such stress relief is possible and the former contention of the rock mechanicists that accepts only positive rock stresses below the earth's surface requires revision. Generally this old view sees the principal rock stresses as oriented vertically and horizontally (ANDERSON, 1951) with all axes positive.

Open fractures can be viewed (in a somewhat idealistic way) as plane, flat cracks. The porosity (Δk_f) and permeability (Δk_f) additions due to such a feature can be readily assessed. Table I shows the permeability increase due to either one, two, or three sets of fractures. The fracture spacing is 1 m (3.3 ft) and the fracture width is 0.1 mm (0.004 in). It is immediately apparent that the effect on permeability is phenomenal. On the other hand it is easily seen that the porosity addition is negligible and varies from Δn_f= 0.0001 to 0.0003 (0.01 to 0.03%) depending on the number of fracture sets. This phenomenon was well known to the early writers on the subject such as REGAN (1953) or HUBBERT & WILLIS (1955). Today virtually nobody associates joints with permeability but rather with porosity. The term fracture porosity is widely used and even found in the "Glossary of Geology" (GARY ET AL., 1972, p.276). This indicates that at this time there is a basic misconception prevailing in regards to the role of fractures in hydrocarbon reservoirs.

Depending on the nature of the stress relief different types of fractured reservoirs can be distinguished. Please note that stress relief must always be complete, and at least one principal stress must be reduced to a zero value. Tensile stresses do not exist in large rock masses since such rock has essentially no tensile strength. It is, however, possible to relieve one, two, or all three principal stresses. This results in three types of fractured reservoirs which are listed in Table II. Under 3-D stress relief the rock is completely relaxed and essentially under stress conditions as found in an outcrop at a roadside. Randomly oriented, open fractures are possible and in fact likely to exist. The one and only way that seems to lead to such a situation in the subsurface is a case of severe geopressures where the overburden is in a state of floatation (TERZAGHI, 1960, p.92). Under these conditions the fracture permeability is enormous and nondirectional (isotropic). The reservoir behaves like a storage tank. In a somewhat less extreme case two principal stresses may be relieved as for instance in a doubly plunging anticline or a domal structure. Such a situation allows two perpendicular (or acute) sets of open fractures to occur, most likely with a vertical orientation. Under such a condition the moving fluid can still reach any point in the reservoir by a continuous fracture path. A weak anisotropy exists such that the vertical permeability is theoretically twice that in the horizontal direction. Considering the high values of fracture permeabilities this difference is irrelevant and such a reservoir still behaves much like a storage tank. Provided the well intersects this fracture system (which is not guaranteed for large fracture spacing) this Type 2 reservoir acts much like the Type 1. It seems thus justified to group both Type 1 and Type 2 reservoirs into Class 1.

There is little doubt that the Type 3 fractured reservoir is the most common one. It requires only 1-D stress relief, a form of stress reduction which is most easily achieved. Bending or buckling are the processes that most conveniently achieve a local stress reduction in one direction (see e.g. FRIEDMAN & SOWERS, 1970, p.488, Fig. 11). Figure 1 shows that this type of stress relief (1-D for one dimensional) can be expected to occur in all geological provinces, be they tectonically relaxed, compressed, or extended. Possible open fractures are not restricted to highly special geological conditions. It must be noted that this type of stress relief, contrary to that produced by geopressures (Type 1), is not universal but both laterally and vertically restricted. It will not prevail throughout the reservoir but be restricted to certain compartments. One must also be fully aware of the fact that a single fracture set only enhances the permeability in the fracture plane (e.g. the x-z plane) and will have no effect on the direction perpendicular to this plane (the y-direction). This type of stress relief is THE major contributor to reservoir heterogeneity and anisotropy.

The reader is further reminded that the ideal models of the Engineers, the sugar cube (Type 1), the square column (Type 2), and the sheet stack (Type 3), all assume infinite length of the fractures, insofar as they traverse the model from end to end (see e.g. REISS, 1980, p.20, fig.12). For 3-D and 2-D stress relief this assumption is not critical since the fluid will always be able to transfer between fractures through crosscutting joints. This is not the case for the single joint set. Even in the x-z plane the fluid will have to move from fracture to fracture via the matrix, that is to say the limited length of the fractures becomes crucial. The fracture flow path becomes a pipeline with constrictions (GRETENER & ZHAO, 1988, p.67, Fig.3). As a result the contrast between fracture permeability and matrix permeability is only moderate and rapid flushing of the reservoirs through fracture-only pathways is not possible.

There can be no doubt that these unconventionally fractured reservoirs are far more prevalent than the officially recognized Type 1 and Type 2 reservoirs. As previously mentioned DRUMMOND (1964) estimated that 13% of the known reserves are contained in Class 1 reservoirs. In view that Class 2 reservoirs have a much higher probability of occurrence it is not overly optimistic to state that more than half of the reserves will occur in this class of reservoir. Why have we not recognized this fact for such a long time? There are a number of reasons besides the already mentioned preconceived notion of the rockmechanics community. One is that vertical wells do not investigate vertical fractures. Intersection may or may not take place depending on fracture spacing. Mechanical logs in general respond poorly to vertical fractures and in particular to near-misses. Also for Class 2 reservoirs the permeability contrast between fractures and matrix is only mild and pressure testing may not readily reveal a two-system permeability.

What then is the evidence for the existence of such fractures. Reservoir heterogeneity is today an accepted concept. W. FISHER (pers. com., 1986) commenting on the successful infill drilling program in Texas made the following statement: "In some old and severely drawn-down fields infill wells encountered the initial reservoir pressure". This is tantamount to saying that after 30 years parts of the field didn't know they had been discovered. Amen! This is clear evidence for much higher compartmentalization of reservoirs than was initially anticipated. In addition it has been frequently reported that neighboring wells, in similar structural positions show highly differing production rates for no apparent reasons. A case in point are the reservoirs in the Asmari limestone in the Zagros Mountains of Iran. Varying amounts of fracturing are the most likely cause for the differing production indices of wells in this area (MCQUILLAN, 1985).

The localized occurrence of open joints in most reservoirs is a major factor for both reservoir heterogeneity and anisotropy. The success of the new completion technique of HORIZONTAL DRILLING is largely due to this factor. It is gratifying to see that most authors describing the advantages of horizontal wells mention first and foremost the intersection of possible open cracks. It seems that the industry is coming around to accept the idea that such joints may be more prevalent than previously thought. To drill into a compartment with such cracks improves the near-well permeability in this section and increases the production rate as well as the ultimate recovery. The general recognition of natural reservoir heterogeneity is the silent acceptance of what might be called NATURAL FORMATION DAMAGE. If the concept of open fractures is indeed valid it becomes mandatory to drill the horizontal section of a well parallel to the least principal stress. However, azimuthal orientation is really not very critical—contrary to vertical positioning. In the case of doubt, which more often than not will exist, one may drill at 45° and only increase the fracture spacing by 40%.

A good hydrocarbon reservoir must have both storage volume (porosity) and continuity (permeability). This is shown in Figure 2. If there Is no volume and no continuity we in fact do not have a reservoir but more likely a seal. The result is D & A. There may be rocks with enormous volume but no continuity. A useful conceptual endmember is the "Big Eye Swiss (Cheese) Reservoir". Much empty space but no way to either fill or deplete it. A natural analogue may be the vesicular basalt but then this is rarely encountered by the oil explorationist. A purely fractured reservoir is also not likely to be a great success. Even hairline cracks provide tremendous permeability but only a pitifully small volume. An exception may be the case of 3-D stress relief by extreme geopressures. Provided the reservoir is large and the condition widespread one can expect vertical dilatancy to occur. The result is opening of the fractures to the point where they themselves will provide an adequate storage volume. Such a case seems to exist in the Alborz structure in Central Iran (GRETENER, 1982). A reservoir of this type behaves like storage tank. The geopressures are indicative of complete isolation and a good gas drive is required for commercial production. The case of a 2-D stress relief reservoir is the Beaver River Field in northern British Columbia. This field produced large amounts of gas but declined rapidly after little more than two years production (DAVIDSON and SNOWDON, 1978; GRETENER, 1986; GRETENER & ZHAO, 1988). The reason: significant lateral extension is unlikely even under complete stress relief.

Absolutely tremendous reservoirs are created when early fractured carbonate rocks are subjected to emergence. Under these conditions karst formation takes place and the early fracture system is enlarged by the solution process. Initial fracture systems not only become vuggy they may in fact be enlarged into caverns. It is well known that these systems of underground "storage tanks" are well connected and lined up in preferential directions. The mapped systems in the Pahasapa limestone in the Jewel and Wind Caves in the Black Hills bear witness to this. One such field is the Yates Field in Texas (CRAIG, 1987).

The fields listed in Figure 2 are endmembers. Most existing fields fall somewhere in between. No good commercial field, however, is without volume and continuity. The message in this paper is simply that in many cases continuity is enhanced by the presence of fractures. It is those fractures that affect the fluid flow in the reservoir during the extremely dynamic period of depletion. In many cases it is the presence of those fractures that renders the field commercial. It is important to realize that these fractures are not omnipresent and it will be necessary to find their habitat and bring them in direct contact with the well.

Detailed geological (-engineering) outcrop studies will provide the general picture. BULK RESERVOIR SURVEYING by such geophysical techniques as: 3-D time slicing, tomography, large step-out VSP's, and the new shear wave splitting technology will delineate the subsurface in situ situation. Logging, a digital sampling process, will be found inadequate since it can never properly assess the key question of lateral continuity.

Post Scriptum: A Note on Fracture Dynamics

Fractures do open and close with time. Once open they may also be healed and become impermeable. It is for this reason that geological field studies should also look past the reservoir proper for evidence of RECENT TECTONIC MOVEMENTS. Such movements may be subtle yet sufficient to re-open fracture pathways (pers. com. E. Kempter, 1990). There is ample evidence that fluid motion in the subsurface is episodic indicating that fluid flow paths are opening and closing. The best such indication is the process of salt solution (e.g. OLIVER & COWPER, 1983). Hydrocarbon reservoir depletion is instantaneous from a geological point of view and it is only the present condition that is relevant.

REFERENCES

ANDERSON, E.M., 1951, The Dynamics of Faulting and Dyke Formation; Oliver and Boyd, London, 206 p.

CRAIG, kD.H., 1987, Yates Field, West Texas: Thousands of Caves, Millions of Years, Billions of Barrels; Bull. AAPG, 71/11, p.1439.

DAVIDSON, D.A. and SNOWDON, D.M., 1978, Beaver River Middle Devonian Carbonate Performance Review of a High-Relief, Fractured Gas Reservoir with Water Influx/ JPT 30/12, p.1672-1678.

DRUMMOND, J.M., 1964, An Appraisal of Fracture Porosity; Bull. Can. Petr. Geol., 12/2, p.226-245.

FRIEDMAN, M. and SOWERS, G.M., 1970, Petrofabrics: A Critical Review; Can. J. Earth Sc., 7/2, p. 477-497.

GARY, M., McAFEE, JR., R., and WOLF, C.L., 1972, Glossary of Geology; AGI, 805 p.

GRETENER, P.E., 1987, The National Conference on Earth Science, Banff, September 15-19, 1986: "Geology and Reservoir Management"; Bull. Can. Soc. Petr. Geol., 35/3, p.375-379.

GRETENER, P.E., 1986, Macrofractures and Fluid Flow; Bull. Swiss Assoc. Petr. Geol. & Eng., 53/123, p.59-74.

GRETENER, P.E., 1982, Another Look at Alborz Nr. 5 in Central Iran; Bull. Swiss. Assoc. Petr. Geol. & Eng., 48/114, p.1-8.

GRETENER, P.E. and ZHAO, Guanjun, 1988, More on Fractured Hydrocarbon Reservoirs; Swiss Assoc. Petr. Geol. & Eng., 54/127, p.61-82.

HUBBERT, M.K. and WILLIS, D.G., 1955, Important Fractured Reservoirs in the United States; Proc. 4th World Petr. Congr., Sec. 1, p.57-81.

MCQUILLAN, H., 1985, Fracture-Controlled Production from the Oligo-Miocene Asmari Formation in Gachsaran and Bibi Hakimeh Fields, Southwest Iran; in Carbonate Petroleum Reservoirs, Roehl & Choquette eds., Springer, p. 511-523.

OLIVER, T.A. and COWPER, N.W. 1983, Wabamun Salt Removal and Shale Compaction Effects, Rumsey Area, Alberta; Bull. Can. Soc. Petr. Geol., 31/3, p.161-168.

REGAN, L.J., 1953, Fractured Shale Reservoirs in California; Bull. AAPG, 37/2, p.201-216.

REISS, L.H., 1980, The Reservoir Engineering Aspects of Fractured Formations; Gulf Publ. Co., Houston, TX, 108 p.

SCHEIDEGGER, A.E., 1979, The Enigma of Jointing; Rivista Italiana di Geofisica, Vol.5, p.1-4.

TERZAGHI, K., 1950, Mechanism of Landslides; in Berkey Volume, GSA, p.83-123.

TABLE 1

Permeabilities for a given fracture spacing (f_s) and fracture width (f_w) using the flat crack model.

f_s = 1 m (3.3 ft)		f_w = 0.1mm (0.004 in)
SUGAR CUBE (3-D STRESS RELIEF)	:	$\Delta k_{fx} = \Delta k_{fy} = \Delta k_{fz}$ = 160 mD
SQUARE COLUMN (2-D STRESS RELIEF)	:	$\Delta k_{fx} = \Delta k_{fy} = 1/2\Delta k_{fz}$ = 80 mD
SHEET STACK (1-D STRESS RELIEF)	:	$\Delta k_{fx} = \Delta k_{fz}$ = 80 mD Δk_{fy} = 0 mD

TABLE II

Type of fractured reservoirs and their characteristics

FRACTURED RESERVOIRS			
TYPE	#1	#2	#3
CLASS	#1*	#1*	#2**
MODEL	SUGAR CUBE[1]	SQUARE COLUMNS[1]	SHEET STACK[1]
STRESS RELIEF	3-D	2-D	1-D
EXAMPLE	ALBORZ, IRAN[2]	BEAVER RIVER, B.C.[3]	"ALMOST UNIVERSAL"
FRAC. VOL.	possibly large	small	small
FRAC. PERM.	isotropic	mildly anisotropic	strongly anisotropic
kf/km	very large	very large	moderate

* conventional fractured reservoirs, officially recognized

** unconventional fractured reservoirs, common but not recognized

[1]REISS, 1980, p.20 [2]GRETENER, 1982 [3]DAVIDSON & SNOWDON, 1978

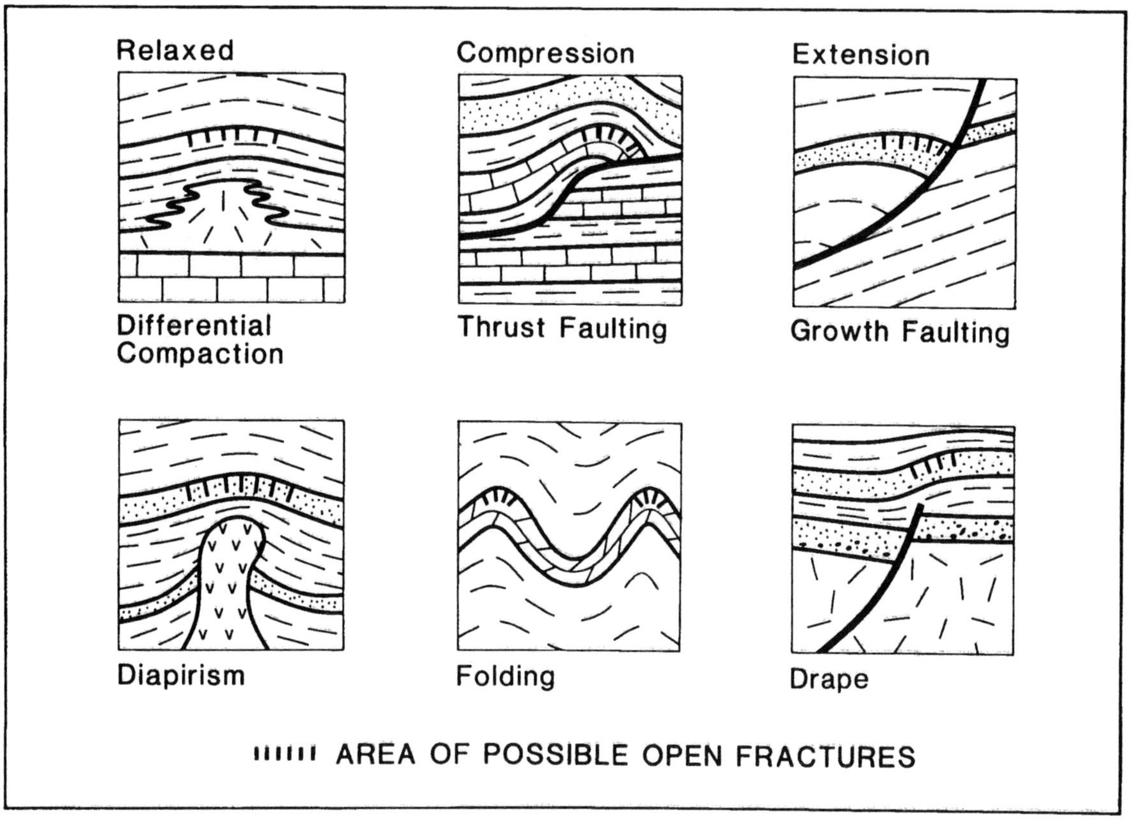

Figure 1—The possibility of open fractures is universal and not restricted to one or a few special geological situations. (from GRETENER & ZHAO, 1988)

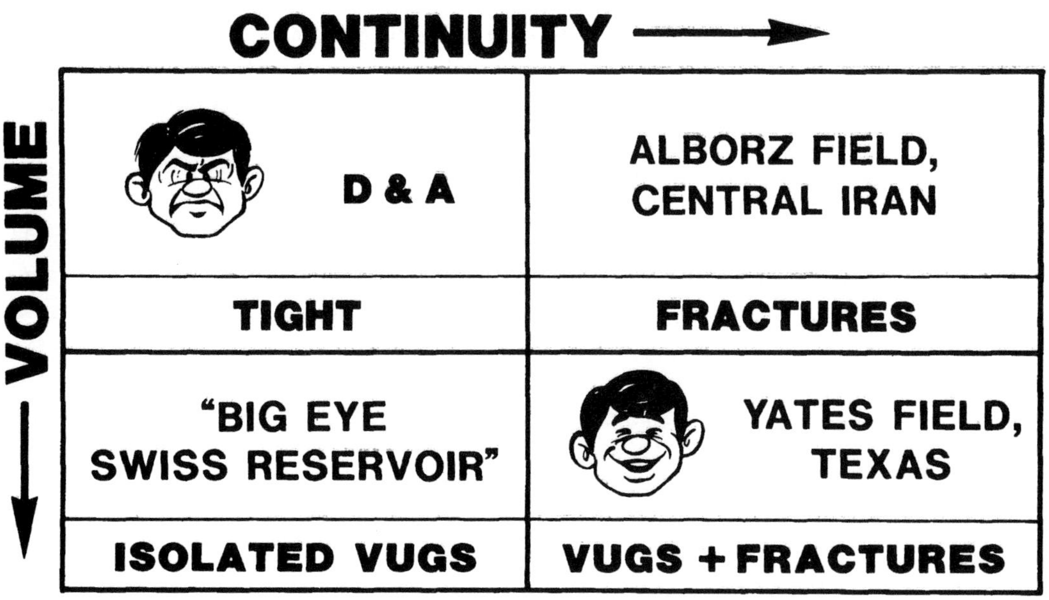

Figure 2—The spectrum of hydrocarbon reservoirs: From rags to riches. (From GRETENER, 1987)

Determination of Hydraulic Fracture Orientation in the Kuparuk River Field, Alaska

J. S. Blundell and R. J. Hallam

The Kuparuk River field on the North Slope of Alaska (Figure 1) is the second largest producing oil field in North America. Currently the production from the Lower Cretaceous Kuparuk formation exceeds 300,000 barrels of oil per day under waterflooding. This reservoir is overlain by the Colville, West Sak, and Ugnu reservoirs which contain an estimated 20 billion barrels of oil in place. These formations are unconsolidated, have widely varying fluid and rock properties, and will require waterflooding and enhanced oil recovery processes. Development options for all of these reservoirs include hydraulic fracturing of the injection and production wells; hence, characterization of the in-situ stress field is critical for optimizing field performance and recovery.

The regional crustal stress field on the North Slope is extensional with maximum principal horizontal stress oriented northwest-southeast. Previous work on fracture direction in Kuparuk, however, indicated that the in-situ stress field was more complex in its orientation. The Kuparuk reservoir occurs within a broad northwest to southeast-trending anticline which plunges to the southeast. Normal fault patterns within the Kuparuk River field show two dominant strike trends: (1) northwest-southeast and (2) north-south.

In this study the hydraulic fracture direction, at both shallow and deep horizons, was determined by integrating geologic, engineering, petrophysical and geophysical data. Dipmeter logs were processed and interpreted to determine wellbore breakout directions for both shallow and deep horizons. Formation microscanner (FMS) images were used to discriminate between incipient wellbore breakout zones and mechanical fracturing from drilling. Hydraulic fracture screenout data were correlated with wellbore azimuth, and a tiltmeter survey was conducted in the northern sector of the field to further refine predictions of hydraulic fracture direction.

The results of all these analyses indicated a variation in fracture direction across the Kuparuk River field. In the northeast sector of the field a northwest-southeast (NW-SE) fracture direction (Figure 2) was determined from dipmeter data in ten wells and confirmed with fracture screen-out data (Figure 3). This direction, which is N39°W with a standard deviation of 26° azimuth, is consistent with the regional orientation of maximum principal stress across the North Slope of Alaska. In the northern sector of the field, a NNW-SSE fracture direction was determined based on dipmeter data from 2 wells, FMS log images (Figure 4), a tiltmeter survey and fracture screen-out data. This direction, which is N10°W ± 10°, is aligned subparallel to the strike orientation of the major faults in that area. In the southwest sector of the field, the fracture direction was determined to be northeast-southwest (NE-SW). The N46°E ± 21° fracture direction is interpreted to result from the influence of a paleostress regime linked to early deposition of the Kuparuk formation in a broad, northeasterly-trending structural trough that existed prior to structural inversion.

The results of this study have indicated three distinct fracture directions for the Kuparuk River field. This variation is interpreted to result from the influence of faults and reservoir structure on a near isotropic horizontal stress field. Recognition of the change in orientation of hydraulic fractures across the anticlinal crest of the Kuparuk structure was accomplished by integrating reservoir engineering data, geophysics, geology, petrophysics and rock mechanics. This paper has shown how interdisciplinary work can lead to improved reservoir description and modelling which will ultimately improve resource exploitation.

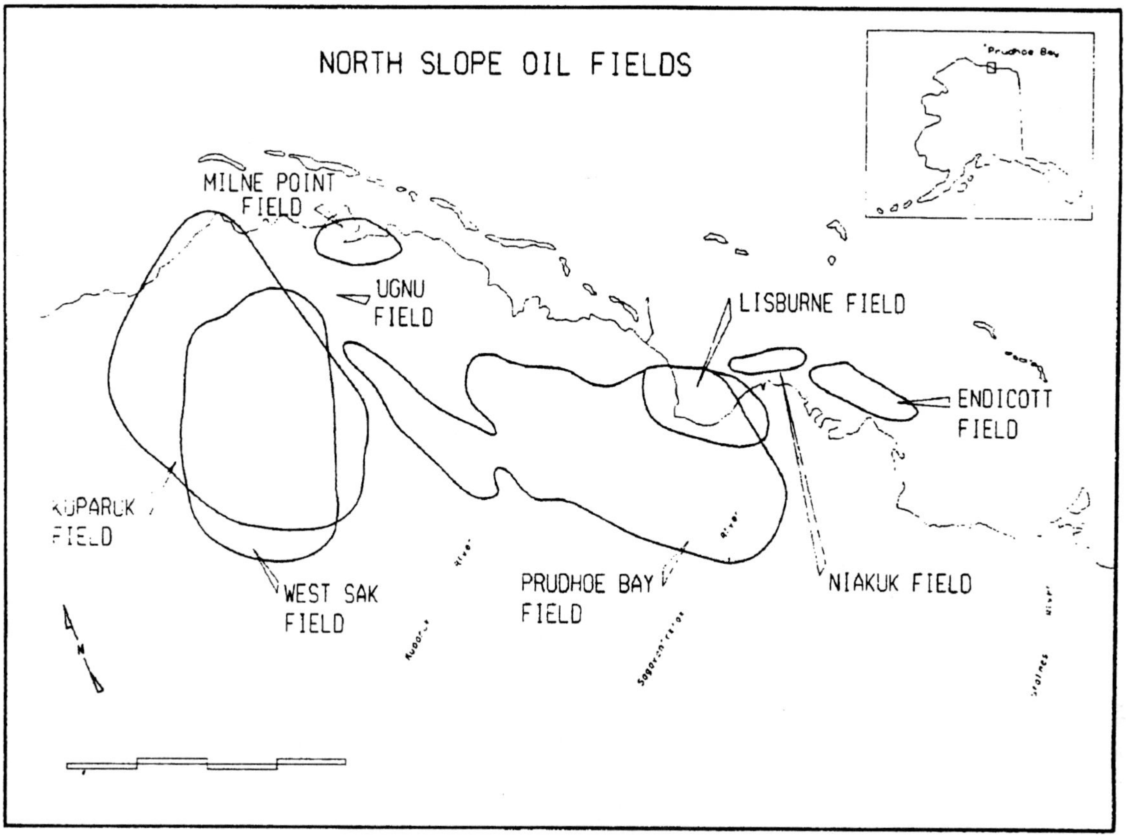

FIGURE 1. LOCATION MAP SHOWING THE OILFIELDS OF THE NORTH SLOPE, ALASKA

AN EVOLVING DESCRIPTION OF A FRACTURED CARBONATE RESERVOIR: THE LISBURNE FIELD, PRUDHOE BAY, ALASKA

R. A. Missman
J. Jameson

ABSTRACT

This paper traces the evolution in models of the Lisburne carbonate reservoir from early wildcat days, through the decision to fund primary development, to waterflood assessment. This work is an integration of geologic and reservoir studies based on nearly ten years of extensive data collection (well test, core, log, pressure and field performance data). Early reservoir descriptions assumed uniform enhancement of continuous pay layers resulting from small-scale fractures. Additional recovery over solution gas drive was expected by gravity drainage and waterflood. Integration of performance and geological data after field start-up reveal greater heterogeneity than originally thought: faults and irregularly distributed fractures dominate performance; much of the porosity is ineffective; and the producing oil-water contact (OWC) varies by as much as 300 ft [90 m]. The evolution in simulation techniques reflects the evolution in geologic models toward increased recognition of heterogeneities.

The complexity of the Lisburne reservoir is a result of faulting, two periods of uplift and erosion, and three distinct phases of porosity development. Particularly important among these diagenetic events is the youngest phase of porosity development, which post-dates significant faulting. As a result, many faults are open and conductive. An additional complexity is a high permeability layer created by a laterally extensive set of small open fractures beneath an unconformity at the top of the reservoir.

Initial development was based on models with limited data which understated the impact of these heterogeneities. In retrospect, several important clues and their implications were not fully appreciated. Despite extensive data collection, the unpredictable influence of fractures still creates uncertainties in reservoir description (e.g. well performance cannot be accurately predicted pre-drill). Recommendations for evaluating potentially complex carbonates include: give full weight to known uncertainties, conduct inter-well tests early in the evaluation, and minimize risk by proceeding gradually with field development.

INTRODUCTION

The Lisburne field is a complex, faulted carbonate reservoir within the Lisburne Participating Area in the Prudhoe Bay Unit on the north slope of Alaska[1] (Fig. 1). The principal productive area lies underneath and adjacent to Prudhoe Bay and is located 100 to 150 ft [30 to 46 m] below the main Ivishak reservoir at Prudhoe Bay (Fig. 2). The development area is 10 mi by 6 mi [16 km × 10 km] and as of January 1991, contains 78 wells, most on 320 acre [130 ha] spacing, from 6 drillsites. Other attributes are listed in Fig. 3. The field is very young with only 4 years of production history. Production has averaged 40 to 50 MBOPD [6400 to 8000 m^3/d]; cumulative production is 68 MMstb [10.8 × 10^6 m^3] including natural gas liquids. Equity in the field is Arco 40%, Exxon 40% and BP Exploration (BPX) 20%.

Lisburne was the principal objective[1] for the 1967/68 Prudhoe Bay State #1 well, which was the discovery well for both the Prudhoe Bay and Lisburne fields. As one would expect, Lisburne reservoir descriptions of the three co-owners have evolved considerably from the early assessment days.

Current published reserve estimates are significantly less than early estimates.[1,2] The factors leading to this change are addressed in this paper. In summary, the principal reservoir unit, the Wahoo formation, is a typical cyclic Pennsylvanian shelfal carbonate[3] (Fig. 2). The complexities in Lisburne reservoir behavior arise from the interactions between the matrix, major faults and a horizon of small-scale fractures beneath an unconformity. The current

References and illustrations at end of paper

geological model consists of four major elements shown in Fig. 5: (1) depositional layering of pay units; (2) major faults and megafractures (faults below seismic resolution); (3) the Subunconformity Alteration Zone, or SAZ; and (4) small-scale, stratabound fractures. The change between early and present models (Figs. 4,5) yielded large changes in field performance expectations. Likewise, performance strongly guided model evolution.

This paper compares "early" and "present" reservoir descriptions. "Early" refers to the assessment stage (including exploration, delineation and the 1984 project funding); "present" refers to the current status after three years of production history and completion of a significant portion of the currently envisioned development drilling. This paper is a snapshot early in the life of a complex reservoir, not the final answer. Our purpose is to record how we became aware of reservoir complexities and their implications, and to indicate that much still remains unknown.

Comparison of our early and present models reveals the importance of understanding the full implications of all the available data. We tended to make conclusions on the basis of data that made 'sense' and to discount those observations that did not fit into a recognizable pattern. In retrospect, we found that several data in the later category provided indications of heterogeneities that were to emerge during field development. In general, discounted Lisburne data can be classified into one of three categories.

1. INCONCLUSIVE DATA: Inconclusive drillstem tests (DSTs) were not used in defining the oil-water contacts, yet they are currently interpreted to indicate that a thick transition zone is present. The full implications of drilling breaks (indicators of thief zones, conductive faults) were only appreciated after extensive testing and field performance.

2. EXOTIC DATA: Rare events were not given enough weight or risk. For example, megafractures were perceived to be rare, yet they are now seen to have a large effect on reservoir performance.

3. ANOMALOUS DATA: Data which did not fit into existing paradigms were not used. The large difference noted between air permeability and overburden permeability to liquids was considered suspect but has a large impact on pay definition. A bore hole gravimeter survey indicated the possibility of large open faults. The implications of these data were not appreciated until reservoir performance offered additional insights.

STRATIGRAPHY / CYCLICITY / FACIES

The Lisburne group is an extensive Carboniferous age carbonate, approximately 2000 ft [600 m] thick. Biostratigraphy, seismic expression and cyclicity form the basis for formal subdivision into the Wahoo and Alapah formations (Fig. 2). The Wahoo formation is further subdivided into seven zones, with the top four zones further subdivided into 16 depositional cycles which provide a precise stratigraphy throughout the productive area. Cycle boundaries, which are often thin shales, are considered as time lines. Individual cycles shallow upwards and are typically 30 to 50 ft [10 to 15 m] thick.

The upper Wahoo productive interval consists of a mosaic of open-to-restricted lagoonal and shoal facies originally deposited on a southward dipping ramp.[3,4] 80% of the succession consists of grain-supported textures with relatively little mud. The Wahoo is a highly stratified carbonate. The productive interval is 30% dolomite, 55% limestone and 15% non-carbonate (shale, silt, chert). The style of Wahoo cyclicity and sedimentation is typical of middle Pennsylvanian shelf carbonates.

STRUCTURE

The Prudhoe Bay structure consists of a faulted monocline defined by: (1) a major boundary fault on the north, downthrown to the north (Fig. 1), (2) the Pre-Upper Permian Unconformity (PUPU) which follows a gentle 1 to 2° structural dip to the southwest, and (3) the Lower Cretaceous Unconformity (LCU) to the east. The PUPU extends over most of the field and is generally stratiform but has as much as 150 ft [46 m] of local relief. The LCU is much younger and steeply truncates the eastern side of the field.

Seismic data reveal two major fault trends. The dominant sense is east-west while a minor set trends NNW-SSE. Detailed interpretation of 3D seismic reveals many more faults than shown in Fig. 1. While there is considerable variation across the field, the small faults have a roughly orthogonal pattern with a fault spacing of ≈1000 feet [300 m].

POROSITY DEVELOPMENT

EARLY POROSITY MODELS recognized that all Lisburne porosity was diagenetic, related to either dissolution or dolomitization. While the presence of a variety of intercrystalline, moldic and micromoldic pores was recognized, models for the origins of porosity were less certain. Lacking genetic models, work focused on relating porosity to facies trends, which could be mapped. This method was fairly accurate for predicting pay thickness but, in retrospect, was an inadequate measure of productivity.

Early work employed a 5% porosity cutoff and mapped facies for each depositional cycle. This produced a high net-to-gross and thick pay layers that extended across most of the field (Fig. 6). Layers of up to 50 ft [15 m] allowed the possibility of gravity drainage. The potential for waterflooding was suggested by high mapped lateral and vertical continuity.

PRESENT POROSITY MODEL—The present model, developed from a more extensive data base, recognizes three distinct episodes of dissolution and/or dolomitization[3,4] (Fig. 7). The earliest stage of porosity is moldic, associated with localized subaerial exposure during deposition in the Pennsylvanian. The second event is Permo-Triassic age, related to burial expulsion of fluids from the overlying Kavik shale to form ferroan dolomites. The third and youngest event is Cretaceous-Tertiary moldic porosity which postdates major faulting.

The present model of porosity formation had several major impacts on reservoir description. Porosity and pore type distribution were changed to reflect the unconformities as conduits for burial diagenetic fluids. One key piece of data that indicated more than one stage of porosity formation was microprobe data which showed a gradient of decreasing iron content in dolomites away from the PUPU. These data complemented results of mapping hot (radioactive) dolomites, which also decreased in intensity away from the PUPU. In addition, petrographic and isotopic data revealed dolomitization post-dated extensive compaction and pressure solution, thus suggesting a shallow burial origin and not a near surface process.

Geochemical observations together with petrographic and performance data were also key in identification of the SAZ, a zone of high permeability underlying and associated with the PUPU. Early production profiles had revealed a disproportionate amount of production from the uppermost 10 to 20 ft [3 to 6 m] of the Wahoo. Later interference tests[5] also showed rapid interwell communication attributed to some sort of fracture system. Geochemical data revealed a unique horizon of ankerite (high-iron dolomite) along the PUPU surface. Associated with ankerite replacement and overgrowth was complete dissolution of fracture filling calcite. Thus the SAZ resulted from a combination of two major factors: (1) fractures with greater vertical extent than elsewhere,[6] and (2) nearly complete dissolution of fracture filling calcite.[3] When these geological observations were combined with performance data (to be discussed later), the SAZ became a major element of the reservoir description.

Harris and Kendall[7] were the first to suggest that some moldic porosity was of late, relatively deep burial origin based largely on cross cutting relationships between fracture filling cements, intercrystalline porosity in dolomite, and moldic pores. This hypothesis was later confirmed by Jameson[4] with core data that was unavailable early in the field history. Late porosity development is attributed to fluid flow along the LCU and/or large E-W trending faults. As a result of this work, the eastern and northern parts of the field have been identified as having significant limestone microporosity in Zone 5 and more pore volume than originally mapped.

Data on the timing and stages of porosity development also provided an explanation of why Lisburne faults were open. The youngest stage of porosity development is particularly important because it is younger than most faulting and provides a mechanism for dissolving out fractures and faults. Petrographic data on cored faults clearly indicated that faults were open due to partial dissolution rather than incomplete cementation.[4] These observations complement the performance data, which together reveal that many faults provide vertical and lateral communication across many parts of the field.

In summary, early models of porosity adopted an empirical, facies-based approach because the geological models were not precise enough for mapping purposes. The microscopic size of most Lisburne pores made it difficult to find evidence for their origin. Present models are based on a considerably larger core sample, with extensive geochemical and performance data. Breakthroughs in our understanding came from combining matrix, fracture and performance data into a unified model.

FLUID CONTACTS

Early DSTs yielded a reasonably consistent gas/oil contact (GOC) but considerable variation in OWC (Figs. 8,9). Many tests were inconclusive, i.e. fluid identification was uncertain—little more than the drillpipe cushion was produced back. It was unclear whether this was due to lack of acidization, lack of means to artificially lift, or simply extremely tight rock. In retrospect, it is likely that some intervals would have flowed had they been acidized. Because of variable results from DSTs in the exploration wells[8] (Figs. 10,11), original estimates of the OWC ranged from –8975 to –9300 ft subsea [–2736 to –2835 m]. Despite an extensive logging suite at each well, it proved difficult to determine fluid contacts and water saturation from openhole logs. Little or no correlation was found between water saturations from openhole logs and oil-based mud cores. This was attributed to two factors: (1) induction logs used with oil-based mud had poor thin-bed resolution, yet rock properties varied over inches; (2) log calculations showed that S_w varied in an unpredictable manner either due to varying R_w[9] or pore geometry effects. Furthermore, gas could not be identified on the logs; this is attributed to near-well flushing. Thus, contacts were based on fluid production in DSTs.

In preparation for funding, four delineation wells were drilled, cored, and extensively tested (L2-06, L3-08, L4-15 and L5-13, indicated by "D" in Fig. 1). The first three indicated a producing OWC of –9147 ft [–2788 m]. The fourth was upstructure; its logs and core indicated a very low OWC (≈–9300 ft [–2835 m]), but none of the prospective intervals below –9044 ft [–2757 m] would sustain flow. Tests from the downstructure delineation wells were considered to be more valid because they were conclusive and yielded a consistent contact. Log and core data were not helpful in defining the OWC. Without clear evidence to the contrary, a planar producing OWC was used: the reservoir would produce oil above –9147 ft [–2788 m], wherever the reservoir was porous and permeable.

We now realize that the apparent early consistency in OWC was because the three key delineation wells were drilled at a similar structural position and all had permeable rock near the –9147 ft [–2788 m] depth. Subsequent development drilling has shown that the oil/water transition zone originally noted below that depth now extends well above that marker in the northern regions of the field. Defining the contact is hampered by poor pay development which makes fluid collection difficult. Mobile oil indicators[10] (resistivity and electromagnetic propagation logs), improved log water saturation calculations, and well production data have shown that the base of producible oil may be as much as 300 ft [90 m] above our original contact (Fig. 11). [The lower limit of the transition in Fig 10 is figurative, for comparison with Fig. 8].

The cause for the variable OWC is not fully understood, however the effect is to reduce original oil in place (OOIP) estimates. Possible causes are tilting late in field history and variation in permeability.

In retrospect, had we been able to explain why many of the early tests were inconclusive, three important points could have been deduced: (1) a thick oil/water transition zone is present; (2) lower porosity rock is not consistently productive; and (3) a large range of flow rates reflects variable influence of fracturing. Perhaps inconclusive data should have indicated the need for additional delineation drilling. This example illustrates that the tendency to discount inconclusive data can lead to simplistic assumptions.

PAY CRITERIA

Lisburne pay cutoffs are based on log porosities calibrated with core. The porosity cutoff was originally assumed to be 5%. In contrast, cutoffs for primary depletion now range from 8% to 12%, and up to 22% for waterflood screening (to be discussed later).

The original 5% cutoff was based on analogy with the porosity cutoffs used in Permian Basin carbonates. Porosity cutoffs in those fields are commonly based on a 0.1 md permeability cutoff,[11] selected to give agreement between material balance and volumetric OOIP with 20- and 40-acre [8 to 16 ha] Permian Basin wells. It is now apparent that a higher porosity cutoff is appropriate for Lisburne.

The currently accepted porosity cutoff range of 8% to 12% is supported by material balance, capillary pressure and Formation Micro Scanner (FMS)* log interpretation[12] work. However, these are only *approximate* indicators. Production logs are consistent with, but do not confirm, the cutoff.

Unlike many sandstones, Lisburne has a large gray area between what is clearly pay and what is clearly nonpay. A pore volume versus cutoff plot (Fig. 12) shows that the effective pore volume drops by ≈30% with a 10% rather than a 5% cutoff.

Producibility of tight layers is a key question in the Lisburne. A large difference was noted six years ago between permeabilities from conventional porosity and permeability (P&P) data (air permeabilities measured at bench conditions) and very low permeabilities to fluids at reservoir overburden pressure (Fig. 13). The discrepancy was so large (up to two orders of magnitude) that the data was considered suspect. Low in-situ permeability is now seen as one of main limitations on waterflood performance. It also further accentuates the difference between matrix permeabilities and measured permeabilities from buildup and flow tests. Work of Juhasz[13] suggests that permeability reductions of this magnitude are to be expected in such low permeability rock.

Fig. 13 is a plot of conventional P&P data which were available at project funding in 1984. Permeabilities are the conventional permeabilities to air at nominal confining pressure. Note the large scatter. The high-permeability, low-porosity points result from fractures in the plug samples. The "bench" line shows our current best-fit correlation for conventional P&P data for dolomites. Other lithologies are not markedly different. The correlation lines are based on a total plug database of over 7000 observations—seven times the data shown here. The correlations exclude fracture data. The "OB" line of Fig. 13 shows air permeabilities with overburden; note that the lowest porosities have the largest change in permeability. The "Kocw" line shows the correlation for permeability to oil at connate water versus porosity. There is little difference between the curves at the upper end, but at lower porosities the permeabilities can differ by two orders of magnitude. For example, 0.1 md permeability to air at bench conditions could equate to 0.04 md for oil flow in the reservoir.

Fieldwide material balance appears too imprecise at this point to be able to define the cutoff. The uncertainty is a result of the difficulty in obtaining a single representative reservoir pressure—the high permeability SAZ may be 1000 psi [7 mPa] or more below the pressure of the tighter matrix.

Fig. 14 shows another line of evidence used to support an 8% to 12% porosity cutoff. The plot of mercury saturation vs. permeability shows how much mercury could be injected at (for example) 500 psi [3450 kPa]. Little invasion occurs below 0.1 md (air permeability, no overburden). Such rock has extremely small pore throats (≈0.1 μm). As seen in Fig. 13, 0.1 md correlates to 10% porosity. [The original 1984 cutoff of 5% was also based on 0.1 md but the correlation may have included some fractured core data].

Attempts to define pay cutoff using production logs appear to be completely infeasible. Synthetic production logs based on k/ϕ transforms of log porosity showed no measurable difference in profiles between a 1 md and a 0.1 md cutoff. Typically, k/ϕ transforms are linear correlations between $\log_{10}(k)$ and ϕ, that is, k is an exponential function of ϕ. Calculated and actual profiles are dominated by high-porosity streaks (and the matrix profile is itself dominated by megafractures or SAZ if present). Low porosity values have extremely low permeabilities with unmeasureable contribution to the flow profile. Currier[14] concluded that low permeability strata are depleted predominantly by crossflow to adjacent higher-permeability layers rather than by radial flow to the wellbore.

The concept of "pay cutoff" may be misleading in this type of rock. Perhaps a better approach would be a gradational weighting such as George and Stiles.[11]

*Mark of Schlumberger

FRACTURE SYSTEM—EARLY VIEW

EARLY CORE WORK recognized four categories of fractures: (1) seismically resolvable faults; (2) megafractures (tectonic faults below seismic resolution); (3) small, strata-bound fractures; and (4) microfractures.[15] Of these four fracture types, megafractures were recognized as the most difficult to predict because they could be found only by drilling. While the potential influence of faults and megafractures was recognized, no attempt was made to model their behavior because they were thought to be infrequent and because of a lack of data to describe them. Early fracture work included structural analysis of seismic data but most descriptive effort was focused on extensive description of small-scale fractures (micro- and stratabound fractures) in core.

Seismic analyses resulted in identification of structural 'provinces,' areas that were expected to have similar fracture properties. Outside of deformation envelopes around major faults, much of the field fell within a single, large province.

Early exploration wells had irregular core coverage, often covering only a fraction of tested intervals. Before funding, a large effort was made to describe small-scale fractures in core. Acquiring fracture data was one of the main justifications for extensive coring. The emphasis was on relating fracture attributes (spacing, orientation, etc.) to the mechanical properties of the host rock.

In our early view, virtually all pay layers were considered to be uniformly enhanced by small-scale stratabound fracture systems of variable lateral extent. This model accounted for high well productivities by pervasive networks of small-scale, poorly-connected stratabound fractures. Fieldwide density of small-scale fractures was 0.69 fractures/m; fracture porosity made up only an estimated 0.4% of OOIP. Most of the fractures observed in core are less than 1 mm wide, calcite-filled, and confined to centimeter-thick strata. Very few fractures cross stylolites, grain-size or lithological changes.

Results of fracture studies indicated that the mechanical properties of a rock exerted a strong influence on small-scale fracture frequency. However, it was also apparent that *open* small fractures were probably too infrequent to account for the observed well performance. Small fractures were ubiquitous; the problem was that 90% were filled with calcite. One attempt to reconcile high test rates (below) with our early fracture model was communication of the partially cemented fractures in three dimensions.

EARLY WELL TESTS—High initial rates from DSTs of low permeability rock provided the earliest indication that Lisburne was a fractured reservoir (e.g. Gull Island State #1 DST #2 with 3125 BOPD [497 m^3/D] from rock with a buildup permeability of 10 md). Core from L1-21, a 1978 wildcat, showed an open megafracture in an interval that produced 2959 BOPD [470 m^3/D] during a 61 hr, openhole DST from an interval with approximately 4 md buildup permeability. Core recovery in the megafracture was poor, and the importance of the megafracture was not appreciated.

T. D. Streltsova interpreted the L1-21 tests in 1981 and, consistent with the early fracture model from core, hypothesized[16] that the reservoir had a `limited connectivity' fracture system. Fig. 16 is a conceptual illustration comparing a limited connectivity fracture system in which the matrix is continuous with the conventional Warren and Root[17] model in which the fracture system is continuous. The limited connectivity fracture system was expected to act like a homogeneous matrix with enhanced permeability.[18] Fig. 17 shows the basis for the limited-connectivity concept: a continually increasing slope on a conventional Horner[19] plot indicates continually diminishing permeability as the pressure pulse investigates deeper into the reservoir. Many of the buildups (including this one) show linear flow characteristics, which could be interpreted as (1) linear flow to a planar fracture, (2) flow to an irregular fracture system, (3) a very stimulated well in a tight homogeneous matrix, or (4) a limited connectivity fracture system. We favored the last explanation based on fracture descriptions of core. In addition, multiple tests on the same interval often show decreasing permeability with increasing flow time and depth of investigation. This was seen as another indication that the fracture system had limits. Some long tests (two months) showed a roll-over at the top which was attributed to either formation of a secondary gas cap or stresssensitive fractures (i.e. drawdown would cause fractures to close near the well, yet remain open deeper in the reservoir away from the drawdown).

DELINEATION DRILLING: Entirely by coincidence, the first of four delineation wells (L2-06, Fig. 1) cored down the trace of an un-mapped megafracture. Core recovery was nearly complete except at one point where over 7000 bbls [1100 m^3] of mud were lost. Core data from this well provided some astoundingly high smallscale fracture frequencies which have not been observed in any other core since. Over half the middle Wahoo in this well had open fracture densities of 10 to 50 fractures/m. Some fieldwide averages of fracture frequency included this outlier and thus provided a misleading impression of the importance of small-scale fractures.

Two-month flow tests were conducted in each of the delineation wells. The L4-15 two-month test yielded up to 1500 ppm H_2S in the produced gas, which was a considerable surprise since it was the first appreciable H_2S detected in the reservoir (and the first appreciable concentration in any commercially significant North Slope reservoir). This indicated that the southeastern region (L4) is poorly connected to the rest of the reservoir. It was consistent with the limited connectivity idea.

Understandably, the early model was based on readily quantifiable data (e.g. seismic and core)*. Core data had particular utility and could be statistically manipulated. Although megafractures were found in two of the first ten wells, they were considered rare and their influence unknown—they conceivably could act simply as barriers or as hydraulic fractures rather than conduits. Difficulties in modeling open faults and megafractures may have delayed recognition of their significance.

FRACTURE SYSTEM—CURRENT VIEW

Drilling, coring and production behavior to date have added a new feature, the SAZ, to the four features recognized in early work (faults, megafractures, stratabound fractures and microfractures). The SAZ, faults and megafractures are now considered to dominate reservoir performance where present. Stratabound fractures and microfractures are believed to have only a minor role in field performance.

Adding to Currier's description,[14] we now think that the flow path to the well can have up to four steps (Fig. 18): (1) tight strata crossflow to adjacent higher permeability strata; (2) the higher permeability strata feed into open, conductive faults and megafractures; (3) the faults and megafractures feed vertically into the SAZ; and (4) within the SAZ, horizontal flow to the well. Faults and megafractures can also provide communication in areas of the field where the SAZ is absent. Comparisons of actual flow profiles from production logs with expected flow profile from core and log data indicate that fieldwide, only ≈30% flows into the well directly from the matrix.

In early wells completed to the top of the Wahoo, production profiles typically showed that most of the influx (and most of the gas) came in at the very top. Fig. 15 shows porosity and perforations vs. depth, and three production profiles. Initially, it was hypothesized that gas segregated in megafractures and percolated to the top of the reservoir before being produced out the perforations. Our present model places more dominance on high permeability within the SAZ, which is fed by fractures draining underlying support pay. Note the dominance of the SAZ in the earliest of the three production profiles in Fig. 15. The SAZS becomes even more dominant with time as tighter layers decline. In this pre-startup example (no drainage limits), the SAZ was fed by crossflow from a large supporting OOIP and so depleted more slowly than lower permeability matrix that fed directly to the well by radial flow.

EARLY TESTS OF FRACTURES—We were concerned that the two-month and extended tests might prove deceptively optimistic because these wells effectively had unbounded drainage radii. That is, the tests were only measuring properties of the fracture system. A multiwell test was proposed to identify the contribution of the matrix and to prepare for a pilot waterflood test. Five wells were drilled and cored at L2 on nominal 80 acre [32 ha] well spacing (pattern `A' in Fig. 21). Interference tests[5] showed rapid inter-well communication of as little as a few hours with 1500 to 2000 ft [460 to 600 m] well spacing. Streltsova[20] (Table 1) concluded that the southern half of the pilot area had significantly higher storativity than the northern half, so much so that the time lag was 11 to 12 hours compared to ≈3 hours in northern wells despite HIGHER permeabilities. Currier[14] attributed the high storativity to crossflow from surrounding layers. In such a complex system, permeability was found to be a function of the way it was measured and the volume of investigation. Single-well short tests yielded higher permeabilities than longer tests because they were more influenced by local fracture enhancement. Interwell permeabilities were lower because the flow path was more tortuous with less probability for the limited-connectivity fracture clusters to connect. Pulse test data gave very high, fracture (SAZ) dominated, permeability.

EFFECTS OF SAZ—Dramatic pilot data (pressure pulses between wells within hours, and an 8-day breakthrough in pilot waterflood tests) triggered a search for the thief zone. Calculations suggested that the observed performance could be explained by as few as one or two open fractures in a very conductive fracture system. Cores in all 5 wells showed only a few small lowconductivity fractures and no clear evidence for a thief zone. Several opportunities to discover the SAZ were missed by tagging the top of the

*Lockman, personal communication

Wahoo with the bit before coring. Even looking back at complete cores, it is easy to miss the significance of one or two open fractures (the visible manifestation of the SAZ).

The SAZ is a key control on reservoir performance and is one reason for the current high field GOR of 8000 scf/STB [1400 m^3/m^3]. Fig. 19 is a production plot of a typical well which shows rapid buildup in GOR from solution ratio of 836 scf/STB [149 m^3/m^3] to over 12,000 scf/STB [2100 m^3/m^3]. The limited drainage test showed a dramatic rise in GORs; analysis by Currier[14] showed that performance is consistent with a stratified model with SAZ and matrix layers.

Fig. 20 shows the impact of the SAZ/fault system on waterflood responses at Well L3-05. Oil rate is plotted versus time; the expected decline trend is indicated by the smooth curve. The first departure from the curve occurred in 1987, just after a one month injection test began in L2-28, one mile [0.6 km] to the west (Fig. 21). The rate benefit is likely due to pressure support rather than formation of an oil bank.[21] The next departure occurred shortly after L2-18 was placed on sustained injection. The third departure is due to L2-32: water broke through and the oil rate dropped severely. Both are 0.7 mi [1.1 km] from L3-05. Waterflood ceased at the end of 1989 and the well has resumed it's previous decline trend. Interference effects are also common for primary depletion, e.g. changes in decline rate and GOR trend.

EVIDENCE OF OPEN FAULTS—While the possibility that faults might be open was acknowledged, this was not incorporated in early modeling work. It was not until a year after field startup that it became clear that faults formed an open system with vertical and lateral communication throughout much of the field. Key evidence came from Repeat Formation Tester (RFT*) surveys which indicated drawdown in zones which had never been produced. Drawdown had occurred by vertical communication between zones. Cross sections showed a highly stratified system which lacks vertical communication except at faults.

Other evidence followed. (1) Core, mud log, drilling breaks and missing section revealed 30% of all wells crossed faults or megafractures. (2) Gas injector LGI-10 was tagged with SF_6 tracer in 1987. SF_6 has been detected in a number of wells surrounding the gas cap (Fig. 22). (3) Native H_2S, produced from the east side of the field and reinjected into the gas cap, also traced gas cap movement and showed the same pattern (Fig. 22). (4) Analysis of pilot waterflood volumes concluded that large volumes of water were lost to the faults and SAZ. (5) RFT surveys from 20 new wells showed unexpectedly high drawdown (50 to 500 psi [340 to 3400 kPa]) in low permeability rock ($\approx$0.1 md), often $\approx$1 mi [1.6 km] from the nearest offset production. Low matrix permeabilities and the high degree of stratification preclude the possibility that these cases of rapid vertical and lateral communication could occur through the matrix, so the most likely explanation is depletion of the matrix by nearby faults or megafractures.

MODELING MEGAFRACTURES—Distance to the fault was estimated using the RFT* data referred to above and some simple assumptions on pressure history in the fault. Solution of the linear diffusivity equation yielded order-of-magnitude distances of $\approx$0.1 mi [$\approx$200 m] from wellbore to fault (Fig. 24). These data, together with mapped faults, provide input for reservoir simulation.

The preceding data indicate local and regional continuity of the fracture/fault/SAZ system. Other data suggests local and regional discontinuities. For example, Wells L3-15 and L2-08 are 0.7 mile [1.1 km] apart. Severe lost circulation in each well precluded running casing and the wells were left as openhole completions. A 300 psi [2000 kPa] difference in static pressures between the two wells suggests that the transmissibility between the two (by SAZ, megafracture or other route) is not large.

Small-scale fractures, the original basis for the limited connectivity fracture model, are now thought to have only minor significance. This is shown by the close agreement between log porosity and RFT pressure drawdown within a pay layer. Fig. 23 shows that pressure drawdown within a pay layer is proportional to its log porosity. Superimposed are open hole log porosity and RFT drawdown from original pressure. Note that the higher porosity part of a layer is more drawn down than the lower porosity part. This relationship has been observed in four of the five available examples. If small-scale fractures were very effective, pressures within a single pay stratum would be more uniform.

IMPACT OF PRESENT FAULT MODEL—The present model of SAZ and faulting has had both positive and negative impacts on field development. Above all, faulting has added greater uncertainty to reservoir management than originally perceived. For example, no correlation is seen between amount of pay in a well and well rate or cumulative production. Thus, net pay maps are not reliable indicators for planning well locations. Three years of production history reveal that faulted wells are some of the best producers in the field. Some wells have produced over 0.5 MMSTB [80,000 m^3] from parts of the field with virtually no pay development near the wellbore. In one region, faults supply pressure support to lower zones. Recognition of the open fault drainage system has also lead to a reduction in ultimate well count.

On the negative side, waterflood plans have been deferred until a better understanding of the fault/SAZ system emerges. In addition, it has proven impossible to deplete all zones uniformly. The SAZ and Zone 7 are drawn down much more quickly and extensively than the tighter lower zones.

In summary, comparison of fault models reveals that four of the presently recognized scales of fracturing were recognized early (microfractures, strata-bound fractures, megafractures and faults, but not the SAZ). Before field startup, difficulties arose in determining which of the various scales of fracturing were likely to be significant. The tendency was to focus on those scales that could readily be analyzed (e.g. small-scale fractures and seismologically defined faults) rather than the harder to assess megafractures. More extensive production logging, pulse/interference testing, and coring the top foot of the reservoir could have aided in early recognition of the importance of faults and discovery of the SAZ.

PRIMARY DEPLETION

The development plan for primary depletion as envisioned in 1984 was for over 200 wells on 160-acre [65 ha] spacing[1] to be drilled with an aggressive 3-to-4 rig program. Ultimate development to 80 acres [32 ha] was anticipated with waterflood. In 1986, a decision was made to restrict drilling to a one-rig program in order to allow more time to understand the reservoir.

SIMULATION: The 1984 project funding decision was based in part on a single-porosity cross-section model which yielded recoveries of 5 to 8% of OOIP downdip and over 25% underneath the gas cap. 90% of produced gas was reinjected. Recovery estimates were based on abandonment rates of $\approx$100 B/D [$\approx$16 m^3/d] per well. The high

*Mark of Schlumberger

upstructure recoveries were from a combination of gas displacement and gravity segregation (gravity drainage). The low downdip recoveries reflected solution gas drive far from gas cap support. This model used a uniform 4 md enhanced matrix permeability and pseudo-relative permeability curves to incorporate fracture effects and reproduce the observed high GOR behavior. The model also had uniform matrix porosity, but partially isolating shales provided some stratification. Overall recovery was 16% of OOIP in rock with $\phi \geq 5\%$. Notwithstanding the relative permeability adjustments, we recognized that the model was too homogeneous and thus optimistic. A risked recovery of 12% was used for evaluation work.

More recent simulation with highly stratified singleand dual-porosity models yielded average primary recoveries of 7 to 10% with gas cap support. Gravity drainage is now expected to provide only minor contribution because the reservoir is more stratified than previously believed.

DECLINE TRENDS: Shortly after field startup, rapid well decline rates were noted (e.g. 50%/year) which yielded pessimistic reserve estimates. RFTs showed drawdowns of up to 1000 psi [6900 kPa] in the L2 test region; fieldwide pressures dropped at 290 psi/yr [2000 kPa/yr], over twice the expected amount. Material balance analysis indicated that contributing OOIP might be half of volumetric OOIP. Since then, decline rates have flattened dramatically in some parts of the field. A number of wells producing near the gas cap show low (or no) decline rates and moderate (considering the location) GORs (5000 to 20,000 scf/STB [890 to 3600 m^3/m^3]). Some wells show apparent hyperbolic decline exponents substantially over $h = 1$.

Fetkovich[22] identifies two major decline regimes: *transient flow* during the infinite-acting period, and *depletion* once drainage boundaries are felt. During depletion, a well producing under solution gas drive would be expected to follow one of the hyperbolic decline depletion tails (with $0 \leq h \leq 1$). Fetkovich et al.[23] state that transient data can yield apparent (false) decline exponents over $h = 1$. A tight, highly stimulated well will follow the transient part of the curve (high apparent h) for a considerable time before it ever starts feeling its drainage boundaries. Interestingly, a number of the better wells near the gas cap also remain on the transient curve for extended periods (up to 3 years), perhaps a result of gas cap support providing a constant potential boundary. This is consistent with material balance work[24] which indicates gas cap drive in the northern part of the field and solution gas drive in the southern part. Upturns in decline trends in several downdip wells, together with produced water trends, indicate partial aquifer support.

In summary, original performance and recovery estimates were optimistic, but correctly predicted the dominant drive mechanisms. Additional production will be needed to ascertain the ultimate benefit of gas cap support and the contribution of tighter rock.

WATERFLOOD

Early assessments, based on historical performance of other carbonate waterfloods, indicated the potential for doubling ultimate reserves by waterflooding. However, low matrix permeability, high matrix stratification and shunting of flow by faults and SAZ suggests limited waterflood potential, at least by conventional methods based upon large well spacing.

WATERFLOOD PILOT TESTING began five months after the field started production. The first injection attempts (Fig. 21, pattern 'A') showed breakthrough to the center well in as little as 8 days. The result was not surprising since these were the same wells which saw pressure responses in a few hours on interference test.[5] These results lead to drilling a second pilot pattern (Fig. 21, pattern 'B'). The newer wells were completed away from the SAZ. The five injectors shown in Fig. 21 had been on line for up to two years. Breakthrough times have ranged from one month to one year, with the earliest occurring in older producers completed in the SAZ. Results have been mixed, as indicated here and on Fig. 20 discussed earlier. A full account of the waterflood tests is outside the scope of this paper.

WETTABILITY/IMBIBITION: Special core analysis indicated mixed wettability.[25] Spontaneous imbibition tests on bland and oil-based mud core gave inconsistent results with 1 to 10% PV water imbibition and 0% to 25% PV oil imbibition. Performance predictions based on scale-up of the most optimistic of these data indicated prohibitively long times (e.g. hundreds of years) for imbibition distances much over 10 ft [3 m].

An imbibition test using partitioning tracers was conducted on L2-28. Exxon's analysis[26] concluded that water did imbibe, although others felt the test was ambiguous. The results likely apply only to the SAZ, where most of the injected water went.

ANALYTICAL PREDICTION METHODS: Our first approach was the Stiles method.[27] A layer cake approach seemed appropriate since there was excellent continuity of strata. The Dykstra-Parsons[28] coefficient for core is $V = 0.8$ as conventionally defined, however the data is not log-normally distributed (power law exponent[29] $p \approx -0.1$). We used the Stiles method with a modified $V = 0.95$ to represent the most permeable, floodable rock (Appendix A). This yielded incremental recoveries of 5 to 8%. The method may be pessimistic since it does not model crossflow, but optimistic in this application since it doesn't include the effects of fractures.

SIMULATION: Waterflood simulation work began with unfractured pattern models. Results were highly sensitive to stratification: using $V = 0.80$ rather than $V = 0.95$ raised recovery by a factor of three.[30] An unresolved question is the relation between core V and simulator V (Appendix A).

Fig. 25 shows a "megafracture" model of one quadrant of an inverted 9-spot flood pattern with 160-acre well spacing. This model was one attempt at representing the extreme heterogeneity indicated by field data. The darker lines are high-permeability cells which represent conductive megafractures or faults. The top layer, shown enlarged for clarity, has high permeability (600 md) and represents the SAZ. The matrix is low permeability (0.16 to 71 md) and highly stratified ($V = 0.94$). Megafracture permeability was assumed to be 100 times matrix permeability. As one would expect, incremental waterflood recovery was lower with increased heterogeneity. This particular model yielded an incremental waterflood recovery of ≈3% OOIP and showed poor sweep in the tighter layers.

PAY CUTOFF affects waterflood more than primary. Fig. 26 left side shows pay with a 10% porosity cutoff; the right side with a 22% cutoff. Analytical calculations (Appendix B) show that the only matrix that may be flooded in a reasonable field producing lifetime may be high permeability rock (with corresponding porosity of over 20%). The higher cutoff reduces both pay volume and pay continuity. Also, field tests have shown that it is easy to communicate through the fault/SAZ system and thus bypass most of the oil in place in the matrix. Fig. 26 shows that there may not be much floodable matrix target even if water could be kept away from faults and megafractures by judicious drilling and completion techniques.

The potential waterflood target is now seen as (at best) a very small segment of the reservoir. In light of currently perceived risks, the owners have decided to suspend waterflood testing.

CONCLUSIONS / RECOMMENDATIONS

Notwithstanding the large amount of data collected, many key attributes of the field still remain unknown. Many of the questions raised over 40 years ago in another fractured reservoir[31] are still unanswered in the Lisburne, e.g. fracture continuity, floodability and reservoir pressure. The reservoir description presented here still needs further refinement.

In summary, early models tended toward uniformity, homogeneity and averages. Although heterogeneities were noted, their implications were not fully recognized. The most important clue that should alert others to the presence of heterogeneities are well test rates well above those sustainable by the matrix. As we have shown, Lisburne heterogeneities have both beneficial and negative consequences. While the SAZ and faults have had a negative impact on waterflood, they also act as major collectors of oil from tight rock, oil that might otherwise be economically unrecoverable. Field development would clearly be most efficient if hetrogeneities were identified early so that development plans could be altered accordingly. Such responsiveness requires complete integration of geology and engineering data.

Specific recommendations for assessing other potentially complex carbonates include: (1) Collect complete suites of data: e.g., openhole logs, core, tests, production logs and full core coverage from the same intervals to better discriminate between matrix and fracture effects. (2) Conduct inter-well tests early in the evaluation to identify fracture system characteristics which may be nearly invisible in single-well tests. (3) Give full weight to uncertainties. (4) Recognize that the data gained during early development can have significant impacts on the reservoir model and hence on the development plan. Development schedules should recognize this and provide flexibility in the development plan.

NOMENCLATURE:

A Pattern area (acres)
c_{tm} Total compressibility of matrix (1/psi)
d Distance between well rows (ft)
E_a Areal sweep efficiency
h Hyperbolic decline exponent
h_l Layer thickness (ft)
i Injection rate (RB/D)
k Permeability (md) [μm^2]
k_o Permeability to oil (md)
M Mobility ratio
p Power law coefficient (probability distribution function parameter[29])
Δp Pressure drawdown (psi)
R Ratio of producing rates of corner well to side well
R_w Resistivity, water (ohm•m)
r Distance between wells (ft) [m]
r_{wa} Wellbore radius, apparent = $r_w \bullet e^{-s}$
s Skin factor
S_w Saturation, water
ΔS_w $1 - S_{or} - S_{wi} = 40\%$
T Transmissibility = kh/μ (md ft/cp)
t_b Time to breakthrough (yrs)
t_L Time lag, hours
V Coefficient of permeability variation
η Diffusivity (md psi/cp)
μ_o Viscosity (cp)
ϕ Porosity
ϕ_m Matrix porosity

ACKNOWLEDGMENTS

We wish to thank the managements of Exxon Company U.S.A., Arco Alaska Inc., and BP Exploration (Alaska) Inc. for permission to publish this paper. The techniques and/or conclusions are those of the authoring company and may not be shared by other Working Interest Owners. We acknowledge and commend the spirit of teamwork between geologists and engineers of the three companies. Within Exxon, we acknowledge the contributions of D.M. Advocate, C.A. Alba, M.L. Belcher, J.R. Bragg, E.M. Braun, J.A. Davis, J.E. Farr, R.A. Humphrey, D.F. Lockman, B.V. Liddell, M.E. Lukk, R.M. McKinley, D.M. Neuhart, L.H. Stiles, T.D. Streltsova and J.J. Woodbury. We also acknowledge the contributions of G.L. Bartholomew, J.D. Moody, J.G. Richardson and C.E. Wanorue.

REFERENCES

1 Paige, K.T. and Dayton, J.S., "Development of the Lisburne Reservoir", paper 16845 presented at the 1987 SPE Annual Technical Conference and Exhibition, Dallas, Sept 27-30, 107-112.

2 "Estimates Cut for Recoverable Reserves at Lisburne on ANS; Expenditures Reduced", *Oilgram News* (August 24, 1989) 5.

3 Jameson, J., Neuhart, D., and Wendlandt, R.: "Multiple Stages of Porosity Development in Wahoo Fm., Prudhoe Bay, Alaska', Annu SEPM Mid Year Meeting (Columbus, Ohio) 1988 Abstr 5, 27-28.

4 Jameson, J.: "Interactions between early and late diagenesis in reservoir formation, Lisburne Field, Prudhoe Bay, Alaska". Annu AAPG-SEPM-EMD-DPA Conv (San Antonio, Texas) 1989 AAPG Bull 73, No. 3, 367-368.

5 Sampson, L.E. and Marcou, J.A.: "Interference Test in a Fractured Carbonate: A Lisburne Case History", paper SPE 18138 presented at the 1988 Annual Technical Conference and Exhibition, Houston, Oct 2-5, 303-317.

6 Belfield, W.C.: "Characterization of a Naturally Fractured Carbonate Reservoir: Lisburne Field, Prudhoe Bay, Alaska", paper SPE 18174 presented at the 1988 Annual Technical Conference and Exhibition, Houston, Oct 2-5, 677-685.

7 Harris, D.C., and Kendall, A.C., "Secondary Porosity in Upper Lisburne Group Carbonates (Wahoo Limestone; Lower Pennsylvanian, North Slope, Alaska)", 1986 SEPM Annual Mid-Year Meeting, Raleigh, N.C.

8 Fisher, D.: "Testing and Evaluation of an Arctic Fractured Carbonate: The Lisburne Formation, Prudhoe Bay", paper 15100 presented at the 1986 SPE California Regional Meeting, Oakland, April 2-4.

9 Owen, L: "Lisburne Water Composition Analysis", Terra-Tek report TR 90-61, Nov 1989.

10 Michel, C.M.: "Well Plans—Producibility/Pay Criteria", handout at Lisburne Operations Subcommittee, Jan 16,

1987.
11 George, C.J. and Stiles, L.H.: "Improved Techniques for Evaluating Carbonate Waterfloods in West Texas", JPT, Nov. 1978, 1547-1554.
12 Casarta, L.J., McNaughton, D.A., Bornemann, E, and Bettis, F.E.: "Fracture Identification and Matrix Characterization Using A New Wellbore Imaging Device in the Lisburne Carbonate, Prudhoe Bay, Alaska" paper presented at SPWLA in Denver, June 12, 1989.
13 Juhasz, I.: "Conversion of Routine Air-Permeability Data Into Stressed Brine-Permeability Data", 10th European Evaluation Symposium, Aberdeen 4-22-86.
14 Currier, B.H. "Lisburne Reservoir Limited Drainage Test: A Pilot Test Case History", SPEFE (Dec. 1990) 337-343; from paper SPE 18277 presented at the 1988 Annual Technical Conference and Exhibition, Houston, Oct 2-5, 169-176.
15 Lockman, D.F. and Schwalbach, J.: "Geologic Characterization in Relationships to Fractures in Outcrop: Pismo Formation, California", SPE Forum on Naturally Fractured Reservoirs (Crested Butte, CO), 1989.
16 Streltsova, T.D.: "Long Term Production Performance of West Bay State No. 1", letter from C. C. Mattax to G. S. Goss, Apr 6, 1983.
17 Warren, J.E. and Root, P.J.: "The Behavior of Naturally Fractured Reservoirs", SPEJ (Sept. 1963) 245-55; Trans., AIME, 228.
18 Chirlin, G.R.: "Flow Through a Porous Medium With Periodic Barriers or Fractures", SPEJ, June 1985, 358-362.
19 Horner, D.R.: "Pressure Build-Up in Wells," Proc. Third World Pet. Cong., The Hague (1951) Sec. II, 503-523.
20 Streltsova, T.D.: "Interpretation of Pressure Data from Lisburne Interference Test", Exxon Production Research Co. report EPR.83PS.86, October 1986.
21 Nelson, M.A. and Gobran, B.D. "Observed L3-05 Gas/Oil Ratio Response", memo to R. G. M. Oba, March 23, 1989.
22 Fetkovich, M.J.: "Decline-Curve Analysis Using Type Curves", JPT, June 1980, 1065-1077.
23 Fetkovich, M.J.; Vienot, M.E.; Bradley, M.D. and Kiesow, U.G.: "Decline-Curve Analysis Using Type Curves—Case Histories", SPEFE, Dec 1987, 637-656.
24 "Areal Reservoir Mechanisms Studies", ARCO presentation, Lisburne Reservoir Task Group meeting, Aug 4, 1988.
25 Humphrey, R.A., Hagan, H.W., Lozano, J.A., Jameson, J., Neuhart, D.: "Lisburne Core Analysis Report", Exxon Production Research Co. report EPR.96PS.86, Dec 1986.
26 Bragg, J.R.: "Summary of Tracer Analyses for Imbibition Test Conducted in Lisburne Well L2-28", Exxon Production Research Co. unnumbered report, April 7, 1988.
27 Stiles, W.E.: "Use of Permeability Distribution in Water Flood Calculations", Trans. AIME (1949) 186, 9-13.
28 Dykstra, H. and Parsons, R.L.: "The Prediction of Oil Recovery By Water Flood", Secondary Recovery of Oil in the United States, 2nd ed., API (1950) 160-174.
29 Lake, L.W. and Jensen, J.L.: "The influence of Sample Size and Permeability Distribution on Heterogeneity Measures", (Tech. Paper SPE 15434) SPERE May, 88 629-637.
30 Bolkovatz, M.J.: "Waterflood Simulation Comparisons—Summary Observations", Lisburne Reservoir Task Group meeting, May 4, 1989.
31 Littlefield, M., Gray, L.L., and Godbold, A.C.: "A Reservoir Study of 174 the West Edmond Hunton Pool, Oklahoma", Trans., AIME (1948) 174 131-155, with discussion by L.E.Elkins, Jack Tarner and L.F.Elkins.
32 Craig, F.F.: The Reservoir Engineering Aspects of Waterflooding, Monograph Series, SPE, 1971, 3.
33 Begg, S.H., Carter, R.R., and Dranfield, P.:"Assigning Effective Values to Simulator Gridblock Parameters for Heterogeneous Reservoirs", SPERE, November 1989, 455-463.
34 Hantush, M.: "Modification of the Theory of Leaky Aquifers", J. Geophys. Res., 1960, 65(11) 3713-3725.
35 Li Yun-Shan: "A Method for Determining the Coefficient of Diffusivity from the Maximum Pressure Response in an Observation Well", (Translated from Russian), Proc, Podzemnaya Gidrodinamica, Tr. MINKh and GP, Gostoptechizdat, 1961 33.

SI Metric Conversion Factors

acre × 4.046 856	E-01 = ha
bbl × 1.589 873	E-01 = m^3
cp × 1.0*	E-03 = Pa•s
ft × 3.048*	E-01 = m
ft^3 × 2.831 685	E-02 = m^3
md × 9.869 233	E-04 = μm^2
psi × 6.894 757	E+00 = kPa

*Conversion factor is exact.

APPENDIX A: SELECTION OF HETEROGENEITY MEASURES

Our Lisburne waterflood simulation work began with Stiles[27] calculations and evolved to simple `layer cake' models. These models allowed crossflow between layers but had no areal variation of properties. Ultimate recovery was highly sensitive to layer heterogeneity.[30] The megafracture model discussed earlier (Fig. 25 in the main paper) was a modification of a layer cake model and is beyond the scope of this analysis. The objective of this analysis is to identify properties of floodable matrix. Therefore, it is assumed that SAZ, faults and megafractures do not interfere with the flood process. This might be conceivable by massive in situ polymerization of the faults or by designing the flood pattern to use the faults as injection or collection surfaces.

The sensitivity of recovery to layer heterogeneity is shown in Fig. 8.3 of Ref. 32: A seemingly small change in V from 0.8 to 0.95 can make up to a three-fold difference in recovery.

The conventional Dykstra-Parsons coefficient of permeability variation is $V = 0.8$ using permeabilities at the 50th and 84th percentiles (median and +1 standard deviation). Our work used a modified Dykstra-Parsons coefficient $V2 = 0.95$ (Fig. 27). $V2$ is calculated the same way as V, but uses permeabilities at the 2nd and 16th percentiles (–2 and –1 standard deviations). Both V and $V2$ honor the core data—the core data does not have a strictly log-normal distribution and thus plots as a curve.

We have attempted to match both V and $V2$ in our models, with emphasis on $V2$ because it honors the important high-permeability data. $V2$ reflects properties of the most permeable and thus most floodable matrix (≈1 to 10 md), whereas V uses the median core plug permeability of 0.05 md which is below our nominal pay cutoff. All core data was used in the Dykstra-Parsons plot because it was found that deleting permeabilities below a nominal cutoff made the remaining population even less log-normal and caused even greater curvature in the plot (Fig. 28).

Other matrix heterogeneity issues which may tend to increase recovery estimates have not been addressed: (1) Lateral changes in permeability may reduce cycling in high-permeability layers and improve vertical sweep efficiency. (2) Matching simulator block heterogeneity to plug heterogeneity could be too pessimistic. Small-scale variations in permeability seen in plug data within a single pay packet may to some extent be averaged out by crossflow. That is, the pay packet may act more homogeneous than the individual plugs within it. However, simple averaging of core data within each simulator block is inappropriate since it would average out the observed variations and yield excessively optimistic reservoir performance.[33]

APPENDIX B: CUTOFF CRITERIA FOR WATERFLOOD SCREENING

The high variability in permeability of the Lisburne carbonate results in a large volume of the rock being in a `gray area' between what is clearly contributing pay and what is clearly non-pay (Fig. 12 of main paper). Waterfloodable pay volume, pay continuity and development limits are highly sensitive to assumed pay cutoff. Since solution gas drive depends upon pressure drawdown and waterflood depends upon fluid displacement between wells, it seems plausible that the difference in drive mechanisms would indicate a need for different pay criteria. Commonly assumed cutoffs for primary depletion range from 8% to 12%, roughly equivalent to 0.1 md. We originally used a pay cutoff of 0.1 md for waterflood based on the primary cutoff and the precedent set by many other fields.[11] However, those other fields have much closer well spacing and we realized that this cutoff value may be inapplicable for Lisburne which has wide well spacing (160 and 320 acres [65 and 130 ha]), low matrix permeability and high abandonment rates (30 to 100 B/D/well [5 to 16 m^3/d]).

This appendix develops an analytical method to define a representative pay cutoff for screening and mapping work to identify the prospective waterflood areas and zones. The method addresses concerns that for the same mapped pay, one region may be economically floodable if its pay is concentrated in easily floodable high-permeability layers whereas another may not be if it has low-permeability pay distributed through a large gross thickness.

Lisburne matrix is highly stratified—individual layers can be correlated over large regions. With little imbibition expected, waterflood recovery depends upon the ability to sweep the matrix. As in the previous section, it is assumed that non-matrix permeability (SAZ and faults) can be controlled.

For this analysis, floodable conduits are considered to be higher permeability strata which can achieve breakthrough in 20 years or less. This timing was selected because most field waterflood flowstream projections to date show a field life of 20 to 30 years. If breakthrough occurs significantly later, the field offtake rate may be so low at that time that it may not be economic to recover much of the remaining potential reserves.

The calculation methodology (example below) makes very simple, idealized assumptions to obtain a figure-of-merit for pay cutoff. Piston-like displacement is assumed through an ideal layer of uniform properties which is continuous throughout the pattern area. No crossflow from adjacent layers is assumed. The flood rate is assumed to be governed by the injection rate, which is calculated as a function of permeability. Breakthrough time is calculated from displaceable pore volume (expressed as a function of porosity), areal sweep efficiency at breakthrough, and flood rate. These equations are combined to yield an equation for breakthrough time as a function of porosity and permeability. This equation is solved for ϕ/k combinations required to achieve a breakthrough time of 20 years or less:

EXAMPLE CALCULATIONS FOR WATERFLOOD PAY CUTOFF-:

Injection rate: (Ref 32, Table 5.12 for inverted 9-spot pattern)

$$i = \frac{0.001538\, k_o\, h_l\, \Delta p}{\mu_o[(1+R)/(2+R)][\log_{10}(d/r_{wa}) - 0.1183]} = 5.2 \text{ B/D/md} * k_o \qquad (1)$$

where k_o = independent variable (unknown)
Δp = 4000 psi
h_l = 1 ft
μ_o = 0.7 cp
R = 1
d = 2640 ft (for 160-ac well spacing)
r_{wa} = 5.85 ft for 7" hole and s = –3

Displaceable pore volume:

$$\text{DPV} = 7758\, A\, h_l\, \phi\, \Delta S_w \qquad (2)$$

where $\Delta S_w = 1 - S_{or} - S_{wi} = 40\%$
A = 160 acres

Areal sweep at breakthrough (Ref 32 Fig. 5.8):

$$E_a = 0.54 \text{ @ M=1} \qquad (3)$$

Displacement volume to breakthrough, from (2) & (3):

$$\text{DPV}\,|_b = 0.54 * \text{DPV} = 1.07 \text{ MRB} * \phi \qquad (4)$$

Time to breakthrough, from (1) & (4):

$$t_b = \frac{\text{DPV}\,|_b}{i} = 565\, \phi/k_o \text{ years} \bullet \text{md} \qquad (5)$$

For $t_b \leq 20$ years,

$$\phi/k_o \leq 0.0354/\text{md} \qquad (6)$$

The resulting cutoff curve (6) is plotted as the solid line on Fig. 29.

For assumptions shown, the pay cutoff is k = 6 md or ϕ = 22%. Sensitivities with widely differing assumptions showed ranges in k from 0.3 to 28 md, with the corresponding porosity ranging from 19 to 24%. Results were only slightly sensitive to overburden and two-phase effects (Fig. 13) because those effects are most pronounced at lower porosities. Probability approaches[11] yielded similar results.

We found that breakthrough times predicted by the above analytical method were in reasonable agreement with simulation results. Differences are attributed to crossflow in the simulator which are not addressed in these simple analytical calculations.

This exercise shows that economically floodable pay may require a 20+% porosity cutoff for a matrix-only system. The method has proved very useful for screening work to identify what the waterflood target looks like, however this method is not recommend as a `stand-alone' definition of pay.

TABLE 1. INTERPRETATION OF INTERFERENCE TEST[20]

Single Well Test Analysis: (modified Horner analysis)

Well	T md-ft/cp	k md	s
L2-26	3589	19.3	–5.5

Inter-well Interference Analysis (L2-26 observing): (Hantush[34])

Well	r ft	T md-ft/cp	η md-psi/cp	k md	$\phi_m c_{tm}$ 1/psi
L2-28	1601	1442	2.2E9	8	5.3E-7
L2-30	1589	375	21.5E9	2.2	1.1E-7
L2-20	2093	1355	10.3E9	7.3	2.0E-6

Inter-well Pulse Analysis (L2-26 observing): (Li Yuh-Shan method[35])

Well	t_L hrs	Δp psi	T md-ft/cp	η md-psi/cp	k md	$\phi_m c_{tm}$ 1/psi
L2-28	2.8	125	6761	1.622E8	36.4	3.2E-7
L2-30	3.2	171	4775	1.433E8	25.7	2.6E-7
L2-20	10.8	24.4	23472	0.958E8	126.4	18.8E-7
L2-24	12	14.3	40912	0.801E8	220.3	39.3E-7

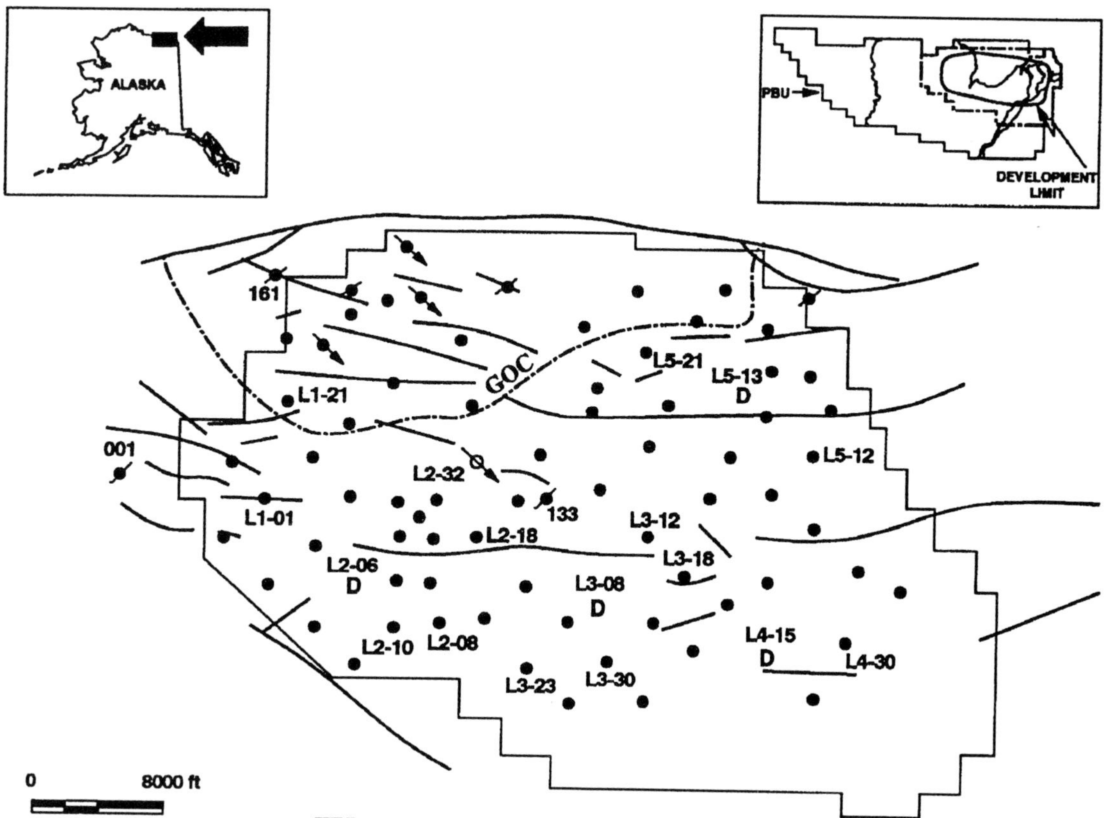

Fig. 1 - Location

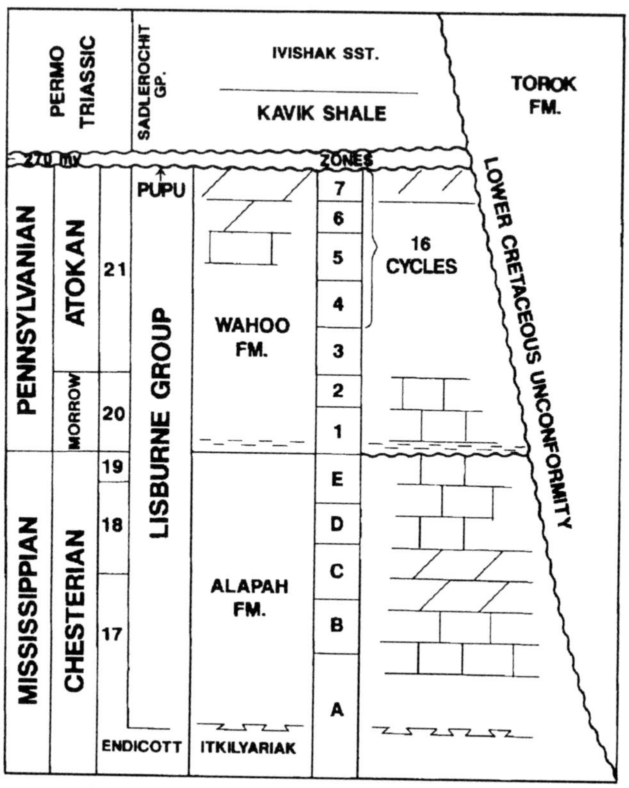

Fig. 2 - Lisburne stratigraphy

	EARLY	PRESENT
OOIP (MMSTB)	3000	2000
RECOVERY FACTOR	12%	7-10%
RESERVES (PRIMARY C&C, MMSTB)	250-350[1]	150-200[2]
WELLS	210	<100
PAY CUTOFF (POR UNITS)	5%	10%
OWC (ft)	-9150	-8900-9150 VARIABLE

Fig. 3 - Comparison of reservoir parameters

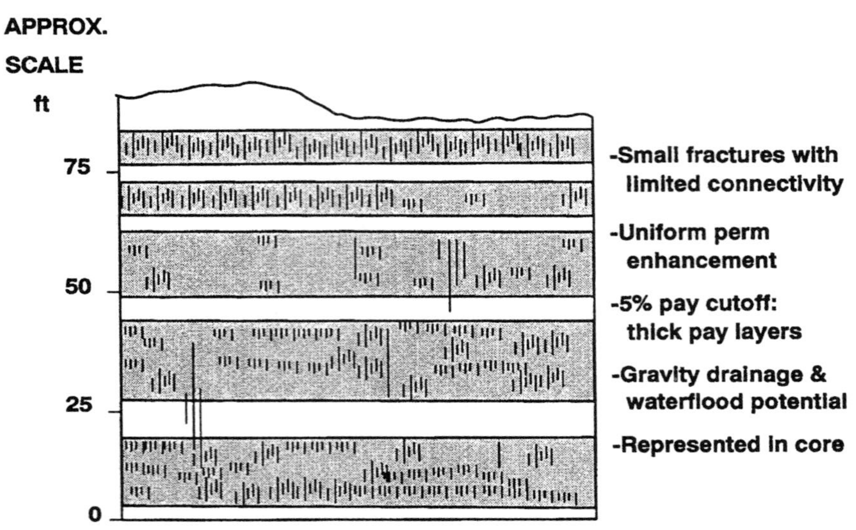

Fig. 4 - Early fracture model

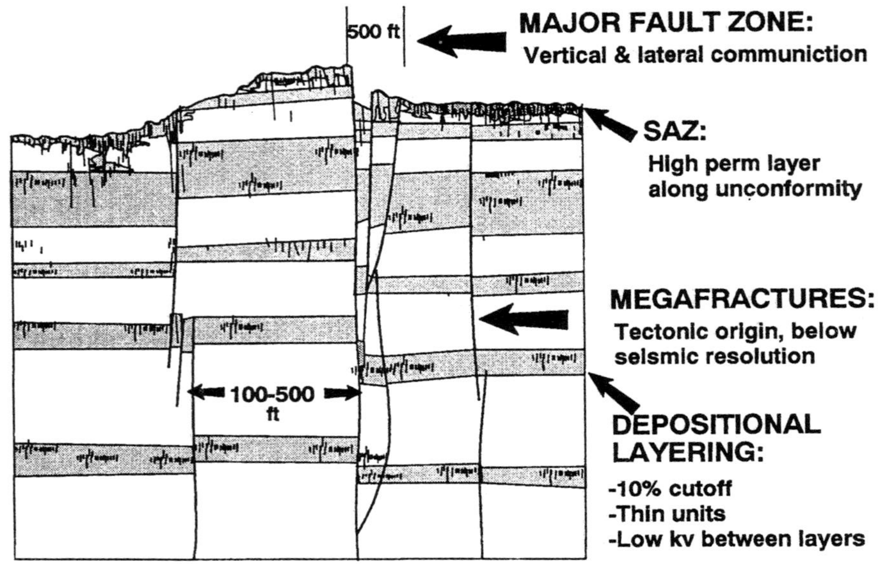

Fig. 5 - Present fracture model

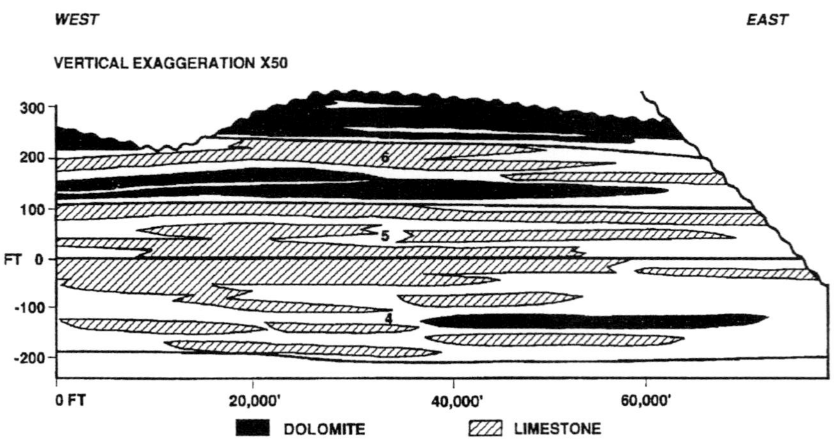

Fig. 6- Early model of pay distribution

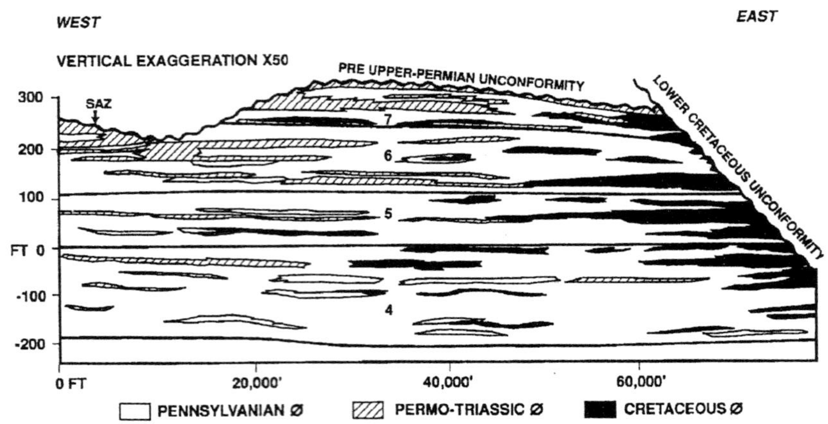

Fig. 7- Present model of pay distribution

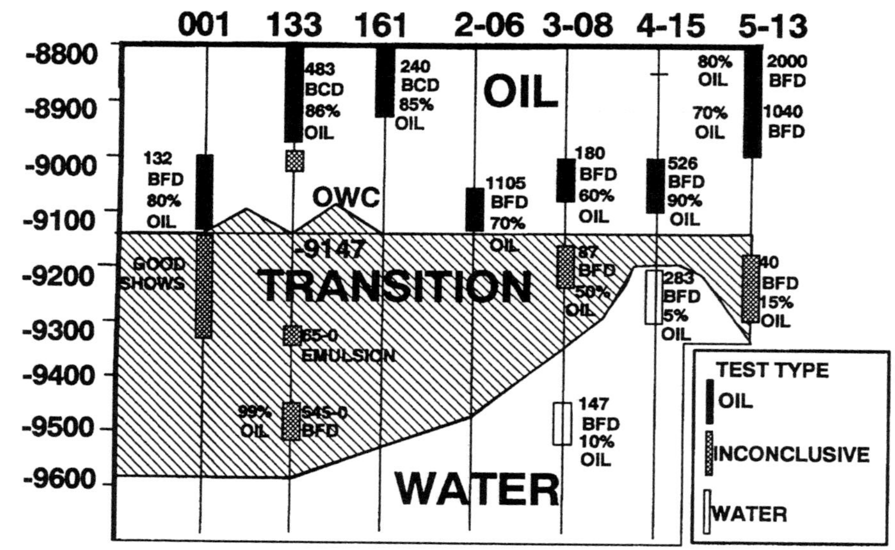

Fig. 8 - Early well data

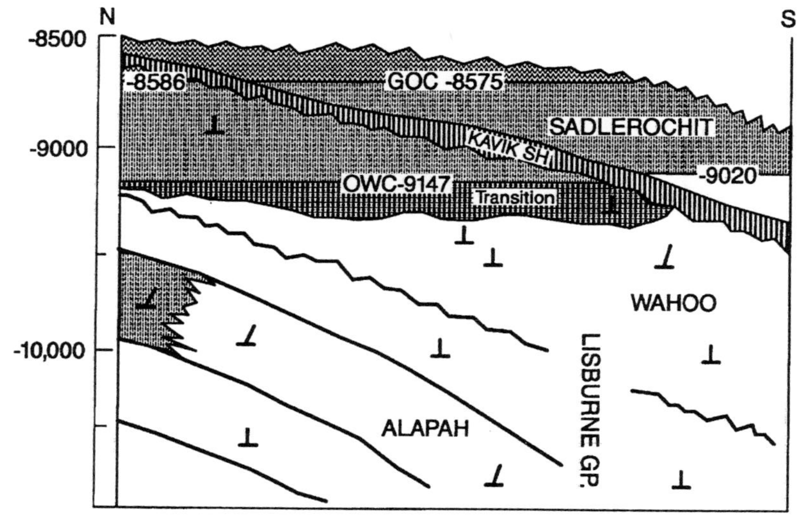

Fig. 9 - Fluid Contacts - Early

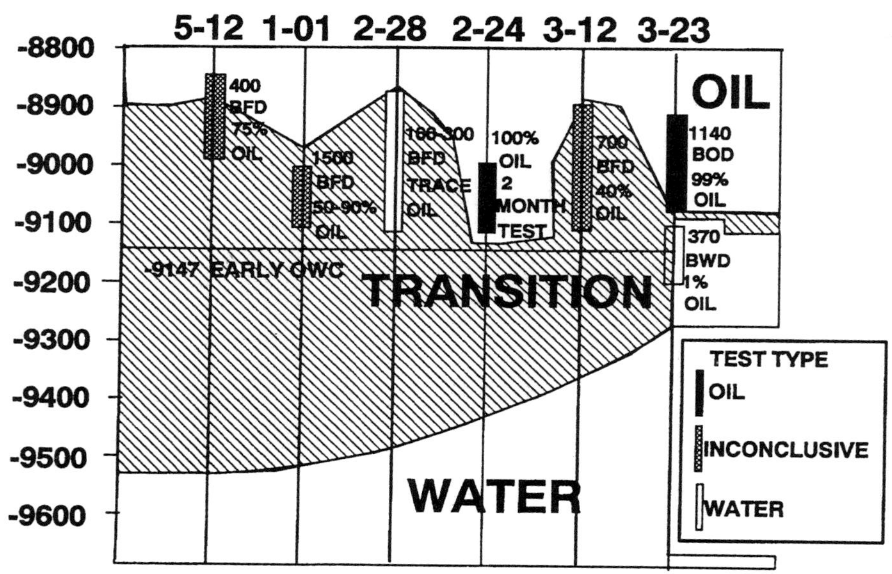

Fig. 10 - Present well data

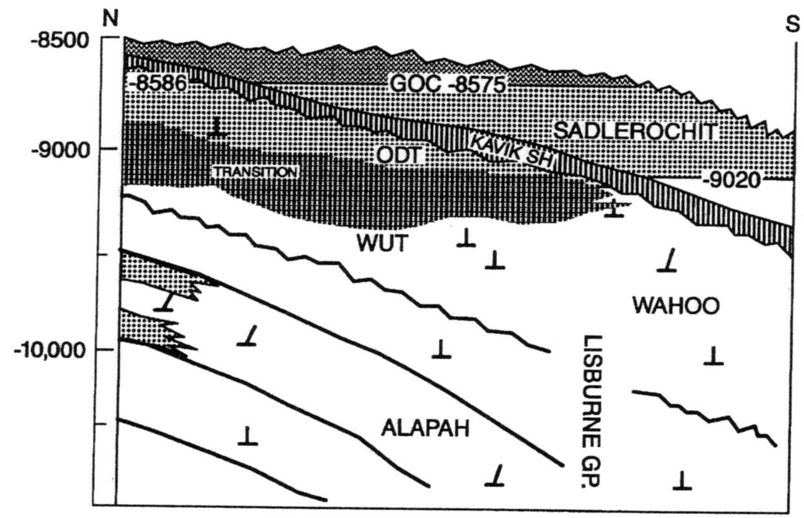

Fig. 11 - Fluid Contacts - Present

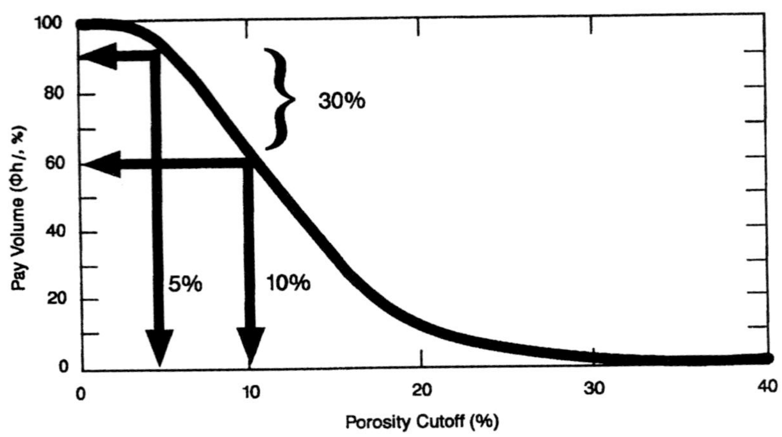

Fig. 12 - Pay Volume Vs. Porosity Cutoff

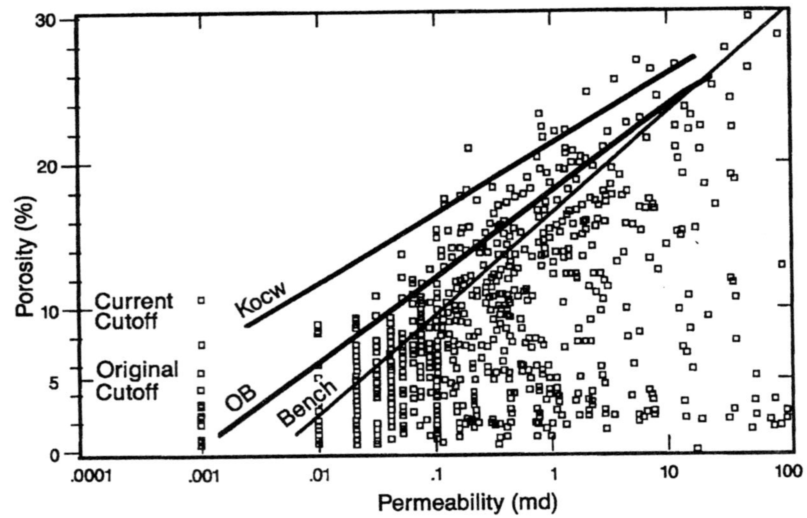

Fig. 13 - Pay Cutoff

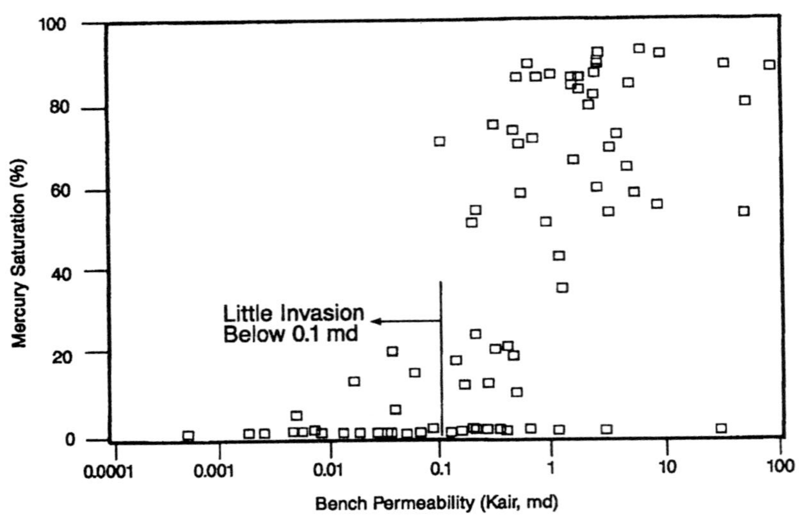

Fig. 14 - Mercury Saturation At Pc=500 psi

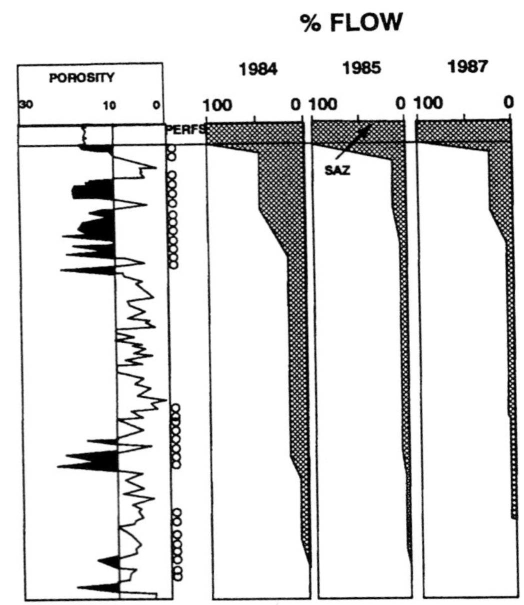

Fig. 15 - Flow profiles L3-08

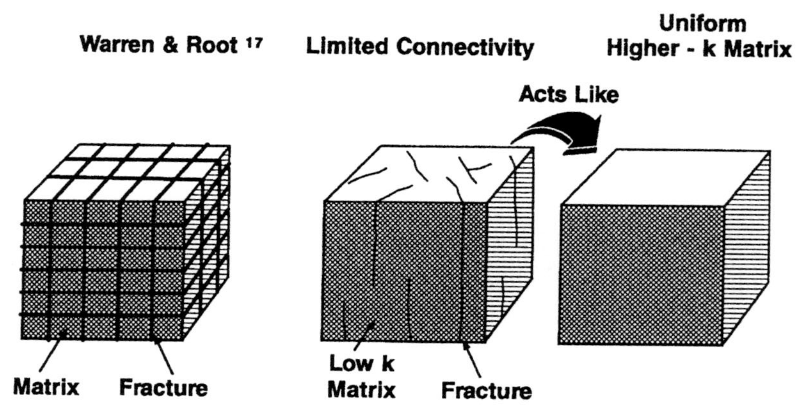

Fig. 16 - Fractures - Early Model

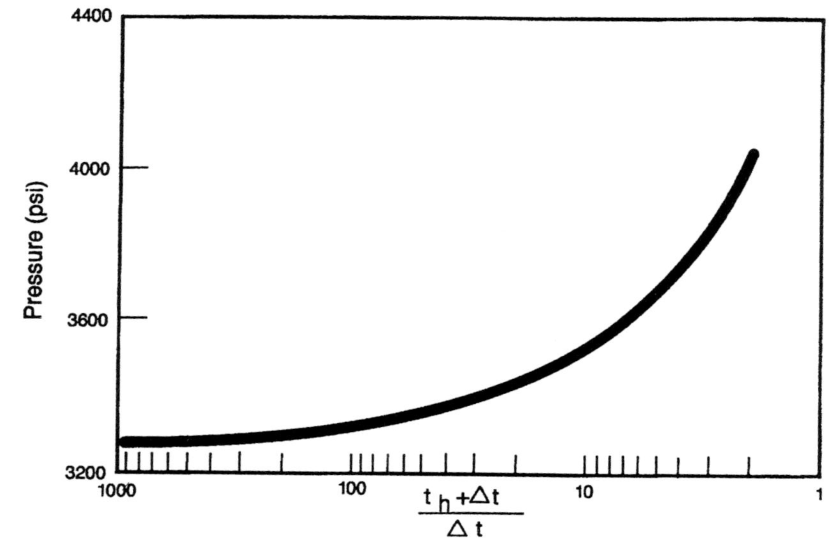

Fig. 17 - Limited Connectivity Fractures Inferred From Buildup Tests

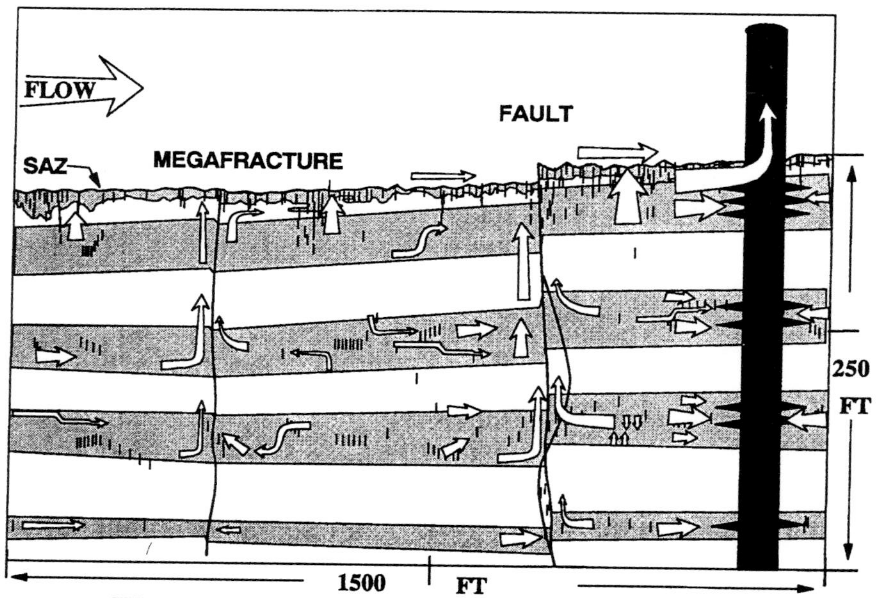

Fig. 18 Fractures - Present Model

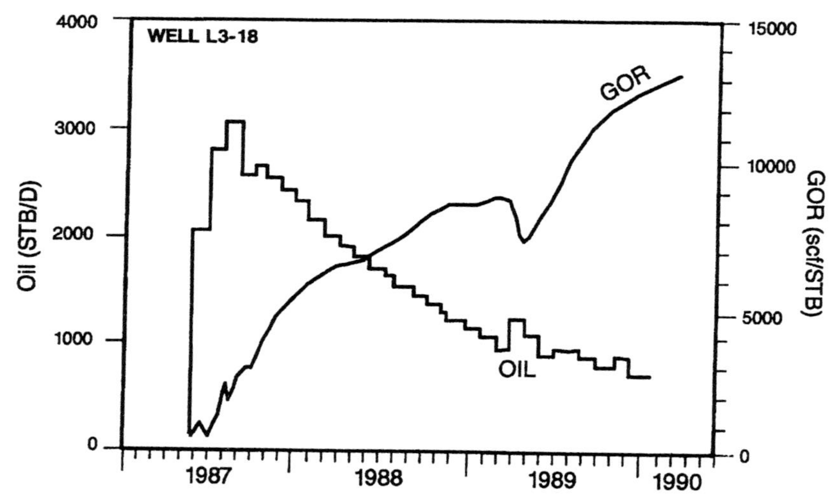

Fig. 19 - Well Production Trend

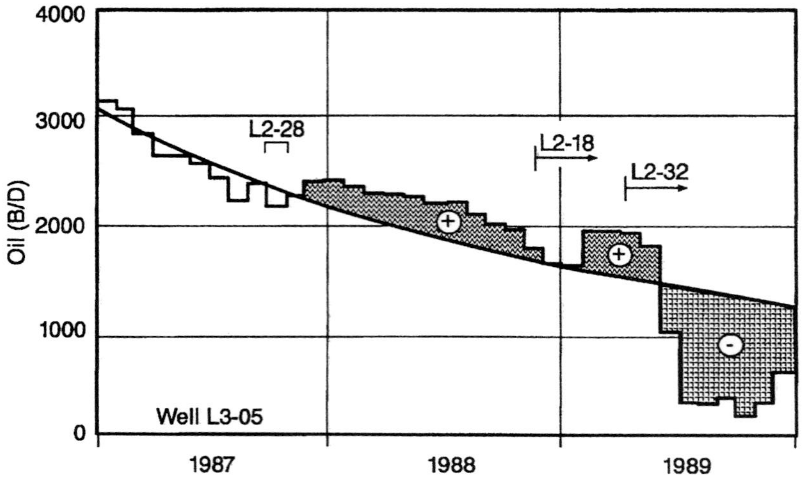

Fig. 20 - Waterflood Response

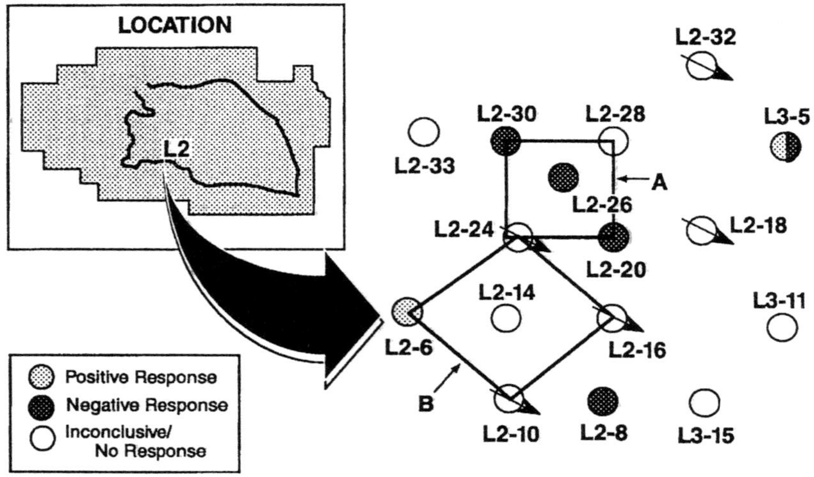

Fig. 21 - Waterflood Pilot Area

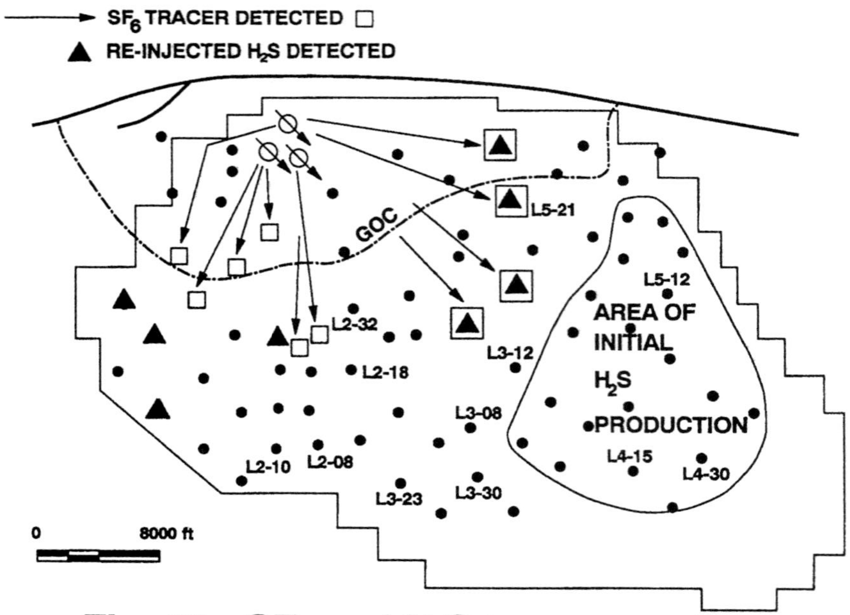

Fig. 22 - SF_6 and H_2S tracer results

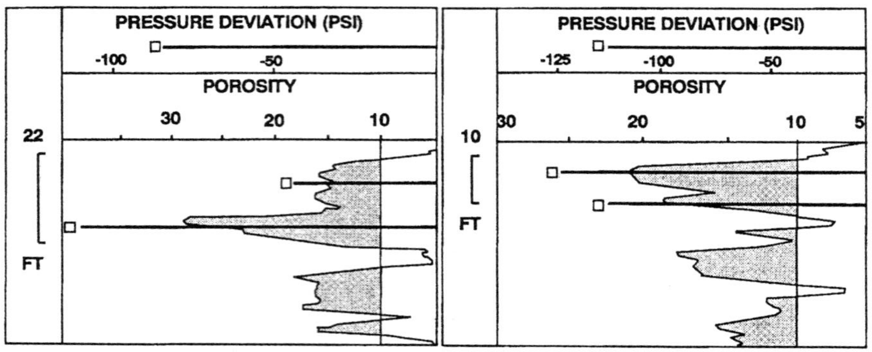

Fig. 23 - Pressure drawdown proportional to porosity

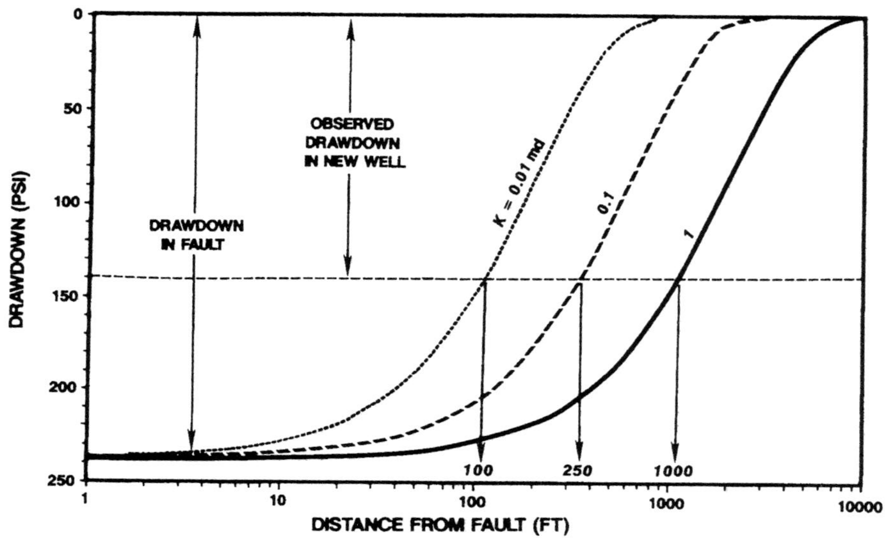

Fig. 24 - Use of RFT data to infer distance to fault

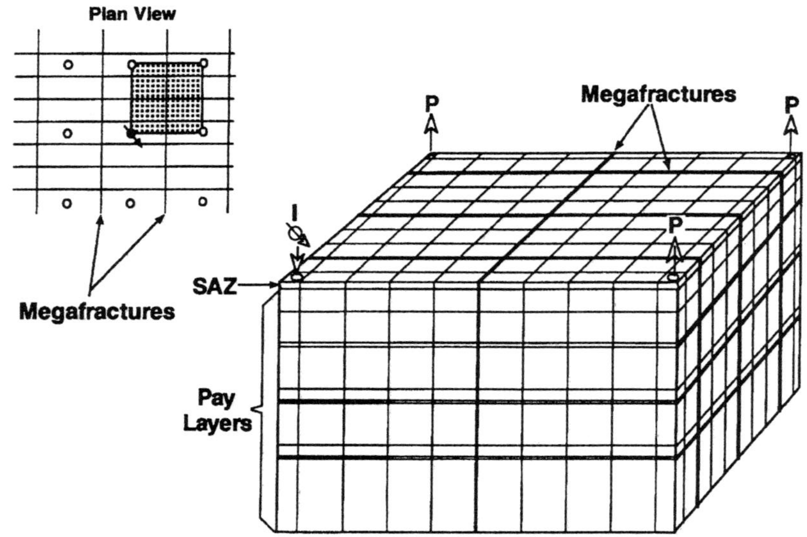

Fig. 25 - Waterflood Simulation

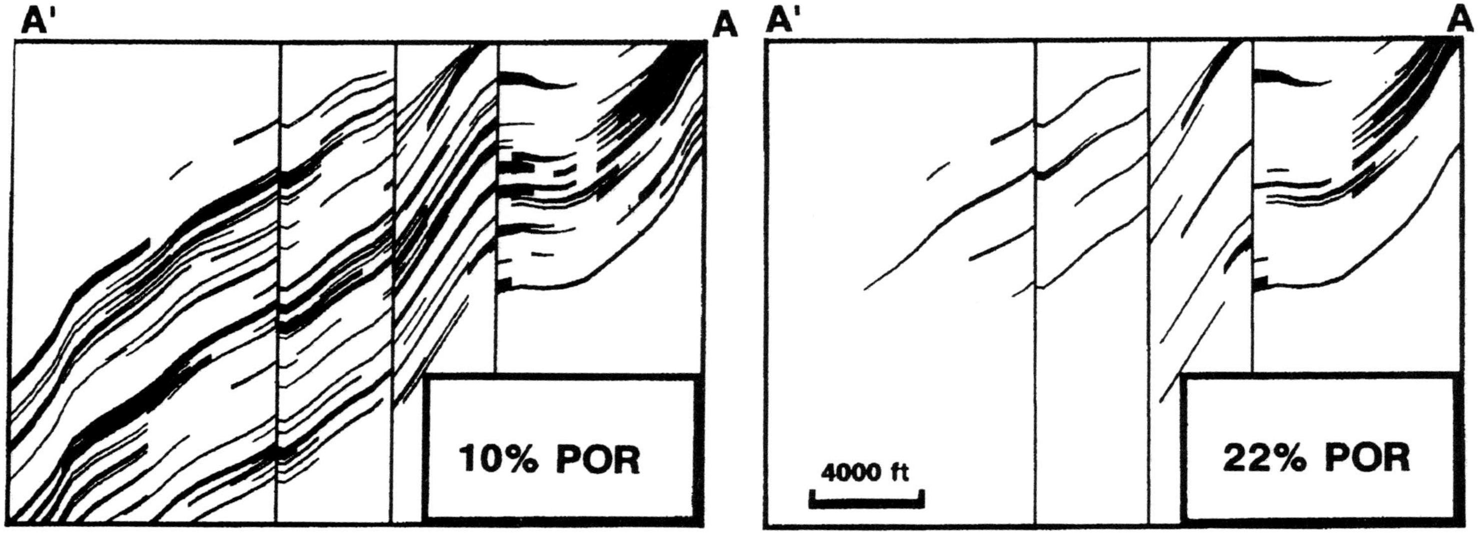

Fig. 26 - Comparison of pay distribution at different cutoffs

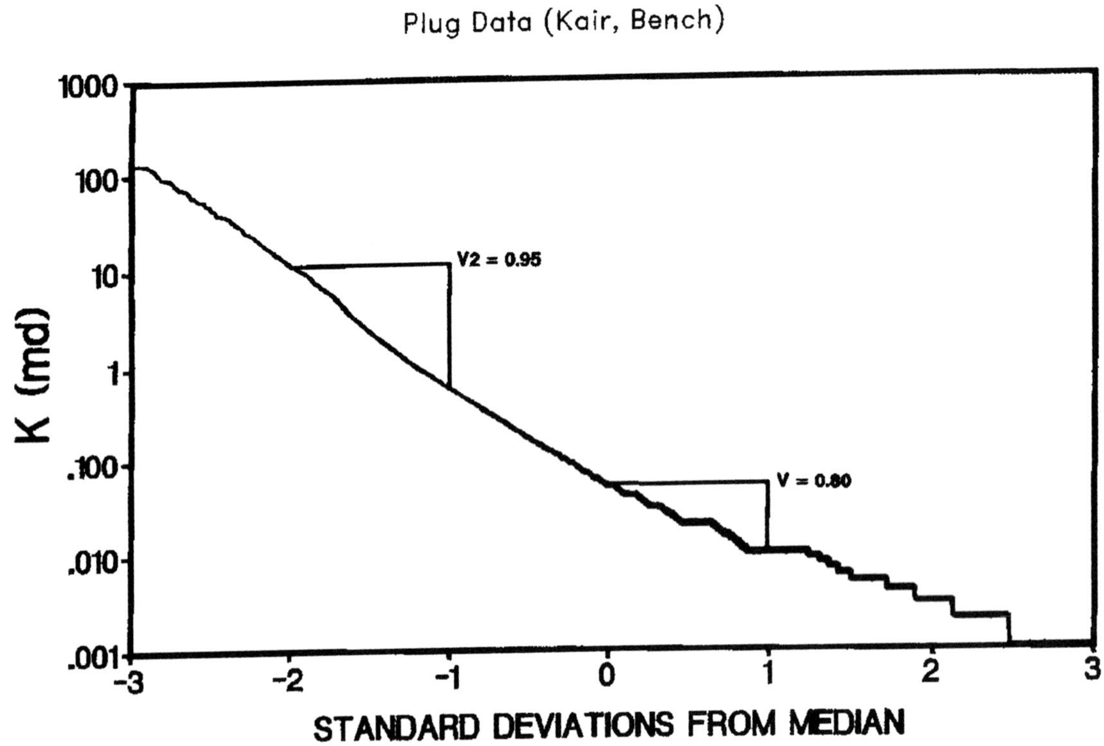

Fig. 27 - Permeability Variation

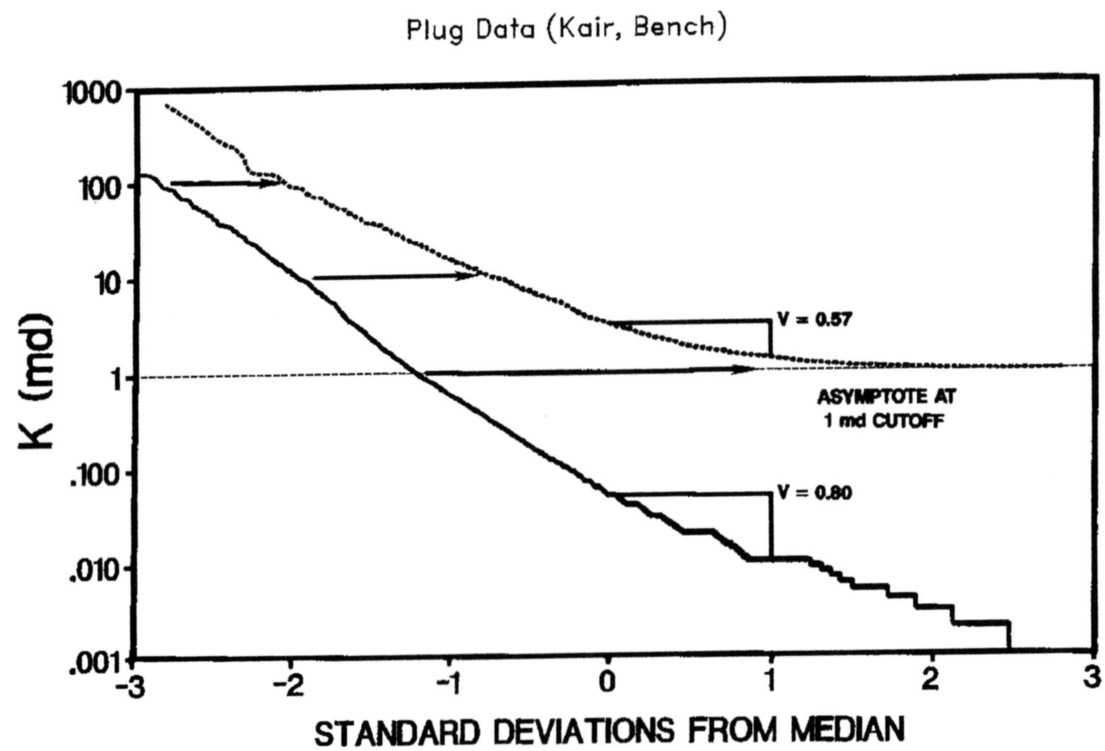

Fig. 28 - Effect of Cutoff on Permeability Variation

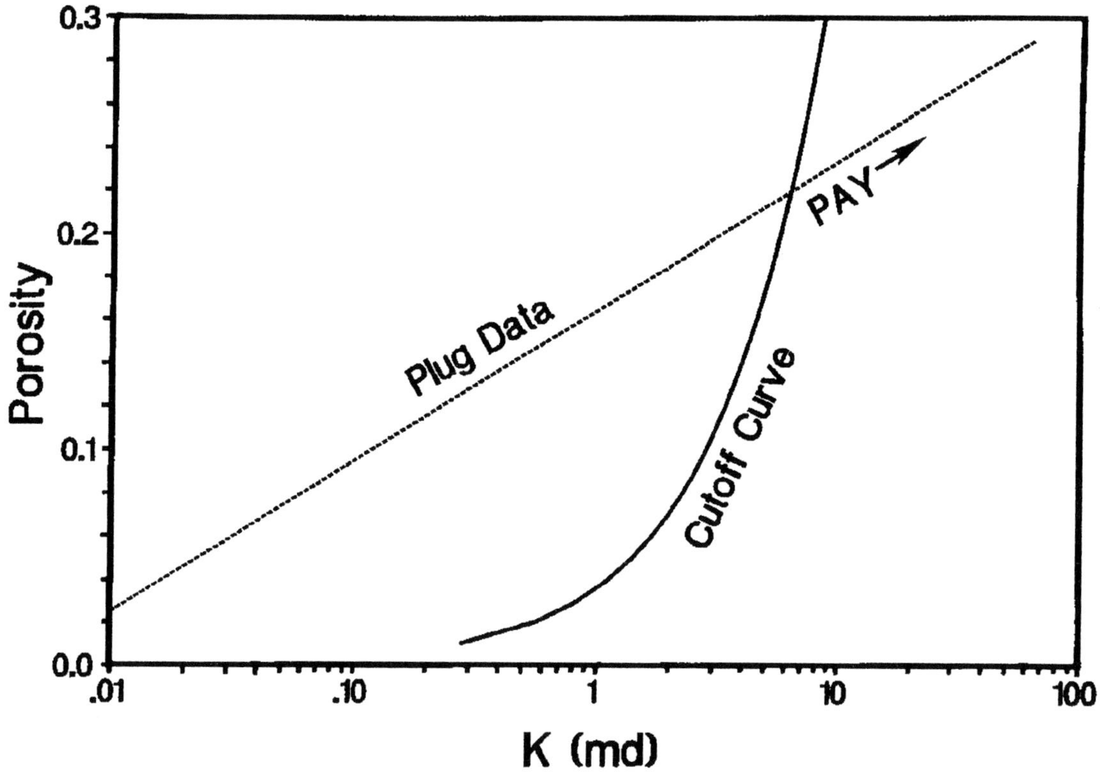

Fig. 29 - Waterflood Pay Cutoff

Fault Interpretation During Seismic Interpretation and Reservoir Evaluation

M.E. Badley,
B. Freeman
A.M. Roberts
J.S.Thatcher
Badley, Ashton & Associates Ltd
Winceby House, Winceby, Horncastle
Lincolnshire, LN9 6PB, United Kingdom

J. Walsh
J. Watterson
Fault Analysis Group, Dept. of Earth Sciences
University of Liverpool, Brownlow Street
Liverpool, L69 3BX, UK.

G. Yielding
Badley, Ashton & Associates Ltd
Winceby House, Winceby, Horncastle
Lincolnshire, LN9 6PB, United Kingdom

ABSTRACT

This paper describes the methodology and application of a number of new structural-geological techniques for analysing faults and fault systems. These are: firstly, fault-displacement analysis, which is used to establish the common fault-patterns for multiple horizons; secondly, analysis of horizon separations on correlated faults to evaluate reservoir connectivity; and thirdly, prediction of the fault-displacement population below the limit of seismic resolution. There exists a substantial methodology for achieving objectivity in the identification and mapping of geological horizons on seismic sections. By contrast, however, the mapping of faults and fault systems remains comparatively subjective with no firm methodology for deriving a 'correct' fault-map. Recent research has demonstrated that displacement varies systematically on fault surfaces and that by an analysis of displacement patterns on putative fault-correlations objectivity and methodology are introduced into fault correlation. We analyse displacement patterns on faults using interactive-graphics software. Once a consistent fault-pattern is established for all mapped horizons, reservoir connectivity throughout the area can be evaluated from horizon-separation diagrams which are produced automatically by our fault-analysis software for any selected fault. Further recent research has shown that prediction of the displacement population of faults and fractures below the limit of seismic resolution can be made from the seismically-resolved, fault-displacement population. Additional calibration of the fault-displacement population can be obtained from measurements made on cores from wells. The integration of these techniques enables a detailed analysis and description of faults and fault systems during seismic interpretation and reservoir evaluation.

INTRODUCTION

The correct interpretation of faults and fault patterns can be crucial to the viability of a prospect or to the development plan for a field compartmentalized by faults. This paper synthesizes much of our recent work on fault interpretation and presents a new methodology for the analysis of faults and fault systems, across a range of scales.

At the largest scale faults may play a central role in creating a hydrocarbon trap. Medium-size faults, or even small faults, may compartmentalize a field into discrete accumulations, and small faults may cause widespread disruption of fine-scale lithological layering. Wherever faults are present, they occur across a broad range of size and there are always many more small faults than there are large faults.

As continued improvements are made in seismic acquisition and processing techniques, it is increasingly common for fault planes to be imaged on seismic sections, enabling detailed and precise interpretation. However, while relatively-unequivocal interpretations of faults and fault relationships can be made on individual seismic sections, the correlation of faults between seismic sections is comparatively subjective. An objective methodology for the routine correlation of faults is essential in order to establish the 'correct' fault-pattern. This is especially important where fault surfaces act, or have acted, as either barriers or as conduits for fluid flow.

Even when the fault system has been mapped correctly there remains the problem of seismic resolution. Fault throws of one quarter of the seismic-wavelet wavelength are difficult to resolve on seismic-reflection data. Depending on the particular interval velocity and seismic-wavelet frequency, one quarter wavelength will be in the range of a few metres to several tens of metres, given typical exploration and production targets. Faults with throws smaller than this threshold, however, may provide critical controls on reservoir heterogeneity.

The new interpretation methodology that is described below comprises:

(1) fault-displacement analysis, used to correlate faults and establish the common fault-pattern for multiple horizons;
(2) analysis of horizon separations on correlated faults, to evaluate reservoir connectivity;
(3) prediction from seismic data of the displacement population of faults below the limits of seismic resolution.

FAULT CORRELATION

There exists a substantial methodology for the interpretation of geological horizons and faults on seismic sections. By contrast there is no widely-used, or consistent, methodology for deriving a 'geologically-correct' fault-map. It is a common observation that fault patterns produced by interpreters can be as characteristic as their handwriting.

The fault-correlation problem is shown schematically in Figure 1. The upper part of the figure shows the conventional views that are made of the sub-surface; seismic sections, interpreted to define the positions of horizons and faults, plus maps of each horizon showing the position of the fault picks on the seismic lines for each horizon. Before a contour map of an horizon can be made, the faults picked on the various seismic sections must be correlated, in order to establish the fault pattern for the horizon. The lower part of Figure 1 shows how arbitrary the fault-correlation process can be. Three of the several possible fault-patterns for the middle horizon which might be made from a limited number of seismic lines are shown. Each of the three correlations is plausible. Which, if any, of the three fault-patterns is correct? While the correlation of the large fault to the west, based on its sense of throw and larger displacement, might be made with confidence, correlation of the smaller faults is equivocal. It is at this point that the correlation process becomes arbitrary. The fault pattern that looks the most pleasing (for example, minor faults striking sub-parallel with major faults) can become the dominant criterion for fault correlation, ie. the apparent fault-pattern in the final map is very influenced by interpreter bias The problem is one of spatial aliasing because not enough data are available to reveal the true correlation. Although fault-correlation problems are most obvious on 2D data, the problem also occurs in 3D datasets where faulting is dense or where data quality is poor.

Can the correlation procedure be made more objective? The conscientious seismic interpreter will not simply join up the fault cuts posted on one horizon map, as illustrated in the top left of Figure 1. Instead, the interpreter will visually compare the faults picked on adjacent seismic lines, and base the correlation on the degree of similarity between them; a large fault on one line would probably be correlated with a large fault in a similar location on the next line. The unstated assumption here is that the throw (displacement) should vary systematically from one seismic line to the next. As is shown below, provided that the seismic line spacing is less than the actual lateral extent of the fault, this assumption is in fact well founded. However, visual comparison of faults from line to line is an inefficient way of assessing such displacement variation. A better way of analysing fault displacements and of using them to assess possible fault-correlations is to plot a strike projection of the fault.

Strike Projections

Several recent papers by the Liverpool University Fault Analysis Group have described in detail the three-dimensional displacement-fields associated with both isolated faults and fault networks (e.g. Watterson 1986, Barnett et al. 1987, Walsh & Watterson 1987, 1988, 1989, 1991, Gibson et al. 1989). The most fundamental outcome of this work, which has used both detailed mine plans and seismic-reflection surveys, is that the displacement on a fault plane varies in systematic manner. Typically, for an individual, isolated normal fault, displacement varies in a near-linear fashion from a maximum at the fault centre to zero at the fault 'tip-line'

A convenient way to combine all of the horizon and cross-section data for a particular fault is to display this information on a strike projection. The relationships between map, cross-section and strike projection are shown in Figure 2. For a steeply-dipping normal fault the displacement values are projected onto a vertical plane orientated sub-parallel with the strike of the fault.

Barnett et al. (1987) showed that small, planar normal faults typically show a pattern of sub-elliptical-shaped contours on strike projections, demonstrating that displacement decreases not only laterally and upwards, but also downwards. This observation has subsequently been documented many times (e.g. Bouvier et al., 1989) and forms the basis for an objective method for correlating faults.

Figure 3 illustrates schematically the commonly-observed, lateral variation in displacement for an isolated normal fault offsetting a single horizon and the corresponding strike-projection . The separation between the upthrown (footwall) and downthrown (hangingwall) cutoffs (i.e. the lines at which the horizon meets the fault) decreases from the centre, becoming zero at the 'lateral tips' of the fault. The vertical axis of the strike projection may be two-way time or depth, and the horizontal axis is distance along the fault. In this projection, the separation of the footwall and hangingwall cutoffs indicates the throw on the fault. If the fault is crossed by a number of seismic lines, the throw at this horizon can be measured and posted onto the strike projection in two-way time or depth.

Since a single fault will usually offset more than one picked horizon, throws and cutoff points from a number of horizons can be displayed on the strike projection for one fault. This is shown schematically in Figure 4, where throws and cutoffs are plotted for three horizons, A, B and C. Contouring the throw values reveals the vertical as well as the lateral variations in throw. Figure 5 shows throw decreasing both vertically and laterally from a maximum, on horizon C, to zero at the tip-line. The systematic variation of displacement shown both by the contours and the horizon separations in the strike projection would give confidence to the interpreter that the fault seen on the individual seismic lines is indeed the same fault.

A variety of patterns of displacement variation on fault surfaces has been observed for planar-normal, listric-normal, oblique-slip, reverse and reactivated faults (e.g. Rippon 1985, Barnett et al., 1987, Bouvier et al., 1989, Beach & Trayner, 1991, Chapman & Meneilly, 1991). The general characteristic, however, is that displacement varies systematically over the fault plane.

We propose that displacement analysis is the most objective way to assess the integrity of fault correlations between seismic lines. A potential fault-correlation showing a continuously-varying pattern is more likely to be correct than one showing erratic variations from line to line and horizon to horizon (Figure 6). An informed decision can therefore be made about the plausibility of one fault correlation in preference to others. A major advantage of strike projections is that all of the displacement data for a given fault are combined in one display.

Application of strike projections to fault correlation (FAPS)

While it is possible to construct strike projections manually, this is not a practical proposition for the evaluation of a fault network comprising more than a few faults. A computerised system where data access and display are both very rapid is required. The Fault Analysis Projection System (FAPS), developed by ourselves (Freeman et al. 1990), is such a system. FAPS, which works on UNIX-based workstations running the X-windows graphics-system, is an interactive set of programs for the analysis of horizon and fault data imported from a seismic-interpretation system (eg. Landmark, Charisma, Geoquest, etc) or digitised from paper sections. The system allows the interpreter to view, using separate windows on the graphics screen, cross-sections displaying the horizon and fault picks, plus map views of the fault cuts at each horizon. Possible fault-correlations are performed interactively by selecting, in map view, fault cuts on the screen. A strike projection, showing horizon separations and contours of displacement for a chosen fault-correlation, is displayed in a new window. The interpreter judges whether the correlation is admissible on the basis of the continuity in displacement variation.

The displacement pattern may be unsatisfactory for a number of reasons:

(1) An attempt has been made to correlate fault picks which are not part of the same fault. Figure 6 shows a strike projection (redrawn from FAPS) with an irregular displacement-pattern. It is likely that the fault cuts joined for this strike projection are not part of the same fault. A contour map at horizon C (in the vicinity of this fault) would look reasonable whereas maps made at deeper and shallower horizons would be unreasonable. In such a case correlations between other fault cuts should be viewed, in order to establish a more likely fault-correlation.

(2) There may be an incorrect fault pick or horizon pick in the interpretation. In this case the pick can be modified by on-screen graphics editing, after re-examination of the original data.

(3) Seismic data may be absent or of poor quality. In this case horizon and fault picks can be extrapolated into the regions of poor data-quality.

Once a fault correlation is considered acceptable the database is automatically updated with all relevant changes and corrections. The correlation is automatically made at all horizons at which the fault has been picked. Thus the fault pattern defined for any one horizon is known to be consistent with the fault pattern for all other horizons at which the fault occurs. This three-dimensional and structurally-consistent fault-interpretation is the optimal starting point for the production of contoured horizon-maps. On completion of an interpretation the dataset can be exported to a mapping package.

An example using offshore seismic data

The figures and text below describe aspects of fault correlation made during the interpretation of part of an offshore seismic-grid. Figures 7 to 11 are redrawn from FAPS screen images. Figure 7 shows an horizon map for part of the survey, with dip-lines spaced at 500m (strike-lines were used during the seismic interpretation, but are not shown on this figure). Two main features are apparent. Firstly, a major fault, throwing down to the left, has been correlated across all the lines. Secondly, a large number of minor fault-cuts have been posted but not yet correlated by the interpreter.

Figure 8 shows two strike projections for the main fault. North is to the left in each plot, and the vertical axis is two-way time. In Figure 8a, the footwall and hangingwall cutoffs for each horizon have been plotted, together with the measured throws at each line/horizon intersection. Also shown are the upper tips of the faults as seen on the seismic sections. In Figure 8b, the posted throw-values have been contoured. It is apparent from both diagrams that throw increases progressively from north to south on this major fault, and does so fairly systematically. Minor irregularities in some of the contours suggest areas where the interpretation might be re-examined. On the whole, however, the smooth variation in fault throw over the surface of this fault is typical of a 'good', or at least plausible correlation. The steeper displacement-gradient shown on the upper part of the fault is related to the presence of an unconformity between the upper tip-line and the highest picked-horizons (most of the fault displacement pre-dates the unconformity).

Figure 9 illustrates a first attempt at correlating the minor faults in the footwall of the main fault. In map view (Figure 9a), a set of fault cuts has been joined along a trend sub-parallel to the strike of the main fault, on the assumption that large and small faults are likely to have similar orientations. Figure 9b shows, on a strike projection, the throw values for this set of fault cuts. The pattern is highly irregular, with moderate values (20-40ms) on three lines and much smaller values (0-6ms) on the two northern lines. This irregular contour pattern is unlike that seen on well-constrained faults. On the basis of this plot, the fault correlation shown in Figure 9a would be deemed unacceptable.

Figure 10 shows an alternative correlation of fault cuts in this area. The map (Figure 10a) highlights an alignment of picks oriented N-S, at about 45° to the main fault. The strike projection in Figure 10b shows contoured throws for this set of picks. The pattern is broadly elliptical, with a displacement high near the centre and an outwards decrease in values. Observed tips on the northern line strongly suggest that a lateral tip is present immediately to the north. On the basis of this plot of throw contours, we would consider the fault correlation as acceptable, i.e. there is a much greater chance that the correlation shown in Figure 10 is correct rather than that in Figure 9.

However, one major problem remains with the throw-contour plot of Figure 10. On line 9, a large throw (57ms) has been interpreted on the uppermost horizon. This does not accord with the throws seen on adjacent lines, or at deeper horizons on this line. This observation would prompt the interpreter to re-examine the picks on the seismic data. In this example, the 57ms throw value results from a mis-pick on the downthrown side of the fault. This mis-pick had not been identified by standard loop-typing because it occurred so locally (over a few tens of traces adjacent to the fault). Figure 11 shows the result of re-picking this fault cut and then re-analysing the variation in throw over the whole fault. The isolated displacement-high has been removed, and a simple pattern of variation in throw is established.

Therefore, in this brief example we have:

(1) confirmed the correlation of the main fault, and the integrity of the horizon picks around it,
(2) tested two possible fault-correlations amongst the small fault-cuts, and identified one correlation as being much more likely than the other,
(3) identified a probable mis-pick on one horizon on one line, and
(4) progressed towards an objective map of the horizons, incorporating fault correlations and corrected horizon picks.

The example above illustrates the basic methodology involved in the use of strike projections for fault correlation. By sequential iterative evaluation of strike projections within a seismic grid the most 'correct' fault-pattern is gradually assembled.

Fault patterns and fault linkage

Strike projections constrain correlations between fault picks on adjacent seismic lines and allow the prediction of fault throw at locations between seismic lines. They can also be used to extrapolate displacements in the vertical direction. This can be useful if poor data-quality requires prediction of whether a fault picked high on a section, extends down into a reservoir and of what its throw might be at this level (Figure 12).

Additional structural-geological information can be used to aid the interpretation of strike projections and assess the likely lateral extent of faults. There are considerable data, summarized in Figure 13 for a wide range of fault sizes, on the relationship between the mapped strike-length of faults and the maximum observed displacement. Figure 13 shows a well-defined band of data points. The width of the band spans about 1 order of magnitude of displacement range, and the slope is ca. 1.5. The width of the band is mainly a function of shear modulus (rock strength) at the time of faulting; soft rocks can accommodate a given displacement on a shorter length of fault plane than can hard rocks. These relationships can be used to predict the minimum, lateral extent of a fault for a given displacement, i.e. whether or not the observed fault-throw on one seismic line is sufficiently large for the fault to extend to and intersect a neighbouring seismic line?

During fault correlation the problem of how neighbouring faults relate to each other in map view must be considered. Are the faults isolated dislocations or are they linked in a network? Faults formed contemporaneously will be linked kinematically, but need not link physically to each other. Figure 14 shows conceptually how the uniform, bulk deformation of a block of material is accommodated by the non-uniform pattern of isolated, en-echelon and physically-linked faults. The linkage can be either hard (faults join physically) or soft (displacement on adjacent faults is linked by a combination of ductile strain and isolated small faults, many of which will be below the limit of seismic resolution).

Figure 15 is a cartoon illustrating hard linkage. The strike projection of Fault 2, which trends NE to SW, indicates that if the observed displacement-gradient is extrapolated to the west, the fault should extend at least to X on the illustrated horizon-map. No fault cut is observed near location X. However, the N-S Fault 1 has been correlated to the immediate west of the last fault-cut observed for Fault 2 but to the east of location X. The displacement pattern on Fault 1 also shows a strong asymmetry, with a marked reduction in displacement to the north of the point where the extrapolated westward extension of Fault 2 would meet Fault 1. The combination of a converging trend, a strike-length for Fault 2 much less than that predicted, and an asymmetry of displacement on Fault 1 about the projected intersection point of the two faults, indicates that the faults physically join and are hard linked. Summing the displacements of Fault 2 with those of Fault 1 produces a more symmetric displacement pattern.

Figure 16 shows a sub-surface example of soft linkage. A large, N-S-trending fault (Fault A) has been correlated through a seismic grid. The strike projection for Fault A shows a regular pattern of displacement for values of less than 100ms but a shortfall in displacement at higher values occurs below 1.75s on seismic lines 8 to 20. In the hangingwall, on lines 8 to 20, a small, synthetic fault (Fault B) occurs, which is mapped as trending parallel to Fault A, but not observed to join it physically. Summation of the displacements on Faults A and B produces a more regular displacement pattern on the strike projection (Figure 16d). This regularity can be interpreted to indicate that, although the two faults do not appear physically to join, they are nevertheless kinematically linked and behave as a single fault. Zones of soft linkage are zones of locally-high strain and small-scale fracturing. Recognition of soft linkage between faults may therefore be important for the evaluation of reservoir properties in the intervening rock volume.

FAULT SEAL AND ACROSS-FAULT CONNECTIVITY

Once the fault pattern has been established, strike projections can be used to evaluate potential reservoir-connectivity across faults or the likelihood of particular faults providing a reservoir seal. Superposition of the footwall and hangingwall cutoffs converts the strike projection into a horizon-separation diagram and reveals juxtaposition relationships across the fault. Preparation of horizon-separation diagrams for fault planes is a common technique in prospect evaluation but, done manually, is time consuming. Production of such diagrams is routine, however, using a system such as FAPS. Figure 17 shows an example of an horizon-separation diagram and its application to the evaluation of across-fault connectivity.

Allan (1989) described the general application of horizon-separation diagrams to the evaluation of migration and entrapment within faulted structures. However, he did not include displacement variation across the fault surface in his illustrations or discussion. Figure 18 illustrates how displacement variations may be a primary control on across-fault communication. Bouvier et al. (1989) presented example horizon-separation diagrams for a fault in the Nun River Field, Nigeria, across which multiple, water-bearing and hydrocarbon-bearing, sandstone reservoirs are juxtaposed. Additionally they calculated the potential for clays to be smeared along the fault plane, thus extending the fault's sealing characteristics beyond simple sand to shale juxtaposition. Bouvier et al. (1989) empirically derived the variation in clay-smear potential across the fault surface by relating the clay source-bed thicknesses and position of clay source-beds to the displacement at a given point on the fault surface. The incorporation of varying clay-smear potential and stratigraphy into strike projections provides a powerful, new technique for the evaluation of fault-sealing potential. Field studies indicate that complete smearing on fault surfaces may extend down fault planes from shale footwall-cutoffs for distances equivalent to at least six times the thickness of the shale source-layer (Lindsay et al., in review).

PREDICTING THE NUMBERS OF FAULTS BELOW THE LIMIT OF SEISMIC RESOLUTION

Seismic data provide information on faults where displacements are above the limit of seismic resolution. It is widely accepted that this resolutional limit approximates to one quarter of the wavelength of the dominant, seismic frequency. This translates to fault throws in the range of 10-40m for commonly-encountered interval-velocities and seismic-wavelet frequencies. While seismic data provide information on the spatial distribution of larger faults, well data provide spatially-isolated, essentially-one-dimensional information on the density of fractures encountered in cores. Reservoir performance may be significantly affected by areally-distributed small faults, unresolvable on seismic data.

The effect of faults on reservoirs is twofold. Firstly, the bulk permeability of a reservoir layer may be impaired by the offsets at faults, or it may be enhanced if depositionally-isolated, reservoir units are juxtaposed. Secondly, the faults themselves are likely to have permeabilities different to those of their hosting rocks and therefore may act as either conduits or barriers to fluid movement. To understand the impact of faults on reservoirs requires that the fault network can be fully characterised, and an important first step in this process is knowing how many faults of different sizes are present in the rock volume of interest, i.e. what is the fault-displacement population? However, seismic-reflection data only resolve the larger faults (offsets > ca10-40m), and well cores generally provide information on only the smallest features. The information gap between the scales represented by seismic data, on the one hand, and well core on the other represents a potentially-critical weakness in the assessment of many reservoirs.

Fault-displacement populations

Can more information about the unimaged faults be extracted from the seismic and well data? Several lines of evidence indicate that this may be possible.

The systematic and fractal nature of fault populations has been the subject of recent interest by ourselves and others (King, 1983, Hirata, 1989, Childs et al., 1990, Heffer & Bevan, 1990, Sassi et al., in press, Walsh & Watterson, in review, Walsh et al., in review, and Marrett & Allmendinger, in review). These studies have indicated that the displacement population of smaller faults can be predicted from the displacement population of larger faults.

Figure 18 shows a typical example of a log-log plot showing the cumulative frequency of fault displacements with throws greater than a particular size, as imaged on a grid of seismic data. This particular plot has been generated by measuring all of the throws observed on sample lines at a single horizon within the grid. This information can either be measured directly on seismic lines, derived from a map, or be extracted from a FAPS or workstation file. The cumulative-frequency data have been normalised to numbers of fault displacements per unit-length of line. Accounting for fault strike allows the population-distribution curve to be expressed in terms of the fault-trace-length per unit area of horizon, for different values of displacement. A similar type of distribution can be generated for several horizons within a seismic dataset with cumulative frequency expressed in terms of fault-surface-area per unit-volume of rock, again for different values of displacement.

The curve in Figure 18 has three distinct segments:

(1) A steep, right-hand portion reflecting the multiple-sample-line nature of the population. For single sample-lines the population curve is ideally straight over its entire length with no right-hand segment (Walsh, et al., in review).

(2) A flat, left-hand segment consisting of values below the limit of resolution of the dataset. Below the limit of seismic resolution fault-throws will not be fully recorded, with the degree of under recording becoming more severe with decreasing fault-throw.

(3) A straight, central segment, which shows the slope of the displacement population. The slope of the displacement-population curve is a key feature as it defines the relative frequency of faults of different size. The position of the curve expresses the absolute frequency of all faults. A low slope, e.g. -0.5, indicates that the increase in number of fault displacements as their size decreases is rather modest, whereas a high slope, e.g. -1.0, indicates a very rapid increase in number with decrease in size.

The displacement population of faults below the limit of seismic resolution can be predicted by extrapolating the population slope towards lower values of displacement. The number of observed shear fractures measured in a well drilled on the same structure as the seismic data is also plotted in Figure 18. The projected population-curve, extrapolated over three orders of magnitude of fault-displacement size, agrees closely with the observed shear-fracture distribution seen in the well. Our research on numerous datasets has shown that, providing there is at least one order of magnitude variation in fault values of throw (e.g. 20ms to more than 200ms) on faults within an area, the fault-displacement population below the limit of seismic resolution, can be predicted down to displacements of centimetres.

Factors affecting fault displacement populations

Fractures and joints

An important distinction must be made between faults (i.e. shear fractures with perceptible offset) and joints (fractures with no perceptible shear-offset, probably of tensile origin). In favourable circumstances (i.e. with fine-scale lamination in the host rock), it might be practical to identify faults with displacement as little as 1mm. In less-favourable circumstances (i.e. more-massive rocks), faults with substantially-larger displacements than 1mm would be indistinguishable from joints. There is no reason that faults and joints should be part of the same displacement population, since they have different origins. Consequently all predictions of fault-displacement populations relate to shear fractures only.

Sampling effects

A fundamental requirement of the sampling is that a large-enough range of fault displacements is included. As a general rule, a range of at least one order of magnitude of displacements must be sampled in order to achieve an estimate of the displacement-population characteristics. For example, if a seismic dataset has an effective limit of resolution (for fault offsets) of 20ms, fault displacements at least as large as 200ms are needed. In some circumstances this requirement will mean that the seismically-resolved, fault-displacement population cannot be fully characterised. However, in general the condition can be met by making full use of the available data, e.g. interpreting downthrown terraces as well as crestal areas of a structure. Even if such interpretations are not constrained by well ties they provide data which greatly improves the fault-population characterisation.

The effects of lithology

We have found that where a number of pre-faulting horizons in the same dataset have been analysed, their individual, displacement-population curves tend to be very similar with minor but real differences in slope. Because seismically-imaged faults are usually quite large compared to the scale of lithological layering that may be mapped, any lithological effect governing the fault-displacement population is likely to derive from the bulk properties of the sequence in which the faults are present. For example, displacement populations of faults cutting an alternating sequence of sandstone units and shale units will be controlled by the bulk properties of the succession rather than exhibiting different displacement populations in the sandstones and shales. However, as the fault size decreases the influence of lithology becomes more important. The displacement population of small faults will be lithologically controlled where the faults occur within thick, ductile units (e.g. salt, overpressured shale).

Applications

Mapping spatial variations in the size and number of faults below the limit of seismic resolution

In general, fault-displacement populations are not distributed homogeneously through the rock volume. Seismological and geological studies suggest that the fault distribution is hierarchical, at least to a certain degree. Thus smaller faults relieve the strain discontinuities around larger faults, at all scales. The spatial distribution of faults is also likely to have a fractal character, in addition to the size distributions described above.

Spatial variations in the distribution of faults can be examined in two ways. Firstly, the population parameters can be determined on a sample-line by sample-line basis, throughout the dataset. Secondly, the dataset can be divided into subareas each containing a small number of lines or portions of lines, and population parameters determined for each subarea. The choice between these two methods involves a trade-off between spatial resolution and population definition: the sample-line method may divide the area into smaller samples, but if these samples have too few data-points then the population characteristics may be difficult to define.

Variations in both population slope and absolute fault-density may be significant. On individual sample lines, population slopes are determined by least-squares regression to all data-points above the effective limit of seismic resolution. The spatial variation in slopes can then be contoured and, when combined with cumulative-frequency data, may be used to map the lateral variation in the numbers of faults of different sizes (Figure 19). Figure 19(a) shows the main faults observed in an offshore area. A c.E-W-trending transfer-zone is interpreted to divide a more-complexly-faulted area to the north from a less-complexly-faulted area to the south. In order to analyse the spatial variation in numbers of faults the map has been divided into subareas. Figure 19(b) shows the contoured values of the cumulative number of fault displacements, per kilometre of sample line, above the limit of seismic resolution (20ms) in the area covered by the map in Figure 19(a). To use this information to predict the density of faults below the limit of seismic resolution a contour map (Figure 19(c)) showing the variation in population slopes throughout the area is also required. The contour patterns in Figures 19(b) and (c) show similar, overall patterns with steeper population-slopes and greater cumulative-frequencies to the north of the interpreted transfer-zone. The population slopes and cumulative-frequency data define the power-law relationship which can be used to map the spatial variation in the density of fault displacements within a given range of fault throws.

The strike and orientation of the faults with throws below seismic resolution are not predicted and must be deduced using standard structural-geological methods.

Horizontal wells

A prediction of the number of faults of a given size likely to be encountered by a horizontal well may be crucial to well planning. An understanding of the fault-displacement population encountered by a horizontal well may allow better interpretation of production results. If a horizontal well is designed to follow, for example, a thin, high-permeability layer, then knowing the number of faults which the well is likely to encounter will determine the practicality of drilling the well to specification. The number of faults with throws less than a given value, including those with throws less than the limit of seismic resolution, which are likely to be encountered by a horizontal well can be calculated from fault-displacement-population data.

Reservoir connectivity and reservoir-permeability impairment

The contribution of faults to permeability enhancement and impairment can be estimated from fault-displacement-population data. Heffer and Bevan, (1990) discussed how the juxtaposition across non-sealing faults, of thin-bedded sands, in an otherwise vertically-isolated sequence, may have led to enhanced permeability and productivity in a

North Sea reservoir. The use of fault displacement population data combined with information on sand-body dimensions and net:gross ratios enabled the prediction of connectivity within the reservoir. Using similar methodology it may be possible to calculate the likely reservoir impairment resulting from permeability reduction in fault zones.

CONCLUSIONS

We have presented a new and integrated methodology for the analysis of faults (across a range of scales) during seismic interpretation and reservoir delineation. An important feature of the methodology is that it uses existing and available data and does not require significant modification or additions to existing interpretation practices. In common with other recent developments in exploration and production, however, the methodology does require the extensive use of computer-based, interpretational techniques. The methodology may be formalised as follows:

Step 1: Establish fault correlations and fault pattern

Establish the most likely fault-correlations using displacement patterns on strike projections. This procedure yields an internally-consistent fault-pattern for all mapped horizons. The methodology enables comparisons between different interpretations of the same data. It is easy to document the reasons why one fault correlation is favoured over another.

Step 2: Assess reservoir continuity across faults

Evaluate reservoir continuity across the established fault-pattern using horizon-separation displays.

Identify areas where faults may be kinematically related, but not necessarily hard-linked; such relationships may affect the reservoir properties of the enclosing sediments.

Step 3: Predict fault-displacement populations below the limit of seismic resolution.

Extract the observed fault-displacement population.

Using the observed fault-displacement population derive the displacement population of faults with throws below the limit of seismic resolution.

ACKNOWLEDGMENTS

We thank David Kemp for drafting the figures.

REFERENCES

Allan, U.S., 1989, Model for hydrocarbon migration and entrapment within faulted structures: AAPG Bulletin, v. 73, p. 803-811.

Barnett, J.A.M., J. Mortimer, J.H. Rippon, J.J., Walsh and J. Watterson, 1987, Displacement geometry in the volume containing a single normal fault: AAPG Bulletin, v. 71, p. 925-937.

Beach, A. and P. Trayner, 1991, The geometry of normal faults in a sector of the offshore Nile Delta, Eygpt, in A.M. Roberts, G. Yielding and B. Freeman, eds, The Geometry of Normal Faults: Special Publication of the Geological Society, London, v. 56.

Bouvier, J.D., C.H. Kaars-Sijpesteijn, D.F. Kluesner, C.C. Onejejekwe and R.C.van der Pal, 1989, Three-dimensional seismic interpretation and fault sealing investigations, Nun River Field, Nigeria: AAPG Bulletin, v. 73, p. 1397-1414.

Chapman, T.J. and A.W. Meneilly, 1990, Fault displacement analysis in seismic exploration: First Break, v. 8, p. 11-22.

Childs, C., J.J. Walsh and J. Watterson, 1990, A method for estimation of the density of fault displacements below the limits of seismic resolution in reservoir formations, in North Sea Oil and Gas Reservoirs - II, Graham & Trotman, p. 309-318.

Freeman, B., G. Yielding and M.E. Badley, 1990, Fault correlation during seismic interpretation: First Break, v. 8, p. 87-95.

Gibson, J.R., J.J. Walsh and J. Watterson, 1989, Modelling of bed contours and cross-sections adjacent to planar normal faults: Journal of Structural Geology, v. 11, p. 317-328.

Heffer, K., and T. Bevan, 1990, Scaling relationships in natural fractures - data, theory and applications: Society of Petroleum Engineers, Paper No. 20981.

Hirata, T., 1989, Fractal dimension of fault systems in Japan: fractal structure of rock geometry at various scales, in C.H. Scholz and B.B. Mandelbrot, eds, Fractals in Geophysics: Birkhäuser, Basel, p. 157-170.

Lindsay, N.G., F. Murphy, J.J. Walsh and J Watterson, in review, Outcrop studies of shale smears o fault surfaces: submitted to Sedimentology.

King, G.C.P., 1983, The accommodation of large strains in the upper lithosphere of the earth and other solids by self-similar fault systems: the geometrical origin of b-value: Pageoph, v. 121, p. 761-815.

Marret, R., and R. Allmendinger, in review, Estimates of extension from balance cross-sections: are small faults important? Submitted to Geology.

Rippon, J.H., 1985, Contoured patterns of the throw and heave of normal faults in the Coal Measures (Westphalian) of north-east Derbyshire: Proceedings of the Yorkshire Geological Society, v. 45, p. 147-161.

Sassi, W., S.E. Livera and B.P.R. Caline, in press, Quantification of the impact of sub-seismic scale faults in Cormorant Block IV, UK Northern North Sea, in: Structural and Tectonic Modelling and its Applications to Petroleum Geology,ed. Norsk Petroleums Forening, Graham & Trotman.

Walsh, J.J. and J. Watterson, 1987, Distribution of cumulative displacement and seismic slip on a single normal fault surface: Journal of Structural Geology, v. 9, p. 1039-1046.

Walsh, J.J. and J. Watterson, 1988, Analysis of the relationship between displacements and dimensions of faults: Journal of Structural Geology, v. 10, p. 239-247.

Walsh, J.J. and J. Watterson, 1989, Displacement gradients on fault surfaces: Journal of Structural Geology, v. 11, p. 307-316.

Walsh, J.J. and J. Watterson, 1991, Geometric and kinematic coherence and scale effects in normal fault systems, in: A.M. Roberts, G. Yielding and B. Freeman, eds, The Geometry of Normal Faults: Special Publication of the Geological Society, London, v. 56, p. 193-203.

Walsh, J.J. and J. Watterson, in review, Populations of faults and fault displacements and their effects on estimates of fault related regional extension: submitted to Journal of Structural Geology.

Walsh, J.J., J. Watterson and G. Yielding, in review, The importance of small-scale faulting in regional extension: submitted to Nature.

Watterson, J., 1986, Fault dimensions, displacements and growth, Pageoph, v. 124, p. 365-372.

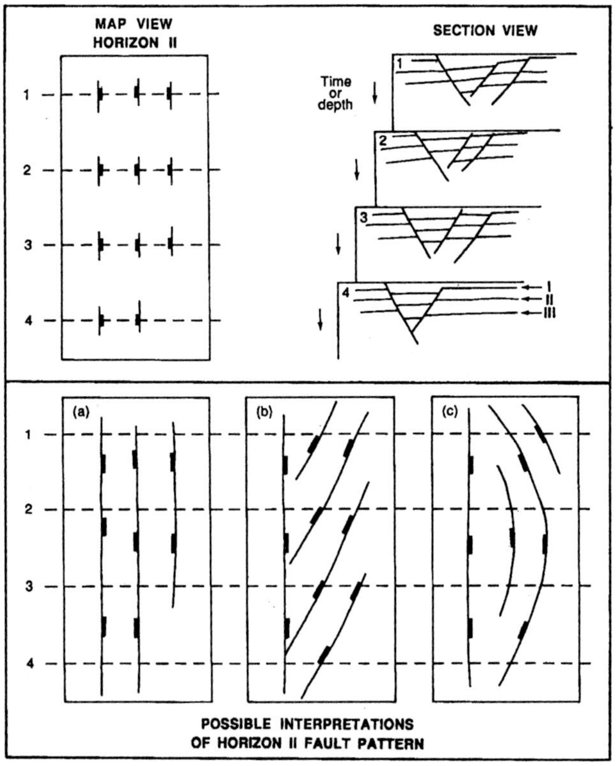

Figure 1—Cartoon illustrating that the correlation of faults seen on seismic sections is not necessarily simple or straightforward.

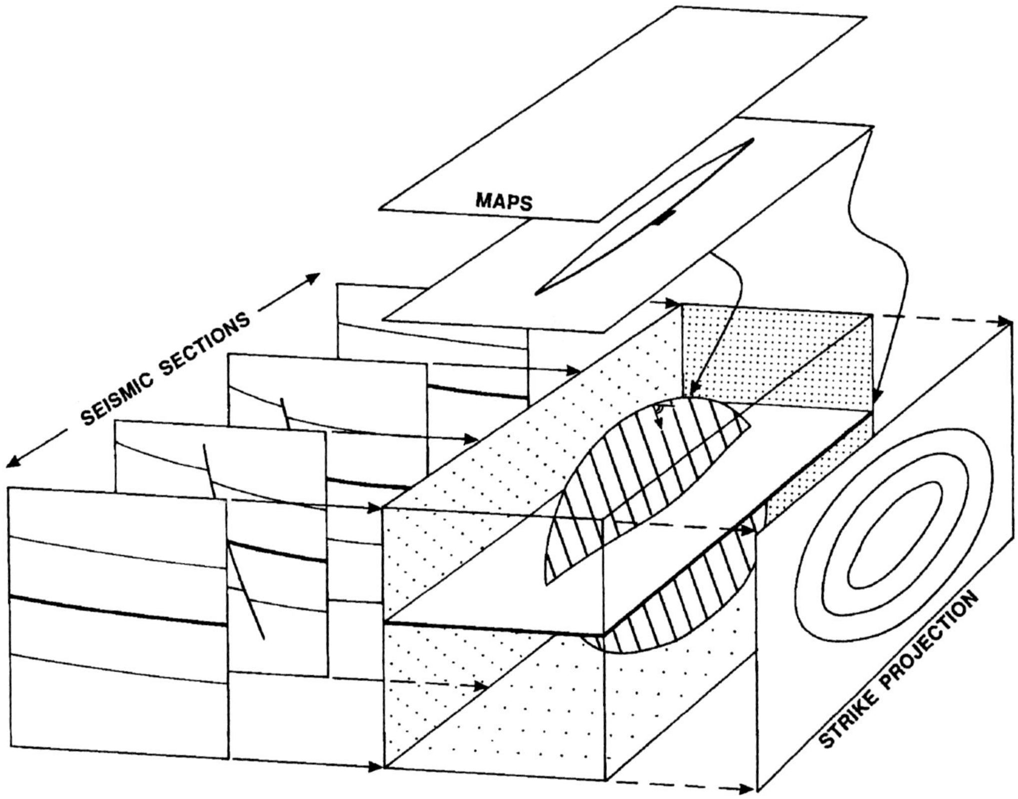

Figure 2—Cartoon showing the 3-dimensional aspects of faults and the relationships between the different views of faults provided by seismic sections, maps and strike-projections.

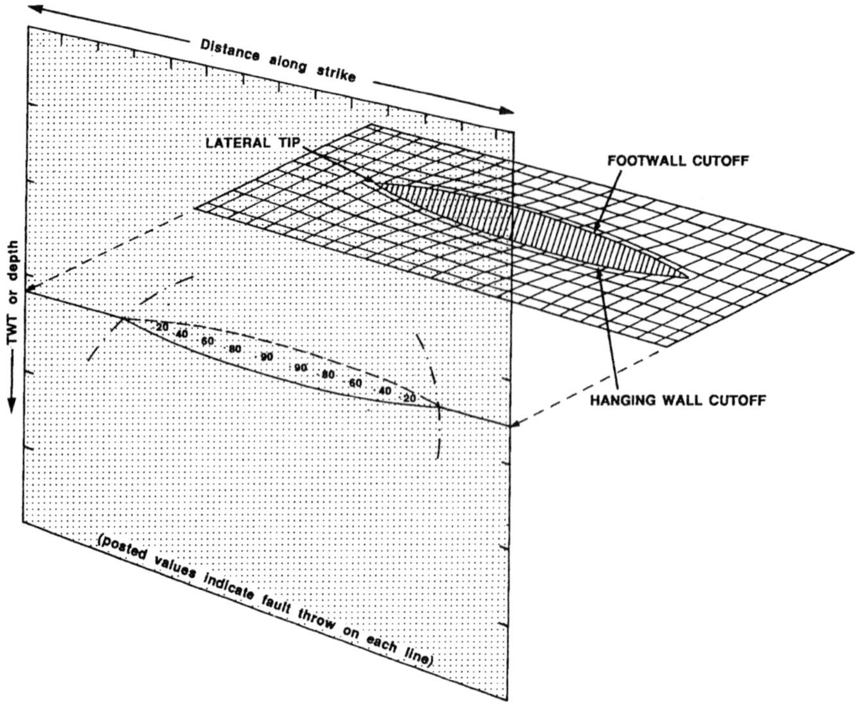

Figure 3—Diagram showing a single normal fault cutting a single horizon and the vertical strike-projection of the fault cut. The posted values indicate the throws observed on successive seismic sections crossing the fault.

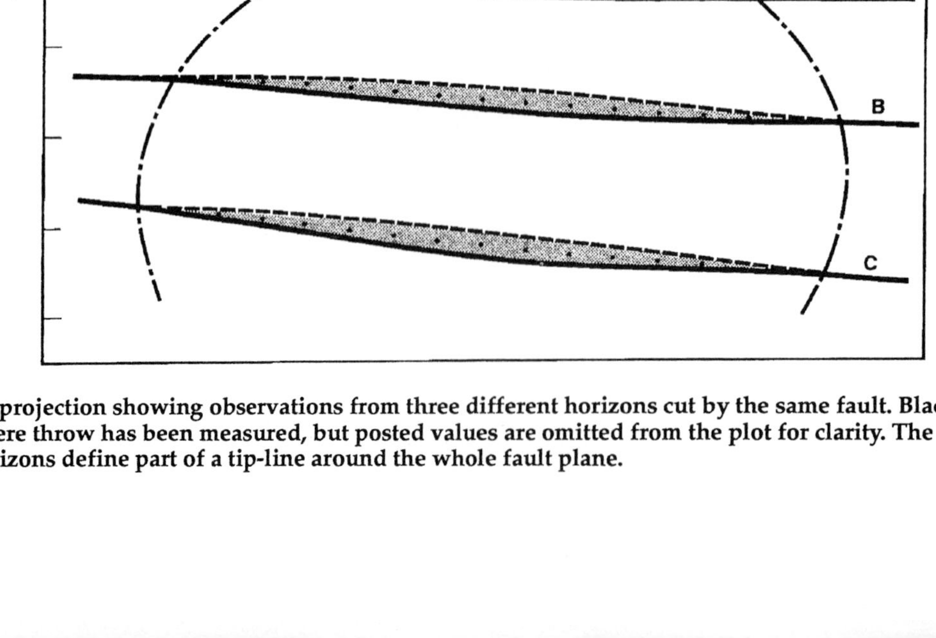

Figure 4—Strike-projection showing observations from three different horizons cut by the same fault. Black dots indicate the positions where throw has been measured, but posted values are omitted from the plot for clarity. The lateral tips seen at each of the horizons define part of a tip-line around the whole fault plane.

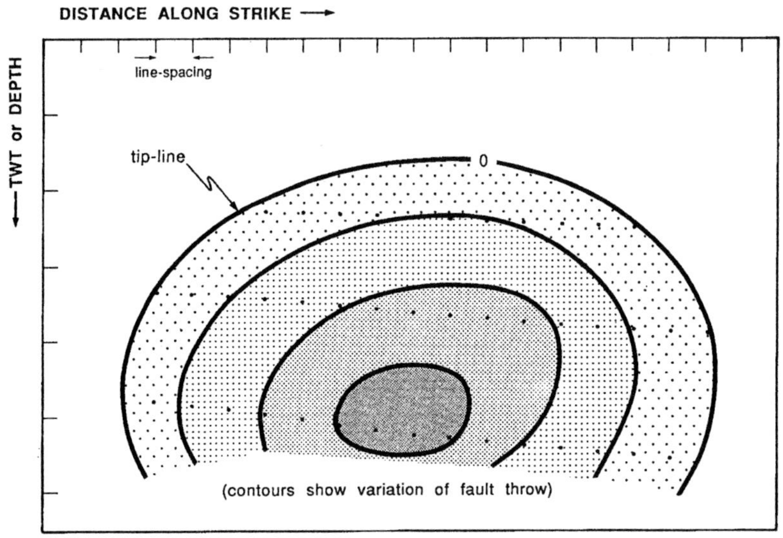

Figure 5—Strike projection based on Figure 4, but showing contours of fault throw (based on the posted throws). The contour values decrease outwards from a maximum near the centre of the fault to zero at the tip-line, i.e. displacement varies laterally and vertically. This broadly-elliptical contour pattern is typical of that observed on well-constrained datasets (Walsh and Watterson 1987, 1988, 1990).

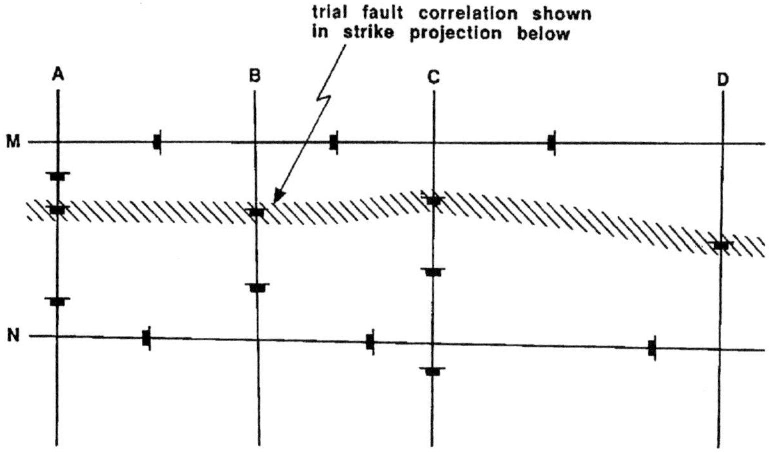

Fault-cut map for Horizon C

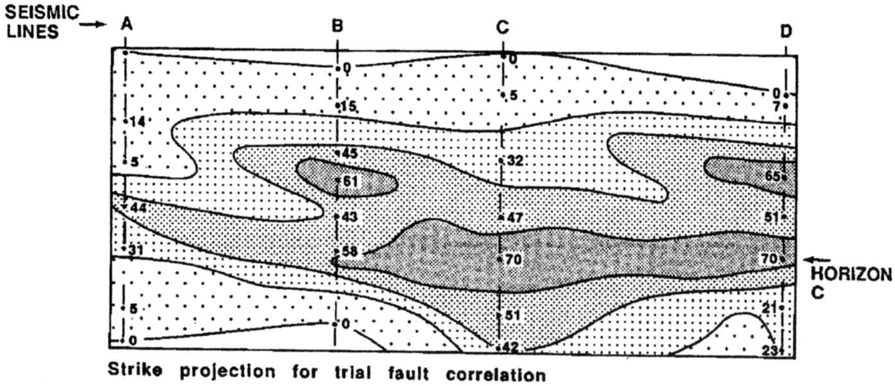

Strike projection for trial fault correlation

Figure 6—Example of fault correlation which may be incorrect. The strike projection (redrawn from FAPS) shows contoured values of throw for the trial fault-correlation shown in the map. The irregular displacement-pattern indicates that the fault correlation may be incorrect and that other possibilities should be evaluated.

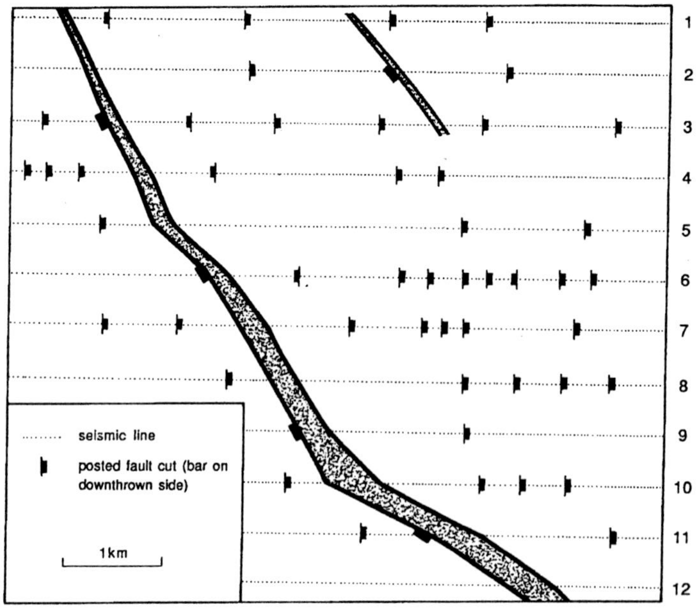

Figure 7—Map of a faulted horizon from a partially-interpreted area of an offshore survey. A major fault has been correlated across the survey, but minor fault-cuts remain uncorrelated.

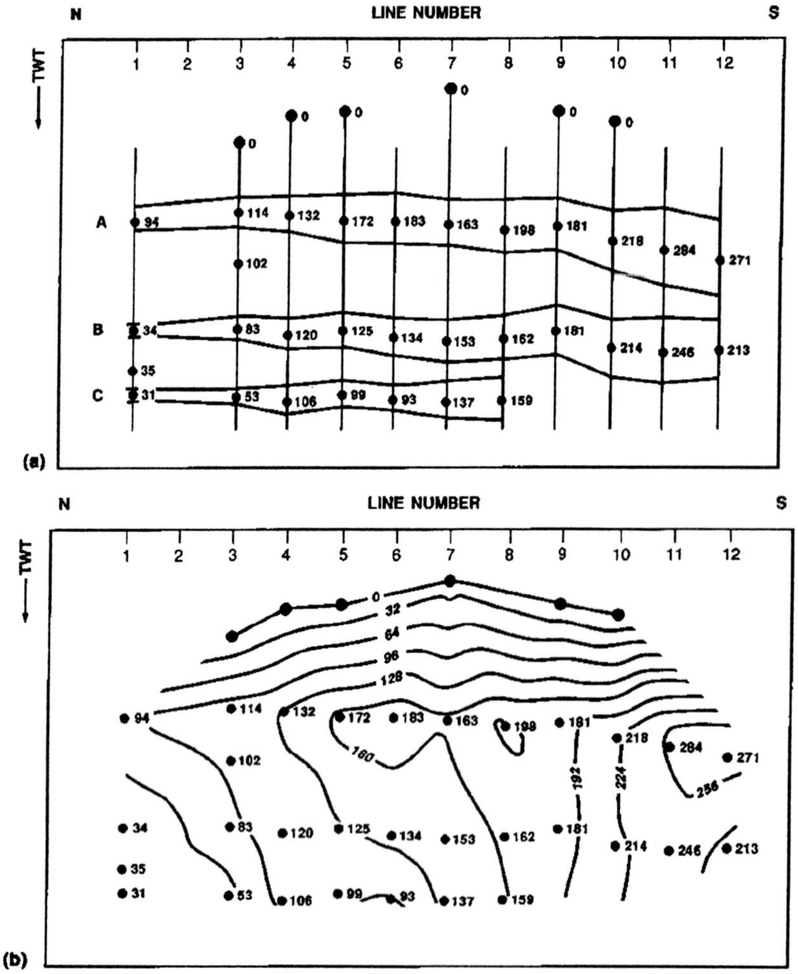

Figure 8—Strike projection displaying displacement data for the major fault in Figure 7. (a) Separation diagram showing footwall and hangingwall cutoffs, posted throw values and fault traces. Note the general increase in displacement from the north (left) to the south (right). (b) Contour plot based on the posted values of throw in (a). Note the smooth variation in fault throw along the correlated fault-surface.

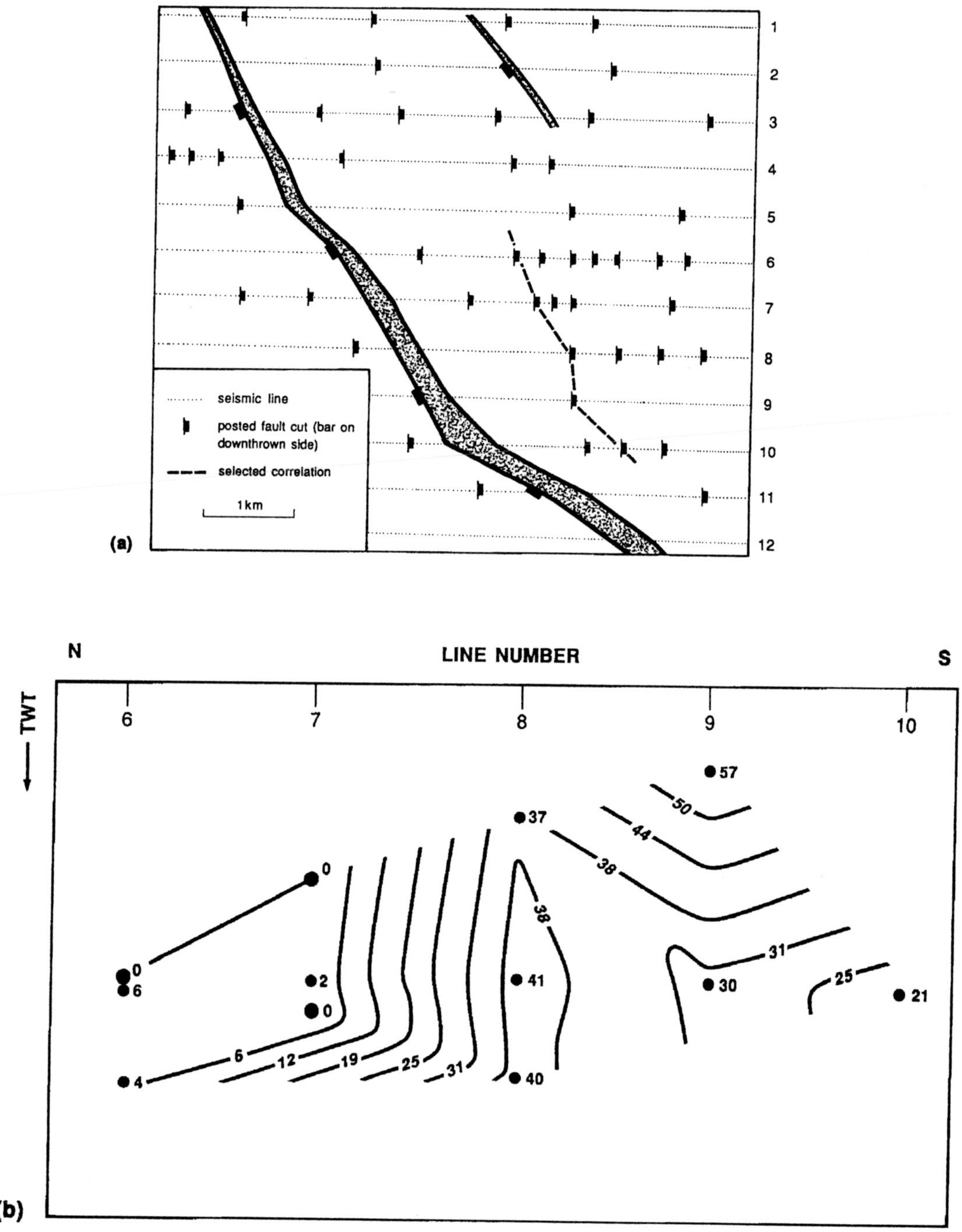

Figure 9—First attempt at correlating minor faults from section to section. (a) A possible correlation, sub-parallel to the main fault. (b) A strike projection of these correlated fault-cuts with throws posted and contoured. The irregular pattern indicates that this is unlikely to be a correct fault-correlation.

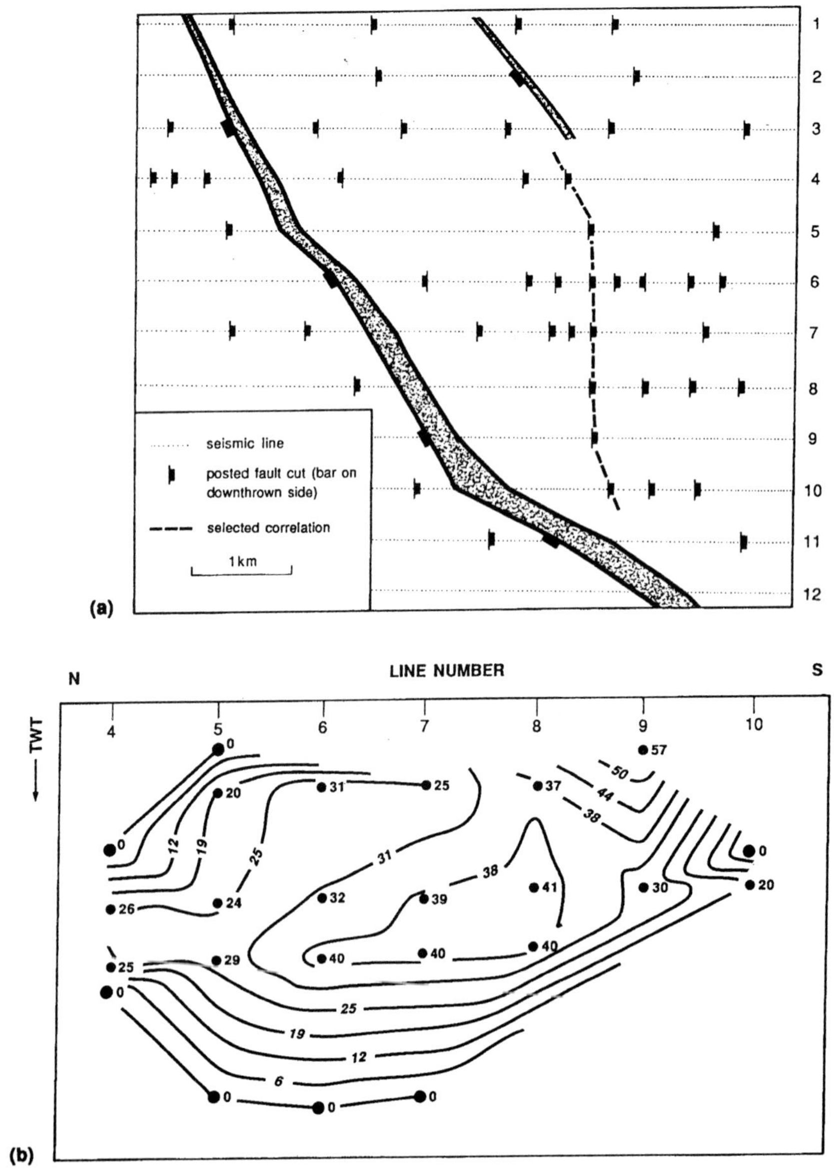

Figure 10—Second attempt at correlating the minor faults. (a) The potential fault-correlation in map view. (b) The strike projection of contoured throws. The overall, semi-elliptical pattern suggests that this fault correlation is plausible. However, a separate contour-high exists near the right-hand end because of a local mis-pick.

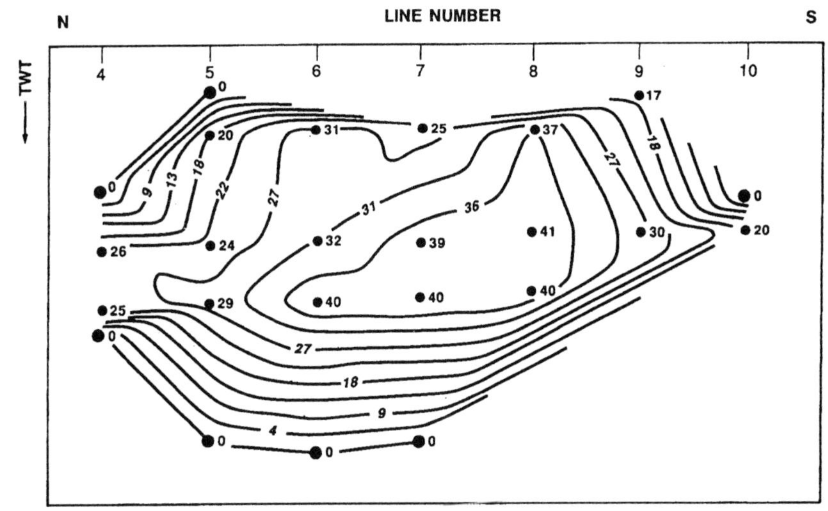

Figure 11—Re-display of the strike projection for the fault in Figure 10, after correction of the mis-pick on one horizon of one line.

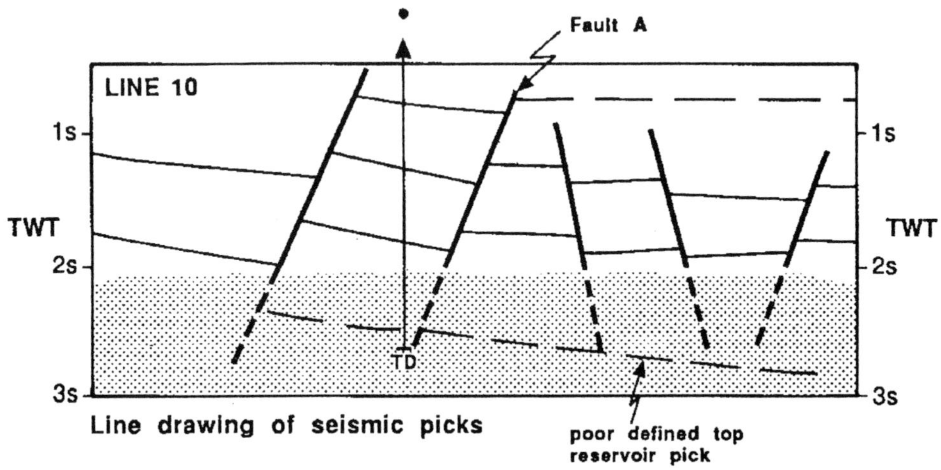

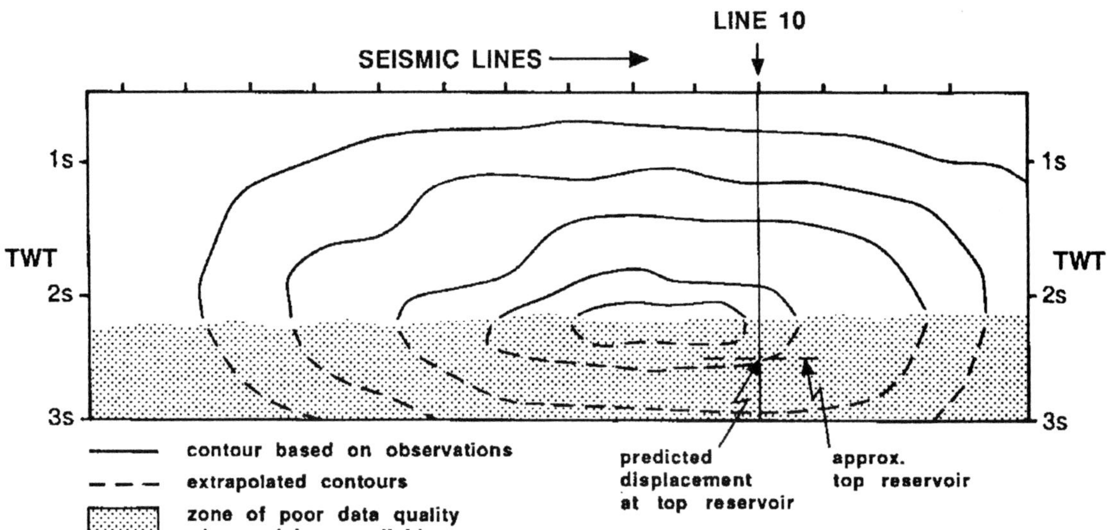

Figure 12—Cartoon showing application of strike projections to predict fault throw at a location where poor data quality does not allow direct picking of the horizon or fault.

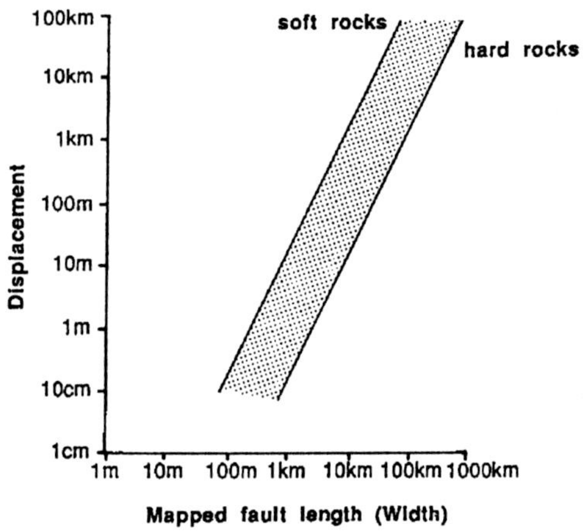

Figure 13—Summary of observations showing the relationship between fault strike-length and displacement (after Walsh and Watterson, 1988).

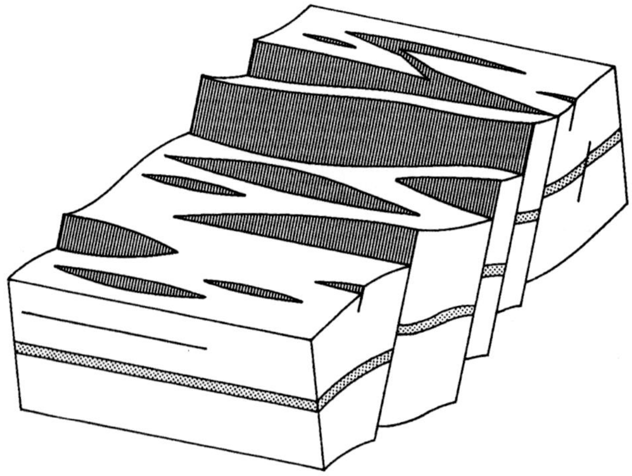

Figure 14—Conceptual diagram showing how uniform, bulk deformation is achieved by a non-uniform pattern of isolated, en-echelon and physically-linked faults (after Walsh and Watterson, 1991).

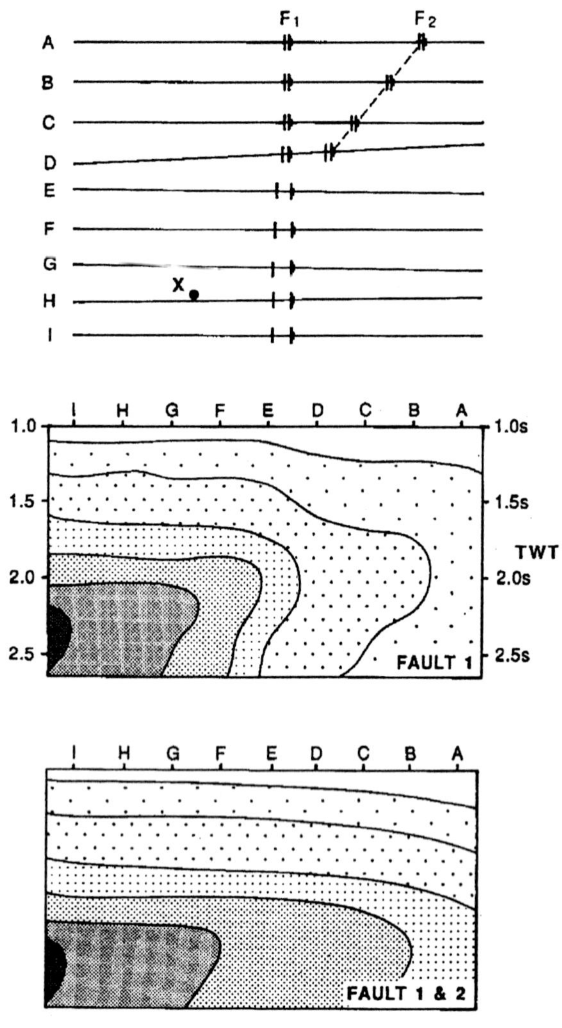

Figure 15—Cartoon showing how fault-length/displacement relationships and fault-displacement summation may be used to predict hard linkage.

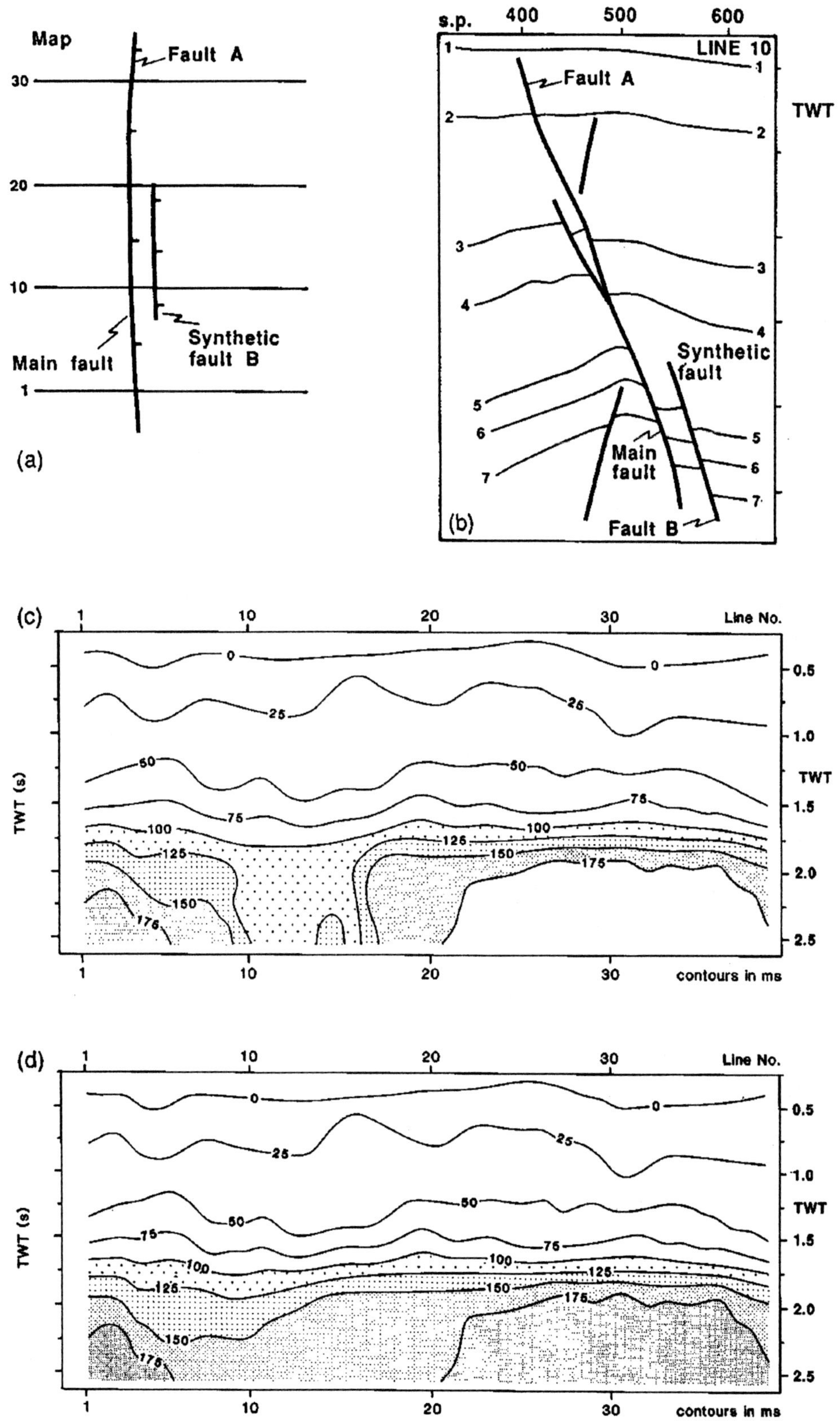

Figure 16—An example of soft linkage (after Walsh and Watterson, 1991). (a) Map showing location of Fault A and synthetic Fault B at horizon 6. (b) Line drawing of interpreted seismic picks on line 10. (c) Contoured strike-projection for Fault A. Note consistency of displacement contours in upper part of fault, but zone of reduced displacement on lower part between seismic lines 8 and 20. This zone corresponds to the presence of synthetic Fault B. (d) Contoured strike projection for the summed displacements of both Faults A and B. Note that the pattern of displacement is much smoother than in (c), indicating that the displacement is partitioned between the two fault strands.

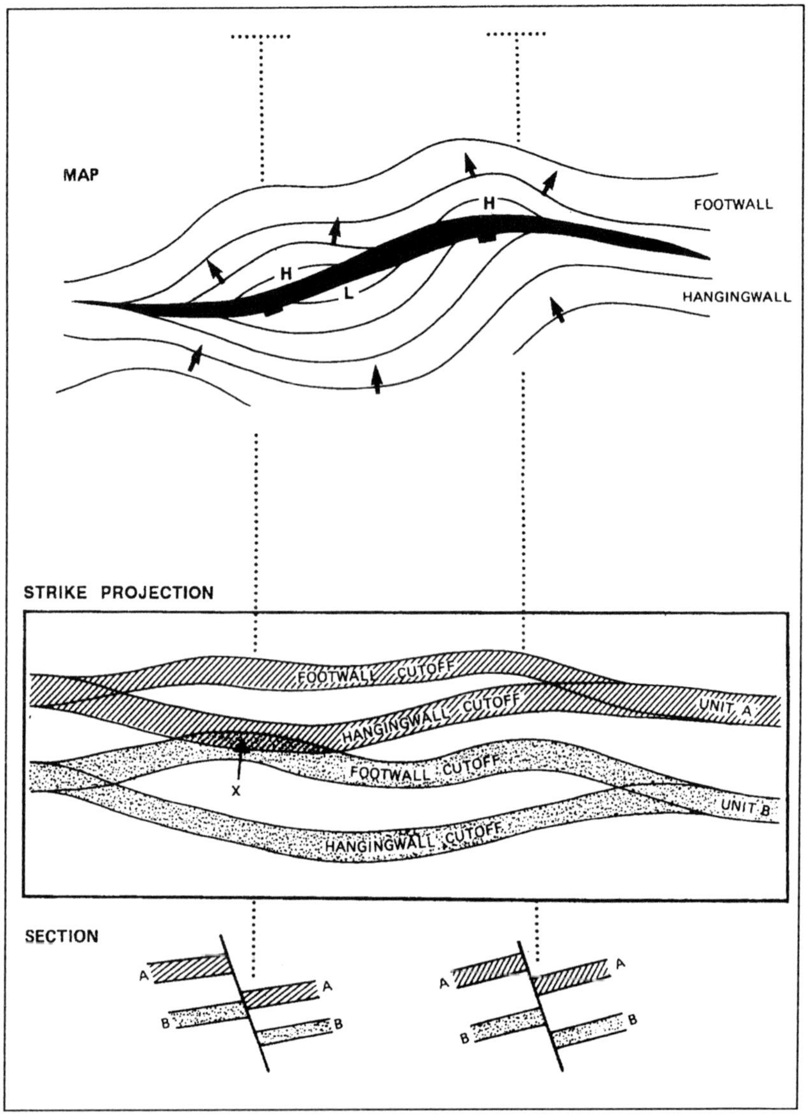

Figure 17—Cartoon showing the application of horizon-separation diagrams to the evaluation of fault seal (after Chapman and Meneilly, 1990). Two sandstone units, A and B, in a shale sequence are offset by a normal fault. Variation of displacement on the fault causes sands A and B to be juxtaposed across the fault at X. This is potentially an area of no seal across the fault.

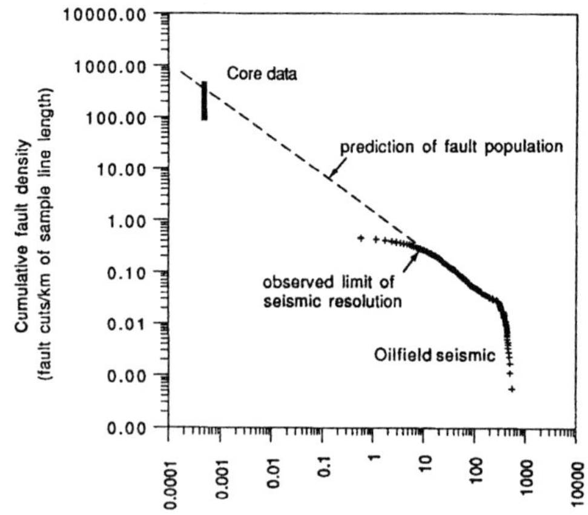

Figure 18—Fault-displacement-population plot for seismic and well data from an offshore oilfield. The plot shows good agreement between the predicted fault-displacement population below the limit of seismic resolution and the observed fault-displacement population derived from core data.

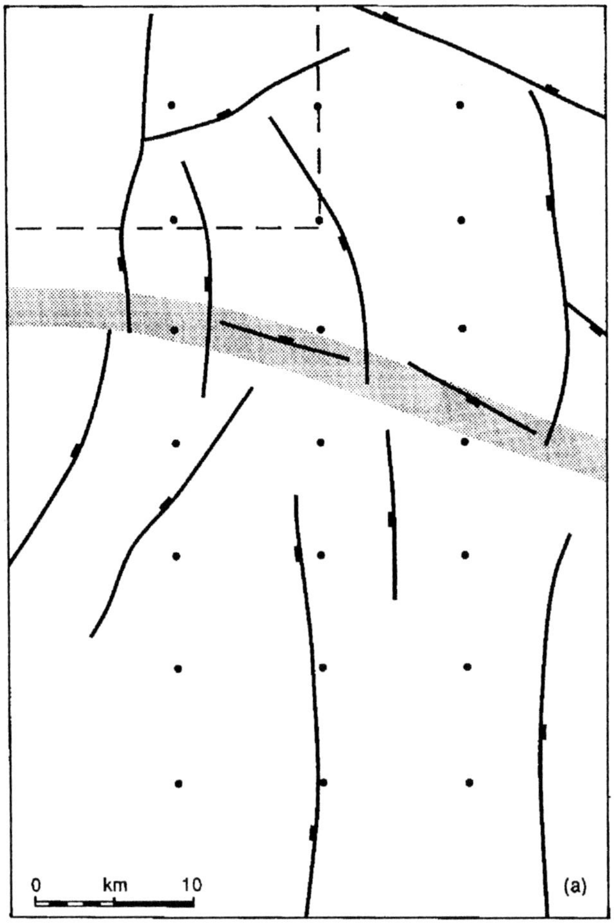

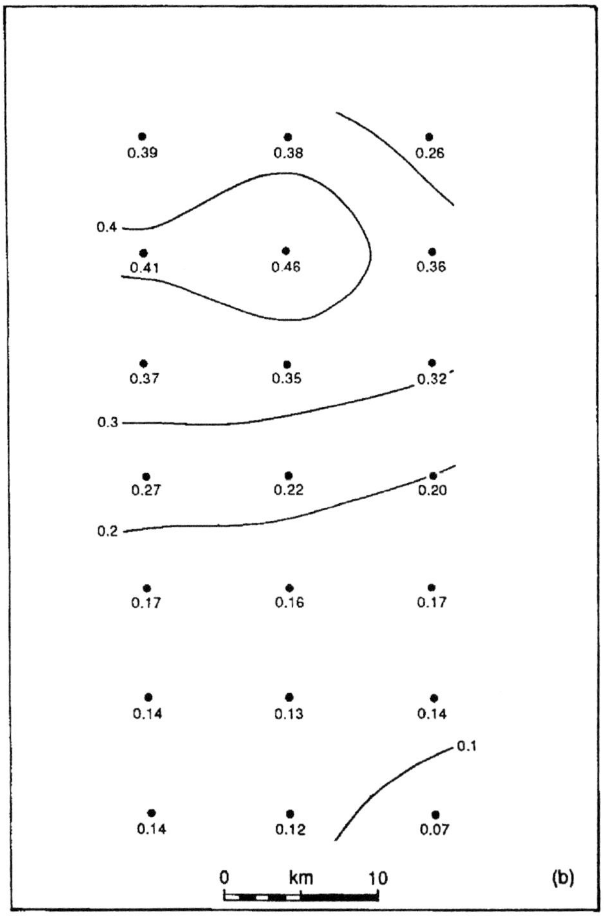

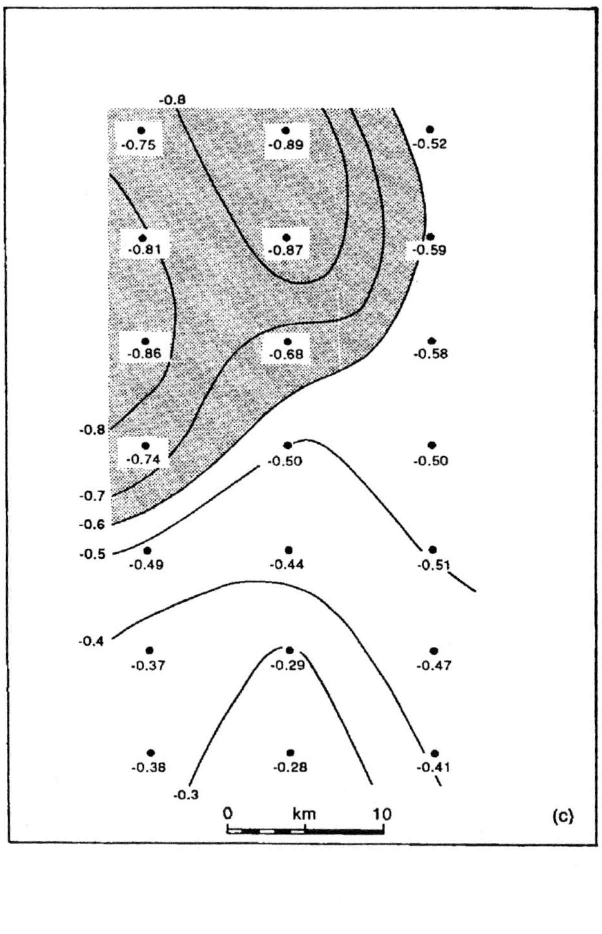

Figure 19—Figure showing the use of fault-displacement populations in predicting the numbers of faults below the limit of seismic resolution. (a) Map showing the main faults observed in an offshore area. Shaded line shows location of interpreted transfer zone which separates the area into a more-highly-faulted area to the north and a less-faulted area to the south. Dots indicate centres of sub-areas used for sampling the fault-displacement population. Dashed line in NW of map indicates size of the sub-areas. (b) Contoured values of the cumulative number of faults (per kilometre of sample-line length) with displacements greater that 20ms, derived from the observed fault-displacement population in the area covered by map (a). The cumulative fault-density is greatest in the northern part of the area. (c) Contoured slope values of single, normalised, displacement-population curves for the subareas from (a). The more-highly-faulted area is associated with steeper population-slopes.

Fracture Evaluation of Block P-0315, Point Arguello Field, Offshore California, Using Core, Outcrop, Seismic Data and Curved Space Analysis

M. J. Padgett
D. C. Nester

ABSTRACT

The fracture system in the Miocene Monterey section at Point Arguello Field, was studied to find the relationship between fracture intensity and local tectonic structure. The analysis included five studies: seismic structural mapping, seismic amplitude, outcrop, core, and surface curvature. The goal was to develop a predictive and quantifying methodology for the exploitation and exploration of fractured reservoirs of the type found in the Point Arguello field.

A detailed structure map on the Monterey chert member was constructed from a 3D seismic survey on a Landmark workstation. The time structure map was depth converted with an accuracy of ±100 feet to the lithologic Monterey chert unit within existing well control. Seismic amplitude studies were performed to relate fracture intensity to a reduction in seismic amplitude across the field. An outcrop study was performed at Lion's Head fault to determine the relationship between the distance to a large fault and fracture density. A statistical evaluation of core analyses for 6 wells by Terra Tek (Salt Lake City) was performed to evaluate the variability of fracture intensity as a function of structure.

The surface curvature analysis consisted of treating each fault block as a 2 dimensional curved space (closed-compact 2D manifold in R^3). At each point the distance to the boundary (faults) was determined as well as measures of the local curvature. The Gaussian curvature, ellipsoidal curvature and second derivative were used to characterize the curvature. The distances to faults and curvature measures were correlated to core analyses and to each other. Results are consistent with the outcrop and core studies, and indicate an ability to use the topology of a seismically derived structure map to predict fractured reservoir parameters.

INTRODUCTION

The Point Arguello Field, Offshore California, consists of Blocks P-0315, P-0316, P-0464 and P-0465. Block P-0315 was leased by Texaco, Pennzoil, Sun and Koch in 1979. Initial exploratory drilling established production from fractured cherts and dolomites of the Monterey interval. Initial flow rates in the discovery well exceeded 20,000 BOPD from a reservoir rock whose unfractured permeability is approximately 0.1 millidarcies. The results of eight production tests available for analysis in this study indicated that oil producibility varied substantially across the field. The objective of this study was to determine if fracture intensity and hence producibility was related to the shape of the producing interval (local structure) at Point Arguello. To accomplish this, core data, outcrop observations and seismic data were related to measures of the curvature of the Monterey section and to distances to faults.

Various aspects of the Point Arguello Field have been studied previously by numerous investigators. Review articles by Belfield, et. al. (1984) and Crain, et. al. (1985) discuss general fracturing in the Monterey and the Point Arguello discovery in particular. Van Bloemen Waanders and Litvak (1990) discuss interference testing and their extensive reservoir modeling of the field. Sadler (1990) and Narr and Suppe (1989) discuss the fractures found in the cores at Point Arguello.

SEISMIC INTERPRETATION

In order to compute curvatures or distances to faults, it is necessary to first produce a depth map on the stratigraphic unit of interest. For this study, a 3D interpretation was performed using a Landmark workstation with all relevant wells and petrophysical curves loaded. To ensure that the top of the Monterey chert was properly interpreted, synthetic seismograms were constructed for several wells across the field. Figure 1 demonstrates the seismic character of the of the chert section. Arbitrary well-tie seismic lines were constructed for proper integration of the stratigraphic tops and faulting identified in the wells.

Once the basic correlation of well tops and faults was made on the workstation, every inline from the data volume was interpreted to ensure the most accurate structural interpretation possible. Care was taken to identify and correlate all seismically resolvable faulting for later use in correlation with core data. The final fault pattern with a basic time map is illustrated in Figure 2. Time-to-depth conversion was accomplished using a best fit statistical checkshot constructed from five checkshot surveys datumed to the water bottom. Adding a digital bathymetry map to the depth converted horizon resulted in a depth map (Figure 3) with an estimated accuracy of ±100 feet at the Monterey chert level.

As seen in the final depth map (Figure 3), the Point Arguello Field is an elongated, thrusted anticline with an east to west trending crest located in Block P-0315. The primary thrust, located south of the structure, bifurcates and dies out to the west as the structure turns northward. To the west, the thrust turns southward and dies out into Block P-0316. A large antithetic thrust forms the northern boundary of the structure and parallels the primary fault. A release fault connecting the primary and antithetic fault is present near the border between blocks P-0315 and P-0316. Two examples of interpreted lines from the seismic data volume are shown in Figures 4 and 5. The compressional structure between the primary and antithetic thrusts is demonstrated in line 75 (Figure 4). Seismic line 105 (Figure 5) crosses the western portion of the field and shows the character correlation of the interpreted horizons in an area of less structural deformation. In these figures, the chert horizon is colored yellow, the base of the overlying Sisquoc formation blue, and the base of the Monterey carbonate member orange.

SEISMIC AMPLITUDE STUDY

In addition to normal structural mapping, seismic amplitude studies were performed to determine if observable amplitude changes across the crest of the structure could be related to fracture intensity. The calculated synthetic seismic response of the chert as seen in Figure 1 was consistent for wells across the field, but a loss in character and amplitude of the correlating seismic data was observed in areas of structural flexure. In Figure 1, a decrease in amplitude is seen within the chert member as it crosses the gentle roll-over on this seismic line. This degradation is believed to be attributed to the presence of tectonically induced fractures in the chert member.

The seismic amplitude analysis involved first a detailed autopicking of the trough below the top of the Monterey chert member (Figure 6) using the Landmark "ZAP!" program. After picking the trough, the Landmark "Stratamp" program was used to compute the rms amplitude of the trough (red) in the interval between the red and yellow markers in Figure 6. The resulting amplitude map and a lineament interpretation are contained in Figures 7 and 8. The lineaments seen in Figure 7 are trends of low seismic amplitude.

Most of the lineaments parallel the structure's axis and appear to be tectonically induced. All the interpreted faults can be identified in the lineament trends. In addition, this display appears to highlight areas of faulting not recognized in vertical seismic displays. Lineament trends appear to intersect where the structure bends to the north into Block P-0450. A similar intersecting lineament pattern is seen to the east where the structure bends southward into Block P-0316. It is conceivable that some of the amplitude lineaments may in fact represent zones of joint sets or fracture swarms similar to those with surface exposure at the Teton anticline structure in Montana.

Using amplitude studies to perform lineament analyses is becoming commonplace in the petroleum industry today (Dalley, 1989 and Ruljtenberg, 1990). These analyses are most reliable when applied to high amplitude events since the signal can be expected to dominate both random noise and processing artifacts. At the Monterey level in this data volume, the seismic reflector is weak across the top of the field. Here the signature of major fracturing is a loss of signal. While a loss of signal is the desired result, the analysis of Figures 7 and 8 must be evaluated in light of the influence of noise and processing artifacts (migration effects) which can be present in low signal data. With respect to this 3D data volume, extensive time migration studies were performed prior to final migration in order to reduce "migration swings" at the Monterey level. Cross line 320 (Figure 9) shows the level of migration swing contamination at the Monterey level. Corresponding events in Figures 6 and 9 have been marked with arrows. We have, however, not performed depth migration studies or re-interpreted the data using a depth migrated 3D data volume.

OUTCROP STUDIES

It has been noted (George, 1988) that fracture density increases rapidly when a major fault is approached. In order to observe and verify this phenomena, an outcrop study was performed at Lion's Head Fault on Vandenburg Air Force Base with J. Crouch of Crouch, Bachman and Associates (Figures 10-18). The Monterey section exposed at this beach outcrop extends from within the upper siliceous member to the middle phosphatic member. The Lion's head Fault brings Jurassic ophiolites into contact with the middle phosphatic member.

The outcrop was measured using the Lion's Head Fault as a datum. The strike and dip of bedding was recorded at 20 stations marking changes in lithologies (Figure 11). Fracture intensity and the location of chert layers were also recorded. As a result several minor adjustment faults were identified. In addition a large fault was detected within an erosional valley in the sea cliffs. The relationship reported by George et.al. was observed with differing degrees at all faulting recognized in the outcrop. A key field observation made was that local fracture intensity is most sensitive to the location of the nearest fault regardless of size and distance from the major Lion's Head Fault. Small faulting caused a large local increase in fracturing. In particular, the chert units at 270-310 meters (Figures 11 and 16) were most strongly influenced by the larger fault at 230-260 meters.

CORE ANALYSIS

Cores from six wells were included in this study. The location of these cores are noted on the structure map in Figure 3. The cores were analyzed by Terra Tek (Salt Lake City, Utah) for fracture intensity (fractures per foot), fracture length, width, type, staining, filling, and orientation (Figures 19 and 20). The cores were analyzed for non-penetrating versus penetrating (cross-cutting) fractures but fracture terminations were not counted. The method of George, et. al. to infer actual mean fracture length could not be used since it requires knowledge of terminating versus non-terminating fracture counts.

For the sake of this report, we will focus on the mean fracture intensity as measured in core to study the variability of intensity as a function of structure. The mean fracture intensities in this report are simple uncorrected means of the fractures from cores in the chert member. All stylolites, induced fractures and bedding plane fractures have been deleted prior to averaging. A reciprocal of the cosine of the apparent dip correction has not been applied because large numbers of 90 degree apparent dips were logged. In attempting to correct the data for apparent dip, it was found that imposing an arbitrary cut-off of 88 or 89 degrees caused a corresponding change in the means computed due to the large number of apparent high angle fractures. Also, the fracture spacings encountered in the more highly fractured cores at Point Arguello are often much less than the core diameter, implying an adequately sampled system. These two considerations have led us to forego the normal reciprocal cosine correction.

CURVATURE ANALYSES

Any surface in space is composed of 2 parts, the interior points of the surface and the boundaries. For a geologic surface, the boundaries are faults, erosional truncations, depositional terminations or a change in facies. In studying fracturing, the faults are crucial in terms of both the breaking of the rock and the reduction of stress in a fault block after faulting. The truncations (erosional and stratigraphic) determine the mechanical properties of the unit which is being fractured. In this study, we have attempted to study the bulk properties of the Monterey unit in each fault block. This philosophy determined the mathematics and mapping procedures which were employed.

A traditional method of studying the distortion of a geologic surface as related to fracturing is by using the radius of curvature (Murray, 1968). In Murray's original paper,the radius of curvature is approximated using the second derivative. Along a line of section, the second derivative of a horizon will approximate the curvature if the dip is not too large. Note that the radius of curvature is the reciprocal of the curvature. It should be noted that the error in using the second derivative to estimate radius of curvature increases as the dip increases. Results deteriorate rapidly for dips greater than a few degrees. In spite of this, some form of the second derivative is normally used since it is fast and available in most mapping packages (Figure 21).

A modest improvement to the second derivative can be made by using a local "ellipsoidal curvature." The local curvature is found in the dip direction and normal to the local dip (in the tangent plane). This allows two independent curvatures to be computed which define an ellipsoid of revolution. For example, if the dip were in the x-direction the radius of curvature used is given by (Burrington, 1973):

$$\text{Radius of Curvature (x direction)} = \frac{(1 + (dz/dx)^2)^{3/2}}{d^2z/dx^2}$$

This radius along with the corresponding curavature in the y-direction would be used to define a local ellipse, in this approximation. Note that if dz/dx is small this reduces to the reciprocal of the second derivative. To compare with the other curvature measures we use a maximum local ellipsoidal curvature (and hence the minimum radius) as seen in Figure 22.

As noted above second derivative methods are normally applied for fracture analyses. Various combinations of sums of second derivatives, gradients of gradients, and other quantities are often applied to geologic surfaces. For this reason, it is important to determine what precise calculation an investigator is making when a second derivative based method is used. In addition, these methods can depend on the orientation of the grid (i.e., the details of the coordinate system). These problems are addressed by treating a geologic surface as a 2D Riemannian manifold in R3 with metric topology. This allows the computation of curvature tensors which are both rigorously defined and which embody invariants which do not depend on the choice of coordinate system.

For a 2D manifold, the Riemann-Christoffel curvature tensor (Rtuvw) contains only 1 algebraically independent component, eg. R1212 (Weinberg, p.142-147). For a 2D surface, the curvature tensor, Ricci tensor (R_{uv}) and gaussian curvature (K) depend only on this single constant and on the metric (g_{uv} with $g=\det(g_{uv})$):

$$R_{tuvw} = (g_{tv}g_{uw} - g_{tw}g_{uv}) * R_{1212}/g$$

$$R_{uv} = g_{uv} * R_{1212}/g$$

$$K = -R_{1212}/g.$$

$$\text{Gaussian Radius of curvature} = 1/(|K|)^{1/2}$$

Measures of the state of distortion of a surface can thus be obtained by computing the metric and Gaussian curvature (or R_{1212}), Figure 23.

To compute the Gaussian radius of curvature, the z-values at the grid nodes were used to compute the metric first and then the relevant derivatives of the metric. To compute a metric at a point requires a minimum of 4 adjacent grid points. To compute the curvature tensors requires that the metric be defined on at least a 3X3 grid about the point of interest. One must have the metric defined on at least a 4X4 grid of contiguous points to compute a curvature tensor. This means that the curvature calculation is invalid within several grid nodes of any fault. For areas near faults the curvature is extrapolated using CPS-1 gridding and contouring procedures. This limitation applies to each of the curvature measures studied here.

The three curvature maps in Figures 21-23 are qualitatively similar but quantitatively different. The correlations between each of these measures and mean fracture intensity in core are contained in Figures 24-26. Each of the measures correlates well with the mean intensity at the core locations. In reviewing the maps, however, some differences are seen in the extent to which low radii areas extend away from the main fault system. For all three maps, the lowest radii are seen near the crest of the structure and near the faults. The second derivative indicates the most rapid transition to high radii in moving away from the main thrust fault. The gaussian curvature shows the largest area of low radii, especially near the A2 well. In addition, the gaussian curvature measures low radii across most of the structure. All three methods give similar results off structure. Other methods beyond core analysis will have to be applied to determine which of the three maps is most accurate.

In addition to studying measures of curvature, the distances to both the major thrust fault and the nearest seismically mappable fault were studied. These studies were included based on our experience at Lion's Head and because we had no reason a priori to know whether curvature or faults would be most important. The correlations between the mean fracture intensities (in core) and radii of curvatures and distance to faults are demonstrated in Figures 24-28. These plots demonstrate that both the radii of curvatures and the distance to the nearest fault correlate well with the fractures as seen in core. The correlation with the distance to the main thrust fault is poor for mean intensity.

The fact that both the radii of curvature and the distance to the nearest fault correlate with the fracture intensity in core raises a question of independence. In Figure 29, the second derivative based radius of curvature is plotted against the distance to the nearest fault. The correlation is quite good and the fit is roughly linear. The radii of curvature maps (Figures 21-23) demonstrate this correlation in the way that the low radii areas track the faults and the radii increase fairly consistently away from the faults. These observations indicate that for the Point Arguello field mean intensity can be inferred using either radius of curvature or the distance to the nearest fault and that these two parameters are not independent.

CONCLUSIONS

After completing the mapping, seismic studies, curvature computations, and fault proximity studies, a strong correlation is seen between both radius of curvature and distance to the nearest fault and mean fracture intensity in cores. These two correlation parameters are however not independent due to the geometry of this thrusted anticline. In addition, each of the curvature measures employed produced very good correlations. The seismic amplitude studies show large decreases in amplitude in the more highly deformed areas of the field as well as lineations, some of which are known faults. This structural study was successful in determining the spatial distribution of fracture intensities and areas of highest oil producibility within the Point Arguello Field.

From this study, a four step methodology is developed that can be used as a tool in the exploration and exploitation of fractured reservoirs. The four steps are listed below:

STEP 1. CONSTRUCTION OF A DETAILED SURFACE MAP

In this step, detailed depth structure maps are constructed on the stratigraphic unit to be analyzed by integrating existing seismic and well data bases.

STEP 2. SEISMIC AMPLITUDE ANALYSIS

This step may first require the construction of synthetic seismograms and modeling to determine if the stratigraphic unit to be evaluated is seismically resolvable. To properly extract, map and display the amplitude information requires that the data be interpreted on an interactive workstation.

STEP 3. SURFACE CURVATURE AND FAULTING ANALYSIS

The Gaussian and ellipsoidal curvature methods applied at Point Arguello are not presently available in off the shelf mapping packages. The second derivative method available on most mapping packages can be used but with knowledge of its limitations as discussed in this report. Curvature methods should be closely coupled to fault proximity and fault distribution studies in fractured reservoirs.

STEP 4. INTEGRATION OF PREVIOUS STEPS WITH EXISTING WELL DATA

This step requires integrating additional seemingly disparate well and engineering information with the curvature and amplitude results. Included are initial production rates, show reports, driller's logs, and lithologic data.

Through this iterative process, fracture trends and "sweet spots" can be high graded allowing for better risk analysis in areas where tectonically induced compressional fractures constitute the potential reservoir or serve to enhance permeability.

ACKNOWLEDGMENTS

The authors would like to thank Pennzoil for the opportunity to work on this project. Thanks are also offered to David Anspach and Robert Bruce (Pennzoil) for their help in production geology and the outcrop work. James Crouch (Crouch and Associates, Santa Barbara) was instrumental in our field work. Finally, the authors would like to thank Texaco, Oryx Energy Company, Koch Industries, Harvest Corporation, Oxbow Energy, and Simmons Santa Barbara, Ltd., for their permission to publish this work.

REFERENCES

Crouch, J.K., Bachman, S.B., and Shay, J.T., 1984, Post-Miocene compressional tectonics along the central California margin, *in* J.K. Crouch and S.B. Bachman, eds., Tectonics and sedimentation along the California margin: Pac. Sec., SEPM, v. 38, p. 37-54.

Barton, C. and Zoback, M., 1986, Determination of in-situ stress orientation from Stoneley wave polarization in boreholes: Stanford Rock Physics Report, 27, 399-438.

Belfield, W.C., LaPointe, P.R. and Helwig, J.A., 1984, Analysis of fracturing and fluid flow characteristics of the Monterey formation, Santa Barbara Channel, Ca.: SPE 12734, Presented at 1984 California Regional Meeting, Long Beach.

Blangy, J.P., and Zoback, M., 1986, Restoration of fracture orientation in structurally deformed strata: Stanford Rock Physics Report, 27, 467-484.

Burrington, R.S., 1973, Handbook of mathematical tables and formulas, McGraw-Hill, New York, p.54.

Crain, W.E., Mero, W.E. and Patterson, D., 1985, Geology of the Point Arguello Discovery: AAPG Bull., Vol. 69, No. 4, p. 537-545.

Dalley, R. M., Gevers, E.C.A., Stampfi, G.M., Davies, D.J., Gastaldi, C.N., Ruijtenberg, P.A., and Vermeer, G.J.O., 1989, Dip and azimuth displays for 3D interpretation, First Break, Vol. 27, No. 3, p. 86-95.

George, R.P., Hayes, M.J., Lockman, D.F., and Green, K.E., 1988, A new method for logging fractures in core, submitted to AAPG in 1987.

Hestir, K. and Long, J.C.S., 1990, Analytical expressions for the permeability of random two-dimensional poisson fracture networks based on regular lattice percolation and equivalent media theories, JGR, Vol. 95, No. B13, p. 21,565-21,581.

Kazdan, J., 1985, Prescribing the curvature of a Riemannian manifold: CBMS Regional Conference—Polytechnic Institute of New York, January 6-10, 1984, Num. 57, Am. Math. Soc.

Long, J.C.S., 1983, Investigation of equivalent porous medium permeability in networks of discontinuous fractures: Ph.D. dissertation, Univ. California, Berkeley.

Murray, G.H., Jr., 1968, Quantitative fracture study—Sanish Pool, McKenzie County, North Dakota: AAPG Bull., Vol.

52, No. 1, p. 57-65.

Olson, J., 1986, Fracture Mechanics applied to jointing problems in sedimentary rock: Stanford Rock Physics Report, 27, 521-538.

Narr, W. and Suppe, J., 1989, Fracture density in Monterey formation reservoir rocks of the Point Arguello Oil Field, California, Department of Geological and Geophysical Sciences, Princeton University.

Ruijtenberg, P. A., Buchanan, R., and Marke, P., 1990, Three -dimensional data improve reservoir mapping, JPT, 1990, p. 22-25, 59-61.

Sadler, R.K., 1990, The relationship of lithology and tectonics to fracturing in the Monterey formation, Point Arguello Field, offshore California, Core Workshop, AAPG convention, San Francisco, 1990.

Van Bloemen Waanders, B.G., and Litvak, B. L., 1990, Simulation of a naturally fractured reservoir, Point Arguello Field, offshore California, SPE 18745, p. 9-22.

Weinberg, S., 1972, Gravitation and cosmology: principles and applications of the general theory of relativity: John Wiley and Sons, New York, pp. 4-11 and 142-147.

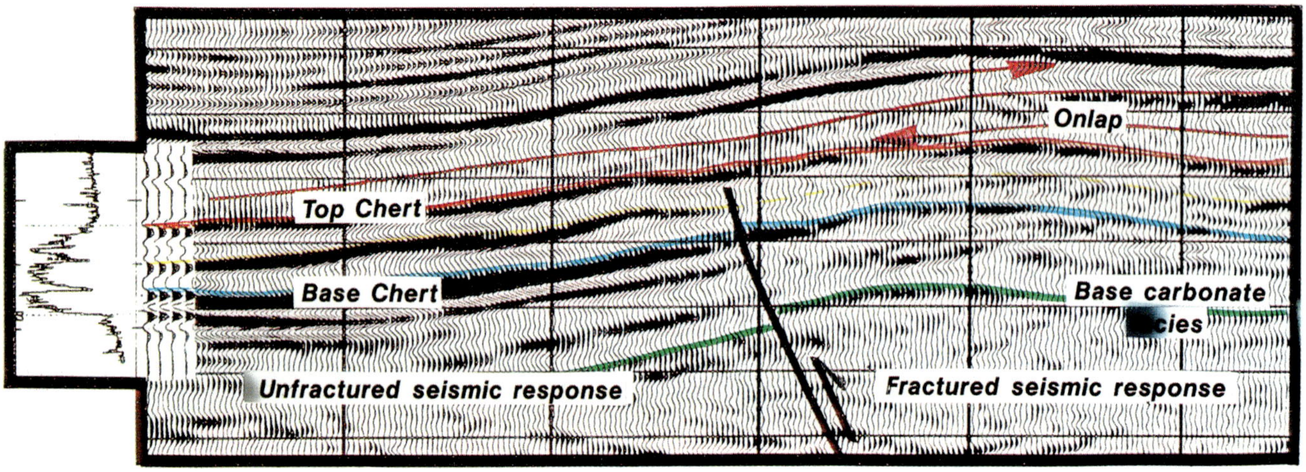

Figure 1—Synthetic tie of Monterey chert member.

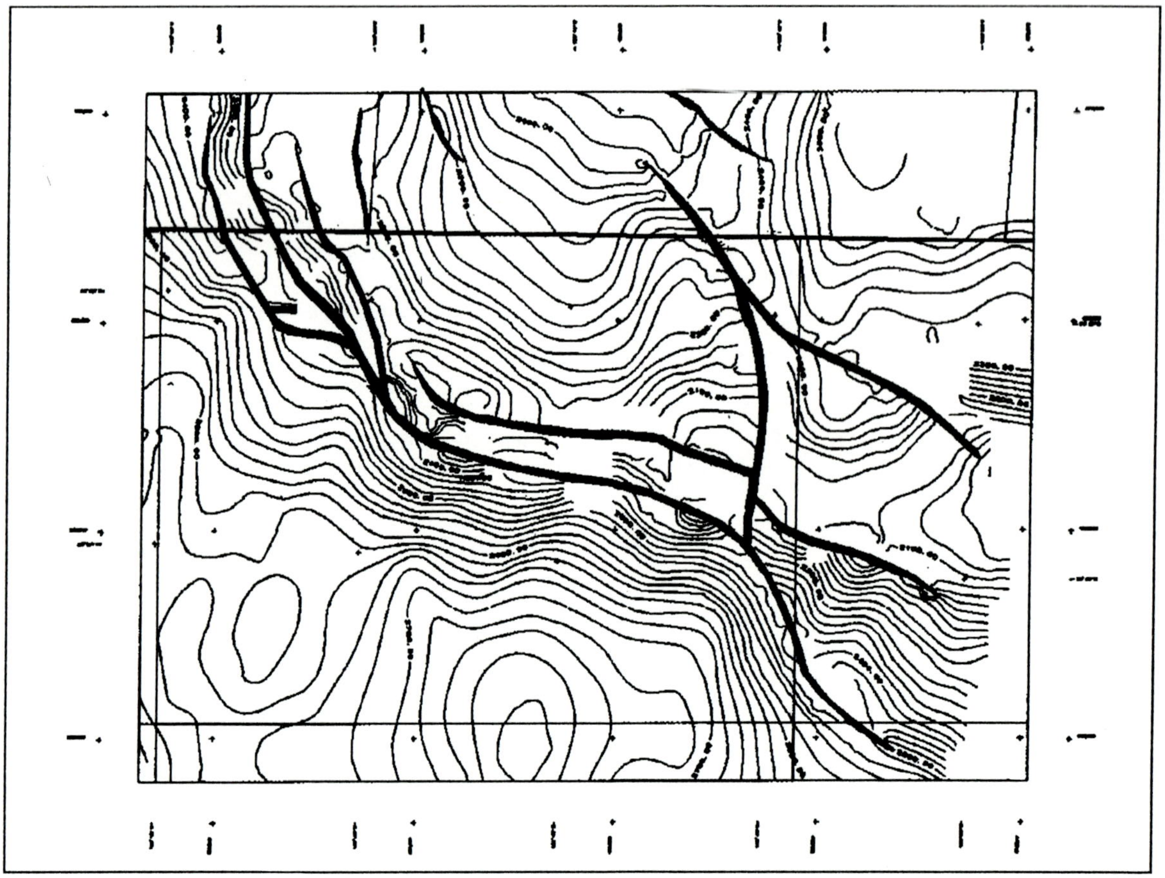

Figure 2—Time structure map at the Monterey (yellow) level. The time is in milliseconds with a contour interval of 20.

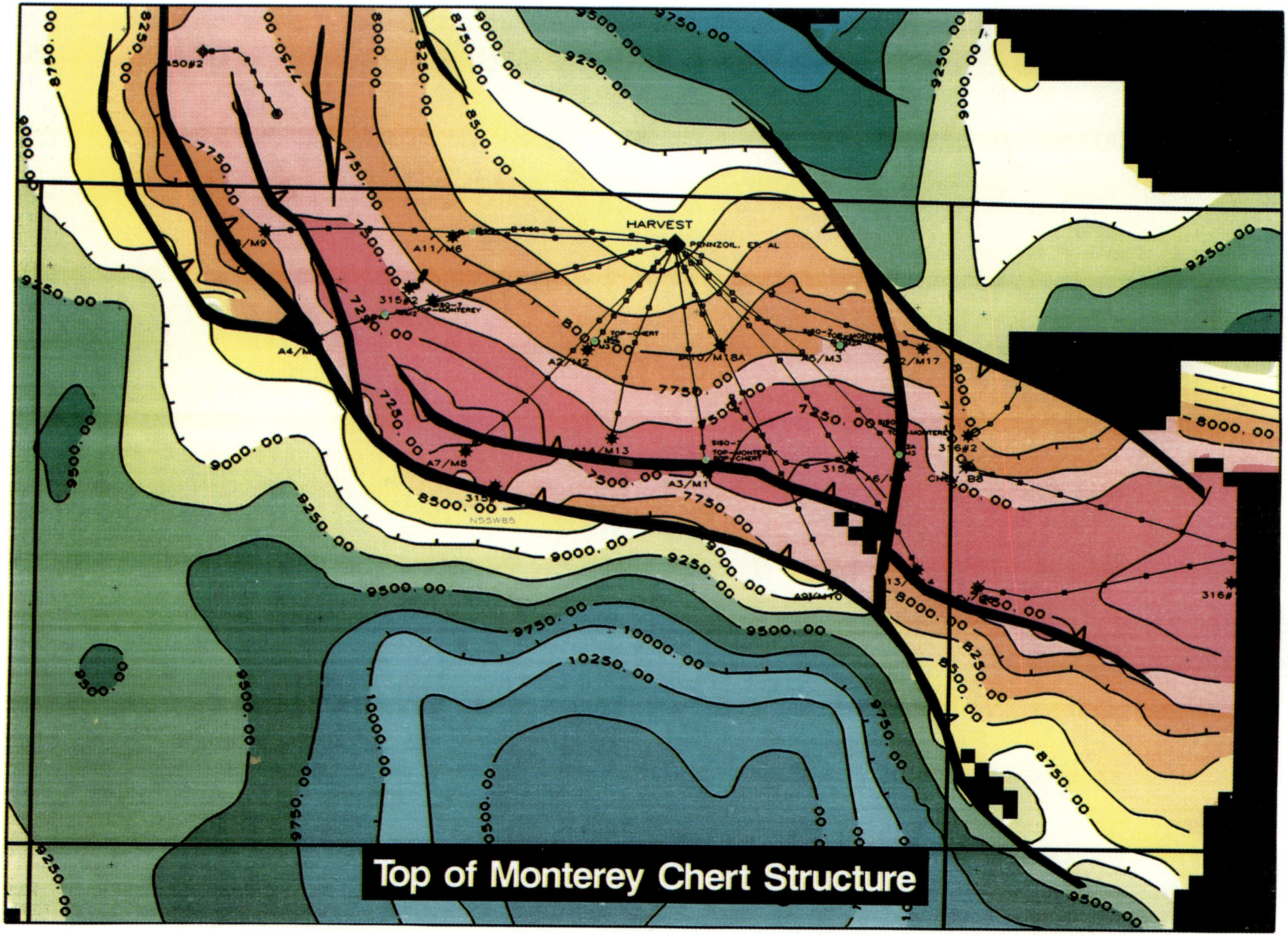

Figure 3—Depth structure map at the Monterey (yellow) level. The depth is in feet and the contour interval is 250 feet.

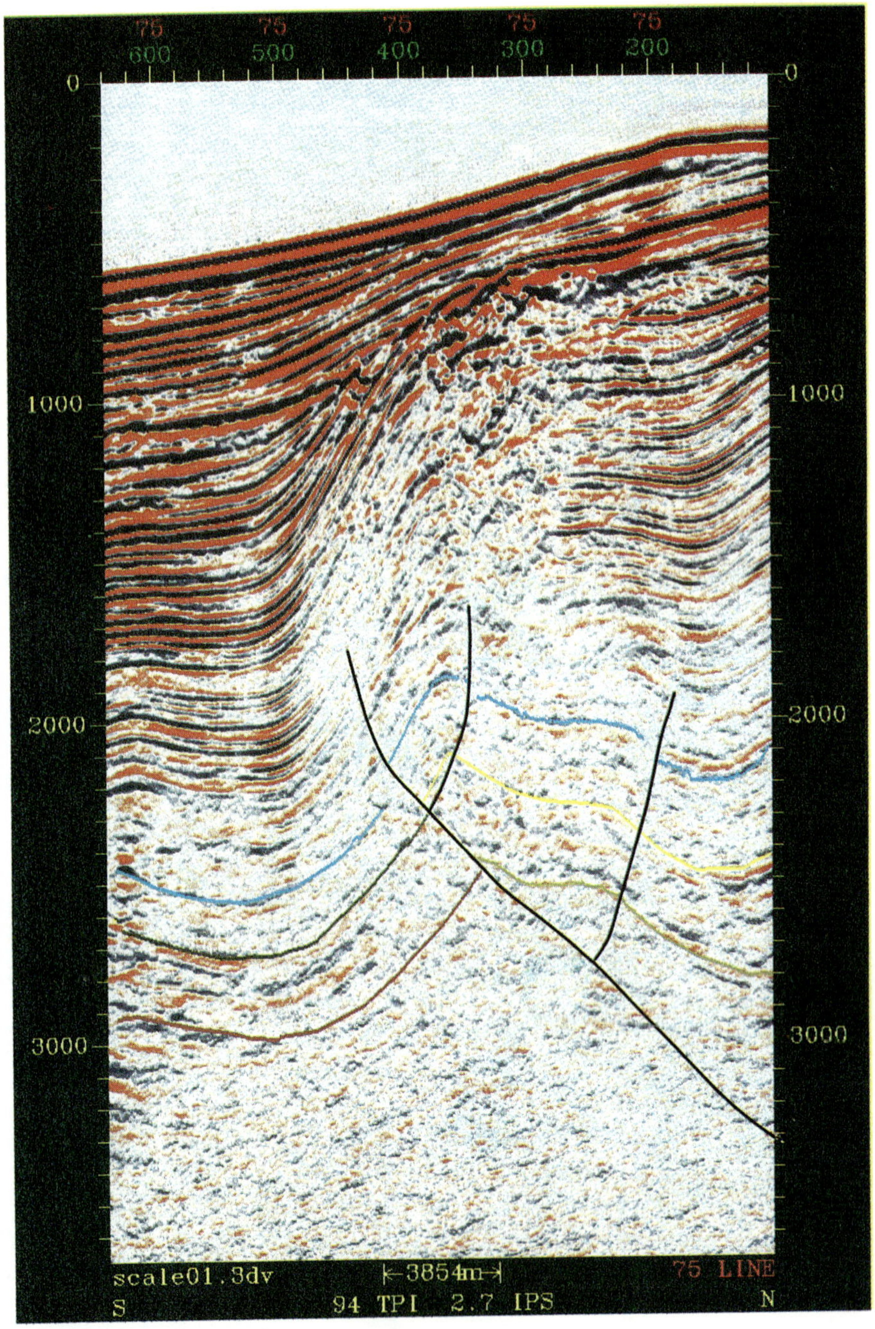

Figure 4—3D seismic line 75 across the Pt. Arguello Field.

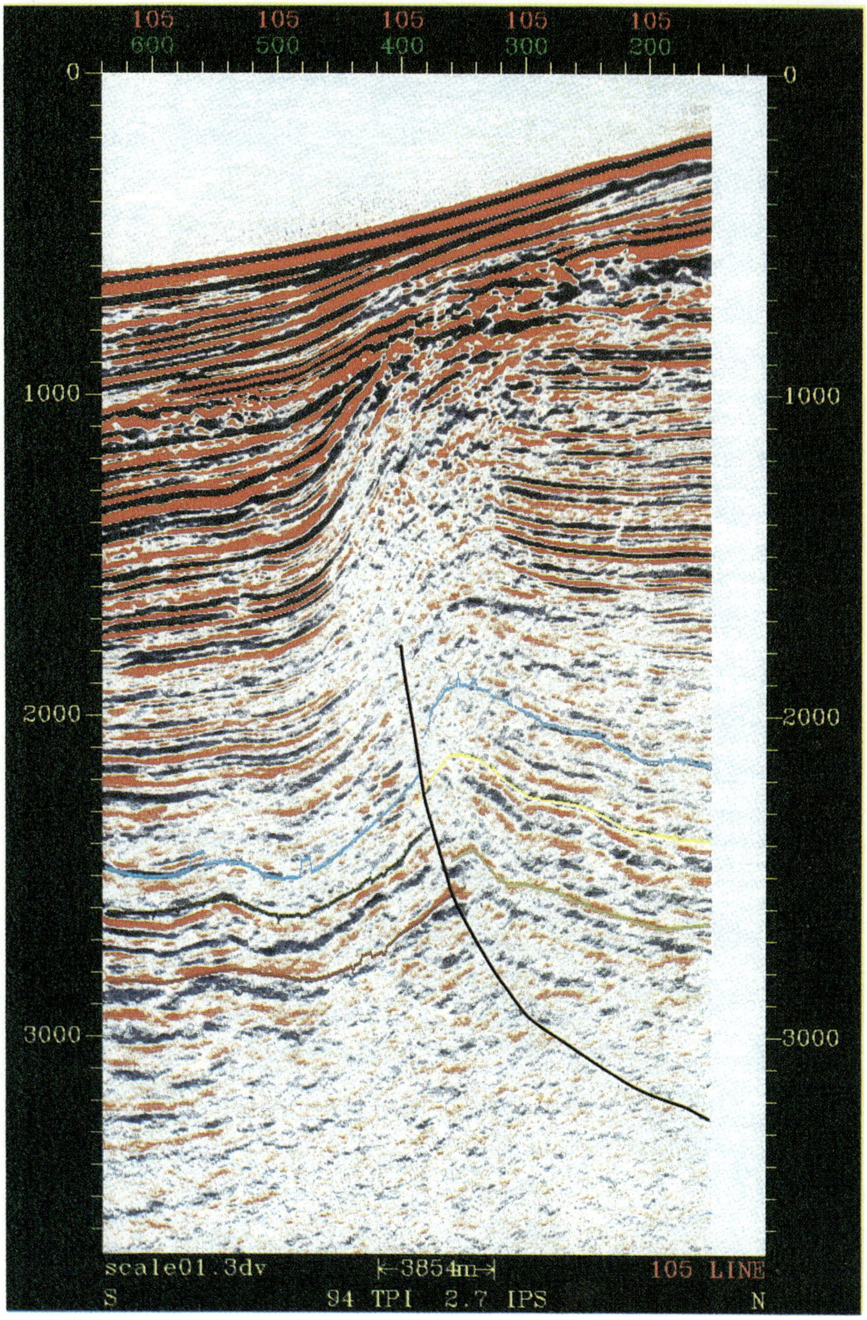

Figure 5—3D seismic line 105 across the Pt. Arguello Field.

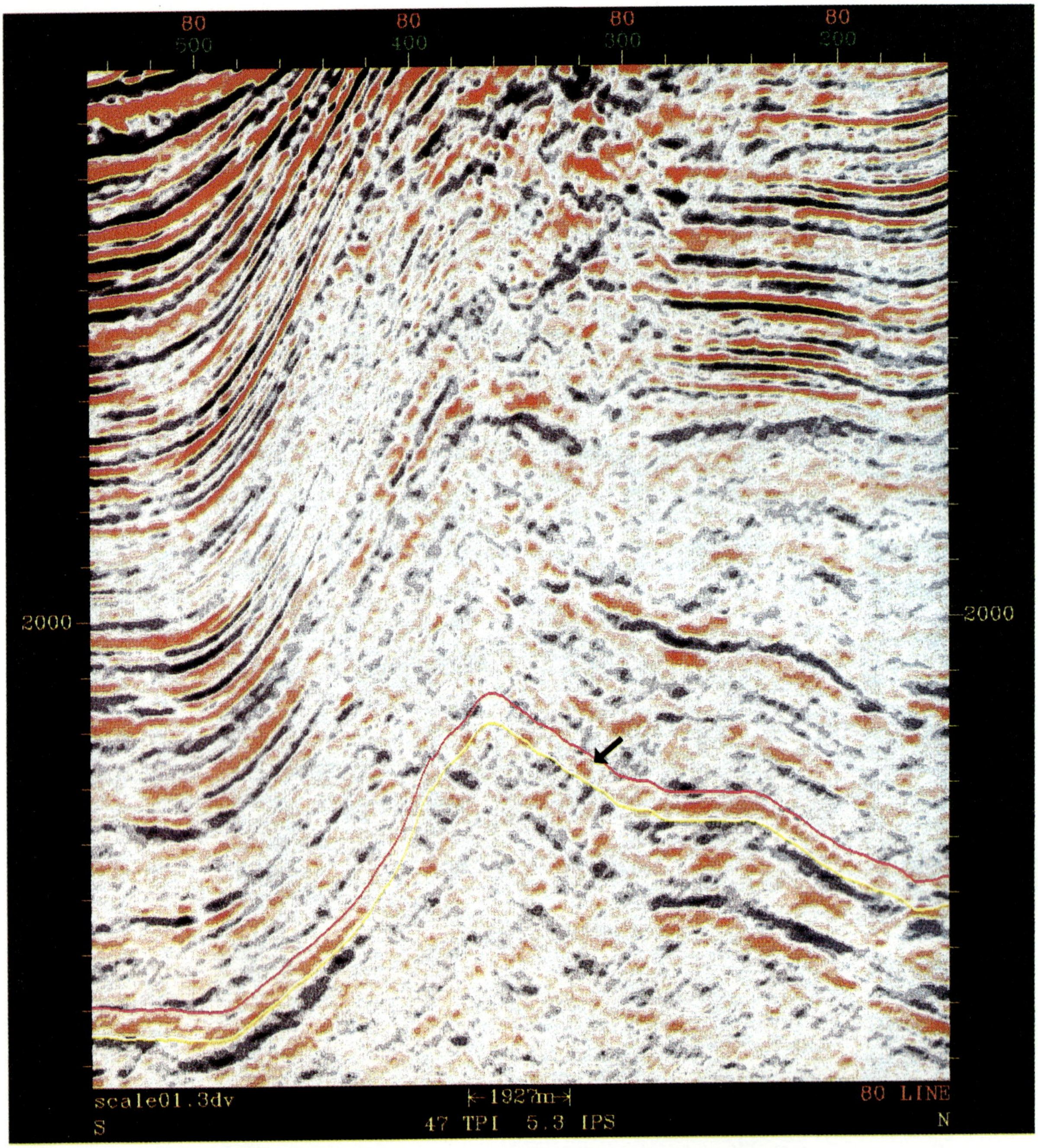

Figure 6—Line 80 demonstrating amplitude analysis window. The Landmark "STRATAMP" program was used to compute the RMS amplitude between the two designated picked horizons.

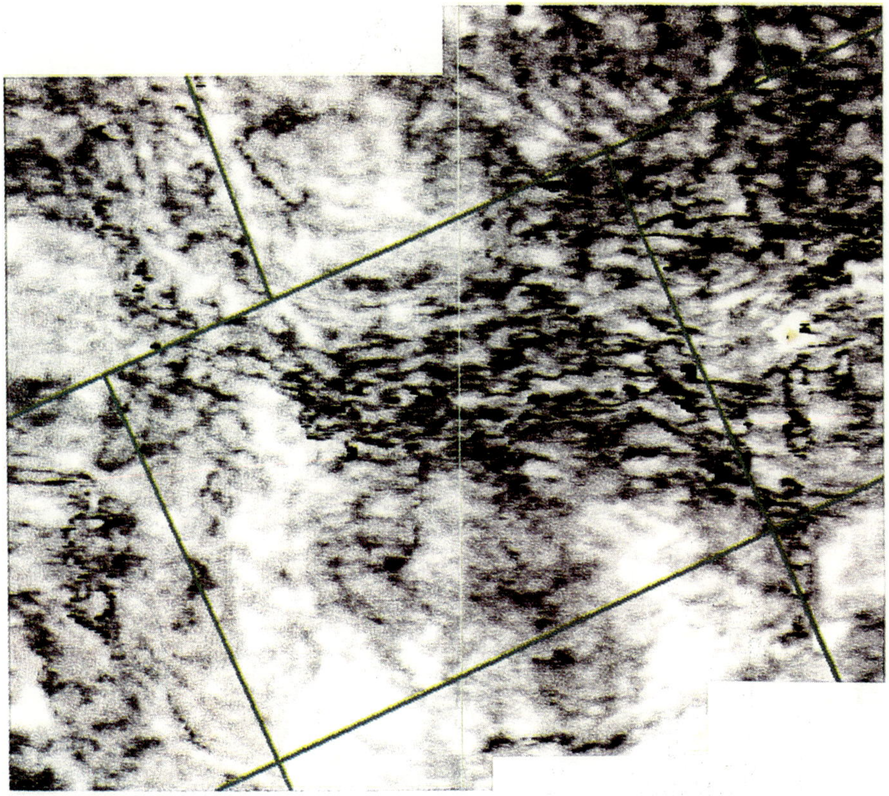

Figure 7—Amplitude map at the Monterey level. Note that darker colors indicate lower amplitudes.

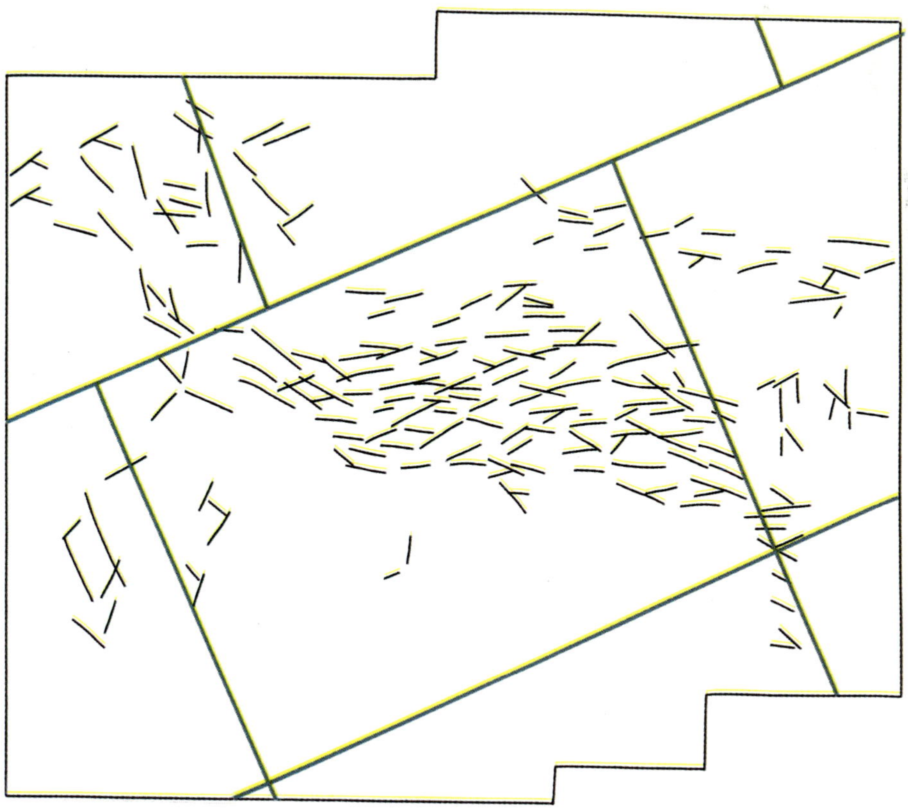

Figure 8—Lineament interpretation of the Monterey amplitude map.

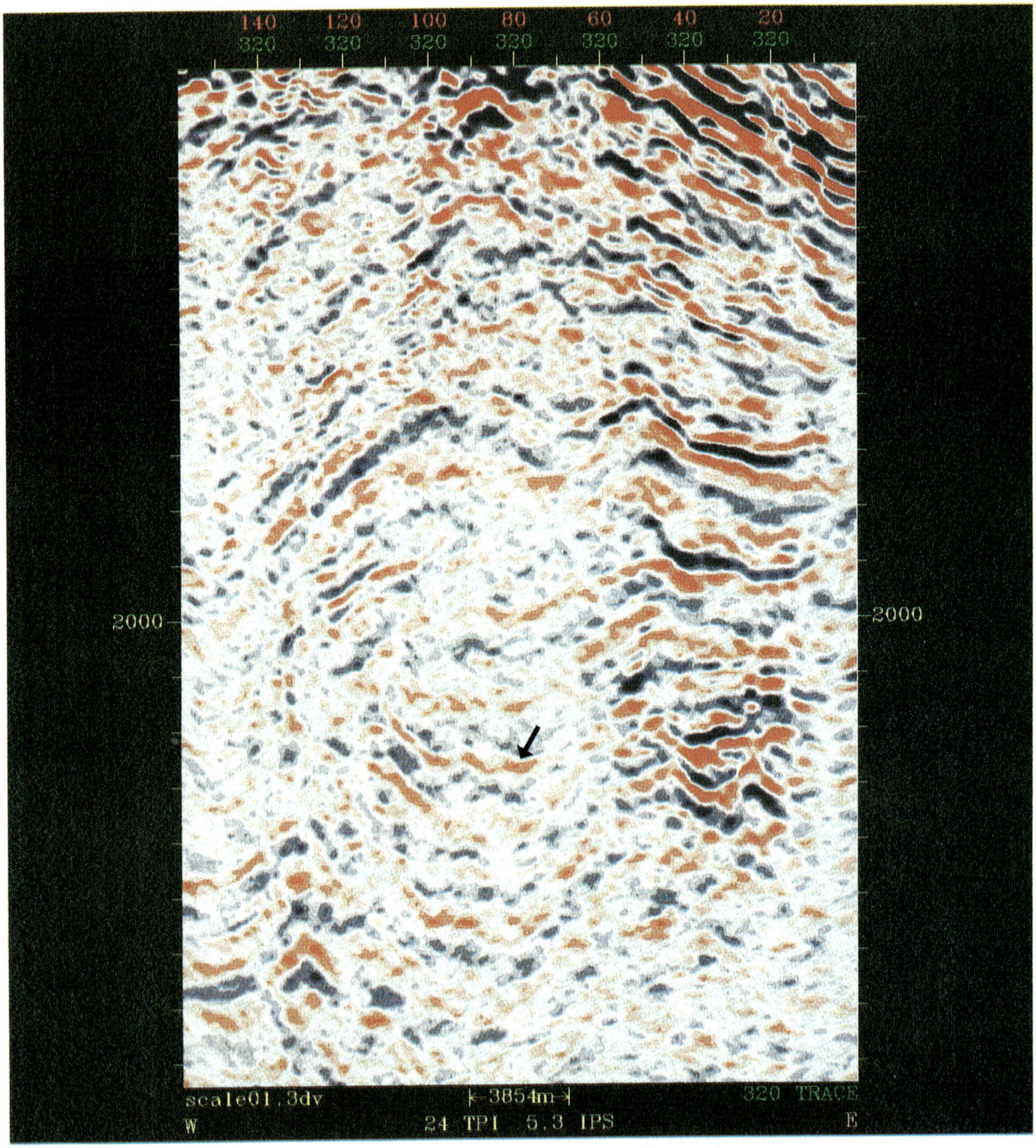

Figure 9—Cross line 320 is intended to help determine the degree of migration swings at the Monterey level. The amplitude marked with the arrow is also marked in Figure 6 with an arrow.

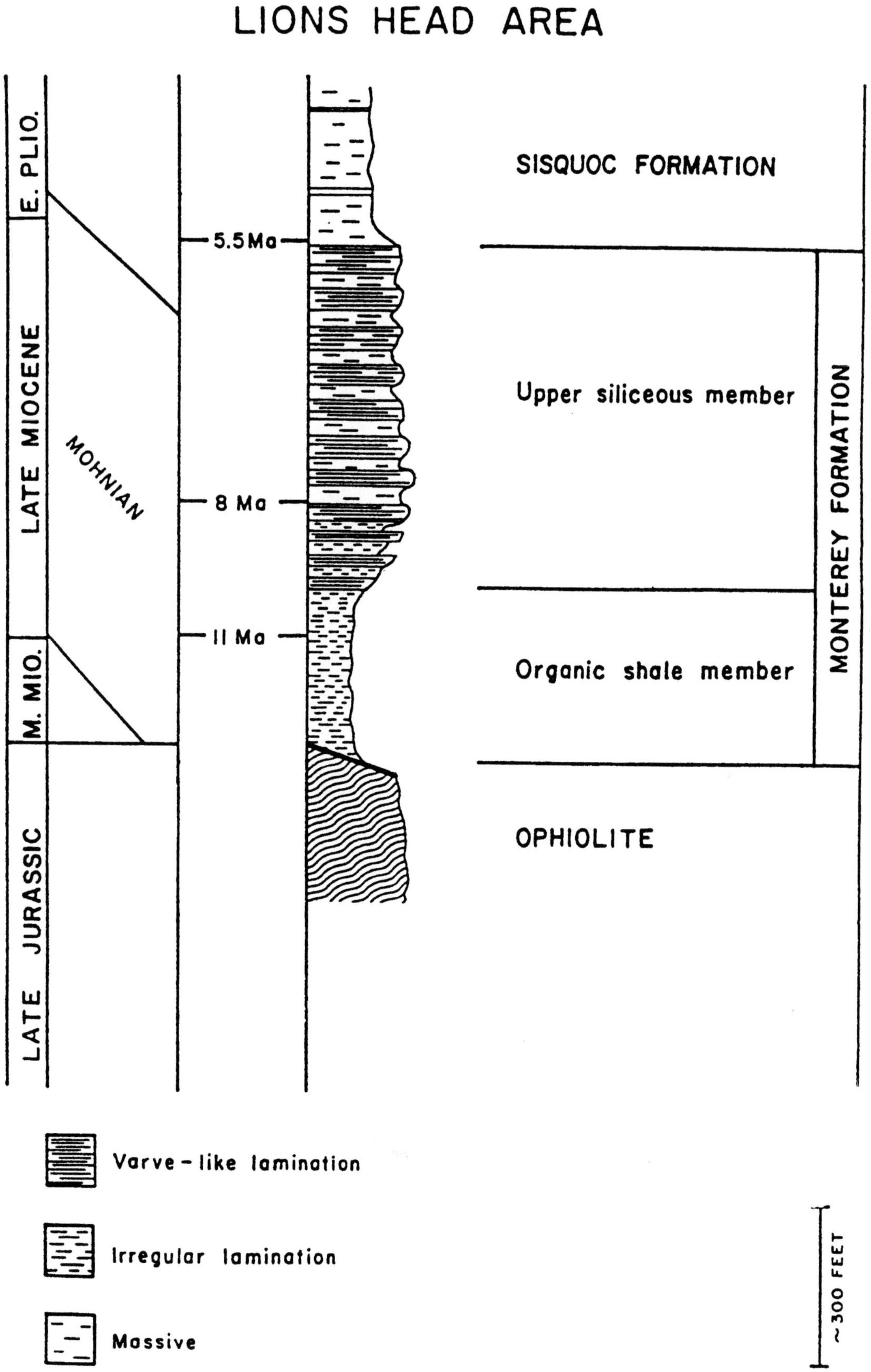

Figure 10—Stratigraphic column at the Lion's head outcrop (from Bachman and Crouch, 1984, with permission of the authors).

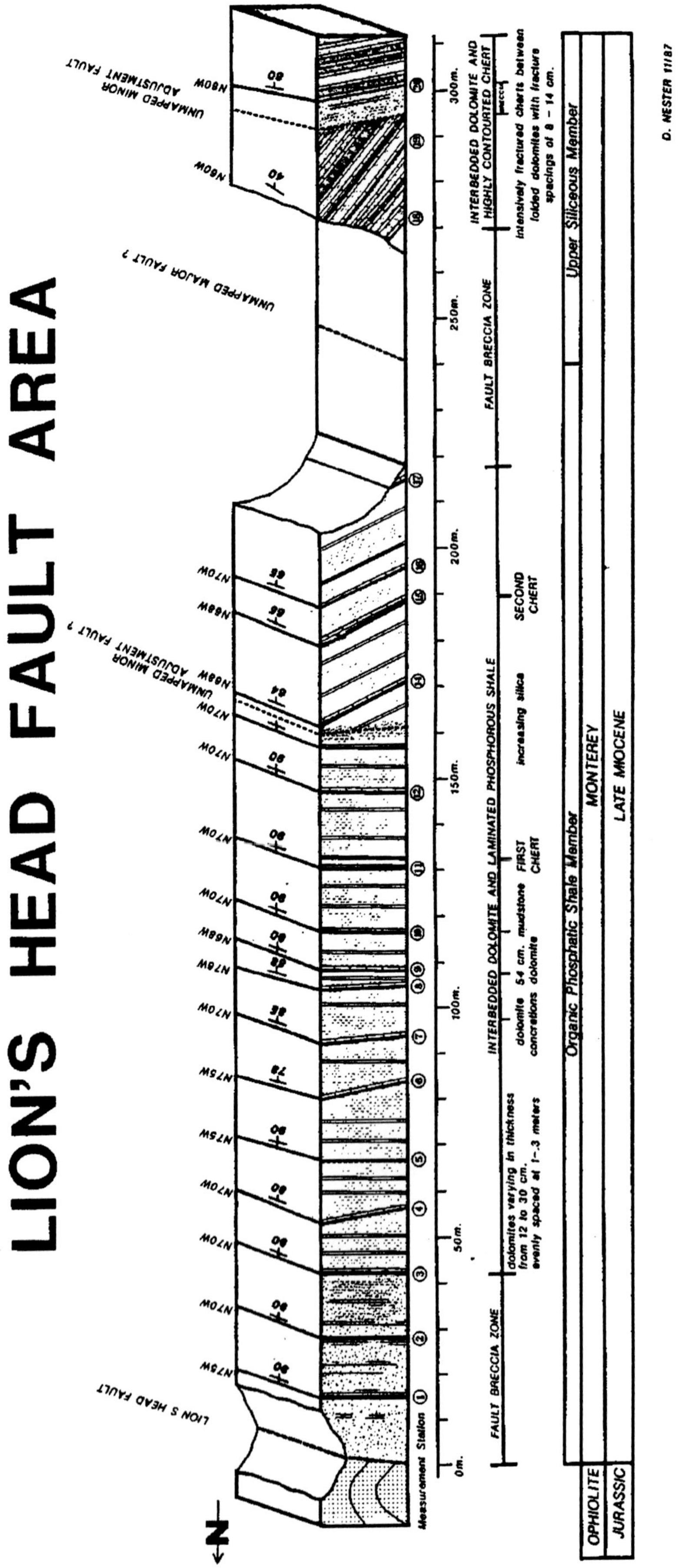

Figure 11—Cross-section of the Lion's head outcrop.

Figure 12—Lion's Head fault (left side of photograph).

Figure 13—Phosphatic member of the upper Monterey.

Figure 14—Interbedded dolomite and low-grade siliceous "shale".

Figure 15—Interbedded chert and dolomite.

Figure 16—Chert outcrops near the fault at 250 meters.

Figure 17—Close-up of chert folding.

Figure 18—Large scale chert outcrop.

Figure 19—Core from the upper Monterey.

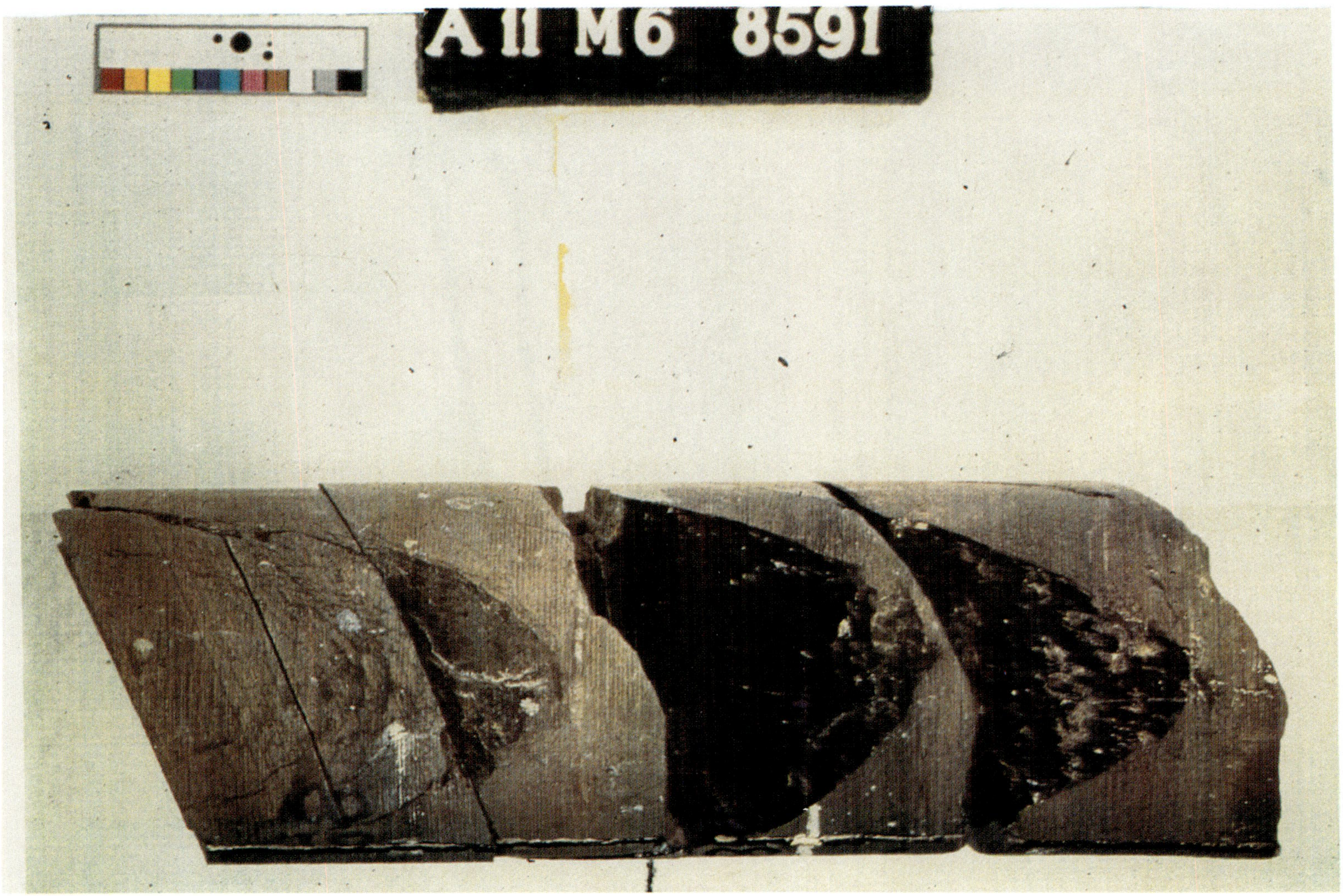

Figure 20—Core from the upper Monterey, re-assembled.

Figure 21—Second derivative radius of curvature map in feet. The radius is given in feet. The contour interval is 1000 feet for labeled contours and 500 feet for changes in color. Core locations are designated by the orange dots.

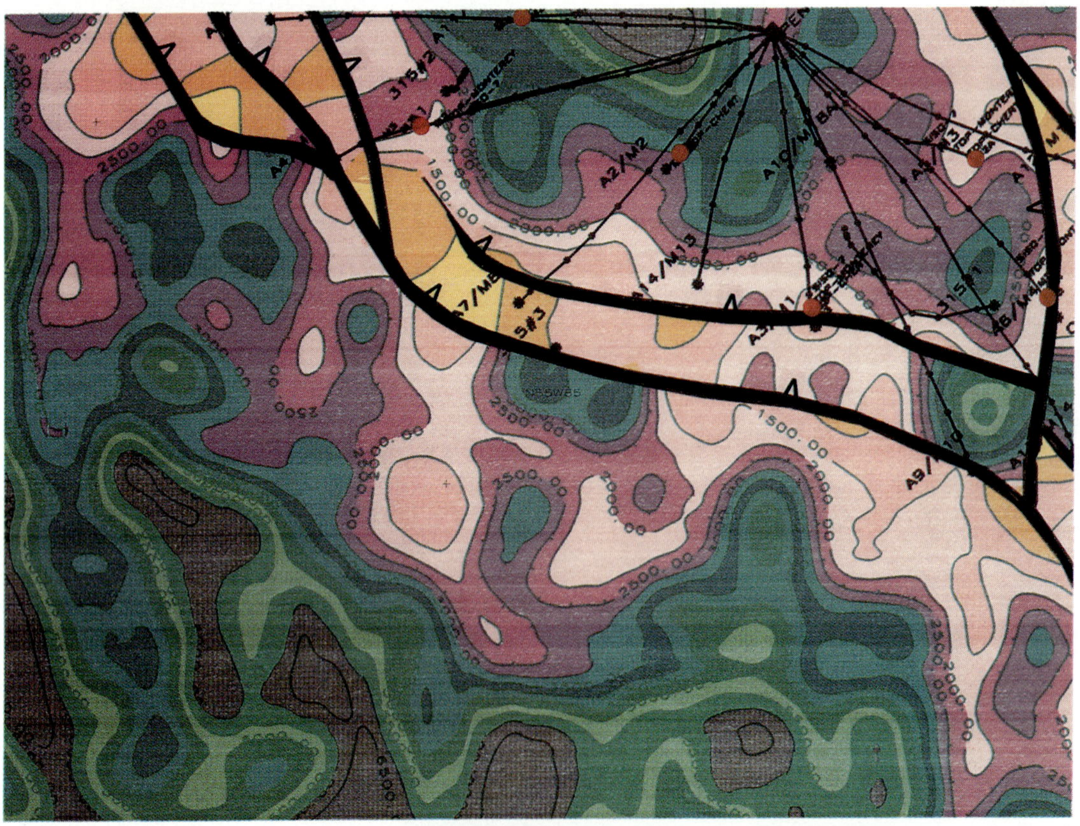

Figure 22—Ellipsoidal radius of curvature map in feet. The radius is given in feet. The contour interval is 1000 feet for labeled contours and 500 feet for changes in color. Core locations are designated by the orange dots.

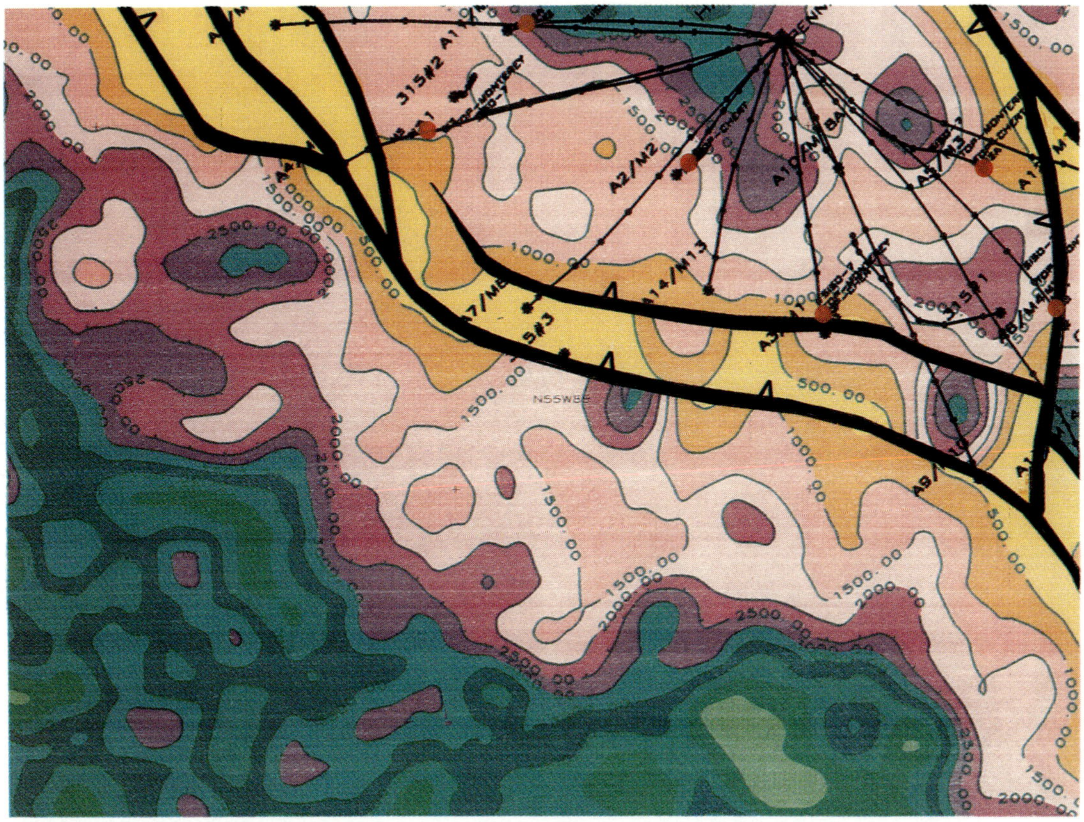

Figure 23—Gaussian radius of curvature map. The radius is given in feet. The contour interval is 1000 feet for labeled contours and 500 feet for changes in color. Core locations are designated by the orange dots.

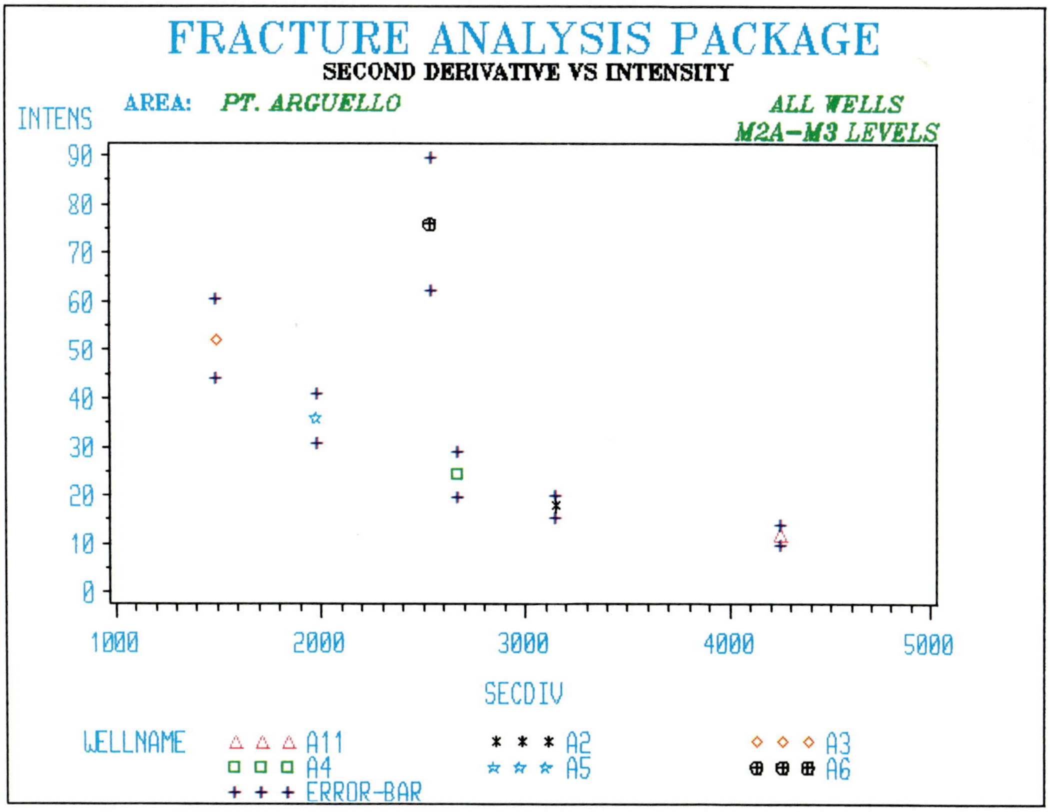

Figure 24—Intensity vs. second derivative radius of curvature. The units are fractures per foot and feet. Note that the A6 well is very close to the fault trending perpendicular to the main thrust faults.

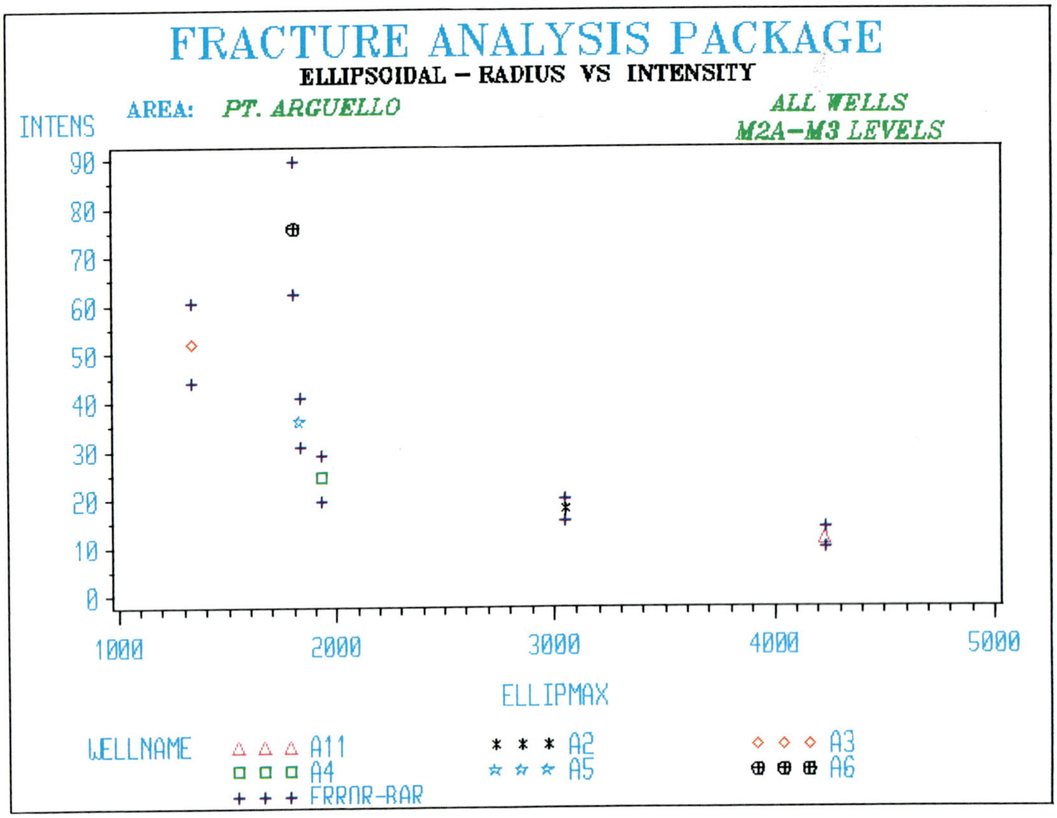

Figure 25—Intensity vs. ellipsoidal radius of curvature. The units are fractures per foot and feet. Note that the A6 well is very close to the fault trending perpendicular to the main thrust faults.

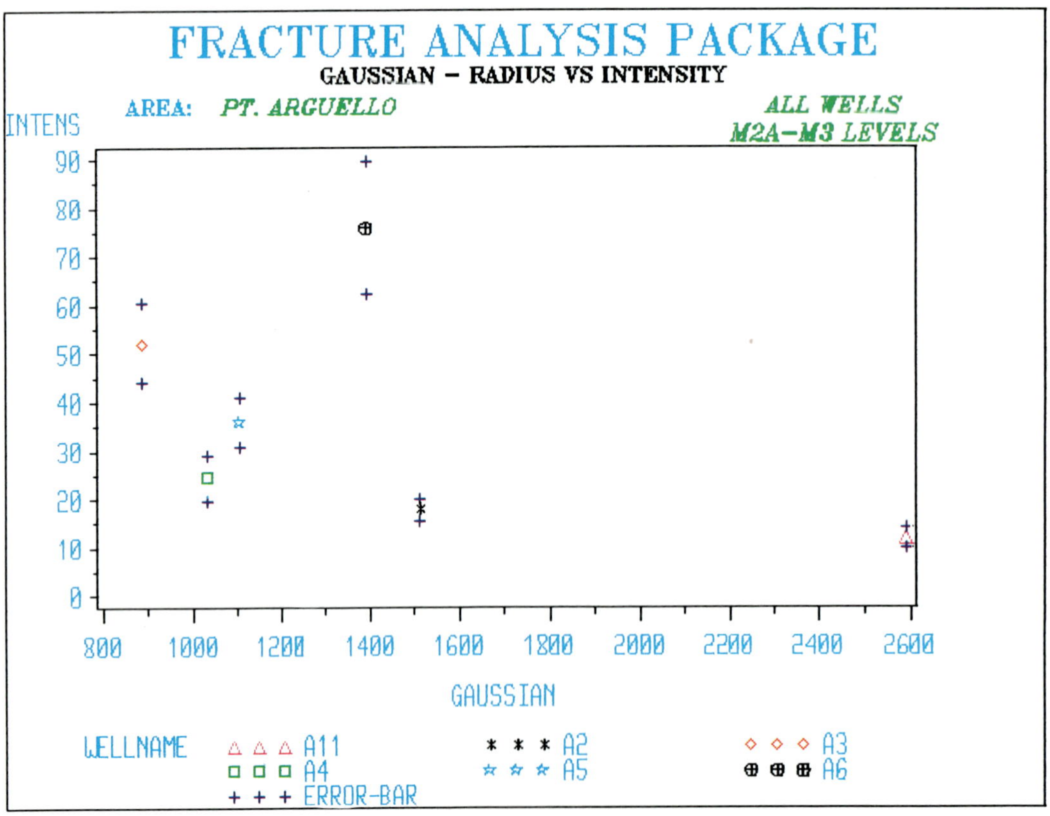

Figure 26—Intensity vs. gaussian radius of curvature. The units are fractures per foot and feet. Note that the A6 well is very close to the fault trending perpendicular to the main thrust faults.

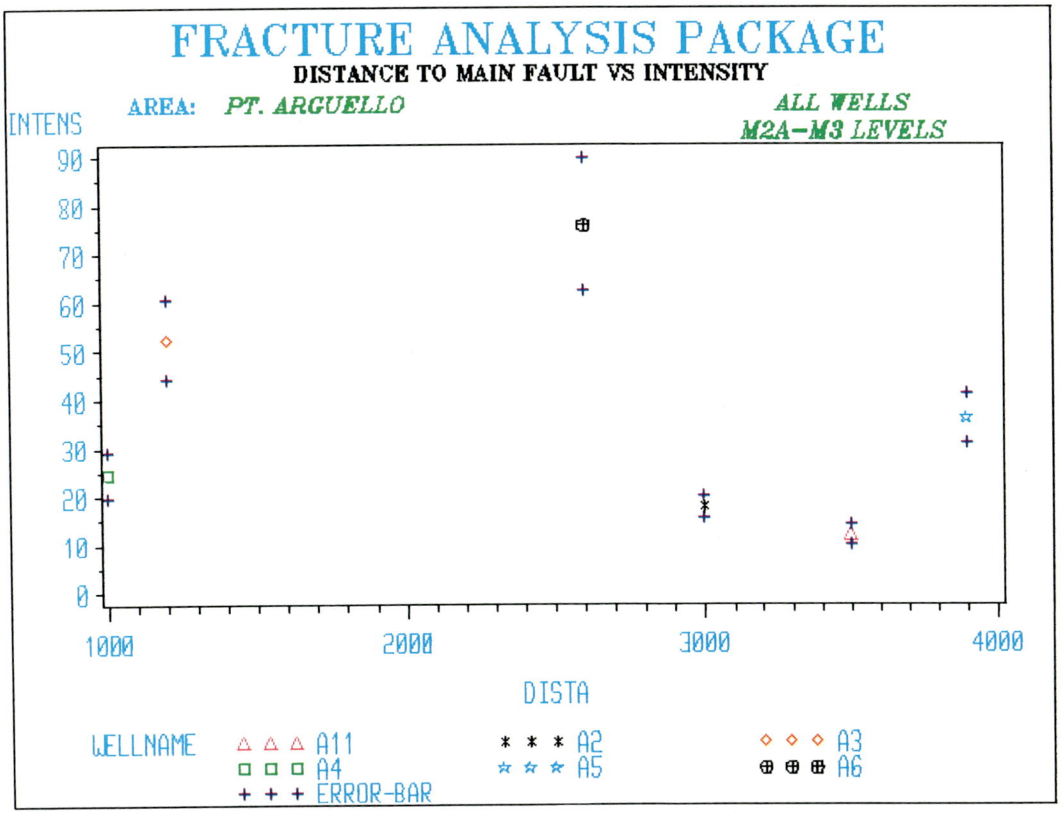

Figure 27—Intensity vs. distance to main fault. The units are fractures per foot and feet.

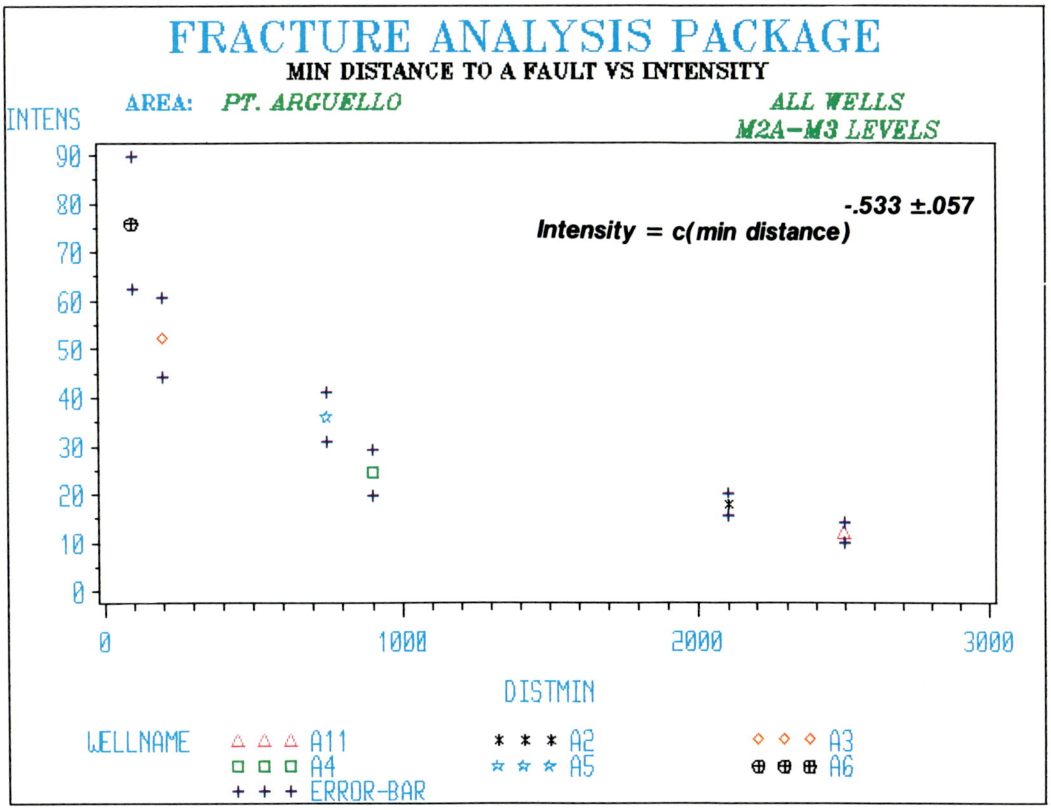

Figure 28—Intensity vs. distance to nearest fault. The units are fractures per foot and feet.

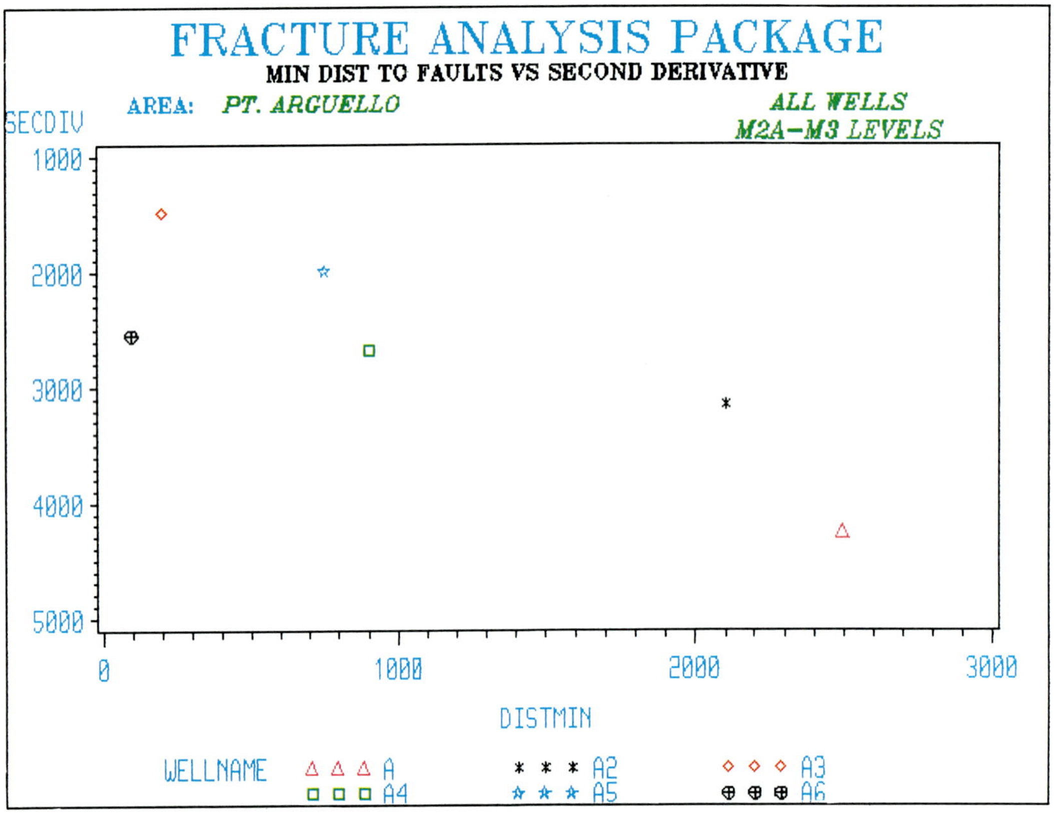

Figure 29—Second derivative radius of curvature versus distance to nearest fault at core locations. Units for both radius and distance to faults is feet.

Prediction versus Production: Waterflood Response of Heterogeneous Reservoirs in an Estuarine Valley-Fill, Little Bow Field, Alberta

John C. Hopkins, James M. Wood, and Federico F. Krause

Upper Mannville G, U and W pools in the Little Bow field are hosted by separate parallel elongate estuarine sandstone bodies within an incised valley fill. Each sandstone body is 3-4 km long, 300-500 m wide, up to 22 m thick, with an average porosity of 22%. Values of horizontal and (vertical) permeability vary widely and average 1324 (125) md in G pool, 2005 (472) md in U pool, and 258 (73) md in W pool.

G pool was discovered in 1972 and placed on primary production. Oil production declined gradually and was accompanied by modestly increasing GOR and WOR. U and W pools were discovered in 1982 and 1983 respectively, and produced by primary methods until initiation of waterflooding in 1985. Response to waterflooding these two pools has been a rise, then decline, in the GOR, followed by rapidly rising WOR, to values much greater than those predicted from reservoir modelling, currently up to 10:1 in wells adjacent to water injectors. Despite the wide variation in permeability values and the different production histories, similar proportions of oil have been produced: 9.2% OOIP in G Pool; 10.5% in U Pool; and 9.3% in W Pool. Production response indicates controls by mesoscale and microscale reservoir heterogeneities.

Mesoscale heterogeneities include permeable sandstone beds several meters thick that are continuous between adjacent wells, and stochastic shale beds up to 80 cm thick. Rapid breakthrough of water has occurred in producing wells adjacent to injectors due to channeling in thick permeable sandstone beds between shales. Microscale heterogeneities are principally mineralogically segregated laminae of cross-stratified sandstones. Core permeability values indicate permeability variations of one order of magnitude among laminae. It appears that water passes preferentially through low-permeability sandstone laminae due to the higher mobility of water compared to oil.

Vertical Seismic Profile Applications for Definition of Reservoir Heterogeneity—South Texas, Vicksburg and Frio Sandstones: A Secondary Gas Recovery Project

S. G. Zinke, L. A. Jirik, R. P. Langford, and R. J. Finley

Approaches to defining the distribution of unrecovered natural gas resources in mature fields are being developed and tested as part of the Secondary Gas Recovery Project, a program supported by the Gas Research Institute, the U.S. Department of Energy, and the State of Texas. The study emphasizes integrated geological, geophysical, petrophysical and engineering assessments of potential reservoir compartmentalization. Selected clastic reservoirs in South Texas have been studied in cooperation with Mobil Exploration and Producing U.S., Inc., Oryx Energy Company, and Shell Oil Company. Vertical seismic profiles (VSP's) have been used to help delineate reservoir sandstones and define structural complexities.

Gas reservoirs of the middle Frio Formation (Oligocene) along the Vicksburg fault zone at Seeligson field (Jim Wells County) consist of fluvial, channel-fill and associated splay sandstones that can be identified using well log correlation and facies analysis. Floodplain and levee mudstones vertically and laterally separate the reservoirs. Production and pressure data confirm lateral reservoir discontinuities. Offset VSP's, having close (30 ft) geophone stations, allow resolution of boundaries of composite channel systems defined using log data. Multi-offset VSP's providing continuous interwell coverage were utilized to interpret fluvial channel boundaries in the 19C-04 reservoir. Forward modelling corroborates the interpretations.

McAllen Ranch field (Hidalgo County) produces from overpressured deltaic reservoirs of the Oligocene Vicksburg Formation. Shingled, southeastward-tapering wedges of the Vicksburg depositionally overlie a basal decollement of the Vicksburg growth fault zone, resulting in complex interplay between sedimentation and tectonism. Gas production is from the shelf-edge, deltaic sandstone packages within the lower Vicksburg. Reservoir heterogeneity is present at four different scales within the five stacked, upward-coarsening progradational intervals of the S reservoir: 1) deltaic lobes, 2) sandstone depocenters within each lobe, 3) distributary sandstones, and 4) diagenetically heterogeneous zones within sandstones. Discrete delta lobes and individual depocenters in the S1, S4, and S5 deltas can be identified and mapped on seismic and overlapping offset VSP data. VSP's also facilitate interpretation of small faults which cut the S reservoir package.

Reservoir Properties Inferred from Seismic Response in Areas with Minimal Log Control

by:

J.D. Loren, J.T. Kulha, K. S. Renbarger, R.M. Sneider

Introduction

The integration of regional petrophysical, geological, and geophysical data permits the detection of reservoir properties and the prediction of seismic response where log control is minimal. Because the necessary acoustic and density logs are frequently unavailable, incomplete or inaccurate, even in mature producing areas, forward modeling to create synthetic seismograms and amplitude versus offset (AVO) response is difficult. Synthetic acoustic and density logs can be created from commonly available resistivity logs and a uniquely derived sand fraction curve. Examination of drill cuttings, sidewall, and whole cores enhances the accuracy of the sand fraction curve, defines reservoir quality and delineates exploration opportunities. Rock, log, and calculated data are transformed into the time domain and integrated into the seismic data by means of the synthetic acoustic transit time.

The process algorithms are project-calibrated with a learning set of diversely representative high quality log data. After verifying algorithm calibration, the synthetic log generation process can then be applied regionally with confidence, even in older wells with "ancient" resistivity logs. The synthetic acoustic and density logs are the basis for models of variable lithology and fluid content, from which synthetic seismograms and AVO response are generated.

Comparison of modeled synthetic seismograms and AVO response to actual seismic data demonstrates a better agreement based upon synthetic acoustic and density logs as opposed to measured acoustic and density, constant density or Gardner's general relationship. The models can be easily modified by varying pore fluids and bed thicknesses to find the range of amplitude changes and AVO effects that can be expected in the seismic data.

Algorithm Calibration

It is rare that complete acoustic and density logs are available throughout the entire depth range in a given well. Reasons for the absence of the logs are many, ranging from being held confidential by the operator to being deleted as a cost savings measure. Even when complete, the density and acoustic logs are often unreliable due to poor borehole conditions, miscalibration or digitizing errors. The commonly run deep-investigating resistivity log is less influenced by borehole alteration, and can be used to generate synthetic acoustic and density logs. This process begins by selecting key wells of high quality log data representative of the complete spectrum of pressure, depth, and geologic conditions. A sand fraction curve (calculated from one of several methods), a fluid pressure gradient curve (obtained from mudweight or pressure data), and a geologic time curve (from paleo or geologic correlations) are created for each well. A composite multi-well file of representative intervals is then resampled to achieve appropriate distribution among variables. A series of multivariable regressions performed upon the resulting data file determines the appropriate coefficients for the algorithms which yield synthetic acoustic and density logs. Comparison of measured acoustic and density logs with the resulting synthetic curves either verifies the algorithm or indicates the need for reprocessing after identifying and deleting from the composite file intervals of questionable data quality.

The sand fraction curve is an important variable since it influences the synthetic curve algorithms and is frequently used either as a discriminator or in the calculation of other parameters. It can be calculated from a shale-baseline shifted SP or normalized SP, the mudweight and borehole-corrected gamma ray, the neutron/density porosity difference or a resistivity curve. The sand fraction is generated by linear interpolation between sand and shale baselines unique to each segment. Provision is made for selecting different methods over different borehole segments (Figure 1). Any one of the five options can provide an excellent sand fraction profile in a typical sand/shale sequence. However, in environments where the standard sand indicators fail, a composite sand fraction curve can be created from a combination of the available methods (Figure 2).

A typical project completed in the middle Texas coast is shown in Figure 3. From an area covering all or parts of seven counties and encompassing approximately 13,000 square kilometers (5,000 square miles), thirty wells were selected for the initial calibration (calibration wells). The resultant algorithm was then applied to an additional twenty-two wells (verification wells). A comparison of the synthetic and measured curves for six of the study wells (both calibration and verification wells) separated up to 160 kilometers (100 miles) and covering 2,600 meters (8,500 feet) of section is shown in Figures 4 and 5. A good fit for both the synthetic acoustic and density is seen through most of the section. Differences are often due to borehole influences such as enlargement, rugosity and probable stress relaxation. Calibration and scaling errors and intervals of obviously erroneous measured data are also common. A calculated resistivity curve representative of wet reservoir rocks drives the process algorithms opposite hydrocarbon intervals. Differences between the calculated and measured resistivities are shaded to flag hydrocarbon intervals such as shown in well 2. The synthetic acoustic and density curves are obviously necessary for the creation of synthetic seismograms where either of the measured logs are totally or partially unavailable, such as in wells 2 through 6; they are usually preferred over the measured logs throughout much of the section, especially in relatively unconsolidated sediments.

Application

The technology demonstrated in the previous examples was applied to a well drilled to 5,200 meters (17,000 feet) in 1961 (Figure 6). Logs available included only an induction, a spontaneous potential, an uncompensated sonic, and an incomplete gamma-ray. Hand edits were applied to the measured acoustic data. A density log was generated from Gardner's relationship using the acoustic log. Below 1,675 meters (5,500 feet), little, if any, definition of sand development is supported by the spontaneous potential.

Drill cuttings sampled at thirty foot increments were obtained and were analyzed using the Sneider rock classification method (Sneider, et al, 1983). Lithology, reservoir quality (permeability), and porosity were estimated and incorporated as continuous curves into the log data file. The sand fraction curve was then calibrated to the rock description and log response. A regional algorithm was applied to generate synthetic acoustic and density logs. Good correlation is seen between measured and synthetic acoustic logs. Major differences are evident between the synthetic and Gardner-derived density; the synthetic density, which correlates with other regional control, is the preferred choice. A synthetic seismogram was created from the measured acoustic log (A). Below 1.7 seconds, a poor correlation in both time and amplitude between the seismogram and seismic section is seen (Figure 7). Good correlation in both time and amplitude from 0.5 to 4.0 seconds is seen when the synthetic density and acoustic curves are used to generate a synthetic seismogram (B). Note the interval below 3.7 seconds (4,900 meters or 16,100 feet) and the improved amplitude and time match. Velocity survey data was incorporated to 3.2 seconds.

The synthetic acoustic transit time was integrated allowing the rock descriptive information and log data to be displayed in time and providing a direct comparison between lithology and the seismic. The amplitude at 3.76 seconds can now be identified as the base of sand development, while poorer quality sands are reflected by the amplitudes between 3.76 and 4.00 seconds.

The algorithm used to generate this example was based on a learning set comprised of 28 wells containing 43,900 meters (144,000 feet) of good quality complete or partially complete density and acoustic data. Within a 3900 square kilometer (1500 square mile) area, the calibrated algorithm was applied to 140 wells in close proximity to seismic control (Figure 8). Of these wells, only a one-inch resistivity log was available in 57 wells; cuttings descriptions were acquired in 19 wells. The synthetic acoustic and density logs were used to generate synthetic seismograms representing wet conditions in all wells. Additionally, modeled lithology or fluid conditions were applied to over half of the wells. Correlation and interpretation of over 1300 kilometers (800 miles) of seismic data in the project area was made possible. This was especially critical in prospective areas where no acoustic or density data was available. A good-to-excellent agreement between the synthetic seismograms and seismic data was established.

Improved Well-Seismic Ties

Figure 9 is a comparison of the well-seismic tie between a model seismogram calculated from measured acoustic and density curves and a model seismogram calculated from synthesized acoustic and density curves. The well from which the logs were taken was drilled directly on the seismic line. A check shot survey was also taken in the well, allowing the models to be referenced to the seismic datum and to apply a drift correction. The four cases shown have all been time referenced to a reflection on the seismic section, for comparison purposes.

The first panel shows the poor agreement between the model seismogram calculated from the measured acoustic and density logs and the seismic section. No check shot correction has been applied. The model seismogram displays a large and variable mis-tie below the reference reflection.

The second panel shows the improved agreement between the model seismogram calculated from the synthesized acoustic and density curves and the seismic section. Again, no check shot corrections have been applied. Only minor time shifting is necessary to make the well-seismic tie.

In the third panel, check shot corrections have been applied to the measured log seismogram. Although a good time tie is achieved, the reflection character still compares poorly to the seismic section.

In the fourth panel, check shot corrections have been applied to the synthesized log seismogram. Not only has a good time tie been achieved, but the reflection character compares favorably with the seismic section. This seismogram allows a confident well-seismic tie to be made.

Acoustic and Density Hydrocarbon Response

Once synthetic acoustic and density logs are generated representing wet reservoir conditions, oil or gas can be substituted and the resultant acoustic and density responses calculated (Figure 10). The in-situ bulk modulus and bulk density of oil and gas at reservoir temperature and pressure is calculated using standard reservoir engineering correlations. The hydrocarbon characteristics shown are for a gas gravity of 0.62 relative to air and an oil gravity of 32o API. A base gas-oil ratio was selected and allowed to increase with depth so that the calculated crude bubble point pressure would never be lower than the formation pressure. As temperature and pressure increase with depth, the differences between oil and gas properties decrease. The greatest impact of fluid substitution, however, is found in the order of magnitude difference between the water and oil or gas bulk modulus.

Gassman's equation and the summation of densities are used, respectively, to arrive at acoustic velocity and bulk density, and in turn, acoustic impedance (Figure 11). The water saturation, Poisson's ratio and matrix bulk modulus used in these calculations are controlled by the sand fraction. The rock matrix bulk modulus varies between the values for quartz and shale, with the shale modulus varying with depth. The reservoir fluid bulk modulus is controlled by the component with the lower bulk modulus (i.e., oil or gas overrides the water). Final confirmation is made by comparing the modeled and measured log response.

Figure 12 is an example of a fluid substitution operation and its effect on the response of the modeled seismogram. Two seismograms were calculated from the synthesized curves for a well drilled to test a high amplitude reflection. This well encountered an eighty foot thick gas sand, which can be seen on the sand fraction curves on the expanded displays. Both water and gas were substituted into the sand, and the resulting model seismograms were compared to the seismic section. Although both seismograms tie the seismic data in time, only the gas case model reproduces the high amplitude reflection seen at 1.95 seconds. By varying the thicknesses of sand layers, as well as their fluid content, suites of simple models can be generated for comparison with seismic data away from a specific well site.

Amplitude Versus Offset

Amplitude versus offset (AVO) analysis of unstacked seismic data can sometimes be used to further discriminate gas from water sands in areas where local rock properties and bed thicknesses can be constrained. The ability to model the expected seismic response of a gas sand is a critical part of such an analysis.

Figure 13 shows two blocked models of a sand body surrounded by sufficient layers to account for interference effects from adjacent boundaries. Layer properties are taken from the synthesized logs, and Poisson's ratio, necessary to calculate seismic amplitudes along non-normal incidence raypaths, is made proportional to the sand fraction curve. In the upper panel, individual layer reflections are displayed for the 854 meter (2,800 foot) offset; the reflections are summed laterally into the offset trace. Modeled sand boundary reflections are highlighted for both the wet and gas cases. Figure 14 shows the full range of response with offset in both cases. The offset ranges from 0 to 2,805 meters (9,200 feet). The single offset model is applied to generate each trace. In this particular model, the acoustic impedance of the sand body is equal to or less than the surrounding shales. Both the wet and gas cases show an amplitude increase with offset for this model.

Figure 15 shows a comparison of unstacked field data from a seismic line acquired over the modeled well site with the theoretical AVO response. The field gathers range in offset from 1,220 to 4,878 meters (4,000 to 16,000 feet), and the AVO response has been modeled over the same distances. The adjacent field gathers show a high amplitude reflection slightly increasing with offset, similar to the gas case. Either case, however, can be supported without the near offset data. Simple modeling shows that variations in bed thickness and rock and fluid properties often lead to non-unique AVO response. Without integration of geologic concepts into forward AVO response models, the usefulness of the technique is even more seriously limited.

Summary

The major points addressed in this work include the following:

1. Accurate and reliable synthetic acoustic and density logs derived from a deep-investigating resistivity log and an interpreted sand fraction curve can be routinely generated provided the process algorithms are properly calibrated with a good quality learning set of data.
2. Rock descriptive data obtained from cores and drill cuttings can be integrated into the log data set in order to better define sand fraction and assess reservoir quality.
3. Synthetic seismograms generated from the synthesized acoustic and density logs routinely provide a good-to-excellent tie to the seismic, even in wells with sparse poor quality data such as older wells where only a correlation-scale electrical log is available.
4. In order to understand amplitude variations observed in the seismic section, pore fluid substitutions are made to vary the rock properties. When gas replaces water in the reservoir rocks, a variety of seismic signatures may be observed, such as amplitude increase, polarity reversal, and amplitude decrease.
5. Amplitude versus offset analysis of unstacked seismic data can sometimes be better understood through forward modeling starting from the synthesized acoustic and density logs. Because variations in bed thickness and rock and fluid properties often lead to non-unique AVO response, geologic concepts must be integrated into the model and interpretive rules of thumb avoided.

REFERENCES CITED

Sneider, Robert M., King, Howard R., Hawkes, H. Earl, Davis, Thomas B., 1983, *Methods for Detection and Characterization of Reservoir Rock, Deep Basin Gas Area, Western Canada*, Journal of Petroleum Technology, v. 35, p. 1725-1734.

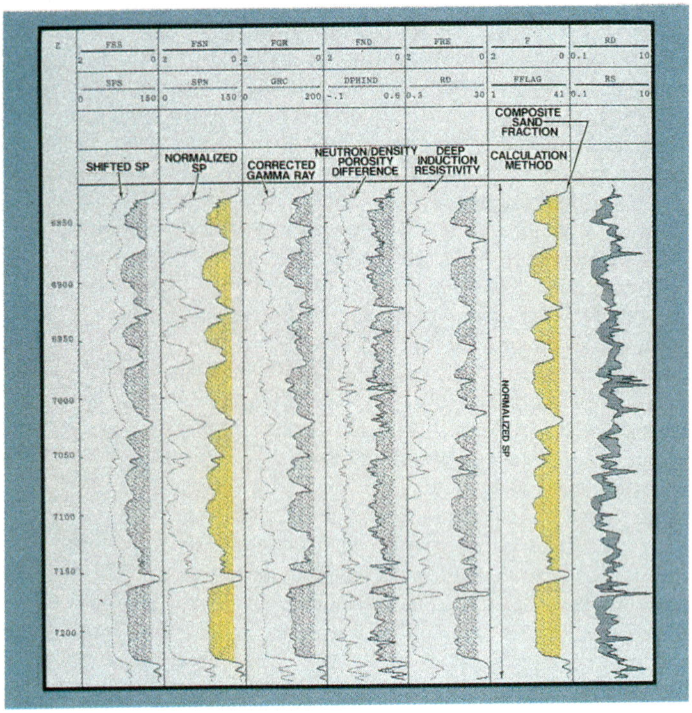

Fig. 1 - Similar Sand Fraction Curves Derived from any Method

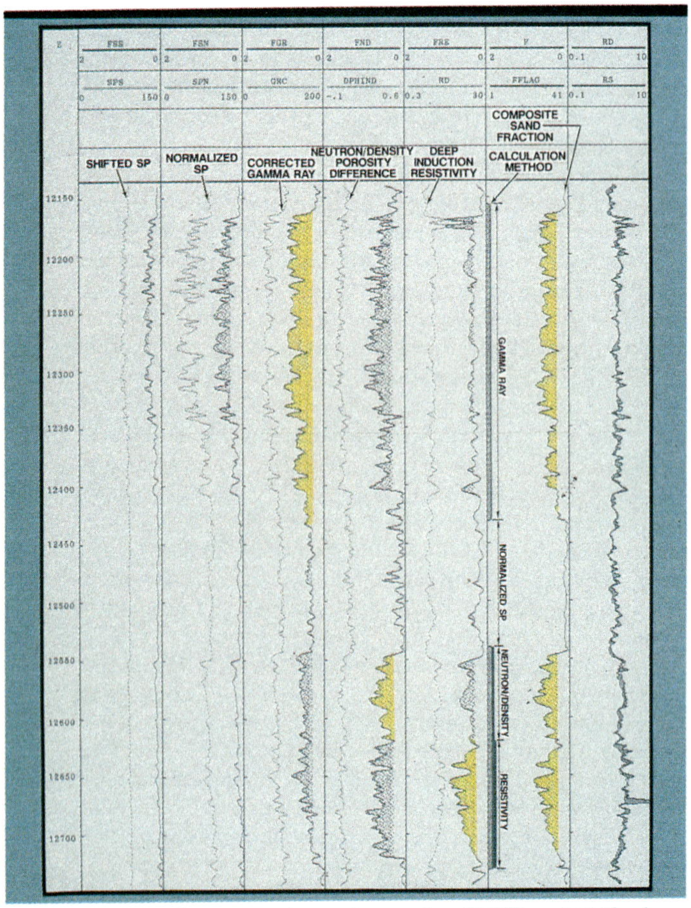

Fig. 2 - Sand Fraction Curve Composite of Different Methods

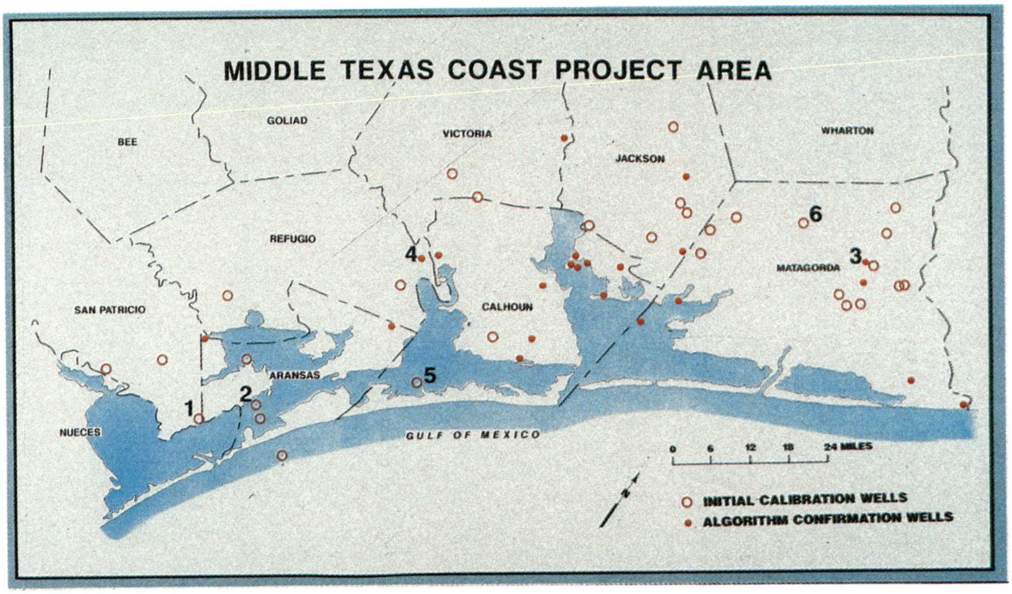

Fig. 3 -Middle Texas Coast Project Area

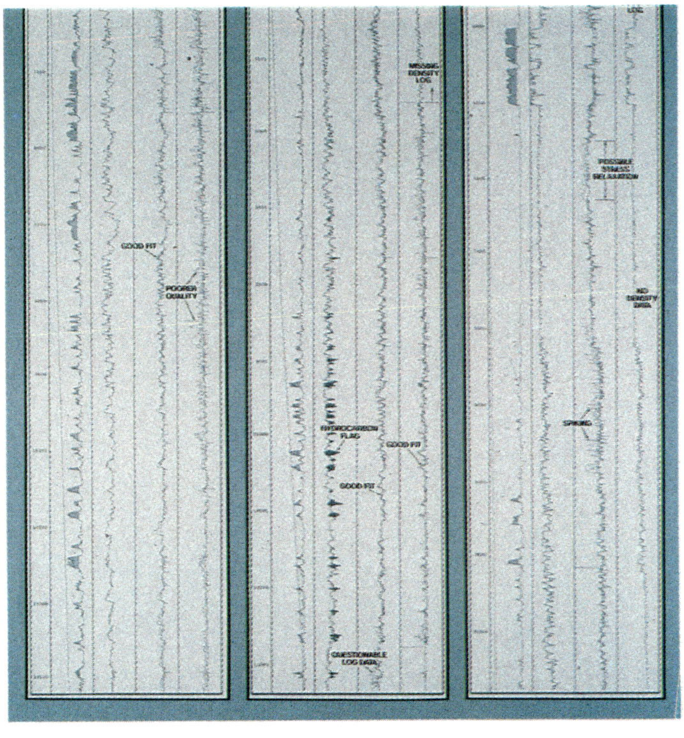

Fig. 4 - Measured and Synthetic Logs Compared - Panel 1

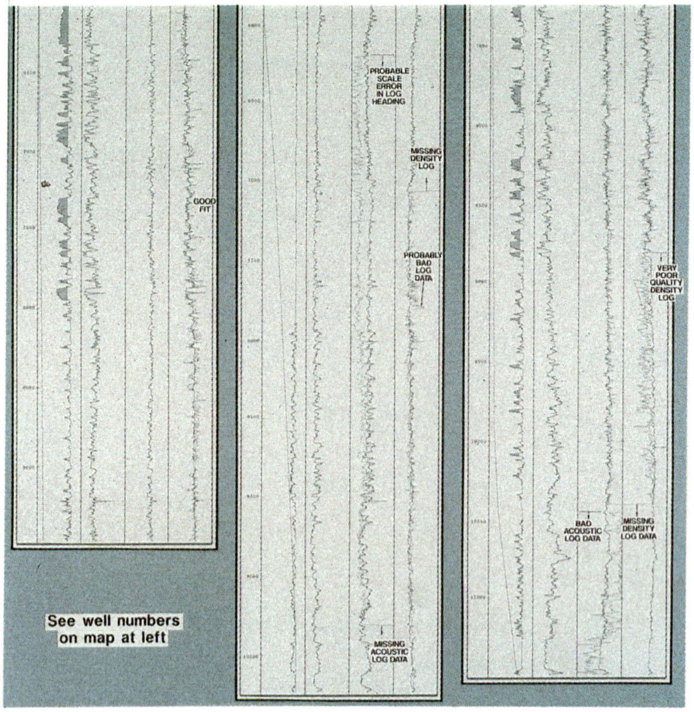

Fig. 5 -Measured and Synthetic Logs Compared - Panel 2

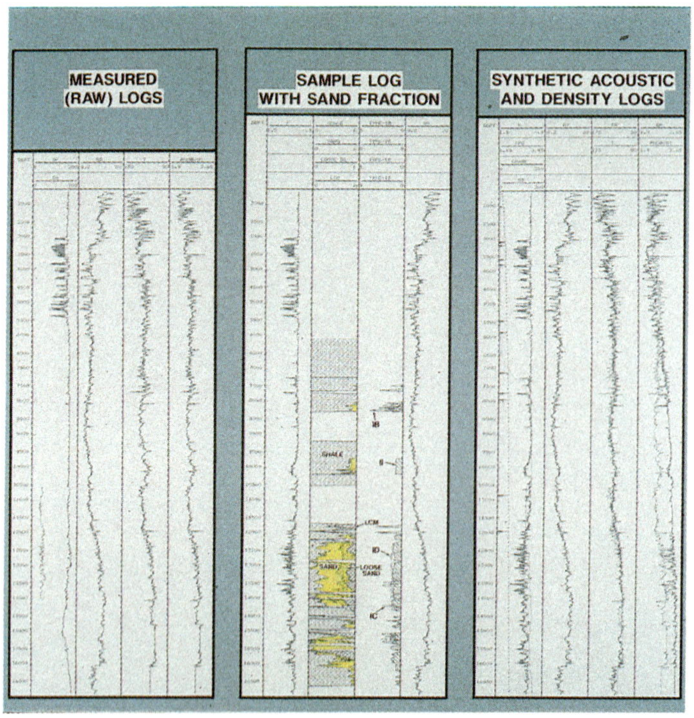

Fig. 6 -Logs and Rock Cuttings Used to Derive Synthetic Acoustic and Density

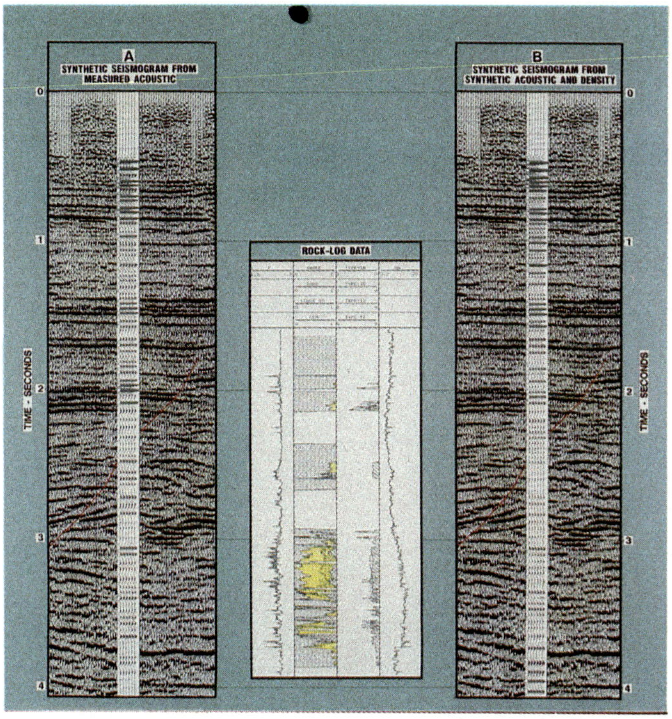

Fig. 7 - Rock-Log Interpretation Improves Synthetic-Seismic Correlation

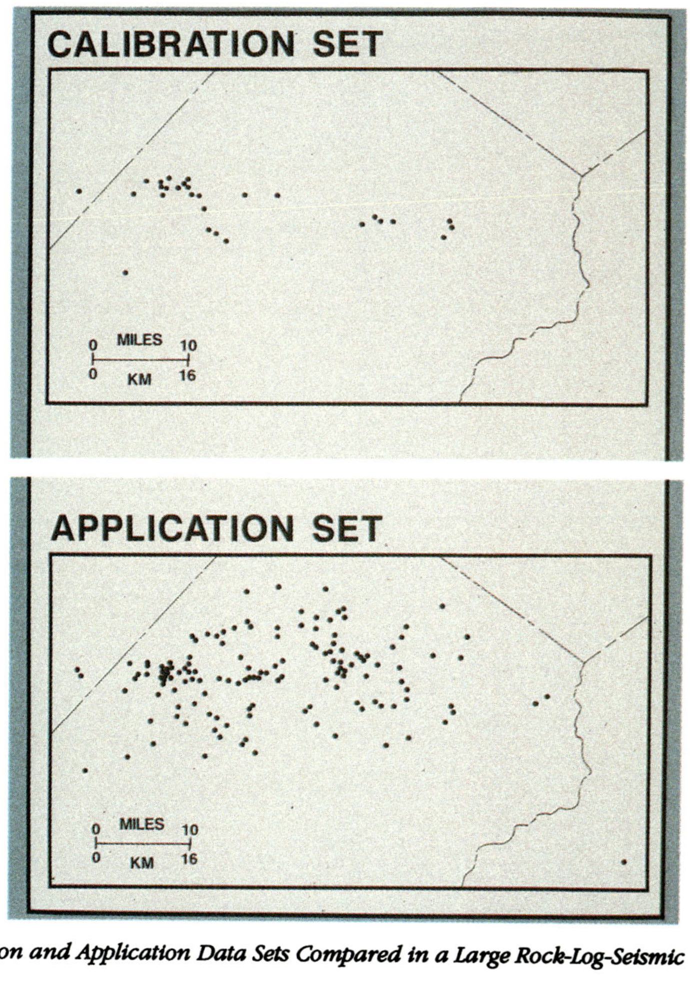

Fig. 8 - Calibration and Application Data Sets Compared in a Large Rock-Log-Seismic Calibration Project

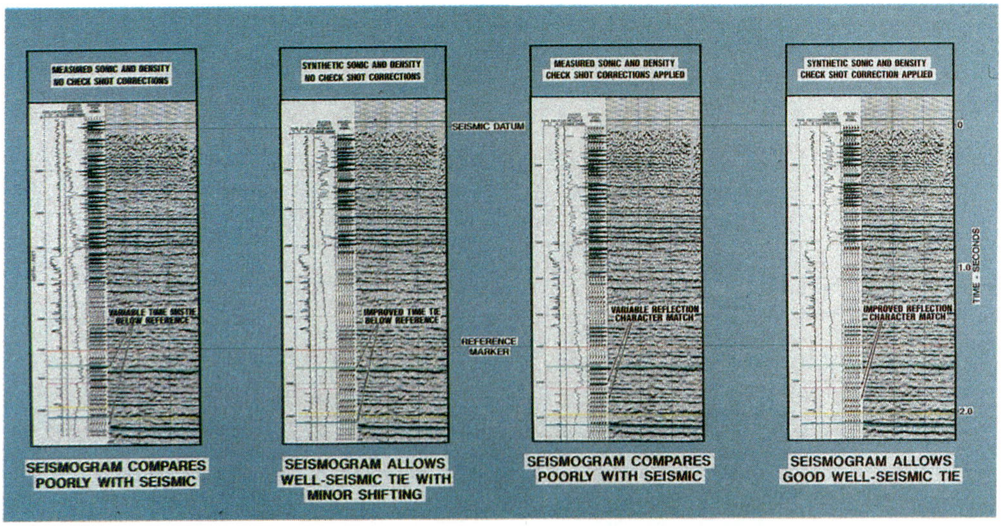

Fig. 9 -Improved Well-Seismic Ties Using Synthetic Log Data

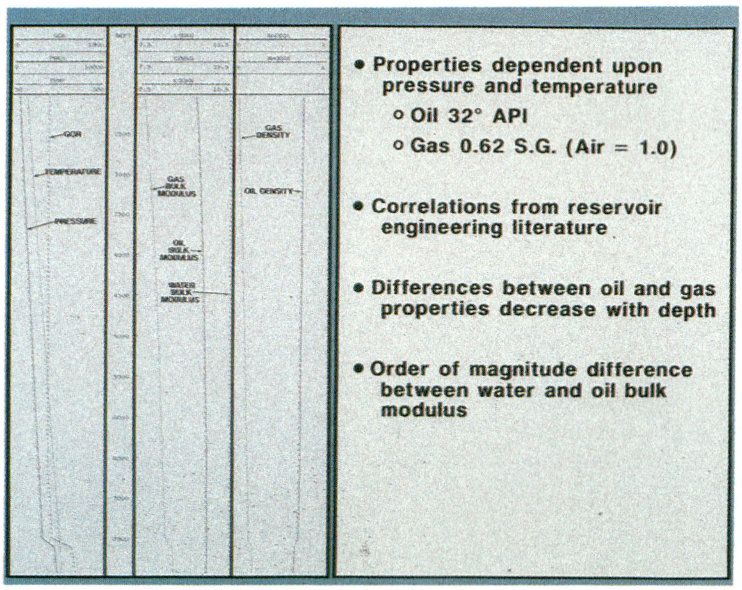

Fig. 10 - Oil and Gas Property Variation with Temperature, Pressure and GOR

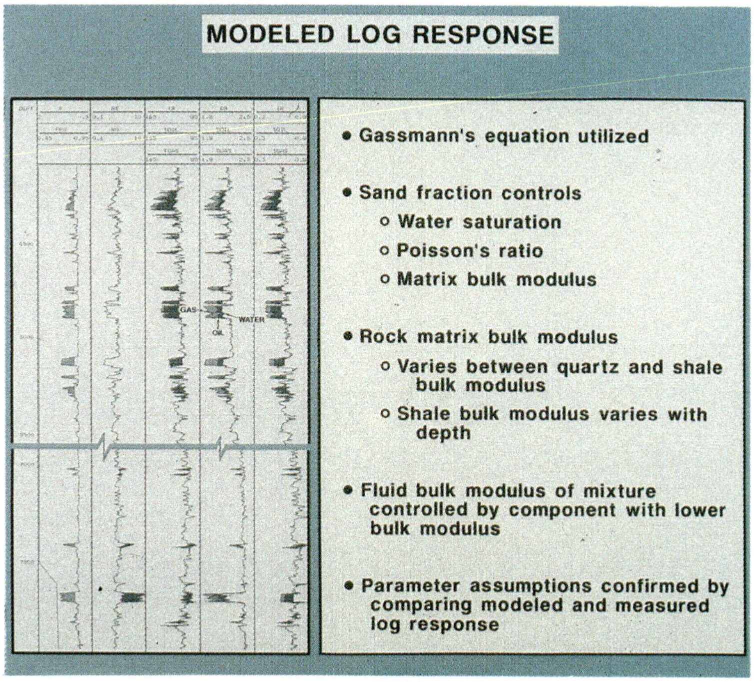

Fig. 11 - Modeled Acoustic and Density response due to hydrocarbons

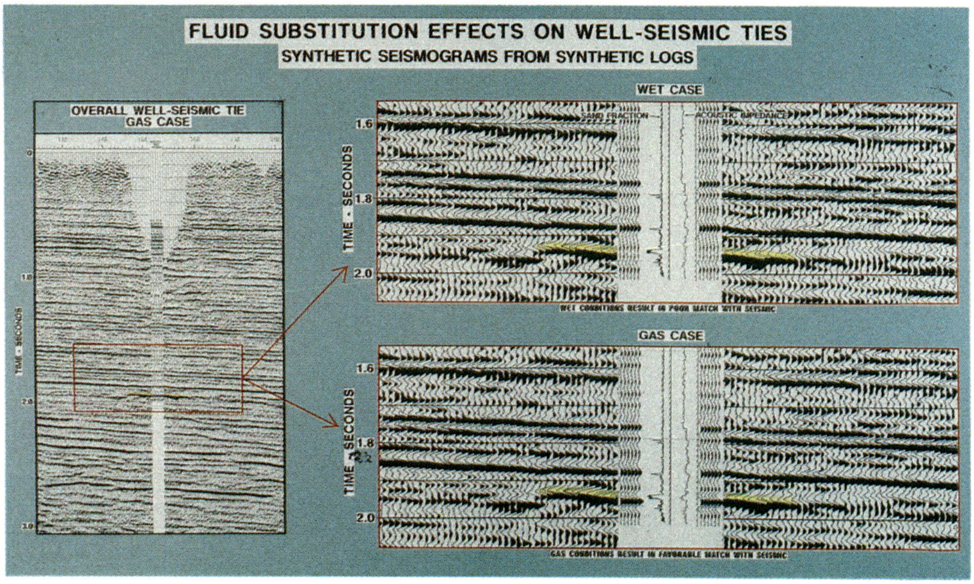

Fig. 12 -Fluid Saturation Effects on Well-Seismic Ties

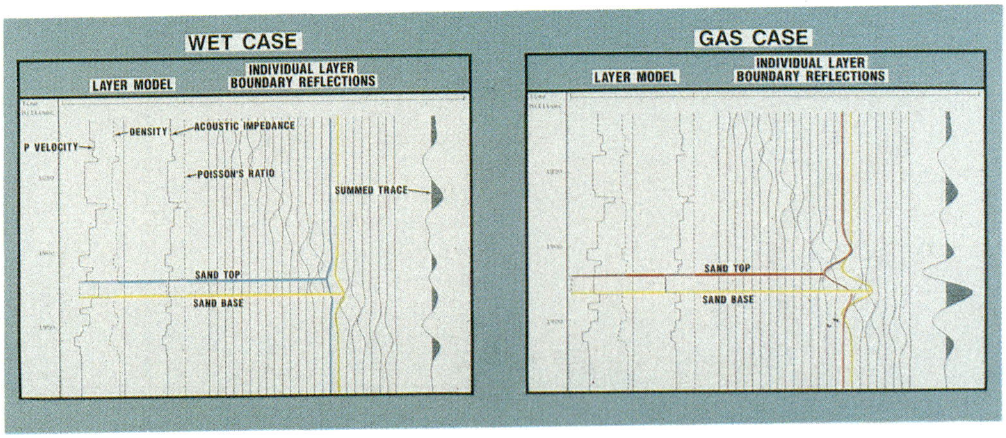

Fig. 13 -Individual Layer Reflections for a 853 Meter (2,800 foot) Single Offset Model

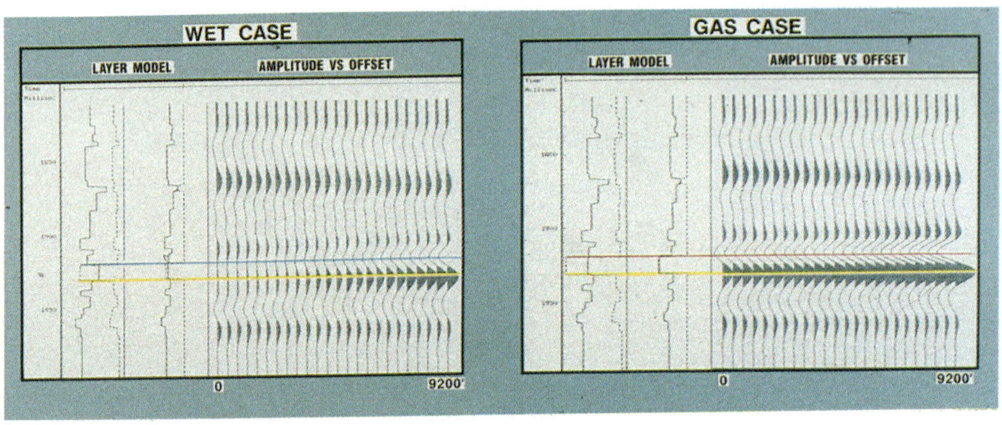

Fig. 14 -Amplitude Variation Versus Offset for a Multiple Offset Model

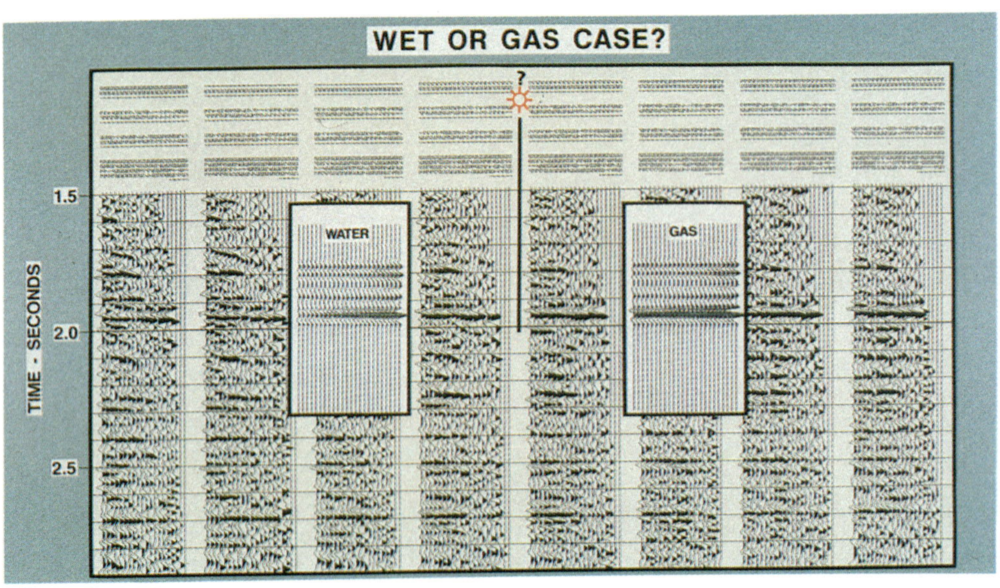

Fig. 15 -Comparison of Multiple Offset Model with Actual Data

COMPUTER-BASED ROCK DESCRIPTION FOR DEVELOPMENT AND EXPLORATION GEOLOGY

Mark H. Scheihing & Gary L. White, ARCO Oil & Gas Company, Exploration and Production Research Center, 2300 West Plano Parkway, Plano, TX 75075

ACKNOWLEDGEMENTS

We thank ARCO Oil and Gas Company for permission to publish this paper.

ABSTRACT

A computer-based rock description system has been developed for on-site description, graphical display and data transfer of cuttings, sidewall cores and conventional core. This system is designed to assist the geologist in efficiently collecting, graphically displaying and building databases of the geological and petrophysical attributes of cores and other types of rock samples. This system is used at the rig or core storage facility to collect rock description for later graphical display and processing in the office.

The rock description software is written as a HyperCard™ stack. This stack contains a wide variety of geological and petrophysical descriptors for siliciclastic, carbonate rocks and some non-sedimentary rocks. The geologist describes the core or sample by selecting description parameters by mouse- or keystroke for a given interval of core or discrete sample. For core description, there are no restrictions placed on the thickness of an interval of description. The description of an entire core or set of samples is stored as a set of "cards," each card containing the geological and petrophysical attributes of a given interval of the core or single sample.

Data entered into the rock description stack can be output as a graphical display, ASCII text file, or spreadsheet-style file. The graphical display uses a series of compiled programs to display the rock description data as lithology, sedimentary structure, and textural and comments columns. Extraction of data from the stack to make text and spreadsheet-style files permits easy transfer of rock description data to other computers and software.

Use of this system has dramatically reduced the time and laboriousness associated with the collection of rock description data. More importantly, this system has reduced the time involved in preparation of graphical displays and data entry for computer applications.

INTRODUCTION

An important responsibility of the petroleum geologist is the description and interpretation of cuttings, sidewall cores and conventional cores. Even with tremendous advances in the imaging capability of some types of wireline logs and resolution of seismic data, rock samples are the most direct indicators of the geologic and petrophysical properties of subsurface formations. Rock samples are the basis for the interpretation of depositional environments and the material upon which permeability, porosity, capillarity and other petrophysical measurements are made.

Whereas wireline log, seismic, and well test data are now routinely collected, processed, stored and manipulated in digital form, rock description data are typically collected manually and often remain in paper copy form. For some studies, all or a subset of the description data may be entered into a "core description" or "core net pay" database. However, this involves a great amount of time-consuming and laborious data recoding and entry. One of the major obstacles to the effective incorporation of rock description data and interpretation into reservoir and exploration studies is the general difficulty in quickly and efficiently translating descriptions and interpretations into a computer-based format that can be manipulated by graphical display, mapping and statistical software already available for these purposes.

This paper discusses considerations for computer-based collection and processing of rock description data, and the implementation of a HyperCard™ application for rock description on a Macintosh computer. We hope to demonstrate that a system can be developed that is acceptable to the wide range of geologists who describe rock samples from the subsurface and that can efficiently record descriptions and store that data in a form usable by a wide variety of programs and computers.

METHODS OF RECORDING ROCK DESCRIPTION DATA

By "rock description" we mean the gathering of data based on visual examination of cuttings, sidewall cores or conventional core using the unaided eye, hand lens and low magnification reflected light microscope rather than the petrographic and scanning electron microscope. These latter techniques fall beyond the scope of the type of data collecting dealt with in this paper.

Geologists describe rock samples for a wide variety of purposes. The scope and detail of description ranges from fairly basic descriptions of lithology to highly detailed sedimentological descriptions for interpretation of depositional environments and reservoir characterization. Table 1 summarizes the major types of data commonly collected during rock description.

Manual Methods

Rock description data, particularly conventional core description data, are generally recorded in three major forms: (1) written descriptions and voice recordings; (2) graphical description forms; and (3) coding sheets. The way in which data are recorded influences what can be done with the data and who can use it. Written descriptions and voice recordings recount the description verbally without the use of vertical profiles. They require no special formats but often suffer from incompleteness and inconsistency from interval to interval and core to core in terms of observations made (and not made).

Graphical description forms are superior to written descriptions in that they categorize the kinds of entries made while allowing a fair degree of flexibility in the amount of detail recorded. Graphical forms are perhaps the most widely used formats for core description by both clastic and carbonate geologists and a variety of standardized forms are available (e.g., Rusnak and Luft, 1963; Boyles, Scott and Rine, 1985; Kersey, 1986).

Customized coding sheets intended for computer entry are sometimes used, especially for large exploration or development projects in which a large quantity of core is available (e.g., Lehmann, 1978; Sneider and King, 1984). These forms are useful when the major description categories of interest are known in a given stratigraphic interval or reservoir but are less useful when starting out in a relatively unknown area. In any event, they require subsequent computer entry.

Computer-based Rock Description

A computer-based rock description system should be able to do all of the following things:

- Record all of the desired data about the sample or cored interval in a flexible order and format with minimum effort and without the use of codes that must be looked up or memorized.

- Store description data in a form that can be easily reviewed and edited and transferred to other programs and computers for database creation, graphical display, or statistical manipulation.

- Be usable by someone with a minimum amount of training.

The requirement for ease and flexibility in data entry is a critical factor in determining the acceptance of a rock description system by geologists. Systems in which data entry is laborious are not likely to be used, even if such systems can demonstrate a downstream saving of time in data handling.

Previous Efforts at Computer-based Rock Description

Much of the software described in the literature is dedicated to the storage, manipulation and graphical display of rock description data rather than data collecting. References to these types of systems are summarized in Ward and Waltho (1988).

Perrin (1984) describes a core description system using the Apple II microcomputer. This system provides for the entry of alphanumeric codes and numeric percentage data into several panels that prompt the user for the data. The system is designed to be operated by a technical assistant who enters descriptions verbally communicated by the geologist. Ward and Waltho (1988) describe a series of BASIC programs used for entry of core description data from coal-bearing strata. This system permits the entry of coded lithological descriptions into a hand-held computer. The system checks and validates the entered codes.

Systems have been developed expressly for the purpose of recording data on modern soil profiles. For example, Clarke, Fisher and Ragg (1986) describe a BASIC program used for recording modern soil profile descriptions in the field. This system uses a hand-held microcomputer for data entry in the field and transfer of the data to a Soil Information System on a mainframe computer.

There is at least one commercial software package for computer-based rock description. The software package called "LogOn™" sold by Rockware, Inc. permits the selection of descriptive terms from a menu to create the lithologic description. These terms can be edited and additional terms added. The text file created by this software is the basis for input into a graphical display package marketed by the same firm.

"HYPERCORE," A HYPERCARD™ APPLICATION

We have developed a rock description system, called HyperCore, running under HyperCard™ on the Apple Macintosh family of computers. The Macintosh computer was used because of its superior graphics and the ease of its user interface. Data entry and on-line help functions were programmed as HyperCard™ stacks because of the ability of HyperCard™ to freely associate graphics, text, and numbers, as well as the ability to handle database management functions required by the application and the relative ease of programming in HyperTalk™, the programming language of HyperCard™. Routines for creating graphical and text file output were programmed in QuickBasic™ and linked to the HyperCard™ data entry stacks.

HyperCore was designed with three things in mind: (1) the utility of the software to geologists describing rocks for a wide variety of purposes; (2) the portability of rock description *data* to other applications and computers; and (3) the ability to expand the system at a later time with minimal reprogramming. We desired a system that would be useful both to the generalist and the specialist in rock description.

Structure of HyperCore

As previously mentioned, the data entry part of HyperCore is a HyperCard™ stack. The physical analogy to HyperCard™ is a stack of cards that can contain data in the form of text, numbers and graphics. Manipulations can be performed upon these data and relationships established among cards using a programming language called HyperTalk™ built into HyperCard™. One of the features of HyperTalk™ are "buttons" whose physical appearance can be customized and which can be programmed to perform some function(s) when clicked by the mouse. HyperCard files are called "stacks" in keeping with the physical analogy. The data entry "stack" of HyperCore consists of four main types of "cards:" (1) the Table of Contents card; (2) the Interval Record card; (3) Selector cards; and (4) the Reports card.

The **Table of Contents** card (Figure 1A) contains basic information about the set of cuttings, sidewall cores or conventional cores such as well name, API number, geologist responsible for the description, notes and listings of depth intervals described and lithofacies defined. This information is entered into designated areas on the card by clicking anywhere within the area and then entering the data into a dialog box that appears. The card also contains buttons that perform various database functions such as indexing and sorting cards. The Reports card is also accessed from this card.

The **Interval Record** card (Figure 1B) contains the summary of geologic attributes that have been chosen for each depth interval described. All depth intervals described have an Interval Record card. The interval is specified by a top and bottom depth. This card is the starting point of description for the depth interval specified by the user. This card also contains buttons that perform other functions such as access to other Interval Record cards, creation of lithofacies and listings of geologic attributes of that interval.

Selector cards (Figure 1C) are cards that contain various classification systems and geologic descriptors from which selections are made or data are entered. Typically, selections are made simply by clicking on the button denoting the selection. Selections or entries made on these cards are stored under the appropriate Interval Record card.

The **Reports** card (Figure 1D) is the card from which output options are executed. These output options include graphical display of the data and creation of comma-delimited text and spreadsheet-style files. Output is generated by QuickBasic™ compiled routines that are launched from this card.

Use of the Data Entry Stack

To record rock description data, there are three major steps that are repeated for each cuttings sample, sidewall core or core interval (Figure 2):

Step 1 - The geologist decides what constitutes a particular "unit of description" or "interval". For cuttings and sidewall cores this "interval" corresponds to the depth range over which the sample is representative of the formation. In cuttings taken every ten feet of drilling, an interval of 10000 to 10010 feet, for example, would be appropriate; for a sidewall core of about one inch diameter, an interval of 10000.0 to 10000.1 feet, for example, would be appropriate.

For conventional core, the thickness and geological significance of an interval are the decisions of the geologist. The interval can correspond to a simple lithology, a "sedimentation unit" as described by Otto (1938), a "genetic unit" and its variants as described by Siemers & Tillman (1981) or a "lithofacies" as defined by Reading (1986). Each interval described in a core can have a different thickness, varying from a fraction of a foot or meter to tens of thousands of feet or meters. In our experience, geologists typically subdivide cores into intervals for description that are between 0.5 and 3 feet thick.

Step 2 - The geologist creates a new Interval Record card and enters the top and bottom depths of the interval.

Step 3 - The geologist makes selections and entries in the appropriate Selector cards to describe the discrete sample or cored interval. These selections and entries are stored under the Interval Record card for that depth interval.

These steps are repeated for the next discrete sample or cored interval. By this procedure, the geologic attributes of a set of samples or conventional core are stored as a "stack" of Interval Record cards in which each card contains the description of one interval.

Types of Selector Cards

There are a great many descriptors and classifications that could be applied to igneous, metamorphic and sedimentary rocks. However, in most cases, a smaller subset suffices, especially since HyperCore is predominantly intended for describing sedimentary rocks. Consequently, Selector cards were developed that represent a wide range of categories and sedimentary rock and sedimentary structure types but generally avoid highly detailed and obscure specialist classifications of sedimentary rocks and detailed classifications and descriptors of igneous and metamorphic rocks. Table 2 lists the Selector cards currently in the system. Selector cards were designed to efficiently record different types of data. In addition, attention was given to the "look and feel" of Selector cards to visually reinforce the geologic descriptors represented.

This approach led to the design of several major types of Selector cards. The most common type is a card that contains descriptors that are selected by clicking the mouse (Figure 3A). Most of these cards permit the creation of additional categories should the desired category not be present on a particular card. Another type (Figure 3B) consists of text or numeric entries that are entered from the keyboard. A third type (Figure 3C) permits the entry of semiquantitative percentage data made from visual estimates. This card keeps a running total of each entry and will not permit totals over 100%.

All Selector cards are accessed either from the Interval Record card or from any other Selector card. In this way, the geologist can enter data in any order that is deemed appropriate or as observations are made. There is no requirement that all or any subset of Selector cards have entries made in them.

The Interval Record card for the described interval displays a black dot next to the Selector card listing if an entry has been made in that particular card. There is also a "Table" button on the Interval Record card that, when clicked-on, displays a summary of the description to date for that particular depth interval. Both these features help the geologist to remember to enter all desired data into the appropriate cards.

Special Functions - Defining and Replaying Lithofacies

Sedimentary sequences can usually be described by a number of lithofacies that are repeated vertically within the sequence. Lithofacies are typically defined on the basis of lithology, grain size, sorting, mineralogy, color, physical structures and other associated attributes. It is advantageous to be able to define and subsequently replay descriptions of lithofacies when they are again encountered in the core. Thus, a function was included to define and store a lithofacies and then replay that lithofacies description for a new interval.

After deciding that a particular described interval is to be identified as a lithofacies, the geologist clicks on the "Make a Facies" button contained on the Interval Record card. A name for the lithofacies of the geologist's choosing is entered into the dialog box that appears. A separate Interval Record card is automatically created that contains the description associated with the *lithofacies name* rather than the depth interval. This lithofacies name is also listed on the Table of Contents card.

In the event that the same lithofacies is again encountered in the core, the geologist can automatically enter its description by clicking on the "Replay a Facies" button on the Interval Record card of the interval currently being described. A scrolling window appears containing a listing of all lithofacies previously defined. The geologist selects the desired lithofacies name and the description associated with that lithofacies name is copied into the new interval. This description can then be modified for this specific depth interval if that is deemed necessary.

Graphical Display and Other Output Options

Various reports routines provide graphical display and other output files of data collected and stored in the HyperCore stack. These routines are compiled QuickBasic™ programs launched from the Reports card. The most fundamental file output is a comma-delimited ASCII text file in which each interval is a record and all descriptors are fields of that record. This file is the basis for all other graphical and file output. The graphics display routines (CoreChart A,B,C) offer two different columnar displays at scales varying from 1 inch = 0.5 foot or meter to 1 inch = 20 feet or meters. An example of CoreChart A is shown in Figure 4. This format displays lithology, grain size profile, cementation, sedimentary structures, grain size range and mode, grain angularity, sorting, hydrocarbon staining and comments. The text translation of the coded description and comments may be printed out either in the "Comments" column next to the graphical display or separately as a text file. CoreChart B contains somewhat different display categories and is intended primarily for display of carbonate core descriptions. CoreChart C is a columnar section for the display of semi-quantitative data collected by HyperCore (Figure 5).

In addition to the comma-delimited ASCII text file and graphics columnar sections, there are two additional output options, a wireline log-style file and a spreadsheet-style file. The wireline log option produces a table of rock description codes on a constant footage basis that can be input into a log analysis program and integrated with the wireline log curves. The spreadsheet-style option produces a tab-delimited table of rock description codes that can be input into a spreadsheet or any other program that uses data in table form. The spreadsheet is the most useful data output format for moving core description data to other Macintosh programs and for transfer to other computers.

THE HELP STACK

A Help stack was created and linked with the HyperCore description stack to provide on-line help in using the rock description system. The stack is designed to function both as an on-line aid and as a stand-alone tutorial. Each Selector card has its own Help card which explains its use and any literature citation associated with the classification system employed in the card. In addition, all functions associated with data storage, manipulation and output are explained in the Help stack.

PHYSICAL SETUP OF SYSTEM

HyperCore is intended to be used on site at the rig or core storage facility. For cuttings and sidewall core description, a Macintosh computer is set up on the work bench or table next to the samples and binocular microscope. For core description, the computer is usually set up on a cart next to the slabbed cored and rolled along the core lay-out benches as description proceeds. The use of a Macintosh computer on site is limited only by the availability of an uninterrupted power supply and environmental conditions within the operating range of the computer. Most of the core description work has been done to date using a Macintosh SE with an accelerator card or SE 30. HyperCore will run on any Macintosh computer but we have found that the response time of Macintosh Plus and SE computers without accelerator cards is usually too slow to be tolerable by the user.

SYSTEM EXPANDABILITY

For most rock description needs, the number and type of Selector cards already in the description stack have been more than adequate but we have encountered several core description projects that required specialized Selector cards. These custom Selector cards were created and integrated into the rock description stack with a relatively minor programming effort. Thus, with the appropriate programming expertise, the rock description stack is readily customized for special or unique description needs.

The use of HyperCard™ as the operating environment for the core description program has the added advantage of providing an excellent programming environment for the creation of geologic tutorial stacks that can be used both as stand-alone tutorials or can be linked to the core description stack itself. Such stacks could be on any subject relating to core description, lithology, reservoir or depositional environments and facies analysis. With such stacks in place, the geologist could access information relating to subjects such as the vertical sequence of lithofacies in strandplain environments or the identification and environmental interpretation of trace fossil types.

CONCLUDING REMARKS

At ARCO a number of clastic and carbonate sedimentologists have used or reviewed the HyperCore system and are favorably disposed toward it. The system has proved especially valuable in quickly providing detailed core description data to be used in a comprehensive well summary chart incorporating well log, core description, and core plug reservoir data that was generated using PETROS™, a wireline log analysis system. In addition, HyperCore has been successfully applied to section measurements of several surface exposures.

Hopefully, we have demonstrated that a relatively easy to use and versatile system can be designed that meets the needs of a variety of individuals who describe rock samples from the subsurface. We hope that the implementation of the core description system will encourage both the description of core by geologists and the incorporation of this information into exploration and reservoir engineering databases where such data are desperately needed.

REFERENCES CITED

Boyles, J.M., A.J. Scott & J.M. Rine, 1986, A logging form for graphic descriptions of core and outcrop: *Journal of Sedimentary Petrology*, v. 56, p. 567-568.

Clarke, S.R., P.F. Fisher, and J.M. Ragg, 1986, Soil profile recorder: a program to enable the recording of soil profile descriptions in the field: *Computers & Geosciences*, v. 12, p. 779-806.

Kersey, D.G., 1986, Coring: Part 9--Geological aspects: *World Oil*, January, p. 103-108.

Lehmann, R.J., 1978, Application of a core logging technique for standardizing descriptions and simplifying computerization of data, *in* H.E. Hodgson, ed., Proceedings of the Second Symposium on the Geology of Rocky Mountain Coal--1977, Colorado Geological Survey, Department of Natural Resources, Resource Series 4, p. 149-152.

Otto, G.H., 1938, The sedimentation unit and its use in field sampling: *Journal of Geology*, v. 46, p. 569-582.

Perrin, J.D., 1984, Use microcomputers for core description entry: *Oil & Gas Journal*, January 2, p. 78-80.

Reading, H.G., ed., 1986, Sedimentary Environments and Facies: 2nd edition, Oxford, Blackwell Scientific, 615 pp.

Rusnak, G.A., & Stanley J. Luft, 1963, A suggested outline for the megascopic description of marine sedimentary cores: Institute of Marine Science, The Marine Laboratory, University of Miami, 13 pp.

Siemers, C.T., & R.W. Tillman, 1981, Recommendations for the proper handling of cores and sedimentological analysis of core sequences, *in* C.T. Siemers, R.W. Tillman & C.R. Williamson, eds.,

Deep-Water Clastic Sediments: A Core Workshop, SEPM Core Workshop No. 2, San Francisco, May 30-31, p. 20-44.

Sneider, R.M., & H.R. King, Integrated Rock-Log Calibration in the Elmworth Field--Alberta, Canada: Part I - Reservoir Rock Detection and Characterization, *in* J.A. Masters, ed., Elmworth--Case Study of a Deep Basin Gas Field: American Association of Petroleum Geologist Memoir 38, p. 205-214.

Ward, C.R., & Waltho, A.E., 1963, A BASIC program for in-field entry of lithologic descriptions in borehole logs to a hand-held portable computer system: *Computers & Geosciences*, v. 14, no. 1, p. 83-97.

Lithology	Sedimentary structures
Grain size	Trace fossils & bioturbation
Grain types	Nature of bedding contacts
Sorting	Body fossils & fragments
Grain shape & angularity	Cementation & dissolution
Color	Concretions
Mineralogy	Fractures & faulting
Porosity amount and types	Hydrocarbon shows

Table 1 - Major classes of data commonly collected during visual rock description.

Siliciclastic Rocks
Sand-Silt-Clay
Conglomerate-Sand-Fines
Arenite-Wacke-Mudstone
Mudrock
Limestone-Sandstone-Mudstone
Dolomite-Sandstone-Mudstone
Carbonate & Evaporitic Rocks
Carbonate Textures
Gypsum Fabrics
Anhydrite Fabrics
Evaporite Structures
Organic Rocks
Coal, Organic-rich Mudrocks
Paleosols
Paleosol Textures
Igneous & Metamorphic Rocks
Igneous
Pyroclastic
Metamorphic
Grain, Cement & Pore Types
Grain Attributes & Color
Particles & Skeletal Fragments
Mineral Constituents
Cement Mineralogy
Cement Texture
Pore Types
Physical & Chemical Diagenesis
Consolidation & Induration
Concretions
Trace Fossils
Burrowing Intensity
Trace Fossils
Sedimentary Structures
Heterolithic Structures
Ripple Cross Lamination
Parallel & Cross Lamination
Graded & Massive Bedding
Soft Sediment Deformation
Bedding & Lamination
Paleocurrent Data
Reservoir
Net Pay
Plug Data
Staining & Show Analysis
Other
Fractures & Faults
Samples
Comments
Lithofacies
Core Missing?
Contact

Table 2 - Listing of Selector cards currently in HyperCore data entry stack. Major Selector card categories are shown in bold.

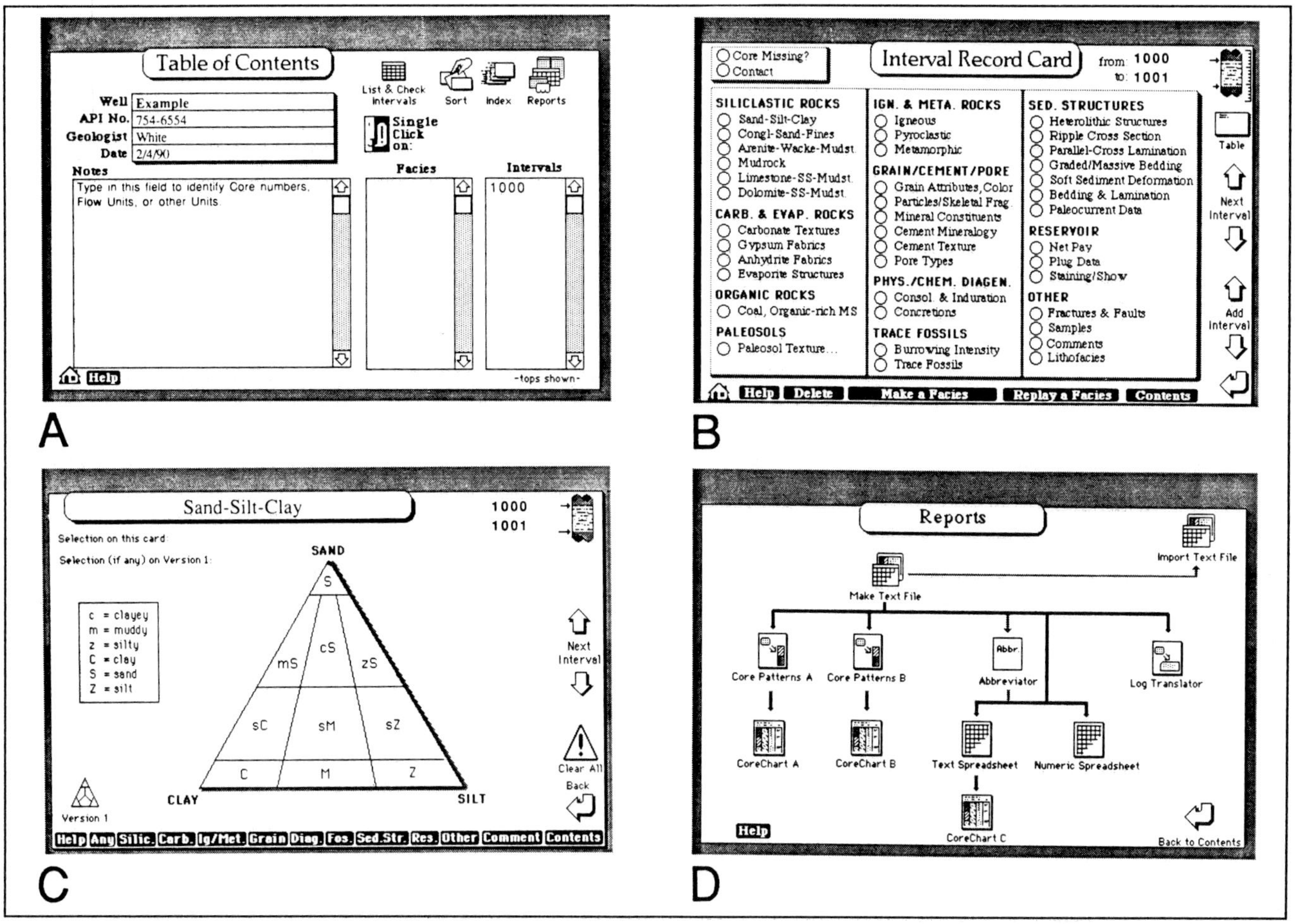

Figure 1 - Examples of major types of cards in the HyperCore data entry stack. (A) Table of Contents card; (B) Interval Record Card; (C) Selector card for classification of rocks using sand-silt-clay triangle; (D) Reports card.

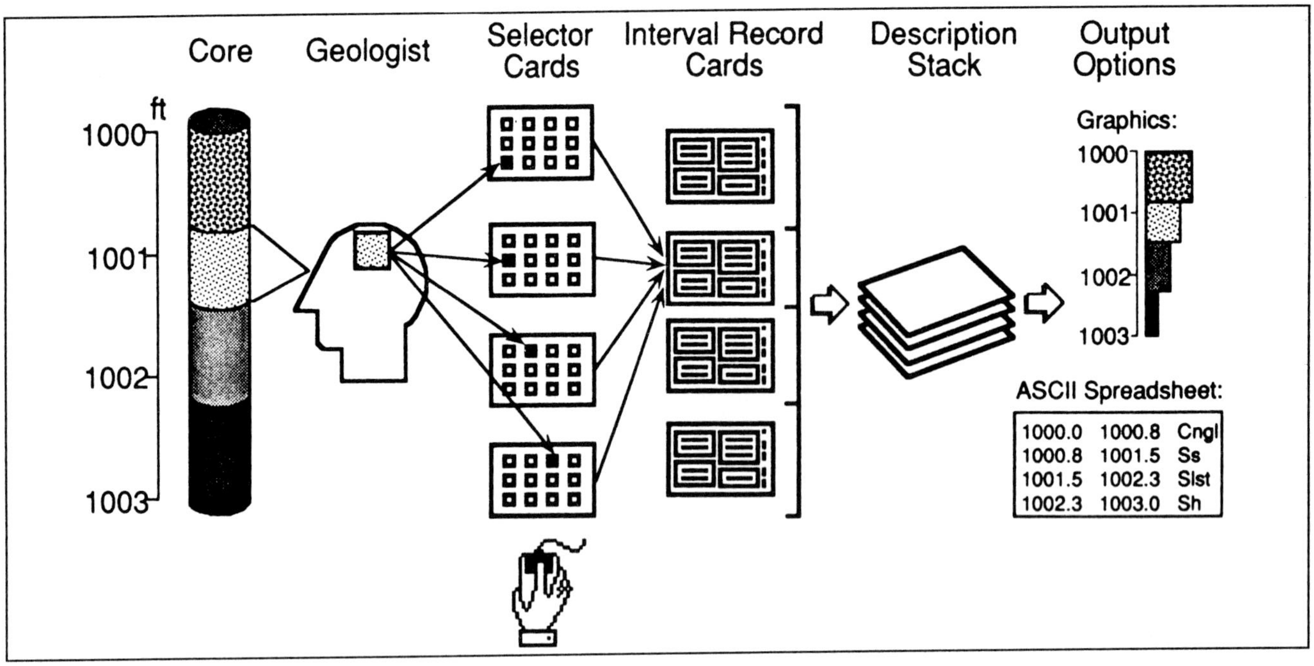

Figure 2 - Cartoon of use of data recording using the HyperCore data entry stack.

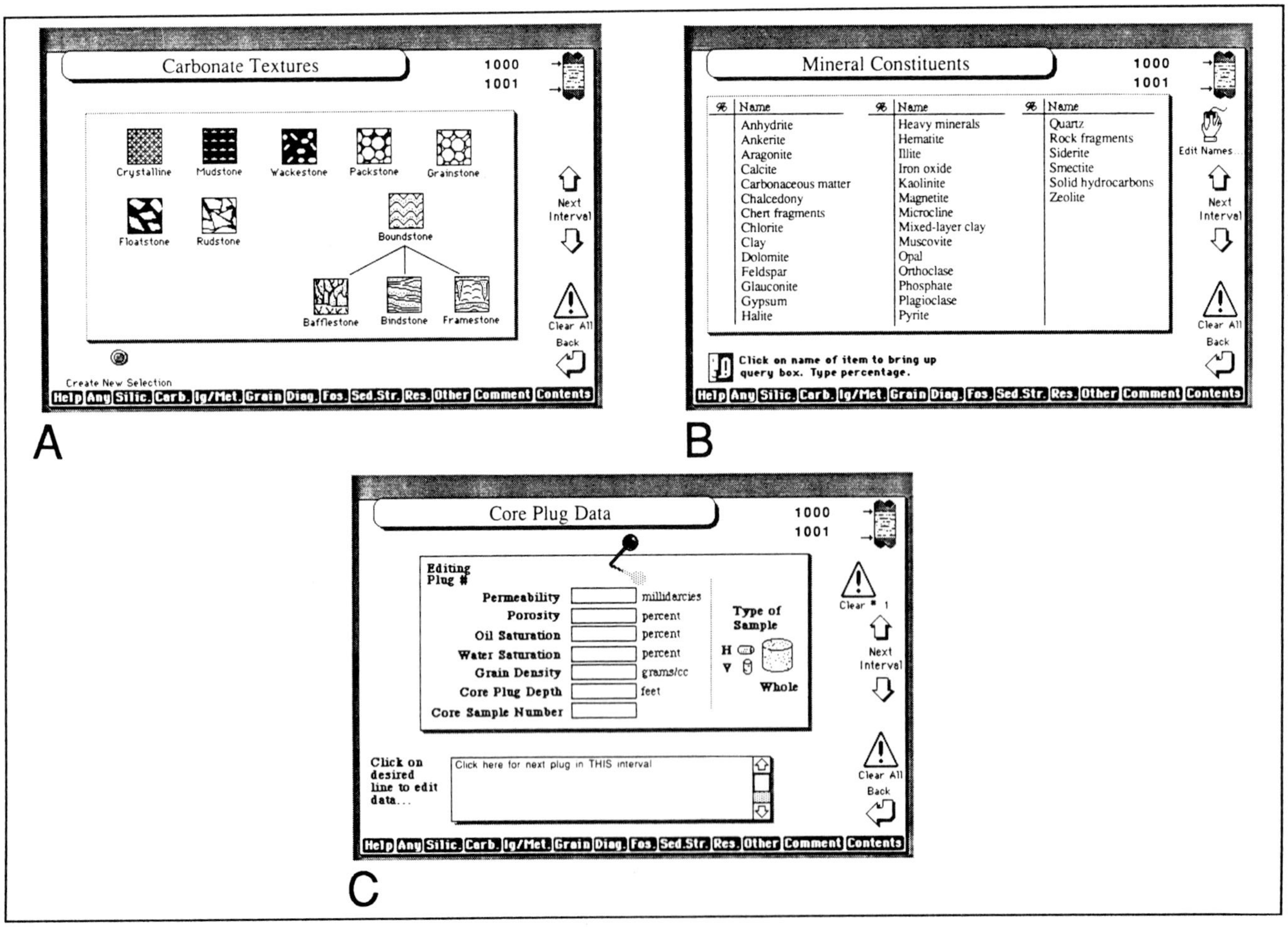

Figure 3 - Examples of major types of Selector cards in the HyperCore data entry stack. (A) Carbonate Textures card; (B) Mineral Constituents card; (C) Core Plug data card.

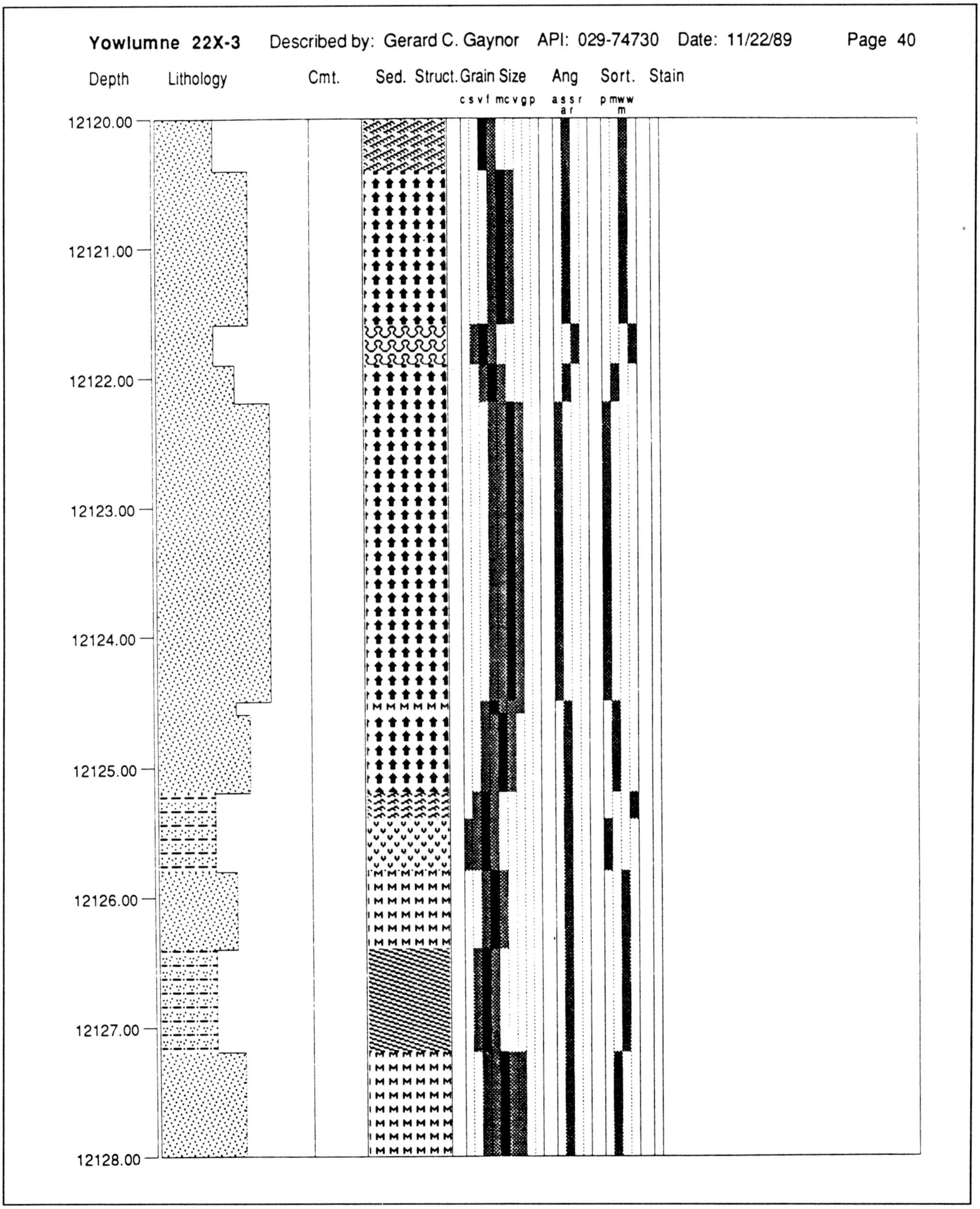

Figure 4 - Example output using CoreChart A routine. Grain size profile in "Lithology" column is generated using modal grain size entered into data entry stack. In "Grain Size" column, black bar represents modal grain size and gray bars on either side represent grain size range. Patterns in "Lithology" and "Sed. Struct." columns are explained in a separate sheet output by the ChoreChart A routine.

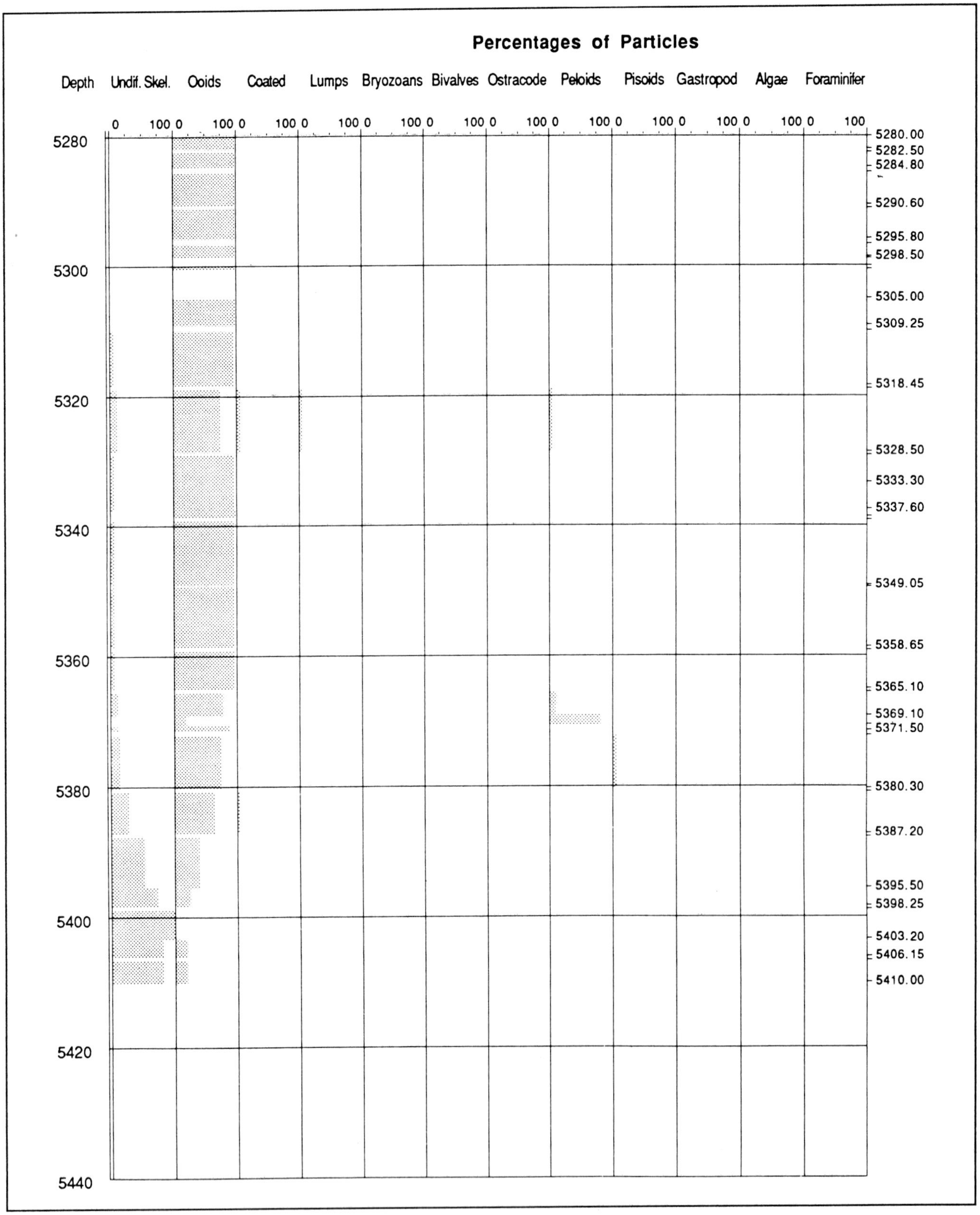

Figure 5 - Example output using CoreChart C routine. Semi-quantitative visual estimates of carbonate particle types are expressed as vertical profiles. This particular example shows the upward transition from undifferentiated skeletal fragments into carbonate ooids in a grainstone shoal.

Multidisciplinary Reservoir Description, Walker Creek Field, Columbia and Lafayette Counties, Arkansas

D. M. Bliefnick
J. G. Kaldi
ARCO Oil & Gas Co., Plano, TX

S. K. Bissmeyer
ARCO Oil & Gas Co., Lafayette, LA

T. T. Dang
ARCO Oil & Gas Co., Plano, Tx

INTRODUCTION

Objectives

An integrated reservoir description of the Smackover Formation in Walker Creek Field in southern Arkansas was conducted in order to evaluate the field's hydrocarbon potential and to determine the best method of increasing recovery. Geologic, petrophysical and engineering studies comprise the multi-disciplinary reservoir description of Walker Creek Field. Evaluating the field's potential included optimizing current operations, proposing options that accounted for reservoir complexities, and planning future operations that would maximize the unit's economic value. The specific aspects of the reservoir description consist of:

- Determining the depositional and diagenetic controls on reservoir lithologies and the resulting reservoir geometry;
- Establishing the internal geometry of the reservoir for the purpose of 1) zonation of reservoir flow units, 2) discrimination of pay from non-pay intervals and 3) evaluation of the continuity of different pay zones and of barriers to flow;
- Relating facies to reservoir properties, including 1) petrographic evaluation of pore types, 2) relating depositional and diagenetic textures to capillary pressure (Pc) trends and 3) recognizing log responses associated with reservoir layers;
- Providing a depositional and diagenetic model of the occurrence and distribution of reservoir and non-reservoir units in Walker Creek Field, which may have applications to exploration efforts elsewhere in the Smackover Formation.

Field Location and Background

Walker Creek Field is located north of the Arkansas-Louisiana border in southern Lafayette and Columbia Counties, Arkansas (Figure 1). It is one of the largest stratigraphic traps discovered along the State Line Trend and is nearly 30 square miles in area (Becher and Moore, 1979). The field is bounded to the south by a down-to-the-south, east-west trending normal fault of the Buckner-Haynesville graben system (Bishop, 1973). Hydrocarbon production (oil, gas and condensate) is from a 100 foot thick section of the ooid grainstone facies of the Upper Jurassic Smackover Formation. Walker Creek has an estimated 95 MMB of oil-equivalent originally in place in the field.

Field History—The Smackover Formation in southern Arkansas and northern Louisiana has been the objective of exploratory drilling for over five decades. Early exploration centered on high-relief, salt-cored anticlines which were easily located by geophysical methods. Subsequent drilling away from these large structures has led to the discovery of several significant stratigraphic traps. Walker Creek Field, one of the largest of these stratigraphic traps, was discovered in March, 1968 with the drilling of the Chapman #1 Helms (Section 29—T19S—R23W, Lafayette County, Arkansas). The well flowed 534 BOPD and 1.1 MMCFD with a gas/oil ratio of 2015 SCF/STB (Brock, 1980). The Arkansas Oil and Gas Commission established 320 acre drilling units, each with an allowable of 320 BOPD and a limiting GOR of 2000 SCF/STB (Chimene, 1976). The field was unitized in 1974, and Pennzoil Production Company was named operator. There are 59 wells within the boundary of the field (Figure 2), of which 37 have produced hydrocarbons at economic rates, 21 are still producing, and 6 are being used for gas injection. Since 1975, the field has been under partial pressure maintenance by reinjection of produced gas at the crest of the structure. As of January, 1988, 36 MMB of oil have been produced from the estimated 95 MMB of OOIP.

Reservoir Fluid Data—Walker Creek crude is considered a high shrinkage crude or volatile oil, characterized by an oil gravity of 49° API, an initial solution gas/oil ratio of 1840 SCF/STB and an oil formation volume factor of 2.035 RVB/STB. Initial reservoir pressure was 5300 PSIA, with an average reservoir temperature of 250°F. At reservoir temperatures, brine density is 1.126 gm/cm^3, oil density is 0.515 gm/cm^3, and the interfacial tension is 16 dynes/cm. The average depth to the top of the Smackover Formation varies from North to South from -10,590′ to -11,140′ subsea. The original oil-water contact varied from North to South, from -10,720′ to -10,733′ subsea (Chimene, 1976). A characteristic of reservoirs producing volatile oil is that the reservoir gas phase carries a significant amount of oil, which is recoverable as stock-tank liquid and should be included in estimates of original-oil-in-place.

Production History—Until 1975, when gas re-injection commenced, condensate production contributed 18% of the total liquid rate, but became increasingly dominant until it presently comprises 100% of the total liquid rate. Similarly, solution gas production was initially large (200,000 MCF/month), but has diminished to very low rates, indicating that essentially all gas produced today is non-associated gas.

The current producing mechanism at Walker Creek is primarily a reservoir liquid vaporization process, or gas stripping. When lean gas is injected into a volatile oil reservoir, a chemical equilibrium between the injected gas and the reservoir oil will be established. Initially, the most volatile components (propane, butane and pentane) will account for most of the material transferred from the oil phase to the gas phase. As the partially stripped oil phase is contacted with additional dry injection gas, heavier components (hexane, heptane and octane) will gradually transfer to the gas phase in increasing amounts.

The original-oil-in-place for the major producing intervals (Zones 2 and 4), was determined by material balance calculations to be 50MMBO, and 35MMBO for Zones 2 and 4 respectively. Based on these values, a total of 16% of OOIP had been recovered from Zone 2 and 13% of OOIP had been recovered from Zone 4 at the end of primary recovery. After 14 years of gas injection, the total recovery to date for Zone 2 is 45% of OOIP and that from Zone 4 is 35% of OOIP.

Each reservoir zone was originally in pressure isolation. Plots of pressure vs. time show a relatively uniform pressure decrease within each reservoir zone with no breaks to suggest smaller divisions within that reservoir zone. The pressure data were used to assign producing stratigraphic intervals to each of five reservoir zones and to correlate these zones across the field. Correlation of reservoir zones based on log character or lithology alone would have been nearly impossible.

Geologic Setting—During rifting along the southern margin of the U.S., clastics of the Late Triassic Eagle Mills Formation and Early-Middle Jurassic Werner Formation were deposited in continental settings. These are overlain by the Middle Jurassic Louann Salt, which was deposited in a series of evaporative lagoons and shallow basins around the Gulf Coast. The Louann is unconformably overlain by the Late Jurassic Norphlet Formation. The Late Jurassic Smackover Formation conformably overlies the Norphlet and is, in turn, conformably overlain by the Buckner Member of the Late Jurassic Haynesville Formation (Figure 3). In Late Jurassic time, the site of Walker Creek Field was a gently dipping shelf (ramp) between a narrow open-marine basin to the south (north Louisiana salt basin) and a broad coastal shelf to the north (Chimene, 1976; Becher and Moore, 1976).

The effects of contemporaneous salt tectonics on the deposition of Smackover and younger sediments is reflected in thinning of Upper Jurassic sediments over salt-cored anticlines (Dickinson, 1968 and Hughes 1968) and increases in grain size, better sorting and higher porosity in the Smackover over these anticlines (Akin and Graves, 1969). A structural contour map on the top of the Smackover in Walker Creek (Figure 4) shows regional southward dip, no structural closure and three, subtle, positive features, which are most likely the result of flowage in the Louann Salt, as is the case in many other Smackover fields (Dickinson, 1968; Hughes, 1968; Bishop, 1971; Keyes, 1971). These positive structural features affected the distribution of Upper Smackover facies at Walker Creek by localizing turbulent conditions. Thus, the highest energy facies are best developed on or near the crests of the highs. Salt movement apparently ceased by the end of Buckner deposition, as structure contour maps on the top of the Buckner show only regional dip to the south (Brock, 1980; Becher and Moore, 1976).

In southern Arkansas, the Smackover Formation is composed of three members (Dickinson, 1968). Only the middle and upper members are penetrated at Walker Creek. The middle member comprises 65% to 75% of the total thickness of the formation, and is composed of brown, dense limestone that may contain anhydrite and dolomite. Deposition of this member probably provided the thickness and weight of sediment to initiate movement of the underlying Louann Salt.

The upper Smackover, also known as the Reynolds Oolite (Weeks, 1938), ranges between 100 and 200 feet thick and varies from a mudstone and pellet packstone updip to oolitic grainstones with intraclasts, oncolites and peloids in a downdip position. It is these oolitic grainstones that produce hydrocarbons at Walker Creek Field and throughout the upper Gulf Coast province.

There is a sharp break between the Smackover Formation and the overlying Buckner Member of the Haynesville Formation, as indicated by sharp deflections to the left on the SP log and to the right on the resistivity log (Figure 5). The Buckner is 200 -300 feet thick and consists of anhydrite, red shale, and dolomite. In the field area, the Buckner is predominantly interbedded red and green shales. The Bossier Shale is the basinward equivalent of the Upper Smackover and Haynesville. Along with the lower Smackover, it may be a source rock for the hydrocarbons now being produced in the field area (Brock, 1980).

Walker Creek is interpreted as a stratigraphic trap with the reservoir limited in an updip direction by diagenesis and/or a lack of reservoir facies (Brock and Moore, 1982; Brock, 1980; Becher and Moore, 1976; Chimene, 1976). The interaction of shallow marine currents with early, subtle salt structures resulted in deposition of high energy ooid grainstones over the axis of growth, with lower energy sands (pellets and oncolites) deposited on the flanks. The incorporation of grains of various types and sizes within the ooid shoals reflects the migration of active shoal lobes over lower energy inter-shoal areas. This continued movement resulted in a widespread sand sheet dominated by ooids but with recurring intervals containing abundant non-ooid grains. Zones 2 and 4 represent two major phases of ooid shoal development in a regressive offlap relationship (Figure 6). Subsequent burial and diagenesis resulted in porosity preservation in much of the ooid facies, and porosity destruction by compaction and cementation in the lower energy facies. The resulting trap has no structural closure, is influenced by depositional and diagenetic controls, and may exhibit multiple stacked reservoirs.

DEPOSITIONAL AND DIAGENETIC HISTORY OF THE UPPER SMACKOVER FORMATION IN WALKER CREEK FIELD

Methods

Samples from eleven wells (Figure 2) were examined for this study. Table 1 lists each well in the field (well name and tract designation) and those wells from which samples were obtained. Three cores and over 150 thin sections were examined. Routine core analyses (porosity, permeability, water saturation, and oil saturation) were obtained on all cored wells in the field. Core porosity was corrected for the effects of net overburden stress with the percent reduction ranging between 2.4 and 3.8% from ambient to 6000 PSIA simulated reservoir stress.

Digitized log curves of all wells in the database were compared to blue-line logs to ensure that the digitized data were properly on depth. In wells where samples were available, a combination of petrophysical and petrographic criteria were used to select lithologic units and calibrate their characteristic log responses. From log responses and production data (initial formation pressure values and decline curves), it was possible to extrapolate unit tops and bases to wells where samples were not available. Representative samples of reservoir and non-reservoir rock types were analyzed by mercury injection/capillary pressure tests.

Samples were analyzed petrographically to determine texture, grain types, cement types and composition, amount of compaction, and relative abundance and distribution of pore types. Porosity and permeability vary widely (1% to 24% and .01 to >5000 md). To determine pore geometry, thin sections and pore casts from three typical wells were examined, and mercury injection capillary pressure analyses were run on nine samples from those wells.

Pore casts were prepared by impregnating samples with epoxy. The rock material was then gently dissolved in 1-2% HCl, leaving the epoxy within the pore space standing out in relief. Pore casts were examined in the scanning electron microscope (SEM) to determine the abundance and distribution of pore types, the sizes and shapes of pore throats, and to determine relationships between the pore geometry and rock type.

Facies of the Upper Smackover Formation

Rocks of the Upper Smackover were divided into four facies based on a combination of primary depositional features and secondary diagenetic textures. All samples had an original grain-supported depositional fabric, and most are essentially free of silt and clay-size material. Lithologic subdivisions are based on the grain types present, the textural maturity of the samples and the amount of cement present. The grain types and fabrics of each facies are interpreted to reflect a specific depositional environment and energy range. Energy conditions during deposition played a major role in determining primary porosity by controlling grain types, sizes and sorting.

Shallow Shelf Facies.—This is the deepest facies identified and occurs below all reservoir intervals. It is composed of fine-to medium-grained pellet packstones with 10-25% quartz and minor peloids, ooids and skeletal fragments (bivalves, echinoderms and algae) (Figure 7A). The facies is argillaceous, in part, and grades to quartz-rich crystalline limestone in places. The low porosity (<1%) is due to a combination of pore-filling micrite, argillaceous material and calcite cement.

Mixed Grain Shoal Facies.—This facies occurs above the shallow shelf facies and is dominantly a grainstone of coarse to very coarse oncoids, intraclasts and pisoids in a 'matrix' of fine to coarse peloids, ooids and pellets (Figure 7B). The texture grades to packstone in places where it is poorly sorted. Calcite and baroque dolomite cements partially fill pore spaces. Porosity (2-8%) and permeability (0.1-20 md) are low due to poor sorting and cementation.

Cemented Ooid/Mixed Shoal Facies.—This facies occurs below reservoir Zone 2, between Zones 2 and 4 and above reservoir Zone 4. This facies also occurs within reservoir zones (especially Zone 2) where it is predominantly a cemented ooid grainstone. Generally, the texture is a mixed packstone/grainstone and is locally argillaceous, indicating energy conditions lower than those producing the grainstones of the reservoir intervals. The main rock type is a fine-to medium-grained, moderately sorted ooid-peloid packstone to grainstone with 10-20% coarse to very coarse intraclasts, oncoids, peloids and skeletal fragments (bivalves, echinoderms, forams, gastropods, algae and rare coral) (Figure 7C). This mixture results in overall poor sorting. The facies contains fine pellets and quartz and grades to a pellet packstone/grainstone or calcite-cemented quartz sandstone in places. Compaction is indicated by sutured grain contacts and microstylolites. Coarsely crystalline calcite and baroque dolomite fill intergranular pore spaces (Figure 7D). These processes result in porosities of 2-5% and permeabilities of 0.1-0.5 md. The grain components of this facies and of the reservoir intervals are very similar and it is predominantly diagenesis, in the form of compaction and cementation, that has differentiated the two.

Porous Ooid/Mixed Shoal Facies.—This facies comprises the majority of the two major reservoir zones (Zones 2 and 4). The dominant texture is a clean grainstone composed of fine-to medium- (locally coarse) grained, well sorted ooids (Figure 7E) indicating a high energy environment of deposition. Porosity and permeability values range up to 22% and 5 Darcies. Present within the ooid grainstones are intervals of 5-25% coarse to very coarse oncoids, intraclasts, peloids, multiple or compound ooids, algal plates and skeletal fragments (bivalves, echinoderms, forams, gastropods, coral) (Figure 7F). Very fine-to fine-grained pellets and ooids 'fill in' around the larger grains. Oncolites are more abundant toward the base of the zones. Intervals where oncolites or pellets are abundant have lower porosity and pemeability values than intervals dominated by ooid grainstones. This is due to a combination of compaction of the softer grains, infill of finer grains between larger ones, and calcite cement.

Diagenetic History

Diagenesis is the most important control over the present porosity distribution. The major diagenetic products that reduce porosity include fibrous calcite rim cement, microstylolites and coarse sparry calcite cement.

Fibrous calcite rim cement.—Fibrous calcite rim cement occurs in the ooid grainstone facies in the Chapman #1 Helms (1-1), the Woods Stuart #3 (15-1) and the ARCO Bodcaw #1 (28-1). Two of these wells (Stuart and Bodcaw) occur along one of the structural highs in the field and the third occurs on a structural nose. The cement forms isopachous crusts 30μ-95μ in thickness with the long axis of the crystals perpendicular to the grain surfaces (Figure 8A). The fibrous crust is interpreted as an early post-depositional, shallow marine, phreatic cement of original aragonite or magnesian calcite mineralogy (Becher and Moore, 1976; Moore and Druckman, 1981). This early cement provided rigidity to the rock, which reduced the amount of compaction in the ooid grainstone facies (Figure 8A).

Fine Equant Calcite.—Fine equant calcite cement occurs locally in the upper member of the Smackover Formation. It occurs as pore filling cement on surfaces of grains and in internal cavities. The distribution is not evenly distributed around grains, but tends to be concentrated at grain contacts and within the smaller pore spaces (Figure 8B). This type of cement has been termed "meniscus cement" by Dunham (1971) and represents cementation in a meteoric vadose environment.

Compaction.—With progressive burial, compaction became the major diagenetic process. Compaction resulted in both physical (e.g., grain breakage) and chemical (e.g., stylolitization and pressure-solution at grain contacts) alteration. In facies dominated by pellets and oncolites with no early cementation, porosity was totally destroyed by compaction (Figure 8C). The effects of compaction were substantially reduced where early cement was present. Porous and nonporous zones are thinly interbedded in the Upper Smackover. The pores in the latter are ineffective due to a combination of compaction plus cementation.

Sparry Calcite Cement.—Following compaction, coarse calcite spar cement (calcspar) precipitated and further reduced porosity (Figure 8D). This cement occurs in nearly all samples, comprising up to 90% of the cement in Walker Creek (Brock, 1980). Evidence from two-phase fluid inclusions, stable oxygen isotopes and the presence of hydrocarbon inclusions in the cement crystals led to the interpretation of a late, deep origin for this calcspar (Brock, 1980; Moore and Druckman, 1981). The cement was precipitated in the phreatic zone, but from a fluid that was either a connate marine fluid or a brine similar to present Smackover formation water, or a mixture of the two (Carpenter and Trout, 1978).

Dolomite.—Dolomite pore-filling cement is a minor component, but it is widely distributed at Walker Creek. The equant blocky crystals exhibit the curved crystal faces and undulatory extinction of "baroque dolomite" (Folk and Assereto, 1974). Physical character and chemical composition parallel that of the associated calcspar, and the dolomite formed concurrently with the calcite in a similar setting (Moore and Druckman, 1981).

Dissolution.—Following compaction and late-stage cementation, secondary enlargement of pores occurred through partial dissolution of grains and cement. As hydrocarbon maturation proceeded in the lower Smackover and in its basinal equivalent, fluids associated with hydrocarbon migration and charged with CO_2 and/or H_2S corroded the carbonate constituents. Significant dissolution occurred along pathways of preserved primary porosity and some of the present pore interconnection is due to this late dissolution event. The migration and trapping of hydrocarbons marked the end of major diagenetic activity in the upper Smackover at Walker Creek (Figure 9).

PORE GEOMETRY

Pore Types

Diagenesis has modified depositional pore types and has resulted in distinctive pore geometries at Walker Creek. Two pore types, interparticle and microporosity, account for most of the porosity in the reservoir.

Interparticle Porosity.—Interparticle porosity is the most important pore type and comprises the majority of the effective porosity in the Walker Creek reservoir. By definition, this type of porosity occurs between grains. It may form by one of two processes: 1) preservation of pore space between the grains from the time of deposition, or 2) dissolution of matrix and/or cement between grains to form secondary interparticle porosity. This latter process, if present, is minor and nearly all interparticle porosity in the reservoir is interpreted as preserved primary porosity. Good to excellent permeabilities are frequently associated with this pore type and the majority of the production is from zones with this pore geometry.

Interparticle porosity exists where calcite cement has incompletely filled the pore spaces between grains (Figure 10A). The sizes and shapes of the pores are largely governed by the sizes and shapes of the grains. A more poorly sorted sample will exhibit a wider variety of pore sizes and shapes than a better sorted sample. In general, the better sorted the sediment and the larger the pore sizes, the more effective is the connection between them. Interparticle porosity is uniformly distributed within the various generations of ooid grain shoals and extends across the field within the reservoir facies. This type of porosity does not occur in cemented streaks within the reservoir facies, nor does it occur in the rocks between the ooid shoal facies because of a combination of cementation and presence of mud matrix in these lower energy packstones.

Microporosity.—Microporosity is a common although ineffective pore type in the Walker Creek reservoir. It ranges from nearly 50% to <5% of measured porosity. The upper size limit used here for micropores is a 5μ radius. This is a commonly used, albeit somewhat arbitrary, cutoff, and it corresponds approximately to the 'blue haze' optical characteristic of micropores seen at low magnifications (5x-10x) in standard petrographic thin sections. However, SEM photomicrographs clearly show the size, shape and distribution of micropores (Figure 10B,C).

The micropores occur within carbonate grains, predominantly ooids, which are in facies with preserved intergranular porosity (Figure 10B). The microporosity is ineffective, because the pores are very small and have very high capillary pressures. Thus, this pore type would require hydrocarbon columns far in excess of that present at the field to be charged. Electrical properties testing indicates that the fluid in the micropores is bound formation brine and not immobile hydrocarbons. Although microporosity contributes to total porosity as seen on porosity logs and measured by routine core analysis, it does not contribute significantly to oil production.

Microporosity is present in varying amounts in the reservoir facies across the field. It is present in very small amounts where coarse calcite cement or compaction has completely occluded the pore space, or where micrite or argillaceous material are present. The microporosity was most likely formed early by micron-scale dissolution of unstable components of the ooids. Following minor early meteoric cementation and compaction, the grainstone facies was variably cemented by coarse calcite in a burial setting. Where this cementation was incomplete, late solution enhancement of preserved interparticle porosity occurred. The same solution event also enhanced the microporosity (Figure 10D). Portions of the grainstone facies that were completely cemented provided no conduits for the dissolving fluids, and as a result, neither interparticle porosity nor microporosity are present in appreciable amounts.

Mercury Injection Capillary Pressure (MICP) Analyses

Mercury injection capillary pressure (MICP) data were acquired for nine samples from three wells (Table 2). The analyses were performed by injecting mercury into evacuated, cleaned and extracted core plugs. Mercury injection pressure was increased in a stepwise manner and the percentage of rock pore volume saturated by mercury at each step was recorded. Sufficient time was allowed for equilibrium to be reached between successive pressure increments.

Capillary pressure data were used to help distinguish reservoir from non-reservoir rock (Table 3) following the recommendations of Sneider (1987). The shapes of the curves of mercury saturation vs injection pressure vary as a function of the distribution of the effective pore throat sizes, as derived from the capillary pressure equation:

$$P_c = \frac{2\,\sigma \cos\theta}{r_c}$$

(where P_c = capillary pressure, r_c = capillary radius, θ = contact angle, and σ = interfacial tension). Samples dominated by throats of similar size (well sorted) have broad, flat plateaus. The plateau is steeper for more poorly sorted samples, and is absent (the slope approaches 45°) for samples with unsorted pore throats. Samples with more than one population of pore throat sizes (e.g., bimodal size distribution) have curves with two or more plateaus.

There are three distinct groups of capillary pressure curves (Figure 11). The reservoir rocks all show similar characteristics. Samples 1, 2, 4, and 6 have large, well-connected pores and pore throats as seen in thin section and pore casts (Figure 12A, B). On the MICP curves, the large pore throats are indicated by the low displacement pressures (<1 PSIA). There are broad, relatively flat plateaus between 50 and 100% wetting phase saturation, suggesting that about 50% of the pore volume is accessed by throats with effective radii greater than 1 micron. Also, the wetting phase saturation is low at low injection pressures for all these samples (50% saturation at 20 PSIA), indicating excellent reservoir quality.

The presence of two plateaus, one at about 5 PSIA and another at about 500 PSIA in curves 1, 2, 4, and 6 (Figure 11), reflects a bimodal pore throat size distribution in these samples. The mercury entered the larger interparticle pores between the ooids at 5 PSI (first plateau), and then entered the smaller pores of the microporosity network within the ooids at 500 PSI (second plateau). Examples of thin sections and pore casts of this pore type are presented in Figure 12A and B.

The capillary pressure curves for the marginal reservoir rocks (samples 7 and 9) illustrate that the pore throats are smaller and the intergranular pores are less well connected than in the reservoir rocks. Also, displacement pressures are higher and the wetting phase saturation is greater at any given injection pressure. The slope of the curve (approximating 45°) reflects the very poorly sorted distribution of the pore throats. The visible variation in pore throat size distribution in these samples is exemplified in Figure 12C and D.

The capillary pressure curves for the non-reservoir rocks (samples 3, 5, and 8) show more variation than the other groups. Displacement pressures are high (>200 PSIA), and reflect the fine pore throats that provide access to most of the pore volume in these samples. Note that sample #3 has a bimodal pore throat size distribution as reflected in the two plateaus in the curve. In this case, the relatively larger pore throats (around 0.05μ effective radius) that the mercury first enters are comprised of very fine rims of porosity around individual ooid grains (Figure 13A, B). The original interparticle pore space is now filled with calcite cement. The second plateau represents the very fine (effective pore throat radius ~ 0.001μ) intercrystalline pore network between the calcite cement crystals.

Sample 8 contains abundant micritic and argillaceous material (Figure 13C, D). A high percentage of small and poorly connected pores and pore throats are indicated by the steep slope of the curve; less than 10% of the total pore volume is accessed by pore throats of 0.0002μ effective radius. Even at 30,000 PSIA, the wetting phase saturation remains about 90%.

Sample 5 is a rock with high porosity (9.5%) and low permeability (0.1 md), that has been classed as non-reservoir. Calculations converting MICP to height above free water level (FWL) and pore volume to bulk volume (Sneider, 1987), suggest that this rock type would be reservoir at a height of 75 feet above FWL (Figure 14A). However, mathematically derived relative permeability curves (Figure 14B), constructed from the capillary pressure analysis (Purcell, 1949; Brooks and Corey, 1964, 1966) of this sample, indicate that at a sufficient height to be reservoir, this rock would have a relative permeability to oil (k_{ro}) of only 0.04 md. Hence, though this rock type may be contributing to pay, it is classed as conditional pay (based on the assumption that other rock types contribute much greater volumes to total production).

Correlation With Facies

The mercury injection capillary pressure curves were used to relate reservoir quality to facies (Table 2). The porous ooid/mixed grain shoal facies comprises most of the reservoir rock with abundant interparticle porosity. This reservoir rock is characterized by samples 1, 2, 4, and 6 on Figure 9. These samples are well-sorted ooid grainstones with 10—20% total porosity and up to 5 Darcies permeability (Figure 12A, B). Rocks from different depositional environments with different diagenetic histories can have similar capillary pressure curves and reservoir quality. This is exemplified by samples 7 and 9. Sample 9 is from the porous ooid/mixed grain shoal facies (Figure 12C, D), but is of marginal reservoir quality due to poor sorting of different grain types (finer grained ooids and pellets 'fill in' the pore spaces between larger oncoids and intraclasts), and partial infill of pore spaces by calcite cement. Sample 7 occurs between porous ooid shoals and is in the cemented ooid/mixed grain shoal facies. Here, calcite cement has largely filled interparticle pore spaces resulting in marginal reservoir quality similar to that of sample 9.

Another example of rocks from different depositional settings with similar capillary pressure curves and reservoir quality is samples 5 and 3. Sample 5 characterizes a cemented interval within the porous shoal facies where the interparticle pores have been filled with calcite cement. Sample 3 characterizes an ooid-rich portion of the cemented ooid/mixed grain shoal facies where calcite cement has also filled interparticle pore spaces (Figure 13A, B). The presence of abundant calcite cement results in the occurrence of non-reservoir rocks both within and between porous ooid shoals. Sample 8 was deposited in a lower energy setting between high energy, active shoals. Pores and pore throats are extremely small due to the abundance of micritic and argillaceous material that accumulated in this lower energy environment (Figure 13C, D).

Reservoir Implications

The capillary pressure data demonstrate the relationship of pore geometry to facies and to reservoir quality. The depositional environment controlled what grain types were deposited and what textures developed at any given interval in the field. Diagenesis modified depositional textures and resulted in the present pore geometries. Each reservoir zone contains within it reservoir, marginal reservoir and non-reservoir rock types. Each of these types plots as distinctive capillary pressure curves. The curves reflect the diagenetically modified pore systems of lithologic facies that originally ranged from nearly identical to totally different. Capillary pressure analysis of an unknown sample would place it within the reservoir, marginal reservoir or non-reservoir category, and provide guidance on such questions as perforations, testing and production. New porosity (5PU) and permeability (0.6 md) cutoffs for net pay were suggested by the results of log analysis used in conjunction with capillary pressure work. Old cutoffs did not take into account the importance and ineffectiveness of microporosity, which is shown so clearly in the pore casts and capillary pressure curves.

Cross Sections—The new cutoffs and designation of pay and non-pay were used in the construction of strike- and dip-oriented cross sections across the field. The strike-oriented sections illustrate the continuity of facies in an east-west direction, parallel to the paleo-shoreline. The north-south dip-oriented sections show the progradational nature of the reservoir zones. A representative dip cross section from the middle of the field is shown in Figure 5. The datum is the first SP and resistivity marker going up section in the overlying Buckner shales. The gray shaded areas indicate non-porous intervals. Zone 2 is yellow and Zone 4 is pink. Their relationship illustrates the vertical stacking and lateral progradation (more open marine is to the south) of the two main generations of ooid shoal development. This relationship can also be seen in a map of the distribution of major and minor zones of ooid shoal development across the field (Figure 15).

Parameter Maps—With these zone definitions, average parameters, such as porosity, net pay and porosity X thickness (Øh), were computed for each well on a zone by zone basis. The included figures are for Zones 2 and 4 only.

Before porosity maps were constructed, a grainstone isopach map was made for Zones 2 and 4 (Figures 16 and 17). The amount of grainstone was determined from the SP log where a strong deflection to the left indicates the presence of grainstone. The top and bottom of the interval for each reservoir zone was chosen to be where the SP curve crossed the 50% scale line. These maps show the strong east-west orientation of the shoal complex, which parallels the paleo-shoreline to the north. Localized thicks and thins represent individual ooid shoals and the tidal channels between them, which can be at any angle to the trend of the shoal complex. The thickest part of Zone 4 is located seaward of the thickest part of Zone 2, reflecting the prograding nature of the ooid shoals.

On the porosity isopach maps (Figures 18 and 19), the contours end at 5% which represents the new log-determined cut-off. The porosity contours show an east-west trend with thicks and thins along strike. The development and preservation of porosity has followed the trend of the ooid shoals and exhibits roughly the same distribution pattern.

The net pay (Figures 20 and 21) and Øh maps (Figures 22 and 23) all show the east-west orientation of the shoal complex. The net pay maps include only those portions of the reservoir with porosity greater than 5% and at least 0.6 md permeability; the zero contour line corresponds to the 5% porosity contour. The values on the Øh maps were obtained by multiplying the average porosity values in the reservoir zone by the cumulative net pay feet. Not surprisingly, both of these maps reflect the distribution of carbonate grain shoals and development of porosity within them. The shape and location of the individual shoals and channels may vary but the overall pattern is very similar. These maps were also used in the volumetric evaluation of the reservoir for comparison with material balance and for input into the simulation model.

CONCLUSIONS

1) The reservoir at Walker Creek is a clean ooid grainstone with variable amounts of peloids, oncoids, intraclasts and skeletal fragments. This high energy facies was localized on a broad carbonate ramp by low relief salt structures.

2) There are two major and three minor generations of shoal development. Seaward migration of these shoal complexes during the Late Jurassic sea level high stand resulted in a regressive offlap relationship.

3) The most abundant and effective porosity is preserved primary intergranular. Microporosity, created by secondary dissolution, is common, but does not contribute to reserves or production.

4) Non-pay zones, both within and between the generations of shoal development, are caused by a combination of compaction and cementation. The most abundant cement is a coarse, pore-filling calcite which precipitated in a burial setting late in the diagenetic history. The non-pay rocks between ooid shoal facies are less well sorted, mixed-grain, packstone/grainstone that is somewhat argillaceous.

5) The relationship of pore geometry to facies and to reservoir quality is strongly indicated by the capillary pressure work. Samples from reservoir, marginal reservoir and non-reservoir intervals plot as distinct groups on mercury injection capillary pressure curves and are derived from different lithologic facies and their diagenetically modified pore systems.

6) Field maps of grainstone thickness show the east-west trend of all shoal complexes. Localized thicks and thins represent individual ooid shoals and the tidal channels between them. Porosity, gross pay, net pay and Øh maps all reflect the trend of the ooid shoals and the development of porosity within them.

ACKNOWLEDGEMENTS

We would like to thank Susan Bissmeyer of ARCO Oil and Gas Co., Lafayette, LA, for her help and enthusiasm and cheerful willingness to answer all the questions that arose during the course of this project; and Chuck Vavra of ARCO Oil and Gas Co., Plano, Tx for plotting Figure 14 as well as helpful discussion.

REFERENCES

Aiken, R. K., Jr., and R. W. Graves, 1969, Reynolds oolite of southern Arkansas: AAPG Bull., v. 53, p. 1909-1922.

Baria, L. R., D. L. Stoudt, P. M. Harris and P. D. Crevello, 1982, Upper Jurassic reefs of the Smackover Formation, U.S. Gulf Coast: AAPG Bull., v. 66, p. 1449-1482.

Bathurst, R. G. C., 1975, Carbonate sediments and their diagenesis (Second Edition): Developments in Sedimentology, 12, Elsevier Scientific Publishing Co., Amsterdam, 620 pp.

Becher, J. W. and C. H. Moore, 1976, The Walker Creek Field: a Smackover diagenetic trap: Gulf Coast Assoc. Geol. Soc. Trans., v. 26, p. 34-56.

Bishop, W. F., 1971, Stratigraphic control of production from

Jurassic calcarenites, Red Rock Field, Webster Parish, Louisiana: Gulf Coast Assoc. Geol. Soc. Trans., v. 21, p. 125-137.

Bishop, W. F., 1973, Late Jurassic contemporaneous faults in NorthLouisiana and South Arkansas: AAPG Bull., v. 57, p. 858-877.

Brock, F. C., 1980, Walker Creek revisited: a reinterpretation of the diagenesis of the Smackover Formation of Walker Creek Field,Arkansas: unpublished master's thesis, Louisiana StateUniversity, 78 pp.

Brock, F C. and C. H. Moore, 1982, Walker Creek revisited: a re-interpretation of the diagenesis of the Smackover Formation of Walker Creek Field, Arkansas: Gulf Coast Assoc. Geol. Soc. Trans., v. 31, p. 49-58.

Brooks, R.H. and Corey, A.T., 1964; Hydraulic properties of porous media. Hydrology Papers of the Colorado State University, Ft. Collins, CO., pp 1-27.

Brooks, R.H. and Corey, A.T., 1966; Properties of porous media affecting fluid flow. J. of Irrigation and Drainage Div. Proceedings of the Am. Soc. Civil Engineers, June, 1966, pp 61-88

Carpenter, A. B. and M. L. Trout, 1978, Geochemistry of bromide-rich brines of the Dead Sea and southern Arkansas: Oklahoma Geol. Survey Circ. 79, p. 78-88.

Chimene, C. A., 1976, Upper Smackover reservoirs, Walker Creek Field area, Lafayette and Columbia Counties, Arkansas, *in* North American Oil and Gas Fields: AAPG Memoir 24, p. 177-204.

Crevello, P. D. and P. M. Harris, 1984, Depositional models for Jurassic reefal buildups, *in* The Jurassic of the Gulf Rim, Gulf Coast Section, SEPM, 3rd Ann. Research Symp., Program and Abstracts, p. 57-102.

Dickinson, K. A., 1968, Upper Jurassic stratigraphy of some adjacent parts of Texas, Louisiana and Arkansas: U.S. Geol. Survey Prof.Paper 594E, 25 pp.

Dunham, R. J., 1971, Meniscus cement: Studies in Geology, no. 19, Johns Hopkins Univ., p. 297-300.

Folk, R. L. and R. Assereto, 1974, Giant aragonite rays and baroque white dolomite in tepee-fillings, Triassic of Lombardy, Italy: AAPG-SEPM Ann. Mtg. Abs., v. 1, p. 34-35.

Hughes, D. J., 1968, Salt tectonics as related to several Smackover fields along the northwest rim of the Gulf of Mexico Basin: Gulf Coast Assoc. Geol. Soc. Trans., v. 18, p. 320-330.

Keyes, P.L., 1971, Geology of the Jurassic, Flomaton-Jay area, Alabama and Florida (abstr.): AAPG, v. 55, p. 1695.

Moore, C. H., 1984, The upper Smackover of the Gulf rim: depositional systems, diagenesis, porosity evolution and hydrocarbon production, *in* The Jurassic of the Gulf Rim, Gulf Coast Section, SEPM, 3rd Ann. Research Symp., Program and Abstracts,p. 283-307.

Moore, C. H. and Y. Druckman, 1981, Burial diagenesis and porosity evolution, Upper Jurassic Smackover, Arkansas and Louisiana: AAPG Bull., v. 65, p. 597-628.

Purcell, W.R., 1949; Capillary pressures—their measurement using mercury and the calculation of permeability therefrom. Petroleum Transactions, AIME, 186, 39-48.

Shinn, E. A., 1975, Polygonal cement structures from the Holocene: a clue to recognition of submarine diagenesis: AAPG-SEPM Ann.Mtg. Abstr., v. 2, p. 68.

Sneider, R. M., 1987, Practical petrophysics for exploration and development: AAPG Education Dept. Short Course Notes, Houston, TX, October 8-9.

Todd, R. G. and R. M. Mitchum, 1977, Seismic stratigraphy and global changes of sea level, Part 8: identification of upper Triassic, Jurassic and lower Cretaceous seismic sequences in the Gulf of Mexico and offshore West Africa: in Seismic stratigraphy—applications to hydrocarbon exploration, AAPG Memoir 26, p. 145-163.

Weeks, W. B., 1938, South Arkansas stratigraphy with emphasis on the older coastal plain beds: AAPG Bull., v. 22, p. 953-983.

Woods, D. J., 1971, Smackover's significant Walker Creek (Arkansas), AAPG Midcontinent Sect. Mtg., Abstr., AAPG Bull., v.55, p.1689.

TABLE 1. WELL NAME/TRACT DESIGNATION CROSS REFERENCE LIST

TRACT		OPERATOR/LEASE/WELL NUMBER		
Tract 1	No. 1	Chapman Helms	No. 1	S
Tract 2	No. 1	Chapman Whitehead	No. 1	
Tract 3	No. 1	Chapman Burton	No. 1	
Tract 4	No. 1	Bodcaw Kosek	No. 1	
Tract 4	No. 2	Pennzoil Tract 4	No. 2	
Tract s	No. 1	Chapman Hughes	No. 1	
Tract 6	No. 1	Oakland Umphries	No. 1	
Tract 6	No. 2	Pennzoil Tract 6	No. 2	S
Tract 7	No. 1	Bodcaw Bodcaw	No. 2	
Tract 8	No. 1	Bodcaw Bodcaw	No. 3	
Tract 9	No. 1	Woods Jack	No. 1	
Tract 9	No. 2	Woods Blackwell	No. 1	
Tract 9	No. 3	Woods McDole	No. 1	
Tract 10	No. 1	Woods Blackwell	No. 2	
Tract 11	No. 1	Woods Stuart	No. 1	
Tract 12	No. 1	KWB IPCO	No. 2	
Tract 12	No. 2	KWB IPCO	No. 1	
Tract 13	No. 1	Woods Powell	No. 1	
Tract 13	No. 2	Pennzoil Tract 13	No. 2	S
Tract 14	No. 1	Woods Stuart	No. 2	
Tract 15	No. 1	Woods Stuart	No. 3	S
Tract 16	No. 1	Woods Knight	No. 1	
Tract 17	No. 1	KWB Boyette	No. 1	
Tract 18	No. 1	Bodcaw Burton Est.	No. 1	
Tract 19	No. 1	Chapman Brigham	No. 1	
Tract 19	No. 2	Partee-Chapman E Kosek	No. 1	
Tract 20	No. 1	Bodcaw Pickler	No. 1	S
Tract 21	No. 1	Marathon Henderson	No. 1	
Tract 22	No. 1	Pennzoil Nations	No. 1	S
Tract 22	No. 2	Pennzoil Tract 22	No. 2	S
Tract 22	No. 3	Pennzoil Tract 22	No. 3	S
Tract 23	No. 1	Woods Teutsch	No. 1	
Tract 24	No. 1	Bodcaw E Kosek	No. 1	
Tract 25	No. 1	Bodcaw Bodcaw	No. 5	
Tract 26	No. 1	Pennzoil Pickler	No. 1	S
Tract 27	No. 1	Woods Stuart	No. 4	
Tract 28	No. 1	ARCO Bodcaw	No. 1	S
Tract 29	No. 1	Pennzoil E Kosek	No. 1	
Tract 30	No. 1	Bodcaw Bodcaw	No. A-9	
Tract 30	No. 2	Bodcaw Bodcaw	No. 9	
Tract 31	No. 1	ARCO Cabe	No. 1	
Tract 32	No. 1	Pennzoil Pickler	No. 2	
Tract 33	No. 1	Pennzoil Bodcaw	No. A-1	
Tract 34	No. 1	Bodcaw Jack	No. 2	
Tract 35	No. 1	Marathon Boucher	No. 1	
Tract 36	No. 1	Hunt Ham	No. 1	
Tract 37	No. 1	Bodcaw Bodcaw	No. 12	
Tract 38	No. 1	Austral Shirey	No. 1	
Tract 38	No. 2	Stanolind Bodcaw	No. 1	
Tract 39	No. 1	Clegg Stuart-Ham	No. 1	
Tract 40	No. 1	Partee Disotell	No. 1	
Tract 41	No. 1	Bodcaw Bodcaw	No. 1	
Tract 42	No. 1	ARCO McFadden	No. 1	
Tract 43	No. 1	Bodcaw Bodcaw	No. 1	
Tract 43	No. 2	Bodcaw Bodcaw	No. A-1	
Tract 45	No. 1	Bodcaw Bodcaw	No. 10	
Tract 46	No. 1	Bodcaw Bodcaw	No. 11	
Tract 49	No. 1	ARKLA Burton	No. 1	
Tract 49	No. 2	Lyons Hendron	No. 1	

S: Well with available samples

TABLE 2. MERCURY INJECTION CAPILLARY PRESSURE SAMPLES

Sample #	Well	Core Depth (ft)	Ø (%)	K (md)
1	1-1	10912	8.5	31
2	1-1	10918	15.1	210
3	15-1	10691	3.4	<0.1
4	15-1	10717	22.8	960
5	26-1	10851	9.5	0.1
6	26-1	10867	16.3	133
7	26-1	10881	1.1	<0.1
8	26-1	10900	0.7	<0.1
9	26-1	10916	5.8	0.5

TABLE 3. SNEIDER'S MERCURY INJECTION CAPILLARY PRESSURE RESERVOIR/NON-RESERVOIR ROCK CRITERIA

	RESERVOIR	NON-RESERVOIR
P_c where mercury occupies 1% bulk volume	< 300 psi	> 500 psi
Mercury Saturation at 1000 psi	> 3% bulk vol.	≤ 2% bulk vol.
Mercury Saturation at 2000 psi	> 3 % bulk vol., and > 50 % pore vol.	≤ 3 % bulk vol., and ≤ 50 % pore vol.
Displacement pressure	< 100 psi	> 100 psi

Modified after Sneider, 1987

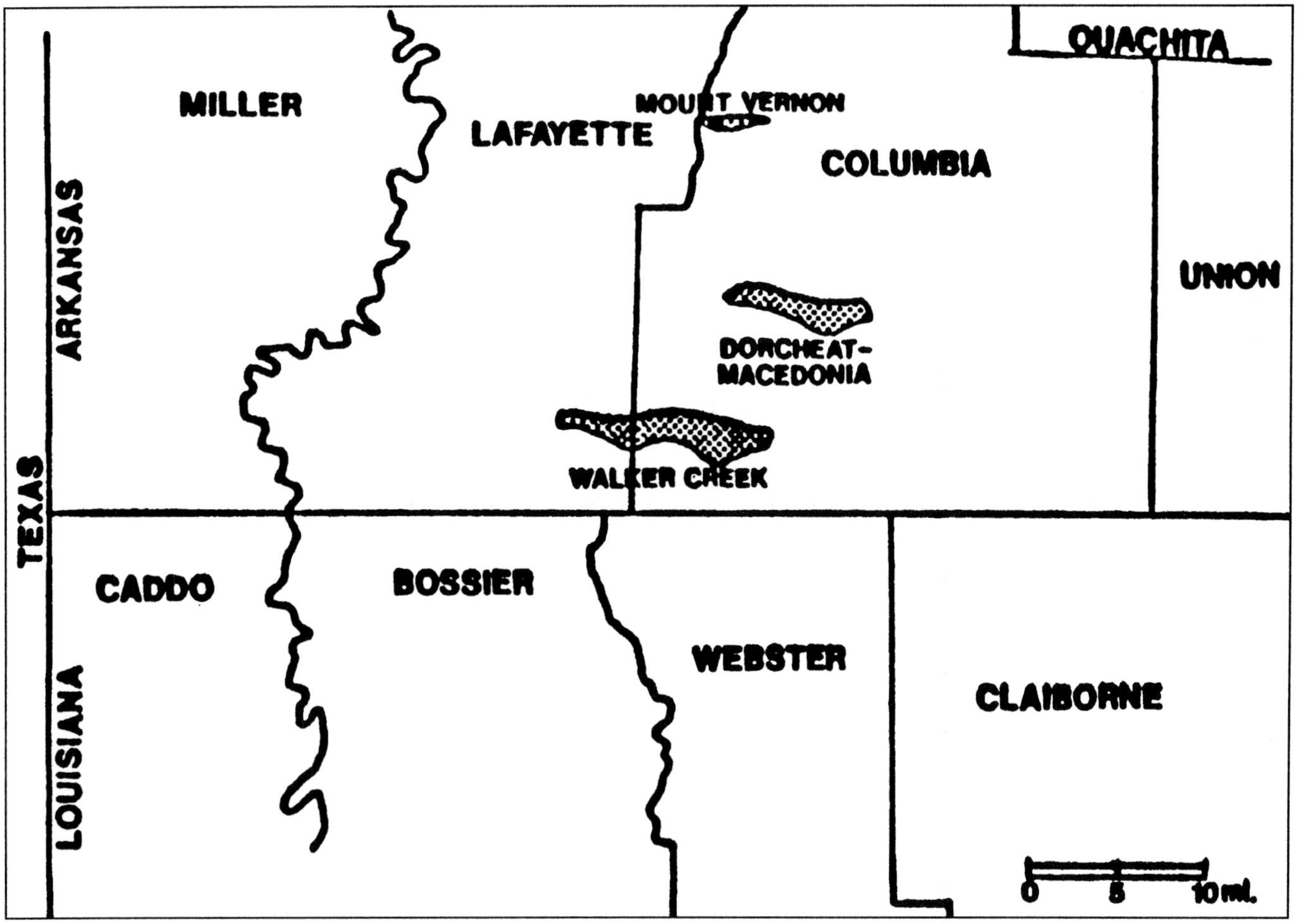

Figure 1. Location of Walker Creek Field

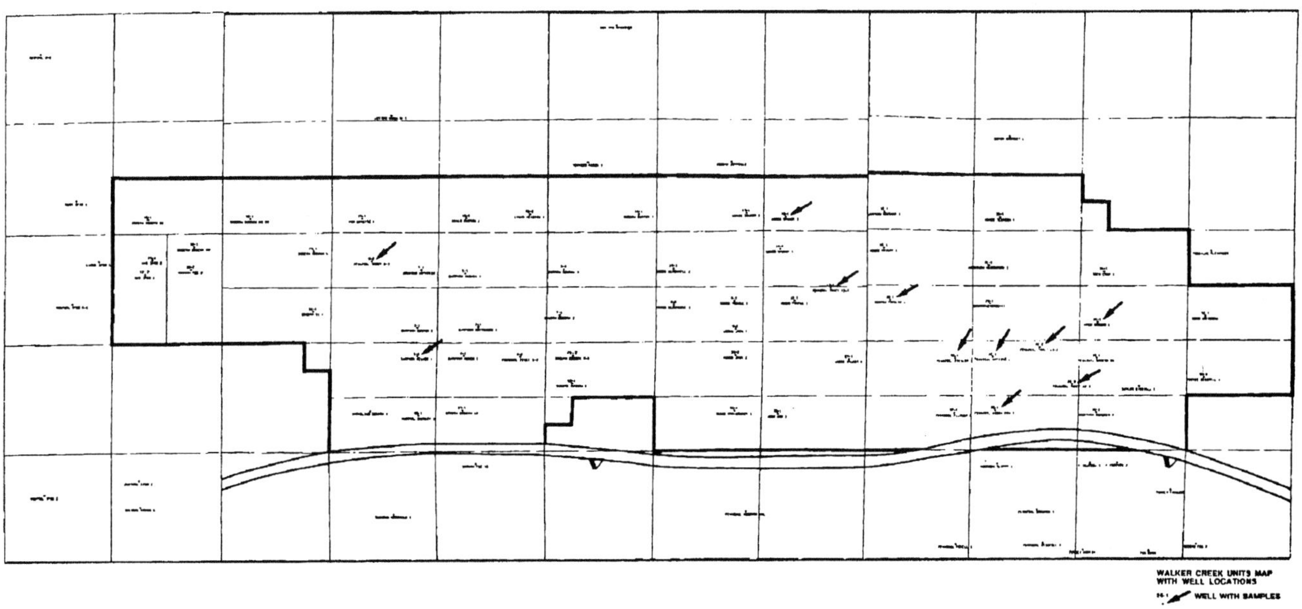

Figure 2. Walker Creek well locations.

	STAGE	SUBSTAGE	FORMATION	MEMBER
JURASSIC SYSTEM	Kimmerid-gian	Havrian	Haynesville	
		Sequanian		Buckner
	Oxfordian	Argovian	Smackover	upper
				middle
				lower
			Norphlet	
	Pre-Oxfordian		Louann	
			Werner	

Figure 3. Generalized Stratigraphy at Walker Creek. Dashed lines indicate unconformities.

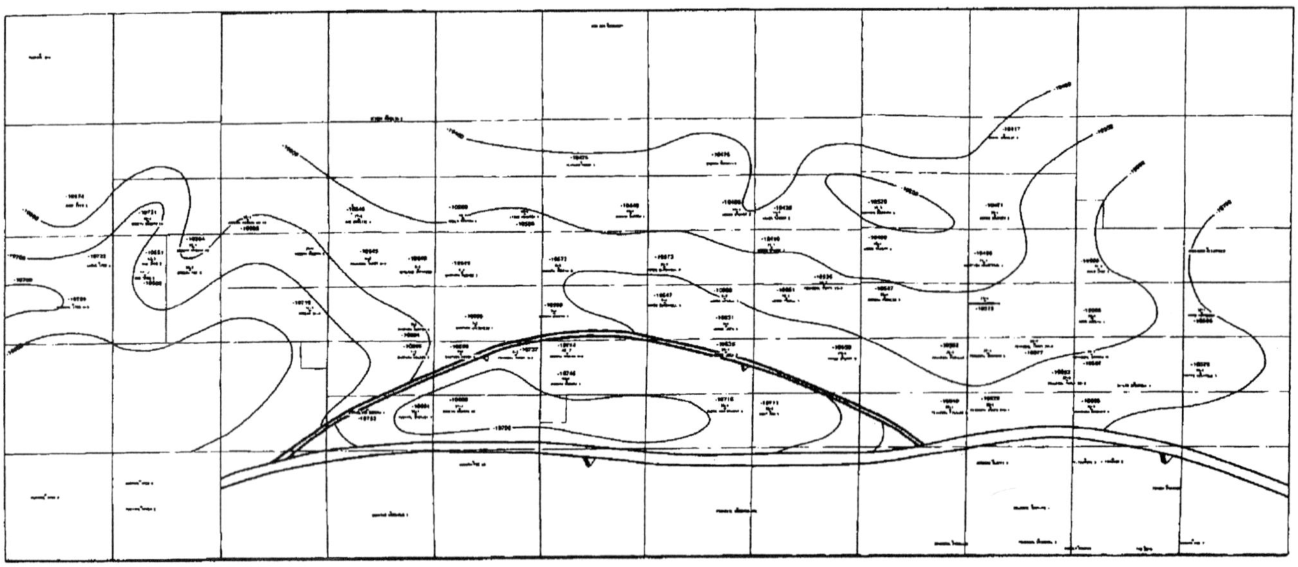

Figure 4. Top of Smackover structure map.

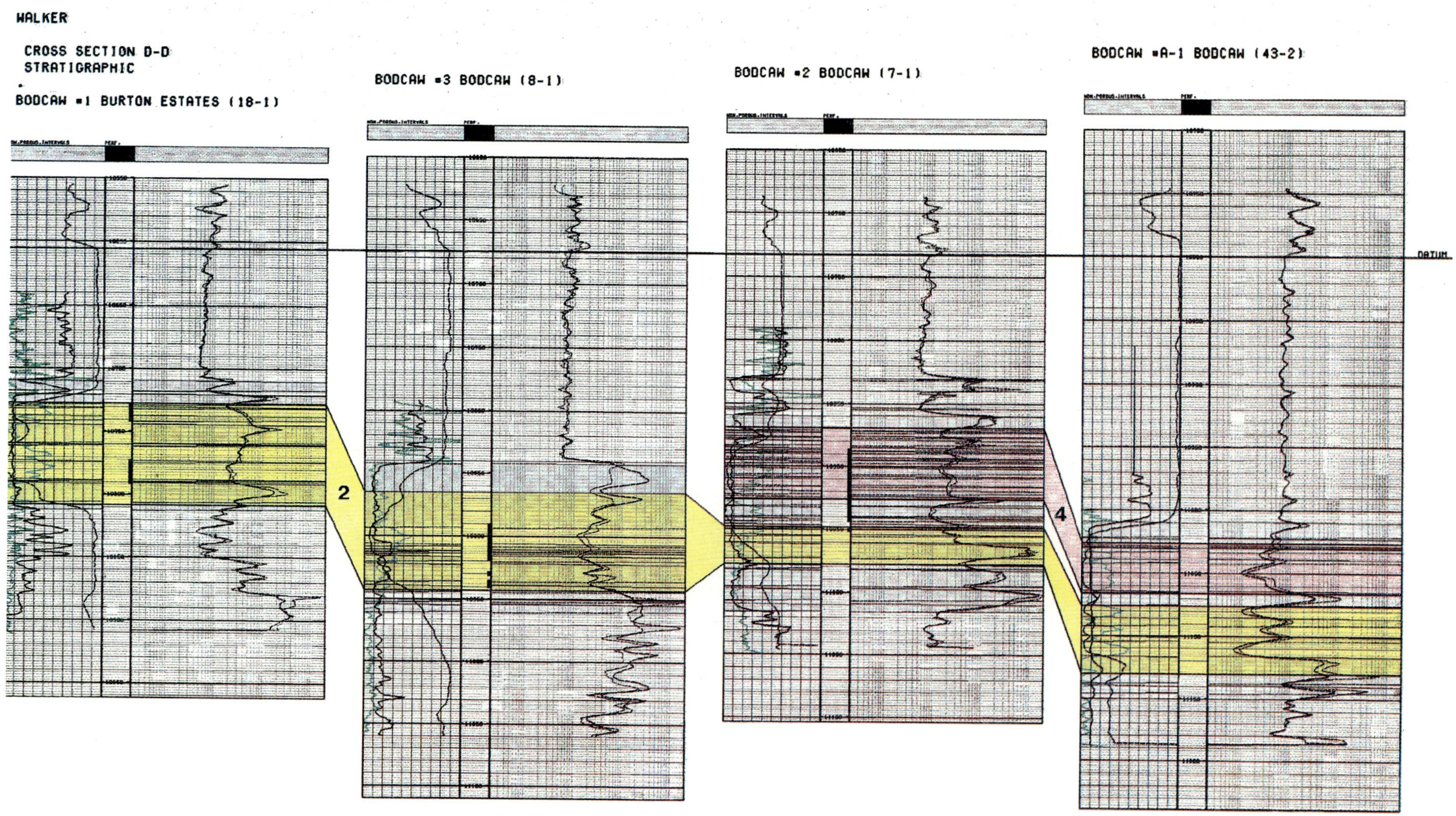

Figure 5. Stratigraphic cross section from the middle part of the field. 31

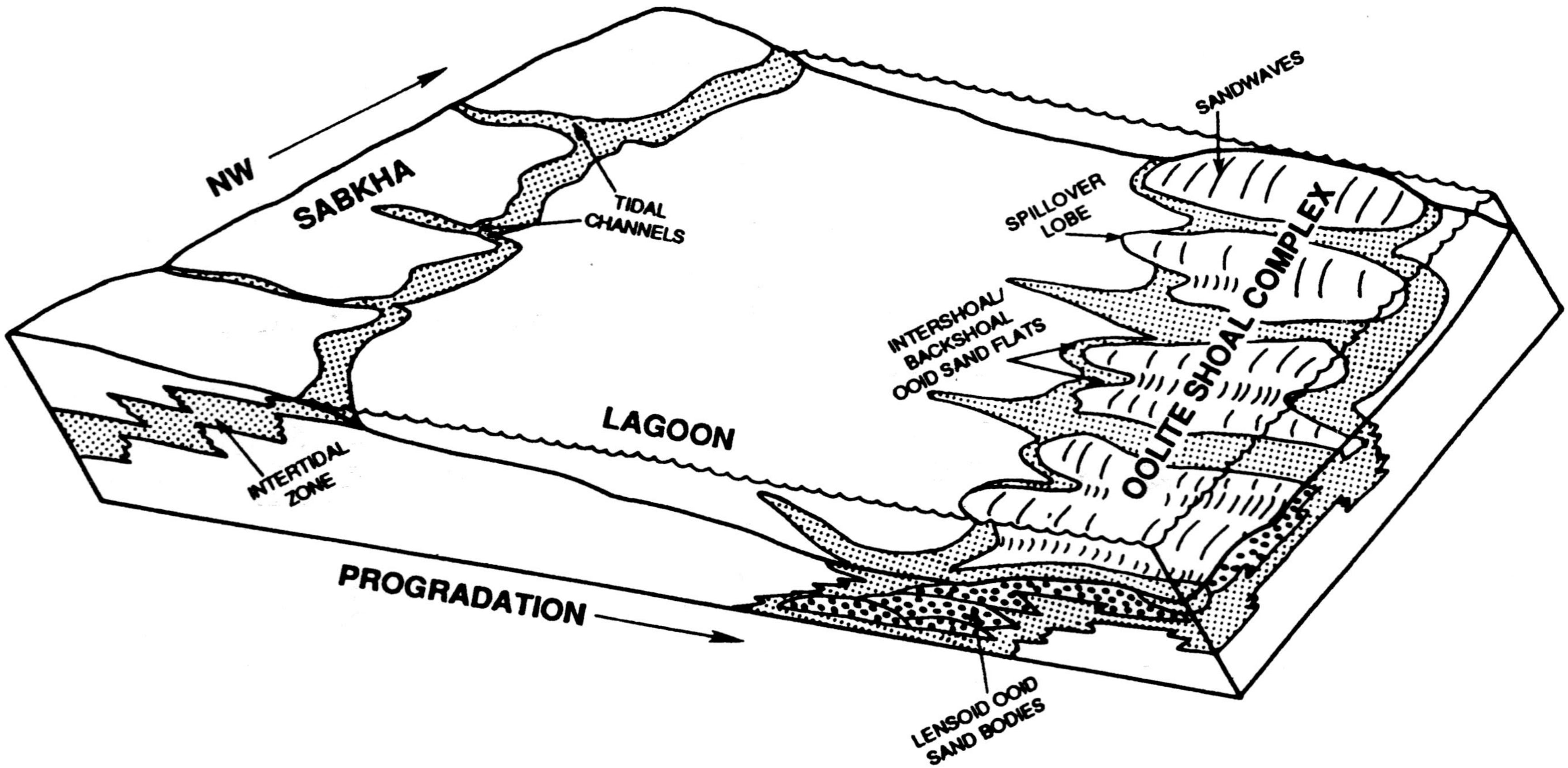

Figure 6. Paleogeographic Reconstruction of Depositional Environments at Walker Creek

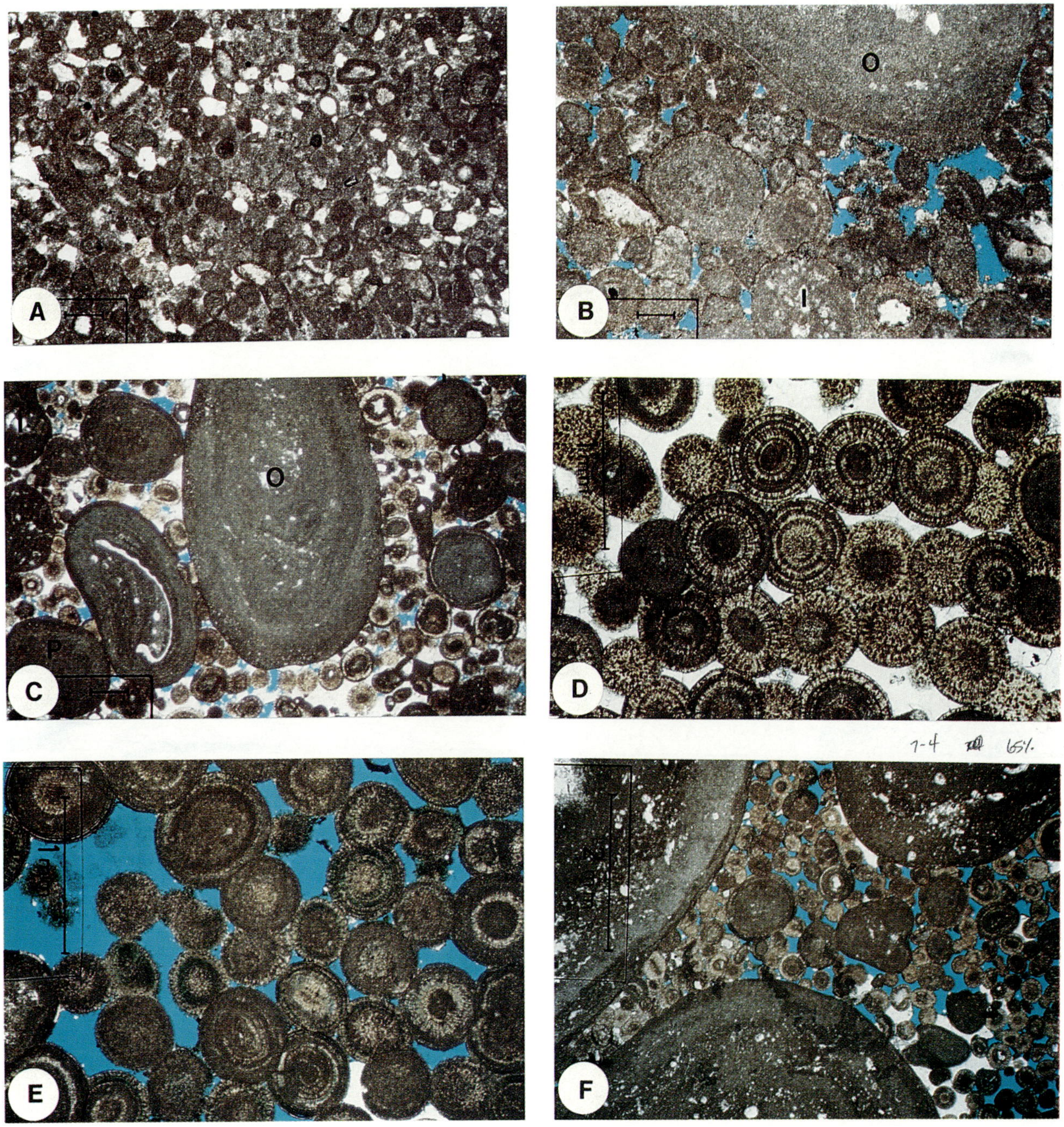

Figure 7. Depositional Facies
A. Pennzoil 13-2, 10927 ft.
Thin section photomicrograph of quartz-rich peloid packstone from the shallow shelf facies. Pellets, peloids of various origins, quartz grains and skeletal fragments (mainly bivalves) occur in a micrite matrix indicating lower energy conditions. Scale bar is 0.2mm in A and B.
B. Woods Stuart #3 (15-1), 10779 ft.
Poorly sorted grainstone from the mixed grain shoal facies composed of coarse to very coarse oncolites (0) and intraclasts (I) in a 'matrix' of fine to coarse peloids and ooids. Support is provided by compaction and minor calcite cement. Porosity is 7.9% and permeability is 6.6 md.
C. Pennzoil Kosek #1 (29-1), 10923 ft.
Poorly sorted grainstone from the cemented ooid/mixed shoal facies composed of coarse to very coarse intraclasts (I), oncolites (0), peloids (P) and coated skeletal fragments in a finer matrix of ooids and pellets. Scale bar is 0.4mm.
D. Woods Stuart #3 (15-1), 10691 ft. (Sample 3 in Table 2)
Well sorted ooid grainstone from between Zones 2 and 4. Intergranular porosity is occluded by coarsely crystalline calcite cement. Porosity is 3.4% and permeability is < 0.1 md.
E. Woods Stuart #3 (15-1), 10717 ft. (Sample 4 in Table 2)
Fine-to-medium grained ooid grainstone from the porous ooid/mixed shoal facies in Zone 2. Porosity and permeability are high (22.8% and 960 md) in this high energy rock type from the heart of the ooid shoal complex.
F. Chapman Helms #1 (1-1), 10880 ft.
Porous grainstone from Zone 4 with bimodal size distribution. Very coarse oncolites 'float' in a sea of finer grained ooids, multiple ooids and peloids. Porosity and permeability are 8.1% and 68 md.

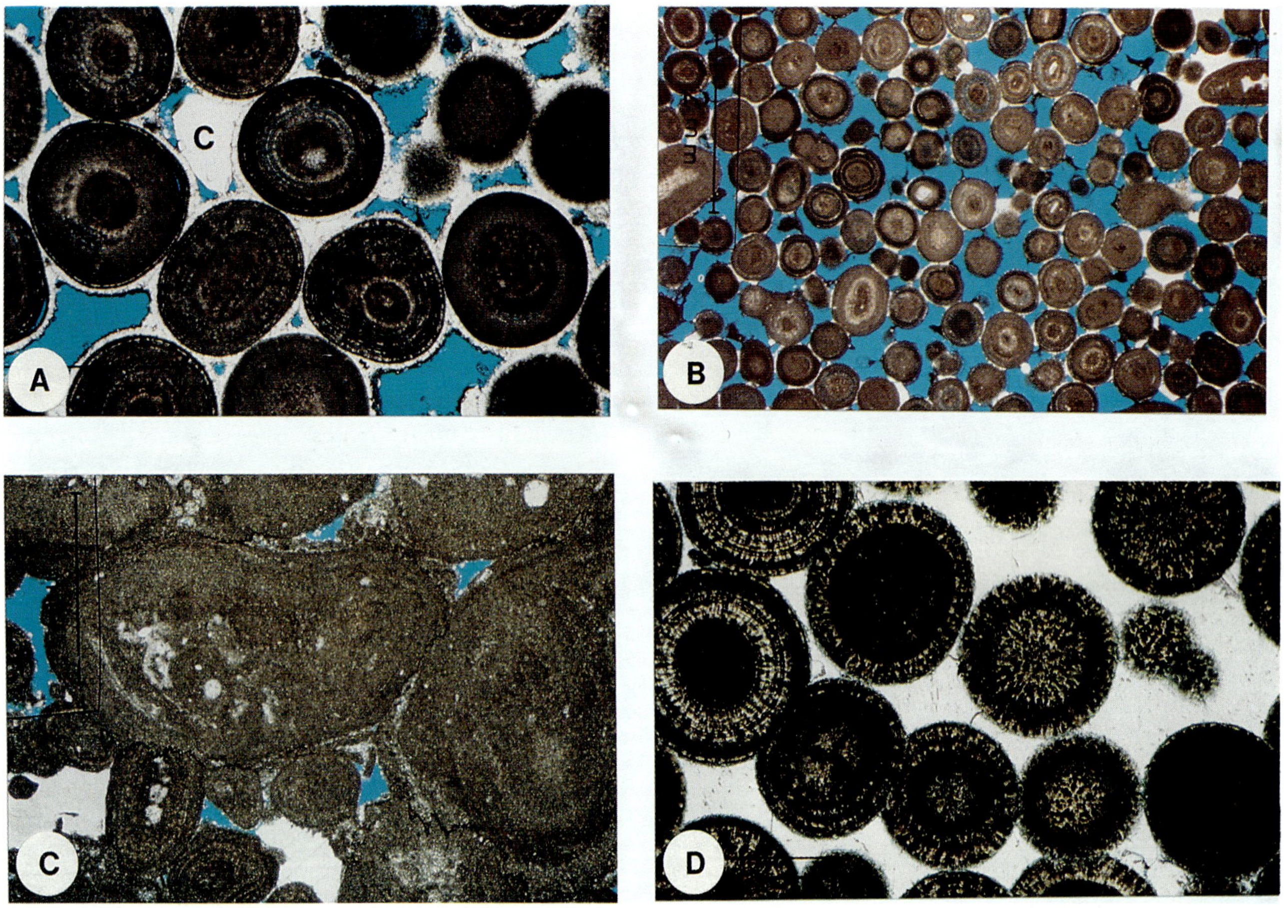

Figure 8. Diagenetic Textures

A. Woods Stuart #3 (15-1), 10710 ft.
Thin section photomicrograph of a porous ooid grainstone with fibrous calcite rim cement isopachously lining the grains. Coarse calcite of a later, burial origin (C) partially fills an intergranular pore. Intergranular porosity is well preserved and microporosity is also present (21.4% total porosity with 778 md permeability). Bar scale is 0.2mm.
B. Woods Stuart #3 (15-1), 10713 ft.
Thin section photomicrograph of a porous ooid grainstone with fine equant calcite cement concentrated at grain contacts. Porosity and permeability values are high (21.8% and 5400 md) and this rock type provides some of the best production in the field.

C. Pennzoil Pickler #1 (26-1), 10916 ft. (Sample 9 in Table 2)
Thin section photomicrograph of a sample dominated by oncolites and ooids with an infill of finer grained pellets. Compaction has resulted in pressure solution and microstylolites at grain contacts with a significant loss of porosity.
D. Woods Stuart #3 (15-1), 10691 ft. (Sample 3 in Table 2)
Thin section photomicrograph of an ooid grainstone with intergranular pore spaces filled by coarsely crystalline calcite spar cement. The occurrence, morphology and geochemistry of this cement indicate it precipitated in a deep burial setting. Porosity is 2.8% and permeability is <0.1 md. Bar scale is 0.2 mm.

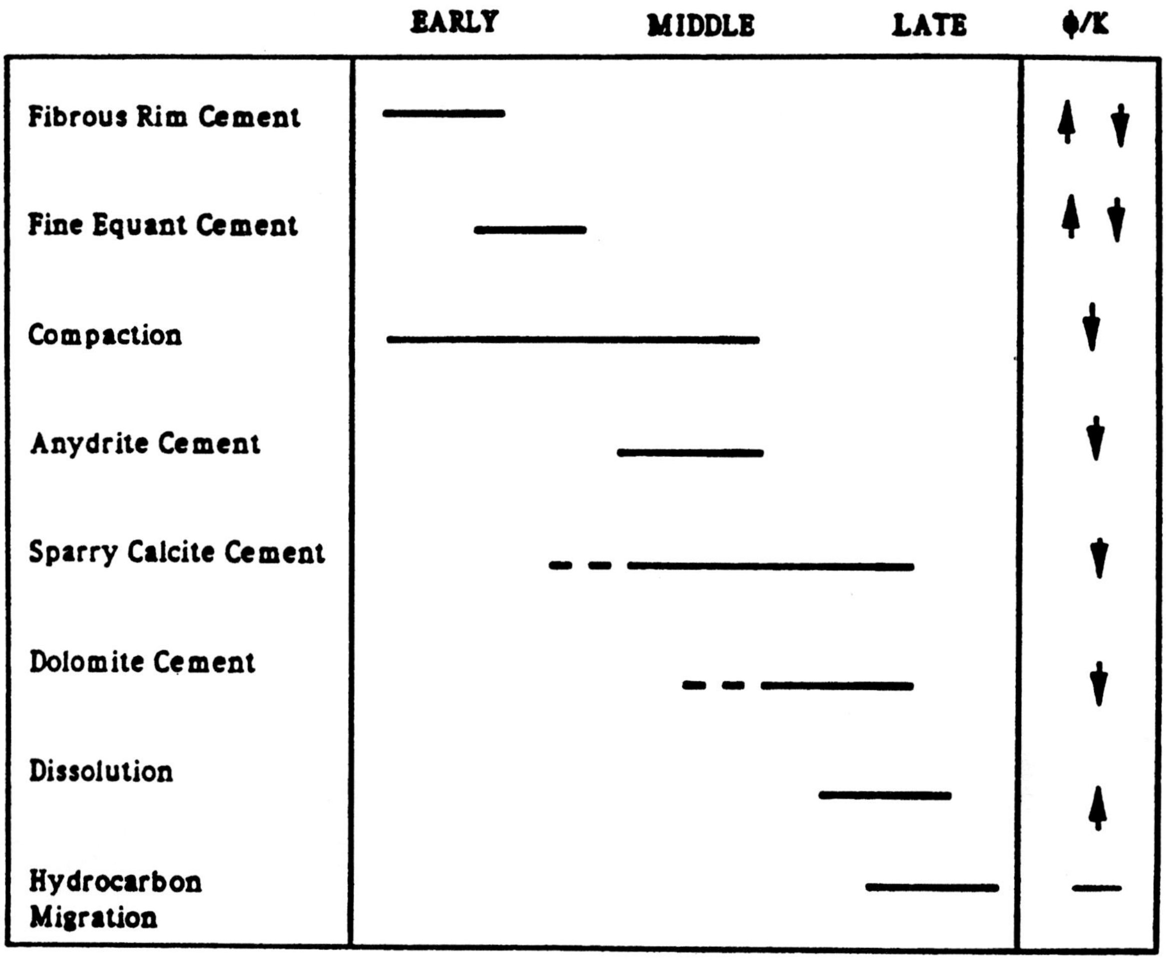

Figure 9. Paragenetic Sequence at Walker Creek and the Effect on Porosity and Permeability of the Diagenetic Events

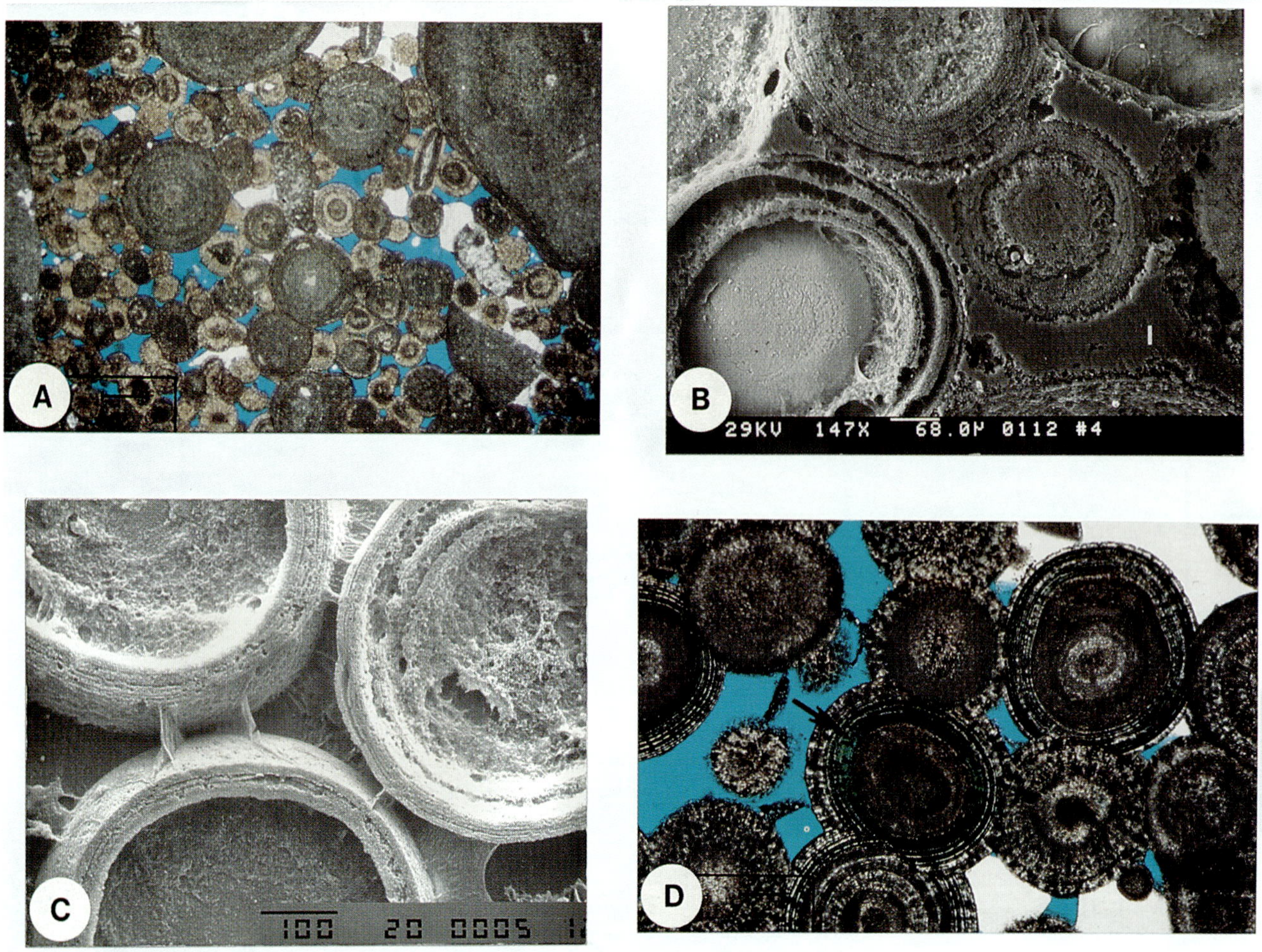

Figure 10. Pore Types

A. Chapman Helms #1 (1-1), 10877 ft.
Thin section photomicrograph of a porous ooid/mixed grain grainstone with preservation of primary interparticle porosity. Sparry calcite cement has only partially filled the pore spaces between ooids and oncolites. Porosity is 8.8% and permeability is 103 md. Bar scale is 0.4mm.
B. Woods Stuart #3 (15-1), 10717 ft. (Sample 4 in Table 2)
SEM photomicrograph of a pore cast of an ooid grainstone. Epoxy was used to impregnate the sample which was then etched with hydrochloric acid. Porosity now stands out in positive relief. In addition to preserved intergranular porosity (I), micron-scale dissolution within the ooids has resulted in layers of microporosity. Although most ooids have abundant microporosity, it is poorly connected making it ineffective. Total porosity is 22.8% with 960 md permeability. The bar scale is 68µ.
C. Pennzoil Pickler #1 (26-1), 10851 ft. (Sample 5 in Table 2)
SEM photomicrograph of a pore cast of an ooid grainstone. Porosity is dominantly microporosity within the ooids with minor intercrystalline porosity (porosity between calcite cement crystals, now represented by the very fine 'bridges' between ooids). Porosity is 9.5% with only 0.1 md permeability. The bar scale is 100µ.
D. Woods Stuart #3 (15-1), 10699 ft
Microporosity (arrow) is present in the ooids of a porous grainstone. Late stage solution enhancement of preserved interparticle porosity likely also improved the microporosity. Total porosity is 15.8% and permeability is 80md. Bar scale is 0.2 mm.

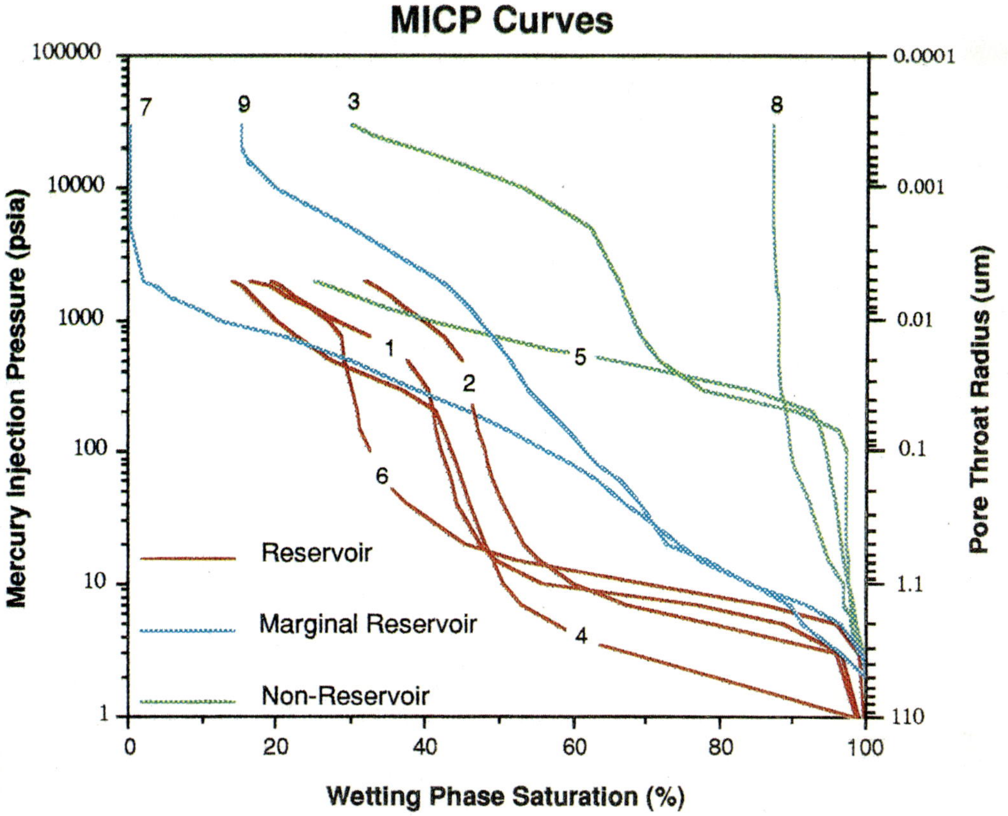

Figure 11. Mercury injection capillary pressure curves for nine samples from three reservoir categories. The large, well-connected pores and pore throats of the reservoir samples (1, 2, 4, and 6) are indicated by low displacement (<1 PSIA) pressures (pressure at which mercury enters the pore system) and low wetting phase saturation at low injection pressures (50% at 20 PSIA). The broad, relatively flat plateaus between 50 and 100% wetting phase saturation suggest that about 50% of the pore volume is accessed by throats with effective radii greater than 1 micron. The presence of two plateaus, at 5 PSIA and 500 PSIA, reflects the bimodal pore throat size distribution of the samples. The mercury entered the large interparticle pores between the ooids (first plateau), and then entered the smaller pores of the microporosity network (second plateau). Curves for samples of marginal reservoir quality (7 and 9) show higher displacement pressures and higher wetting phase saturation at any given pressure indicating pore throats are smaller and pores are less well connected. The slope of the curve (approximating 45°) reflects the very poorly sorted distribution of the pore throats. Curves for the non-reservoir rocks show high displacement pressures (>200 PSIA) which reflect the fine pore throats of the samples. The first plateau in sample 3 represents very fine (around 0.05μ) rim porosity around the ooids. The second plateau represents the even finer pore throats (~0.001μ effective radii) that occur between pore-filling calcite cement crystals. In sample 8, the high percentage of small pore throats and small, poorly connected pores are indicated by the steep slope of the curve; less than 10% of the total pore volume is accessed by pore throats greater than 0.0002μ effective radius. Even at 30,000 PSIA, the wetting phase saturation remains about 90%.

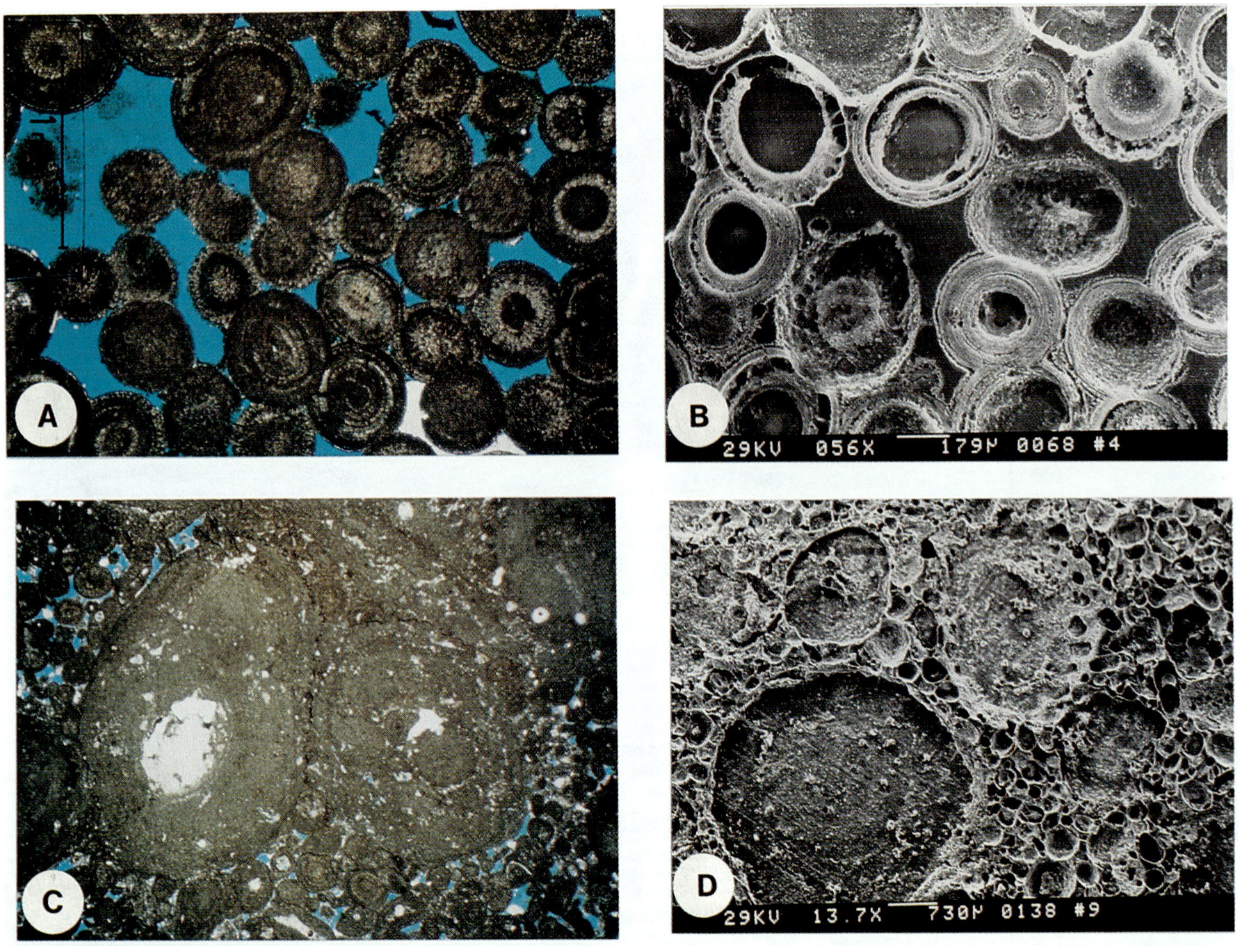

Figure 12. Pore Geometries

A. Woods Stuart #3 (15-1), 10717 ft. (Sample 4 in Table 2)
Photomicrograph of well sorted, ooid grainstone from the porous ooid/mixed shoal facies in Zone 2. Intergranular porosity is the dominant type, although microporosity is also present. Permeability values (960 md) reflect the abundant, well connected pores (22.8%).
B. Woods Stuart #3 (15-1), 10717 ft. (Sample 4 in Table 2
SEM photomicrograph of a pore cast of A. Intergranular pores stand out as smooth areas of positive relief. Microporosity occurs as more irregular positive relief areas within the ooids. Pores are large and well-connected by large pore throats. Scale is 179µ.
C. Pennzoil Pickler #1 (26-1), 10916 ft. (Sample 9 in Table 2)
Photomicrograph of a poorly sorted grainstone from the porous ooid/mixed shoal facies in Zone 2. The sample is composed of coarse to very coarse oncolites, intraclasts and peloids, with an infill of finer grained ooids and pellets. Porosity is 5.8% and permeability is 0.5 md. Width of field of view is 5 mm.
D. Pennzoil Pickler #1 (26-1), 10916 ft. (Sample 9 in Table 2)
SEM photomicrograph of a pore cast of C. The bimodal grain size distribution shows up clearly. Intergranular and microporosity are visible in positive relief. The pore throats are smaller and less well connected than the sample in B. The permeability cut-off of 0.6 md makes this sample of marginal reservoir quality. Scale is 730µ.

Figure 13. Pore Geometries
A. Woods Stuart #3 (15-1), 10691 ft. (Sample 3 in Table 2)
Photomicrograph of a well sorted, ooid grainstone from the cemented ooid/mixed shoal facies. The low porosity (3.4%) and permeability (<0.1 md) values reflect the filling of intergranular pores by coarsely crystalline calcite cement.
B. Woods Stúart #3 (15-1), 10691 ft. (Sample 3 in Table 2)
SEM photomicrograph of a pore cast of A. Microporosity occurs as thin rims around the ooids. These pores are connected by even finer intercrystalline pore throats. Porosity is 3.4%, permeability is <0.1 md. Scale is 1000μ.
C. Pennzoil Pickler #1 (26-1), 10900 ft. (Sample 8 in Table 2)
Photomicrograph of a poorly sorted argillaceous wackestone with pellets and skeletal fragments from between reservoir Zones 2 and 4. Slightly lower energy conditions resulted in deposition of micritic and argillaceous material between the grains. Porosity and permeability are correspondingly lowered (0.7% and <0.1 md).
D. Pennzoil Pickler #1 (26-1), 10900 ft. (Sample 8 in Table 2)
SEM photomicrograph of a pore cast of C. The jagged texture of the intergranular pores is a consequence of the presence of abundant micritic and argillaceous material. Pores are small, irregular and poorly connected by very small pore throats. Scale is 1000μ.

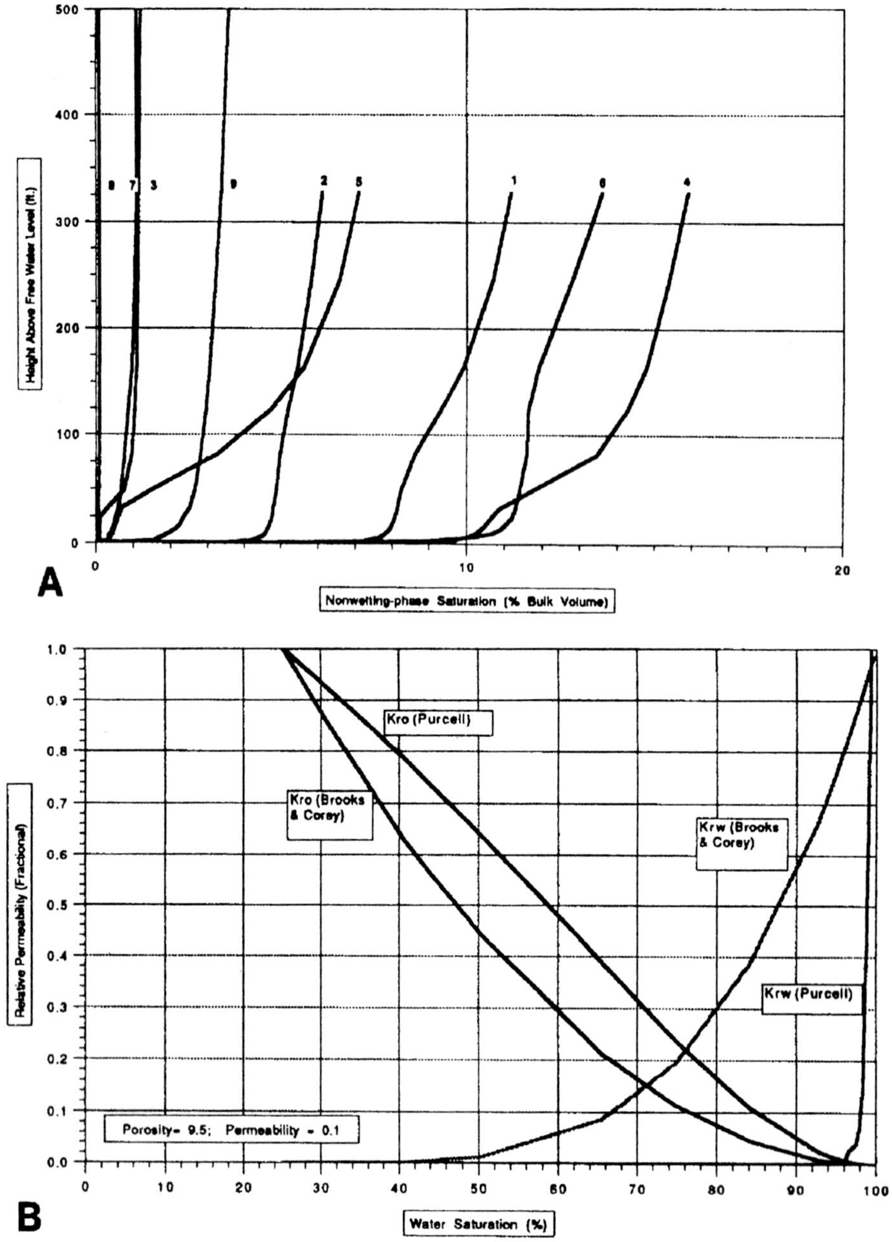

Figure 14
A. Plot of samples showing relationship of non-wetting phase saturation (as a bulk volume percentage) to height above Free Water Level (FWL). Calculation is made by first converting P_c air/mercury to P_c brine/oil of the reservoir:

$$P_{Cb/o} = P_{Ca/m} \times \frac{\sigma_{b/o}\cos\theta_{b/o}}{\sigma_{a/m}\cos\theta_{a/m}}$$

where $\sigma_{b/o}$= reservoir brine/oil interfacial tension = 16 dynes/cm

$\sigma_{a/m}$ = air mercury interfacial tension = 480 dynes/cm

$\theta_{b/o}$ = reservoir brine/oil contact angle $\approx 0^\circ$.

$\theta_{a/m}$ = air/mercury contact angle = 140°

Next, the relationship of the reservoir brine/oil capillary pressure to height above FWL can be determined:

$$P_{Cb/o} = h(\rho_b - \rho_o) \times 0.433$$

where ρ_b = reservoir brine density = 1.126 gm/cm³

ρ_o = reservoir oil density = 0.515 gm/cm³

h = height above FWL

Pore volume to bulk volume conversion is made by:

$$PV = BV \times \frac{100}{\text{Ø}}$$

where PV = pore volume occupied (in percent), BV = bulk volume occupied (in percent), and Ø = total porosity of the sample (in percent)

B. Relative permeability curves constructed from the capillary pressure analysis of sample 5. Both Purcell (1949), and Brooks and Corey (1964, 1966) calculation methods indicate that at a sufficient height to be reservoir, this rock would have, at best, a relative permeability to oil (k_{ro}) of only 0.04 md.

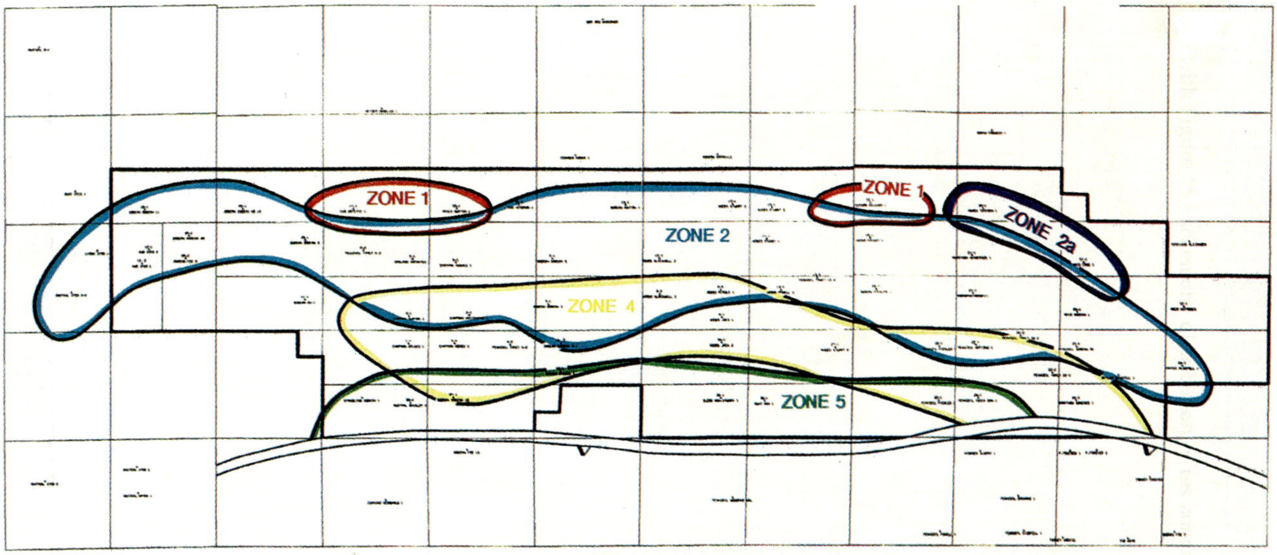

Figure 15. Schematic distribution of reservoir zones corresponding to generations of ooid shoal development.

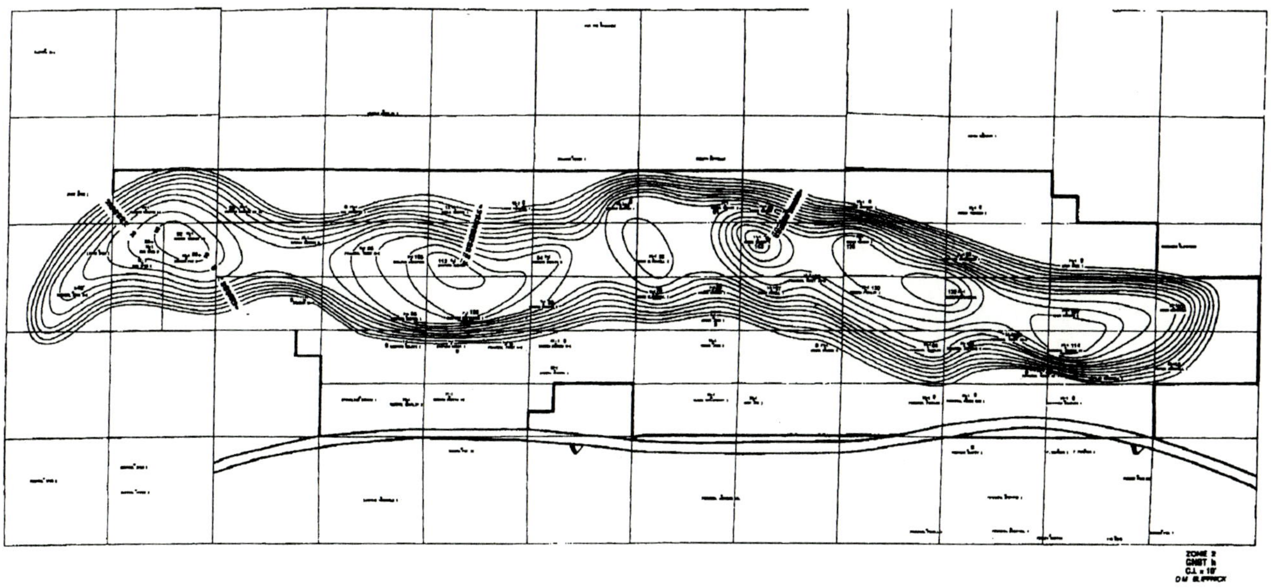

Figure 16. Ooid grainstone isopach for Zone 2.

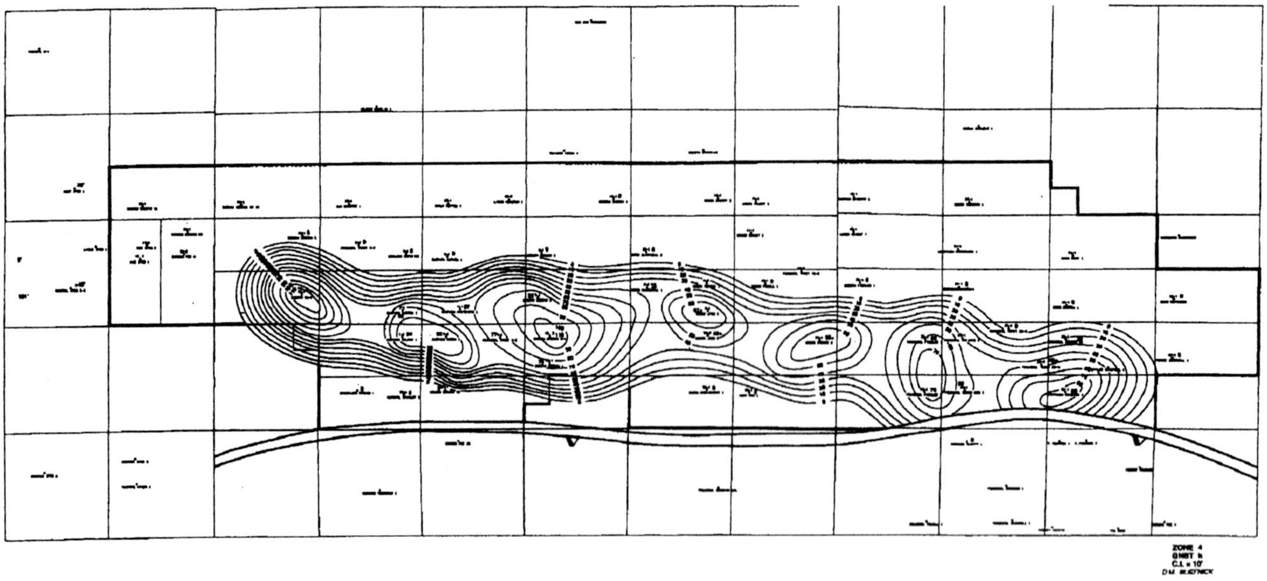

Figure 17. Ooid grainstone isopach for Zone 4.

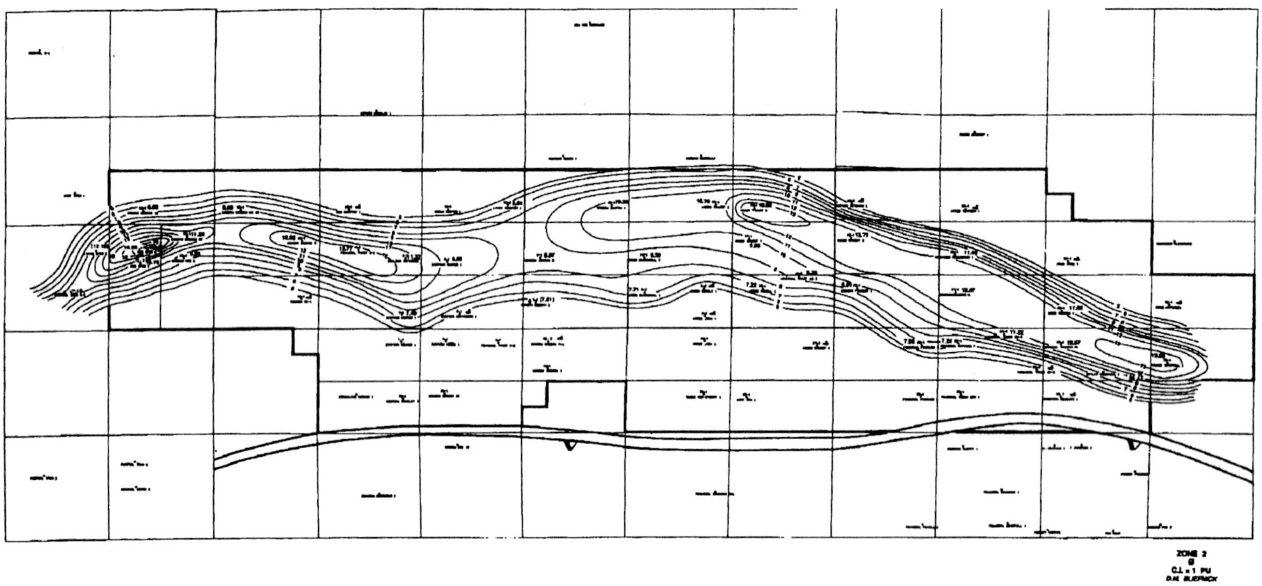

Figure 18. Porosity isopach for Zone 2.

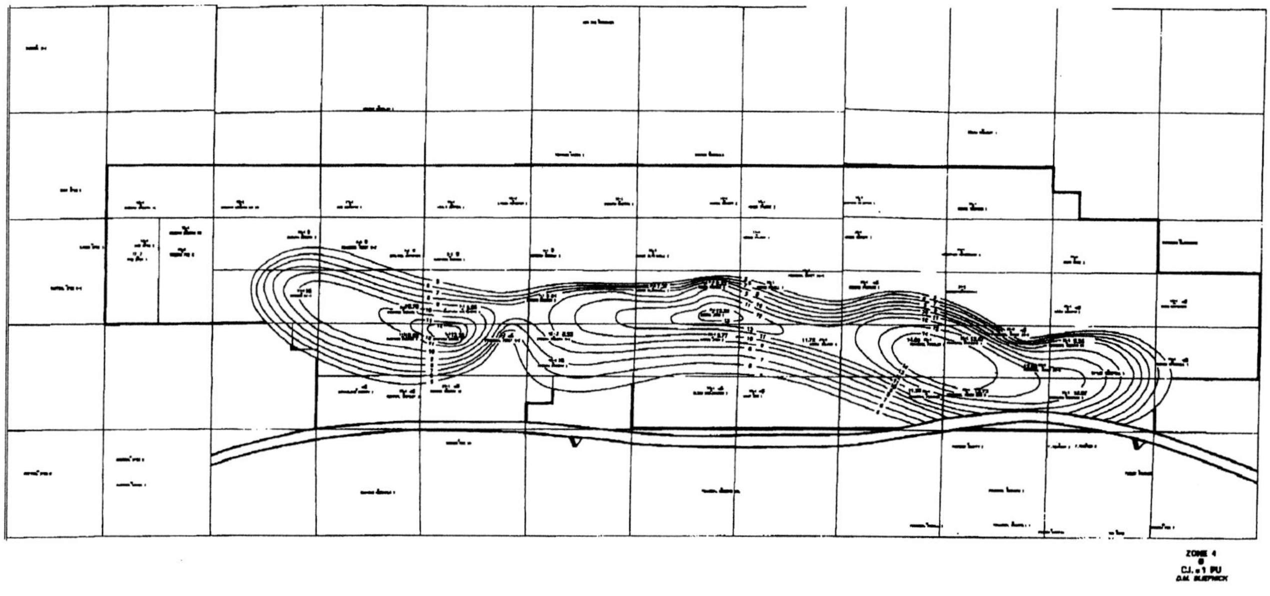

Figure 19. Porosity isopach for Zone 4.

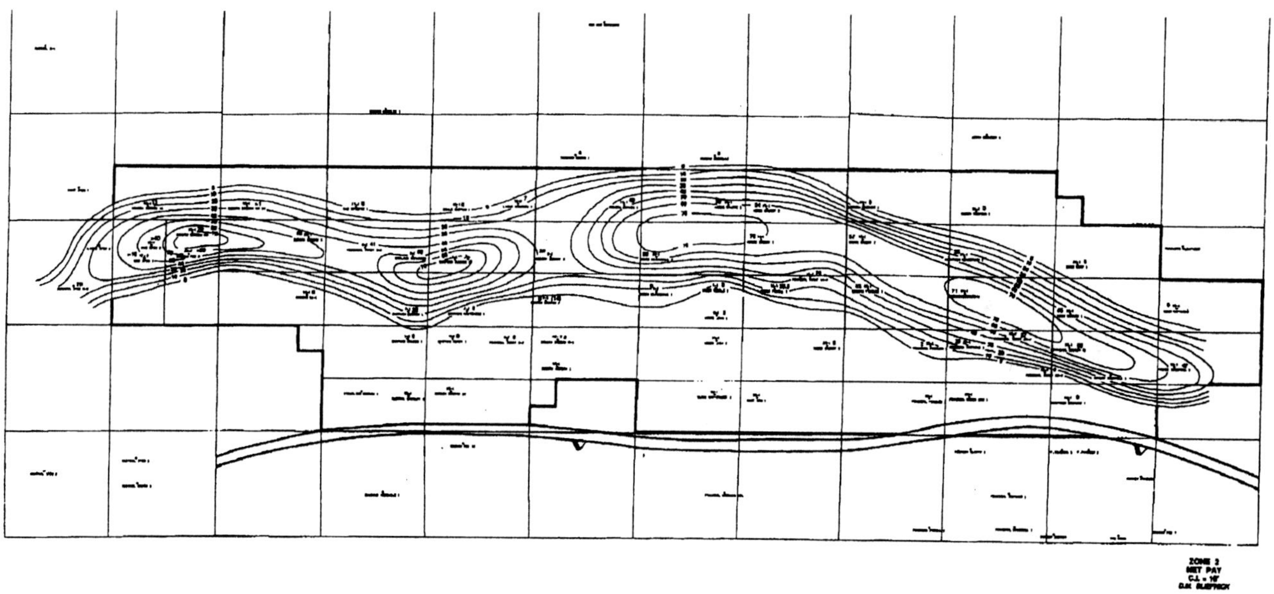

Figure 20. Net pay isopach for Zone 2.

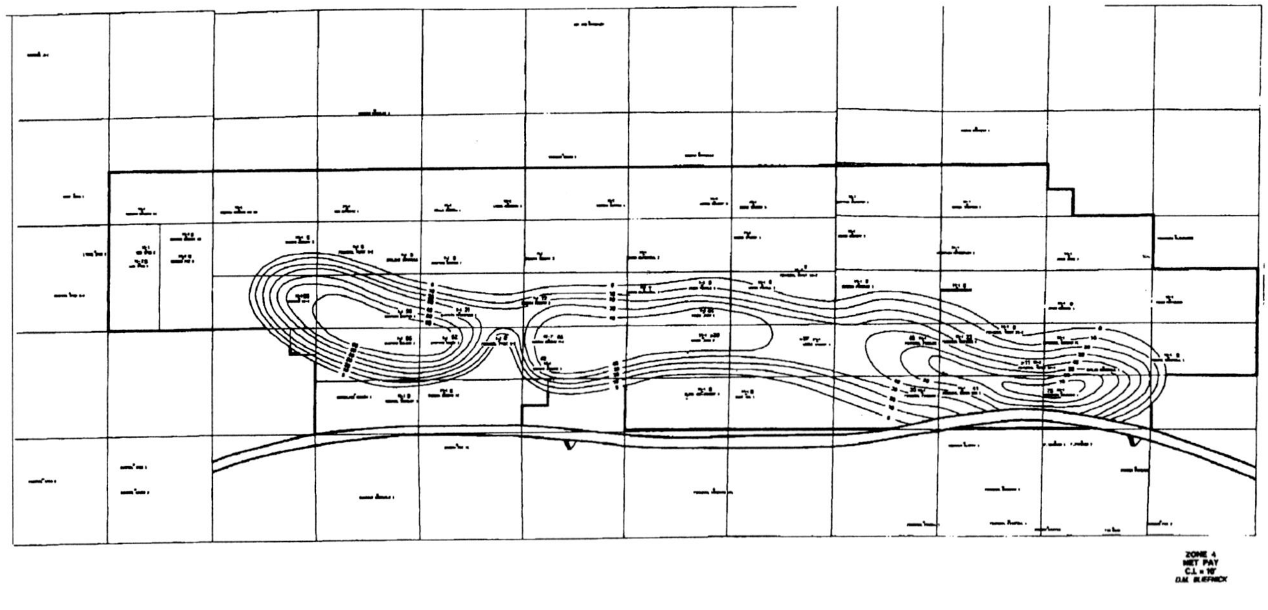

Figure 21. Net pay isopach for Zone 4.

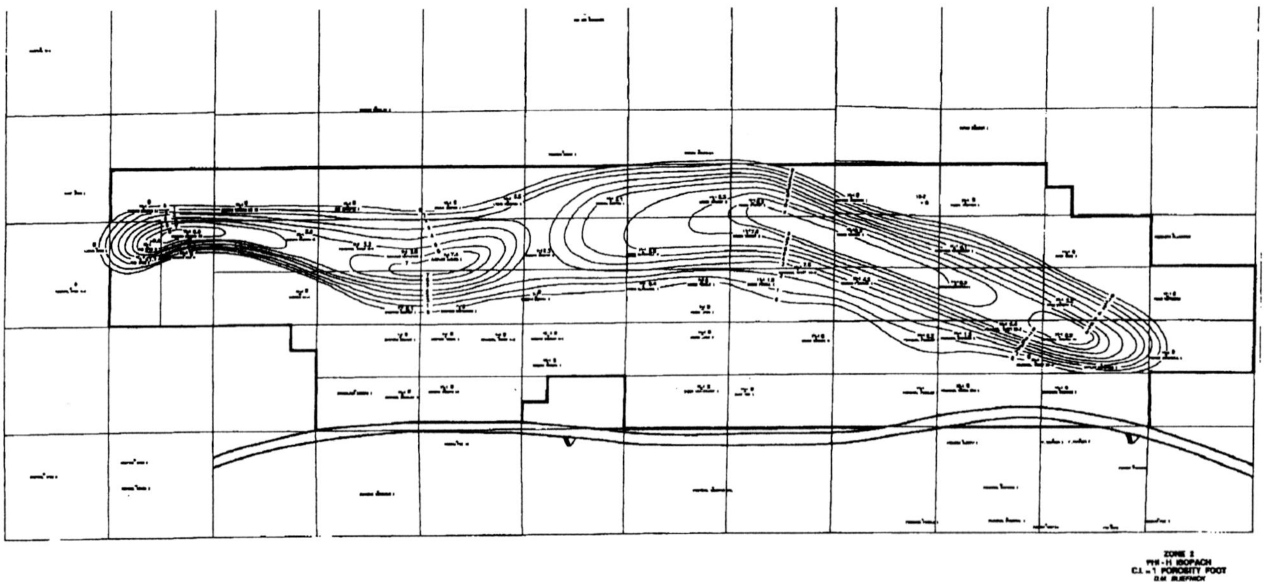

Figure 22. Øh isopach for Zone 2.

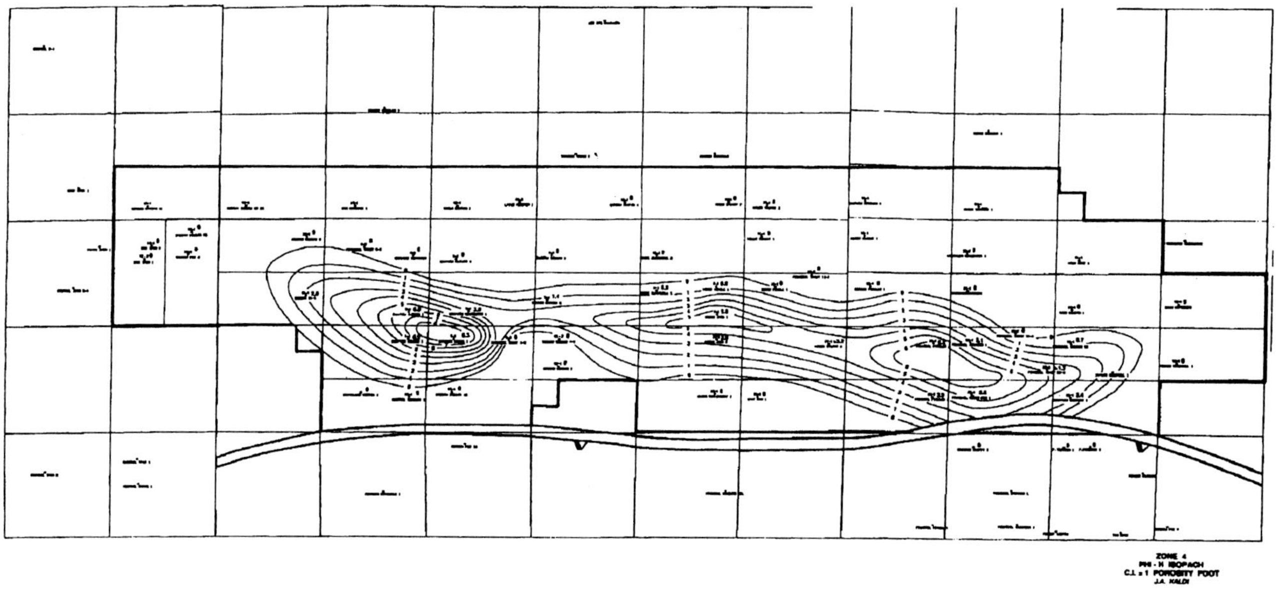

Figure 23. Øh isopach for Zone 4.

Batanga Sandstone Reservoir: A Turbidite Stratigraphic Trap

D. R. Smith, S. B. Desantis, L. A. Dunne, B. L. Faulkner and R. E. West

A synergistic approach was instrumental in conducting an exploitation study of the Upper Cretaceous Batanga reservoir of Gabon. The Batanga reservoir is an oil productive, turbidite channel sequence deformed by syn- and post-depositional salt piercement and intrusion. This study illustrates how several techniques derived from geological, geophysical, petrophysical, and reservoir engineering disciplines were integrated into a comprehensive description of this reservoir.

These Oguendjo "B" and "C" fields were discovered in 1981. Nineteen wells have produced 38 MMBO from stratigraphic traps associated with salt diapirism. Oil column heights and productive acreage average 165 feet and 650 acres, respectively, in these two fields.

Regional and local depositional trends were interpreted from isopach maps derived from three dimensional seismic data and from true stratigraphic thickness (TST) log cross sections. Salt uplift contemporaneous with deposition resulted in thicker sands on the flanks of salt piercements. Five separate depositional sequences were identified in each field and are composed of laterally discontinuous proximal channel, medial levee, and distal overbank facies. Using a statistical analysis of wireline log data, the oil productive channel and levee facies were discriminated from non-reservoir rock.

Log analysis and net pay identification of thinly bedded turbidites was accomplished using the "Dual-Water" model for water saturation calculations and the M-N crossplot technique for V-shale calculations. Routine and special core analyses further refined these models. True Vertical Thickness (TVT) logs were used to generate net pay and hydrocarbon pore volume (HPV) estimates for each well.

An estimate of the original oil-in-place based on isochore thickness and log derived HPV calculations were prepared for each field. Verification of HPV-derived oil-in-place was obtained from material balance computations. Reservoir performance and fluid movement was interpreted through water-cut maps, production logs (PLT), thermal decay logs (TDT), and RFT and bottom hole pressure analyses.

The recommendation for additional infill drilling and water shut-off workovers in existing wellbores will provide for the recovery of additional reserves during the next two years.

Managing a Technical Cooperation Team for Hydrocarbon Exploration and Development in the People's Republic of China

Kenneth M. Mallon

Multi-disciplinary teams in the hydrocarbon exploration and development business often work in a super-charged atmosphere, finding exciting solutions to oil field problems. Managing a team involved in the integration of various disciplines to achieve a common goal can be a big challenge. Involve such a team in a Technical Cooperation with Chinese co-workers, and locate them in a small city in the People's Republic of China, an emerging Third World country that has a weak infrastructure outside the major cities, work-related cultural and language barriers, and a bureaucratic oil and gas industry that is a government agency, and the challenge becomes a daily obstacle course with its own special pitfalls. In this environment, maintaining synergism within the team, and between the team and the Chinese co-workers, is a unique and difficult task. The attainment of team goals thus becomes a very real reward!

A four-man team with practical experience in many specific geoscience and related fields was recruited to participate in a one year Technical Cooperation with Chinese geoscientists, engineers and computer professionals at the ZhuoZhou Interpretation Center, a division of the Bureau of Geophysical Prospecting. This Technical Cooperation included the transfer of "USA-style" oil and gas exploration and development techniques, training in the art of seismic interpretation workstation methods, and the interpretation of problem areas by the team.

The main goal of the cooperative effort was to define and initiate a multi-disciplinary approach to exploration and development in the Lang Gu Depression, a mature producing province. This structurally and stratigraphically complex area is located in the Bo Hai Basin of northeast China. Difficulties in interpreting the geology have led to disappointing drilling results: too many small fields and low-productivity wells.

With experience in seismic interpretation and processing, exploration and development geology, log analysis, reservoir characterization, and geoscience programming, the team was well equipped to handle the technical aspects of the Cooperation. There were, however, obstacles to the Cooperation's eventual success. Maintaining a synergism that was sometimes fragile, and overcoming the bureaucracy, the logistics and communications barriers, and the limited availability and usefulness of mountains of existing data were daily adventures in management.

THE ECONOMIC VALUE OF A SYNERGISTIC ORGANIZATION

Robert M. Sneider

Synergy is defined as the "action of discreet agencies so that the total effect is greater than the sum of the effects taken independently". Within the context of the petroleum exploration and production business, synergy means that geologists, geophysicists, petroleum engineers and others work together on a project more effectively and efficiently as a team than working as a group of individuals (Sneider, 1986). The synergistic team approach has been tried by several large and small oil companies in the 1970's and 80's in order to compete more effectively and profitably with fewer staff and managers.

During the past sixteen years, I have had the opportunity to build or help build both small and large synergistic teams and organizations. Although it is relatively easy to measure qualitatively a group's performance, it is very difficult to evaluate quantitatively the technical and economic benefits of a synergistic team and organization versus that of other traditional E & P organizations. Also, it is very difficult to determine what are the critical ingredients that make one organization more effective, productive and profitable than another. The following "experiment" is the only attempt that I know of to help evaluate quantitatively the technical benefits and economic value of the "traditional" versus" synergistic" team approach in exploration and production.

"How can you really prove that synergistic teams are more effective and profitable?" This question was posed to me by a senior director (banker) of a large oil and gas company. I proposed an "experiment" that I thought would unequivocally provide the answer: Form a small synergistic E and P company that would compete on an exactly equal basis with one of the company's E and P divisions.

The Board of Directors agreed to a three to five year test and hired a president to form the new company organized as integrated, multidisciplinary teams. This small company of about thirty-five professionals and support staff competes in the same geographic area with the larger traditional exploration and production division, which has about 175 professionals and support staff. The age distribution and experience of the staffs (Figure 1), the budgets, technical data bases and economic/risk criteria for projects of the small and large groups are essentially identical. The organizational structure (Figure 2) and approach of the two groups are different. The small group is organized by plays or projects and has a very flat organization. The production-operations group is organized into projects like the acquisition group. Most of the staff are members of several teams/projects.

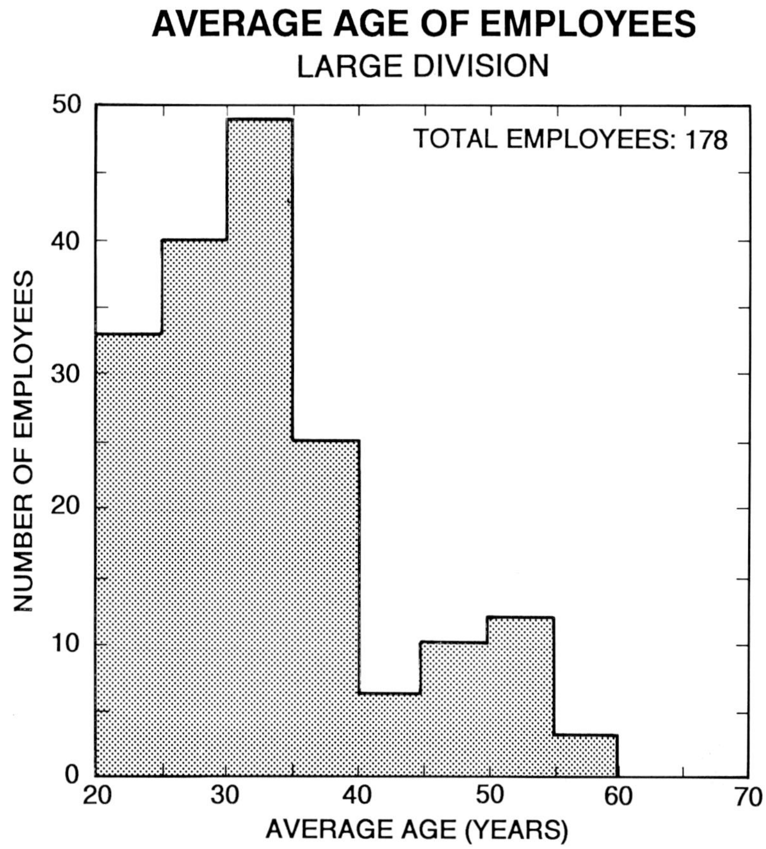

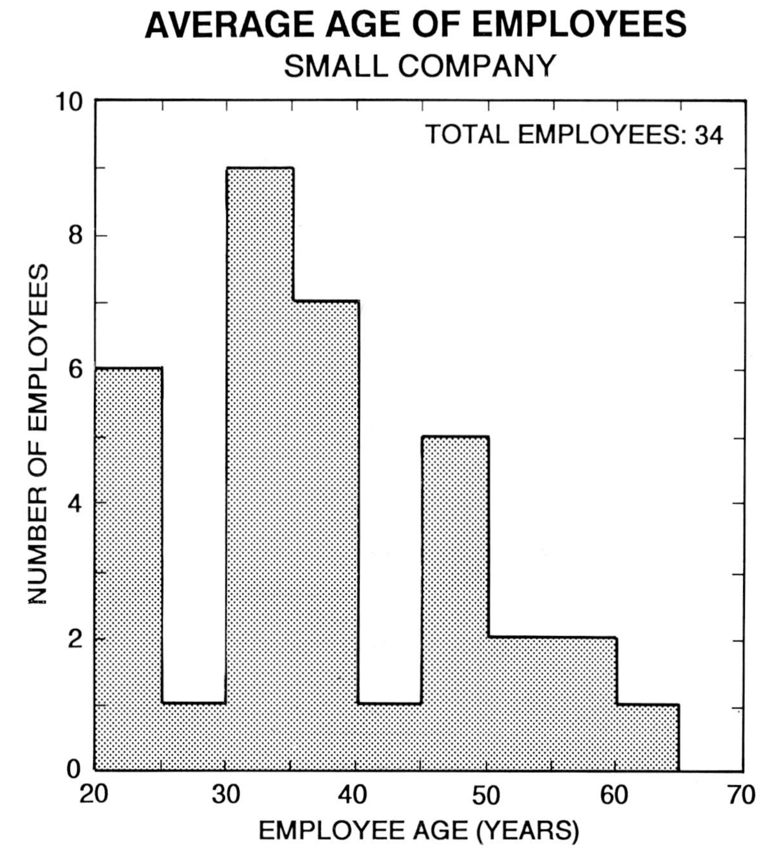

Fig. 1. Comparison of the age distribution of the large company division and the small synergistic company.

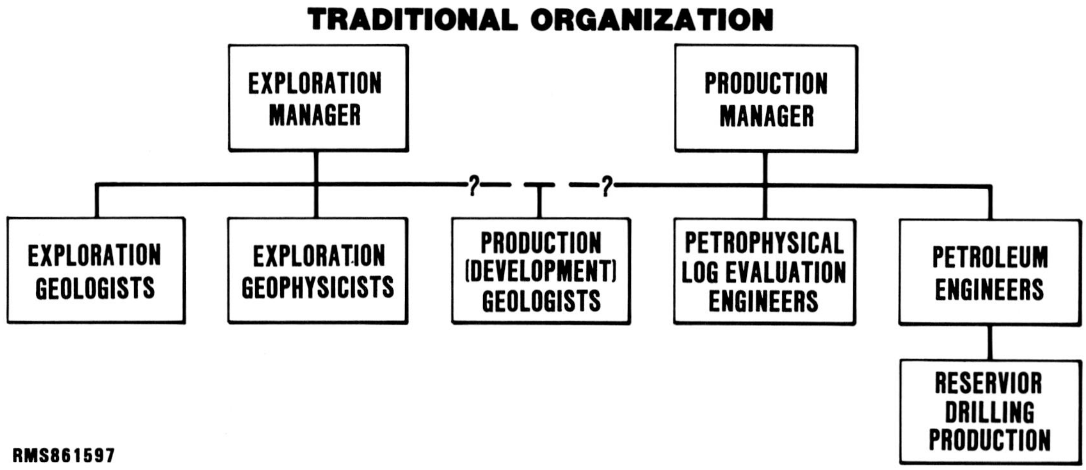

A.

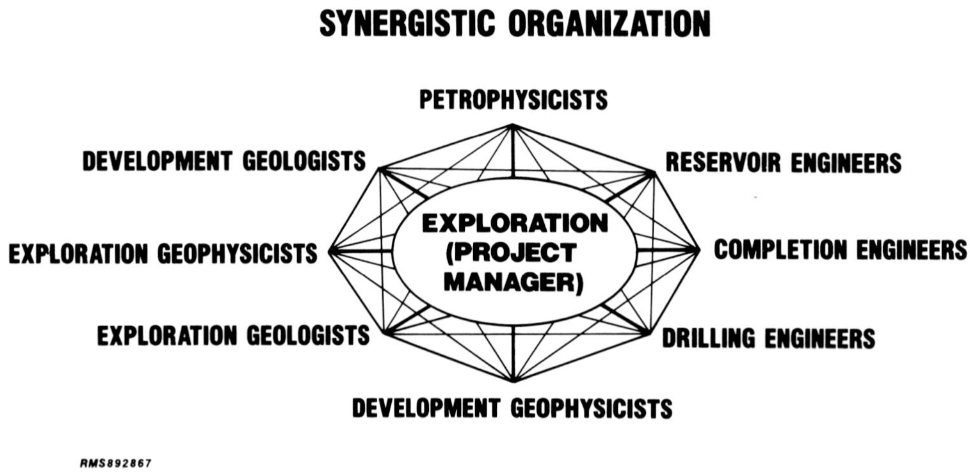

B.

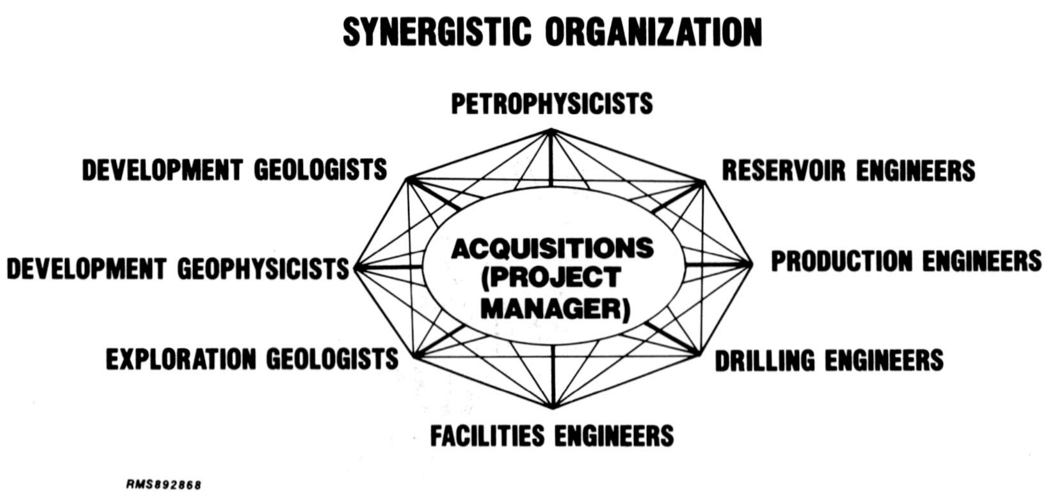

C.

Fig. 2. Comparison of the organization of the traditional large company division (A) and the small synergistic company (B & C). The small company has from four to six teams of exploration and production technical and support staff.

Project/play supervisors report directly to the president (Figure 3). The individual team leaders have a great deal of technical and monetary authority. In contrast, the larger organization is structured as a traditional exploration and production group with four levels of management and review before project/play approval (Figure 3). Figure 3 shows that at the beginning of the "experiment", the large division had four levels of supervision compared with two in the small synergistic company.

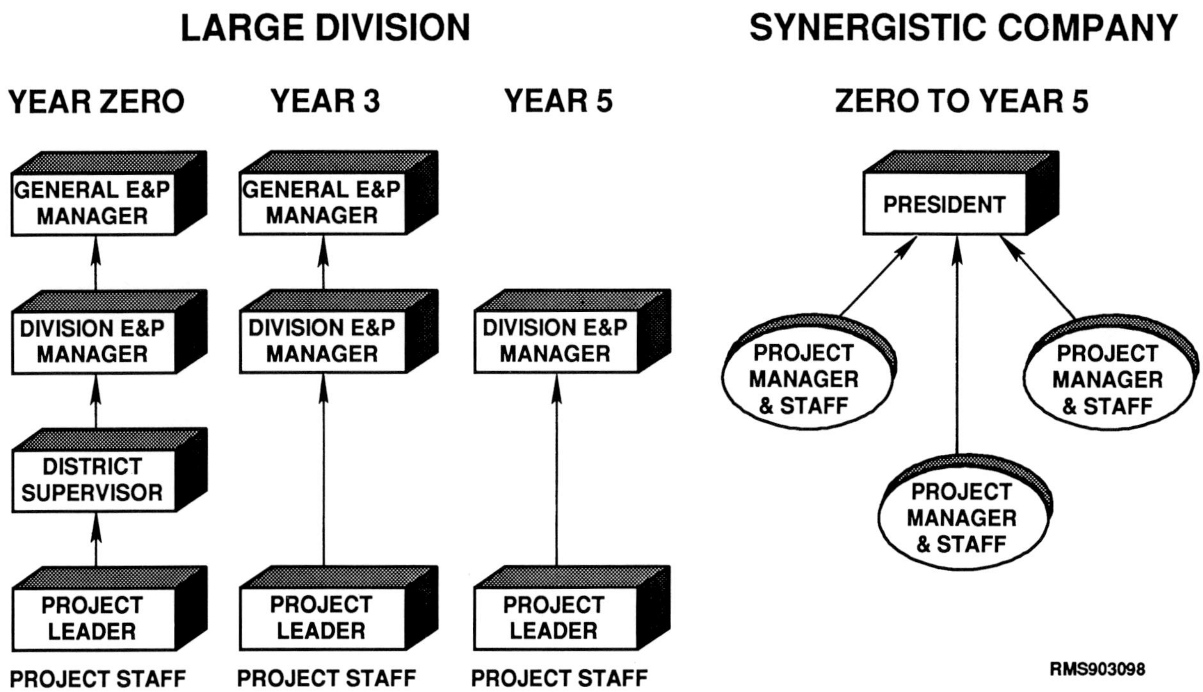

Fig. 3. The management hierarchy of the large division and the small company. At the beginning of the "experiment" the large division had four levels of management and authority. The number of levels were reduced to three in the third year and to two in the fifth year. The synergistic company had two management levels from the beginning to year five.

The two groups have had similar budgets each of the first five years of the "experiment", which ranged from more than 35 to 60 million U.S. dollars per year. The economic criteria and minimum field-size targets for the two groups are identical. The two groups did not know that they were competing with each other.

The results of the first five years of competition are amazing. Figure 4 shows that the small synergistic team found about 2.8 times the reserves at about half the finding costs. The development costs are significantly lower for the smaller group.

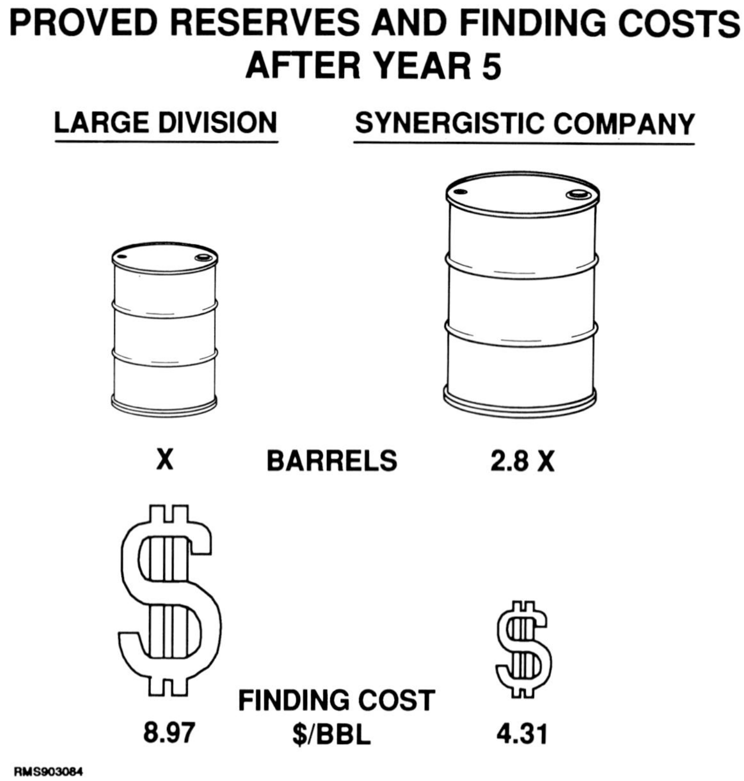

Fig. 4. Comparison of the proved reserves found and the finding cost for the large division and the small company for five years.

Based on the findings of the first three years of the "experiment", the parent company reorganized the larger group more along the lines of the synergistic group. It eliminated one level of management and increased the monetary authority and pushed decision-making to lower management levels. After year five, the management of the larger organization eliminated another level of management and technical supervision in order to approach a more streamlined organization. The Board of Directors was convinced that fewer management levels and increase of authority to lower levels in the larger organization resulted in improved efficiency and profit. Some of the keys to the increased efficiency and profitability of the synergistic company and the "streamlined" large division are (1) fewer meetings and therefore, much more time for work, (2) the delegation of financial responsibility downward, and (3) excellent communication of objectives, goals and results between management and the staff.

For synergistic teams to be a significant force in an organization and for it to be truly effective in exploration and production, all participants must be open to new ideas and both the individuals and management must want and be able to work and share. There are barriers to synergism and the synergistic team.

Some participants have a personality that is not suitable, others lack good communication skills - an essential ingredient, and some fear that by working in a team that their contributions will not be fully recognized. Some managers lack the understanding of the concept or do not wish to share the glory, defeats or problems with other managers. Some organizational structures inhibit synergistic decision-making.

The use of synergistic teams is for both large and small companies. These teams have an easier chance for success in smaller organizations. In larger companies, synergism is and can be employed, although bureaucracy and size often tend to dilute and hinder it, sometimes to the point of ineffectiveness.

Reference Cited

Sneider, R.M., "The Synergistic Team Approach...Its Value to Petroleum Exploration and Development in the Next Decade", Keynote paper presented at the Eighth Australian Geological Convention, Adelaide, Australia, Feb. 16-21, 1986, p. 16-19.

Teamwork and Technology—A Competitive Advantage

John F. Greene

Recent changes in the traditional organization of many oil and gas companies has been caused by several factors. In an era of fluctuating oil and gas prices, an increased emphasis on profit margins, finding and development costs, producing expenses, and optimized exploitation of producing properties have been of major importance. Utilization of integrated data base systems and networked, interactive workstations has provided efficient access to the data needed to effectively problem solve and evaluate projects incorporating geologic, geophysical, production, land and economic inputs.

The utilization of computer enhanced team technology is playing an increasingly important function in providing the efficiencies, quality, risk-reduction and results needed for success in the 1990s. An effective blend of hardware, software, teamwork with skilled professionals and a task focused work plan has proven to be a dynamic combination. The impact and results of this methodology may be documented in many exploration, development, and reservoir management examples.

Removal of traditional organization structures and boundaries facilitate the development of networks that enhance creativity and innovation. Hierarchy, command, and control are replaced by coordination, communication, and co-operation.

The full potential of computer enhanced team technology has yet to be fully realized. Rapid development and growth of this mode of operation has caused ramifications to the management of our human resources. Our industries' existing method of training, career pathing, recognition and compensation require a thorough reevaluation.

A revised definition of "career success" must be established as layers of management are stripped away and interdisciplinary teams proliferate. The new image of success must combine job satisfaction, winning performance, career and personal growth and peer recognition. Recognition of contributions and results on an individual, team and corporate basis is a necessity. Effective use of a performance driven compensation and reward system is key to a strong work ethic and a high energy, high morale environment.

Successful employment of task focused teamwork and technology can provide a company a competitive advantage in the oil and gas industry of the future.

Teamwork

John A. Masters

We began to build exploration teams about 12 years ago. Others are trying to do it now so Bob asked me to describe our procedure to you. He favors my approach because he considers it constructive damage to the status quo. All of you know how teams work technically. Maybe I can tell you something about the sociology. I tell it from the viewpoint of an explorationist—but it will work for us all.

Frankly much of what we do we don't know how to describe. You people blessed with a large family, how actually does it operate? Can you predict and control the response of every member of the family to every new situation which will arise during the week? You can't—and you don't—because you respect each other. Each member is a precious and unique human being. Well, bear with me and see how this analogy relates to life at Canadian Hunter.

The fundamental Concept (Figure 1) is that most worthwhile problems today are complex mixtures of engineering-geology-mathematics-geophysics with variable additions of finance, gas marketing, drilling costs, etc.

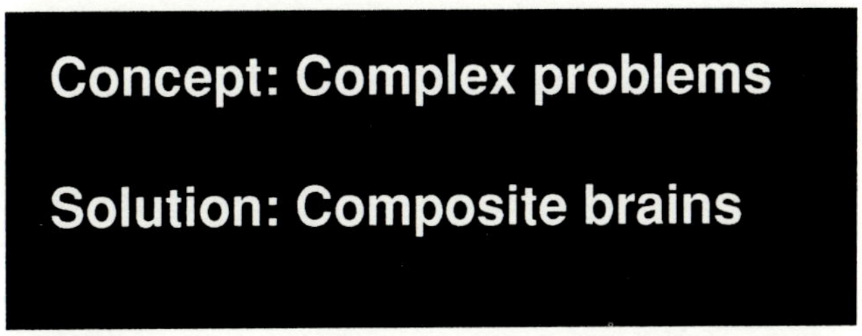

Figure 1

The fundamental Solution (Figure 1) is not to wait for a genius to come along, but to build a genius to solve the problem; In short, a team composed of people with the necessary specialty skills. I call it a form of artificial intelligence.

Subclause A (Figure 2) of the Solution is: Convince everyone that they are there to find oil by solving problems.

Subclause B (Figure 2) is: Give everyone on the team freedom to contribute and communicate.

Subclause C (Figure 2) is: Disband the team when the problem is solved.

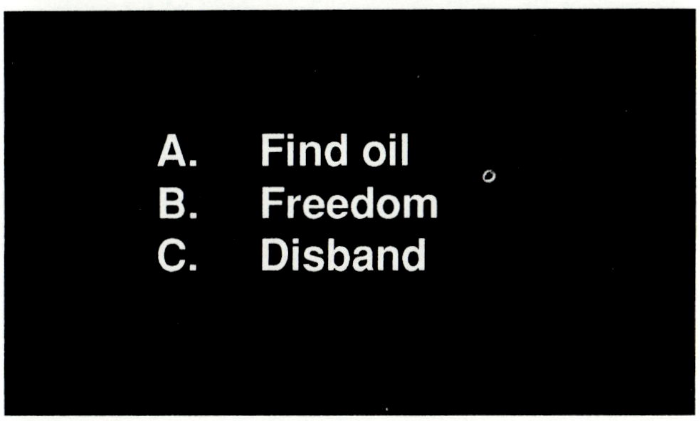

Figure 2

I want to elaborate on the Subclauses.

Subclause A: We come together to find oil. We do it by solving problems. If you want to join us in that demanding, challenging, stimulating quest—welcome aboard. But leave the political warfare behind you. Look, we all know each other here (Figure 3). We all work together. We all trust each other. We have all found oil fields together. You won't get anywhere if you don't hunker down and put your shoulder to our problems. If you're a drilling engineer you're not here to make footage, you're here to drill a hole that can be properly evaluated and optimally completed. Your job isn't done until you've satisfied geologists, electric log analysts, DST specialists, petrographers, frac experts. You're a drilling engineer second. First, you're an oil finder. Do you get the idea? We're all in this together.

Figure 3

Subclause B: freedom to contribute and communicate. Everyone, techs, draftsmen, et al. Here is where your corporate culture starts to be tested. Most of your companies are too authoritarian. People occupy slots and do what they're told. Do you have any idea of the cumulative brain power that is available in your workers, your troops?

People have a way of rising to what you expect of them. Select them very carefully, give them good training, tell them they are good, tell them you are counting on them, then get out of the way. Support them, encourage them, let them experiment and fail, and try again. Above all, TRUST them. Tell them you TRUST them. Brag about how you TRUST them. Go amongst them and show how you TRUST them. All of them. Everyone participates, everyone suggests, communicates, contributes. At first, you will be distressed to find that this spills over. Pretty soon everyone has an opinion about the kind of coffee you have in the kitchen, about working hours, about how their desk is positioned for Chrissake. Do not despair. This is the price you pay for giving people freedom to breath. Some have never experienced it in their working lives and it makes them a little giddy. Don't worry about it. It's just the blood flowing.

Subclause C: Disband the team when it has

accomplished its purpose. Nothing is so deadening as a group that is pretending usefulness. A poison envelopes the members and seeps out into the rest of the organization. It becomes the seed of discontent. Besides, it's dumb. You need those people working on real problems.

I can't take the time to explain all the science involved. I only want to leave the false impression that I understand all the specialty knowledge areas which the ordinary geologist doesn't know but someone has to know in order to make a stratigraphic trap play in obscure, fine grained rocks. Someone has to know structure and stratigraphy, sedimentation, clay mineralogy, geophysics, electric log analysis, DST interpretation, drilling, fracing, and on and on.

Well, a good many years ago we decided it was kind of hopeless to expect one person to do all of this competently. The big answer did not occur to us instantaneously. We got at it by steps. First, we realized that few geologists could do state of the art electric log analysis (Figure 4). We hired a log analyst to do all the critical measurements for our geologists. So, we had four geologists working with one log analyst. Slowly, we began to realize that we actually had four two-man teams in action. So we had expanded our force from 5 to 8. Next thing we did was add a petrographer,to integrate his observations with the log

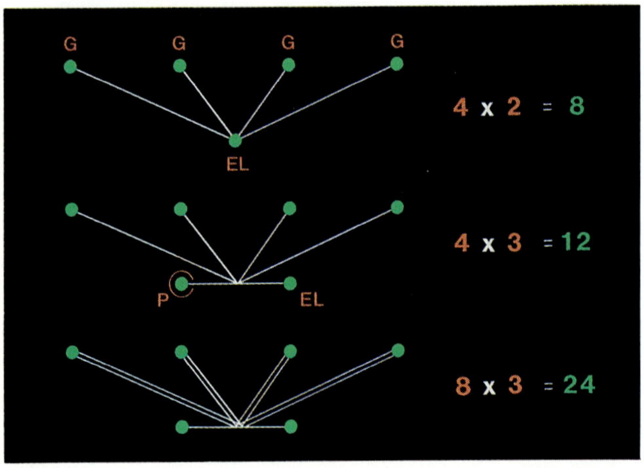

Figure 4

analysis. Then we had four three-man teams. Eventually, the four geologists each had two projects they were working on so we realized that we had eight three-man teams. A staff of 6 people had now divided and expanded itself so that it had become equivalent to 24 people who were working so closely and in such interconnected ways that they were very effectively cross-fertilizing each other with the results of eight separate projects. It looks a little like card tricks but this is a real effect.

Our two or three man partner idea has evolved into a concept that teams of experts should supply all the critical measurements. In exploration our geologists' job should be to study the regional geology, select the depositional or structural model that best fits the data, then insert the measured data into the model and solve for location of the oil or gas. The geologist is the conceptualizer in the first place and then the integrator. It is he who recognizes misties of data, conflicts, and weak spots. He gets the appropriate specialists together and they resolve the problem. The geologist is usually the critical team leader. In the final analysis, his guidance, his lacing together of the many strands of information is what forms the prospect.

The first team was rather neat and carefully formed (Figure 5). The next one was a little different (Figure 6). The third one (Figure 7) became a bit of a blob with some new arrangements out on the east end. Team D (Figure 8) started to experiment with different arrangements in the S.W. corner. Team E (Figure 9) said we're working on something so damn important we want to be all by ourselves; we need engineer and geologist co-leaders, and we need entirely different communication arrangements. No one tried to stop them. It was their problem. The point of these slides is to describe diversity. There is no set way to form these new units. The slides also show how we solve problems. No big

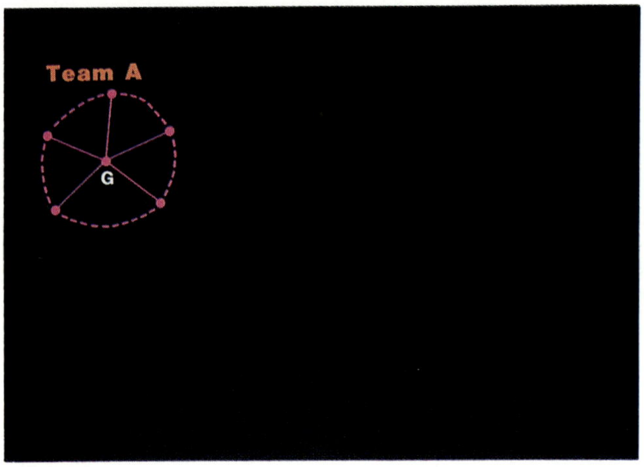

Figure 5

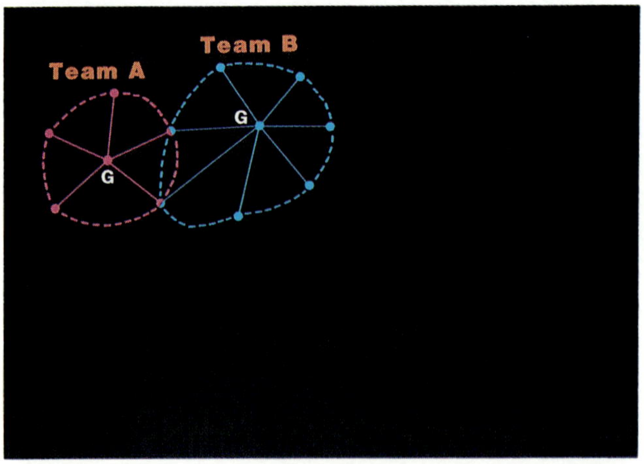

Figure 6

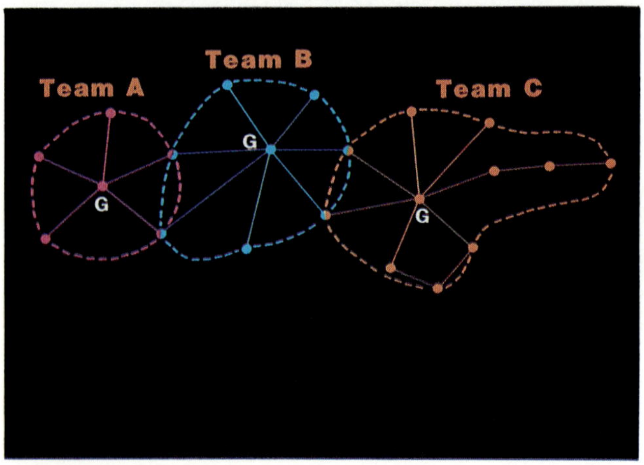

Figure 7

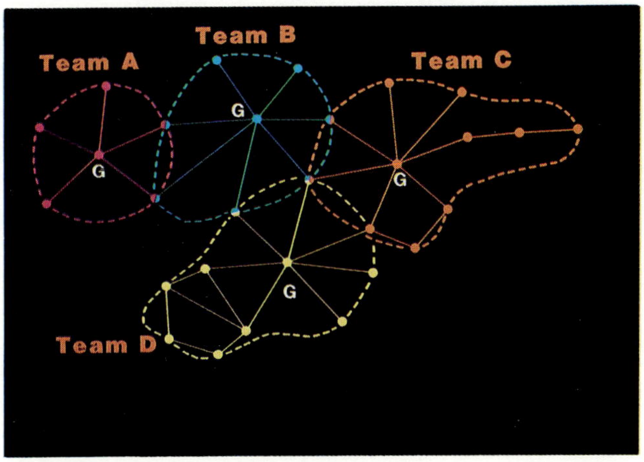

Figure 8

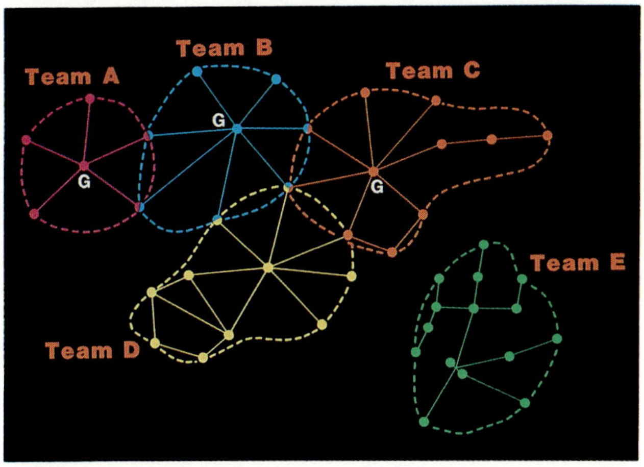

Figure 9

overall corporate plan. Just nibble away at it. Get everyone involved. Let it be theirs.

We believe our high powered teams deliver a better technical product, solve complex problems which defy individual geologists, generate unusual confidence both in the geologists and management, and create a powerful sense of ownership and motivation among the technical specialists. Furthermore, the team is a support system which allows the inevitably necessary leaps of faith in any exploration project to be performed with even more audacity than the single individual can do.

Of course, if it's an engineering problem you have an engineer in the middle. The team system applies throughout the company.

OK, so it doesn't look like an organization chart. No, and it doesn't work like the U.S. Army—more like a pirate ship. In fact, you can't figure out how it works. I can't either. But I know what is crucial. You have to have really good people and you have to let them alone. Turn a planner loose in here, or a Vice President, and in one week they would re-organize Team E and Team D because they didn't look right. In my company we think the guys doing the job know best.

Against all logic, contrary to your best judgement, this works. Like a bumble bee which aeronautical engineers say can't fly.

Look at Figure 9 again. Do you notice there is no hierarchy indicated? That is meaningful. Everyone on this diagram has an important part to play. There is an equality of effort indicated here. Everyone has ownership in these teams. The product has been created by all.

Once you get in the groove, there's something visceral about it. It strums on ancient strings.

I suggest that you are looking at a crude representation of perhaps the most important exploration breakthrough of the next decade. These are integrated technological teams which operate In fluid, flexible, changeable new systems, insulated from the old hierarchical command structures. They provide state of the art scientific and engineering measurements into the center, to the integrator geologist who isn't trained sufficiently to make those specialized measurements. We have proven oil finders only 2 or 3 years out of school. In normal companies, It takes 10 years.

What Canadian Hunter has discovered how to do (Figure 10) is build artificial intelligence with all these technologies making up different lobes of one single, custom-built brain. It sits on a base of fundamental support systems. The brain holds together until the problem is solved and then it self destructs. No residue of managers and supervisors and private parking stalls is left over. Everything goes back into the void but re-appears in new forms, attached to different centers.

This composite brain is necessary to handle many of

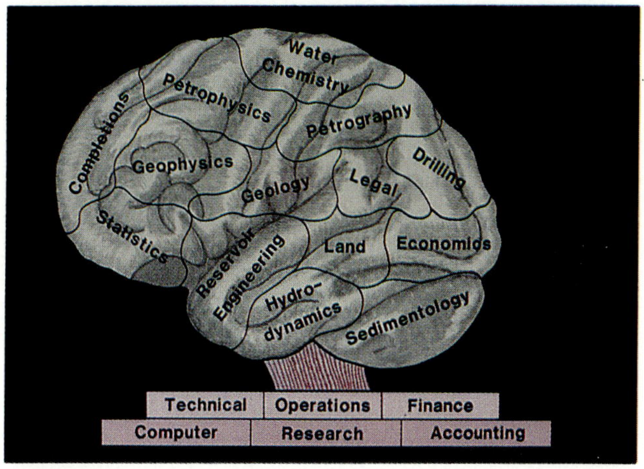

Figure 10

today's complex problems. It will beat almost any single individual brain you can put up against it. The astonishing thing is the parts are simply normal, smart, competent people who have become state-of-the-art experts in one particular specialty. But you can mold the individual, isolated, single gifts of good people with less than genius into a product of genius. Remember, however, a team cannot function unless every person on it is competent. Otherwise, some of the input cannot be trusted. A weak link destroys the chain.

I asked our managers to make a list of all the teams operating at present within their jurisdiction. To simplify, I showed the managers of the various districts and a square for each team working under him and the number of people in each team (Figure 11). Nick has about 30 people working on 18 different problems in various configurations. The 30 people are spread among so many teams, contributing their knowledge to so many different problems, they represent the equivalent of 194 specialists at work under Nick's general guidance. Understand, these 18 teams are led very adequately by interior team leaders. These are the most critical people on the teams, the all-important quarterbacks. Nick doesn't have to lead and integrate the many separate teams. The team leaders have the authority to make most of the decisions, they keep Nick informed, they go to him for advice. Nick acts as a trouble

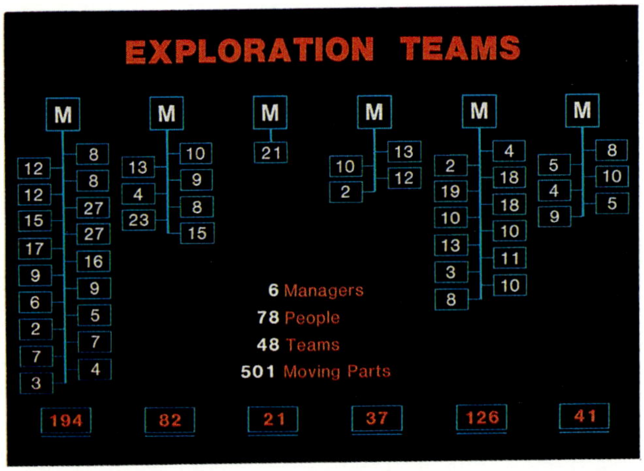

Figure 11

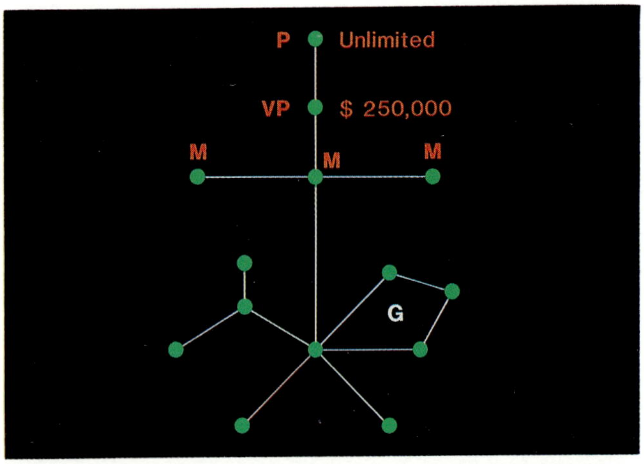

Figure 12

shooter. If there is friction between teams, Nick solves it, not with the heavy hand of authority, but the healing hand of a friend. In total this chart shows 84 people which include 6 exploration managers, 30 geologists. 48 teams and 501 moving parts. That seems to deny some basic physical principle; how do you get 501 worker units out of 84 workers? But it goes a long way toward explaining how Canadian Hunter, about the 20th sized company in terms of numbers of people, carries on the largest exploration program in the entire Canadian industry.

You might sense that being a Canadian Hunter manager is a different kettle of fish than managing in a conventional company. First of all you don't exercise authority in the same way. You don't give orders. You help other people choose the right way. And when it all works, you don't get any glory for it. You make sure the team gets it all. Around Hunter, we say if you need uniforms and titles and gold braid, join the Shriners.

Decision making is pretty streamlined in our company (Figure 12). A favorable recommendation goes to the Manager, then to the Vice President who has $250,000 authority for land and seismic. Larger recommendations and all well proposals go to the President. All homework is thoroughly prepared and decisions rarely require more than one meeting. The President's authority is limited only by total budget and good sense.

What I want to leave you with is not a simple "10 ways to become a new exploration company". I want to let dawn on you a new conception of how to enlist the brain power of your total human resource into the high tech search for scientifically complex stratigraphic traps.

That's about the long and short of it. When you've been doing it pretty successfully for ten years it seems fairly simple. But I know from the questions I am asked that the ordinary corporate wheel dog, trained to the harness and the voice signal, can't clearly comprehend how a reservoir engineer is going to work harmoniously with a geophysicist. I've heard a number of comments like that here. It's kind of like listening to prisoners talking at a rest period.

Here's a Lee Iacocca story. An executive came to him with a typical explanation: "He's a real smart guy, Lee. His only problem is he can't work with people." Iacocca said, "Well, that's too bad because all we've got around here are people. Get rid of him."

If it's important enough to you to build harmonious teams to solve your tough problems, I commend to you this simple personnel policy. It may lack subtlety and nuance, but it doesn't cause much confusion.

Everyone wants to know what our organization chart looks like. It looks like this (Figure 13). Same as everyone else's. We're separated into the conventional oil company divisions with lines connecting the boxes. One difference is we don't pay any attention to the lines, and we tell everyone very specifically, "Go straight to whoever has the information, don't waste time In channels." However, we realize that an oil company has many standard operational problems which are well met by functional divisions. We get most of our wells drilled by our drilling department. No fuss about it. But we have a company ethic and management support for throwing in any number of specialists to make a company drilling brain to work on any particular drilling problem that starts to bite us.

You ask, "Who decides the priorities?" A good answer is, "the appropriate people." The less fuss there is about it, the better organization you have. I've never been called on to settle a dispute. Our guys don't dispute. They just make fair and sensible decisions. If they don't they face a growing back pressure of dis-satisfaction from the rest of the ... "family".

Organizational experts will be horrified at the potential for conflicts and misunderstanding. My rejoinder is: This operation doesn't work on organization. It works on human and social principles which are much more subtle but also more effective than organizational rules.

Each company's system and procedures should be different. If I could explain and chart every interaction at Hunter, which I can't, it wouldn't do you any good anyway. Your company is different.

The biggest wrench in the works is someone who wants to plan everything. Authority. Rules. "I know better than everyone. You guys just do what you're told."

Now I want to talk to you about how we built a different kind of company at Canadian Hunter.

First of all, we started in 1973, not 100 years ago, with just two of us. Very early, we decided we would hire only the best. There wasn't anything particularly noteworthy about that in regard to our professionals but we decided to have the best secretaries and mail clerks, and draftsmen too. We wanted a company of pervasive excellence. We decided on this a long time before we ever heard of Tom Peters. Now, you don't give that job to a Human Resources department. You give it to your department managers and you tell them that's their top priority responsibility: hiring the best people. Their second priority is firing their mistakes. If they do those two parts right the rest of their job will be easy. Everyone who comes into this company is initially selected through contacts as a promising candidate:

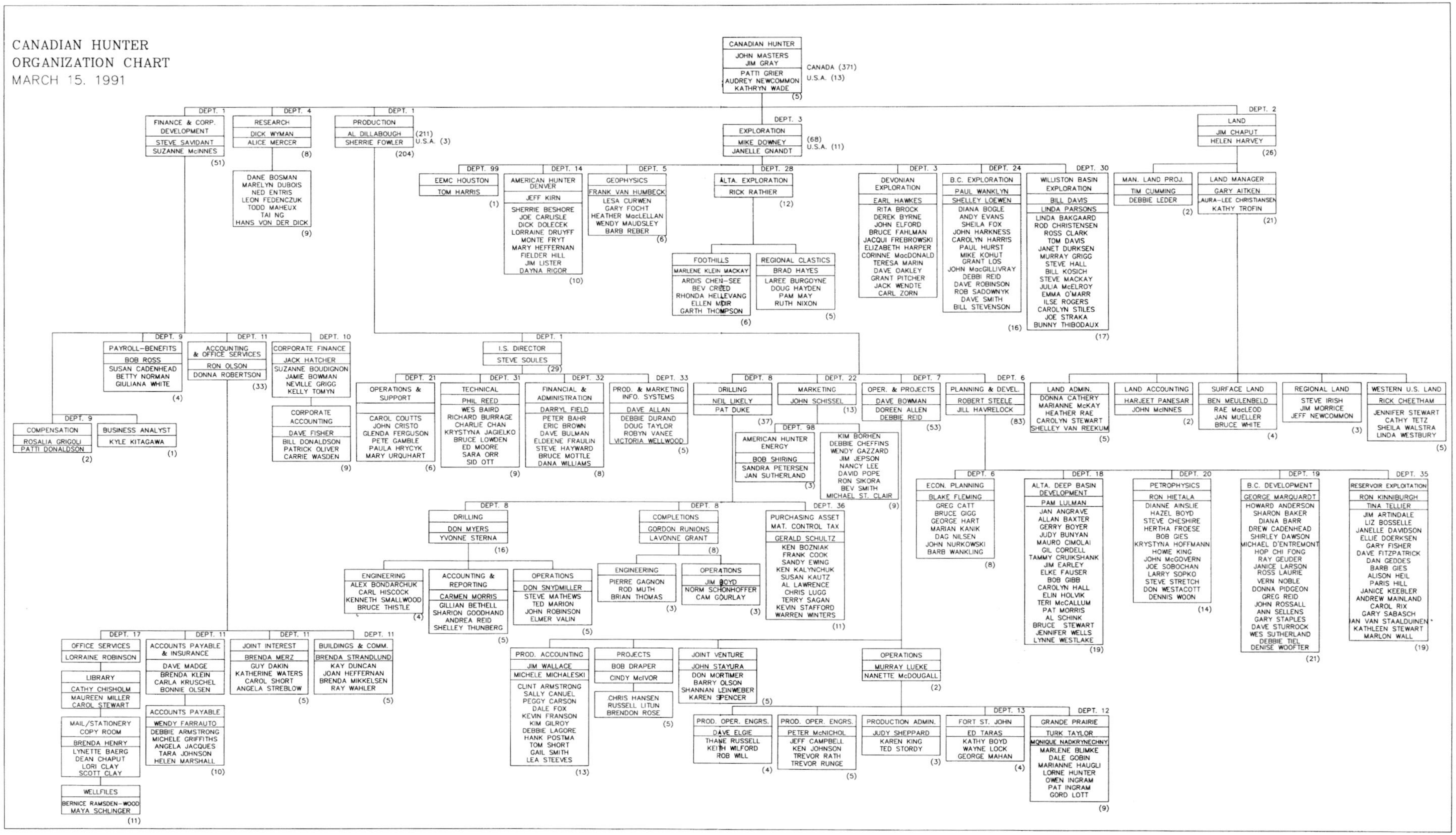
CANADIAN HUNTER
ORGANIZATION CHART
MARCH 15, 1991
CANADIAN HUNTER
JOHN MASTERS
JIM GRAY
PATTI GRIER
AUDREY NEWCOMMON
KATHRYN WADE
(5)
CANADA (371)
U.S.A. (13)
DEPT. 1
FINANCE & CORP. DEVELOPMENT
STEVE SAVIDANT
SUZANNE McINNES
(51)
DEPT. 4
RESEARCH
DICK WYMAN
ALICE MERCER
(8)
DANE BOSMAN
MARELYN DUBOIS
NED ENTRIS
LEON FEDENCZUK
TODD MAHEUX
TAI NG
HANS VON DER DICK
(9)
DEPT. 1
PRODUCTION
AL DILLABOUGH
SHERRIE FOWLER
(211)
U.S.A. (3)
(204)
DEPT. 3
EXPLORATION
MIKE DOWNEY
JANELLE GNANDT
(68)
U.S.A. (11)
DEPT. 2
LAND
JIM CHAPUT
HELEN HARVEY
(26)
DEPT. 99
EEMC HOUSTON
TOM HARRIS
(1)
DEPT. 14
AMERICAN HUNTER DENVER
JEFF KIRN
SHERRIE BESHORE
JOE CARLISLE
DICK DOLECEK
LORRAINE DRUYFF
MONTE FRYT
MARY HEFFERNAN
FIELDER HILL
JIM LISTER
DAYNA RIGOR
(10)
DEPT. 5
GEOPHYSICS
FRANK VAN HUMBECK
LESA CURWEN
GARY FOCHT
HEATHER MacLELLAN
WENDY MAUDSLEY
BARB REBER
(6)
DEPT. 28
ALTA. EXPLORATION
RICK RATHIER
(12)
FOOTHILLS
MARLENE KLEIN MACKAY
ARDIS CHEN-SEE
BEV CREED
RHONDA HELLEVANG
ELLEN MOIR
GARTH THOMPSON
(6)
REGIONAL CLASTICS
BRAD HAYES
LAREE BURGOYNE
DOUG HAYDEN
PAM MAY
RUTH NIXON
(5)
DEPT. 3
DEVONIAN EXPLORATION
EARL HAWKES
RITA BROCK
DEREK BYRNE
JOHN ELFORD
BRUCE FAHLMAN
JACQUI FREBROWSKI
ELIZABETH HARPER
CORINNE MacDONALD
TERESA MARIN
DAVE OAKLEY
GRANT PITCHER
JACK WENDTE
CARL ZORN
DEPT. 24
B.C. EXPLORATION
PAUL WANKLYN
SHELLEY LOEWEN
DIANA BOGLE
ANDY EVANS
SHEILA FOX
JOHN HARKNESS
CAROLYN HARRIS
PAUL HURST
MIKE KOHUT
GRANT LOS
JOHN MacGILLIVRAY
DEBBI REID
DAVE ROBINSON
ROB SADOWNYK
DAVE SMITH
BILL STEVENSON
(16)
DEPT. 30
WILLISTON BASIN EXPLORATION
BILL DAVIS
LINDA PARSONS
LINDA BAKGAARD
ROD CHRISTENSEN
ROSS CLARK
TOM DAVIS
JANET DURKSEN
MURRAY GRIGG
STEVE HALL
BILL KOSICH
STEVE MACKAY
JULIA McELROY
EMMA O'MARR
ILSE ROGERS
CAROLYN STILES
JOE STRAKA
BUNNY THIBODAUX
(17)
MAN. LAND PROJ.
TIM CUMMING
DEBBIE LEDER
(2)
LAND MANAGER
GARY AITKEN
LAURA-LEE CHRISTIANSEN
KATHY TROFIN
(21)
DEPT. 9
PAYROLL-BENEFITS
BOB ROSS
SUSAN CADENHEAD
BETTY NORMAN
GIULIANA WHITE
(4)
DEPT. 11
ACCOUNTING & OFFICE SERVICES
RON OLSON
DONNA ROBERTSON
(33)
DEPT. 10
CORPORATE FINANCE
JACK HATCHER
SUZANNE BOUDIGNON
JAMIE BOWMAN
NEVILLE GRIGG
KELLY TOMYN
CORPORATE ACCOUNTING
DAVE FISHER
BILL DONALDSON
PATRICK OLIVER
CARRIE WASDEN
(9)
DEPT. 9
COMPENSATION
ROSALIA GRIGOLI
PATTI DONALDSON
(2)
BUSINESS ANALYST
KYLE KITAGAWA
(1)
DEPT. 1
I.S. DIRECTOR
STEVE SOULES
(29)
DEPT. 21
OPERATIONS & SUPPORT
CAROL COUTTS
JOHN CRISTO
GLENDA FERGUSON
PETE GAMBLE
PAULA HRYCYK
MARY URQUHART
(6)
DEPT. 31
TECHNICAL
PHIL REED
WES BAIRD
RICHARD BURRAGE
CHARLIE CHAN
KRYSTYNA JAGIELKO
BRUCE LOWDEN
ED MOORE
SARA ORR
SID OTT
(9)
DEPT. 32
FINANCIAL & ADMINISTRATION
DARRYL FIELD
PETER BAHR
ERIC BROWN
DAVE BULMAN
ELDEENE FRAULIN
STEVE HAYWARD
BRUCE MOTTLE
DANA WILLIAMS
(8)
DEPT. 33
PROD. & MARKETING INFO. SYSTEMS
DAVE ALLAN
DEBBIE DURAND
DOUG TAYLOR
ROBYN VANEE
VICTORIA WELLWOOD
(5)
DEPT. 8
DRILLING
NEIL LIKELY
PAT DUKE
(37)
DEPT. 22
MARKETING
JOHN SCHISSEL
(13)
DEPT. 98
AMERICAN HUNTER ENERGY
BOB SHIRING
SANDRA PETERSEN
JAN SUTHERLAND
(3)
KIM BORHEN
DEBBIE CHEFFINS
WENDY GAZZARD
JIM JEPSON
NANCY LEE
DAVID POPE
RON SIKORA
BEV SMITH
MICHAEL ST. CLAIR
(9)
DEPT. 7
OPER. & PROJECTS
DAVE BOWMAN
DOREEN ALLEN
DEBBIE REID
(53)
DEPT. 6
PLANNING & DEVEL.
ROBERT STEELE
JILL HAVELOCK
(83)
LAND ADMIN.
DONNA CATHERY
MARIANNE McKAY
HEATHER RAE
CAROLYN STEWART
SHELLEY VAN REEKUM
(5)
LAND ACCOUNTING
HARJEET PANESAR
JOHN McINNES
(2)
SURFACE LAND
BEN MEULENBELD
RAE MacLEOD
JAN MUELLER
BRUCE WHITE
(4)
REGIONAL LAND
STEVE IRISH
JIM MORRICE
JEFF NEWCOMMON
(3)
WESTERN U.S. LAND
RICK CHEETHAM
JENNIFER STEWART
CATHY TETZ
SHEILA WALSTRA
LINDA WESTBURY
(5)
DEPT. 6
ECON. PLANNING
BLAKE FLEMING
GREG CATT
BRUCE GIGG
GEORGE HART
MARIAN KANIK
DAG NILSEN
JOHN NURKOWSKI
BARB WANKLING
(8)
DEPT. 18
ALTA. DEEP BASIN DEVELOPMENT
PAM LULMAN
JAN ANGRAVE
ALLAN BAXTER
GERRY BOYER
JUDY BUNYAN
MAURO CIMOLAI
GIL CORDELL
TAMMY CRUIKSHANK
JIM EARLEY
ELKE FAUSER
BOB GIBB
CAROLYN HALL
ELIN HOLVIK
TERI McCALLUM
PAT MORRIS
AL SCHINK
BRUCE STEWART
JENNIFER WELLS
LYNNE WESTLAKE
(19)
DEPT. 20
PETROPHYSICS
RON HIETALA
DIANNE AINSLIE
HAZEL BOYD
STEVE CHESHIRE
HERTHA FROESE
BOB GIES
KRYSTYNA HOFFMANN
HOWIE KING
JOHN McGOVERN
JOE SOBOCHAN
LARRY SOPKO
STEVE STRETCH
DON WESTACOTT
DENNIS WOON
(14)
DEPT. 19
B.C. DEVELOPMENT
GEORGE MARQUARDT
HOWARD ANDERSON
SHARON BAKER
DIANA BARR
DREW CADENHEAD
SHIRLEY DAWSON
MICHAEL D'ENTREMONT
HOP CHI FONG
RAY GEUDER
JANICE LARSON
ROSS LAURIE
VERN NOBLE
DONNA PIDGEON
GREG REID
JOHN ROSSALL
ANN SELLENS
GARY STAPLES
DAVE STURROCK
WES SUTHERLAND
DEBBIE TIEL
DENISE WOOFTER
(21)
DEPT. 35
RESERVOIR EXPLOITATION
RON KINNIBURGH
TINA TELLIER
JIM ARTINDALE
LIZ BOSSELLE
JANELLE DAVIDSON
ELLIE DOERKSEN
GARY FISHER
DAVE FITZPATRICK
DAN GEDDES
BARB GIES
ALISON HEIL
PARIS HILL
JANICE KEEBLER
ANDREW MAINLAND
CAROL RIX
GARY SABASCH
JAN VAN STAALDUINEN
KATHLEEN STEWART
MARLON WALL
(19)
DEPT. 8
DRILLING
DON MYERS
YVONNE STERNA
(16)
DEPT. 8
COMPLETIONS
GORDON RUNIONS
LAVONNE GRANT
(8)
DEPT. 36
PURCHASING ASSET MAT. CONTROL TAX
GERALD SCHULTZ
KEN BOZNIAK
FRANK COOK
SANDY EWING
KEN KALYNCHUK
SUSAN KAUTZ
AL LAWRENCE
CHRIS LUGG
TERRY SAGAN
KEVIN STAFFORD
WARREN WINTERS
(11)
ENGINEERING
ALEX BONDARCHUK
CARL HISCOCK
KENNETH SMALLWOOD
BRUCE THISTLE
(4)
ACCOUNTING & REPORTING
CARMEN MORRIS
GILLIAN BETHELL
SHARION GOODHAND
ANDREA REID
SHELLEY THUNBERG
(5)
OPERATIONS
DON SNYDMILLER
STEVE MATHEWS
TED MARION
JOHN ROBINSON
ELMER VALIN
(5)
ENGINEERING
PIERRE GAGNON
ROD MUTH
BRIAN THOMAS
(3)
OPERATIONS
JIM BOYD
NORM SCHONHOFFER
CAM GOURLAY
(3)
DEPT. 17
OFFICE SERVICES
LORRAINE ROBINSON
LIBRARY
CATHY CHISHOLM
MAUREEN MILLER
CAROL STEWART
MAIL/STATIONERY
COPY ROOM
BRENDA HENRY
LYNETTE BAERG
DEAN CHAPUT
LORI CLAY
SCOTT CLAY
WELLFILES
BERNICE RAMSDEN-WOOD
MAYA SCHLINGER
(11)
DEPT. 11
ACCOUNTS PAYABLE & INSURANCE
DAVE MADGE
BRENDA KLEIN
CARLA KRUSCHEL
BONNIE OLSEN
ACCOUNTS PAYABLE
WENDY FARRAUTO
DEBBIE ARMSTRONG
MICHELE GRIFFITHS
ANGELA JACQUES
TARA JOHNSON
HELEN MARSHALL
(10)
DEPT. 11
JOINT INTEREST
BRENDA MERZ
GUY DAKIN
KATHERINE WATERS
CAROL SHORT
ANGELA STREBLOW
(5)
DEPT. 11
BUILDINGS & COMM.
BRENDA STRANDLUND
KAY DUNCAN
JOAN HEFFERNAN
BRENDA MIKKELSEN
RAY WAHLER
(5)
PROD. ACCOUNTING
JIM WALLACE
MICHELE MICHALESKI
CLINT ARMSTRONG
SALLY CANUEL
PEGGY CARSON
DALE FOX
KEVIN FRANSON
KIM GILROY
DEBBIE LAGORE
HANK POSTMA
TOM SHORT
GAIL SMITH
LEA STEEVES
(13)
PROJECTS
BOB DRAPER
CINDY McIVOR
CHRIS HANSEN
RUSSELL LITUN
BRENDON ROSE
(5)
JOINT VENTURE
JOHN STAYURA
DON MORTIMER
BARRY OLSON
SHANNAN LEINWEBER
KAREN SPENCER
(5)
OPERATIONS
MURRAY LUEKE
NANETTE McDOUGALL
(2)
PROD. OPER. ENGRS.
DAVE ELGIE
THANE RUSSELL
KEITH WILFORD
ROB WILL
(4)
PROD. OPER. ENGRS.
PETER McNICHOL
JEFF CAMPBELL
KEN JOHNSON
TREVOR RATH
TREVOR RUNGE
(5)
PRODUCTION ADMIN.
JUDY SHEPPARD
KAREN KING
TED STORDY
(3)
DEPT. 13
FORT ST. JOHN
ED TARAS
KATHY BOYD
WAYNE LOCK
GEORGE MAHAN
(4)
DEPT. 12
GRANDE PRAIRIE
TURK TAYLOR
MONIQUE NADKRYNECHNY
MARLENE BLIMKE
DALE GOBIN
MARIANNE HAUGLI
LORNE HUNTER
OWEN INGRAM
PAT INGRAM
GORD LOTT
(9)

Figure 13

1. Recognition of teams
2. Commitment from top
3. People
4. Trust
5. Friendship
6. Contribute
7. Communicate
8. Achievement
9. Management style
10. Flexible organization
11. Participation
12. Pull together

Figure 14

we don't advertise. They take a personality test and are interviewed by 5 to 10 people over a period of weeks or even months. Professionals go through a two-hour seminar with a dozen of their peers. Many don't make it. One of our old hands said proudly, "It's harder to get into Canadian Hunter than into the F.B.I." Each selectee has a final interview with me. All 395 of our current employees were interviewed by me, are known to me by their first names, and comfortably call me "John."

So, we start with one of the fundamental pre-requisites for team building: consistent high-quality people. You can't have fluid, flexible, changing teams without consistent, interchangeable, high-quality players. This is where most big companies can't cut it. They have lost the discipline and, in many cases, even the legal right to cull out less than excellent people.

Next, the players have to trust each other. They have to be confident of the others' professional abilities and confident of their human values: honesty, trustworthiness, responsibility, persistence. All of these qualities have to be TRUSTED. You don't even want to have to think about the problem. It is too horrible: the work of 10 or 20 people for several months threatened by the weakness or inattention of one person. So we help these people to become friends, because we want these people to depend on each other. We have ski trips and barbecues, dances and birthdays, celebration lunches and even Santa Clauses. Canadian Hunter parties are famous. We have over 95 percent attendance at everything we do. Our parties are fun, clean, and inclusive. Everyone has a share. We invite the spouses. No one has to buy a ticket. Tell me a corporate practice that is more ineffective—buy a ticket to your own Christmas party!

I wonder how many companies make friendship an important part of their internal policy? How many managers would growl at it as artsy fartsy?

Team building is a social activity. It is tribal. It is attitude.

Let me review what you need: (Figure 14)

1. Recognition that many complex problems require the integrated brain power of numerous specialists. This is the "team."

2. Commitment and support from the top for the team concept. What Ron Hietala calls the "receptive environment."

3. Consistently high-quality people.

4. Trust, from the management and amongst themselves.

5. Friendship

6. Freedom to contribute.

7. Freedom to communicate.

8. Job satisfaction by achievement rather than title.

9. Supportive, enabling management style.

10. Flexible, fluid organization. No planners.

11. Once committed, anyone who doesn't participate is OUT. (a vital rule).

12. The incontrovertible ethic: WE ALL PULL TOGETHER.

13. Finally, the guts to stand behind this and make it work. This is not a system for pussycats.

Do you see that what I am talking about is not just team units? It is not new corporate organization so much as a new social contract within the corporation.

I don't have a final solution to all your woes. I think I do know a way that you can run your exploration-production business better and become more competitive in the increasingly complex technological search for oil that we're all involved in.

I suggest that you have a relatively simple choice to make in the new game on the invisible frontiers: either you choose not to play or you integrate into a team system so that you can play. The old style line organization cannot compete with the new teams.

How an Integrated Team Turned a 5 BCF Pool into a 1 TCF Field

Nick Wemyss
Murray Grigg
Janelle Davidson

Ladies and Gentlemen:

Today I'll be discussing the process of discovering the giant Ring/Border field in British Columbia Canada using an integrated team approach to exploration problem solving.

Our exploration team consisted of a geologist, petrographer, log analyst, hydrodynamicist. engineer, landman, and support staff.

I would like to tell you about this group of professionals who each had to work together to understand and support a major play which had previously been missed. This play was identified in an area of modest well control where many wells previously had been logged, tested, cored, completed and even production tested. At the end of this talk I hope to convince you that a leap of faith had occurred on behalf of this group, a leap of faith that no one person could have taken alone—but that this exploration team did—together and turned a 5.0 BCF pool into a 1.0 TCF field.

The field area straddles the border between Northern British Columbia and Alberta (Figure 1). At the time of drilling the closest pipeline tie-in was 25 miles away across swampy muskeg. Canadian Hunter was drilling a well for a deep Devonian gas target. The Devonian target did not materialize but an uphole zone was thought to have marginal potential.

The clock (Figure 2) will keep track of our progress. Time zero was the drilling of the deep Devonian dryhole. This is a story about that marginal uphole zone.

The standard information source, PIX data, described a few uneconomic gas shows. From the initial show map (Figure 3), the red dots are wells we thought could flow in excess of 500 MCF/D when completed; the blue dots show water recoveries on test; and the yellow dots show mud or tight tests. We had 32 control points in this formation. The PIX data listed a small number of cased, shallow gaswells as being from several Cretaceous zones notorious for poor, and discontinuous reservoirs. Cumulative reserves were 5.4 BCF from six individual wells which defined 5 gas pools. These pools are 3,000′ deep and were reported to have 50-60% initial water saturation. The scale of the map is one square mile for each little square.

In the face of this discouraging data our group pressed ahead. We have learned that even with our excellent database in Canada, there is no substitute for a rigorous analysis of the original raw data.

The clock (Figure 4) shows Suitable Environment.

First, let's review the geology.

Within the exploration area, the Cretaceous Gething Unconformably overlies the Triassic Montney Formation (Figure 5). Our analysis showed that the Triassic actually comprises the bulk of the potential reservoir. The Cretaceous/Triassic Montney cores displayed characteristics of three unique depositional environments.

Lowermost Triassic Montney sands were laid down in moderate to low energy lower shoreface or marine shelf environment (Figure 6). Massively bedded, very fine grained sands shaled out marineward to finely interbedded cross stratified sands and shales. According to core measurements the very fine grained and dolomite cemented reservoir retained porosity from 10 to 20% and permeability from 1 to 20 md. Montney sands were truncated and overlain by the Cretaceous Gething floodplain and valley fill units. Figure 7 shows the Cretaceous/Triassic unconformity with the tree root in the Cretaceous section.

Figures 8 and 9 show a Gething channel with high angle x-beds. These highly variable units contain reservoir and tight sands, coals, rip up clasts and shales. The Gething is disconformably overlain by the transgressive lag of the Cretaceous Bluesky.

See Figure 10. This glauconitic lag sand is occasionally of reservoir quality. Note the Bluesky lag showing pebble conglomerates and glauconite sand with carbonate debris. Geometries of the Cretaceous units are highly variable and undependable.

In contrast, the Triassic Montney shoreline is parallel to a well defined basin edge, and is lithologically consistent (Figure 11). This formation is also proven productive several hundred miles southeast in Alberta.

With the knowledge that we had a reservoir unit of good size and consistency along strike in the Triassic Montney our team met to closely examine other available data.

All drill cuttings (Figure 12) were examined for all the old wells in the play area. The very fine sands in this formation can make well site description of these reservoirs difficult.

On the left hand side of Figure 13 you see a conventional photo micrograph at lOX magnification.

Note that the reservoir appears as poor and tight. On the right side of the figure there is an SEM photo of the same well cuttings showing good porosity at magnification of 500 times. Observations such as this showed us that we should not trust historical wellsite reports. This well turned out to be one of the best in the field with an AOF over 4 MMSCFD.

At Canadian Hunter all plays are subject to routine drill cutting examinations. We have established unique criteria that relates the textures and fabrics of cuttings to permeability and other petrophysical properties. With the use of microscopes and SEM photos we routinely estimate porosity and permeability. The chart in Figure 14 depicts rock typing classifications as defined by Sneider/King/Hawkes/Davis in their JPT paper/Sept. 1983.

Another historical well, 14-32, (Figure 15) was drilled to a deep Devonian target through the Triassic/Montney section with water. The hole was completely washed out and the logs were of no value. Using drill cutting information, we saw potential and mapped the permeability trend. The next well was later drilled based on the drill cutting permeability map and flow tested at nearly 4 MMCF/D from the Triassic Montney sands.

In summary, the examination of cuttings from all the wells within the three hundred square mile area independently confirmed the permeable trend of the Triassic shoreline providing a focus for land acquisition (Figure 16). Note the 10–100 md permeability yellow fairway which became apparent by re-examination of this historical data. Drill cutting technology proved its worth on this prospect!

Next, let us review our study of fluid continuity (Figure 17). This is a study of pressures and fluids by our hydrodynamics group. A preliminary scan of PIX cards showed that 32 DSTs and 6 production tests had been run over the Triassic-Montney interval. This play was heavily drill stem tested by other exploration companies prior to discovery by Canadian Hunter. The hydrodynamic group addressed three key concerns for this play:

- What is the trap type?
- Is there reservoir continuity?
- What is the fluid distribution?

Interpretation of the pressures (through the construction of a pressure depth plot, Figure 18) revealed a significantly large gas column, approx. 280 feet (red line on pressure-depth plot), and two regional water systems (2 blue lines on pressure-depth plot). There is a pressure break between the two water systems which identified the potential for a large stratigraphic trap. Gas continuity was demonstrated over six consecutive townships by the pressure-depth relationships based on historical DST data. Two questions had been answered and only one question remained the question posed by several water recoveries on tests within the gas-bearing area. Further investigation showed these wells had been drilled with KCL muds which are notorious for filtrate invasion. In fact, these drill stem tests indicated significant formation damage. Water analysis of the recovered water proved to be high salinity KCl filtrate. In summary, these high salinity tests did not recover formation water.

Figure 19 indicates better completions calculated. Now, I would like to show you a case history DST at b-62-I. This pressure chart from a drill stem test (Figure 20) shows a low gas recovery at 226 MCF/D and high liquid recovery of 404 feet of gas cut KCl mud. The steep shut-in buildup indicates damage, and the slow building flow pressure indicates filtrate invasion flowback. Taking into consideration the effects of relative permeability, the drill stem tests were re-analyzed. On this slide a conventional DST interpretation was 2100 MCF/D after stimulation. Our re-analyses, including relative permeability effects of the flowing filtrate, yielded 3550 MCF/D. The well actually completed for 4164 MCF/D at a later date. This is a new technique as conventional DST analysis does not include relative permeability effects.

Refer to the initial show map in Figure 3. When we re-analyzed all the old DSTs with this new technique in mind, the initial 5 wells which were capable of greater than 500 MCF/D grew to 18 wells with this more sophisticated DST interpretation. Together with the removal of the KCl filtrate wells which had been confused with salt water, the new DST show map is shown in Figure 21.

Note the high occurrence of wells which we thought could complete for more than 500 MCF/D spread out consistently across a large area. This play depended on including a relative permeability effect ln DST analysis, enabling the projection of economic well completion rates.

Figure 22 shows the clock at 6 months.

THE PETROPHYSICAL EVALUATION

Finally one last piece of the puzzle remained. Petrophysically, high water saturations and low reserves were still a problem. Our log analyst worked closely with us.

Figure 23 shows the interbedded Triassic/Montney core. The geological model predicted there would be thin bed shale effects which would reduce formation resistivity. Our hydrodynamic and engineering studies indicated there should be no movable water within the gas cap. Armed with these assumptions the log analyst developed an aggressive interpretation which increased thin sand resistivity to accommodate shale beds and shoulder effects. The log analysts assumed lower water saturations and increased pay but did not have the data to prove it. Later in the play, we did confirm this assumption with modern log technology.

The two major differences between the old and new logs (Figure 24) is higher bed resolution from close tool spacing and "phasor" processing software to resolve the thin bed effects. Note that the water saturations have been reduced from 50% to 23.0%. The pay was improved from 18 feet to 31 feet. The reserves dramatically improved by 150% and crossed the economic threshold. We also checked mercury and brine capillarity testing on core showed these calculations of low water saturation to be correct.

We now had an integrated geological, petrophysical and engineering model which supported an economic gas play. The projected gas rates and reserves were now sufficient to justify a play. We were now 6 months into the play Figure 25).

Management quickly recognized the unique nature of the opportunity. All available government lands were posted for auction. These were purchased at a very low bid price before industry interest raised land costs. Next an aggressive acquisition program began: purchasing existing wells and mineral rights. We achieved a dominant position in the play and were then able to negotiate a 42,000 acre pooling which provided nearly complete coverage of the main potential tracts. (Figure 26). Within 10 months we had a position in 220,000 acres of land, and were the operator for the play.

We had not yet drilled our first well to test the new play concept—the time is 11 months.

Figure 27 is a photo of our first well into the new Triassic/Montney play—during completion testing at b-62-B. It tested 2 MMCF/D.

We have now drilled another 74 wells over an area of 250 sections with a 90% success rate. Presently our gross deliverability is about 110 MMCF/D with about 1000 BOPD. The field will be on stream in 1991 and will ultimately produce 1 TCF of gas. Finding costs calculate at less than 1¢/MCF.

The final map (Figure 28) shows the field as we see it today after development.

In review, we had a cohesive team which supplied answers to various parts of the puzzle. There was the geologist who recognized a continuous Triassic Montney unit which previously had been confused with the discontinuous Bluesky and Gething. There was the petrographer who calibrated to the previously tight rock and mapped subtle permeability variations using drill cutting data with a microscope backed up by SEM studies. There was the hydrodynamicist who verified gas continuity and identified the trap and the log analyst who adjusted pay calculations assuming thin bed shale effects. There was the engineer who analyzed the DSTs to predict expected completions and systematically modified the test analysis to account for invasion fluid flowback (i.e., relative permeability effects) to predict economic well rates. Lastly, there was the management team who recognized this opportunity and acted aggressively.

The competence and confidence of the team members provided the momentum required for this accomplishment. Each discipline depended on the other to make that leap of faith required to get across the holes in the data. All exploration plays require a leap of faith during early phases of technical development. A successful exploration team—both technical and management members—must come together to create an environment of creativity as well as technical excellence where leaps of faith can occur and are understood. The final two figures (Figures 29 and 30) show the land situation map before and after, as 5.0 BCF was turned into 1.5 TCF gas-In-place or 1.0 TCF of recoverable reserves in the course of one year. In retrospect, each individual member's interpretation could have pointed to this play, but only through the balanced integration of all disciplines was the speed and success of a major discovery achieved.

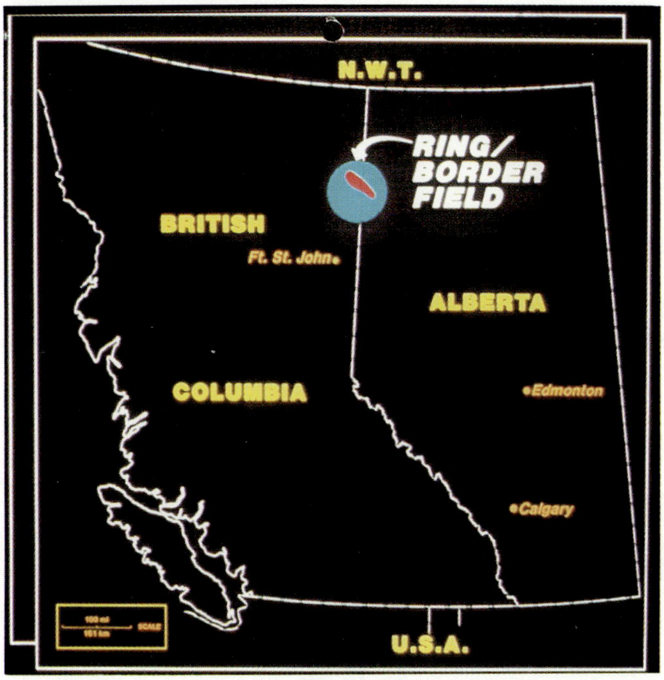

Figure 1

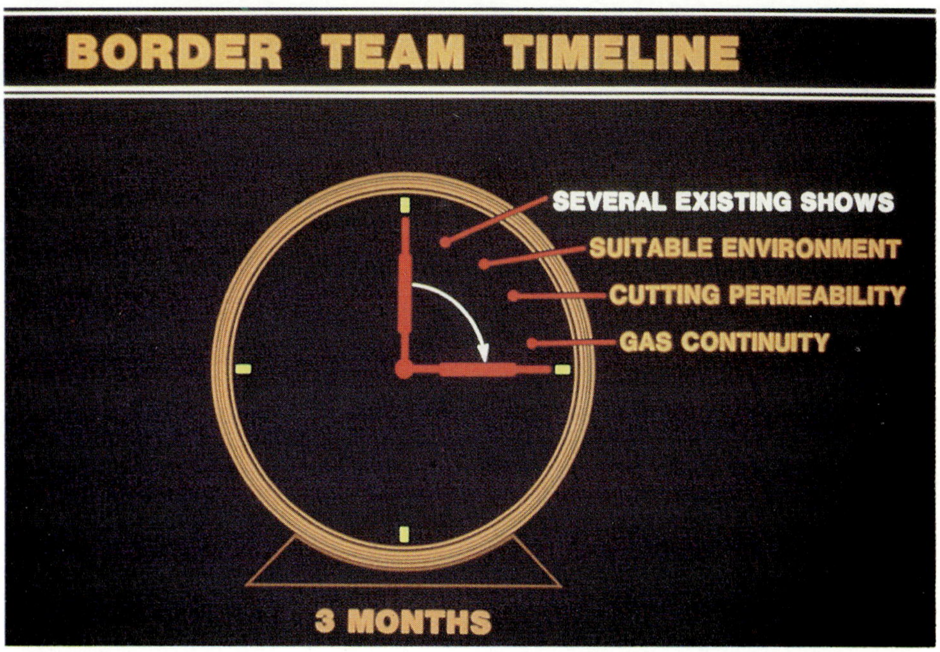

Figure 2

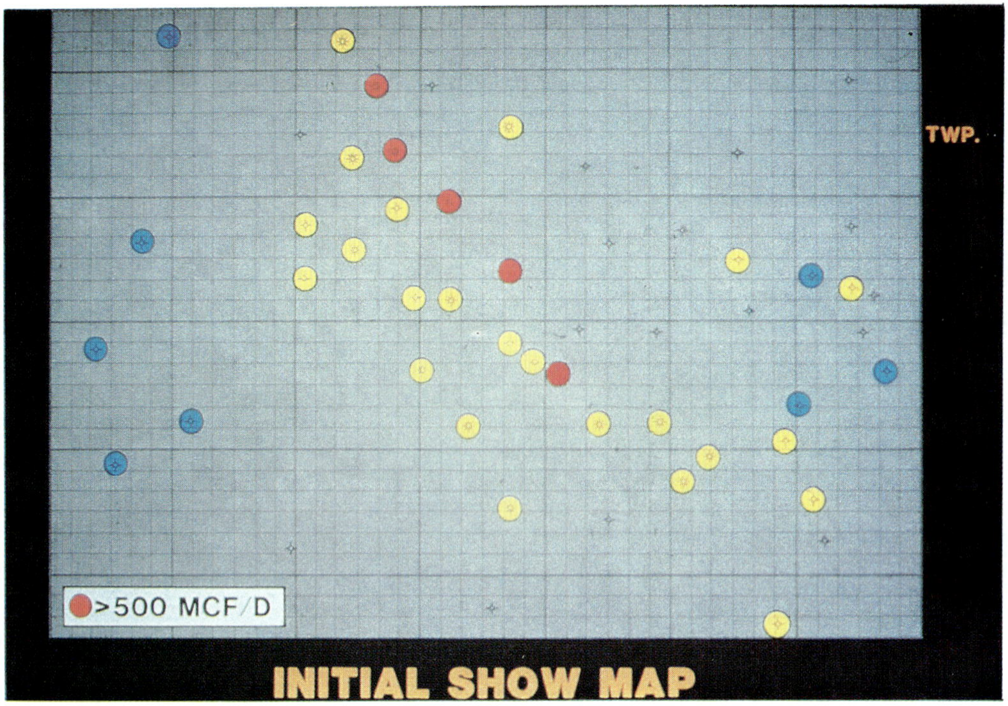

Figure 3

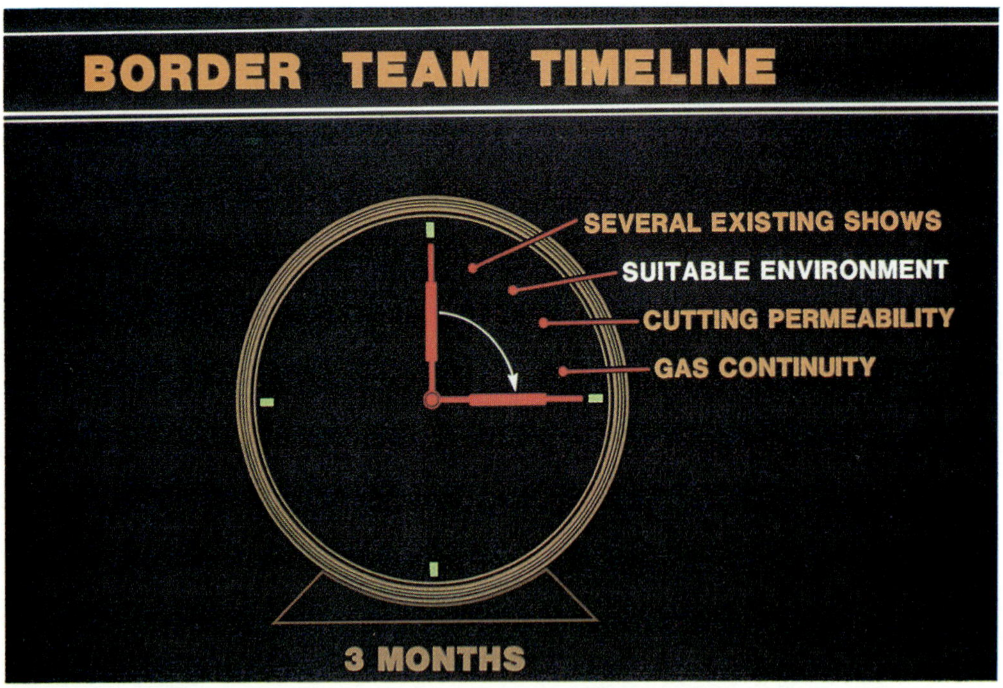

Figure 4

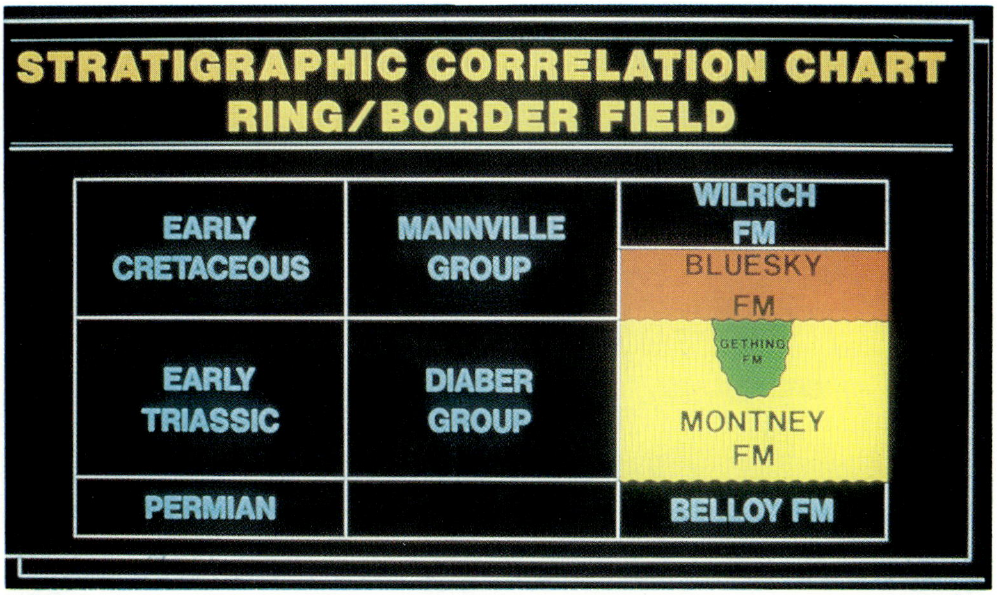

Figure 5

Figure 6

Figure 7

Figure 8

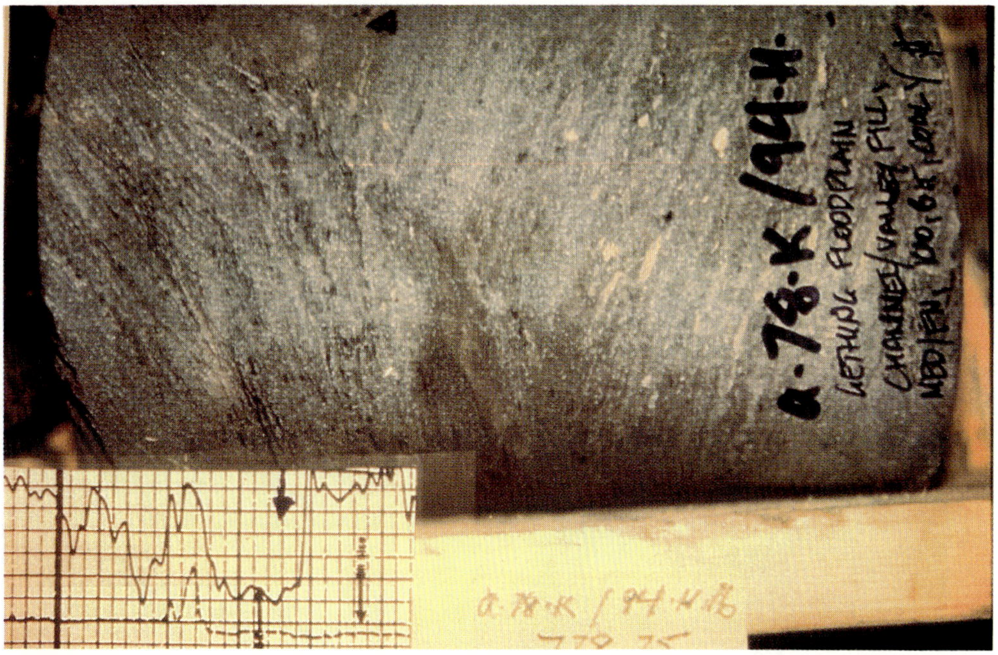

Figure 9

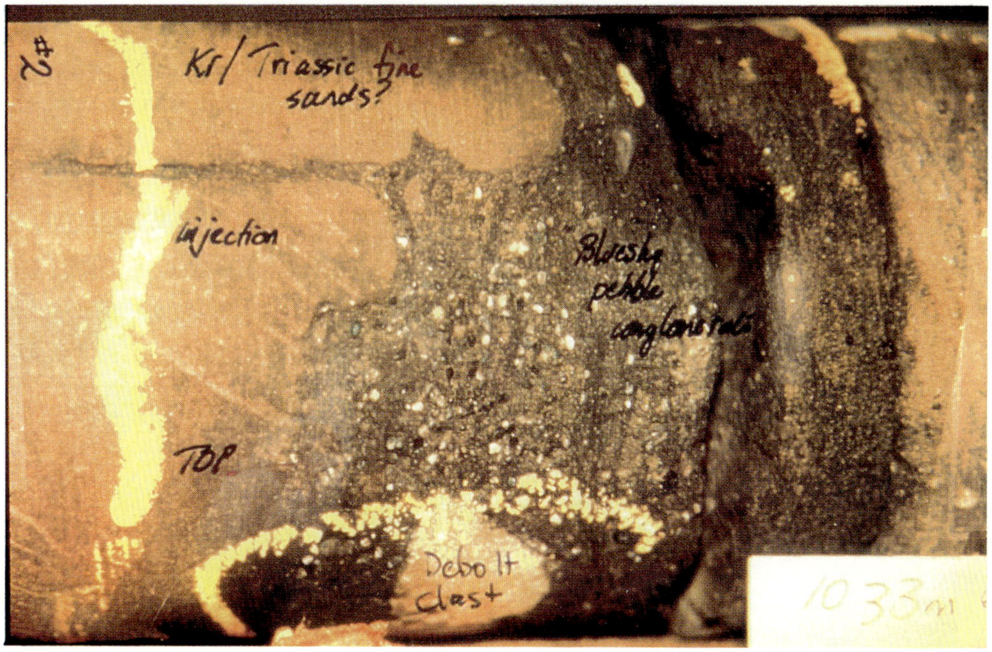

Figure 10

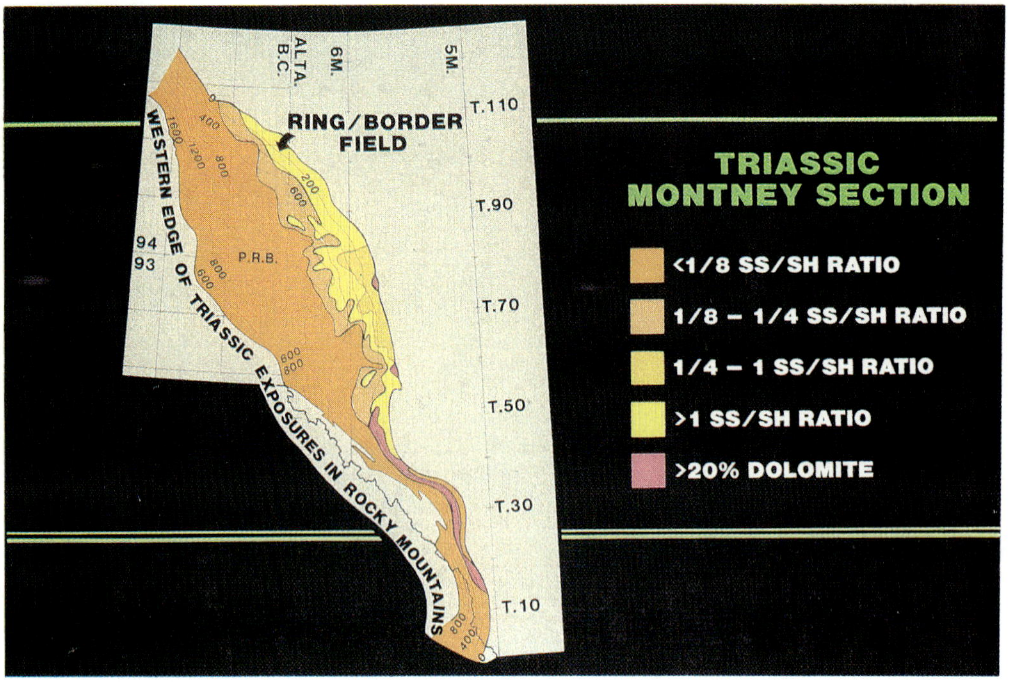

Figure 11

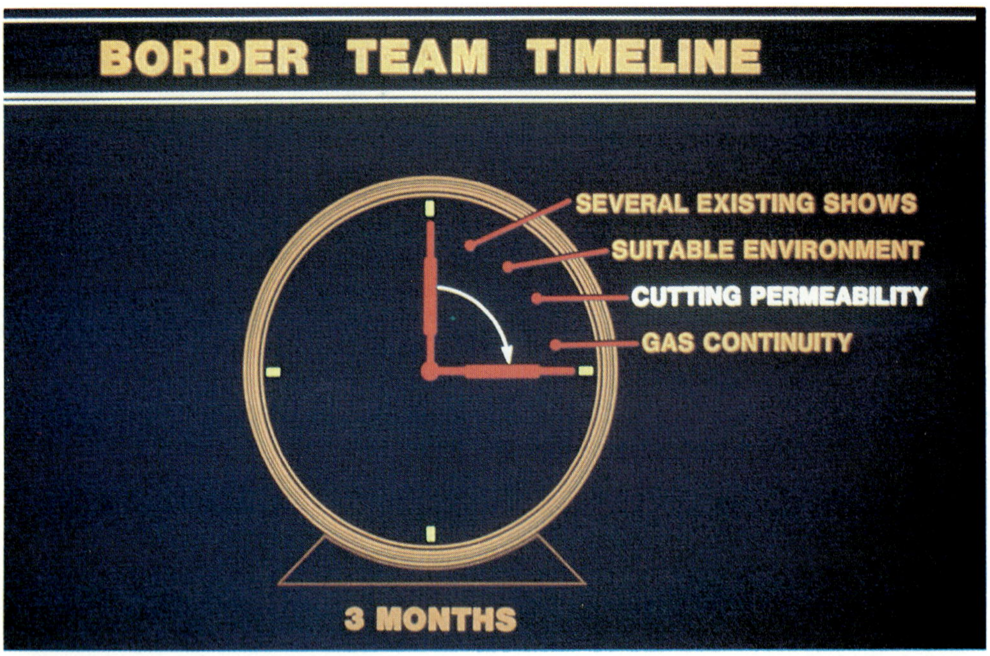

Figure 12

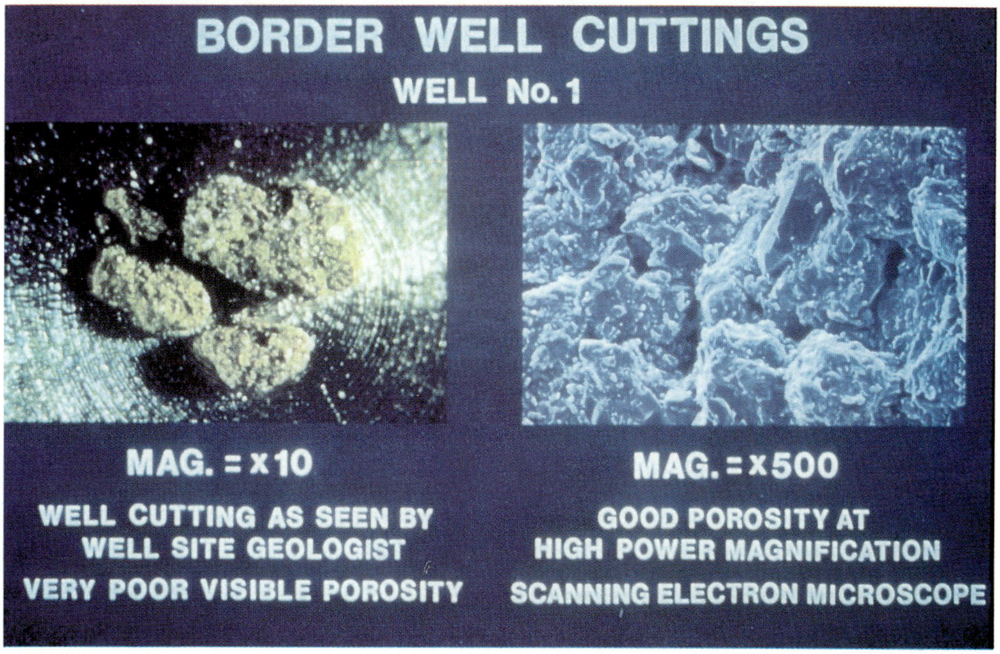

Figure 13

PORE-ROCK TYPES FOR SANDSTONE

TYPE	CHARACTERISTICS OF DRY, FRESHLY BROKEN ROCK SURFACES AT 20X MAGNIFICATION	REMARKS
I	• Visible Ø – Very abundant to common • Pinpoint Ø – Very abundant to common • Pore Interconnection is visible on many pores • Needle Probe can easily dislodge grains from some rock surface and reveal pore	• Reservoir quality rock for gas without natural and/or artificial fracturing (if thick enough) PERMEABILITY (md) IA ⟨ 100 IB 10 – 100 IC 1.0 – 10 ID ±0.5 – 1.0

Figure 14

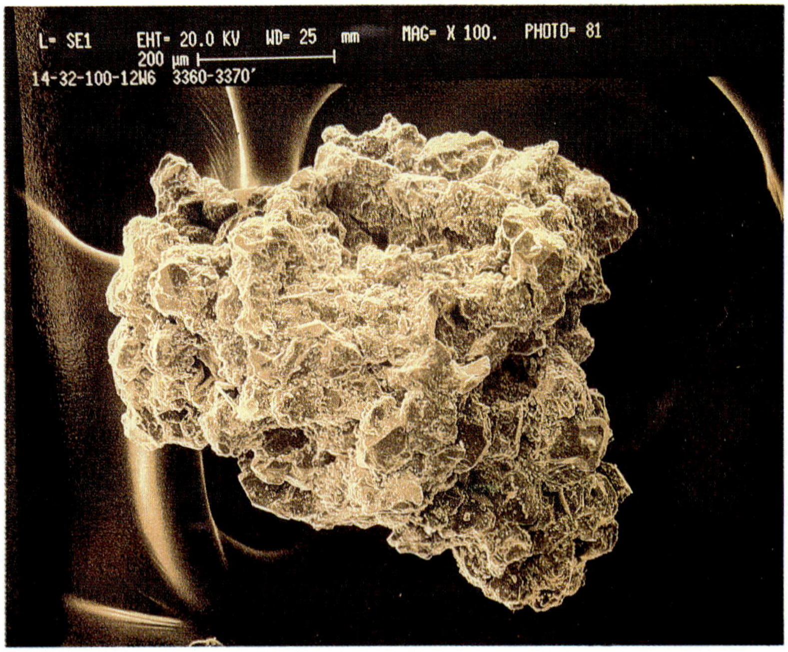

Figure 15

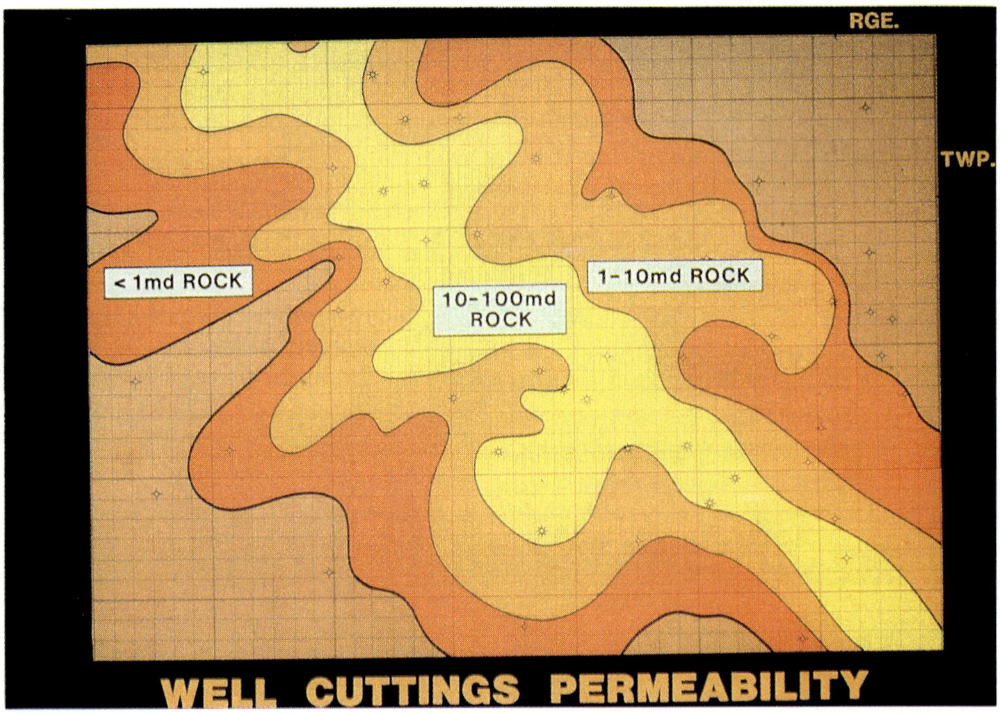

Figure 16

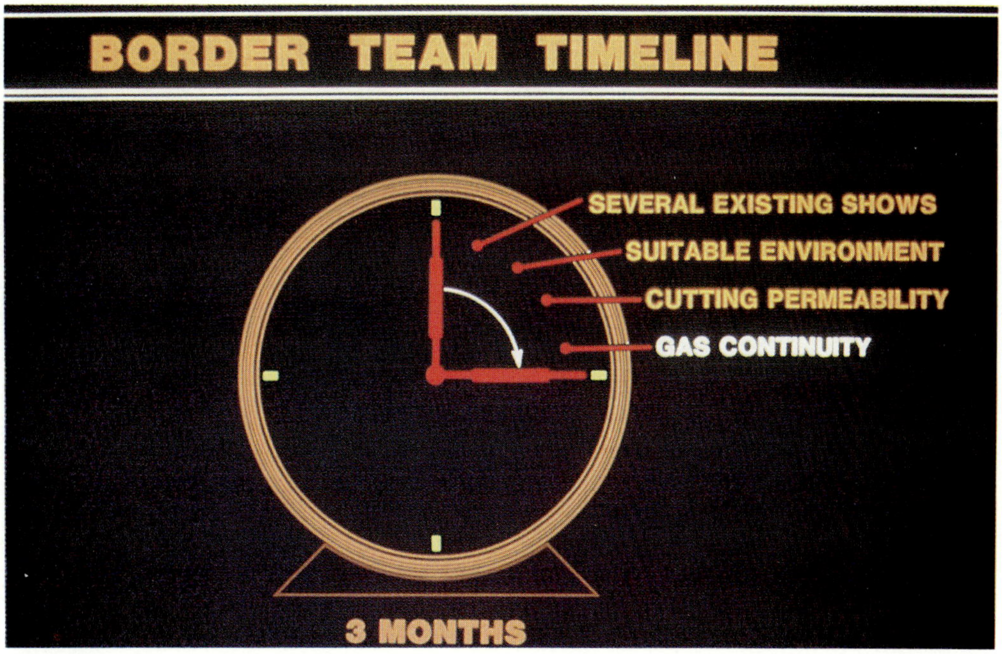

Figure 17

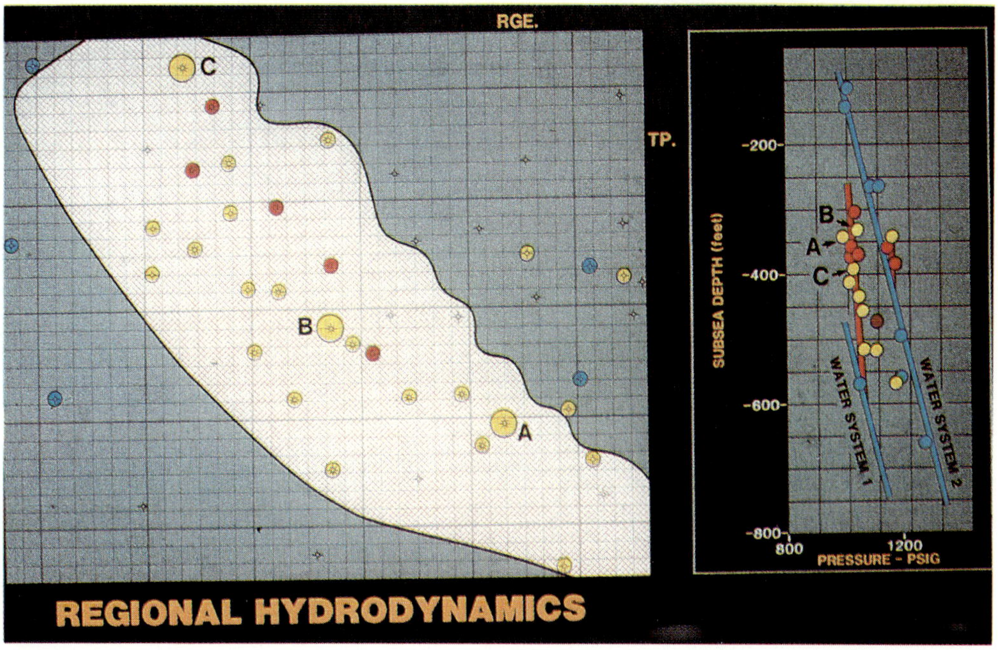

Figure 18

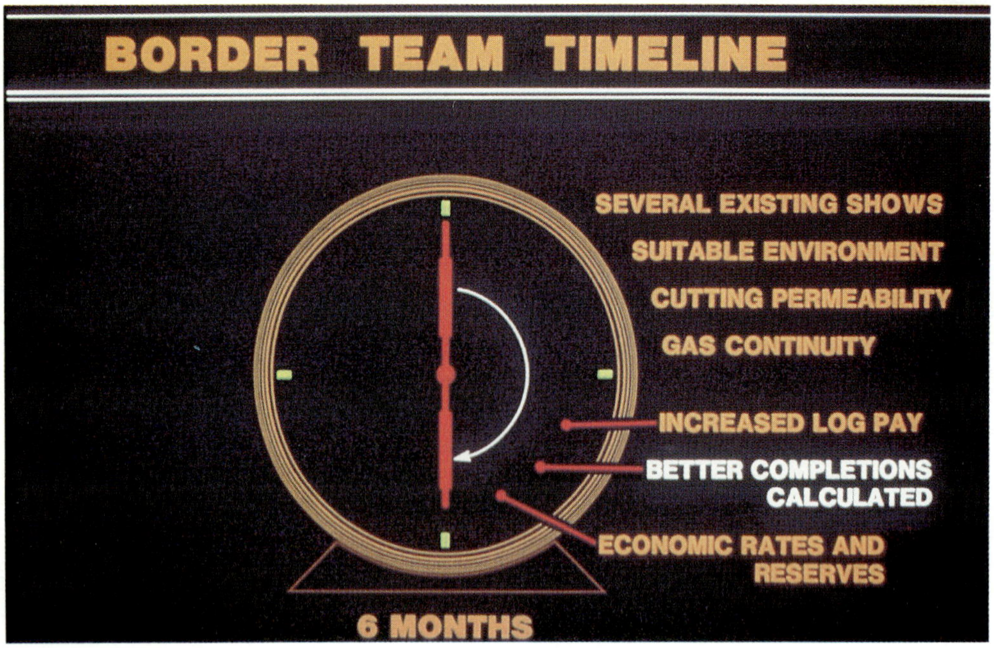

Figure 19

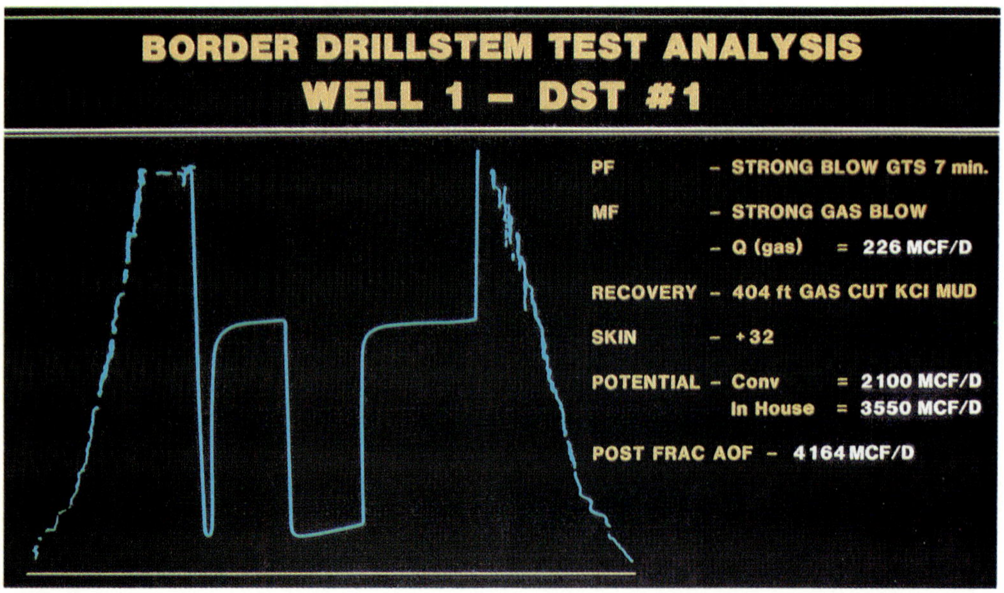

Figure 20

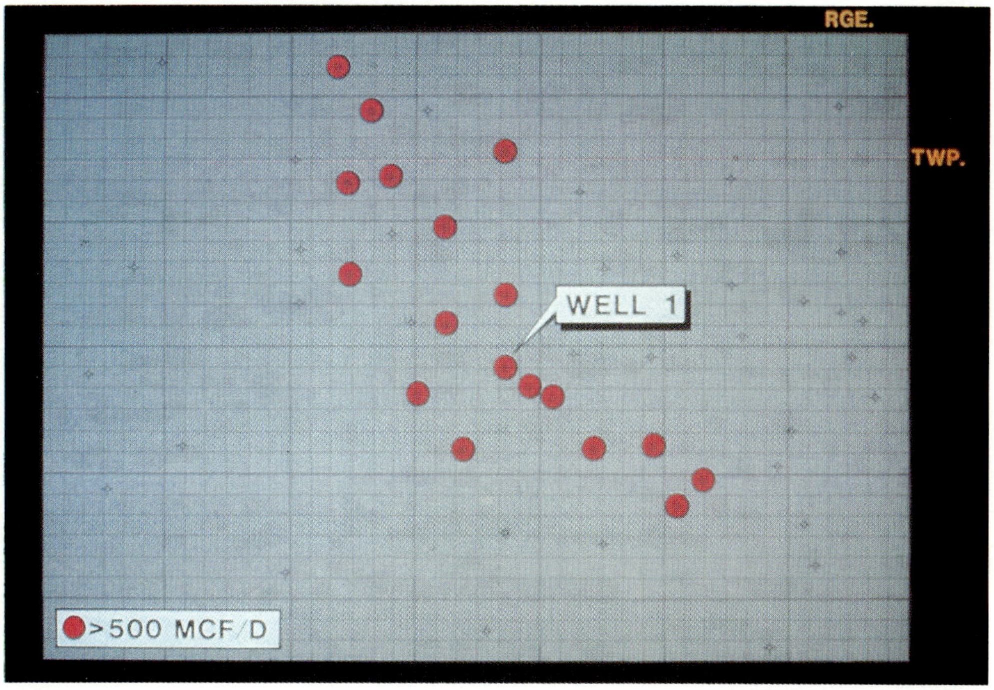

Figure 21

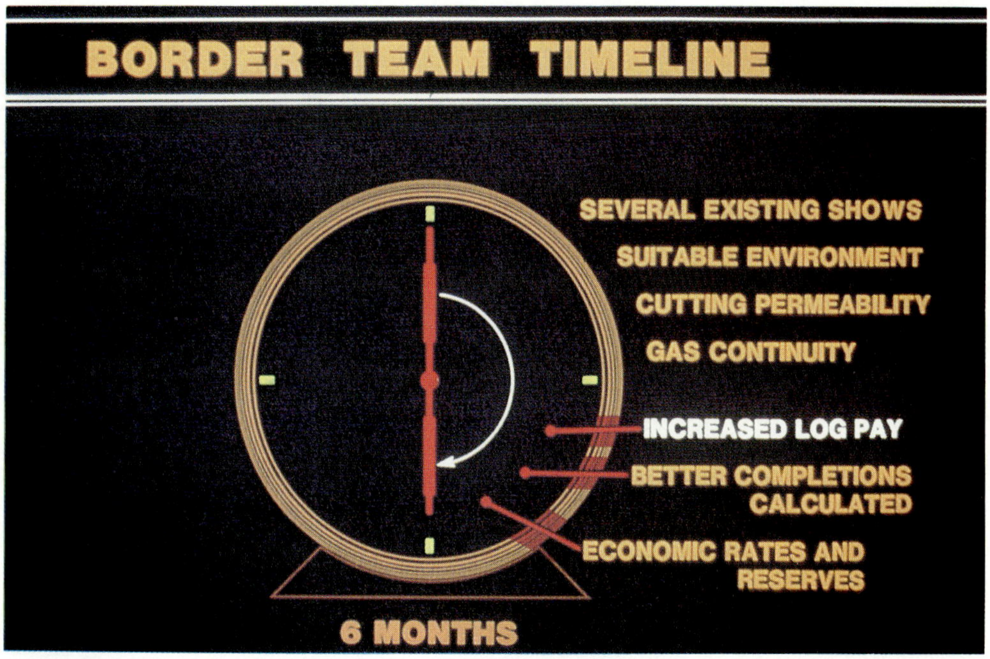

Figure 22

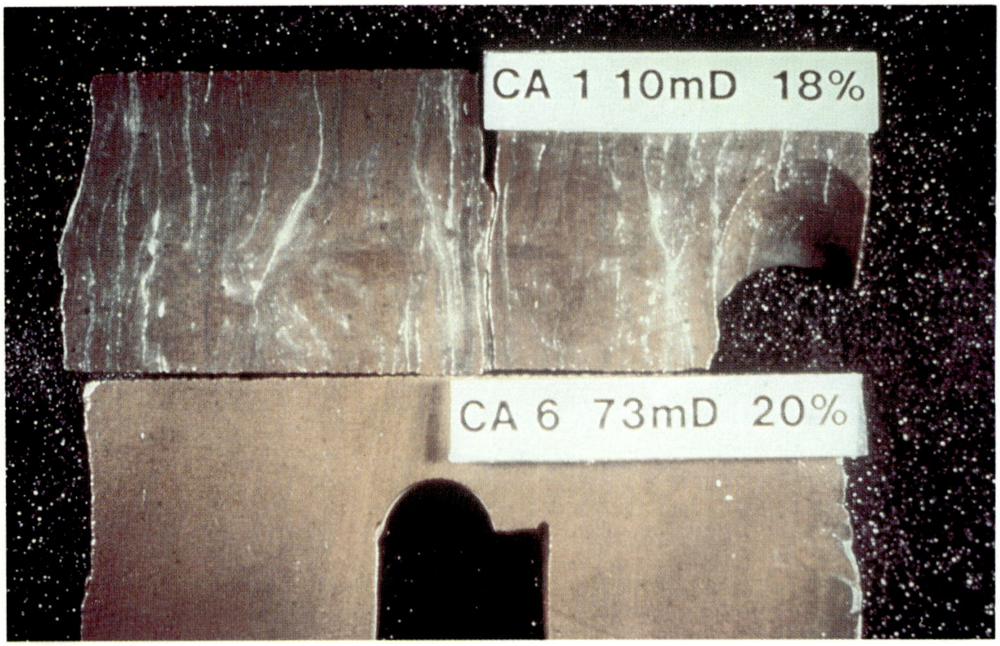

Figure 23

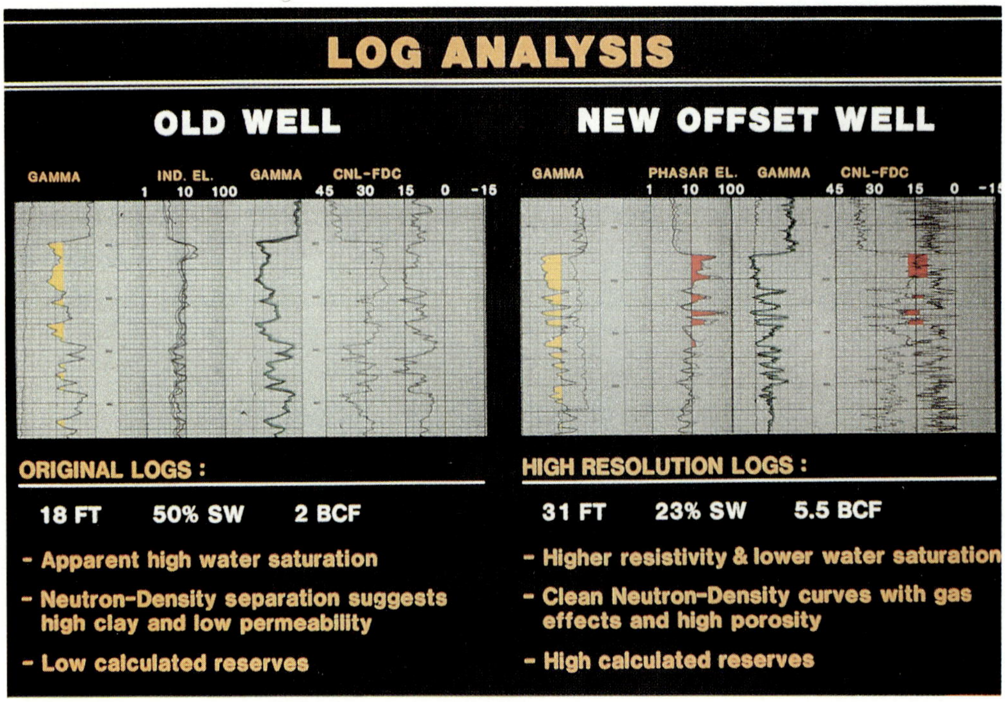

Figure 24

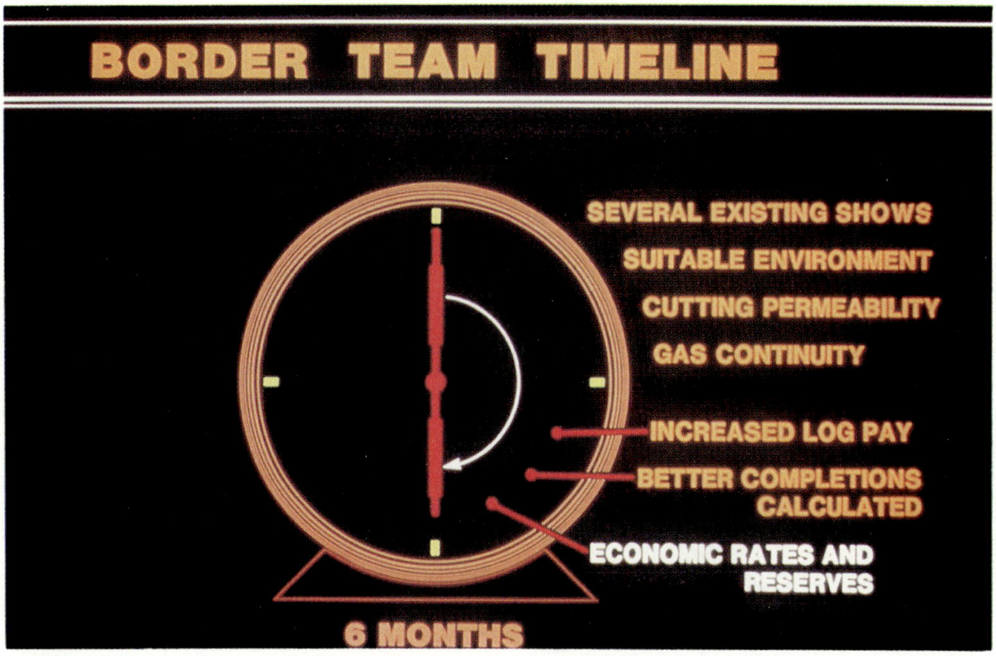

Figure 25

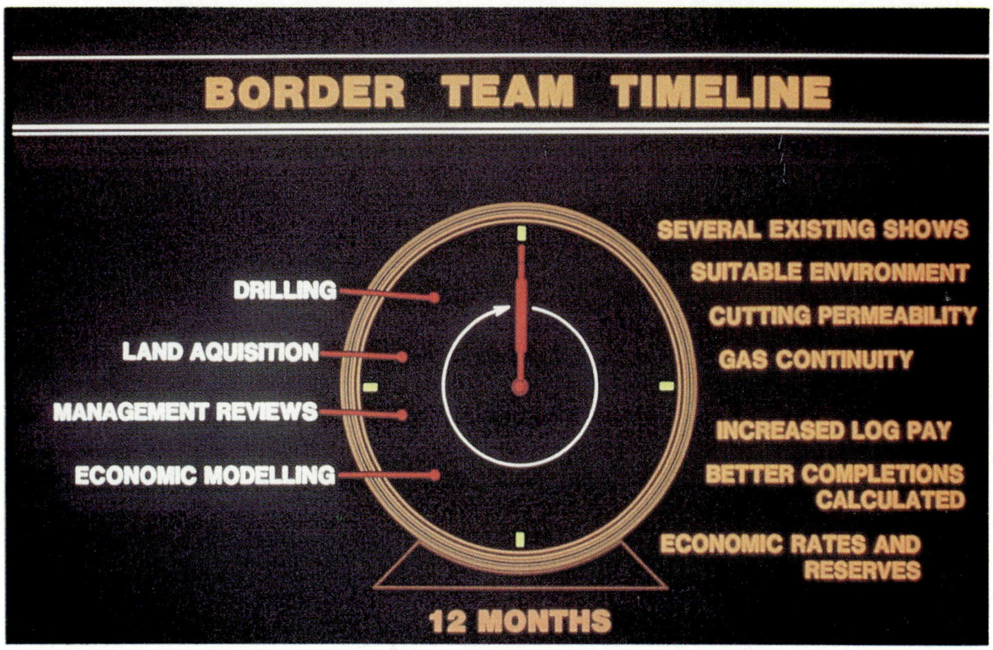

Figure 26

Figure 27

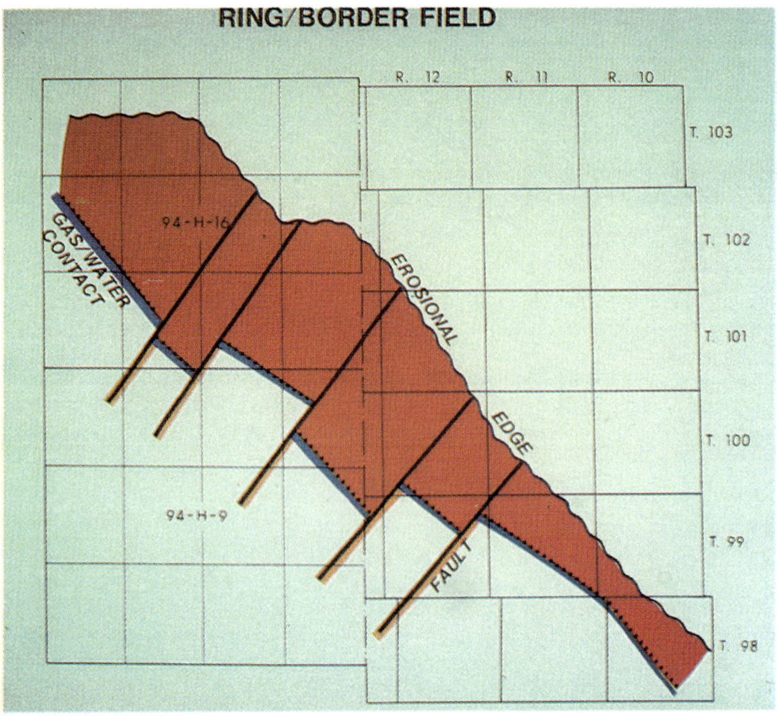

Figure 28

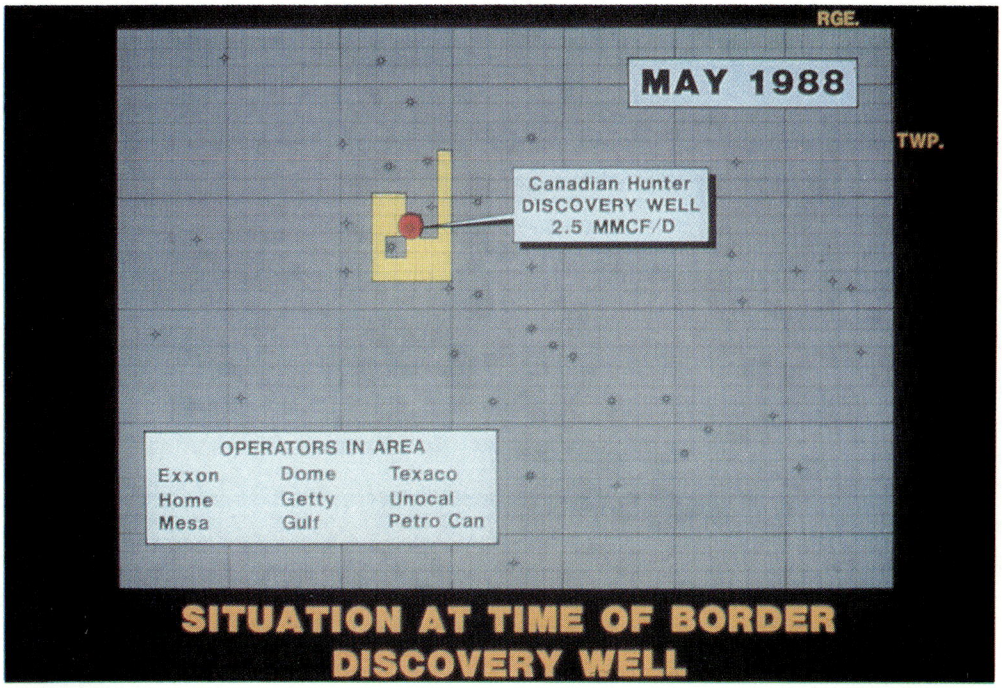

Figure 29

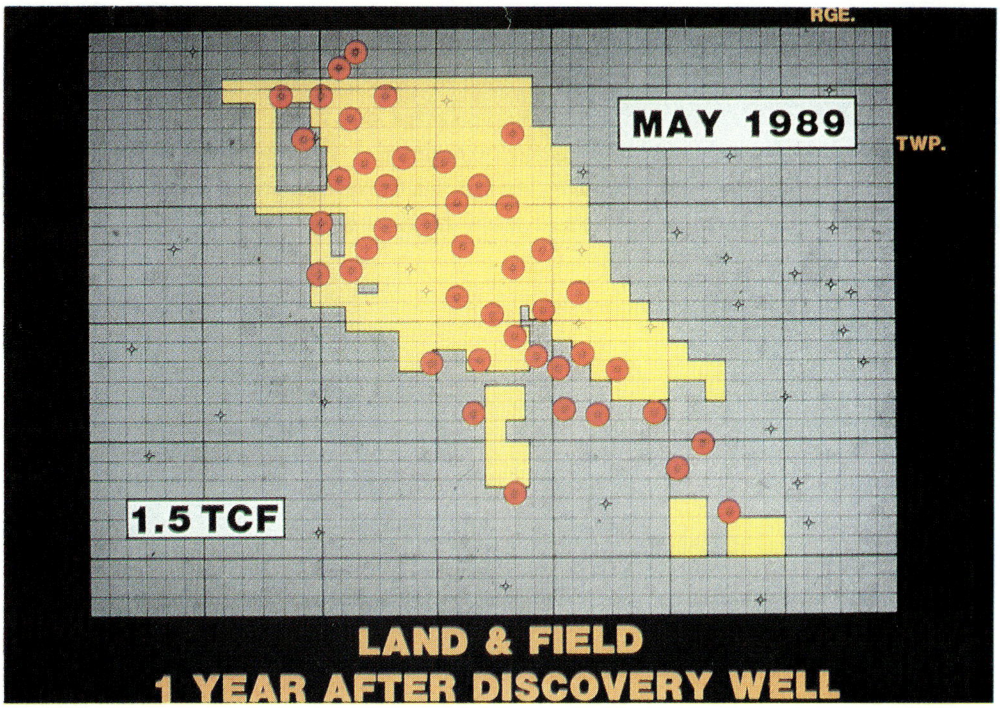

Figure 30

The Computer-Aided Geological Characterization of a Sandstone Reservoir, North Ward Estes Field, Ward and Winkler Counties, Texas

Albert S. Wylie, Jr., E. K. Davidsen, J. D. Gillespie, R. S. Butler, R. G. Stanley, and J. H. Beck

The North Ward Estes Field is located along the western edge of the Central Basin Platform in Ward and Winkler Counties, Texas. The field is part of an Upper Guadalupian productive trend that extends uninterrupted for 90 miles on the edge of the platform. The North Ward Estes Field has produced over 388 MMBO (one-third of the trend's cumulative production) from more than 3300 wells since its discovery in 1929. Production in the field is from back-reef lagoonal sandstones of the Yates, Seven Rivers and Queen Formations.

A correlation scheme was developed for the field based on laterally continuous key dolomites that bracket the productive sands and segment the reservoir into discrete mappable units. Applying this scheme, more than 68,000 correlation markers were selected and loaded into a computer database. Concurrently, 15 million curve feet of log data, 30,000 feet of core analysis data, and 125,000 feet of core lithology data were digitized. Core analyses and lithologies were depth-corrected. Logs were normalized using a 60-foot interval of laterally continuous anhydritic dolomite. Core porosity data were cross-plotted verses bulk density log values to develop equations (transforms) for derivation of porosity. Corrections for hole rugosity, overburden pressure, and lithologic complications were applied to refine the porosity transform. Structure and porosity-feet maps were then merged with fluid contact and water saturation data to calculate volumetrics. Facies and permeability relationships and actual to apparent pay ratios were applied to determine effective hydrocarbon pore volume.

Computer generated net isopach maps of the sands display a north-south strike. The sands pinch out into an evaporite facies updip and a carbonate facies downdip. Dip-oriented cross sections illustrate a systematic migration of the sand thicks through time. The depositional model for the field indicates that the sands were deposited during an overall progradational, shallowing-upward late Permian event that experienced minor transgressive and regressive sea-level cycles.

Design of a Major CO_2 Flood—North Ward Estes Field, Ward County, Texas

R. Winzinger, J. L. Brink, K. S. Patel, C. B. Davenport, Y. R. Patel, and G. C. Thakur

The reservoir engineering aspects of the design of a major West Texas CO_2 flood are presented.

A detailed fieldwide geologic study was conducted to characterize the principal Yates reservoirs which consist of very fine grained sandstones to siltstones separated by dense dolomite beds.

A CO_2 injectivity test was conducted to obtain CO_2 injection rates and to identify any reduction in injectivity either during or following CO_2 injection. Laboratory work included oil-CO_2 phase behavior, slim tube tests with pure and contaminated $C0_2$, and corefloods to determine recovery of waterflood residual oil by $C0_2$ flooding for various WAG ratios.

A comprehensive waterflood evaluation proceeded the selection of average pattern models for reservoir simulation. These three-dimensional models, which have up to eleven layers each, were history matched over the 35-year waterflood period. Predictions were made for continuation of the waterflood and for CO_2 flooding. Scale-up procedures were developed to predict from the average patterns the incremental oil production for the entire project area. Additional reservoir simulation was conducted to study the differences in recovery efficiency between line drives and five spots, the optimum economic slug size and the changes in hydrocarbon gas composition and GOR due to the stripping effect of $C0_2$.

A comparison of the predicted performance with the actual performance since start-up in early 1989 will also be presented.

The Benefits of a Synergistic Approach to Reservoir Characterization and Proration Rose City Prairie Du Chien Gas Field, Ogemaw County, Michigan

C. N. Tinker
L. D. Chambers
H. J. Ritch
C. D. McRae
M. A. Keen

PROLOGUE

Gus Archie stated, in the *AAPG Bulletin* during 1950, "Petrophysics pertains to the physics of particular rock systems whereas geophysics pertains to the physics of larger rock systems. It is the study of the physical properties of rock which are related to the pore and fluid distributions. These properties pertain to the detection and evaluation of hydrocarbon bearing layers."

Twelve years later, during an in-house conference at Shell, Bob Sneider said "The depositional environment is the important clue to the distribution and continuity of porosity, permeability, and barriers to fluid flow ... even though the sand is altered by diagenesis."

During 1977 the Senior Technical Editor of the JPT. Joe Richardson, said "Great strides have been made in studies of how the environment of deposition can effect rock character. Conceptual models based on these studies combined with geophysical as well as engineering data from well tests, cores and logs provide the basis for a modern approach to reservoir description."

As a result of these seminal thoughts, the value of integrating disciplines that focus on the subsurface, namely, geophysics, petrophysics, geology and reservoir engineering, has become more and more apparent over three decades.

Synergy is defined in *Webster* as "The cooperative actions of discrete agencies so that the total effect is greater than the sum of the effects taken independently." How does one integrate independent disciplines into a synergistic team?

Max DePree, in *Leadership Is An Art* provides a clue: "The first responsibility of a leader is to define reality. The last is to say thank you. In between the two the leader must become a servant and debtor."

One must create teams of subsurface scientists and petroleum engineers that are committed to achievement. Each member must be willing to be a leader and define the reality of her/his expertise. Each must be willing to serve and to acknowledge her/his debt to the others. Each must accept the premise that it doesn't matter who gets the credit. With such teams, the benefits of a synergistic approach are bound to accrue, and exemplary reservoir characterization result.

INTRODUCTION

Proration of gas fields in Michigan is regulated by the Michigan Public Service Commission (MPSC). Unlike other states the MPSC determines allowables for the purpose of allocating reserves. Therefore, exemplary reservoir characterization is essential to ensure each party receives, as far as can be practicably determined, an equitable share. SWEPI's Central Division Management recognizes the reality of the Michigan regulatory arena as well as the principles and value of effective leadership and teamwork. Accordingly, to better understand Rose City, a multi-disciplinary team was formed to analyze the extensive database, to prorate the field appropriately and to establish and maintain maximum acceptable production rates.

Field Setting

Rose City Prairie du Chien Gas Field is located in the central portion of the Michigan Basin (Figure 1) and produces from a 250 foot thick sequence of Lower Ordovician sandstones, the Glenwood or "Zone of Unconformity" (ZOU) and the St. Peter or Prairie du Chien (PdC) (Figure 2). This sequence is comprised of interbedded sandstones and carbonates that were deposited in a shallow water marine, progradational shoreface environment.

Rose City Field is developed by five producing wells on 640 acre spacing. The five governmental sections that contain the producing gas wells are located in Township 24 North, Range 2 East, Ogemaw County, Michigan (Figure 3). Also shown in Figure 3 are the seven seismic lines that cross Rose City. Structurally, the Rose City Field is an asymmetrical anticline with a steeply dipping north flank (Figure 4) that lies within a regional northwest-southeast trend of PdC fields (Figure 1).

LITHOLOGY & PETROGRAPHY

A lithofacies model based on detailed lithologic and petrographic analyses of the ZOU and Upper PdC at Rose City Field was developed to improve SWEPI's understanding of reservoir quality and distribution. Core and thin section data from three wells cored through the objective interval were utilized (Figure 4). The first core taken was at the St. Foster 1-28, a dry hole that penetrates the Upper PdC structure in a low flank position. The second core was from the St. Foster 1-20, the discovery well which penetrated the Upper PdC at the structural crest. The third core taken was at the St. Foster 2-28 at a sub-crestal structural position.

Most of the geologic and petrophysical analyses were concentrated on the St. Foster 2-28 core. This core was handled specially; an oil-based mud system was used for coring operations, and the core was frozen on location. Five lithofacies were differentiated based on sedimentary and biogenic features. This work confirmed that the reservoir is comprised of stacked coarsening upward sequences deposited as a shoreface that prograded in a shallow marine setting. Petrographic examination of thin sections showed that reservoir characteristics vary between lithofacies. Finally, correlations based on core data suggest good lateral continuity of each lithofacies and reservoir quality.

Macroscopic Features

Five lithofacies were identified in the ZOU and Upper PdC through macroscopic examination of the cores. Features such as bedding thickness, angularity and contacts as well as the shape, size and orientation of burrows were used. Photographs taken under ultra-violet light enhance these features and also highlight the distribution of permeability in the St. Foster 2-28 core. Light areas correspond to fluorescence of oil-based drilling fluid that has invaded the permeable zones (Figure 5).

Lithofacies 1 (LF 1) is characterized by low to high angle cross-bedding. In places there is excellent invasion of drilling fluid, indicating good permeability. This crossbedded sandstone is commonly interbedded with a Skolithos burrowed sandstone, LF 2. These burrows are elongate vertically oriented traces. Invasion is fairly good, indicating good permeability. In a typical sequence, LF 2 is underlain by a diverse burrowed sandstone, LF 3, which is characterized by fairly abundant, but variably shaped burrows. There are both large and small burrows oriented both vertically and horizontally. Invasion can be good in places, but is mostly poor, indicating variable to poor permeability depending on the degree of bioturbation. LF 4, a churned sandstone, is typically found below LF 3 in sequence. This thoroughly bioturbated sandstone is a mass of small burrows. The originally sandy and shaly layers are thoroughly intermixed. Invasion is generally poor, indicating poor permeability. There is a single 15 foot thick layer at the base of one sequence comprised of parallel-bedded sandstone, LF 5 (Figure 6). This sandstone is comprised of numerous fining upward cycles on the order of one to five inches thick.

Microscopic Features

The microscopic features of LF 1 through 4 indicate an upward coarsening grain size distribution, an upwardly increasing degree of sorting and a uniform sand grain composition of 90 percent quartz and 10 percent potassium feldspar (Figure 7). A scale bar representing 0.5 mm appears on each photograph.

LF 1 is very well sorted and has a low percentage of clay. Large well interconnected pores appear blue due to the thin section epoxy in the photomicrograph. The texture of LF 2 is very similar to LF 1. LF 3 is moderately well to well sorted with a greater percentage of fine grained sand and clay material. There is less visible porosity in LF 3. LF 4 is also moderately well sorted, but has an even greater percentage of fine grains and clay and even less visible porosity. As one proceeds down through a sequence, the core shows decreasing evidence of drilling fluid invasion and thin sections show increasing fine grains and clay with decreasing visible porosity.

LF 5 is comprised of numerous fining upward cycles and has well sorted sands at the base of each cycle that grade into moderately well sorted sandstones at the top (Figure 6). The grain size of this quartz arenite varies from lower fine to upper medium, and within each cycle the volume of very fine grains and clay increases upward.

Depositional Model

These five lithofacies comprise a typical coarsening upward sequence from the parallel-bedded LF 5 at the base to the cross-bedded LF 1 at the top (Figure 8). The thin fining upward cycles of LF 5 are interpreted to be storm deposits in a lower shoreface to offshore setting. LF 3 and 4, bioturbated and shaly sandstones, overlie LF 5 and are interpreted to be low to moderate energy deposits in a lower shoreface environment. At the top of the sequence, LF 1 and 2, well sorted sandstones with very little clay, are interpreted to be high energy deposits in a middle to upper shoreface environment. At Rose City Field, the ZOU and Upper PdC reservoir rocks are comprised of several of these coarsening upward sequences which were stacked together in part and in full as sea level fluctuated.

Diagenesis, Porosity & Reservoir Quality

Petrographic analyses are based on 70 thin section point counts with 300 points each. These data indicate a strong lithofacies related control on diagenesis (Figure 9). The percentage of quartz overgrowth cementation is higher toward the top of each sequence, in LF 1 and 2. The volume of clay is higher toward the base of each sequence, in LF 3, 4 and 5. The percentage of dolomite cement is low in all lithofacies but increases slightly in the most poorly sorted and clay rich lithofacies.

Further analysis of point count and measured core porosity data can be used to calculate the average relative abundance of macroporosity and microporosity in each lithofacies (Figure 10). This distribution of pore sizes controls permeability and ultimately reservoir quality. These data can therefore be used to assess the relationship between reservoir quality and lithofacies.

On average, the clean sandstones, LF 1 and 2, have the highest percentage of macropores, resulting in the highest reservoir quality. The shaly sandstones, LF 3 and 4, are dominated by microporosity and make fair reservoirs. LF 5 has the greatest overall porosity, but a large portion is microporosity, giving it a bimodal pore size distribution. The abundance of microporosity varies significantly within this lithofacies due to the microcycles resulting ln poor to good reservoir quality.

Vertical & Lateral Continuity

A stratigraphic cross section based on log and core data indicates there is significant vertical heterogeneity in the distribution of lithofacies while in a lateral dimension the lithofacies appear very continuous (Figure 11). The same lithofacies were identified at the same stratigraphic position in each of the wells, which are separated by several miles.

The strong correlation between lithofacies and reservoir quality established at the St. Foster 2-28 and the strong lateral continuity established from the lithofacies correlation suggest that reservoir quality is also laterally continuous over the field area. Assuming that permeability is a good indicator of reservoir quality, a cross section correlating permeability measurements from the three cored wells confirms the lateral continuity of the reservoir quality (Figure 12). These data show that permeability streaks are correlative across the field from crestal to flank wells. Nearly identical values were measured at correlative stratigraphic positions. There appears to be very little variation from the diagenetic alteration documented in the St. Foster 2-28 in other parts of Rose City Field.

This understanding of the distribution and types of reservoir rock provided a foundation for the petrophysical engineer to determine the pay, porosity and water saturations.

PETROPHYSICS

The gas productive ZOU and UPdC sandstone formations in the Center Basin in Michigan are difficult to evaluate. The problems are: (l) the thin reservoir quality sandstones are interbedded between lower porosity sandstone and/or dolomite beds; (2) individual reservoir beds range in thickness from less than 6 inches to several feet with resistivities significantly lower than the surrounding non-reservoir quality beds; (3) the average porosity of the gas productive reservoirs is approximately 10% (cut off porosity = 6%) with the reservoir quality sandstones containing varying amounts of silica cement, dolomite cement, and clay; (4) the reservoir quality sandstones have bimodal pore systems resulting ln low "n" saturation exponents (n = 1.20), thus making water saturation calculations above 50% very sensitive to small errors in resistivity and/or porosity (average water saturation for the field is about 50%); and (5) large variations in rock quality based on core and log analysis.

The petrographic study described above, provided the basis for improved petrophysical evaluations. Figure 13 shows the relative reservoir quality characteristics for the five lithofacies. Also, the primary zone(s) of occurrence for each lithofacies are shown which greatly helps the estimation of flow potential. Figure 11 shows the vertical locations of these zones in the three cored wells.

Stressed laboratory measurements were conducted on core material to determine in situ formation properties. In-situ porosity versus in-situ air permeability for all five lithofacies, measured using a differential pressure of 5,200 psig, are shown in Figure 14. LF 1 and 2 (the best reservoir quality sandstones) have better permeability for a given porosity than the other three lithofacies which have poor to good reservoir quality depending on whether they are high quality (HQ) or low quality (LQ).

The differences in reservoir quality are also shown in Figure 16 by stressed mercury-air capillary pressure measurements for all five lithofacies using average porosity samples for each lithofacies. Here again, LF 1 and 2 show the most favorable capillary pressure properties with LF 4 having the least favorable. Average gas column heights for the field are about 100 feet above the free water level for the ZOU and about 60 feet for the UPdC formations. Based on these column heights, LF 4 is almost never commercially productive. When LF 5 is of high quality, then it is usually commercially productive even at very high water saturations.

The Archie relationships shown in Figure 15 are used to calculate water saturations. The petrophysical rock parameters used in the Archie relationships are given in Figure 17. The saturation exponent "n" is the only parameter substantially different than normally found for sandstones. The average saturation exponent "n" is about 1.20 and is due to the bimodal pore size distribution within the ZOU and UPdC sandstones. Figure 18 is a cat scan image of a PdC sandstone. This bimodal pore size distribution is also evident from high pressure (50,000 psia mercury pressure) capillary pressure measurements. The particular lithofacies found in each of the zones shown in Figure 17 are:

ZONE(S)	LITHOFACIES
UZOU1	3 & 4
LZOU1 / ZOU2 / UPdC1	2 & 3 & 4
UPdC2	5
UPdC3	1 & 2

The formation density log is the primary porosity logging tool. The log-measured laterolog resistivities are often too high due to the higher non-reservoir quality surrounding beds; therefore, these resistivities have to be corrected before being used to calculate water saturations.

Formation temperature in the Rose City Field is about 174_F at a measured depth of 10,000 feet. The initial reservoir pressure was 5,150 psia (0.515 psi/ft). The formation waters are saturated at in situ conditions with a large amount of CaCl2 and some NaCl resulting in a measured Rw value of 0.035 r-m. The measured grain density for the reservoir quality sandstones is 2.65 g/cc and does not vary by more than 0.01 g/cc unless the sand becomes very shaly and thus is non-reservoir quality.

The net feet of gas pay were determined for each well and lithofacies. These values were provided to the geological engineer to be incorporated with the structural setting at Rose City for the determination of reservoir volumes.

GEOLOGY & GEOPHYSICS

The structural delineation of Rose City began with a land survey over 38 square miles that was conducted by satellite positioning for selected control points. From these control points, mechanical traverses to each governmental section corner were made to insure base map accuracy. This resulted in an accuracy of 1 foot in 50,000 feet. Within the area, eight deep PdC wells and 80 shallow Dundee (Devonian) wells were located and tied into the survey. Additional data in the area that aided in the structural delineation of Rose City included seven seismic lines, as well as 85 tops for the Dundee, available from wells in the area that were plugged and abandoned.

Seismic Structure

A representative seismic line, shot from the southwest to the northeast, is shown in Figure 19. A one mile scale bar is in the lower left hand corner and closure at the PdC level is indicated by the bar to the right of center. Note the gentle roll to the southwest and the "steep" flank to the northeast. Strong seismic events include the Devonian Traverse and Dundee, Silurian F Salt, A2 Carbonate, Al Carbonate, and Niagaran and the Ordovician Trenton (Figures 2 and 19).

A structure map on the Dundee was created from the

dense shallow well control at Rose City. Very often, due to the tectonic history of the Michigan Basin, deep PdC fields are overlain but offset en echelon by shallow Devonian Dundee fields, as is the case at Rose City. By mapping the Dundee structure based on well control, then tying it into the Dundee event on seismic, much of the substantial geophysical data processing problems caused by glacial drift in the Michigan Basin can be avoided. The Michigan Basin is overlain by fairly thick intervals of glacial drift, which causes very large long-wavelength statics correction problems. In some cases, these static corrections are greater than the structures that are being mapped. By using the Dundee structure in conjunction with the seismic, these corrections can be eliminated. Several other intervals of strong seismic response are mapped.

Using the formation data from the eight deep PdC wells in the area, the seismic response can be tied to the well control. Isopachs between these intervals of strong seismic response were constructed, using the interval transit time and well control (Figure 20). Employing this iterative process, final structure maps of the ZOU and PdC (Figure 4) were constructed for Rose City. The overall structure at Rose City is approximately 3200 acres, with the base of pay at 9057 feet subsea (SS).

Integration of Data

The comprehensive petrographic, lithologic and petrophysical work that was done at Rose City could now be integrated with the accurate structural interpretation to determine the appropriate reservoir characterization and pay distribution. Once the rock data from the three cored wells (Figure 21) were correlated to the electric log data and the depositional model of the field established, lithofacies data were extrapolated to the wells that were not cored (Figure 22). Horizontally, the lithofacies are continuous across the field, while the five lithofacies vary vertically.

SgØH Isopachs & Volumetrics

Rose City Field was divided into three separate intervals to map the pay. Two SgØH (hydrocarbon pore volume) maps were constructed for the ZOU (Figure 23). These two pay maps are divided by the Jordan Shale, a thick, non-producing carbonate interval. This interval is denoted in light blue in Figure 22. The ZOU is dominated by LF 3 and 4 (diverse and churned), which are characteristically fair to poor pay.

The PdC contains the majority of the pay at Rose City (Figure 24). One SgØH map was constructed from the base of pay at 9057 feet SS to the top of the PdC. The PdC is comprised mainly of LF 1 and 2 (cross-bedded and Skolithos burrowed). The only interval of LF 5 (parallel-bedded) is in the PdC and is continuous and approximately the same thickness across the field.

The structure and isopach maps were digitized into a computer mapping program. The shallow and deep well locations, seismic line and shot point locations and land survey data were electronically transferred to the same mapping system. This facilitated data management and editing. Detailed cross sections, Sg_H maps, the underlain volume for each 640 acre drilling unit, as well as the total field volume were drawn and/or calculated by the program. The total field original gas in place was calculated to be 71 BCF from volumetrics.

PRODUCTION TEST PROGRAM

When the Rose City field began producing gas in 1987, it was SWEPI's only PdC field capable of sustained production. The lack of a suitable gas transportation infrastructure prevented several other fields from marketing production. These fields remained shut-in for three years while waiting on the installation of a 125-mile pipeline. Accordingly, a unique opportunity existed at Rose City to utilize sustained production and pressure data to improve the reservoir description, establish an operating policy that maximizes recovery and verify the reliability of the subsurface evaluation parameters used to explore and develop the PdC. A two-year test program was subsequently designed and conducted under the auspices of the MPSC. This necessary and welcome support by a state regulatory agency facilitated the program by permitting temporary increased allowables and ensuring equity would be maintained during the test. Additionally, the sharing of information between both parties provided a common understanding of the reservoir system, which is essential for optimizing future development and depletion of Michigan's PdC gas fields.

Test Results

The ensuing fieldwide production test resulted in an enhanced understanding of the PdC. First, the Team verified the volumetric estimate of original gas in place (OGIP). OGIP, determined independently from both material balance and volumetric calculations, agreed within 10 percent. Second, excellent pressure continuity was observed, which confirmed that the existing spacing of 640 acres/well is appropriate, and that larger spacing may be beneficial in some fields. Finally, the effects of water influx on future performance were quantified. The resulting need for high withdrawal rates from crestal wells was recognized.

All of the foregoing knowledge resulted in better predictions of long-term performance, improved completion strategies and the determination of optimal production rates. An operating policy that maximized ultimate gas recovery and minimized water encroachment was developed and implemented. SWEPI and the MPSC have mutually ensured that production allowables are appropriate, equity is preserved and waste prevented.

Test Operation

The test program was designed to minimize manpower requirements and production curtailment. An emphasis on drawdown testing allowed fieldwide deliverability to exceed plant capacity at all times during the test. Flow rates, tubing-head pressures and bottom-hole pressures in producing wells were measured continuously during depletion using a computerized data gathering system (Figure 25).

A Westinghouse computer control system was used to operate the wells and facility as well as to monitor all temperature-pressure probes and rate meters. A DEC MicroVax computer stored all measurements at any selected sampling frequency. Setpoint's *SETCON* software was used to communicate with the MicroVax from Houston and to program the data collection frequency. Finally, a PC with modem served as the terminal in Houston, where the data was down-loaded and interpreted. The resulting system was fully automated and accurate, provided real-time access and allowed fast interpretation. The commingled production of gas, condensate and water required the use of permanent downhole pressure sensors to improve bottom-hole pressure estimates during drawdown testing. Additional periodic short-term flow tests and pressure surveys were conducted in an observation well to measure reservoir properties remote from producing wells. Pressure profiles were measured during depletion using an RFT to monitor relative contribution of zones.

Data Interpretation

The interpretation techniques utilized include material balance extrapolations, reservoir limits testing, pressure buildup and drawdown analyses and RFT interpretation.

Proper evaluation of the production and pressure data required a knowledge of the size and strength of the aquifer. Regional cross sections showed excellent lateral continuity of sands over a distance of at least 12 miles, with porosity and permeability preservation off structure, indicating a geological setting conducive to water drive. The production performance verified the drive mechanism. Fieldwide water-gas ratio (WGR) increased as a result of water breakthrough in the most permeable pay intervals (Figure 26). In addition, the flank observation well productivity (Figure 27) dropped dramatically (gas rate down 30%, WGR up 400%) as a result of water encroachment.

After verifying the presence of a water drive, an aquifer description was developed for use in modelling the reservoir performance. Figure 28 illustrates a schematic cross section with a random sequence of lithofacies under a single seal. There are often many seals present in one PdC field, as there are at Rose City.

The reservoir/aquifer system is described as infinite-acting and vertically heterogeneous with an edge-water drive. Water drive is expected to be active only in the higher permeability zones (LF 1, LF 2), resulting in low vertical sweep efficiency and minor pressure support overall. This illustrates that the optimal method of depletion is at high rates to reduce abandonment pressure prior to water breakthrough. Additionally, completion strategies require methods to minimize water breakthrough and production.

The anticipated pressure performance was forecast (Figure 29) using material balance techniques. The Carter-Tracy unsteady-state radial flow aquifer equations were used to estimate the rate of water influx at each time step. This pressure performance forecast provided the basis of our ultimate recovery estimates and deliverability forecasts. The pressure support provided by the water influx is shown.

BENEFITS & RESULTS

Proration of Michigan gas fields is intended to: "protect correlative rights; permit a reasonable opportunity to produce recoverable reserves; provide for equitable purchasing and taking of gas from a common source of supply; ensure each producer recovers production in proportion to the recoverable gas under his land; and to prevent waste." As noted earlier, proration demands exemplary reservoir characterization. A thorough reservoir description will lead to better management of the reservoir including predictions of performance, optimal completion strategies and production rates and also facilitate unitization. All of these were accomplished at Rose City.

Historically the MPSC based proration on various combinations of: surface acreage, drilling unit size, underlain area, underlain volume and absolute open flow potential. As a result of its work, the Team was able to convince the MPSC and other working interests to prorate the Rose City Field based solely on underlain hydrocarbon bearing pore volume (ASgOH) and to balance gas volumes for each drilling unit retroactively to the date of first production from the field. In addition, to accelerate gas balancing, maintain optimal production rates and assure that each drilling unit recovers its share regardless of which well is producing, the Team was able to convince the other working interest and royalty owners to voluntarily unitize the field.

Both are "firsts" for Michigan PdC fields.

The reservoir description was comprehensive and involved integration of input from each discipline to better define and illustrate the zonation and heterogeneity of the PdC reservoir system. The work is technically sound and reproducible; it was essential for defining equity during proration and unitization, predicting future performance, and developing an operating policy that will maximize ultimate recovery. This study of the Rose City PdC field benefitted from the synergy generated by a multi-disciplinary team working together in harmony and serves as a model for the exploration and development of other PdC fields.

EPILOGUE

In addition to its primary task, the Team also developed some fundamental understandings about leadership and effective teamwork. The Team now knows that Max DePree knows what he is talking about.

The Team came to realize that it isn't the people you work for but the people who work for you that determine your success. Each member of the Team participates. Each member of the Team is a specialist. Each member of the Team is a leader, but none is a boss. No one drives the Team. No one leads the Team. Everyone serves the Team!

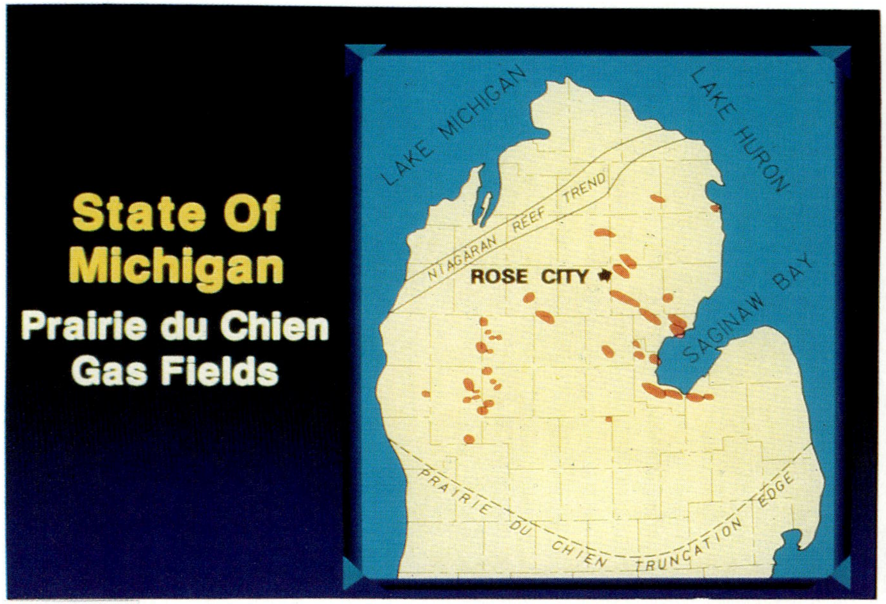

Figure 1

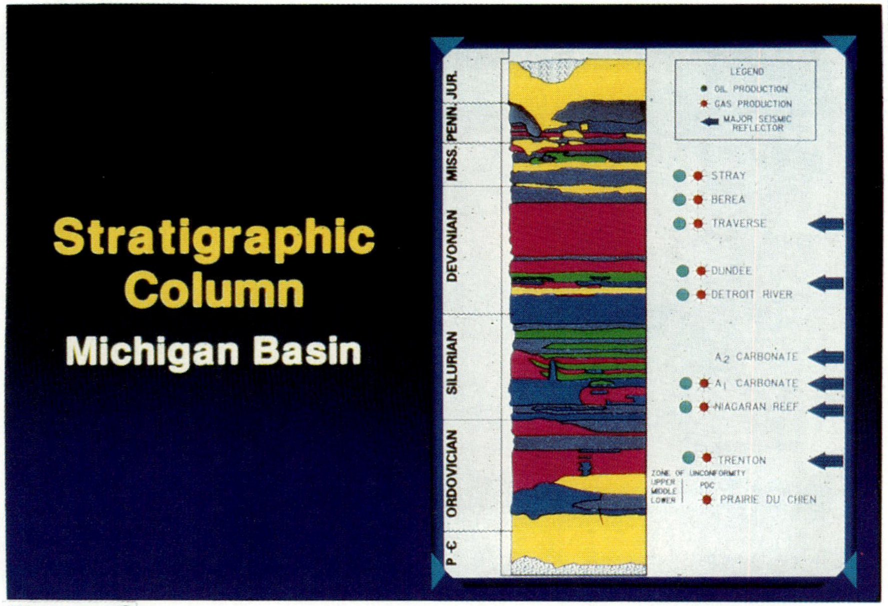

Figure 2

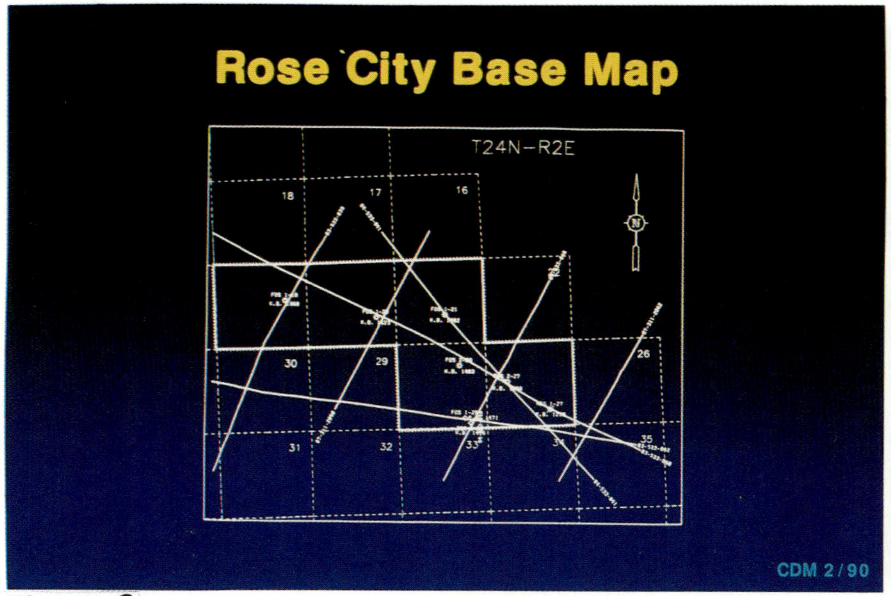

Figure 3

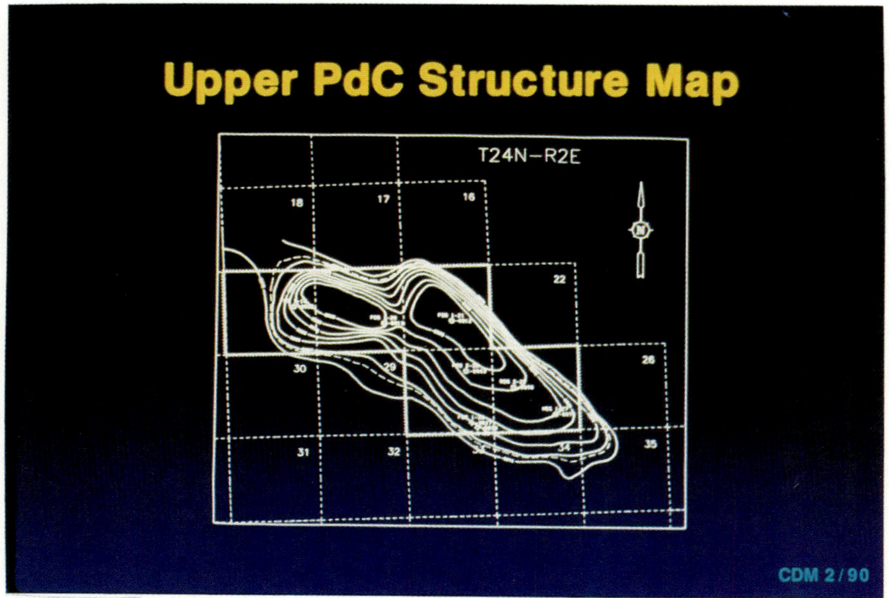

Figure 4

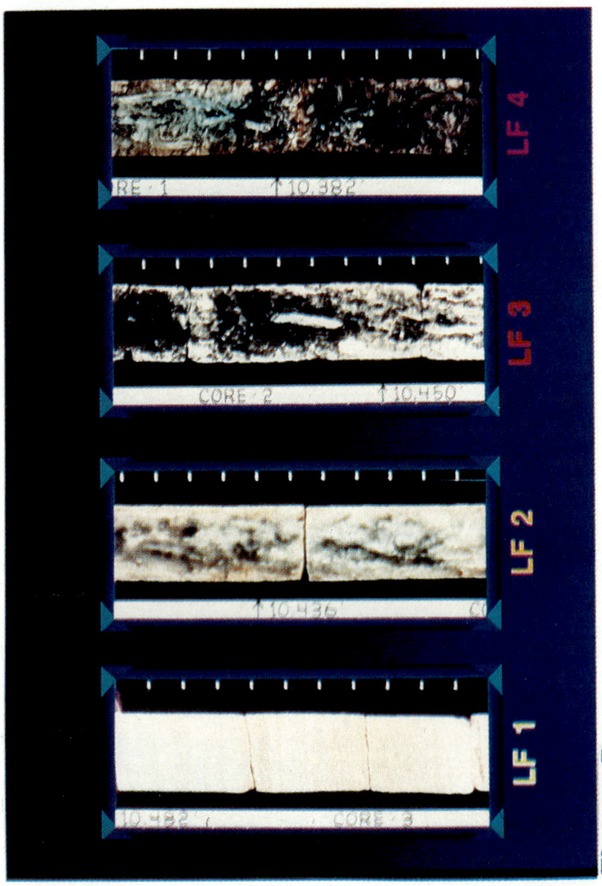

Figure 5

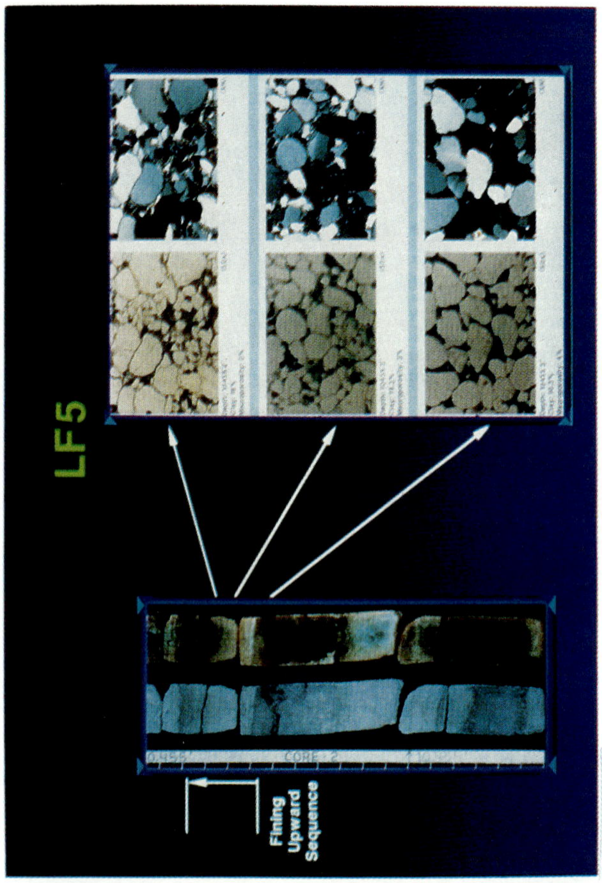

Figure 6

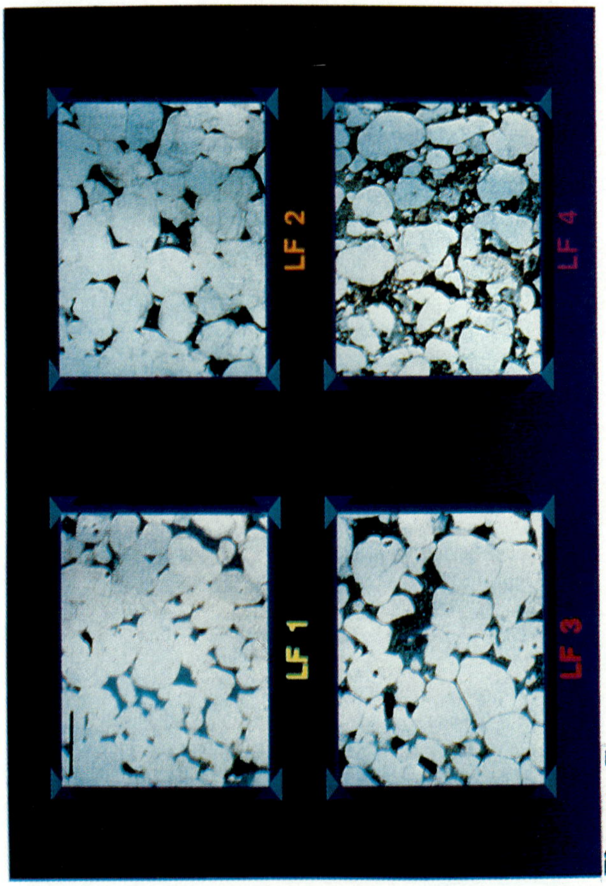

Figure 7

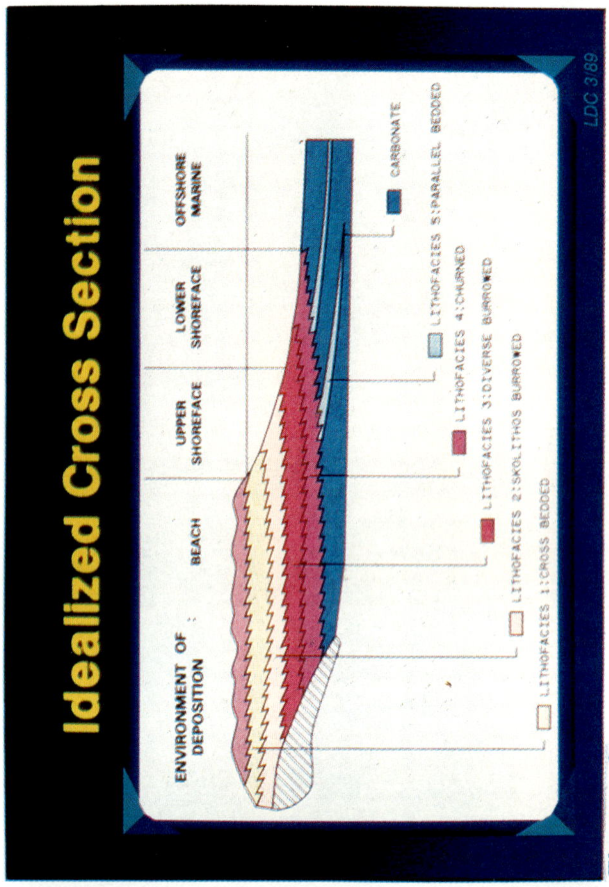

Figure 8

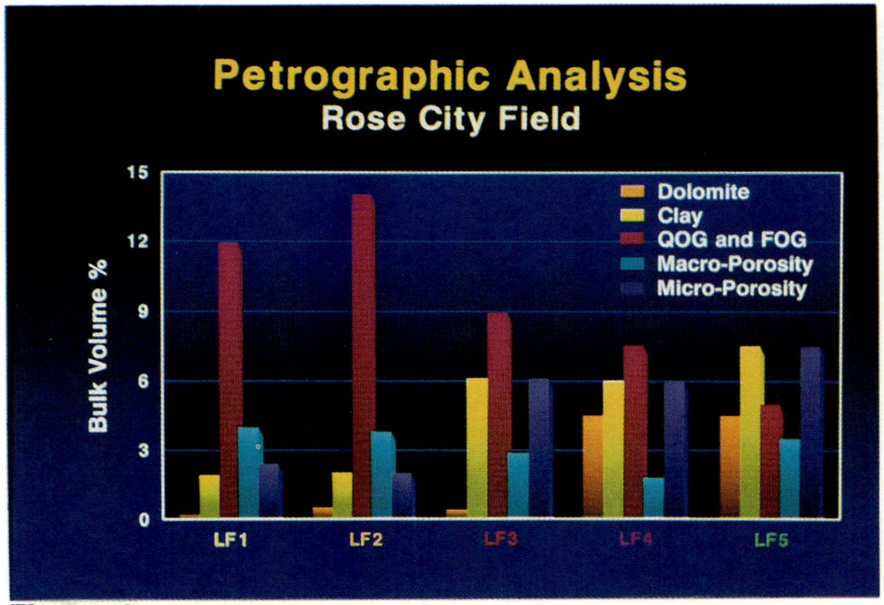

Figure 9

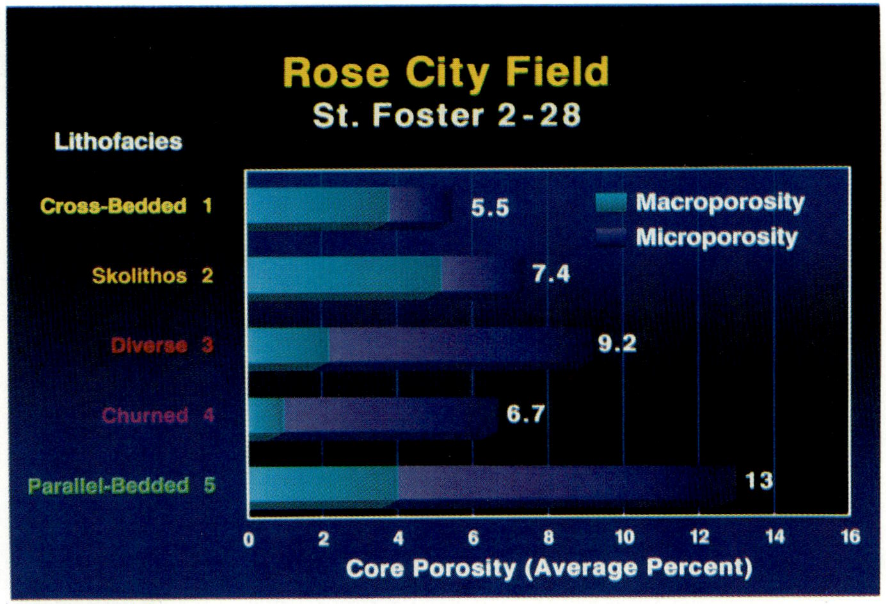

Figure 10

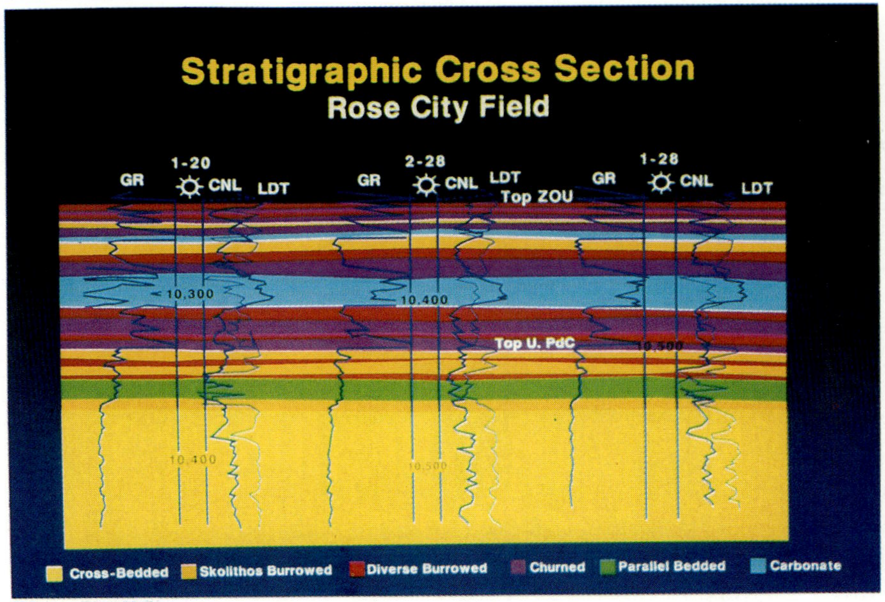

Figure 11

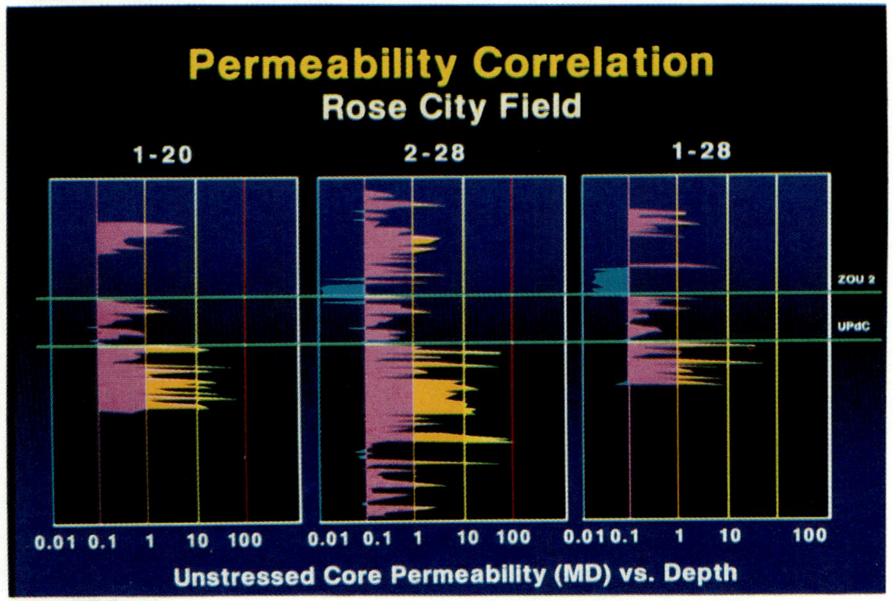

Figure 12

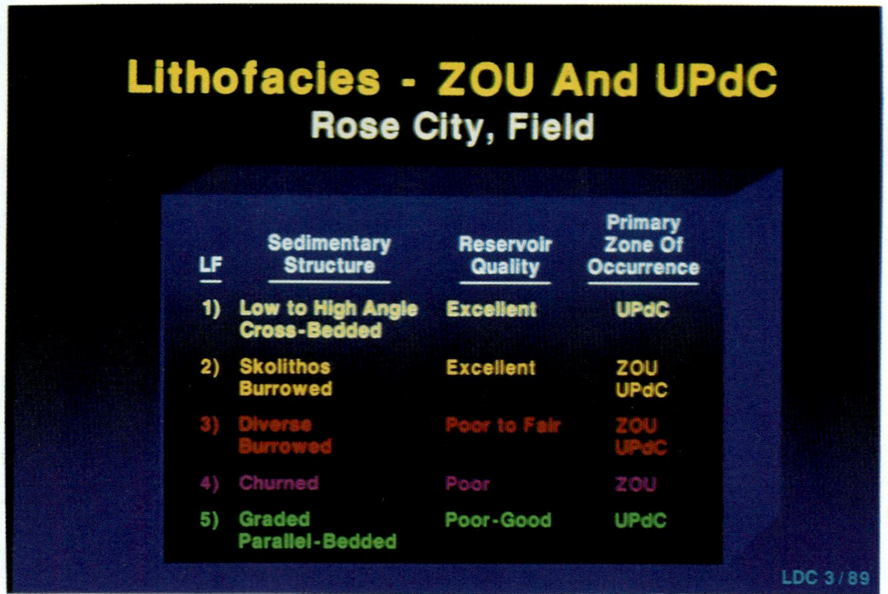

Figure 13

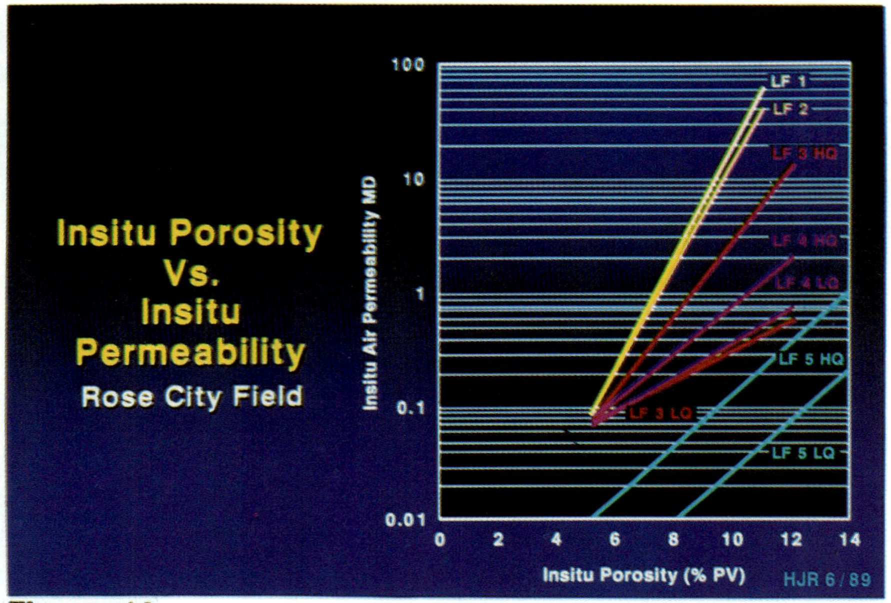

Figure 14

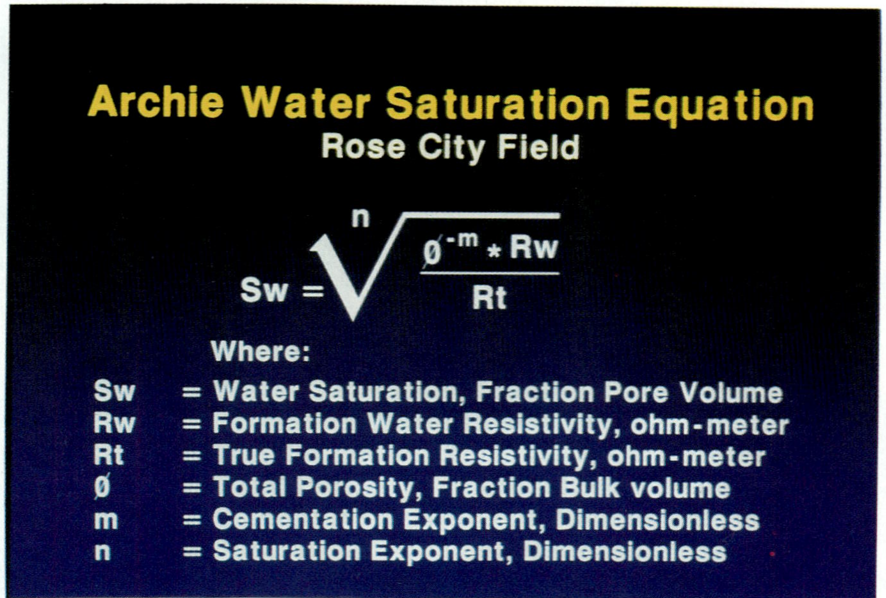

Figure 15

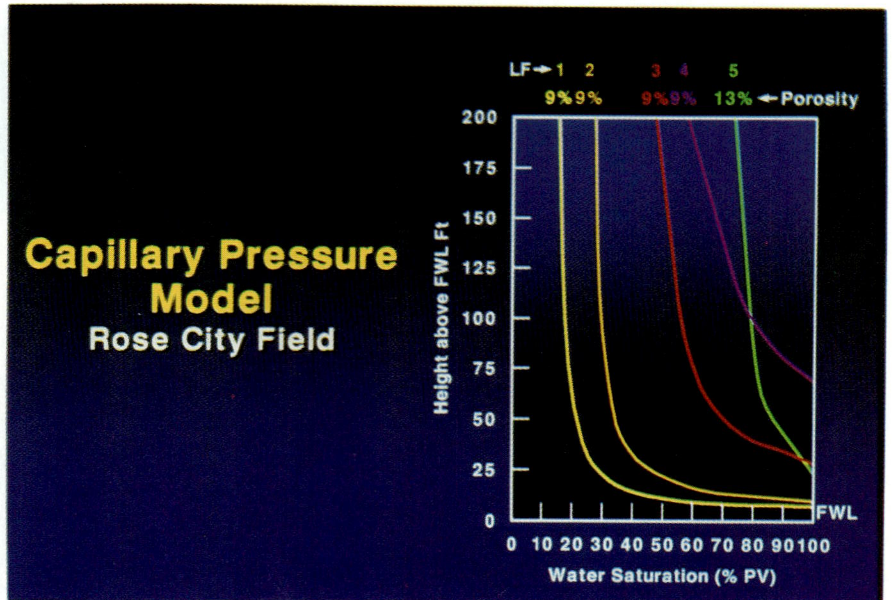

Figure 16

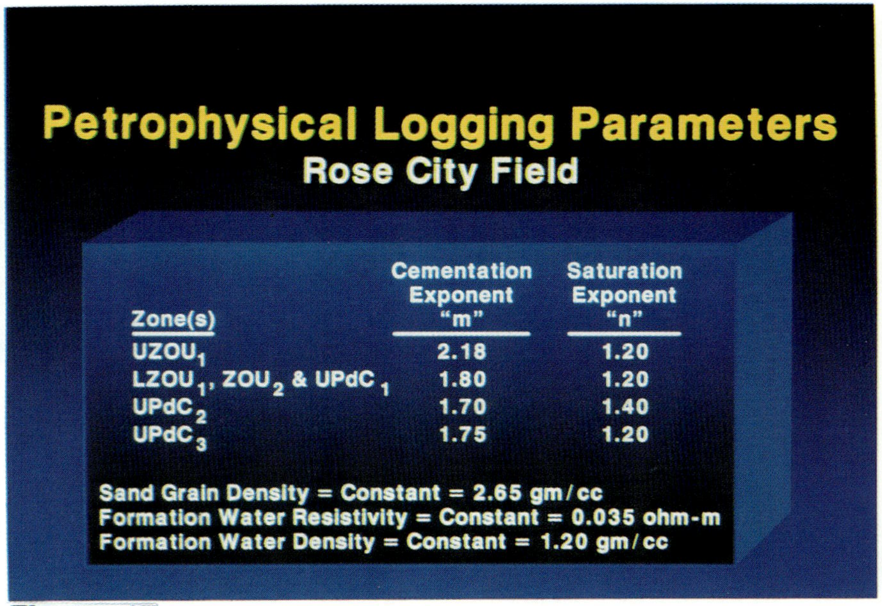

Figure 17

Figure 18

Figure 19

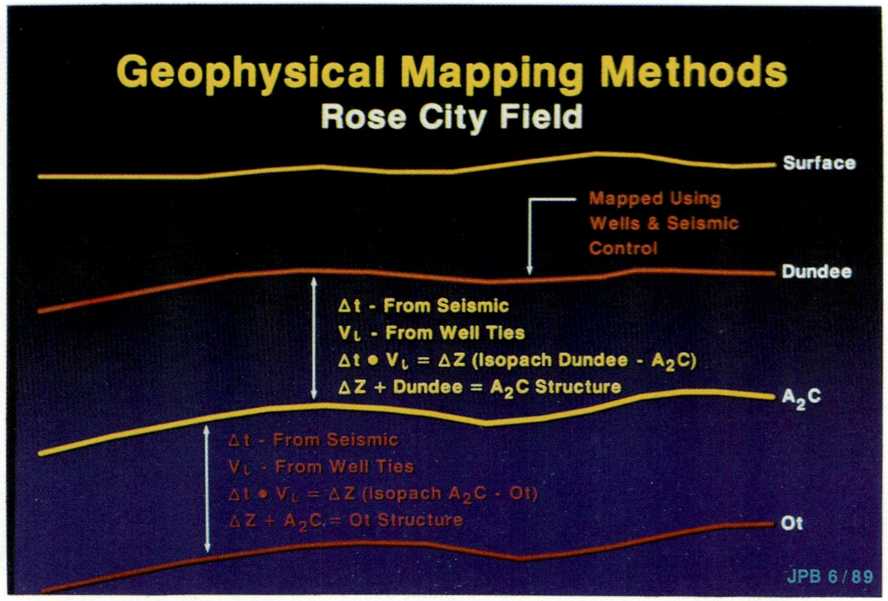

Figure 20

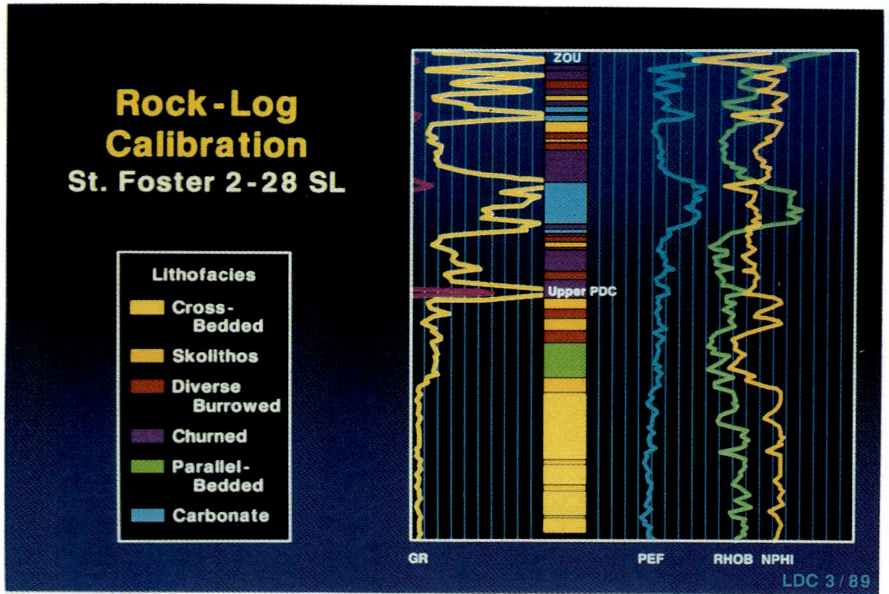

Figure 21

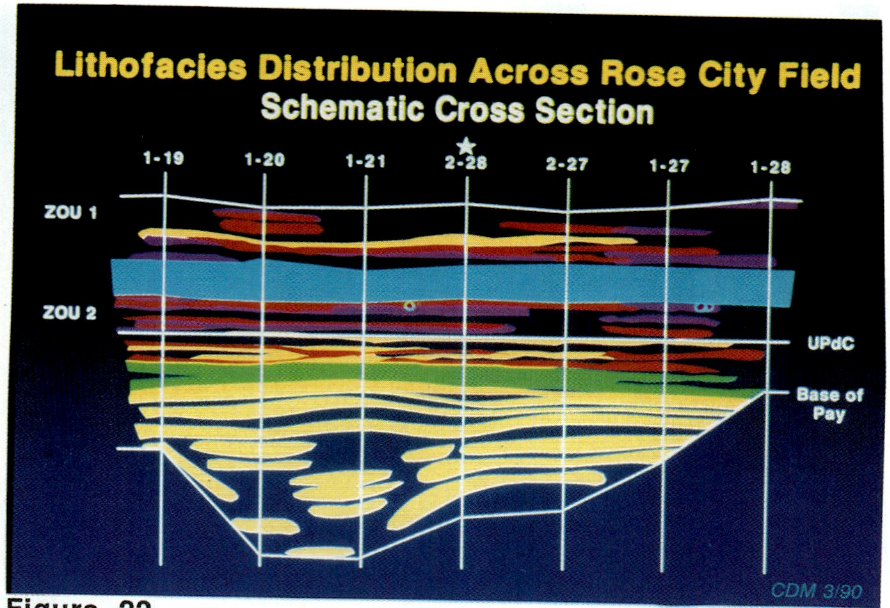

Figure 22

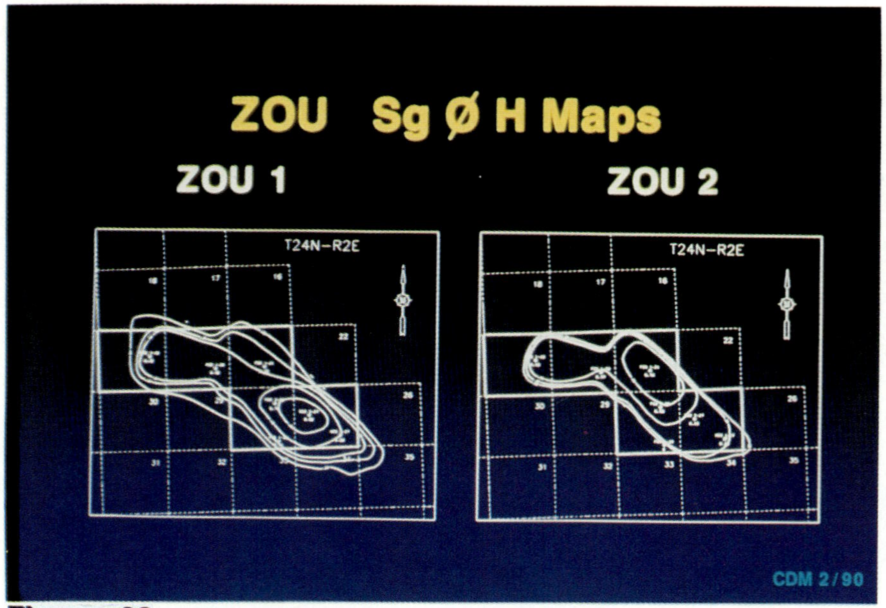

Figure 23

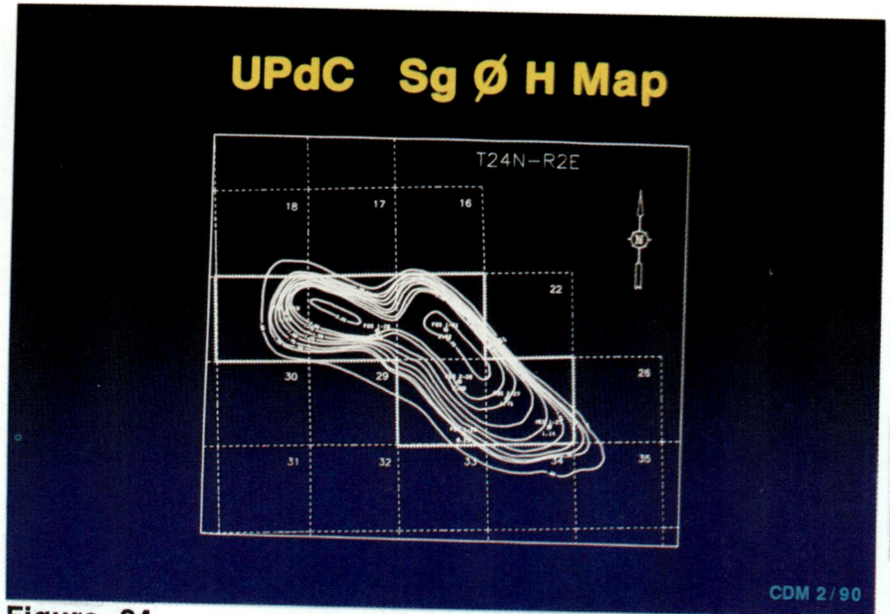

Figure 24

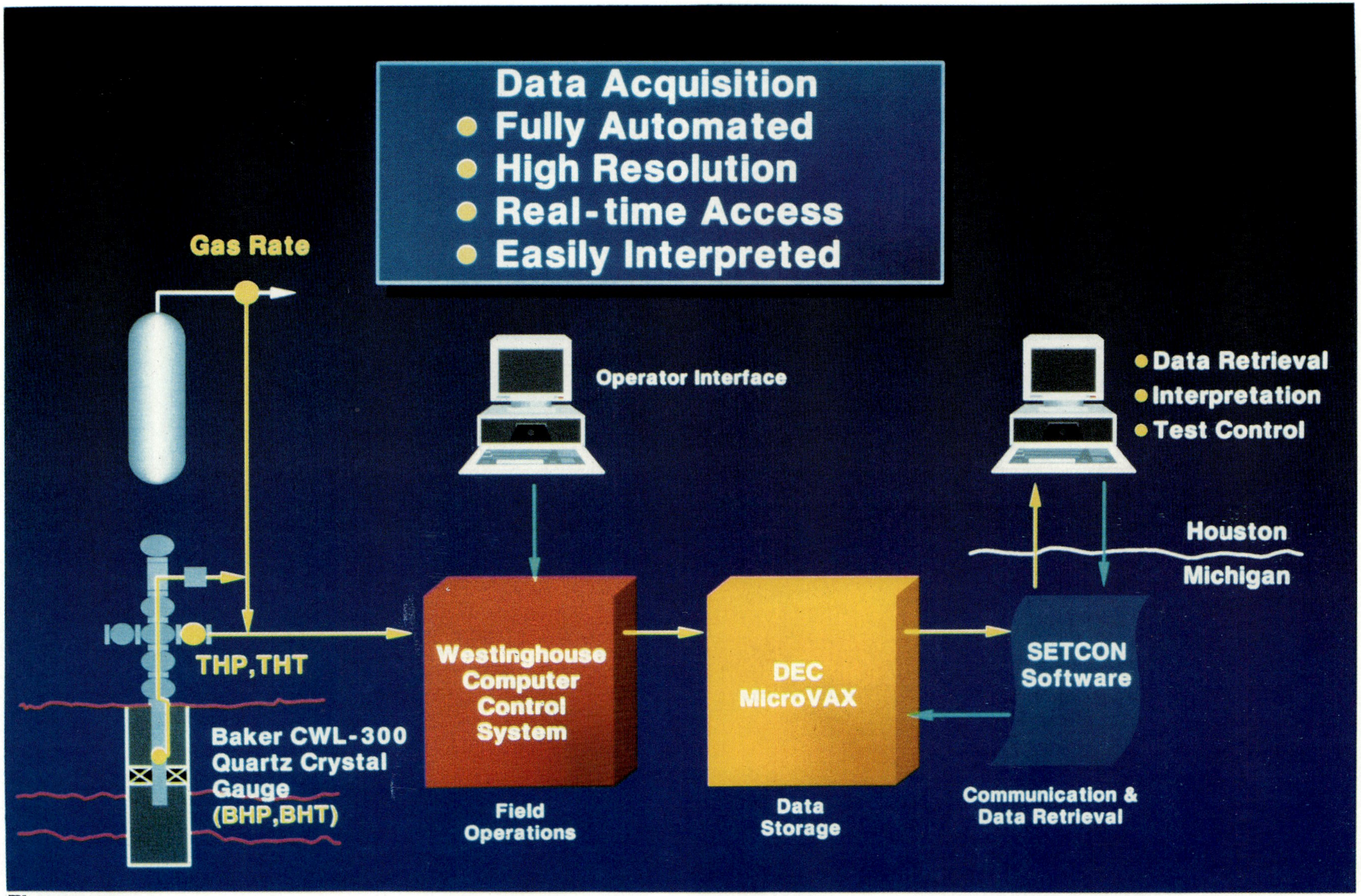
Data Acquisition
Fully Automated
High Resolution
Real-time Access
Easily Interpreted
Gas Rate
THP,THT
Baker CWL-300 Quartz Crystal Gauge (BHP,BHT)
Operator Interface
Westinghouse Computer Control System
Field Operations
DEC MicroVAX
Data Storage
SETCON Software
Communication & Data Retrieval
Data Retrieval
Interpretation
Test Control
Houston
Michigan

Figure 25

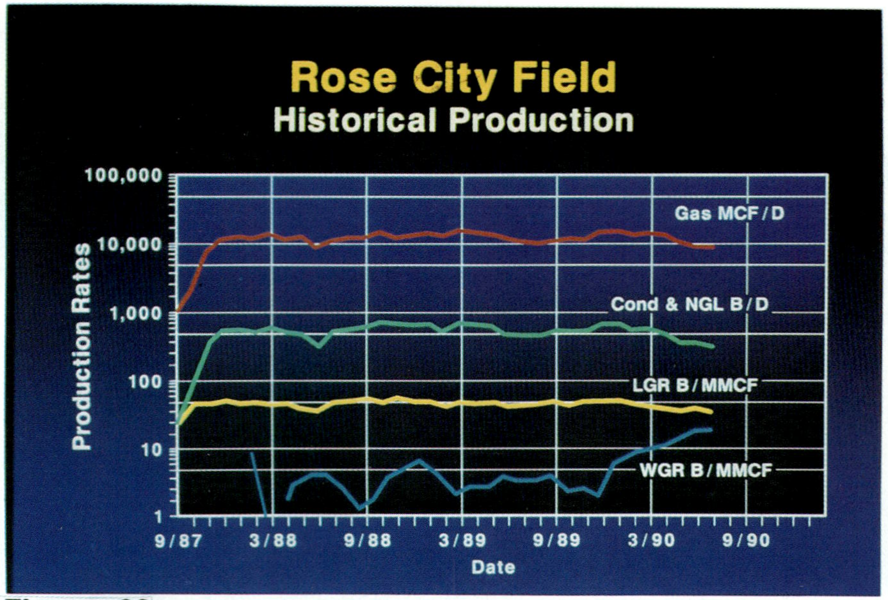

Figure 26

Testing Summary
Foster 1 - 19 Well

	Production Rate MMCF / D	Production Rate BW / D	WGR (BW / MMCF)	FTP (psi)
July 1988	1.20	15	12	290
May 1989	0.85	41	48	338

*Water-gas Ratio Increased 400% in 10 Months

Figure 27

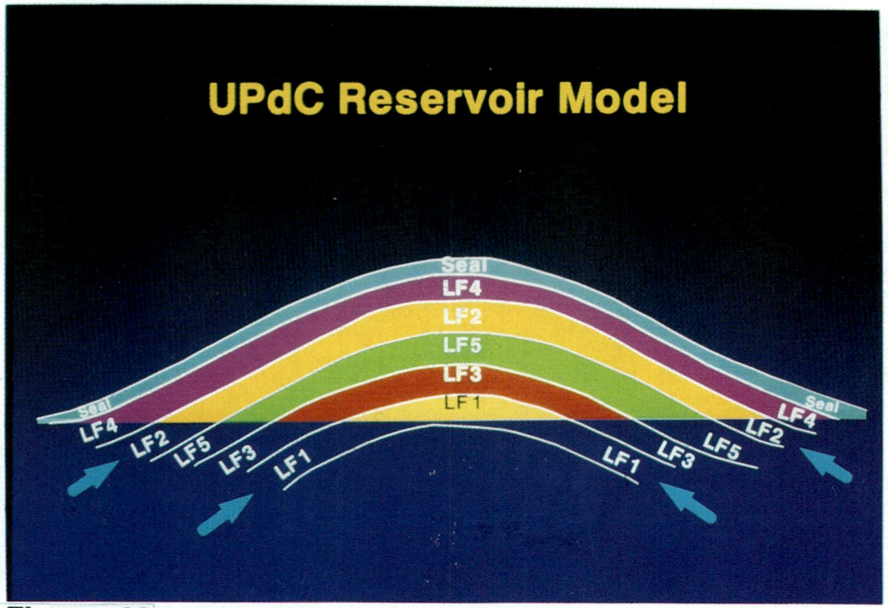

Figure 28

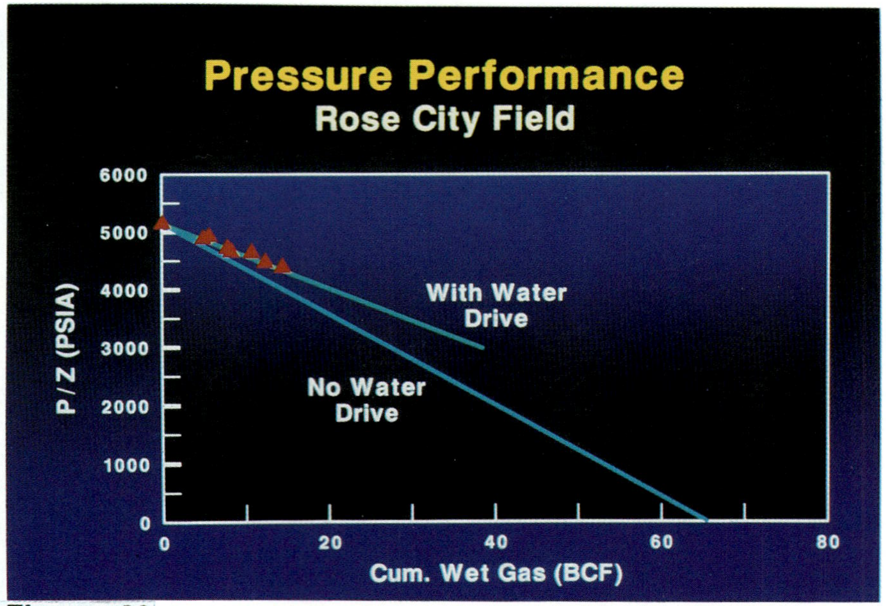

Figure 29

Technological Development Synergy: Kisbey, Saskatchewan—A Case History

David L. Sturrock
Gilbert M. Cordell
Don C. Westacott
David M. Fitzpatrick

ABSTRACT

Synergy between geology, petrophysics, and engineering in the development of the Kisbey Oil Field has led to significant economic benefits in all stages of field exploitation. Detailed integration of advanced petrophysical technology and geological analysis created an enhanced reservoir description which provided input to a synergistic reservoir simulation model. This model fed back into the development drilling strategy and provided a basis for production operations optimization. Data from innovative production techniques was fed back into the reservoir simulation model to create an overall production plan which has led to dramatically lower operating costs.

The Kisbey Oil Field is located in Township 8, Range 6 W2 in the Mississippian "buried hill" trend on the northwest flank of the Williston Basin in southeastern Saskatchewan, Canada. Trapping is caused by a major erosional valley on the Mississippian erosional surface truncating the reservoir units updip and on the flanks of the field. Production is obtained from the Mississippian Kisbey Sandstone with several wells providing additional production from the Mississippian Frobisher and Alida Beds.

Detailed core examination and Mineralog* analysis show the Kisbey to be a complex lithologic mixture. It consists of reservoir quality dolomitic sandstone interbedded with high porosity-low permeability sandy dolomite, both cemented to varying degrees with anhydrite. It is difficult, with conventional log analysis of the Kisbey interval, to determine lithology. In addition, conventional log analysis yields 60 to 80 percent water saturations for intervals producing with a negligible water cut.

Petrographic examination of the Kisbey sandstone reservoir reveals a two-porosity system consisting of intergranular porosity between sand grains and micro-intercrystalline porosity in the dolomite cement. Using this concept together with accurate cementation and saturation exponents determined from core analysis and Electromagnetic Propagation Tool (EPT)** log calculations, a detailed lithologic and pore geometry model for the Kisbey interval was constructed. Recently developed high resolution wireline log measurements processed with thin-bed definition techniques were interpreted using a multi-mineral log analysis program constrained by the lithologic model.

Data from this evaluation was integrated with extensive relative permeability, capillary pressure, and PVT analyses into field-scale reservoir simulations. An interactive approach was employed involving engineers, a geologist, and a petrophysicist to "fine tune" the model to actual field results. The reservoir model recognized a strong aquifer as the dominant depletion mechanism and identified a number of drilling opportunities.

The model also predicted that recovery is not rate sensitive, which led to individual well optimization. Full life forecasts for the field provided the basis for field consolidation and battery construction.

In summary, the integrated approach optimized field development by drilling fewer wells in better places, optimizing individual well production, and planning a production facility designed specifically to the long-term requirements of the field.

* TM Western Atlas International
** TM Schlumberger Well Services

INTRODUCTION

The main purpose of this paper is not only to describe the geology, petrophysics, reservoir, and engineering of the Kisbey Oil Field but also to show how an integrated team of a geologist, a petrophysicist, and reservoir and operations engineers used a synergistic approach to add significant value to a newly-discovered field. It is not hard to find descriptions of new technology in petroleum-related subjects. This paper differs in that it attempts to show how new technology has been utilized within a team environment to efficiently develop an oilfield. The paper is organized by individual disciplines, but we stress that all work was carried out concurrently with a reiterative approach applied amongst geology, petrophysics, and engineering.

LOCATION

The Kisbey Oil Field is located on the northeast flank of the Williston Basin in southeastern Saskatchewan, Canada (Figure 1). The field is situated at the erosional edge of the Mississippian Kisbey Sandstone below the basin-wide unconformity at the top of the Paleozoic (Figure 2).

I. Geology

A structure map on the Mississippian unconformity (Figure 3) illustrates the erosional topography which creates the updip and lateral trap. Kisbey is a "buried hill" or cuesta with a steep erosional edge to the northeast caused by a northwest-southeast trending erosional valley. A dip-oriented cross-section through the field (Figure 4) illustrates the erosional unconformity overlain by impermeable Triassic/Jurassic siltstones. The main reservoir unit, the Kisbey Sandstone, dips gently to the southwest and is truncated updip by the erosional valley at the northeast edge of the field. Production is also obtained from the Glenburn Member of the Alida Beds and from the Frobisher Beds. The oil-water contact corresponds to the deepest downcutting of the erosional valley to the northeast.

The field was discovered using seismic data which clearly shows the Mississippian event dropping into the erosional valley to the northeast (Figure 5). The paleotopographic nature of the structure can be clearly demonstrated by an isopach of the Triassic/Jurassic Spearfish Beds which overlie the unconformity (Figure 6). The rapid thickening on the northeast flank of the structure indicate that the erosional feature is a cuesta. Figure 7 illustrates the structure on the Kisbey Sandstone. The erosional edge and the 600 metre subsea contour define the edges of the field. Intra-Mississippian markers such as the Kisbey Sandstone dip gently to the southwest with only minor structural perturbations.

The density log of 1-13-8-6 W2 (Figure 8) shows a typical succession in which both the Kisbey and Alida are productive. In most wells, the Kisbey is overlain by a thin, impermeable cap of anhydritized Frobisher Beds below the unconformity.

Kisbey Reservoir

The main reservoir lithology in the Kisbey Sandstone is a clean very fine- to fine-grained quartz sandstone with microcrystalline dolomite cement (Figures 9, 10,11). Minor feldspar grains and anhydrite cement are common. There is little or no clay in these reservoirs. High gamma readings on logs throughout these intervals are caused by radioactive dolomite and small volumes of potassium feldspar. Kisbey sandstones are excellent reservoirs with porosities ranging from 8-25 percent and permeability from 50 md to over 800 md with an average of approximately 200 md.

Interbedded with this high quality reservoir is a sandy dolomite with quartz grains floating in a microcrystalline dolomite matrix (Figure 12). Minor amounts of anhydrite cement are common, but again clay volumes are very low with high gamma readings on logs indicating the presence of radioactive dolomite. Porosity values are very similar to the reservoir sandstone (20-26 percent), but permeability is much lower(0.5-2.0 md). Throughout this lithology, microporosity is prevalent and bound water saturations are high. The resistivity log reacts strongly to the bound water with readings often below 1 ohm metre. The two Kisbey lithologies are often interbedded leading many operators to interpret the entire zone to be water-bearing. These zones exist tens of metres above the oil-water contact and produce only small quantities of water.

Alida Reservoir

The Alida reservoir is a coated crinoidal grainstone in which the usual syntaxial overgrowth cement has been suppressed by an oolitic coating (Figures 13, 14, 15). The only cement in this reservoir is syntaxial calcite surrounding an occasional partially-coated crinoid fragment. Most of the porosity reduction has been by compaction as evidenced by fractures in the oolitic coatings (Figure 14). Reservoir quality is excellent with porosity ranging from 15-18 percent and permeability varying from 100 md to over 5 darcies with an average of about 250 md.

This lithology also occurs with an early spar calcite cement filling all the pores with a later stage of leaching which cuts across textural fabrics (Figure 16). These vuggy pores are partially infilled with a late calcite cement. The result is a lithology with identical porosity (15-18 percent) but dramatically lower permeability (less than 1 millidarcy) than the original reservoir. The porosity log, however, cannot differentiate the two Alida reservoirs.

Reservoir Characteristics

The lithological reservoir characteristics are summarized below as a geological reservoir model to integrate with the log analysis:

1. Thin beds, often less than one metre in thickness
2. Complex mineralogy—quartz, dolomite, calcite, anhydrite, and minor clays
3. Four different pore system types, all with different a, m, and n values
4. Bound water in high porosity, low permeability strata

II. Petrophysics

At the time the above geological reservoir model was defined, five wells had been drilled in the field. From this model, a petrophysical program was designed to fully evaluate the reservoir. Each characteristic was examined in light of current petrophysical technology.

1. Thin beds

Modern logging tools can be operated at a slower logging speed and processed to result in far greater resolution compared to standard logs. Figure 17 is an example of a reprocessed 1.2 inch density log compared with a normal 6 inch trace. Note the difference in resolution and the thin high porosity zones which are averaged with the surrounding beds on the normal resolution density trace. A similar significant increase in resolution can be obtained using the Dual Induction Phasor** tool and reprocessing the data. Using these tools, the interbedded sandstone and dolomite in the Kisbey were resolved.

2. Complex mineralogy

The complex mineralogy of the Kisbey was approached using an integration of several techniques. First Mineralog*—a quantitative mineral identification technique based on infrared spectroscopy—was run on the core available at the time. That data, combined with thin section and SEM petrographic data, provided some interesting observations. In any specific rock type, only three minerals were present. We began to build a model for a number of stratigraphically-correlatable rock types. For example, the Kisbey Sandstone is composed entirely of quartz, dolomite, and minor anhydrite. Thus, a matrix was developed for log definition of lithology with only those three minerals which makes the matrix easier to solve with fewer logs. Figure 18 illustrates the accurate mineralogical breakdown of the formation using this technique.

3. Pore system types

Differentiation of pore system types proved to be the most challenging petrophysical problem. It was approached using three different techniques. First, the electrical properties were measured using core samples from each of the different pore system types defined by petrography.

The results confirmed that there was a wide range of "m" and "n" from less than 1.8 in the Kisbey Sandstone to greater than 2.5 in the vuggy Alida limestone. Recognition of this variability led to the use of the EPT** tool.

The EPT** is a logging tool which uses microwaves to directly measure the water saturation in the near wellbore flushed zone. This provides an accurate measurement of water saturation in a zone in which all other variables are known. Thus "m" and "n" can be back-calculated in a continuous function down the borehole. These parameters are used to calculate water saturation in the uninvaded zone away from the wellbore.

The third technique was to cut an oil-based core. As well as providing the reservoir engineers with samples to measure wettability and relative permeability, there was now a measured water saturation to compare with log calculations. Figure 19 illustrates the variable "m" and "n" calculations and the measured and calculated water saturations using the oil-based core.

4. Bound water zones

Identifying high porosity-low permeability zones required an innovative new technique. Using the data at hand, an attempt was made to measure permeability from log data using a variation of the Coates free-fluid model which is based on porosity, hydrocarbon saturation, and irreducible water volumes. The calculated permeability was then matched to permeability measured from core. The low permeability zones which this technique identified correlated well with the high volume dolomite zones recognized using petrography.

A summary log illustrating the geological/petrophysical model shows that using high resolution logging techniques, an accurate match could be obtained with porosity and permeability measured on core and water saturations measured on oil-based core (Figure 20).

At this time, it became apparent that the published government water saturations for the Kisbey Field were far too high, and thus their calculated reserves were too low. In Saskatchewan, the allowable production rate is based on government-calculated reserves. A detailed presentation of the above data and conclusions convinced government geologists that the average water saturation was 26 percent—not the 40 percent carried previously. On that basis, the field allowable for Kisbey was increased by 23 percent—an extra 300 barrels of oil per day.

Using the above techniques, the field was quickly drilled to 80 acre spacing. The reservoir quality was mapped as the drilling proceeded, leading to a success ratio of 88 percent while fully developing the field in 16 months.

Throughout the drilling phase, close communications were maintained amongst all disciplines. Synergy can only be sustained with an iterative approach. When the drilling was completed, the geological reservoir model was used to divide the producing intervals into six flow units on the basis of similar reservoir characteristics. Structure, net pay, porosity, permeability, and water saturation were mapped throughout the field for each reservoir unit using a computer mapping program with considerable geological input to the contouring. This data was transferred directly into the reservoir simulator.

III. Reservoir Engineering

Reservoir

In excess of 60,000 data points were taken from the geologic/petrophysical database and put into the reservoir model. The model itself is a six-layer, three-dimensional, black oil simulator. For the Kisbey field, which measures approximately two miles long by one mile wide, we generated graphical displays in order to verify the accuracy of the model input parameters. Figure 21, the Kisbey unit horizontal oil perm, is an example of this type of output. These representations enabled us to sit down with the geologist and petrophysicist in a round-table fashion, visually verify any known anomalies, and quickly correct any spurious data. The result was that a large quantity of data was processed and quickly verified.

Parameters

The reservoir and operating parameters of the two main producing horizons in the field the Kisbey Sandstone and the Glenburn limestone, are listed as Figure 22. Permeability of the two zones was measured from a combination of core, DST, and oil well buildup tests. The oil gravity ranges from a low of 33° (in the portion of the reservoir near the oil-water contact) to a high of 38°in the crest of the structure. Original reservoir pressure stands at 1625 psi. We are dealing with an undersaturated oil pool as bubble point pressure is 975 psi. The producing gas/oil ratio is 440 scf/bbl. To give you an idea of the size of the pool, we are dealing with 16 MMBO in place, of which only 8.8 MMBO is deemed mobile oil. Approximately 45% of the oil will remain trapped as a residual oil saturation.

Simulation Results

The simulator provided three basic reservoir engineering conclusions, as follows:

1. Aquifer support is the dominant depletion mechanism.
2. There is no need to inject fluid to maintain pressure support.
3. Do not fieldwide infill drill to 40-acre spacing.

Figure 23 depicts individual well DST pressures versus time. A very slow, shallow pressure decline of the field from the original pressure of 1625 psia is apparent. A drop of approximately 30 psia was observed over this three-year period. Figure 24 repeats the same shallow pressure decline and also demonstrates the effect of taking the water aquifer out of the model. Without the pressure support, reservoir pressures by late 1988 would have been more than 250 psia lower. With the higher reservoir pressures observed in the field, we feel confident about the active nature of the downdip aquifer.

In order to accurately quantify the OOIP, particularly in the downdip portion of the pool, sensitivities on the position of the oil-water contact were run. Figure 25 is a plot of watercut versus time for the 1-14-8-6 W2 well. This well has a field watercut of 14%, and our base case model predicted a cut of approximately 17%. If we had placed the OWC 2 m or 6 feet higher, a watercut of 90% would have been predicted. Likewise, if we drop the OWC by 6 feet, the well would not produce any water at all in the simulator. Having matched the field watercut, we felt secure about the model-predicted OOIP and recoverable volumes.

One of the first important pieces of data to come out of the model was a plot of predicted oil rate versus time (Figure 26). It shows production peaking in 1990 at approximately 1000 BOPD and declining rapidly over the next five years at 8% per annum. The total area under the plot or the integrated volume represents a 32% recovery factor.

To address the second reservoir engineering conclusion of the non-necessity of injecting additional fluid, Figures 27 and 28 represent reservoir pressure in the main Kisbey Zone time dated to the years 1994 and 2026. In 1994, the area to the southwest represents reservoir pressures close to the original 1625 psia. The area most depleted by this time (northeast) still has reservoir pressures of approximately 1300 psia.

By the year 2026, which is the end of the economic life of the pool, pressures have dropped to 1100 psia. This is still well above the bubble point pressure of 975—hence, no gas cap.

To support the third reservoir engineering conclusion—that regarding infill drilling—we simulated four 40-acre locations in the top end of the reservoir (Figure 29). The four wells increased oil production approximately 250 BOPD but overall incremental recovery was estimated at 50,000 barrels total (Figure 30). The incremental recovery, coupled with the rate acceleration factor, could not economically justify the drilling of the infill wells.

Two other important pieces of data provided by the simulator were gas and water rates predicted over time (Figures 31 and 32). Gas rates are expected to drop from a current level of 0.5 mmcf/d to 100 mcf/d at the depletion point. Watercuts will climb quickly from the current 30% to in excess of 85% in five years. This data, along with the predicted oil rate curve, was discussed at length with the Production Operations staff. It provided the basis for their field optimization plans.

IV. Production Operations

The Production Operations engineer contributed to the development process in three areas, as follows:

1. By providing model data input and well response rationale;
2. By evaluating and implementing a cost-effective operating scheme; and
3. By optimizing well production as per the model conclusions.

The acquiring of reservoir PVT data for the model is a good example of teamwork. The well sampling program was not only designed jointly by the Operations and Reservoir staff, but also both disciplines were present in the field to conduct the test. This resulted in the acquisition of accurate PVT data in a safe and economical manner.

The Operations staff reviewed all wells for anomalous operating characteristics. Wells with high wax production, emulsion problems, and pumped off or high fluid levels helped explain the differences in well productivities and verify wellbore skin damage.

By January 1989, the field had been delineated, and now the Operations people were charged with optimizing cashflow from the property. At that time, it was understood that additional water did not need to be injected for pressure support; therefore, an injection system did not need to be designed. It was also acknowledged that producing watercut and ultimate oil recoveries were not rate-sensitive parameters.

To optimize the field operation, a fully-consolidated independent process system was installed.

We went from a field of 18 single-well batteries (with all fluid hauled away) to a closed-loop pipeline system that eliminated all fluid hauling (Figure 33). Due to the proximity of the field to the Kisbey townsite, we felt this process would minimize any potential environmental impact.

Armed with the reservoir simulator results, the Operations staff went on a well optimization rampage. A fieldwide dynamometer and fluid level survey was conducted. Through a combination of pumpjack speed increases, stroke length adjustments, and pump changes, oil production across the field was increased 175 BOPD (+17%). Figure 34 depicts the enhanced oil production in excess of the model-predicted oil rate profile.

Timing

Over a 17-month period (between the second quarter of 1987 and the end of 1988) oil production climbed from 25 BOPD to in excess of 1000 BOPD. At the same time, the number of wells on production went from 2 to 18. Figure 35 illustrates the timing of the development process.

Over this 17 month period...

...the first development wells were drilled,

...the geological/petrophysical model was built,

...the pool was mapped and drilled to delineation,

....a reservoir model was built,

...a history match was obtained,

...predictions/sensitivities were evaluated,

...well rates and property cashflows were maximized, and

...partners were successfully AFE'd for consolidation equipment.

We feel that this expeditious development could only have been accomplished by working within a team-oriented group.

Project Economics

The economics of the project are very attractive. As a result of the consolidation, operating costs were cut from \$2.94/bbl to less than half, at \$1.41/bbl. The netback (based on a constant oil price) climbed from \$15.06/bbl to \$16.60/bbl oil. The reduction in economic limit increased the remaining recoverable reserves by 15%. The overall net effect of the incremental reserves and the lower operating costs increased the property present value (disc. @ 20%) by 25%. We now felt that we had come full circle in the development of the field and were realizing full economic value for the integrated effort.

CONCLUSIONS

1. The early development of a geologic/reservoir model spurred a petrophysical evaluation enabling us to characterize this complex field.
2. The integration of geology and reservoir engineering produces greater reservoir simulation accuracy.
3, Inclusion of the production/operations staff in the development plans optimizes reservoir value.
4. Interdisciplinary synergy results in cost-effective, time-efficient exploitation.

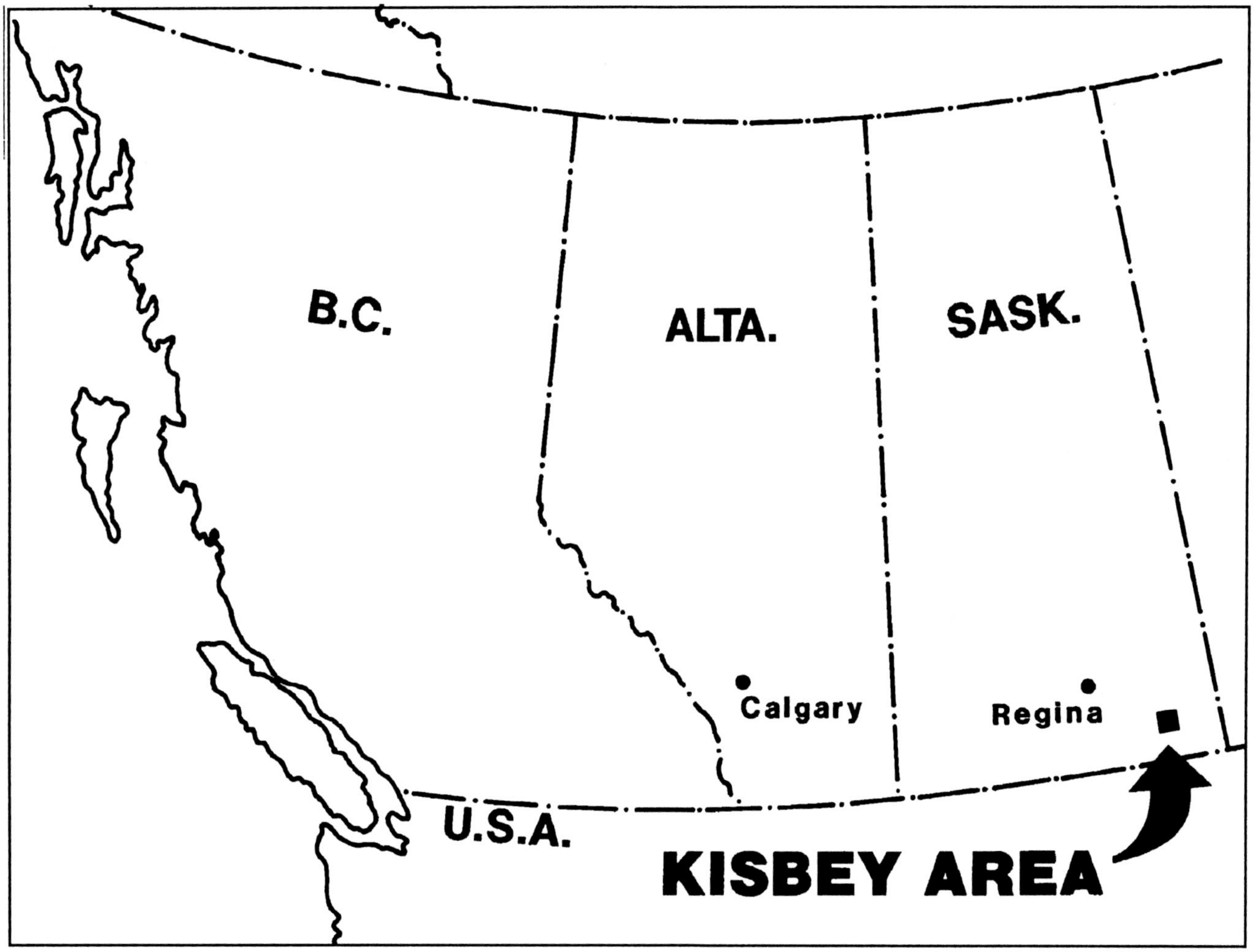

Figure 1—Location of the Kisbey area

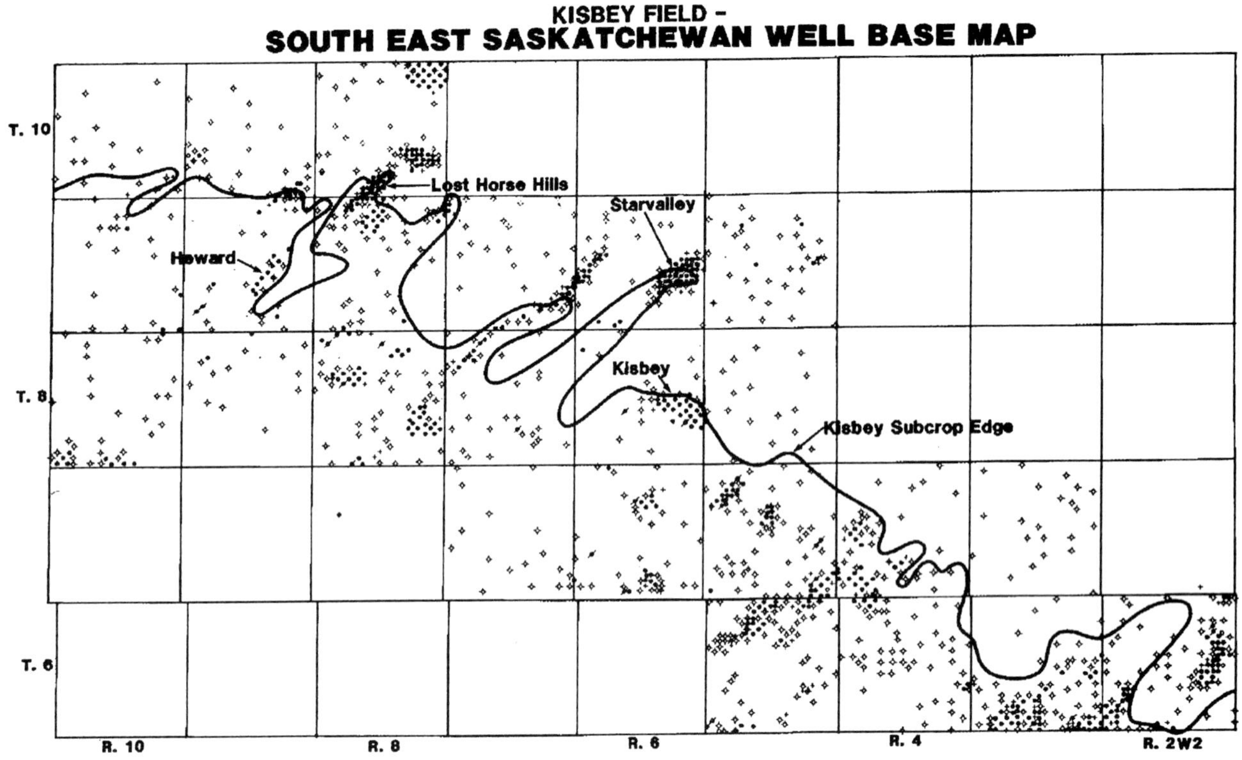

Figure 2—Subcrop map of the Kisbey Sandstone. Note the location of the Kisbey Field and other fields producing from the Kisbey Sandstone.

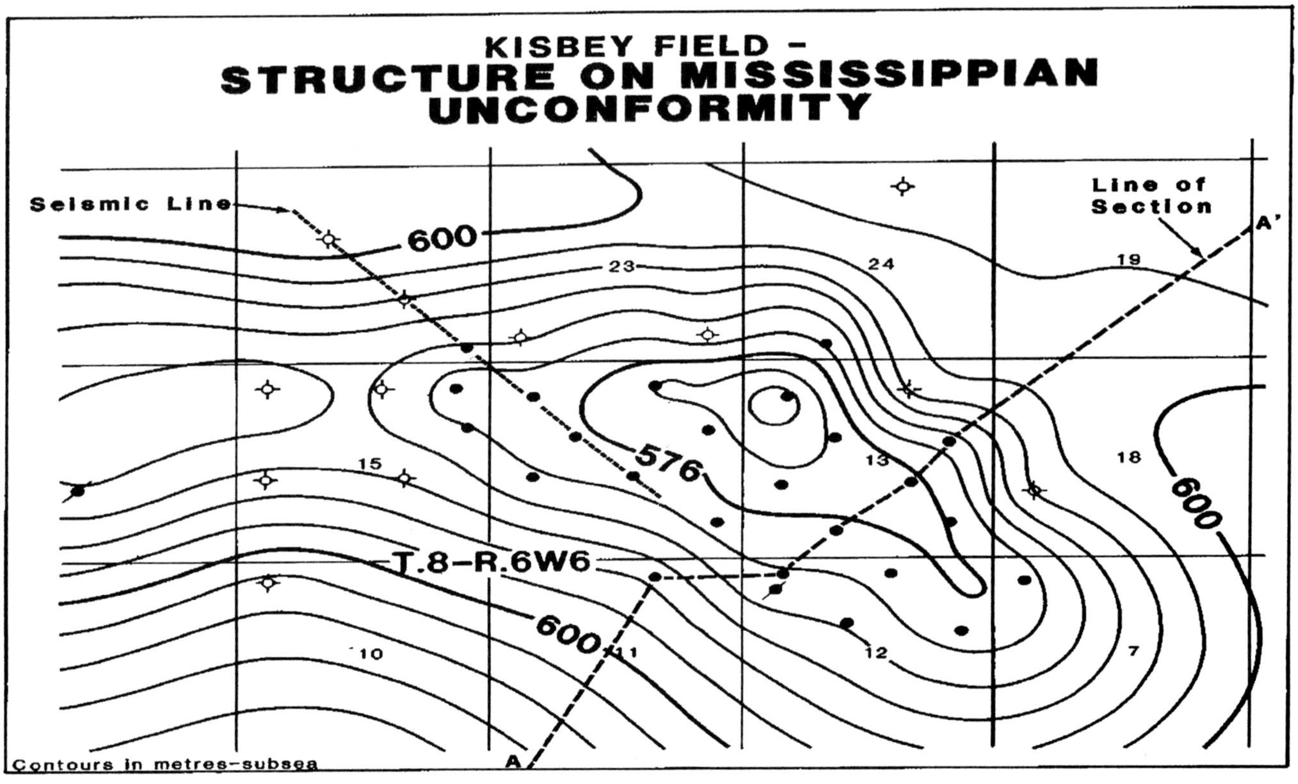

Figure 3—Structure map on the Mississippian unconformity showing the "buried hill" structure. Note the structural low or "valley" northeast of the field.

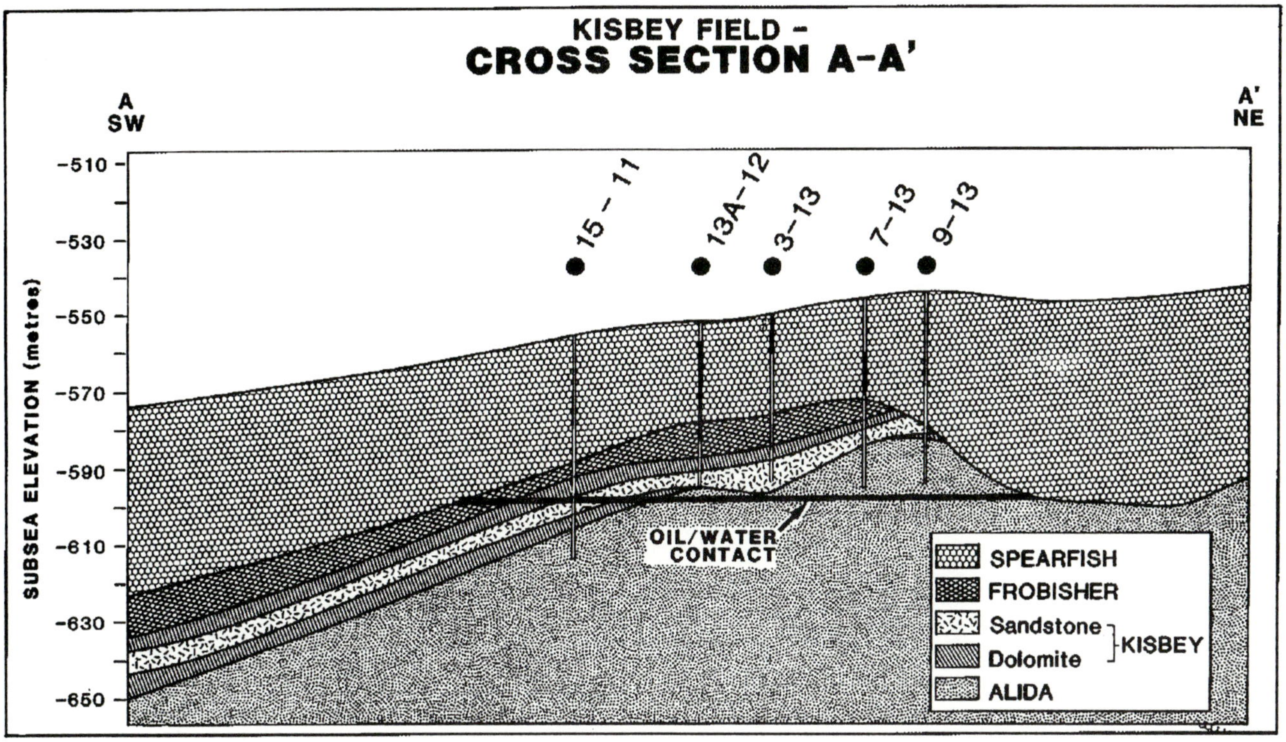

Figure 4—Schematic structural cross-section across the Kisbey Field. For location, see Figure 3. Note the erosional truncation of the Frobisher, Kisbey, and uppermost Alida (Glenburn) into the valley on the northeast end of the section.

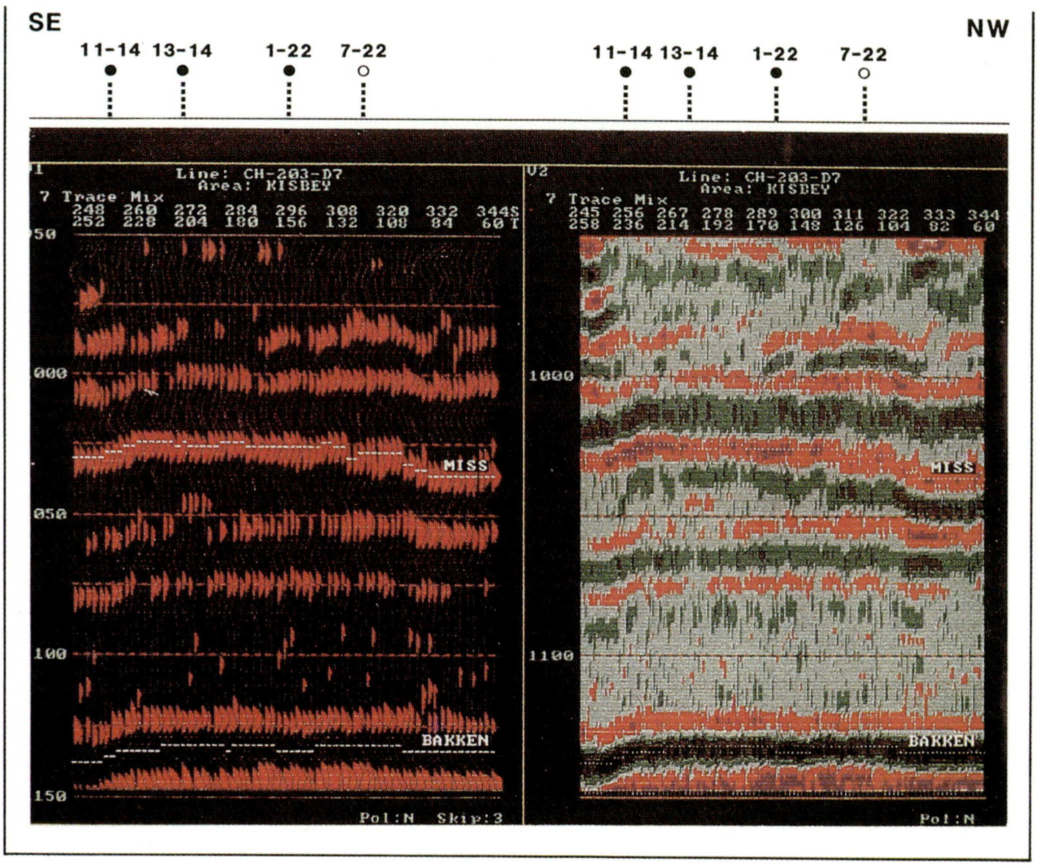

Figure 5—Seismic section across the northwest edge of the Kisbey Field. For location, see Figure 3. The Mississippian reflector drops to the northwest into the erosional valley.

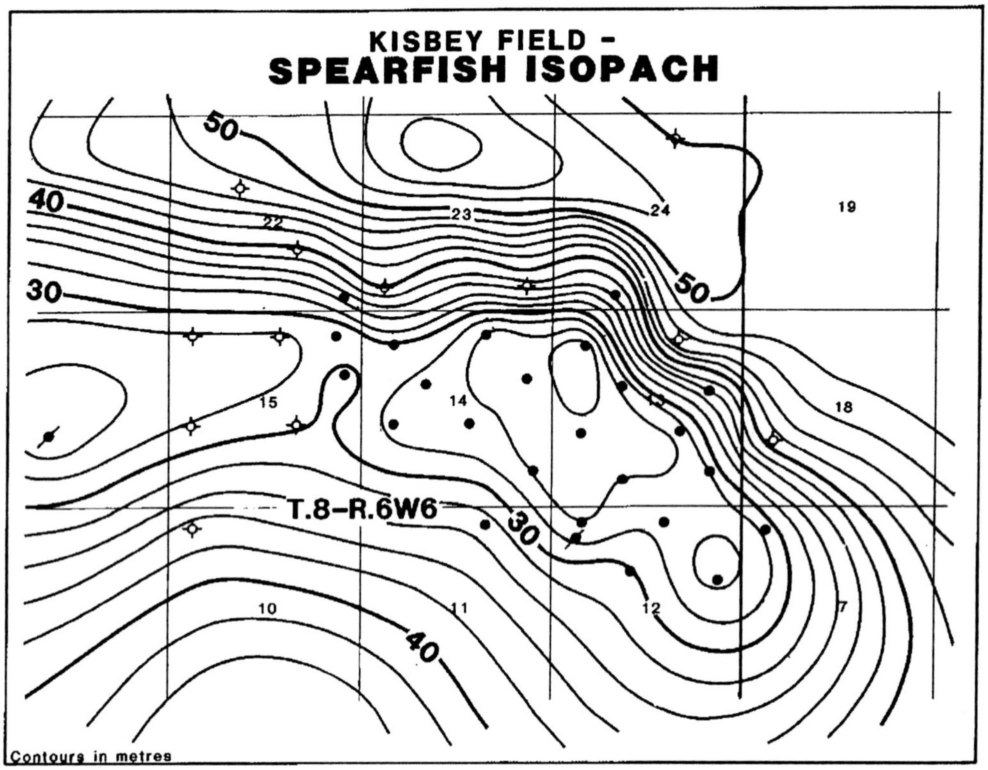

Figure 6—Isopach of the Triassic/Jurassic Spearfish Formation which overlies the Mississippian unconformity. The crest of the buried hill is where the isopach values are thin. The steep north flank and gentle south flank of the hill suggest that the erosion feature is a cuesta.

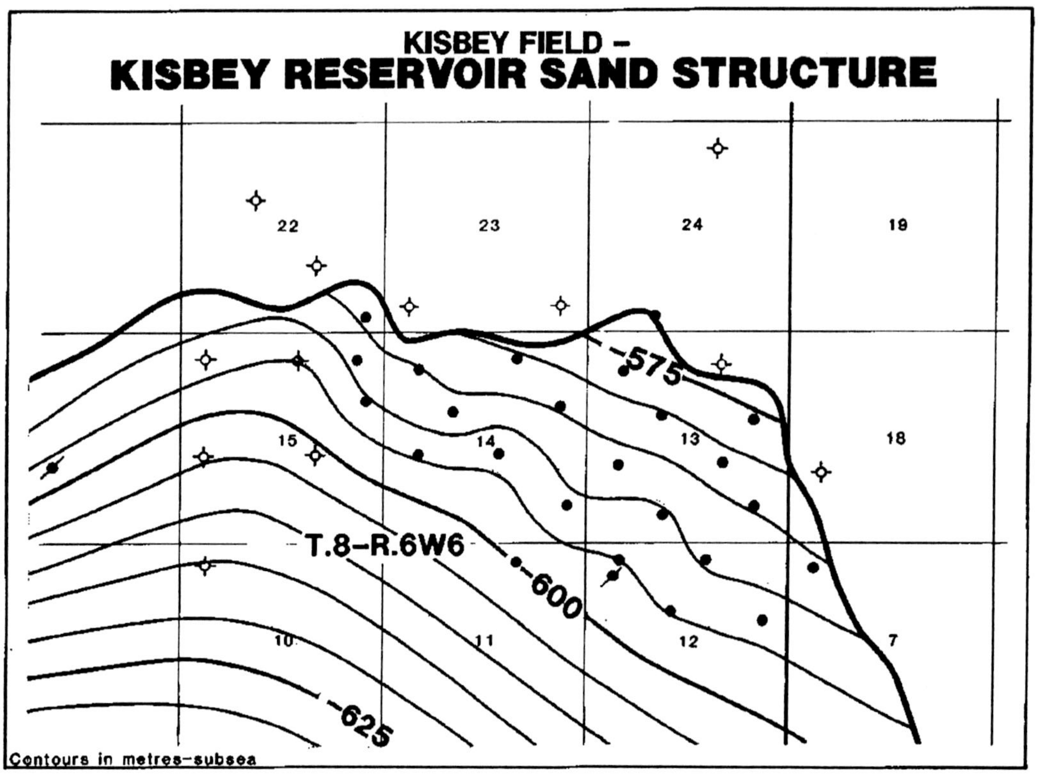

Figure 7—Structure map on the top of the Kisbey Sandstone showing the erosional edge. The contours indicate a gently southwesterly-dipping, relatively unstructured surface.

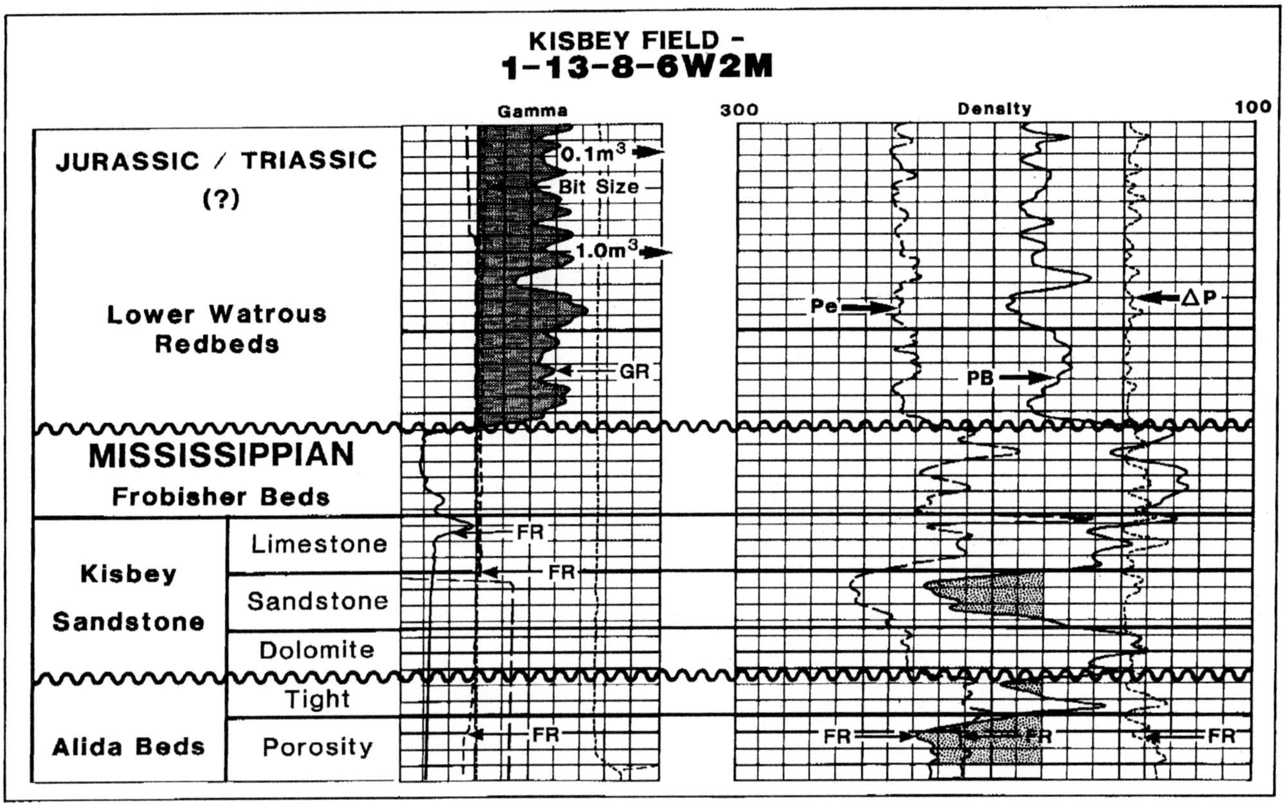

Figure 8—Bulk density-gamma ray log of the well located at 1-13-8-6 W2. This well produces from both the Kisbey Sandstone and the Alida Beds.

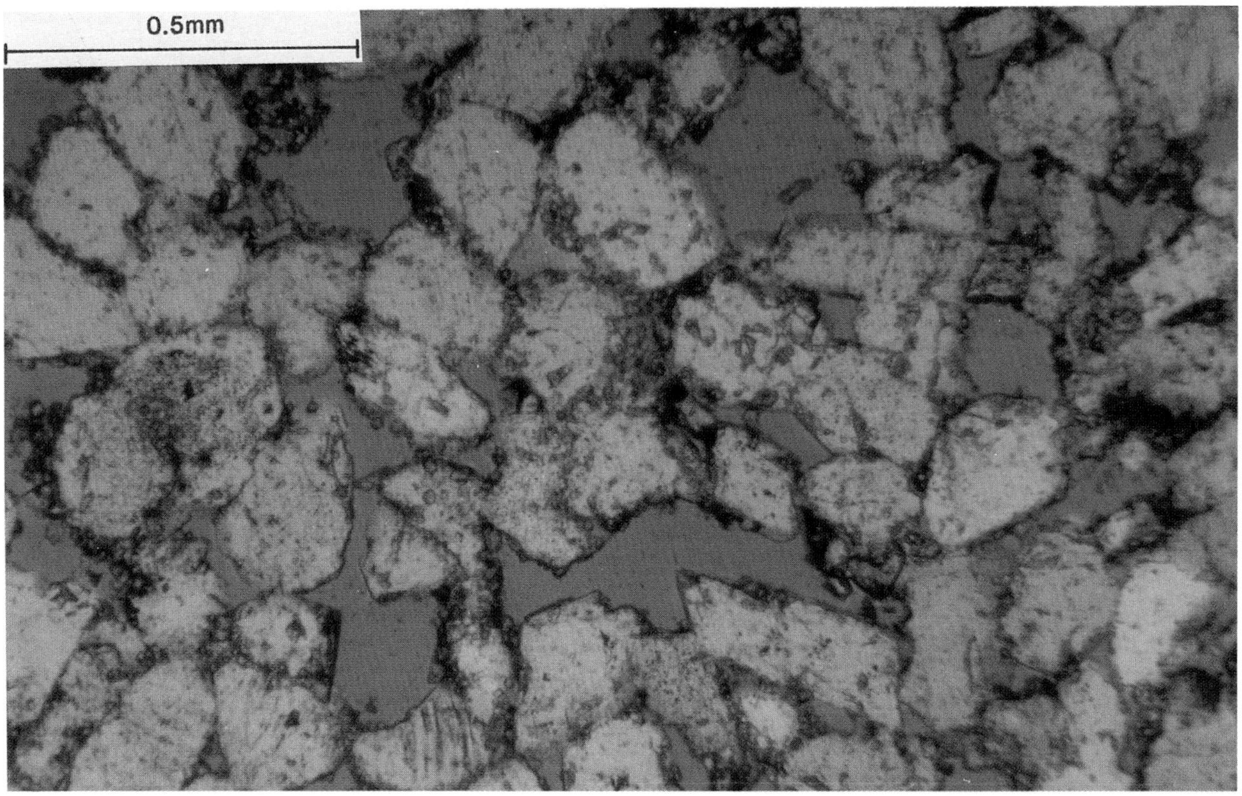

Figure 9—Thin section photomicrograph of the Kisbey sandstone reservoir. Porosity has been impregnated with blue epoxy. The sandstone is composed almost entirely of very fine- to fine-grained quartz with only minor quartz overgrowth cement.

Figure 10—Thin section photomicrograph of the Kisbey sandstone reservoir at greater magnification. Porosity has been impregnated with blue epoxy. Microcrystalline dolomite cement surrounds most of the quartz grains.

Figure 11—Scanning Electron Photomicrograph of the Kisbey sandstone reservoir. The microcrystalline dolomite cement can be clearly seen enveloping most of the quartz grains.

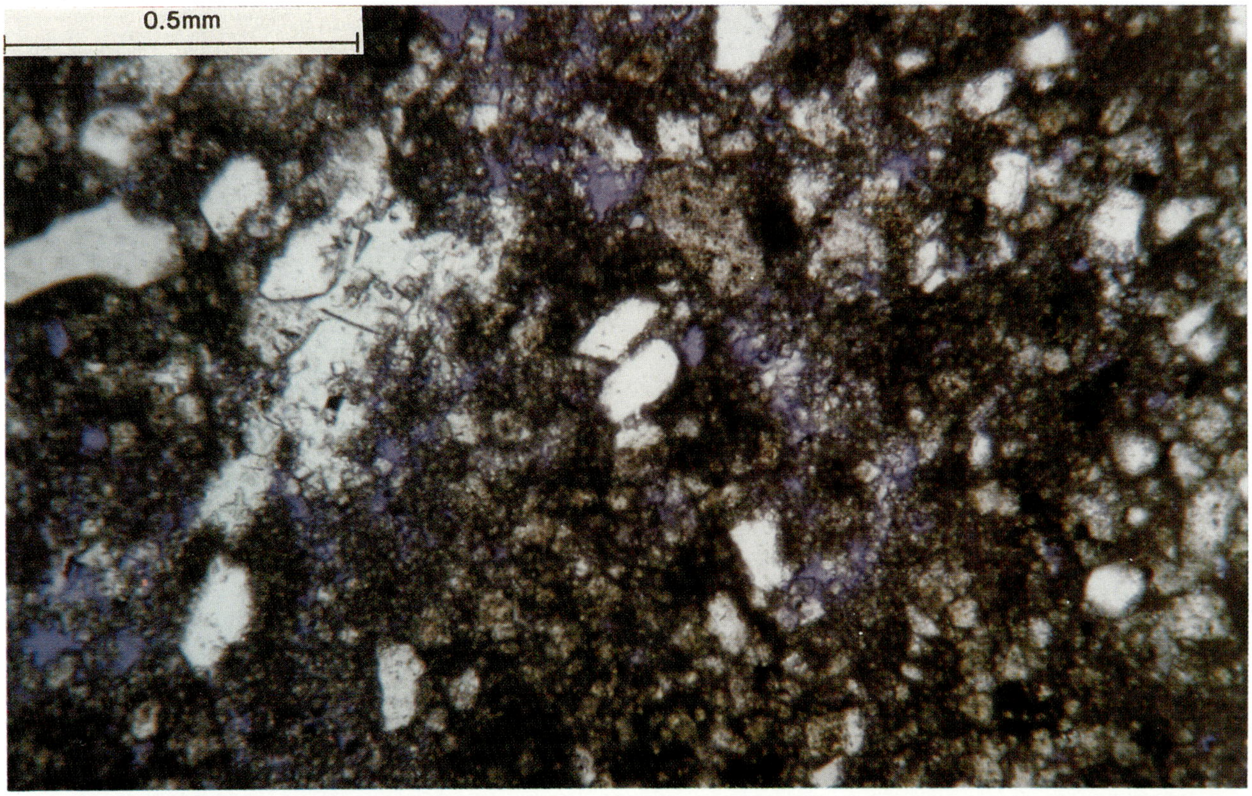

Figure 12—Thin section photomicrograph of the porous microcrystalline dolomite in the Kisbey Sandstone. Porosity has been impregnated with blue epoxy. Very fine grains of quartz float in the dolomite matrix. Minor anhydrite cement (A) has filled some of the porosity.

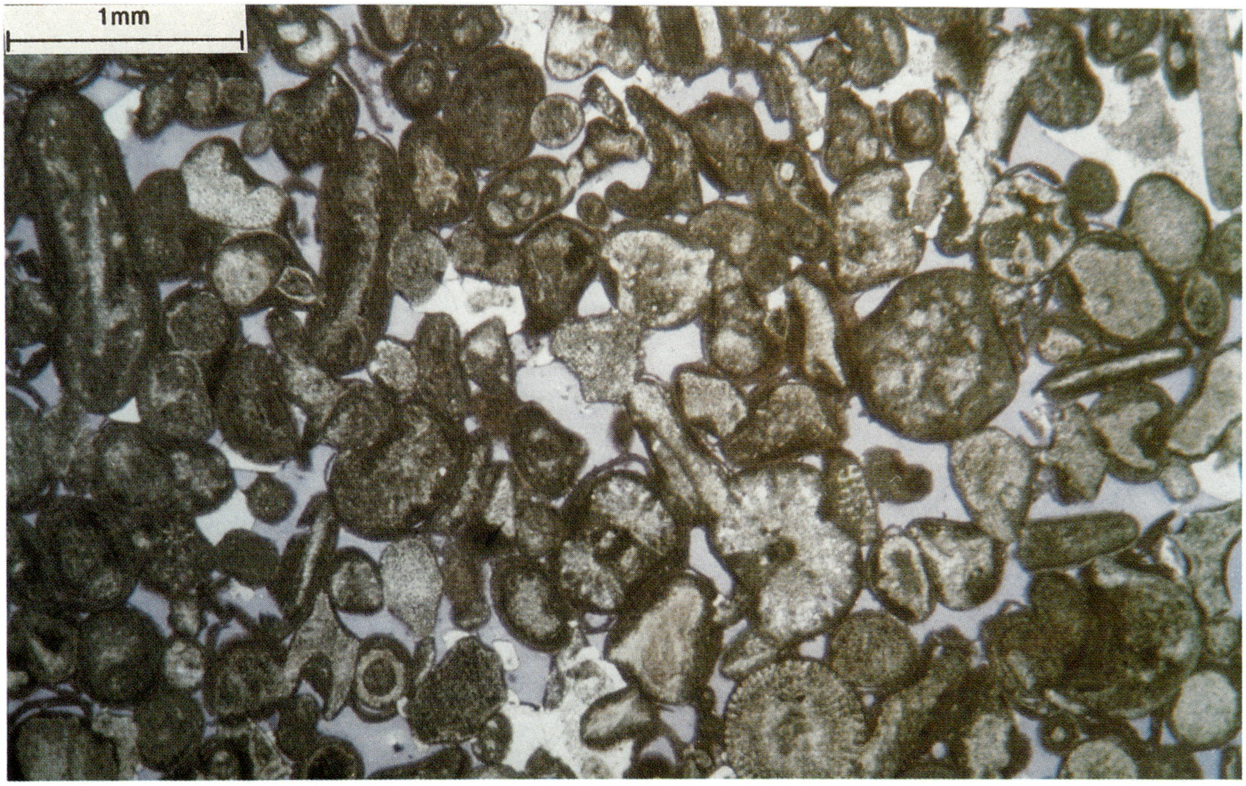

Figure 13—Thin section photomicrograph of the Alida reservoir. Porosity has been impregnated with blue epoxy. Most of the grains are coated crinoid fragments. Only minor calcite cement is present as syntaxial overgrowths of uncoated crinoid fragments.

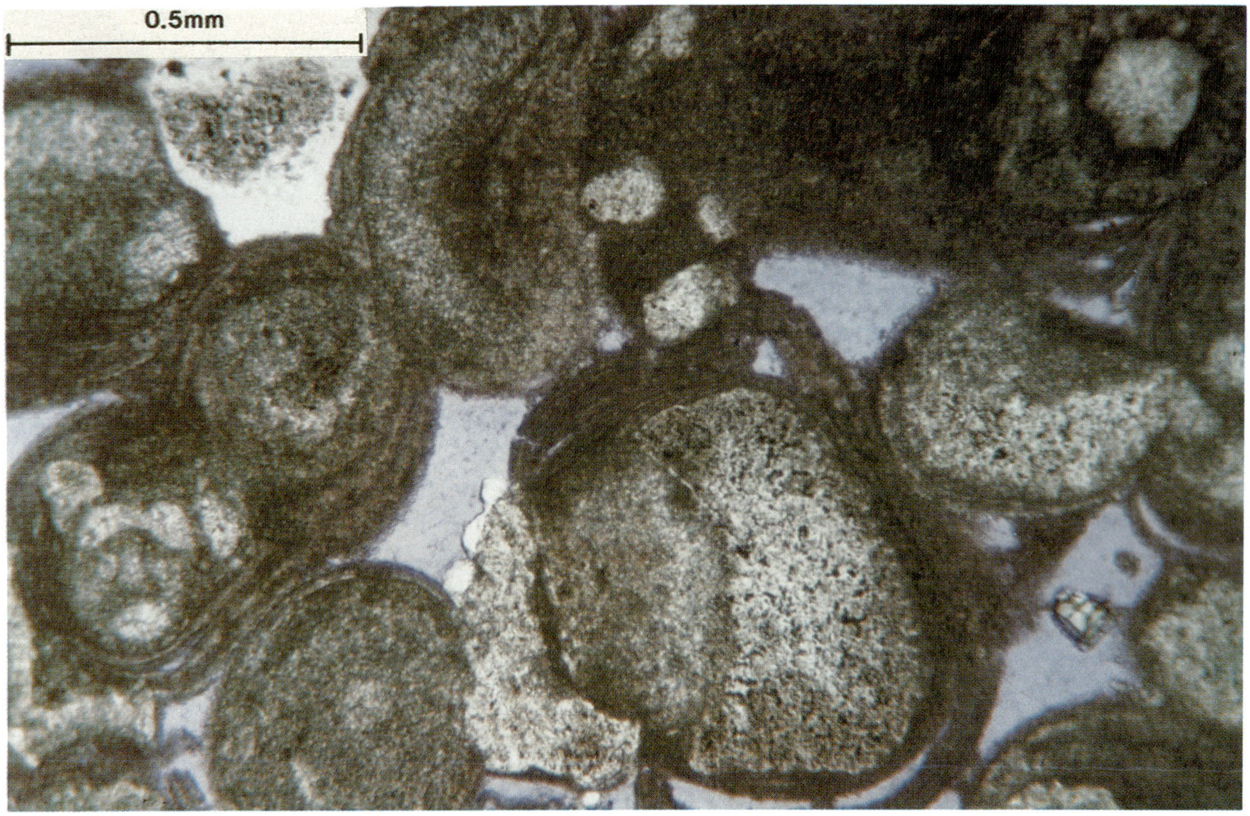

Figure 14—Thin section photomicrograph of the Alida reservoir at greater magnification. Porosity has been impregnated with blue epoxy. Broken oolitic coatings are present, probably due to compaction. Note the syntaxial overgrowth calcite cement.

Figure 15—Scanning Electron Photomicrograph of the Alida reservoir. Note the oolitically-coated crinoid fragments and syntaxial overgrowth calcite cement.

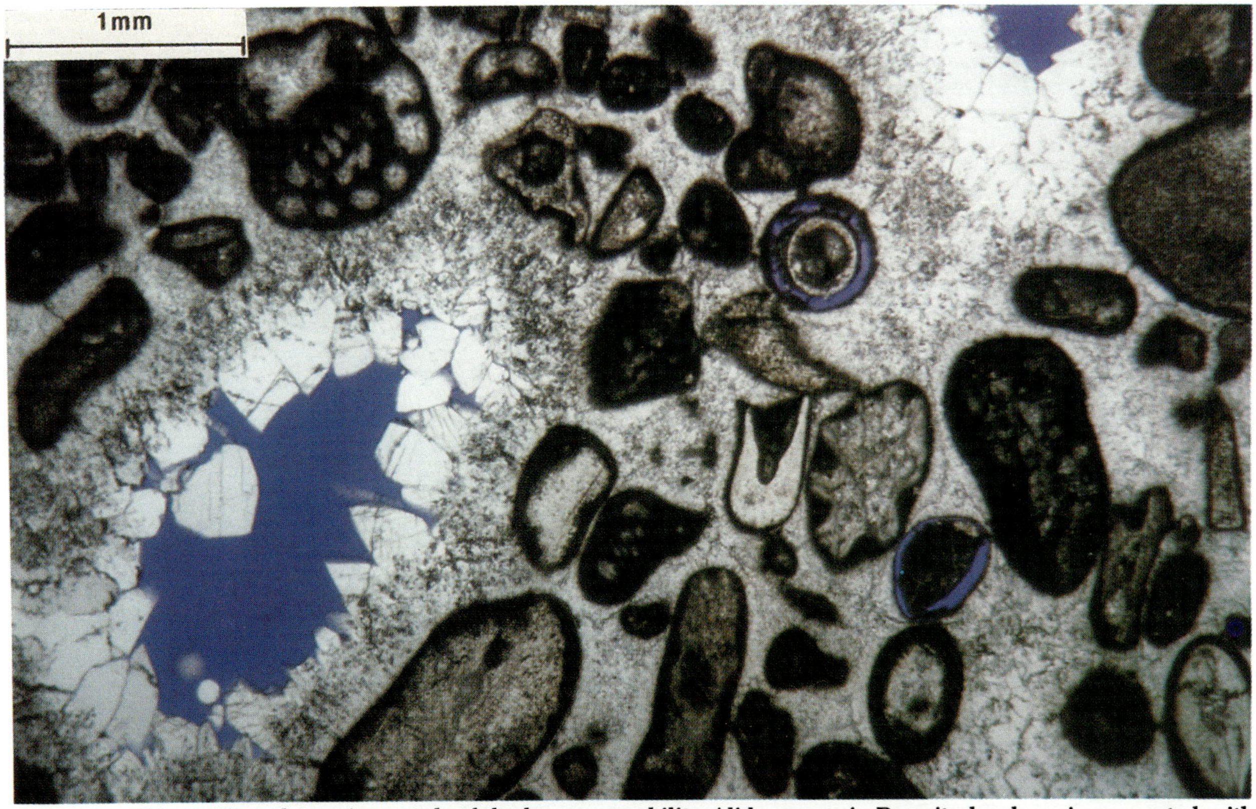

Figure 16—Thin section photomicrograph of the low permeability Alida reservoir. Porosity has been impregnated with blue epoxy. This rock has been cemented relatively early and shows very little compaction compared with the reservoir in Figures 13 and 14. Later leaching has created large vugs which are partially infilled with a later calcite cement.

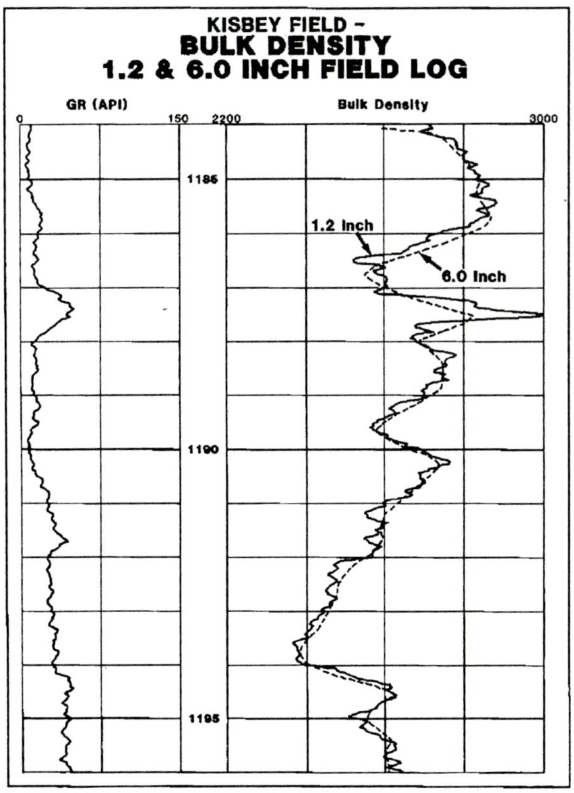

Figure 17—Bulk density-gamma ray log of a well in the Kisbey Field comparing the resolution of the 1.2 inch (high resolution) and 6.0 inch (normal resolution) density tools.

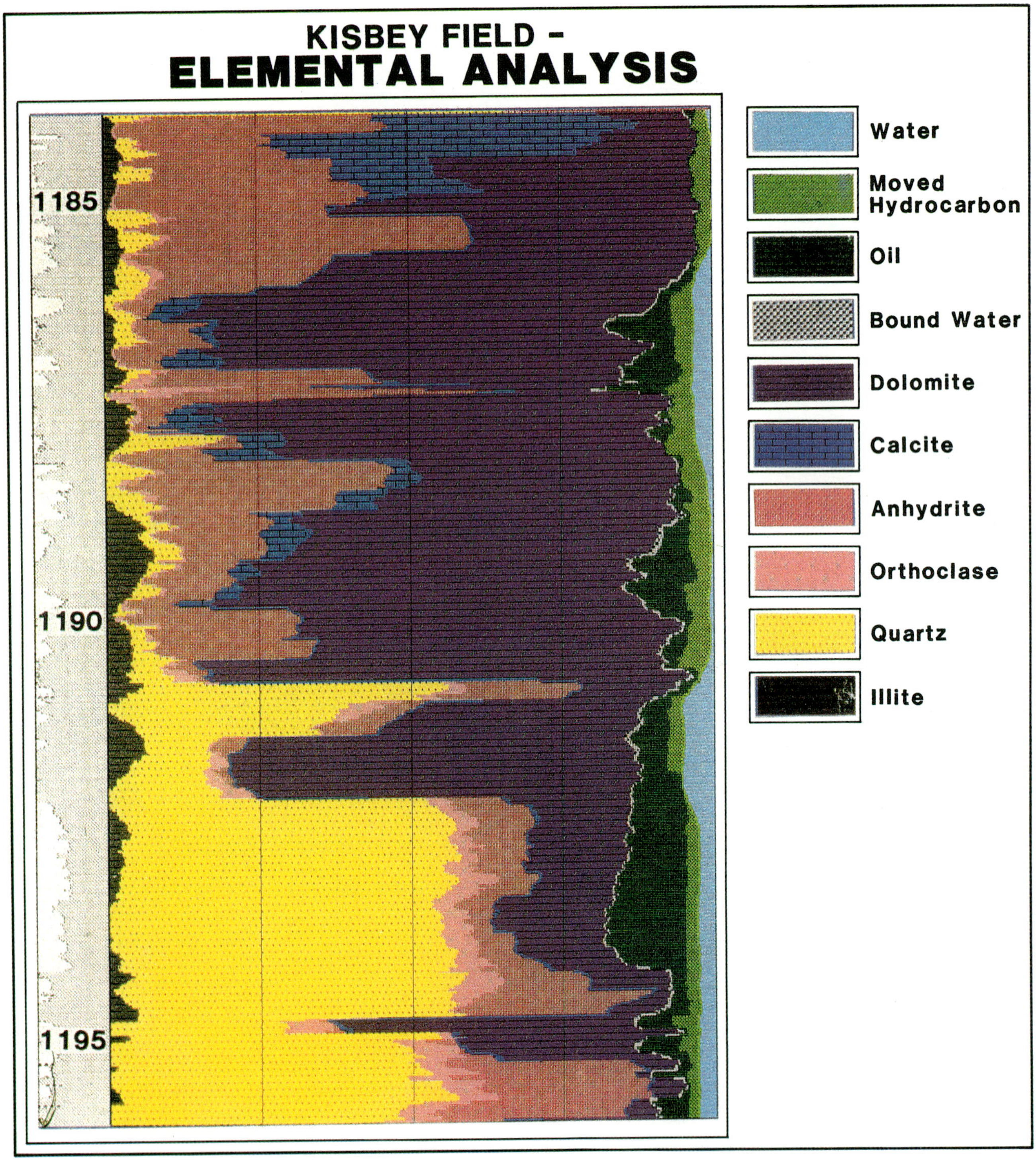

Figure 18—Elemental analysis of a well in the Kisbey Field. The technique breaks down the reservoir into both mineral and fluid components.

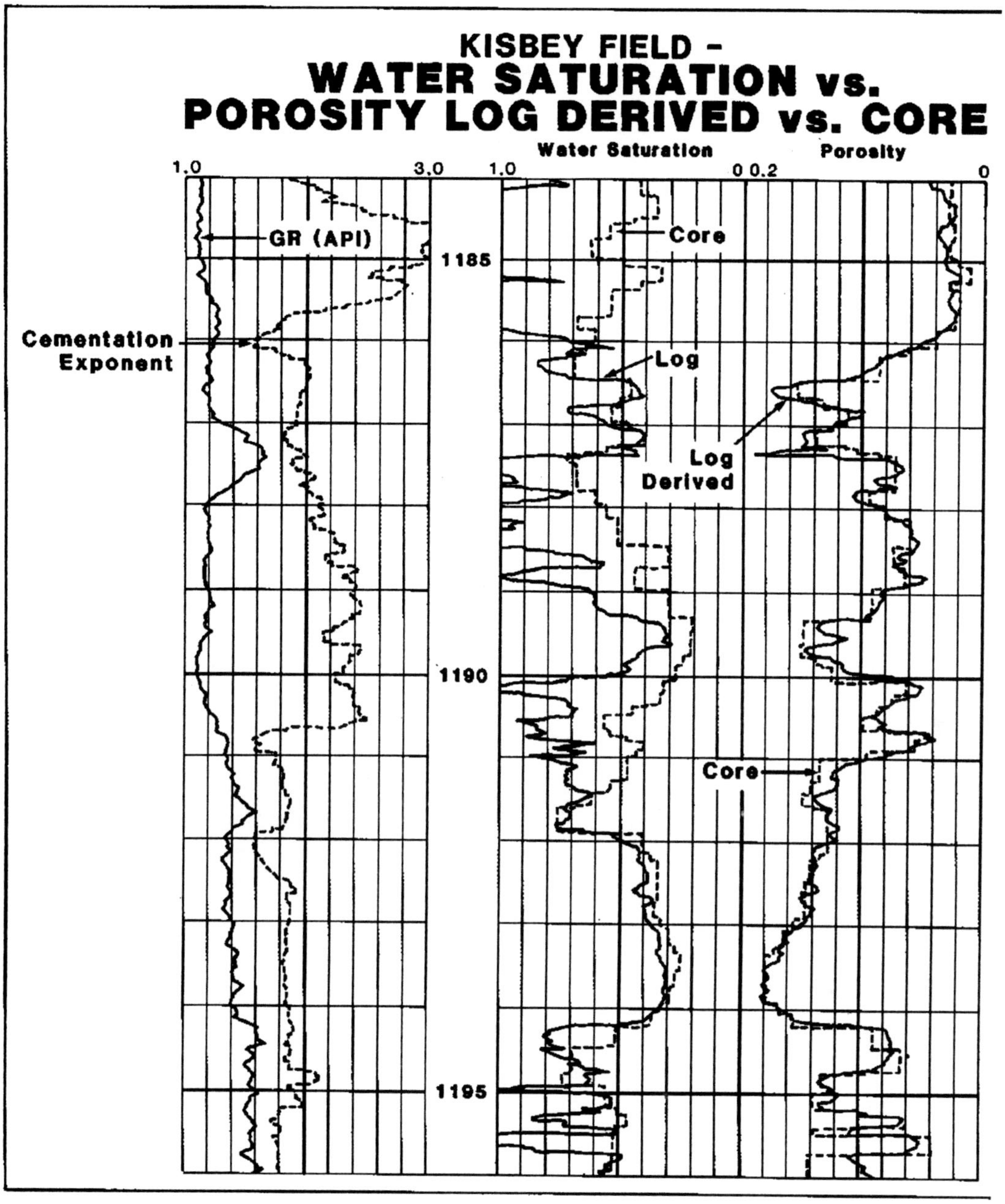

Figure 19—Comparison of log- and core-derived porosity and water saturation using elemental analysis and variable cementation exponent (m) calculated using the EPT log. In the reservoir zones, the match between core and log-derived values is excellent.

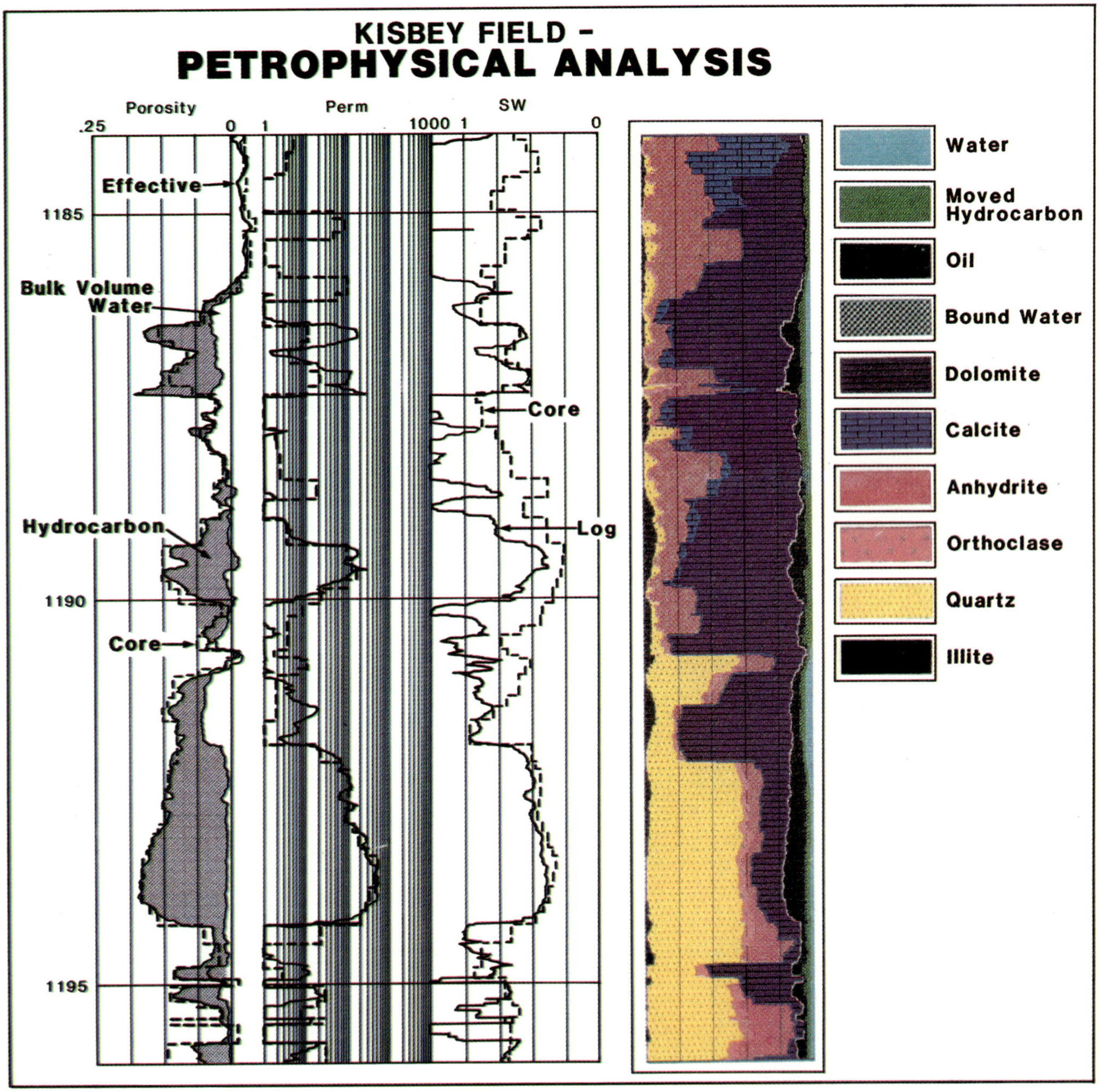

Figure 20—Complete petrophysical analysis of a well in the Kisbey Field. Note the good agreement between core and log-derived permeability.

MAIN KISBEY ZONE HORIZONTAL OIL PERMEABILITY AT IRREDUCIBLE WATER SATURATION

(Color Scale from 0 to 190 md.)

MWAT1. DAT.

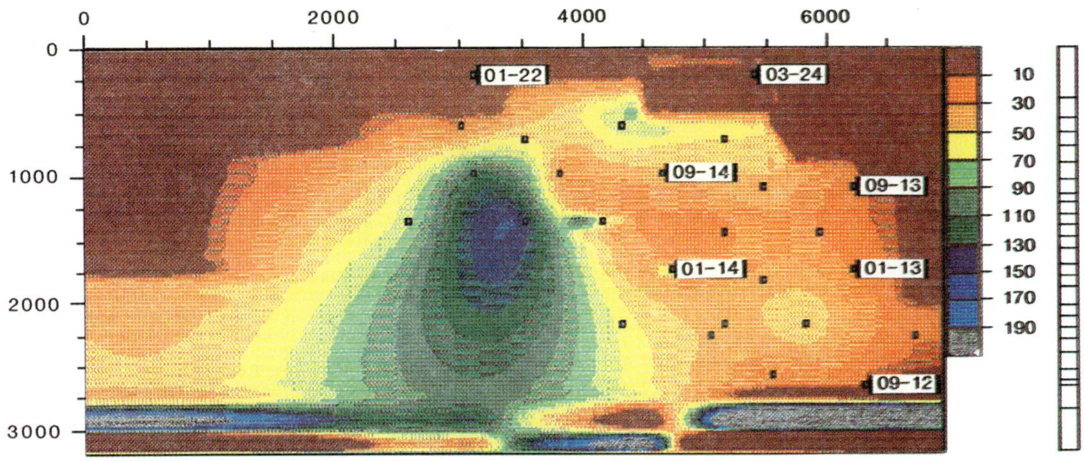

Figure 21—Kisbey unit horizontal oil permeability. Displays such as this were used to check the model input data.

KISBEY POOL – GENERAL RESERVOIR PARAMETERS

	KISBEY FMN.	GLENBURN FMN.
Reservoir Depth (ft. K.B.)	3,900	3,920
Net Reservoir Thickness (ft.)	10	25
Air Permeability (md.)	70	200
Porosity (%)	20	18
Oil Saturation (Pore volume fraction – %)	75	80
Oil Gravity (degrees API)	33 – 38	
Original Reservoir Pressure (psi)	1,625 @ –1,900 ft. subsea	
Bubble Point Pressure (psi)	970	
Original GOR (scf./bbl.)	440	
Original Oil In Place (bbls.)	15,975,000	
Original Mobil Oil In Place (at original conditions) (bbls.)	8,835,000	

Figure 22—Reservoir and operating parameters for the two main producing horizons.

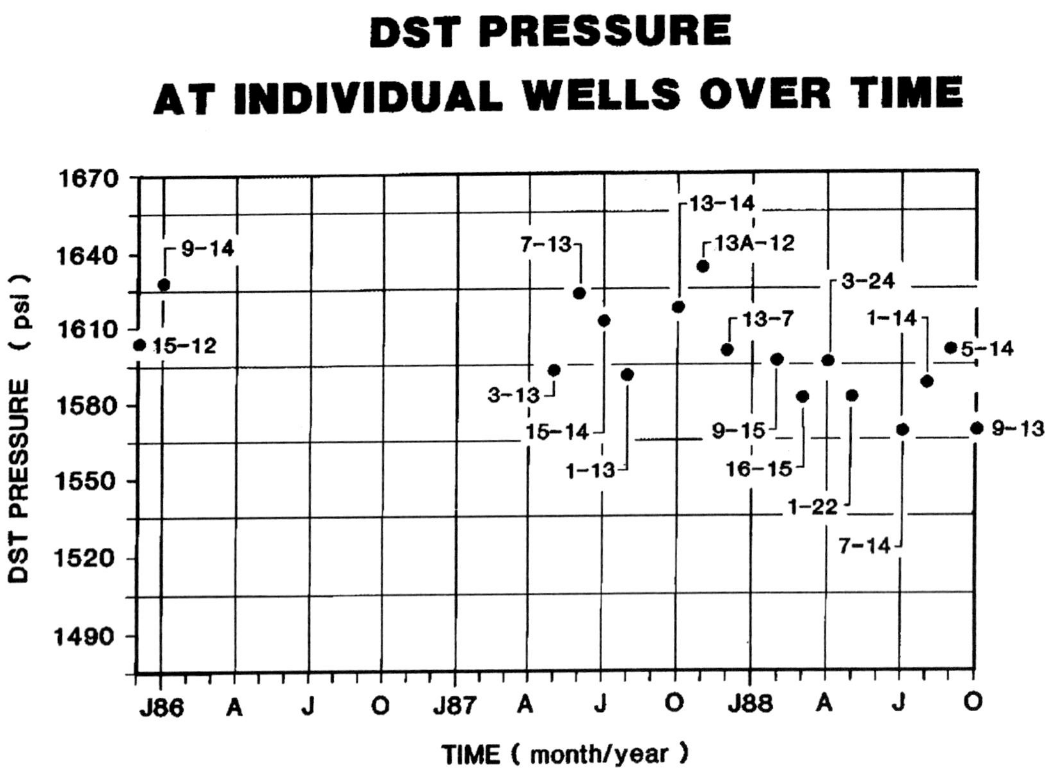

Figure 23—Individual well DST pressures versus time. A modest pressure drop of approximately 30 psia was observed over a three year period.

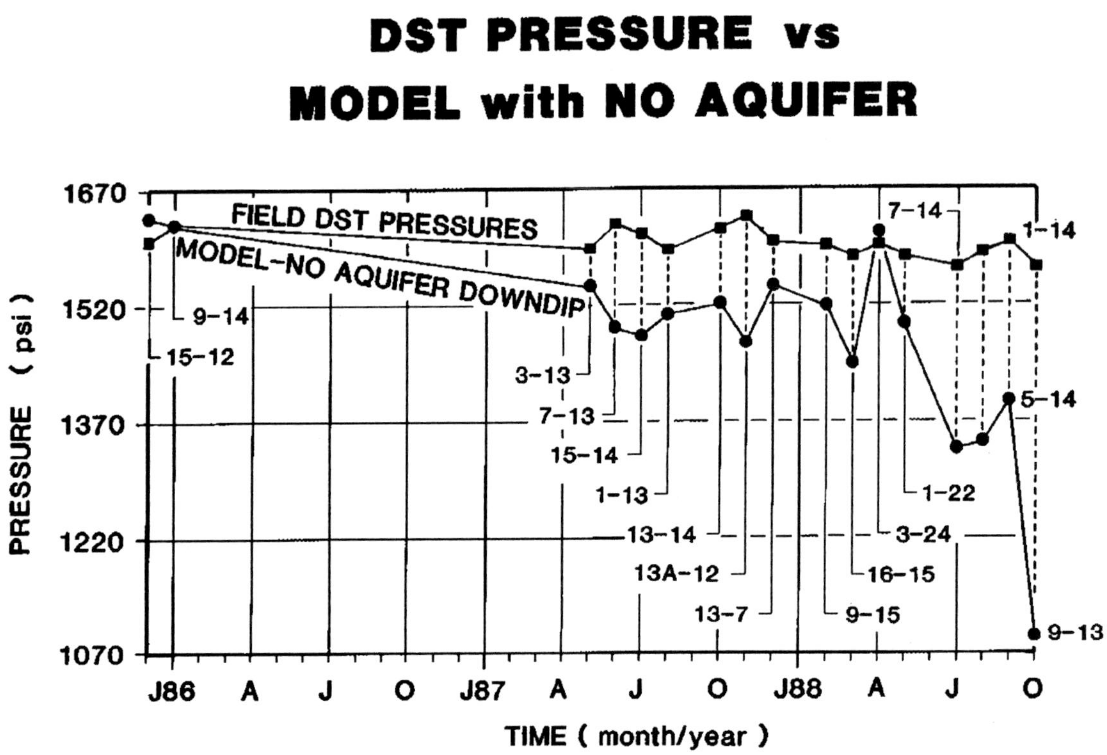

Figure 24—DST pressure versus model with no water aquifer. Without natural pressure support, reservoir pressures would have been significantly lower by late 1988.

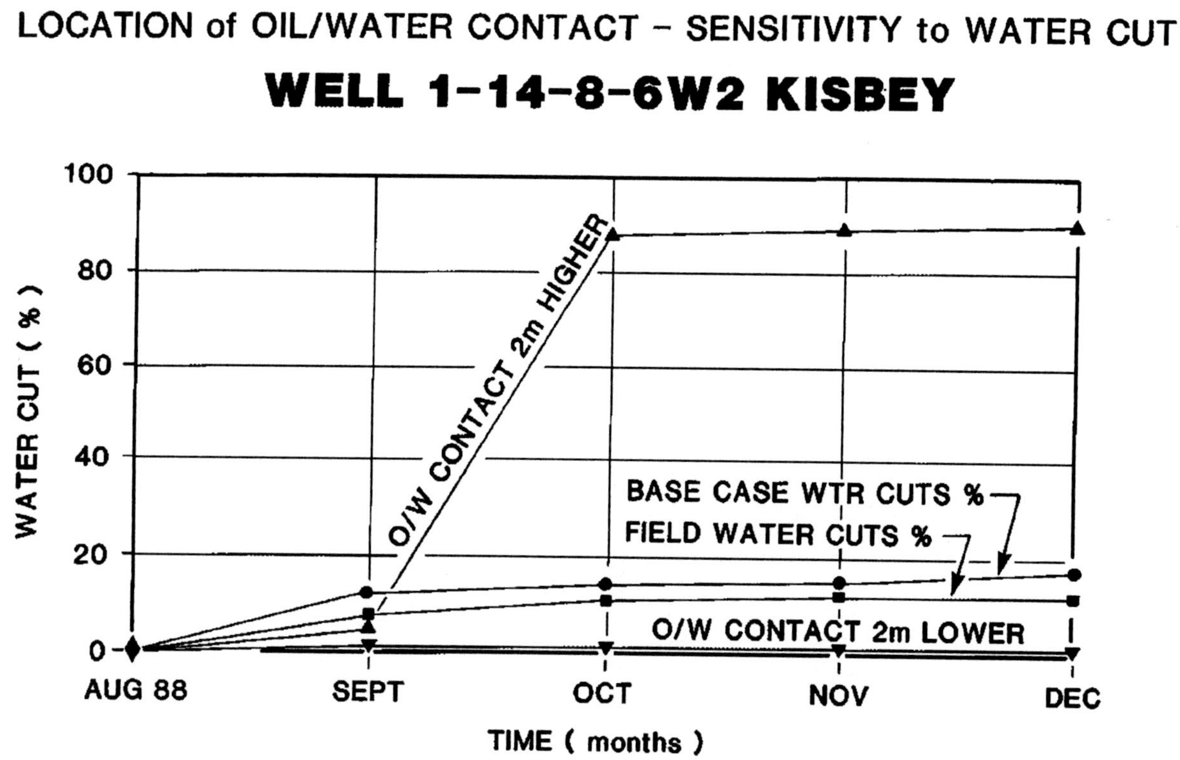

Figure 25—Watercut versus time at 1-14-8-6W2 well. This plot illustrates the sensitivity of producing watercut to the location of the oil/water contact.

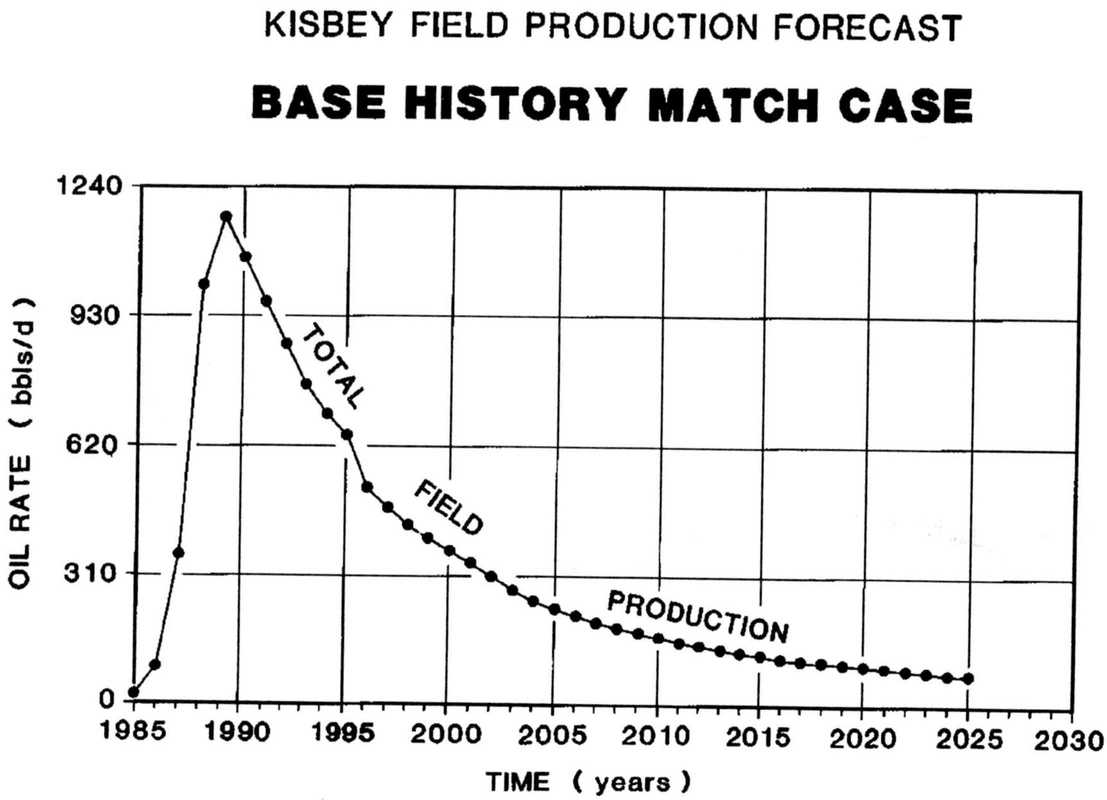

Figure 26—Kisbey field production forecast. Primary recovery is estimated at 32% of OOIP.

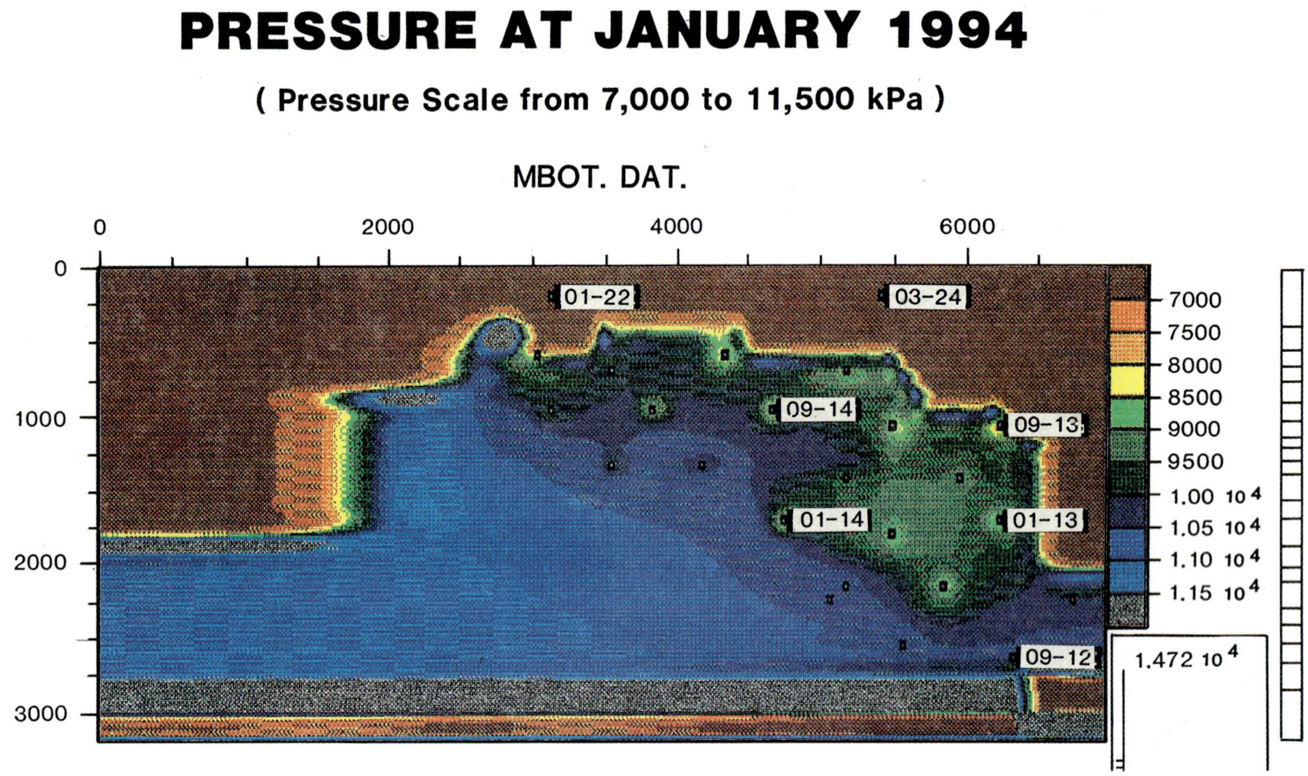

Figure 27—Kisbey reservoir pressure in 1994. The area of highest cumulative withdrawal has a pressure of approximately 1300 psia.

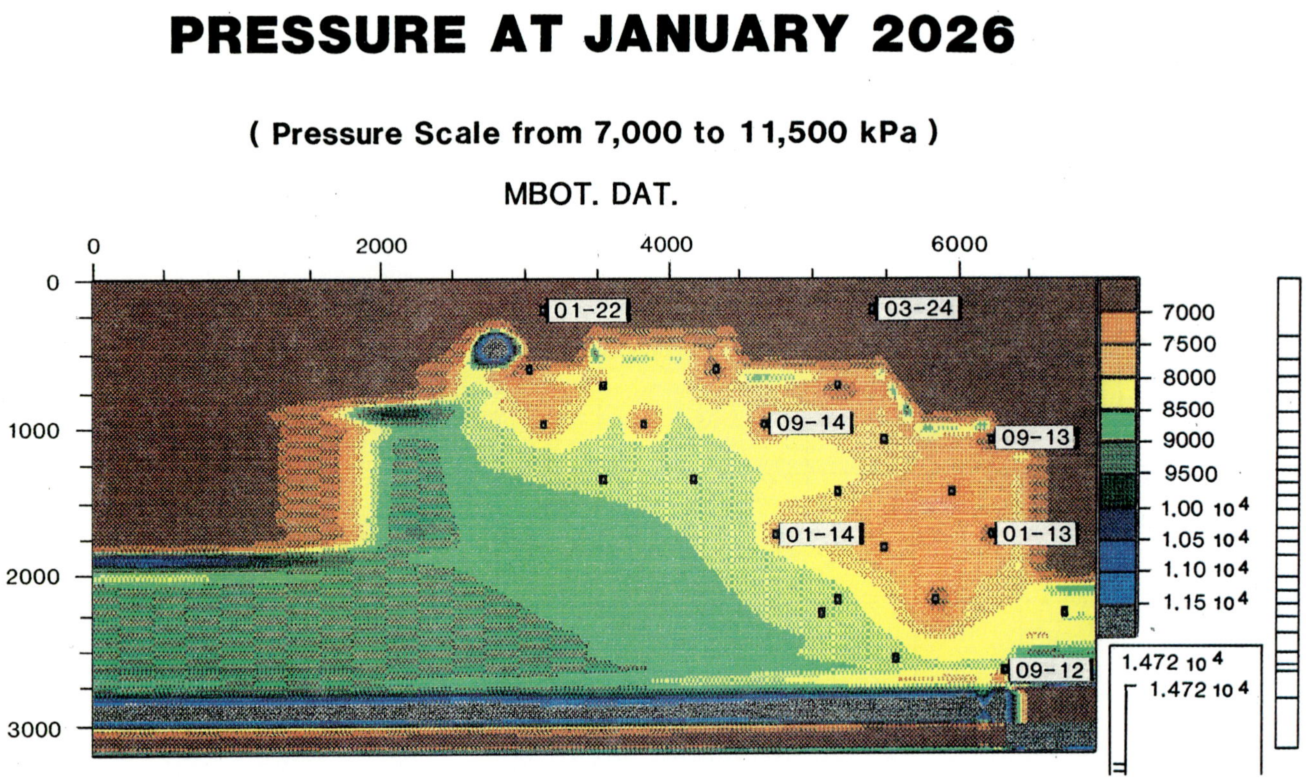

Figure 28—Kisbey reservoir pressure in 2026. No gas cap has formed as reservoir pressure exceeds 1100 psia.

UPDIP EDGE INFILL DRILLING

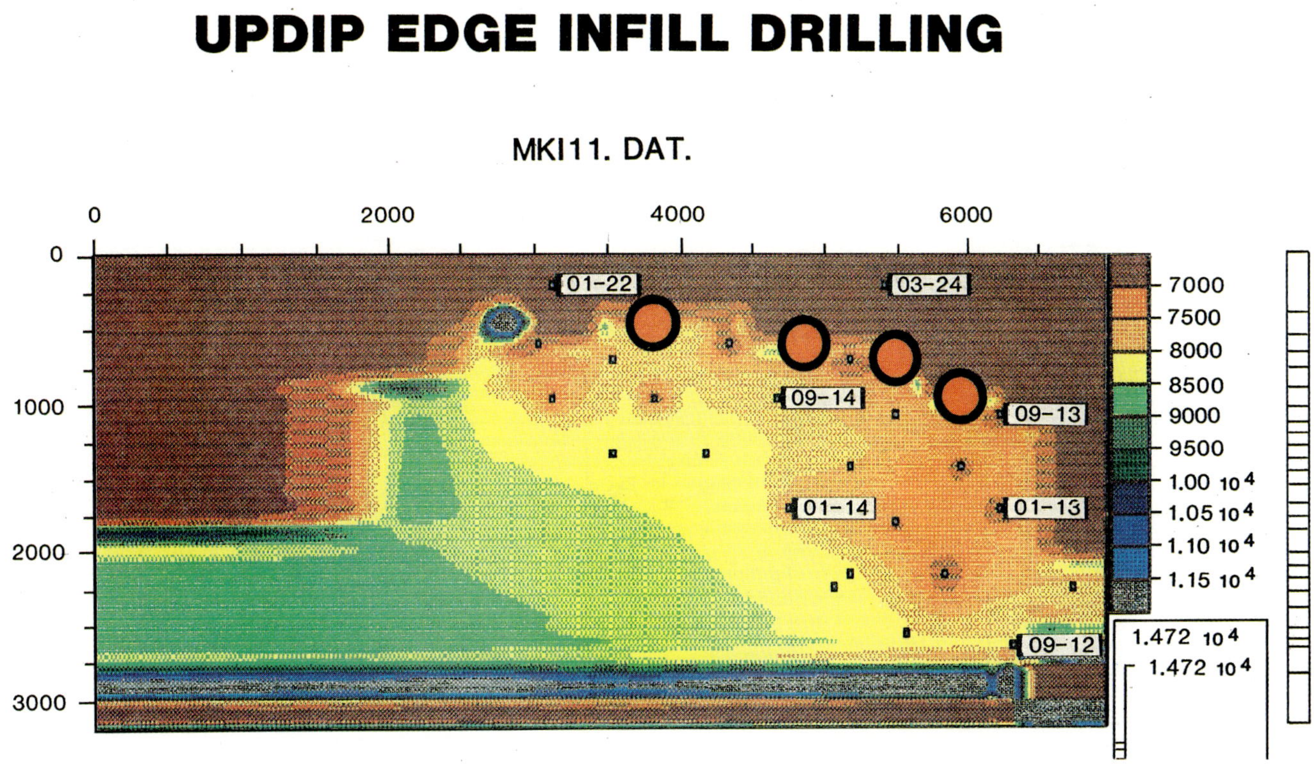

Figure 29—Updip edge infill drilling locations.

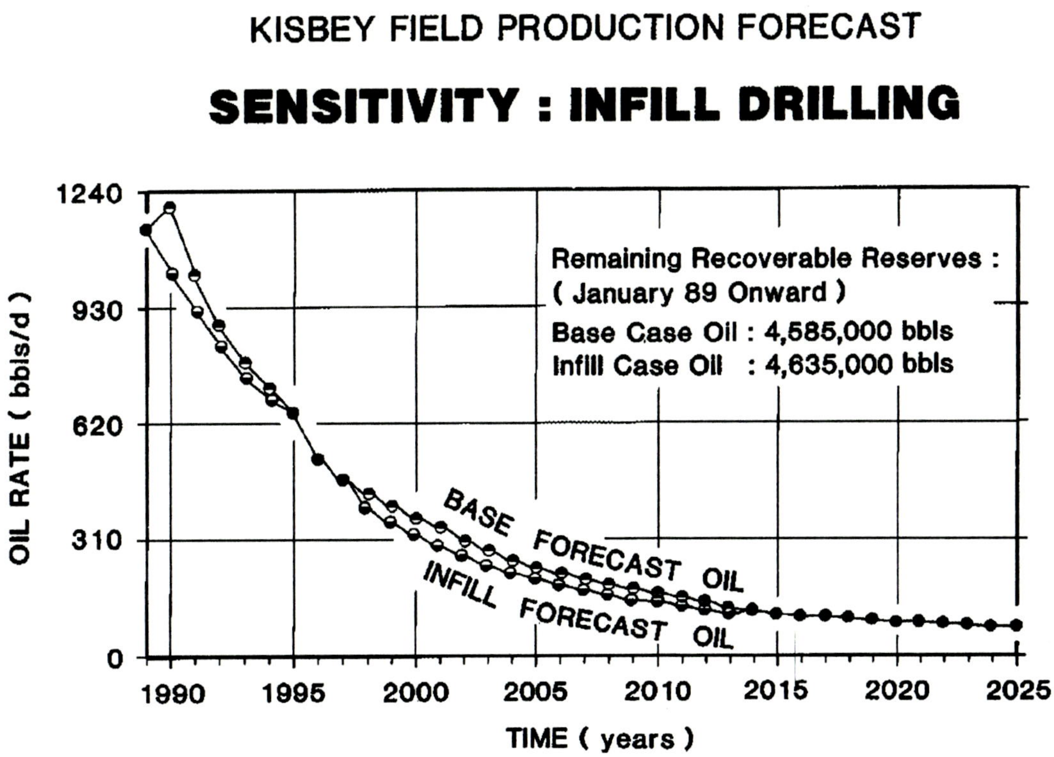

Figure 30—Kisbey field production forecast-sensitivity to infill drilling. The four infill wells recover an incremental 50,000 barrels of oil.

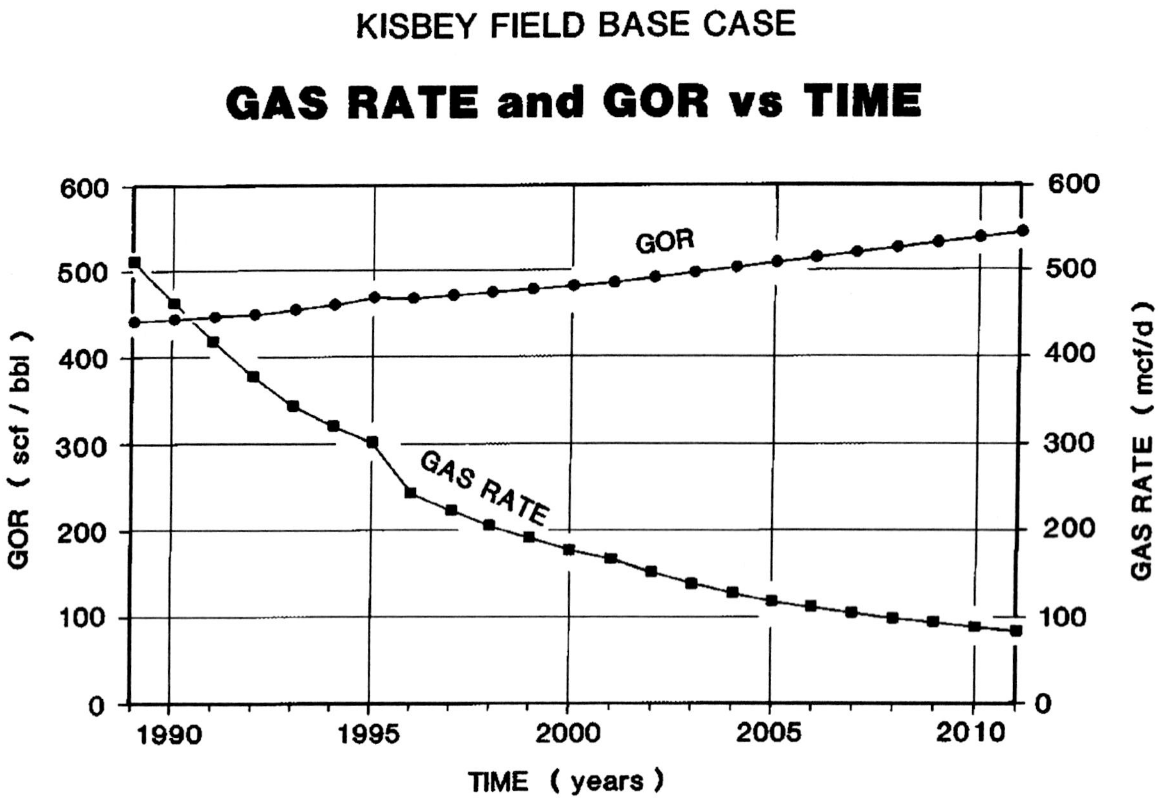

Figure 31—Gas rate and GOR versus time. The GOR is expected to climb very slowly.

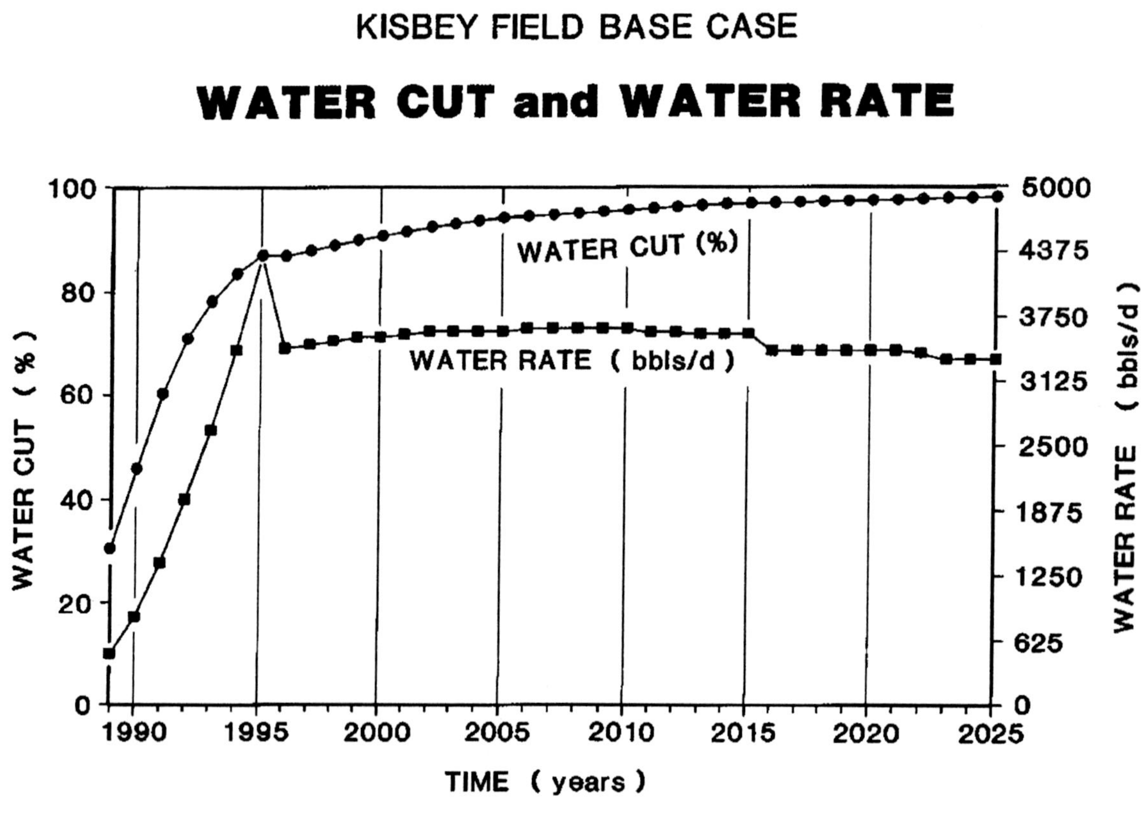

Figure 32—Watercut and water rate versus time.

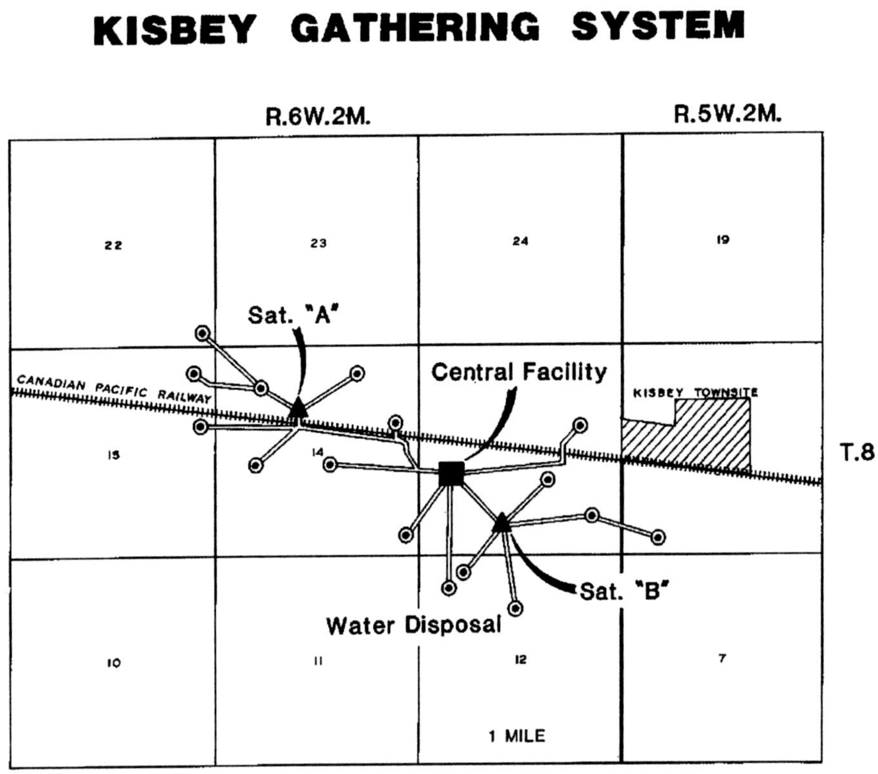

Figure 33—Kisbey gathering system. Trucking of fluid was eliminated as a result of the consolidation.

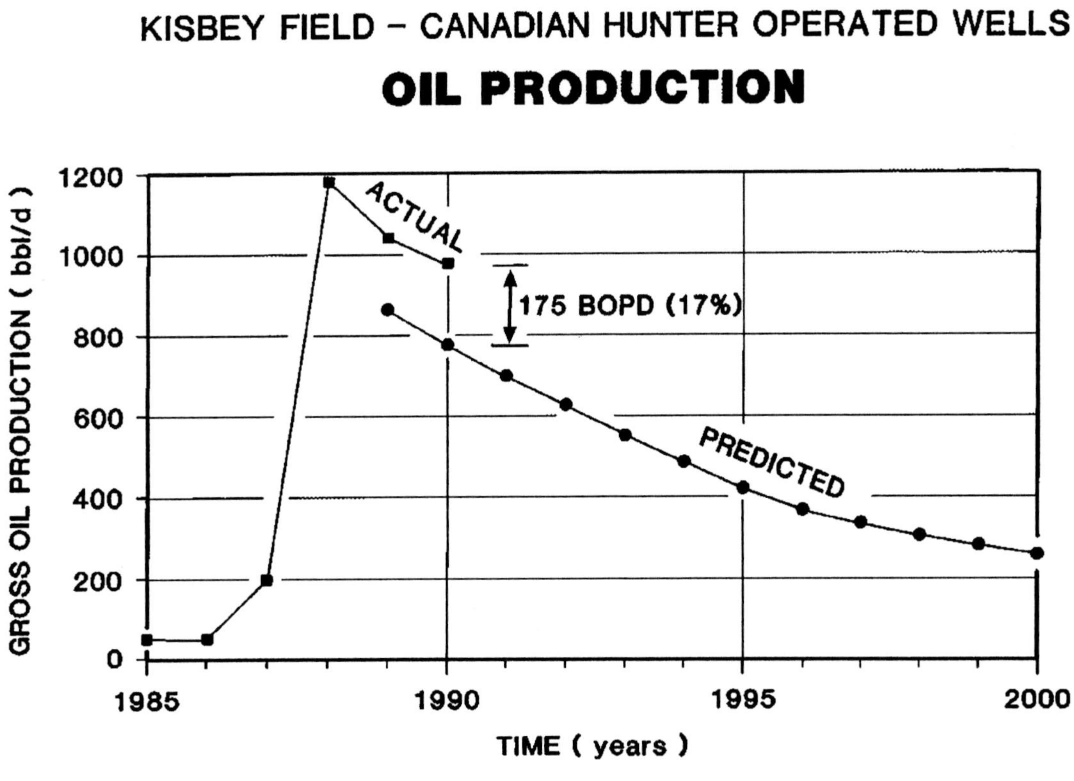

Figure 34—Oil production versus time. Production optimization efforts increased oil production by 17~.

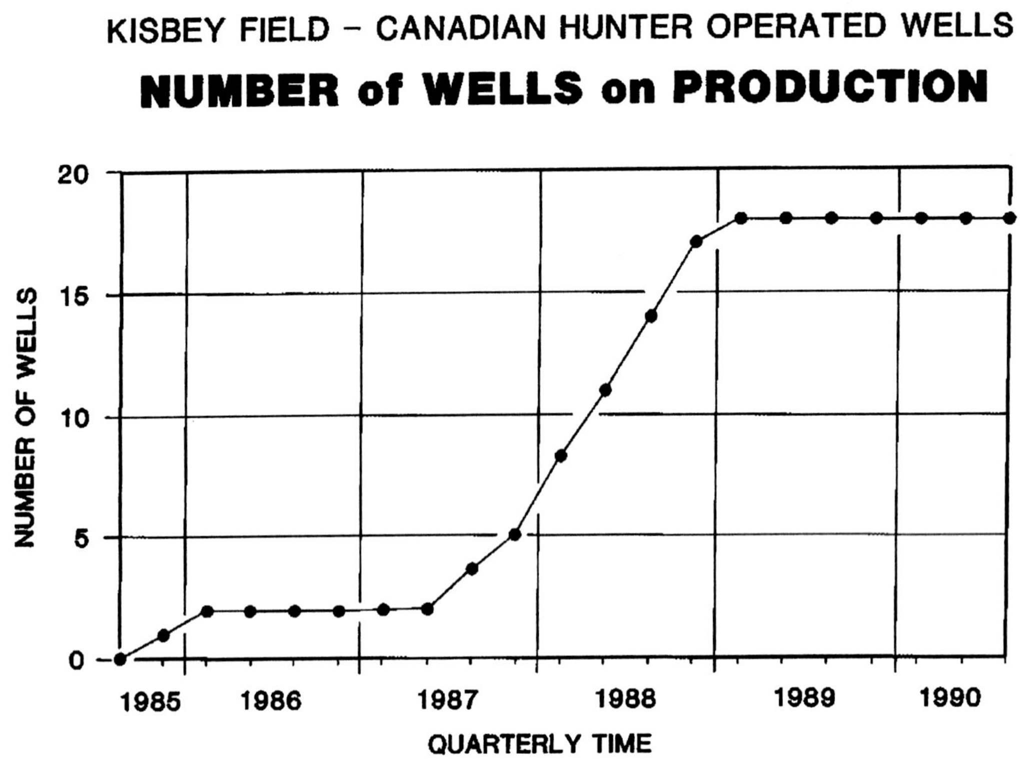

Figure 35—Number of wells on production.

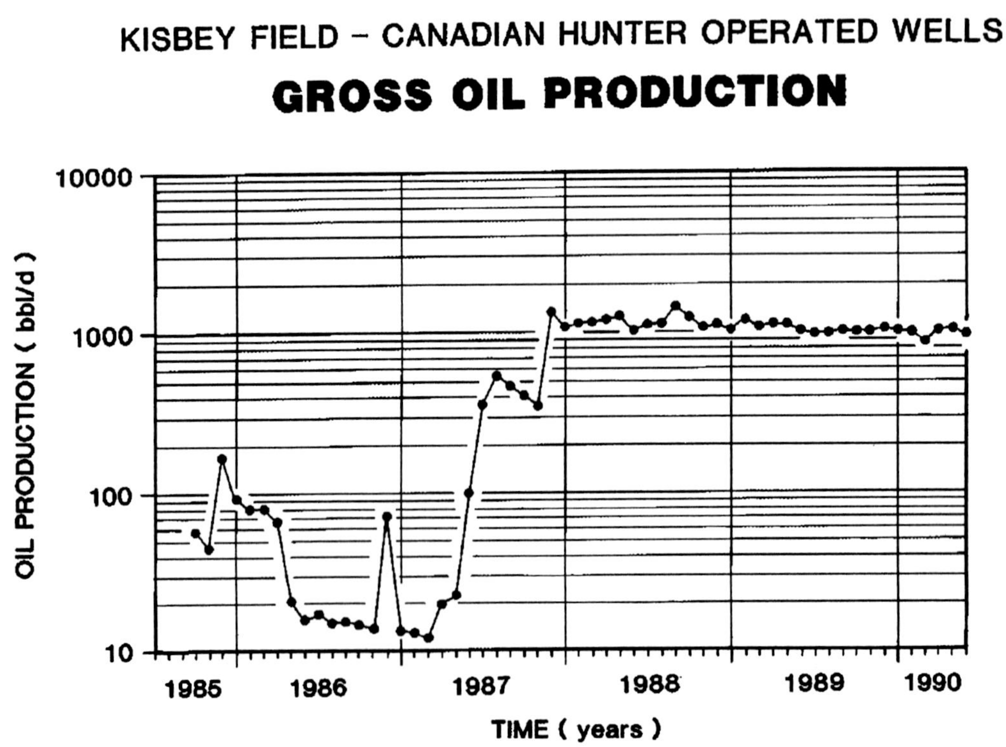

Figure 36—Gross oil production. Figures 35 and 36 illustrate the timing of the development process.

Mid-Dip Tuscaloosa Trend: Analog Geological, Geophysical, and Engineering Data for Prospect Evaluation and Field Development

Roger D. Shew

The Upper Cretaceous Mid-Dip Tuscaloosa trend occurs in southwestern Mississippi and parts of northeastern Louisiana (Figure 1). It is located on the south rim of the Mississippi Salt Basin and occurs updip of the Lower Cretaceous shelf margin. The Mid-Dip Tuscaloosa is primarily composed of terrigenous clastics derived from the Ouachita orogenic belt. Braided and meandering fluvial deposits unconformably overlie the Lower Cretaceous Washita-Fredericksburg Group. These initial progradational deposits were followed by an overall transgressive depositional sequence composed, from base to top, of fluvial, deltaic, and nearshore sediments encased in a dominantly shale-rich interval. These lenticular sandstones, particularly point-bar and channel deposits, are prospective for hydrocarbons in stratigraphic or combined structural/stratigraphic traps. The encasing mudstones and shales are excellent lateral and top seals for the hydrocarbons. Marine shales of the Middle Tuscaloosa cap the transgressive sequence and are interpreted to be the source rock. The mature Mid-Dip Trend has been estimated to have contained greater than one billion barrel equivalents of oil.

Three Shell fields, Little Creek, Olive, and Liberty, provide important analog data for further exploratory and enhanced recovery operation in the trend. Little Creek is by far the largest field with OOIP of 102 MMBO in a combined structural/stratigraphic trap. Olive (8 MMBO) and Liberty (7 MMBO) are typical of the remaining smaller stratigraphically-trapped accumulations. Detailed seismic stratigraphic interpretations, in combination with core and wireline data, are critical to the discovery, delineation, and development of these smaller reservoirs. Mappable seismic anomalies are associated with these porous, "soft" sandstones.

The objective reservoirs (10,000 to 11,000' in depth) are channel and point bar sandstones deposited by highly sinuous meandering streams on the upper to lower delta plain. Individual and composite channel sequences are both prospective. Reservoir quality is high but variable with average porosities and permeabilities of 24% and 100 md, respectively. Lateral continuity is generally good but shale drapes create local barriers to fluid flow. Reduction in vertical to horizontal permeability results from the shale drapes, carbonate cemented layers, and fining upward sequences. However, primary production and water-flood histories indicate these are not continuous barriers. Although several diagenetic events (compaction, rock fragment dissolution, chlorite formation, quartz and carbonate cementation, and formation of kaolinite) are evident in these fields, the most important to reservoir quality and petrophysical evaluation is the abundant grain-coating chlorite. In addition to questionably lessening quartz cementation and possibly compaction, the chlorite contains abundant microporosity which is filled with connate water in these typically thin oil columns. Therefore the effective porosity is less than measured and clean oil production is possible from measured water saturations as high as 75%.

A long history of production from the Little Creek reservoir has provided important information for the development and production of other fields.

Average recovery efficiencies in the three fields have been similar, 24% from primary and 22% from waterflood. Early unitization and waterflooding are usually necessary because of poor aquifer support in these lenticular sand bodies. Based on pilot studies at Little Creek and early field results from Little Creek and Olive, an estimated additional 23% of the OOIP is recoverable using carbon dioxide floods for tertiary recovery.

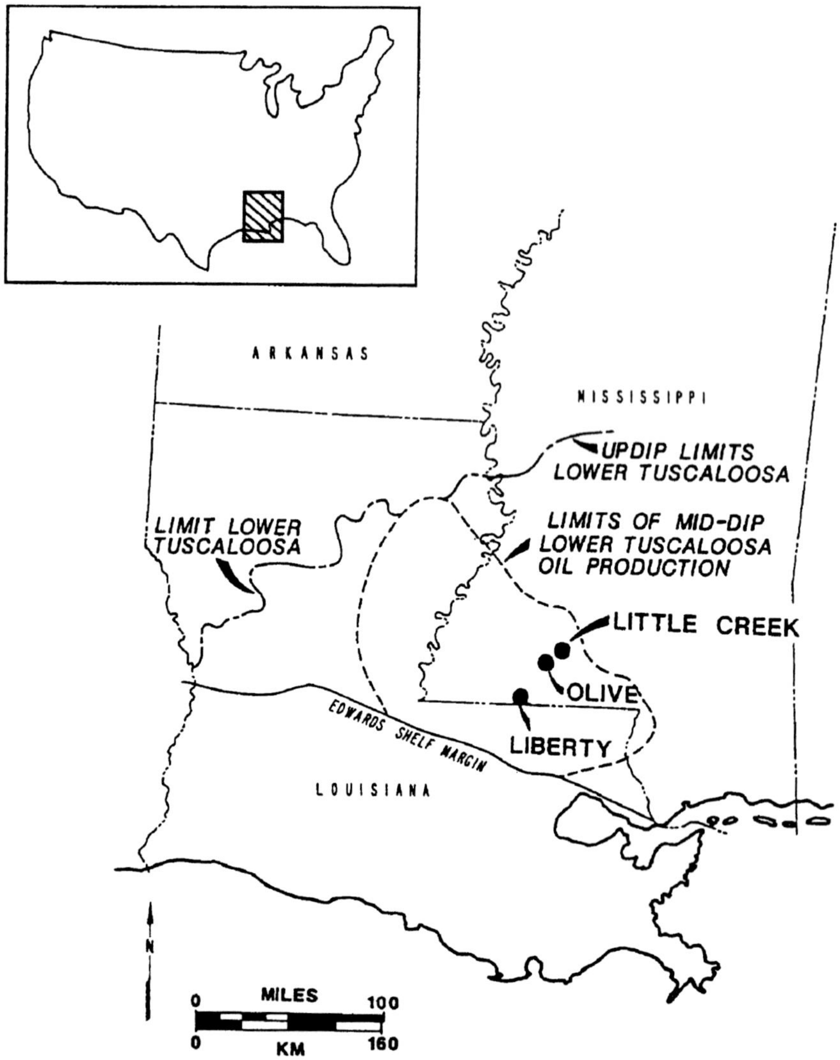

FIG. 1. Location map.

The Challenge of Exploring For and Producing Zone II Sandstone Reservoirs

Norman S. Neidell and Ernest E. Cook

For over 30 years literature describing relations between well log velocities and densities have been only sporadic, uncoordinated and often excluded one or more key parameters such as lithology or geological age. Model relationships are presumed to hold and tested only qualitatively with little to no effort directed toward refining the models or quantitative reconciliation.

In 1986 the concept of the sand/shale acoustic impedance cross-over on Zone II was introduced. Where Zone II exists, it separates strongly contrasting low acoustic impedance sands within shales from the more consolidated but still strongly contrasting high acoustic impedance sands within shales at greater depth. The inconsistent and highly muted reflections characterizing Zone II, and corresponding small fluid signatures make seismic definition especially difficult. Also, familiar tools such as seismogram synthesis and AVO studies give unreliable results in this environment. We also now recognize the relatively widespread occurrence of Zone II.

Important potential for hydrocarbons has been established for Zone II—albeit mostly by accident. It remains largely unrecognized and unexplored except for obvious simple structures. Conscious and detailed seismic procedures in conjunction with appropriately organized cataloging of subsurface parameters offer the most effective methods for defining Zone II to date. Applying such techniques constitutes the Challenge, but the rewards could be great indeed.

Better Decisions Through Quality Information Integration—A Practical Example of a Reservoir Development Situation

Paul D. Vrolijk

For cost effective exploration of hydrocarbon reservoirs it is of critical importance to be able to accurately map lithostratigraphic variations in and around those reservoirs. Sufficient and detailed information on the lithology can be derived from well logs, but this information is, unfortunately, only valid in a small area around the wells.

Although the seismic method can provide the desired areal coverage, in case of thin layering, standard seismic interpretation does not provide sufficient detail. Besides well log and seismic data, the interpreter may have additional knowledge about, among other things, the depositional history of the area.

The purpose of the reservoir case is to combine the available seismic, well and geological information to obtain a detailed lithostratigraphic interpretation for use in further reservoir development.

It will be shown that this objective can be achieved by an interactive and easy-to-use tool for the evaluation of geological hypotheses.

3D Seismic Surveys: Scrimp Now, Pay Later

Christine Warren Brandes

The declining cost and widespread coverage of "recon" Exploration 3D surveys have made 3D almost as available and priced competitively with 2D seismic data. Many operators are using 3D seismic to conduct initial exploration evaluations. There is a growing belief within the Industry that the earlier a 3D survey is gotten, the better. BPX is a technology leader in deep water field development, using 3D seismic data to evaluate 92% of their producing and pre-development fields in the Gulf of Mexico. The savings in time and capital through utilizing an Exploration/Appraisal 3D survey are illustrated with the evaluation history of upper Pleistocene gas zones at BP's Snapper field (East Breaks 165/209).

Post-discovery appraisal may use the same 3D data set as the initial exploration effort. Such data is often cited to support lower development risk estimates in pre-development planning. It may be inappropriate to apply an Exploration-purpose 3D survey to Production development needs. Seismic data acquired "too early" to be appropriate for development use is defined as data that is shot and/or processed without quantifying the critical factors controlling a discovery's commerciality. Beyond structural imaging, these factors usually include continuity of reservoir area, identification of production barriers, and accurate net reservoir thickness. Their resolvability from 3D data are controlled by line orientation, line spacing, shot-receiver offset, wavelet phase and dominant frequency. A number of BPX offshore development examples will show how 3D seismic surveys, designed for exploration/appraisal use, lacked the detailed structural definition and stratigraphic resolution required for optimal placement of production wells.

Reliance on less than optimal 3D data is a false economy. Cost effective alternatives include reprocessing existing 3D data and shooting new 3D data designed to image specific reservoirs. These are not inexpensive options, but the incremental cost can be justified against the savings in "over-engineered" production wells and sidetracks. Additional, though somewhat intangible, benefits arise from optimizing penetration locations for maximum reserve recovery.

Generation of a Field Development Plan, Natih Field, North Oman

D. R. Bostock, S. Adams, C. Mercadier, H. Milatz, H. van der Weerd, and T. Walker

The Natih Field in North Oman is a complex domal fractured limestone structure with a STOIIP of some 500 million m^3. Initially subjected to a depletion drive, rapidly dropping pressures resulted in a switch to water injection and then to gas injection to promote gas oil gravity drainage.

In determining which production mechanism to promote an integrated approach has been taken. Field tests used tracers to investigate the problems associated with water injection; fracture orientation and intensity has been determined using core data including palaeomagnetics, FMS logs and out-crop studies; initial oil saturations have been determined within a given range using resistivity and pulsed neutron logs as well as capillary pressure curves to overcome problems of mixed wettability; to monitor gas saturation development in the matrix recourse has been made to the Borehole Gravity Meter; special core analysis on restored state cores has been utilized to characterize wettability and relative permeability as input to both monitoring the water influx and to reservoir simulation.

Development of a simulator to incorporate all the necessary physics of gravity drainage, including block to block interaction, has been undertaken at the research laboratories in the Netherlands and, using the understanding of the field built up through the above studies, a successful history match was achieved. A field development plan is now emerging requiring a ring of wells to be drilled to a downdip target completion interval to further promote gas oil gravity drainage. The plan will attempt to create a balanced offtake from around the field with GOR control and crestal gas injection to push the fracture oil rim down. Production from the aquifer will also assist in this downward movement.

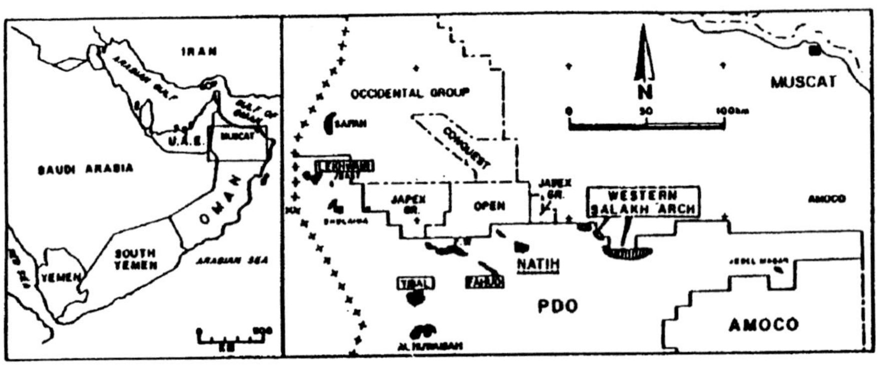

Fig. 1 : LOCATION MAP OF STUDY AREAS

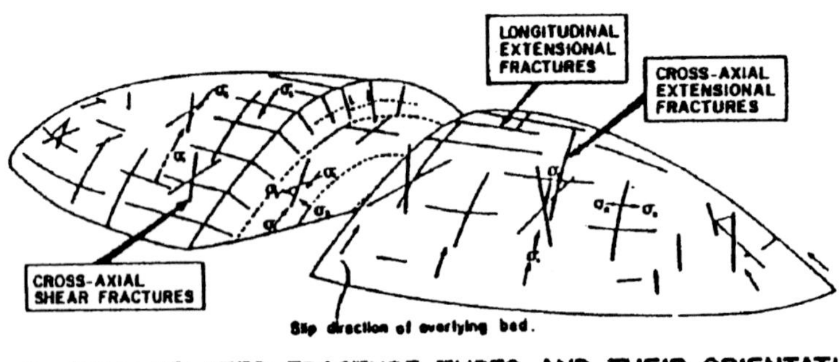

Fig. 2 : FOLD-RELATED FRACTURE TYPES AND THEIR ORIENTATIONS

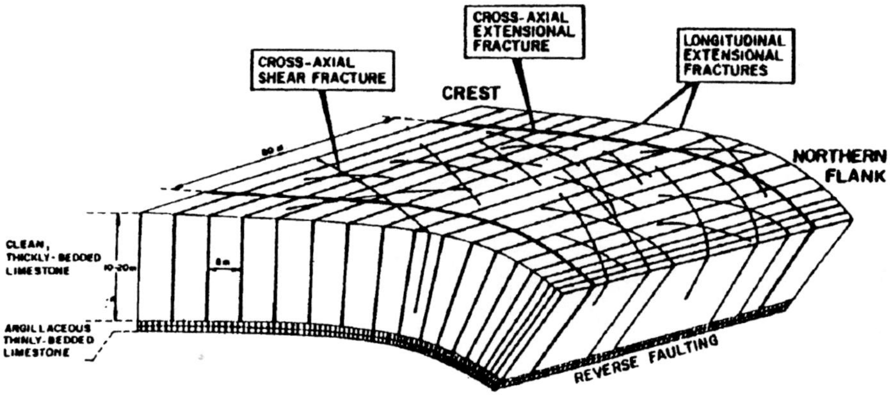

Fig. 3 : CONCEPTUAL FRACTURE MODEL FOR THE NATIH FIELD
(NOT TO SCALE)

NATIH FIELD PERFORMANCE

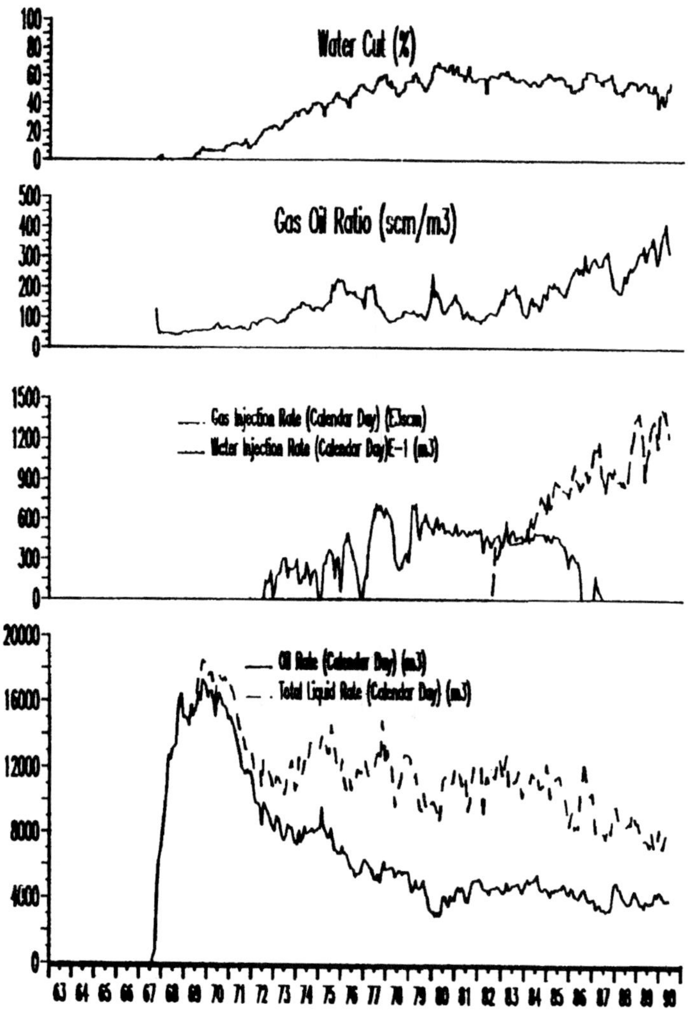

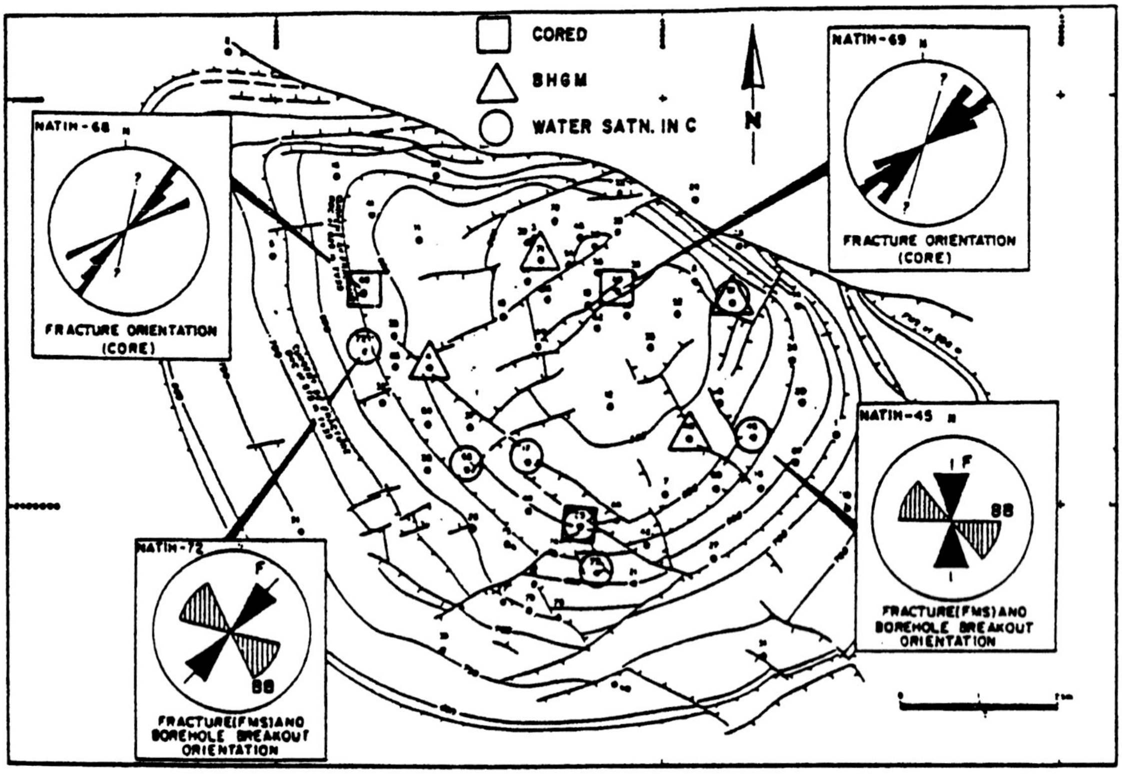

Fig 5 STRUCTURE MAP TOP NATIH FORMATION WITH FRACTURE AND BOREHOLE BREAKOUT ORIENTATIONS

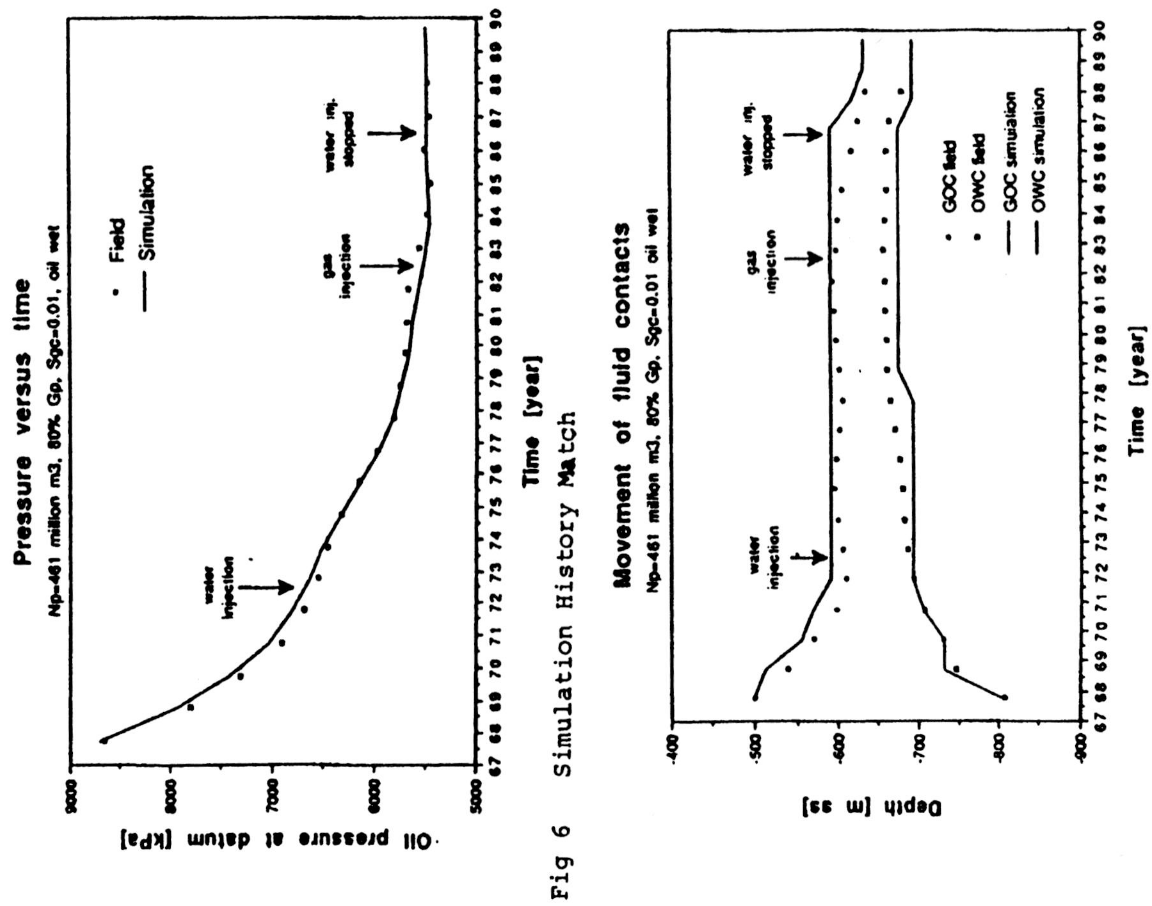

Fig 6 Simulation History Match

INTEGRATED RESERVOIR CHARACTERISATION OF CYCLE III, BRENT GROUP, BRENT FIELD, UK NORTH SEA FOR RESERVOIR MANAGEMENT

by
I.D. Bryant, A.H.M. Paardekam, P. Davies and M.C. Budding

Investigation: 6260E1182

KONINKLIJKE/SHELL EXPLORATIE EN PRODUKTIE LABORATORIUM

RIJSWIJK, THE NETHERLANDS
(Shell Research B.V.)

SUMMARY

Cycle III of the Brent Group, Brent field is a sedimentologically complex interval composed of sands, shales and coals deposited in lower to upper delta plain environments. The traditional method of simulating fluid displacement in a model based on digitised maps of average reservoir properties is not sufficiently detailed to adequately assess the reservoir heterogeneities that affect reservoir drainage at the scale of individual sand bodies. Characterisation of such reservoirs to guide reservoir management requires detailed reservoir models constructed at the scale of these individual genetic units. We have constructed such a model by using workstation-based data integration and modelling techniques. Well data have been used in conjunction with pressure data and historical fluid movements to subdivide the reservoir into ten genetic geological layers that also correspond to the layers within which differential fluid movement takes place during production. Using the same well log-derived data base we have further defined the distribution of geological objects (channel-fill sands, coarsening-upward sands, sheet sands, mudstones and coals) within each of the layers. Each object comprises a number of rectangular cells termed "voxels". Petrophysical evaluations of well log data have been combined with geostatistical techniques to assign porosity, permeability and pore-fill to each of these voxels. Well data and high resolution 3-D seismic data have been used to define the structural configuration of the reservoir. The resultant model functions as a multidisciplinary data base which has been used to determine the distribution of both the original oil in place and the remaining oil volumes within individual correlatable sand bodies. This approach represents an optimal method for data integration and facilitates the transfer of data between geologists and engineers without loss of the relevant geological details.

INTEGRATED RESERVOIR CHARACTERISATION OF CYCLE III, BRENT GROUP, BRENT FIELD, UK NORTH SEA FOR RESERVOIR MANAGEMENT

1. INTRODUCTION

Data transfer between geologists and reservoir engineers has traditionally been by means of maps of reservoir properties averaged over the whole reservoir or over several reservoir intervals. These averaged descriptions of the reservoir are often adequate to guide decisions concerning early field development (e.g. how many platforms are required to develop a given field) but lack sufficient detail to guide later stage reservoir management when much more information is usually available (Slatt & Hopkins 1988; Weber & van Geuns 1989). At this stage of field development a major problem is the time and manpower required to integrate the available data into a detailed reservoir model (Johnson 1989). Even if such a model is successfully created, all too often the detail present in the model is lost during transfer to reservoir engineering simulators.

Recent developments in data management technology have resulted in the creation of a number of detailed, computer-based reservoir modelling systems (Abib **et al**. 1988; Pajot **et al**. 1988; Hastings **et al**. 1989). The reservoir models generated by these systems may be used as data bases that are accessible to a variety of petroleum engineering disciplines and constitute a common, multi-discplinary description of the reservoir that may subsequently be transferred to reservoir simulators. This methodology represents a significant advance from traditional methods since much more information is contained within these models than in maps of average properties.

This paper discusses the use of workstation-based log manipulation and reservoir modelling software to generate a detailed 3-D model of part of the Brent Group in the Brent field and to transfer parts of this model to reservoir engineers for further evaluation. These evaluations have been used to identify the distribution of undrained oil volumes within Cycle III so as to optimise the final stages of field development (Bryant & Livera in press).

2. CYCLE III RESERVOIR DESCRIPTION

The Brent field is a giant oil field located in the East Shetland basin of the northern North Sea (Fig. 1a). Hydrocarbons are found at two stratigraphic levels: in the Lower Jurassic/Triassic Statfjord Formation and in the Middle Jurassic Brent Group. The Brent Group has been divided, for petroleum engineering purposes, into four cycles (Fig. 2). Cycle III represents a complex of lagoonal and coastal plain sediments deposited during progradation and temporary abandonment of small deltas behind the main wave-dominated Brent delta-front.

The present study is confined to the Cycle III interval in the Main Block of the field, which comprises a relatively unfaulted, gently westward-dipping flank separated from a northern accumulation by a series of east-west orientated faults (Fig. 1b). The southern margin of the field is likewise defined by east-west orientated faults, whilst the crest of the structure (to the east) is complicated by slumping and reworking. The average well spacing in the Main Block at Cycle III level is approximately 500 m (Fig. 1b).

3. RESERVOIR SUBDIVISION

Prior to the present study, Cycle III had been subdivided into three reservoir units: 3.1, 3.2 and 3.3, based on the recognition of geologically distinct facies associations within each unit and on the observation of pressure differences developed between the units during production (Fig. 2) (Johnson & Stewart 1985; Livera 1989).

Unit 3.1 represents lagoonal to lower coastal plain facies associations. The top of the unit is taken at the base of the fieldwide lagoonal shale that is often referred to as "the mid-Ness shale". Over most of the Main Block unit 3.1 comprises coarsening-upward sands deposited as lagoonal shoals and lagoonal deltas (Livera 1989). The base of the unit comprises a thin transgressive lagoonal sheet sand and lagoonal shales (Fig. 3).

Unit 3.2 represents an upper coastal plain facies association of channels, crevasse sands, floodplain shales and coals. These coastal plain sediments show significant areal variations in net:gross ratio as a result of localised vertical stacking of channel-fill sand bodies (Fig. 3).

Unit 3.3 represents lagoonal to lower coastal plain facies associations. This interval comprises a series of laterally continuous, stacked, coarsening-upward lagoonal fills (Fig. 3).

Detailed examination of dynamic reservoir data such as pressure and reservoir performance data reveals that fluid movements and pressure changes often occur at a finer scale than that defined by the reservoir units, as a consequence of strong vertical permeability contrasts and local flow barriers formed by laterally extensive coals and shales. Consequently the present study has used the Shell workstation-based log correlation system GEOLOGIX to establish a finer scale subdivision of Cycle III.

GEOLOGIX is a log correlation system with a powerful graphic user interface that allows interactive log subdivision and correlation (Rieuwerts 1989). Most correlation was based on a combination of density and neutron logs but gamma-ray and sonic logs were also used to provide supplementary information. Each well was first subdivided and correlated according to the existing reservoir subdivisions (3.1, 3.2 and 3.3) (Fig. 4a). The "zoom-in" function of GEOLOGIX was then used to establish further subdivision of each of these reservoir units into ten correlatable layers of genetically related sediments (Fig. 3). Consideration of productrion and engineering data indicates that these subdivisions correspond to the layers within which differential fluid displacement has occurred during production (Bryant & Livera in press) and are equivalent to "flow units" (**sensu** Hearn **et al**. 1986; Slatt & Hopkins 1988). Finally each set of logs was subdivided into reservoir elements that represent genetic reservoir units (Fig. 4b). The vertical boundaries of these units were taken to correspond with those defined by petrophysical evaluations of net sand. Whereas the correlatable reservoir layers define the "flow units" within which fluid movements take place, the distribution of reservoir elements within these layers determines reservoir continuity and quality. We have identified five types of reservoir element: channel-fill sandstones, sheet sandstones and coarsening-upward sandstones together with mudstones and coals.

Where geological considerations or production/engineering data indicated that reservoir elements in two adjacent wells are very likely to form part of the same genetic unit they have been correlated within GEOLOGIX, assigned to an element type and given a unique identifier. Non-correlatable reservoir elements were assigned an element type but were not given unique identifiers. Thus named, these elements represent an integration of geological, petrophysical and engineering data. This exercise was carried out for Cycle III in the entire Main Block. The data set created thus represents the best reservoir description that is consistent with the available data and forms the basis of the three-dimensional models that have been generated.

Uncertainty within this description remains as to the exact location of the termination of each element, and in the distribution of non-correlatable elements, within the inter-well volume.

4. RESERVOIR MODELLING

Construction of reservoir models traditionally integrates geological knowledge with reservoir data to provide a qualitative description of the reservoir (Fig. 5). Recently the generation of cell-based reservoir models has been advocated as a means of further quantifying the reservoir description (Haldorsen 1983). These cell-based reservoir models may be constructed using two different philosophies: modelling of continuous variables (e.g. porosity or permeability) or of discrete objects (e.g. pressure zones or channel-fill sand bodies). Attention has often focussed on the use of conditional simulation for modelling of continuous variables and on Boolean methods for modelling discrete objects (Matheron **et al**. 1987; Dubrule 1989). Consequently many modelling systems use either conditional simulation or Boolean methods. However, in order to adequately represent the Cycle III interval, it is necessary to model in three dimensions the distribution of both correlated reservoir elements ("deterministic" layers **sensu** Haldorsen (1983)) and non-correlated reservoir elements to produce a "mixed deterministic/probabilistic model" (**sensu** Weber & van Geuns 1989).

In the Cycle III model, the correlatable reservoir elements have been kriged between the wells. Maps of these elements were used to control the quality of the log correlations and to establish the trends of channel-fill sand bodies (Fig. 6) that were subsequently used in modelling of the non-correlated channel-fill sand bodies. The lateral extent of the latter was modelled probabilistically using an empirical relationship between width and thickness (cf. Fielding & Crane 1987). Workstation-based software was used to generate a 3-D representation of the reservoir from these data (Fig. 7). The relative proportions of the various types of reservoir element in the reservoir volume is close to that obtained by averaging the well data. Thus, if the relationship between channel-fill sand body width and thickness is assumed to be valid, it is suggested that any sand bodies not penetrated at the current well spacing are small and consequently, that the model probably contains all of the volumetrically significant reservoir elements.

The reservoir model comprises volume elements ("voxels") of dimensions 50 m x 50 m horizontally and 1 ft vertically, that are characterised by x, y, z coordinates and a geological designation. In order to be of value to engineering studies it is further necessary to assign petrophysical parameters to the voxels. This has been achieved by averaging the petrophysical evaluation stored in GEOLOGIX for each reservoir element at each well location. These averages have then been kriged into the interwell areas. In this way each voxel is additionally assigned a value of porosity, permeability and hydrocarbon saturation (Fig. 8).

5. UTILISATION OF THE MODEL

The resulting voxel-based reservoir description represents a versatile data base which may be utilised in a variety of different ways. A major advantage of the description is that it may be displayed in a variety of different forms. Thus the geologist may examine maps or 2-D cross-sections, the petrophysicist single well log evaluations or the engineer hydrocarbon volumes within individual drainage units. Revision of the reservoir description by any of these specialists automatically results in a transfer of information between these disciplines.

The emphasis of the present study was to locate and quantify remaining oil volumes within the Cycle III interval. This has been achieved by firstly incorporating the original fluid contacts into the model by assigning each voxel a true vertical subsea depth on the basis of the top of Cycle III as defined by seismic and well data. In this way it is possible to distinguish those voxels which were originally in the oil leg from those in the gas and water zones (Fig. 9a). By incorporating information from reservoir monitoring into the model it was possible to establish more recent fluid contacts within each of the reservoir elements (Fig. 9b). This model indicates that:

a) oil does remain in unit 3.1 due to vertical permeability contrasts but is likely to be unrecoverable by conventional recovery methods due to rapid coning of gas and/or water from adjacent, flooded sands,
b) oil may remain in unit 3.2 in isolated channel-fill sands not connected to the current wells but such accumulations are considered to be uneconomically small, and
c) most of the undrained recoverable oil is expected to occur in unit 3.3 within correlatable sand bodies (Bryant & Livera in press).

Consequently a model of the correlatable sands of unit 3.3 was constructed for the entire Main Block (Fig. 10), such that the model had natural boundaries: the OGOC to the east, OOWC to the west and faults to the north and south (Fig. 1b). By inserting original and currently established fluid contacts within the model (cf. Fig. 9), we have carried out volumetric calculations at the scale of individual genetic sand bodies (Bryant & Livera in press).

When the individual genetic units are relatively homogeneous this geometric approach is adequate and costly simulation is unnecessary. However, where the sand bodies contain significant lateral or vertical variations in petrophysical properties or are in communication with other sand bodies, reservoir simulation of fluid displacement may be required to predict reservoir behaviour. In this case the ability to extract the parts of the reservoir description that are of interest reduces the number of voxels to a manageable number and enables the relevant level of detail to be incorporated into the reservoir simulator.

6. DISCUSSION

Traditional methods of reservoir description are usually, at best, semi-quantitative. These descriptions are often represented by well log correlations (cf. Fig. 4) or schematic sections (cf. Fig. 3). Although these representations of the reservoir may be geologically correct they are often 2-D or pseudo 3-D descriptions of the reservoir and are not in a form which may be easily transferred to other petroleum engineering disciplines. The data are usually used to produce maps of average properties for reservoir intervals that are not necessarily related to the geological subdivision of the reservoir. It is these maps that form the basis of volumetric calculations and any reservoir simulation that is carried out. These maps of average properties are thus "purpose-built" for the reservoir engineer and are not easily usable for other petroleum engineering activities, such as comparison with field seismic data. Neither are the maps easily updated, for instance with revisions to incorporate new well data or to use new petrophysical cut-off values. They thus become static representations of the reservoir at a given moment in time.

The information contained in the maps is often sufficiently detailed to permit bulk hydrocarbon content of a part of the reservoir to be determined with sufficient accuracy to enable early field development decisions to be taken. Very often, however, they lack sufficient spatial resolution to precisely locate these hydrocarbons within the reservoir volume. Neither are they easily modified to perform such a task since they are purposely designed to provide average descriptions.

The method outlined in this paper results in a reservoir description that contains sufficient detail to guide reservoir management on a sand-by-sand basis. Furthermore since it is accessed via an interactive graphic user-interface it can be readily updated or revised. The reservoir description may be produced at a variety of scales: at the Cycle scale (Fig. 7); at the unit scale (Fig. 10) or at the scale of individual genetic geological units (Fig. 8). Such detailed descriptions are often reduced by averaging (Abib **et al**. 1988; Le Loc'H 1989) or by generation of pseudo-values (Lasseter **et al**. 1986) for further evaluation using reservoir simulators. Our approach has been to avoid such reductions in the spatial resolution of the data until reaching a reservoir description at the scale of individual genetic units. At this scale averaging procedures may be used to reduce the number of voxels to an appropriate number to allow transfer to reservoir simulators. Thus, having determined the zones of the reservoir which have the potential to contain remaining oil these may be transferred to reservoir simulators with the relevant detail still present.

These detailed reservoir descriptions provide a challenge to both geologists and petroleum engineers (Haldorsen **et al**. 1987). The quantification involved in the voxel-based description of the reservoir requires a quantitative geological knowledge to guide modelling of sand box size (e.g. Fielding and Crane 1987) and the distribution of rock properties within each sand body. Furthermore the existence of such detailed descriptions justifies the collection and analysis of reservoir engineering data (e.g. pressure measurements) at a finer scale than was hitherto the case.

7. CONCLUSIONS

The key to acharacterising these depositionally complex reservoirs such as Cycle III, lies in a fine-scale, sedimentologically-based correlation of the reservoirs and superimposition of dynamic data collected by reservoir monitoring. This integrated reservoir description is further quantified by generation of voxel-based reservoir models at the scale of genetic geological units. These models assist in reservoir management and represent versatile, dynamic reservoir data bases.

Acknowledgements
The modelling system is based on an original concept of M.C. Budding that was initially realised through the programming efforts of O.R.F. Dubrule, H. Rieuwerts and A.B.G.M. ten Tusscher. We are indebted to Shell Expro, Esso UK, Shell Internationale Petroleum Maatschappij, and Shell Internationale Research Maatschappij B.V. for permission to publish this paper.

References

ABIB, O., MORETTI, F.J., CEN MEI & YANG YUHZE 1988., The application of geological modelling and reservoir simulation to West Saertu area of the Daqing oil field. Society of Petroleum Engineers Paper. 17560.

BRYANT, I.D. & LIVERA, S.E. in press. Identification of unswept oil volumes in a mature field by using integrated data analysis: Ness Formation, Brent field, UK North Sea. **In**: SPENCER, A.M. (ed.), **Generation, accumulation and production of Europe's hydrocarbons**. Proc. 1st EAPG Conf. Berlin. Oxford Univ. Press.

DUBRULE, O.R.F. A review of stochastic models for petroleum reservoirs. **In**: ARMSTRONG, M. (ed.). **Geostatistics**. 2, 493-506.

FIELDING, C.R. & CRANE, R.C. 1987. An application of statistical modelling to the prediction of hydrocarbon recovery factors in fluvial reservoir sequences. **In**: ETHERIDGE, F.G., FLORES, R.M. & HARVEY, M.D. (eds.) **Recent developments in fluvial sedimentology**. Society of Economic Paleontologists and Mineralogists Special Publication, **39**, 321-327.

HALDORSEN, H.H. 1983. **Reservoir characterization procedures for numerical simulation**. Unpub. Ph.D. Thesis, University of Texas at Austin. 556 pp.

HALDORSEN, H.H., CHANG, D.M. & BEGG, S.H. 1987. Discontinuous permeability barriers: a challenge to engineers and geologists. In: KLEPPE, J., BERG, E.W., BULLER, A.T., HJELMELAND, O. & TORSAETER, O. eds) **North Sea Oil and Gas Reservoirs**. Graham and Trotman, London, 127-151.

HASTINGS, A.F., MURPHY, P. & STEWART, L. 1989. Helm reservoir characterisation: the need to use a multidisciplinary approach. (Abs.) EAPG Meeting, Berlin.

HEARN, C.L., HOBSON, J.P. & FOWLER, M.L. 1986. Reservoir characterisation for simulation. Hartztog Draw Field, Wyoming. **In**: LAKE, L.W. & CARROLL, H.B. Jr. (eds.) **Reservoir characterisation**. Academic Press, London, 341-372.

JOHNSON, H.A. 1989. North Sea reservoir simulation: practical considerations. SPE Pat. 19039.

JOHNSON, H.D. & STEWART, D.J. 1985. Role of clastic sedimentology in the exploration and production of oil and gas in the North Sea. **In**: BRENCHLEY, P.J. & WILLIAMS, B.P.J. (eds.) **Sedimentology: recent developments and applied aspects**. Geological Society of London Special Publication, **18**, 249-310.

LASETTER, T.L., WAGGONER, J.R. & LAKE, L.W. 1986. Reservoir heterogeneities and their influence on ultimate recovery. In: LAKE, L.W. & CARROLL, H.B.Jr. (eds.) **Reservoir characterisation**. Academic Press, London, 545-559.

LE LOC'H. 1989. An efficient strategy for combining the permeabilities: practical application on a simulated reservoir. **In**: ARMSTRONG, M. (ed.) **Geostatistics**. **2**, 557-568.

LIVERA, S.E. 1989. Facies associations and sand-body geometries in the Ness Formation of the Brent Group, Brent field. **In**: WHATELEY, M.K.G. & PICKERING, K.T. (eds.) **Deltas: sites and traps for fossil fuels**. Geological Society of London Special Publication, **41**, 269-286.

MATHERON, G., BEUCHER, H., DE FOUQUET, C., GALLI, A., GEURILLOT, D. & RAVENNE, C. 1987. Conditional simulation of the geometry of fluvio-deltaic reservoirs. Society of Petroleum Engineers Paper. 16753.

PAJOT, D., LAFUE, G.M.E. & WINCHESTER, A.J. 1988. Knowledge representation in a workstation for oil reservoir analysis. Paper 3-25, European Association of Exploration Geophysicists Conference, The Hague.

RIEUWERTS, H. in press. GEOLOGIX, an interactive knowledge-based well correlation system. Proceedings of Conference on Artificial Intelligence in Petroleum Exploration and Production, College Station, Texas, 17-19 May, 1989.

SLATT, R.M. & HOPKINS, G.L. 1988. Scales of geological reservoir description for engineering applications: North Sea oilfield example. Society of Petroleum Engineers Paper, 18136.

WEBER, K.J. & VAN GEUNS, L.C. 1989. Framework for constructing clastic reservoir simulation models. Society of Petroleum Engineers Paper, 19582.

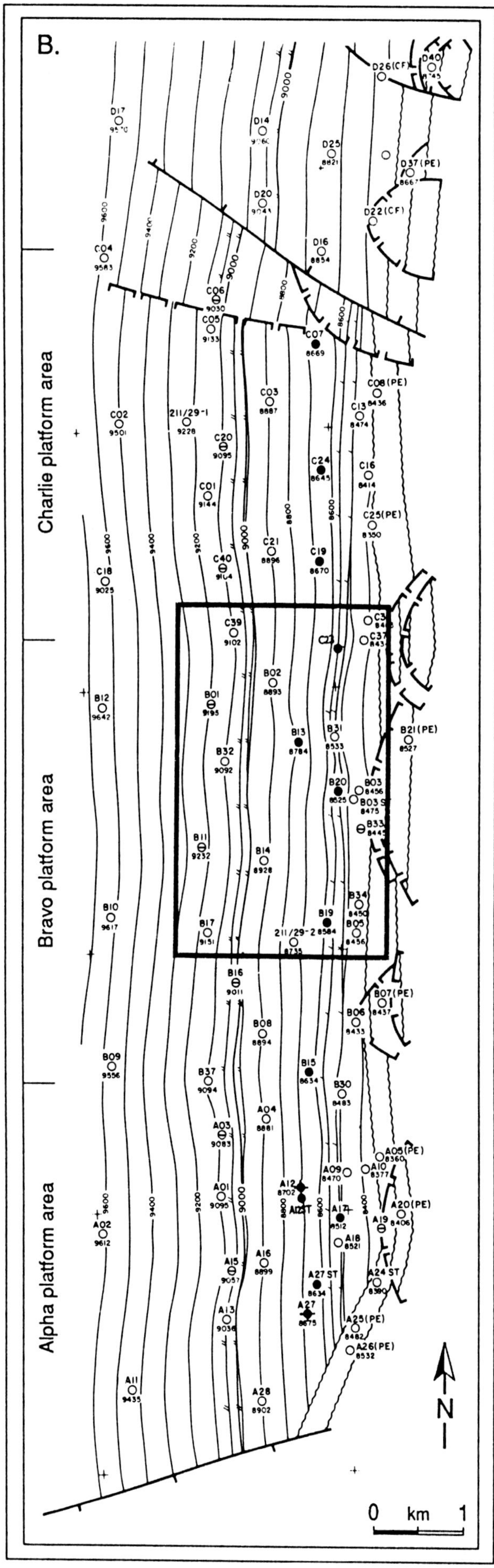

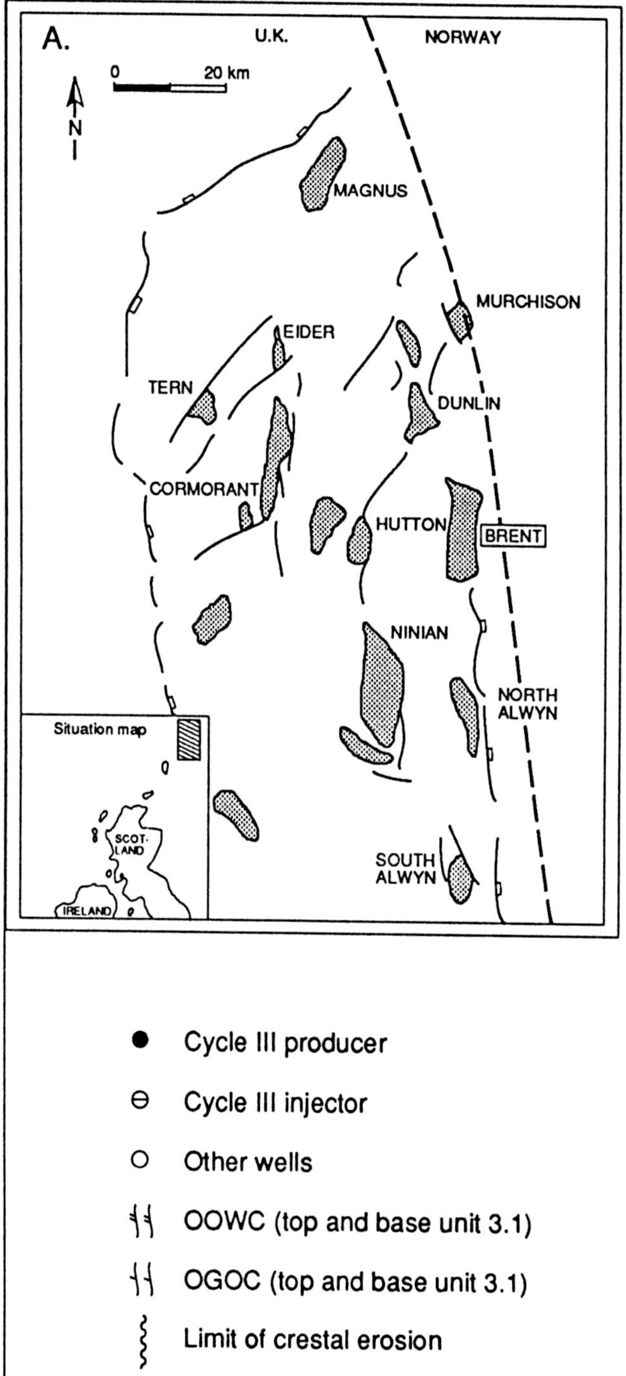

A. Location of the Brent field in the East Shetland basin of the northern North Sea. (Inset shows the location in relation to the British Isles.)

B. Depth contour map (in feet) of the top of unit 3.1. (Inset area shows the location of Figures 6, 7 and 9.)

FIG. 1

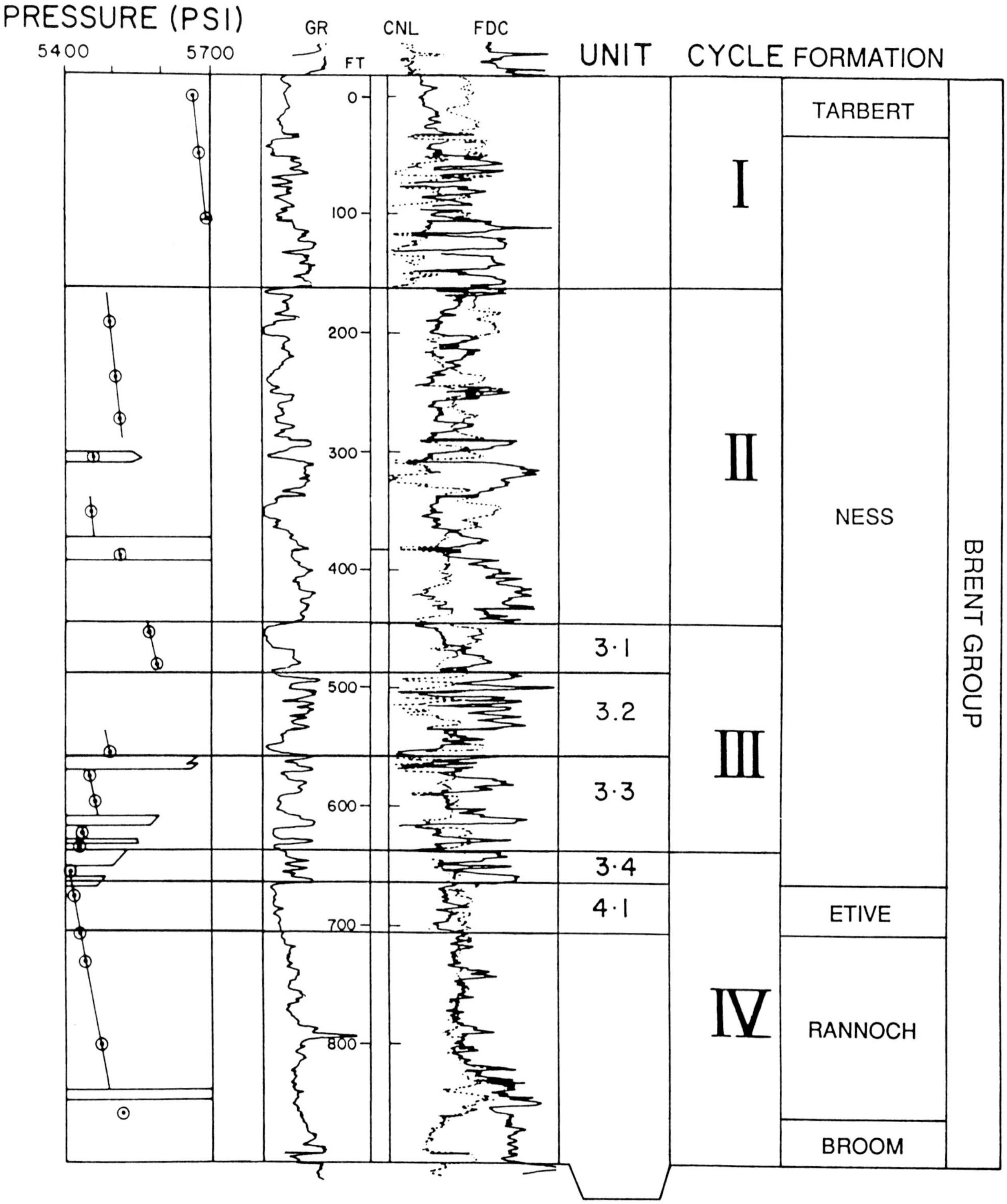

Well logs and pressure data from a well in the Brent field to show the division of the Brent reservoir into geological formations and petroleum engineering "cycles" and "units". The pressure measurements show pressure differences between and within cycles that have developed during production as a consequence of vertical pressure barriers or baffles within the sequence. (Modified from Johnson and Stewart 1985.)

FIG. 2

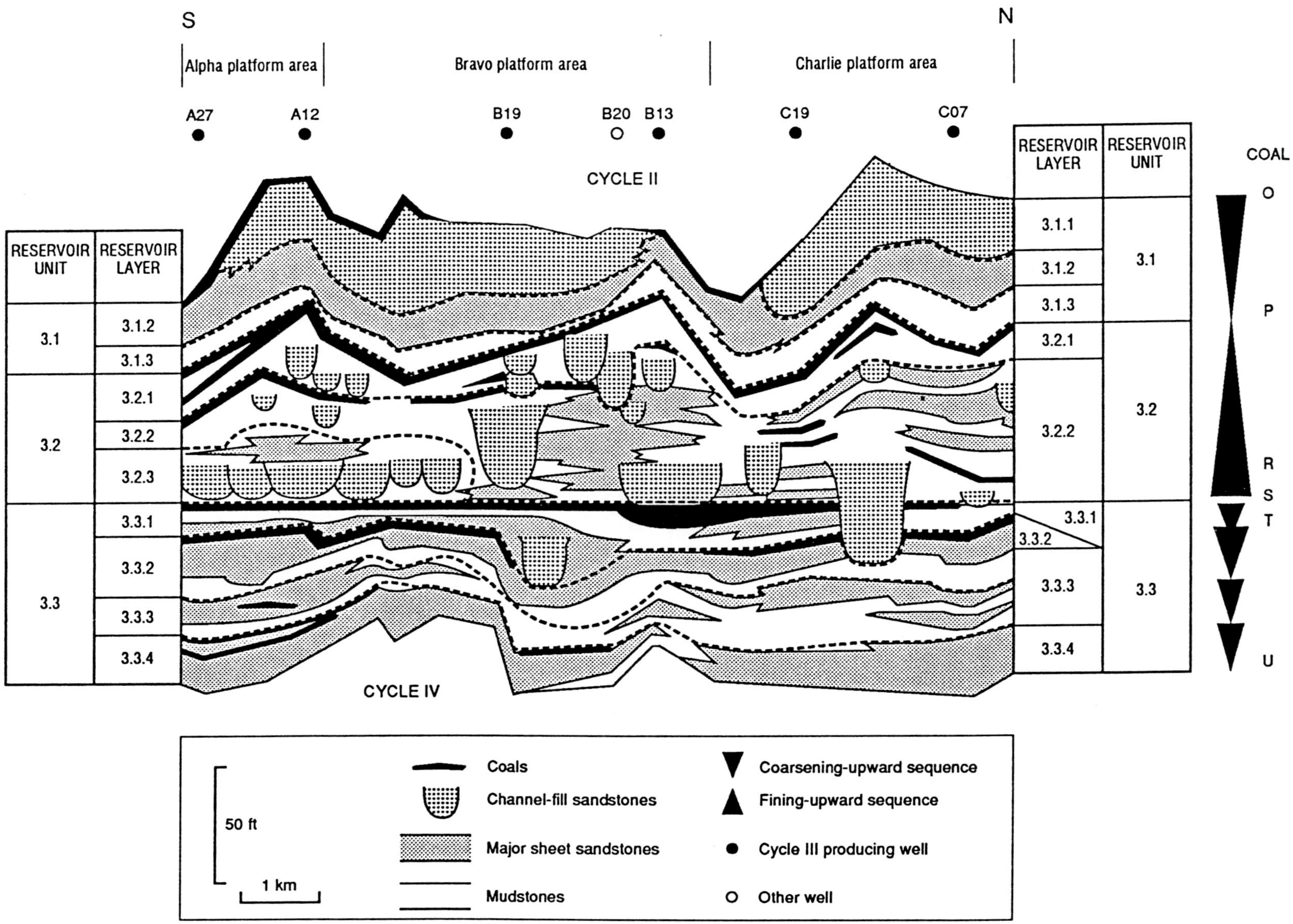

FIG. 3 Schematic N-S cross-section through Cycle III in the Main Block of the Brent field constructed using all of the available well data (only key wells shown). The section schematically illustrates the distribution of sands, shales and coals together with the subdivision of Cycle III into reservoir units and layers.

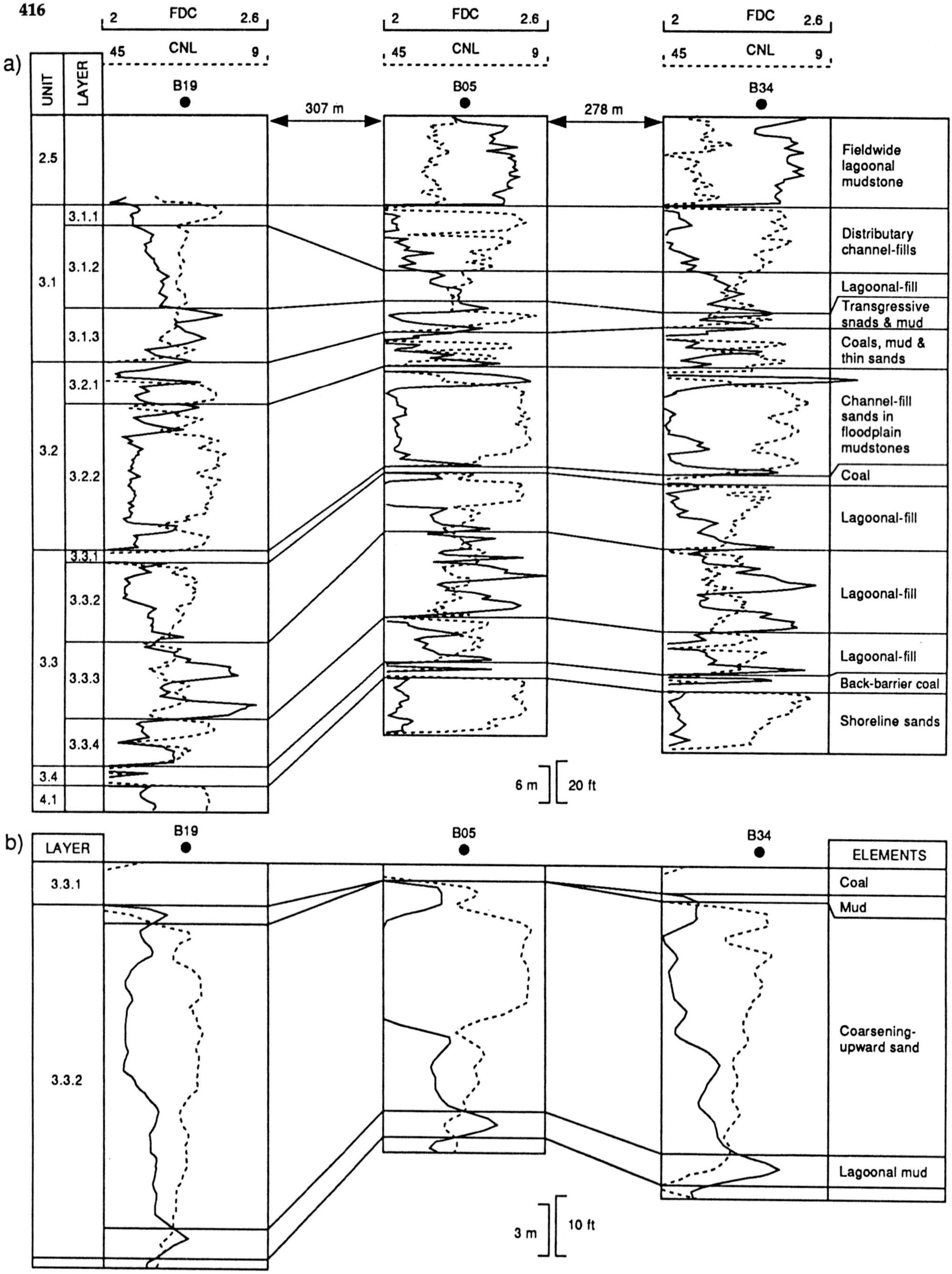

Density and neutron logs from three wells in the Bravo platform area.

a) Logs displayed using the top of unit 3.1 as a datum (base of the "mid-Ness shale") to show the subdivision into reservoir units and layers.

b) An enlargement of part of unit 3.3 to show how layer 3.3.2 comprises up to three reservoir elements.

FIG. 4

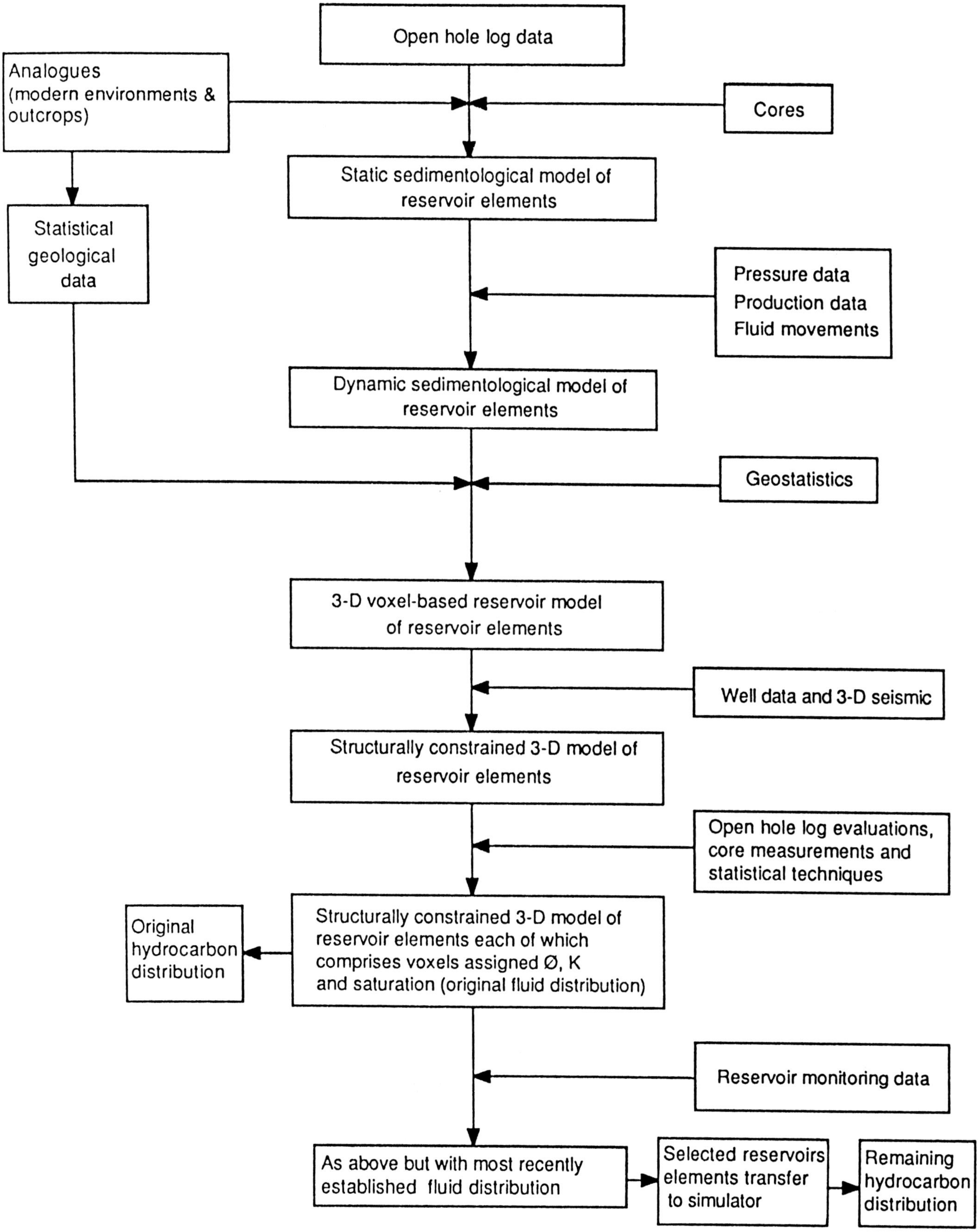

FIG. 5 Integrated reservoir description to locate the remaining oil

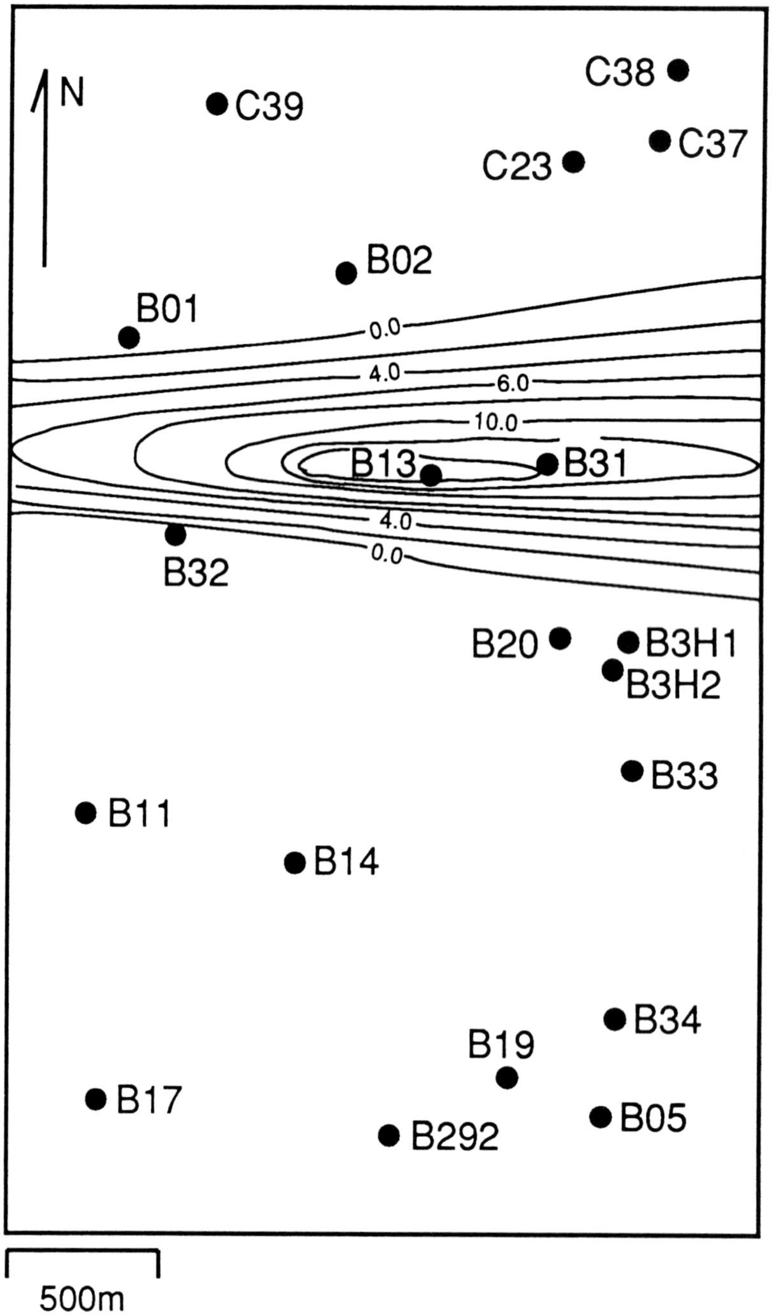

FIG. 6 Thickness maps (in feet) of a channel-fill reservoir element. Such maps have been used to establish the orientation of the channel fill sands.

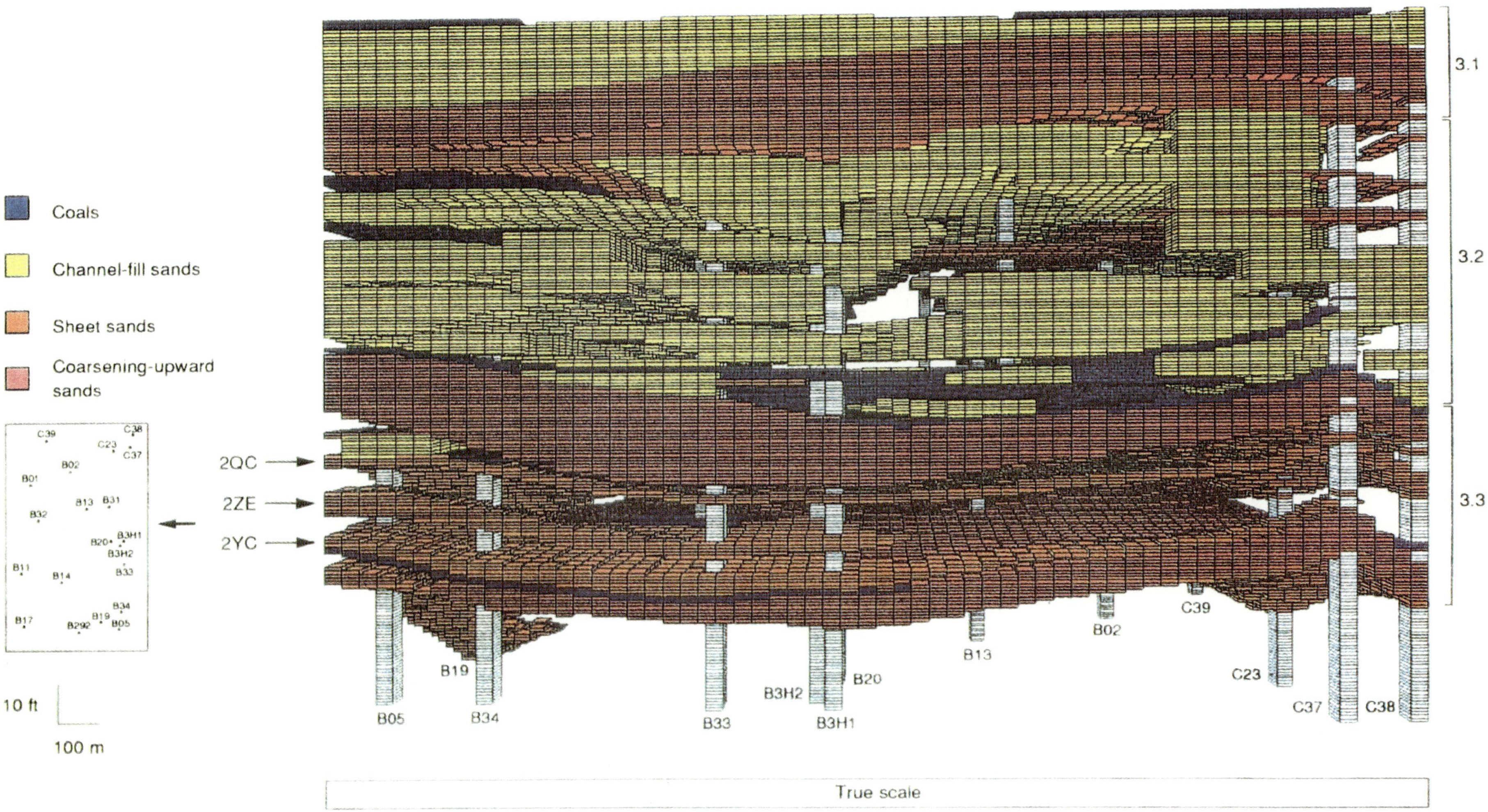

A perspective view, from the east, of a 3-D model of Cycle III in the central part of the Main Block (boxed area of Fig. 1). The model takes the top of Cycle III (base of the "mid-Ness Shale") as datum. The transparent voxels represent shales. (Well positions and viewing direction are shown on the simplified field map to the left of the perspective view.)

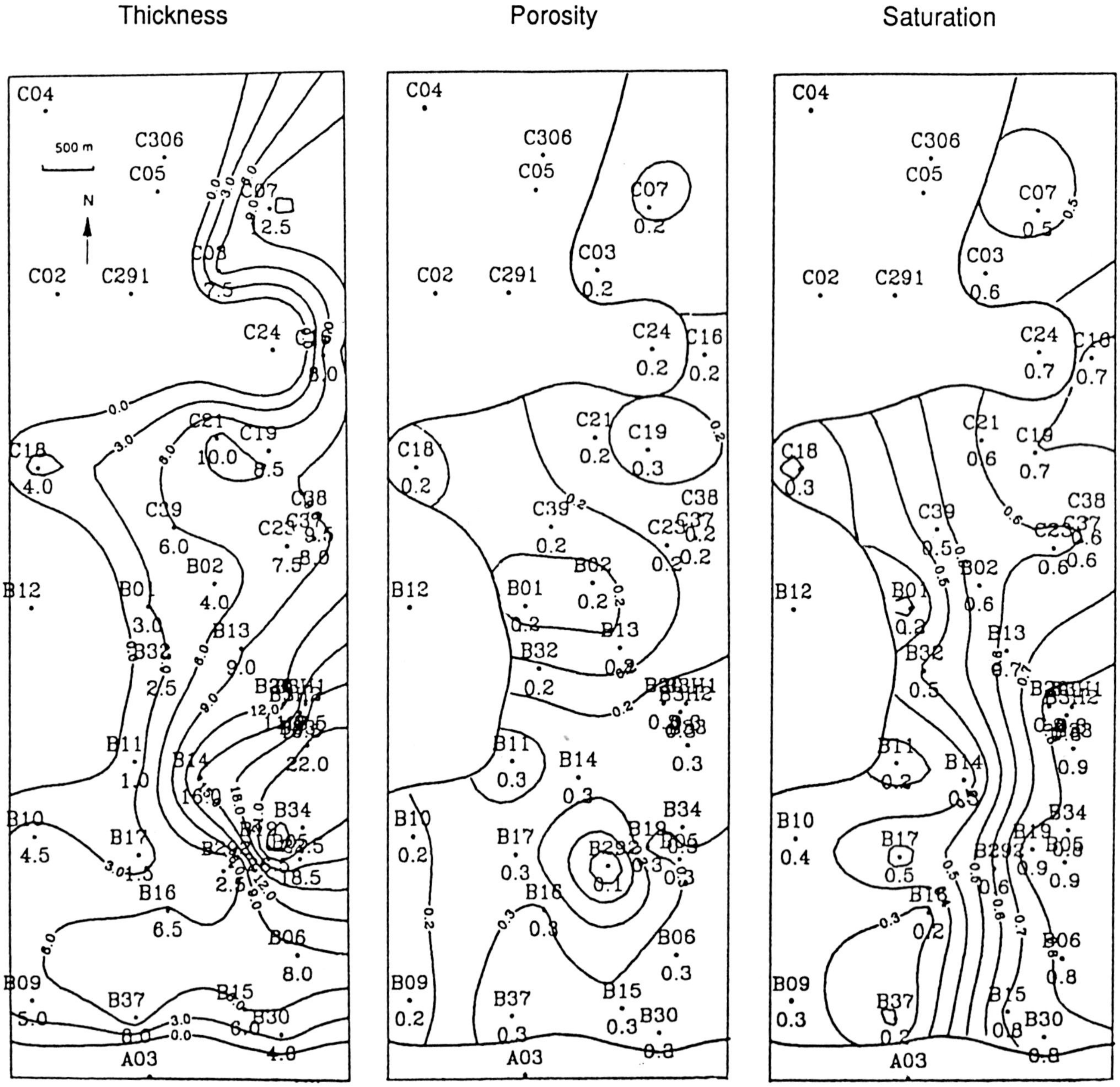

FIG. 8 Maps of thickness (feet), porosity and hydrocarbon saturation for a reservoir element in unit 3.3 (Fig. 10). The information represented on the maps is stored on a voxel basis and indicates that this reservoir element initially contained 39.4 x 10^6 stb of which 6.4 x 10^6 stb of oil potentially remains.

Dr.no. 76888/14

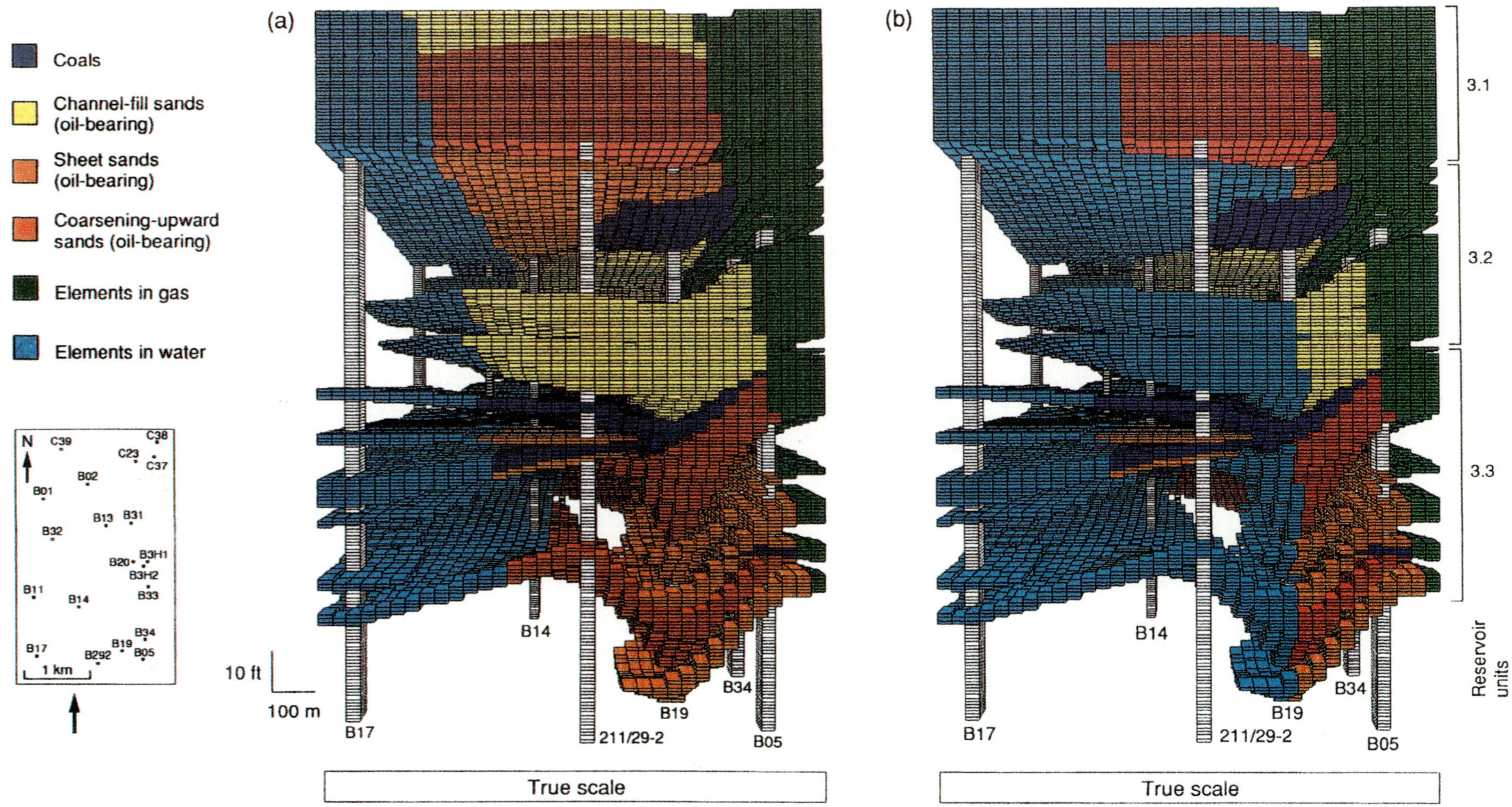

FIG. 9

Perspective views of a 3-D model of correlatable reservoir elements in Cycle III in the central part of the Main Block (boxed area in Fig. 1). The models take the top of Cycle III (base of the "mid-Ness Shale") as datum. The transparent voxels represent shales and non-correlatable reservoir elements. (a) Shows the original oil-water and gas-oil contacts. (b) Shows the currently established fluid contacts on the basis of reservoir monitoring data. (Well positions and viewing direction are shown on the simplified field map to the left of the perspective view.)

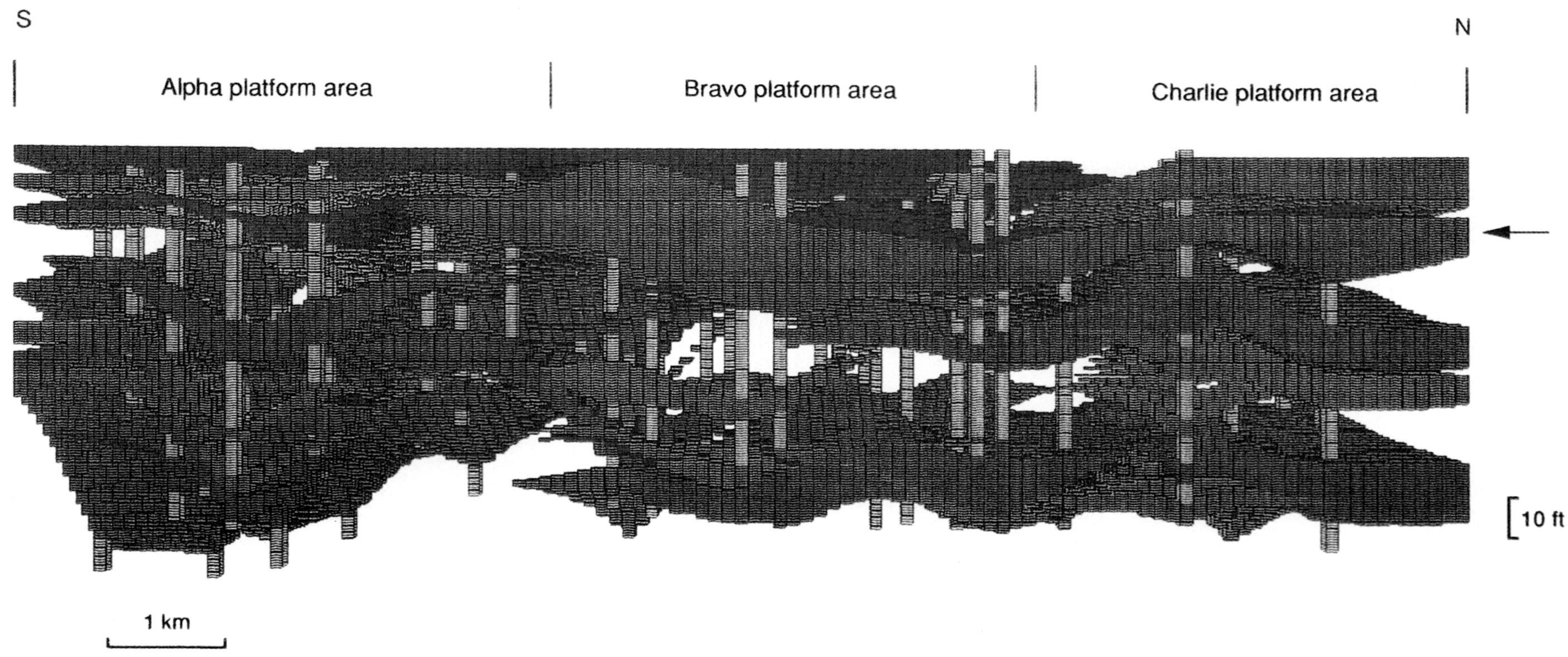

Perspective view of 3-D model of unit 3.3, over the area of the entire Main Block, viewed from the east. The model uses the top of unit 3.3 (top of the S coal over the most of the field) as datum with 430 x vertical exaggeration. The reservoir element shown in Fig. 8 is arrowed.

An Integrated, Multidisciplinary Study of Reservoir Facies and Reservoir Performance of a Multilayer, No-Crossflow Gas Reservoir—Chase Group (Wolfcampian), Guymon-Hugoton Field, Oklahoma

M. J. Fetkovich, W. T. Siemers, J. J. Voelker, J. S. Williams, A. M. Works, and D. J. Ebbs

This paper summarizes a comprehensive study of the depositional, diagenetic, petrophysical, and production characteristics of the Chase Group (Wolfcampian) gas reservoirs, Guymon-Hugoton Field, Oklahoma. Integration of disciplines in geology, petrophysics, and reservoir engineering and management has resulted in accurate and meaningful geological and reservoir performance models for this field.

The Chase Group is composed of cyclical sequences deposited on a gently dipping, low-relief, shallow marine shelf. Each cycle comprises laterally continuous, successively shallower water, marine carbonate (chiefly dolostone), siltstone and sandstone reservoir units capped and separated by shaly redbeds and paleosols. The lateral continuity, low permeability, and high threshold entry pressures of shaly layers produce effective regional barriers that prevent vertical fluid flow and pressure communication between reservoir layers. Although variable, depending on depositional and diagenetic history, porosity is continuous within a reservoir layer because of a well-developed intercrystalline pore network produced by pervasive dolomitization independent of depositional texture and other pore types.

Pressure differences measured between reservoir layers verify predicted reservoir and barrier qualities. A study of 83 replacement wells failed to identify any original field pressures (490 psia), indicating no significant volumes of gas not in communication with producing wells. All replacement well locations, therefore, have contributed to production from the original section well or other offset wells; no new gas-in-place has been found.

Petrophysically unique Archie saturation input parameters were computed based on the abundance of the various lithotypes within each of the reservoir layers. Their incorporation into the log analysis system provided a more accurate determination of effective porosity and water saturation and, ultimately, a meaningful definition of effective pay thickness and hydrocarbon pore volume.

A three-dimensional, multiwell, multilayer, no cross-flow reservoir model was developed which matches 40 years of pressure-cumulative production performance, well deliverability, and current individual layer pressures. The model is used to forecast future reservoir performance and to evaluate possible effects of decreasing well spacing. The model and field performance shows the present 640-acre spacing effectively drains the reservoir. Because of contrasting layer permeabilities, a long producing life (110 yrs) is predicted for the multilayered, no cross-flow reservoir. If market demand is high enough, infill drilling can accelerate production, but infill drilling does not increase ultimate recoverable reserves. At abandonment, infill wells will not deplete high pressure layers any more than will one producing well per 640 acres.

SOLE PIT—IMPROVING THE PERFORMANCE OF A TIGHT GAS RESERVOIR, AN INTEGRATED APPROACH

R. E. Wilson
J. F. R. Branco
B. G. van Heyst
R. H. Turner
Shell Exploration and Production U.K.

ABSTRACT

The initial development plan for Clipper and Barque 'core' areas in the Southern North Sea, Sole Pit Basin called for conventional wells to intersect fissured zones wherever possible and where no open fissures were encountered, connection would be attempted via massive hydraulic, sand propped fracturing.

Since drilling began in 1988 the plan has benefited from a number of new developments:

-Improvements have been made to the design and placement techniques of Massive Hydraulic Stimulation.

-A new Borehole Televiewer that works in oil-based mud is providing information on the fissure type, orientation and density, impacting on the placement and design of later wells.

-Dune slip-face sands have been identified which enable direct production from the matrix, without hydraulic stimulation.

-Production results of highly successful, but now abandoned appraisal wells have been replicated by a technique named 'cloning'.

-The application of horizontal drainage has been very successfully tested and will be extensively utilized in both fields.

-Subtle features identified using 3D seismic derived attribute maps have explained the extent and distribution of the different free water levels.

-Well completion techniques that do not incorporate a cemented liner are being investigated.

These developments have been made possible by close co-operation across the Petroleum Engineering disciplines organized into a small integrated team.

PREPARED FOR THE INAUGURAL ARCHIE CONFERENCE, HOUSTON OCTOBER 1990

A. BACKGROUND

Strategically located in the Southern North Sea within 100 km of south eastern Britain, the southern gas fields have supplied a major part of the United Kingdom's gas requirements over the last 22 years. As the older fields like Leman and Indefatigable have gone into production decline attention has increasingly turned to the smaller accumulations, often found in much tighter sandstone reservoirs.

Shell (as operator) in partnership with Esso, is simultaneously developing two such fields Clipper and Barque, located in the Sole Pit Basin (Fig. 1).

Bringing these fields to development has been a long (and elusive) enterprise. Exploration wells in the 1960s and 1970s confirmed the existence of gas-bearing Permian Rotliegend Leman Sandstone within several structures in the Sole Pit Basin, but well test results were poor and the prospects were deemed subeconomic. In 1982 an appraisal well produced at high rates from fissures and there followed an intensive appraisal drilling campaign and associated studies that culminated in 1988 when project and gas sales agreements were reached.

The Sole Pit Basin trends NW-SE across the western part of the Southern North Sea and is defined by three major tectonic faults. The geological history of the basin is complicated, involving at least two and possibly a third phase of inversion (Glennie and Boegner, 1981). Clipper and Barque are NW-SE orientated horst/domal structures, bounded by major faults on their NE and SW flanks and dip closed on their NW and SE flanks (Fig.l). An overall improvement in reservoir quality occurs from Barque to Clipper due to increasing proximity to the main Rotliegend dune fields in the south east. The Barque field is elongated and is 25 km long and 3 km wide. Host of the gas-bearing Rotliegend reservoir is found between 7500 and 8500 ft. TVSS, the reservoir is 700-800 ft. thick and consists mainly of aeolian quartzose sandstone which crestally, is entirely gas-bearing. The expected reserves in the Barque initial development area* are 375 BCF. The Clipper field is a broad, low relief structure with multiple culminations and differing free water levels (FWLs), it is 12 km long and 5 km wide. In the highest part of the field the aeolian sandstone is fully gas-bearing with a maximum column of 900 ft.; most gas occurs between 7500 and 8500 ft. TVSS. The expected reserves in the Clipper initial development area are 506 BCF. (Farmer and Hillier, 1991).

*The initial development areas (core areas) of the two fields were chosen to enable development from central platforms of the highest and most favorable areas to the formation of natural fissures. In Barque and Clipper this area is defined as a cylinder with radius 11,500 ft. centered on the platform, allowing development of gas reserves by wells with a maximum inclination of 60 degrees (Fig.l).

The Leman Sandstone is divided into 3 zones (Fig. 2). From top to bottom they are:

The A zone is 300-400 ft. thick, generally of poor quality and consists of mixed lithofacies; mainly dune, interdune and sabkha sandstone, usually mantled at the top this zone with thin (50-100 ft.), almost featureless Weissliegend sandstone (Glennie and Buller, 1983; Heward, 1990). In Clipper the average porosity in this zone is 10% and the in-situ effective permeability to gas is extremely low.

Cross-bedded dune sandstones dominate the middle or B zone which is the best quality reservoir rock in the sequence. This zone is 150-300 ft. thick, has an average porosity of 12% in Clipper (as high as 18% ln some slipface sands) and in-situ gas permeabilities up to 10 mD.

The C zone is the poorest of all three zones, consisting mainly of interbedded aeolian and fluvial deposits. These are beneath the GWC over much of the Barque and Clipper core areas and contain only 5% of GIIP.

Conventional core porosity/permeability relationships (corrected for overburden pressure, to Klinkenberg effect and to critical point drying) indicated the dune sands to be the best reservoir.

Palaeoburial of the reservoir to depths in excess of 13000 ft. has resulted in diagenetic changes which reduced the porosity and, more importantly, permeability of the rock by the growth of fibrous illite in the pore throats and by quartz cementation.

Development of Sole Pit would be sub-commercial if wells were dependent on draining the reservoir solely via the tight matrix. However, appraisal (and development) drilling indicates the presence of naturally fissured zones in the fields, and good production tests have been ascribed to the occurrence of a near vertical fissure network that is locally pervasive.

Note: In order to differentiate in the text between hydraulically induced fractures and those naturally occurring, the term 'fracture' has been reserved for hydraulic fractures and the term 'fissure' refers to the in-situ features.

B. INITIAL DEVELOPMENT PLAN

At the end of the appraisal campaign it was clear that substantial volumes of gas were trapped in the Rotliegend in seven structures in the Sole Pit area. It was also evident that the in-situ permeabilities are such that the initial well rates would be low and so would be the recovery. A development could at best be marginally attractive. However, some wells had encountered fissures, which greatly improved their productivity. How extensive would these be? What effect would they have on recovery efficiencies? What was the chance of drilling through more of them? No easy answers were available.

It was therefore decided to first develop only the "core areas" of Clipper and Barque where long gas columns were expected, with a significant volume of the better developed B-sands above the gas water contact and the chance of finding fissures was considered to be high, and where massive hydraulic fraccing could be widely applied to enhance productivity of matrix wells without the risk of fracturing into the aquifer.

It was realized that at the low prevailing permeabilities, recovery is very sensitive to permeability variations, fissure distribution, fissure conductivity and well spacing. In tight gas reservoirs like those in Sole Pit, the reservoir pressure depletes mainly in the neighborhood of producers. Even after many years of production the wells in the lower relief parts where only the poorer quality A-sands are above the GWC, will hardly interfere. Under these circumstances a good geological and fissure model can obviously help to locate wells optimally and maximize economic recovery. But with the information in hand at the time only a semi-statistical description of the fissure model could be made. It could not be predicted whether individual wells would encounter fissures but wells could be grouped in one of three possible categories:

-wells that would intersect fissures (expected to be highly prolific).

-wells that could be "connected" to a fissure system by means of a massive hydraulic frac.

-wells that would only encounter matrix and would therefore, even after massive hydraulic fraccing, make only a marginal contribution.

About forty percent of the wells were expected to be in the third category. Wells in the first two categories were expected to show declines of up 50% over the first winter.

In many Rotliegend offshore developments, pipelines and facilities costs are an order of magnitude higher than the well costs. In Clipper and Barque due to the low recovery per well, a large number of wells are required, therefore well costs formed a substantial part of the overall expenditure. Substantial savings were obtained by omitting one casing string from the well design. A 9 5/8" casing was to be set at the top of the overlying Zechstein Group and a production liner would be set over the Zechstein and Rotliegend together. This necessarily meant drilling the reservoir section with a high mud overbalance in order to keep the Zechstein, which occasionally contains pockets of over-pressured gas, under control. At the time this was not seen as a disadvantage as little matrix production was expected, thus impairment was not considered an issue. Moreover, drilling with a high mud weight was the only means available (when using oil-based mud) to identify when fissures had been encountered, by inducing losses.

The first round of development wells in Clipper found that over substantial parts of the field the GWC was shallower than hitherto assumed, reducing the GIIP by some 30%.

C. MODIFICATIONS TO PLAN

1. Massive Hydraulic Fracturing

In the original field development plan, much emphasis was placed on the ability to create massive hydraulic fractures (MHF) in the development wells. It was assumed that some 80% of the development wells would require MHF stimulation. Typically, job sizes of 1 million lbs of proppant (creating fractures of some 350 to 400 ft. radius) were planned.

Prior to drilling the Sole Pit development wells, experience was gained in MHF design and placement techniques in the Leman Field, to the south. This led to major changes in the fracture design models and material selected to combat the problem of proppant backflow (Clayton and Gordon, 1990). As a result, a sand management system was developed to facilitate rapid clean up of the wells during the commissioning phase and in-line filters were included in the flow lines to combat low level, long term sand production. Resin coated proppant has also been used which, when subjected to reservoir temperature for a short period of time, causes the grains to bond together and thereby resist backflow. A novel frac carrier fluid has been introduced to put the coated proppant in place. In addition, a mechanical down hole proppant exclusion system, called the 'fraction pack' tool (based on gravel pack technology), has been developed and tested.

Although considerable advances in MHF design and placement have been made, the early stages of the field development were characterized by the desire to avoid fracture stimulation (at least in the short term) wherever possible. This represented a basic shift from "if in doubt about productivity, frac the well" to fraccing only as a last resort, thus avoiding proppant back production and reducing costs. As a result, a number of cross discipline studies were undertaken to identify, better understand and exploit the natural productive features in the reservoir and these are discussed in more detail in the following sections.

2. Natural fissure detection

Considerable effort was expended over a number of years in an endeavor to find a fissure detection method that could be relied upon to function in a variety of drilling fluids, including oil-based mud (Frisinger and Gyllensten, 1986). Considerable core data was gathered during the appraisal phase but no single logging tool was found to be effective in the oil-emulsion muds used in Sole Pit. The method adopted to detect fissures in the first development wells was close monitoring of mud losses. As a fissure indicator, measuring drilling fluid losses is expensive in terms of the cost of chemicals, rig time wasted in curing losses and impairment of fissure conductivity. The technique is also crude in that it provides no precise information on the density, distribution, orientation, type or even the exact location of fissures.

It was clear that a better fissure detection method or tool had to be found, with one overriding stipulation; it had to be effective in oil-based mud. An extensive coring programme was considered but besides being expensive, it could not be accommodated within the severely time constrained drilling and well commissioning schedule required to meet first gas.

Attention turned instead to a third generation version of an old logging tool, the Borehole Televiewer. Called CBIL*SM (Circumferential Borehole Imaging Log), this tool produces two acoustic images of the borehole in terms of amplitude and travel time. Unlike its predecessors it can image through an increased range of mud conditions (including OBM). Results of the runs in six wells are very encouraging and as the trials progressed there has been a steady improvement in both image quality and recovery as a percentage of logged interval. Both open and closed fissures have been identified and computations of fissure orientation as well as structural and sedimentary dips are part of the routine image evaluation (Fig. 3). Testing of the logging tool will continue in both fields and a further trial is planned in Clipper where the acoustic image will be compared directly with core. Orientation of the core will also be attempted using the evaluated image.

Adoption of this logging tool as the principal fissure detection device will facilitate a lowering of the mud weight during drilling of the reservoir, thus reducing near-well impairment of matrix and fissures. A major contribution can be made to our understanding of the reservoir through the accumulation of lithofacies and fissure orientation, distribution, and fissure density data without resorting to coring on a large scale. These data will form part of the input to a new geological reservoir model which will lead to further improvements to field development.

*SM Registered service mark of Western Atlas International Inc.

3. Slip-face sands

Slip-face sands (otherwise known as avalanche dune sands or grain flow sands) comprise stacks of fine to medium grained sandflow cross-strata with sedimentary dips of > 20 deg. and < 32 deg. Figure 4 is a cross-plot of poroperm data per genetic type of sediment body and shows that the best reservoir properties are possessed by the thicker slip-face sands. They are identifiable on wireline logs from high porosity, high sedimentary dips and low gamma ray; however, until recently the oil-based mud has prevented the acquisition of dipmeter data. The best developed slip-face sands predominate in the B zone, with a few thin, isolated sands within the A zone. The thickest examples are probably tens to hundreds of feet wide and hundreds to thousands of feet long (Weber, 1987).

The contribution that slip-face sands can make to well productivity was first demonstrated to Shell in the Auk Field where most of the oil produced from the Rotliegend sandstone comes from the more porous slip-face sands (Heward, 1990).

Although better quality dune sands had been incorporated in the original Sole Pit geological model, their effect on the deduced gas recovery efficiency was largely negated by the adoption of geometric permeability zonal averages for the simulation model. The productivity and recovery efficiency benefit that could be gained if wells could connect with these features was therefore not fully taken into account.

Slip-face sands were cored in the second development well, W06 (Fig. 2). After evaluation of the core was complete and with the Auk example in mind, it was decided to complete conventionally across these sands and test their productivity. Similar sands were identified on the logs of well W07 and it too was perforated opposite the sand bodies. Both wells produced on test at close to 30 MMCF/D. No MHF was required immediately and another well category was established— wells that connect with good quality reservoir. Slip-face sands have now been recognized in other Clipper wells, and in the Barque horizontal well B16 much of the gas production is thought to emanate from these 1– 50 milliDarcy range sands. The benefits to the project are: - immediate savings through reduction of costly stimulation; - savings to rig time spent on post stimulation clean up; - no proppant back production problems; - fewer non productive wells; - improved recovery efficiency, dependent on the size and extent of the sand bodies.

4. Well cloning

In the case of new gas field development like Sole Pit, where the petroleum engineers are set a rigid gas delivery target both in terms of quantity and first gas date, there is a strong incentive to drill as many high productivity wells as possible in the early phase of development.

Replicating productive but abandoned exploration/appraisal wells is an obvious first step since the reservoir elevation, quality, fill etc., are known at the well locations but can only be inferred in-between. The variability in Rotliegend reservoir quality is such that drilling a well only a few hundred feet away from a successful appraisal well can lead to very disappointing results as has been experienced elsewhere ln the basin (Hawkings et al., 1990). In the development of Sole Pit replication has been achieved by means of 'cloning'. A very productive appraisal well, 48/19a-2 (43 MMCF/D) has been successfully cloned by carefully drilling directly at the abandoned well and a similar production rate has been achieved. In execution, this technique bears a strong resemblance to blow-out relief drilling and it likewise requires a high level of communication and co-operation between petroleum and drilling engineers. A second clone well has been drilled in the Barque Field and initial results indicate that a similar well productivity to the target well can be expected.

5. Horizontal development

At first sight the Clipper and Barque initial development areas do not look like horizontal hole country. Gas columns of several hundred feet, no problems with fluid coning, and some layering, all suggest massive hydraulic fracturing instead. Theoretically the productivity improvement of a horizontal well over a MHF'd well in a homogeneous reservoir with Sole Pit properties is minimal. However, horizontal holes would require an extra casing string, making them even less attractive. The heterogeneities in the reservoir and their current unpredictability, however, tilt the balance. With the sparse distribution of fissure swarms predicted (possibly some 1000 ft. apart) and the unknown distribution of slipface sands, 40% of the MHF wells are not expected to connect with either fissures or slipface sands and are forecast to produce less than 10 MMCF/D by the end of their first six months of production.

A horizontal well was planned in the Barque field prior to platform installation to increase the chance of intersecting vertical fissures by extending the lateral reservoir penetration of the well. The positive effect of horizontal wells would not be an improvement in production rate so much as a way of increasing the likelihood of any one well connecting with productive features, resulting in a reduced number of non productive wells. As a first approximation, a horizontal well that intersects a fissure swarm is expected to produce at a similar level as a vertical well which hits fissures. The well was drilled successfully, with a 3000 ft. penetration of the reservoir and completed with an uncemented, permeable liner to guarantee later access for logging purposes. The well tested 49 MMCF/D (tubing constrained) for a short period before platform installation. Although indications of semi open and open fissures were found on the CBIL log (Fig. 3), the test interpretation suggests that the flow is originating from circa 1000 ft. well developed dune sand intersected by the horizontal well with porosities which are considerably above average. Unfortunately no spinner survey in the horizontal section has been possible as yet.

Several well test interpretations can be made because of the short duration of the flow period. However all of them suggest a sustained high level of future productivity. Single well simulations indicate that the medium term production of the well will not be determined by the inflow performance, as is usually the case in tight reservoirs, but by the continuity of the dune sand and the amount of recharging of the dune by the surrounding matrix. The transient decline of the wells predicted to be relatively small. This result was reinforced by a horizontal well drilled by another operator in a nearby field which, after six months still produces at rates that are considerably higher than from any other wells and has doubled the deliverability of that field (Hawkings et al., 1990). These two wells have increased the interest in horizontal holes as a means to improve both rates and recovery of the fields.

Core analyses indicate that vertically about 30% of the B sand is occupied by potentially productive dune, although the horizontal distribution and size is not yet known. If, for simplicity, it is assumed that the horizontal percentage is also 30%, then simulation indicates that an increase of 25% recovery could be achieved with 25% fewer wells (Fig. 5) by using horizontal rather than conventional drilling techniques. This result is mainly due to the fact that at the prevailing permeabilities the major part of the pressure drop between reservoir and surface takes place over the dune/matrix interface in the case of the horizontal well, while in the case of the fracced conventional well, the inflow performance of the well is the bottle neck (Fig. 6). The cost difference between vertical and horizontal wells is expected to be minor as the extra casing string and drilling time is largely offset by the considerable savings on well stimulation.

Based on the above results and simulations it is now believed that horizontal wells can:

-substantially reduce the number of non productive wells by intersecting heterogeneities, either fissures or dune sands;

-significantly increase the deliverability of the field in the short term; and

-improve the recovery at reduced cost.

Obviously production performance and a better understanding of the geology will be required to accurately assess the benefits and decide the final well spacing.

Three more horizontal holes are planned in Barque in the near future, also targeted primarily at the B sand. They will be spaced in such a way as to allow a gradual increase in well density, if and when required. In Clipper, the majority of wells needed for B zone development in the core area have already been drilled conventionally with well inclinations in the range 20 – 60 deg. Therefore a horizontal well in Clipper is not expected to have the same spectacular large areas where the 'fissure finder' concept could work. This will be tested in the near future. There is additional scope to apply horizontal drainage in the low relief flanks where fraccing conventional wells would result in communication with the aquifer.

Improved geological models combined with the recently matured horizontal drilling/completion techniques and the quantification of the benefits by reservoir simulation may have opened the way to a much healthier field development.

6. 3D Seismic

Clipper and Barque core areas were covered with separate 3D seismic surveys in 1986 and 1987. Conventional evaluation of the processed data using B seismic workstation demonstrated for the first time that the highly disturbed top reservoir reflector could be traced right across the fields, and as a result Rotliegend maps have been produced that, in terms of the resolution of structural detail, represent a major improvement over the 2D derived maps they replaced. The reliability of the seismic has been put to the test in Clipper with the drilling of the first nine development wells and so far the 3D has proven to be a credible basis for planning field development. Improvements to the structural definition have led to subdivision of the Clipper Field into 4 major blocks rather than three using the 2D seismic. However, after the first 3D analysis was finalized, some problems of partition of different observed free water levels remained unresolved in both fields.

The potential to improve fault resolution using 3D seismic attribute analysis has lately been receiving some coverage in the literature (Dalley et al., 1988; Ruijtenberg et al., 1990; Mondt.,1990). Instantaneous dip and azimuth displays can be used to help identify small faults with offsets so minute that they are normally missed in a conventional interpretation. Dip and azimuth analysis has been applied to the interpreted 3D top reservoir time horizons for both fields and maps have been produced that accentuate minor changes in the magnitude of dip and small variations in dip direction (azimuth) within the horizon. Figure 7 is an azimuth display close to the top of the Barque reservoir. Subtle NNE- SSW and ENE- WSW features (arrowed) are clearly visible and similar features are observed using the same displays of the equivalent Clipper time horizon. These are thought to be fault zones that have resulted from strike-slip movements on reactivated buried faults. Furthermore, there is a growing body of evidence that suggests they coincide with zones of tight reservoir, separating areas with different FWLs. The Barque horizontal well penetrated one of the NNE-SSW features and recorded a sharp and dramatic reduction in porosity at the point of intersection.

7. Further plans

The first phase of Sole Pit development has seen a concentrated effort to achieve the short term goals of getting the project on stream, on time, and within budget whilst maximizing wherever possible, the gathering of reservoir information. Having achieved the first gas goal, the petroleum engineering team has turned its attention to the more fundamental and important objectives of increasing recoverable reserves and reducing capital investment. To this end, a number of additional multi-disciplinary initiatives are being pursued:

- the ability to place MHFs in horizontal well sections and prediction of their performance under the prevailing reservoir conditions.
- follow up well stimulation work (i.e. acidization, MHF of poor producers selective MHF placement). This will largely depend on well performance after first gas.
- completing both conventional and horizontal wells without a cemented liner (barefoot completions). This would allow maximum contact between the wellbore and the reservoir and enable all productive formation features, no matter how thin, to contribute to the well productivity. The low mud weight used during drilling will minimize matrix and fissure impairment through exposure to the drilling fluid and losses control material. Impairment due to liner cementation will be avoided. The confidence to do barefoot completions is based on formation strength analyses carried out on Sole Pit cores, the observed horizontal hole stability in the Barque Field and having a fall back option, in the event of low well productivity, to plug back the open hole section and place a barefoot MHF through the permanent completion string.

D. CONCLUSIONS

1. The incidence of initial well stimulation including massive hydraulic fracturing as envisaged in the development plan is likely to reduce as a result of the realization that the reservoir contains other productive features besides open fissures. The fraccing philosophy has changed from, "frac all wells that have failed to encounter conductive fissures" to "frac only as a last resort".
2. A clearer picture of the orientation and distribution of fissures and their relationship to seismic features is emerging as the fissure data base expands with the continuing use of a new generation borehole imaging log.
3. The existence of better quality reservoir lithofacies in the form of slip-face sands and their proven gas productivity without the aid of MHF has provided added incentive to improve well planning and completions. Wells are planned to maximize the chance of intersecting these sands with minimum impairment and then completed to take full advantage of their productivity potential.
4. The high well test rates of some abandoned Sole Pit appraisal wells have been replicated by drilling development wells directly at the older wells and then completing across the proven prolific interval(s). This technique has been named 'cloning'.
5. Horizontal drainage has been tested in the first Barque development well with very successful results. The technique will be extensively utilized in the development of Sole Pit with an expectation of higher recovery efficiency with possibly fewer wells.
6. 3D seismic has provided a firm and reliable basis for field development in a structurally complex area. New display techniques of 3D seismic attributes have been instrumental in identifying the distribution and partitioning of different FWLs in both fields.
7. Important changes have been made to the project execution by quickly sharing the findings of the PE disciplines through open discussion and then formulating an integrated response. The short time (2 years) between commencement of development and delivery of first gas meant that innovative changes had to be implemented almost immediately and the experience gained was then fed back into a dynamic development plan. Both project and the team members have benefited from this rapid and mostly positive reinforcement of their efforts.

ACKNOWLEDGMENTS

The authors would like to thank the management of Shell UK. Expro, S.I.P.M. and Esso Exploration and Production U.K Ltd. for permitting this material to be published. We wish to acknowledge the contribution made to the development of Clipper and Barque by the Esso seismologists, geologists and reservoir engineers who have also been part of the integrated team.

REFERENCES

1. Clayton, F. M., and GORDON, N. C., 1990. The Leman P and G Developments Obtaining Commercial Production Rates from a Tight Gas Reservoir, European Petroleum Conference Proceedings, Vol. 2, pp 429-440, SPE 20993-

2. Dalley., R., Gevers., E.C.A., Stampfli., G.H., Davies., D.J., Gastaldi., C.N., Ruijtenberg., P.A., and Vermeer., G.J.O. 1989. Dip and azimuth displays for 3D seismic interpretation in First Break, Vol. 7, No. 3, pp 86-95.

3. Farmer., R.T., and Hillier., A.P. (due to be published in 1991). The Clipper Gas Field, in the Geological Society of London Memoir , United Kingdom Oil and Gasfields 25 Years Commemorative Volume.

4. Parmer., R.T., and Hillier., A.P. (due to be published in 1991). The Barque Gas Field, in the Geological Society of London Memoir, United Kingdom Oil and Gasfields 25 Years Commemorative Volume.

5. Frisinger., N.R., and Gyllensten., A. 1986. Fracture Detection in North Sea Reservoirs, Tenth European Formation Evaluation Symposium Proceedings, SPWLA.

6. Glennie., K.W., and Boegner., P.L.E. 1981. Sole Pit inversion tectonics, in Illing., L.V., and Hobson., D. G., Petroleum Geology of the Continental Shelf of North West Europe, Inst. of Petroleum, London, pp 110-120.

7. Glennie., R.U., and Buller., A.T. 1983. The Permian Weissliegend of N.~. Europe: the partial deformation of aeolian dune sands caused by the Zechstein transgression: Sedimentary Geology, V.35, pp 43-81.

8. Hawkings., N., Ramsey., C., Rapach., J. and Seligmann., P. 1990. Results from a Horizontal Well in a Rotliegendes Gas Field, European Petroleum Conference Proceedings, Vol. 2, pp 441-451, SPE 20994.

9. Heward., A.P., (due to be published in 1990). Inside Auk- The anatomy of an aeolian oil reservoir, in Miall., A.D., and Tyler., N., eds., The three-Dimensional Facies Architecture of Terrigenous clastic Sediments and its Implications for Hydrocarbon Discovery and Recovery: SEPM Concepts and Models in Sedimentology and Palaeontology Series.

10. Hondt., J.C. 1990. The use of Dip and Azimuth Horizon Attributes in 3D Seismic Interpretation, European Petroleum Conference Proceedings, Vol. 2, pp 71-78, SPE 20943.

11. Ruijtenberg., P.A., Buchanan., R., and Narke., P. 1990. Three-dimensional data improve reservoir mapping, in Journal of Petroleum Technology, January 1990.

12. Weber., R.J. 1987. Computation of initial well productivities in aeolian sandstone on the basis of a geological model, Leman gas field, U.K., in Tillman., R.E., and Weber., K.J., eds., Reservoir Sedimentology: Society of Economic Palaeontologists and Mineralogists Special Publication No. 40, pp 333-354.

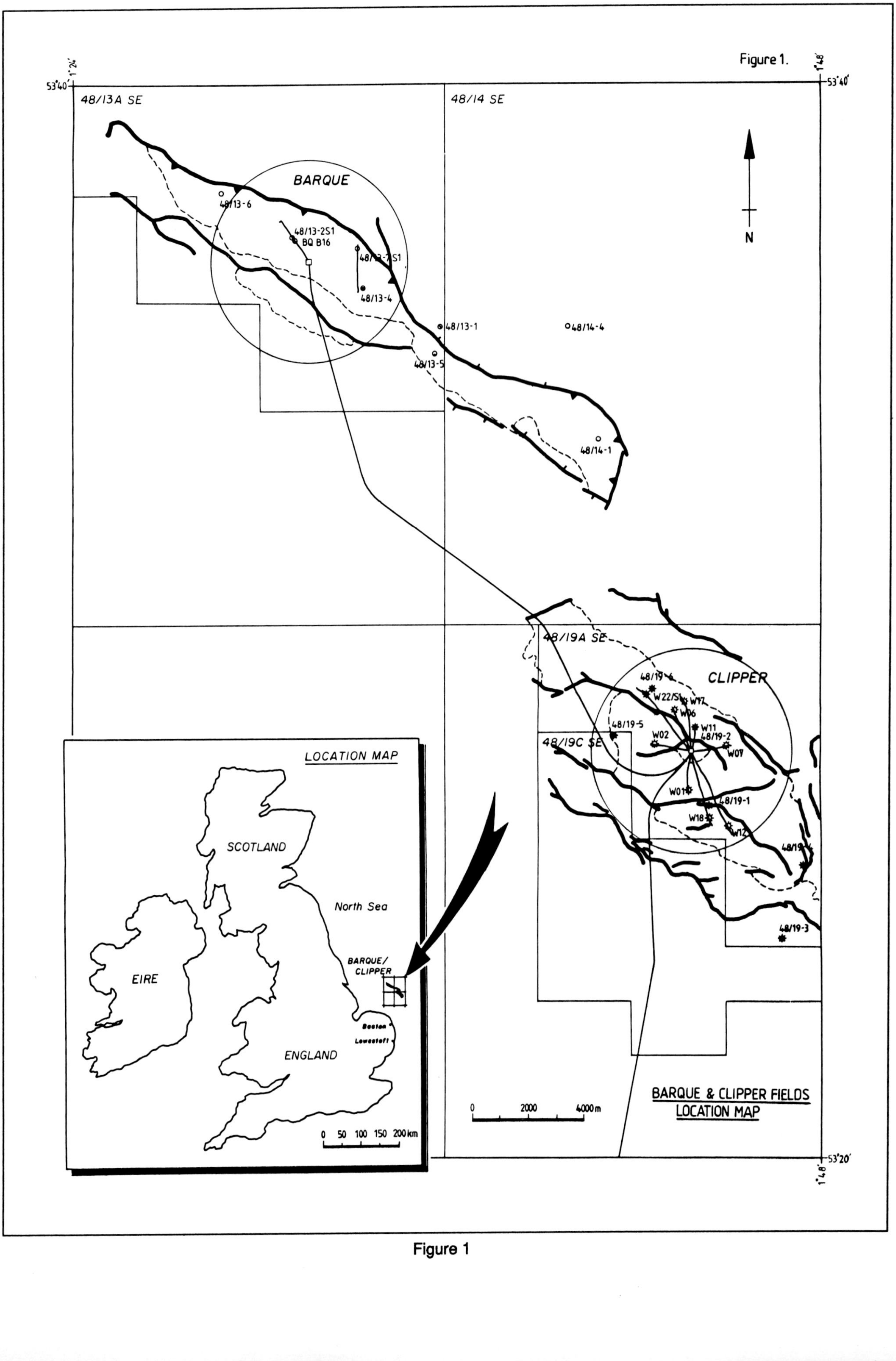
Figure 1.
53°40'
1°24'
1°48'
53°40'
48/13A SE
48/14 SE
BARQUE
48/13-6
48/13-2S1
BQ B16
48/13-4
48/13-1
048/14-4
48/13-5
48/14-1
N
48/19A SE
CLIPPER
48/19-6
W22/S1
W06
48/19-5
W02
W11
48/19-2
48/19C SE
W01
48/19-1
W18
W12
48/19-3
LOCATION MAP
SCOTLAND
North Sea
EIRE
BARQUE/
CLIPPER
Bacton
Lowestoft
ENGLAND
0 50 100 150 200 km
0
2000
4000 m
BARQUE & CLIPPER FIELDS
LOCATION MAP
53°20'
1°48'

Figure 1

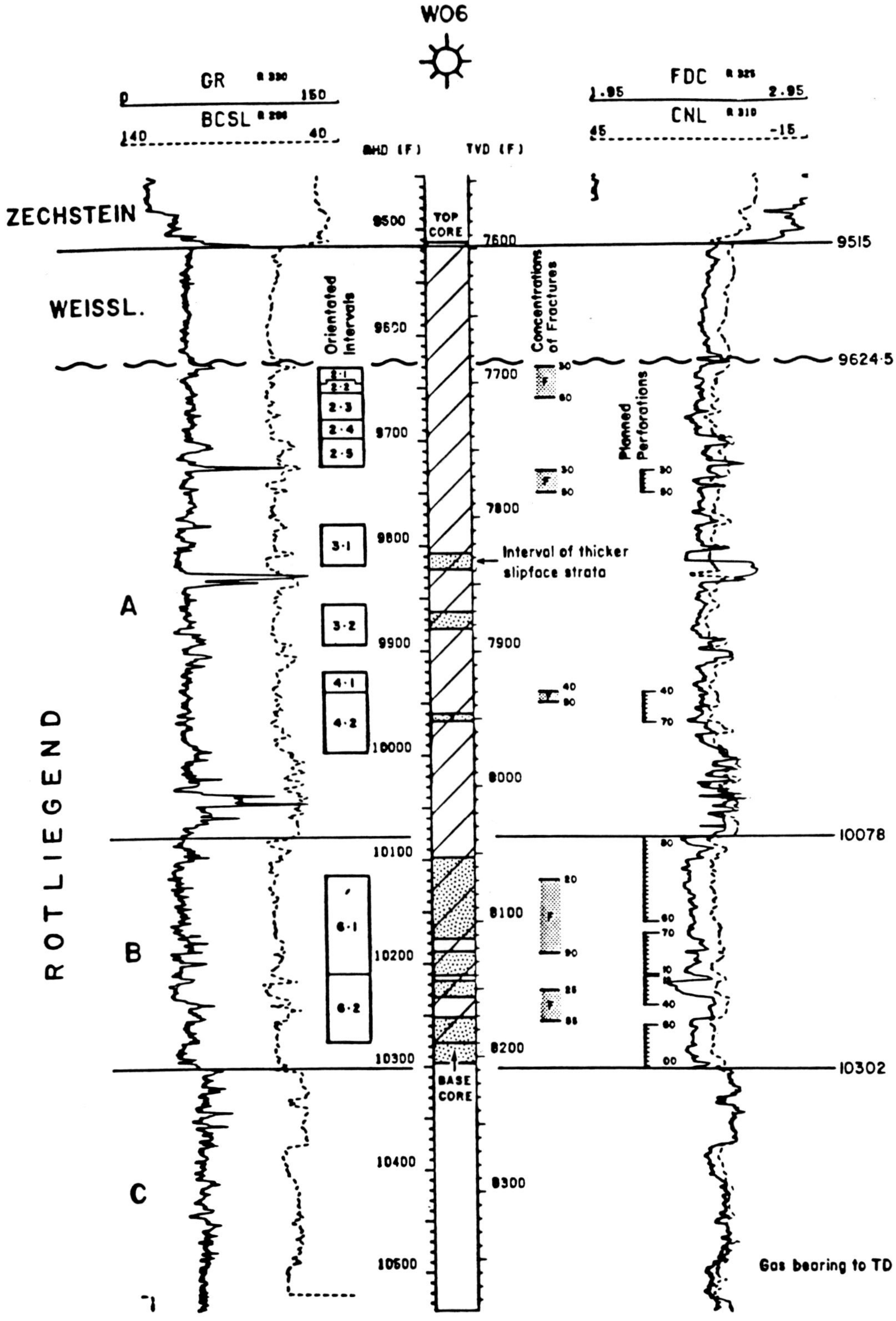

SUMMARY LOG

Figure 2

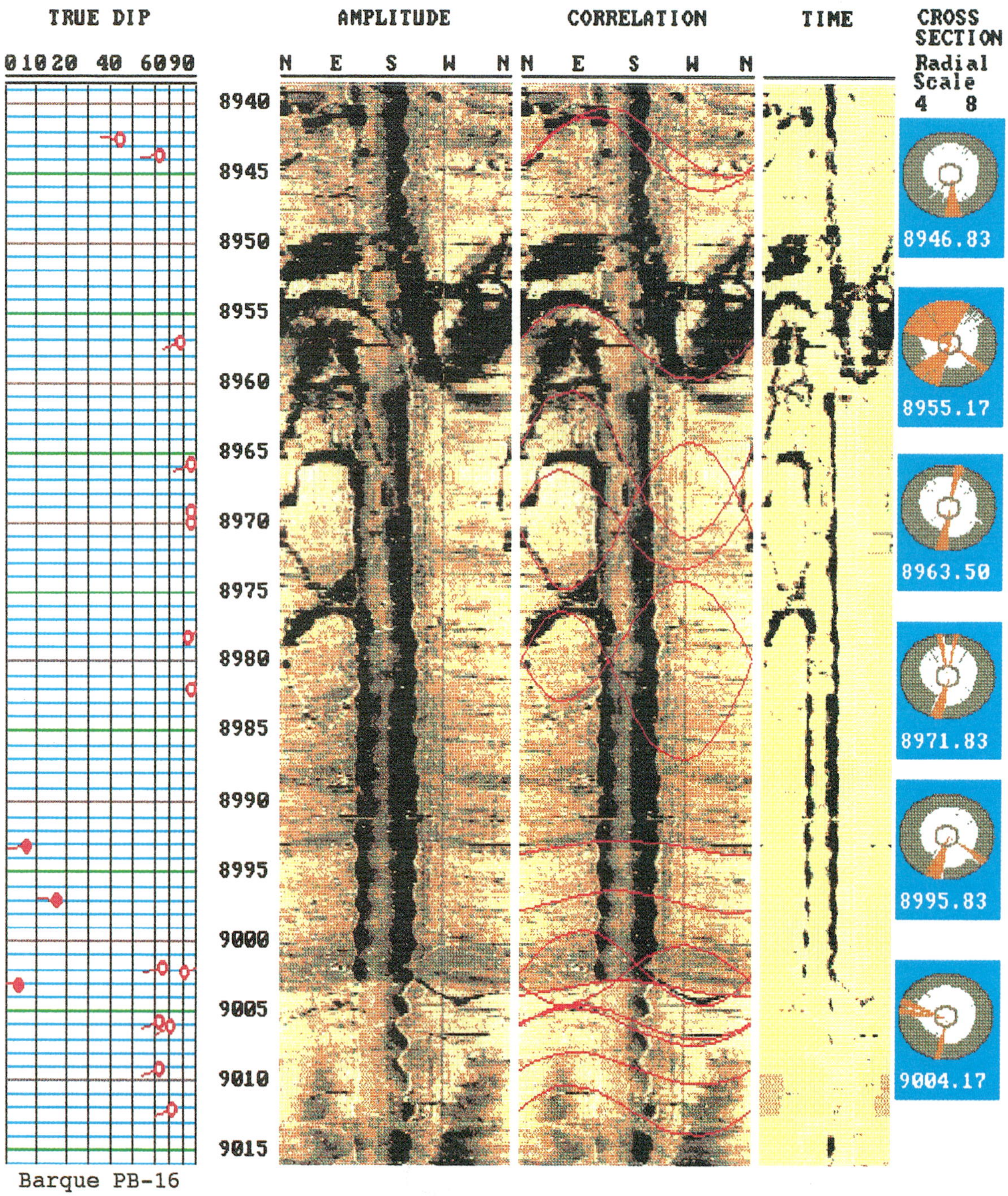

Circumferential Borehole Imaging Log (C.B.I.L.*) , evaluated over a 75′ interval in the build section of the Barque horizontal well,B16.
The image shows a series of steep, well developed open fractures (black) from 8957′ to 8983′ AHBDF, oriented subparallel to the well azimuth (320°), interpreted to be the wing of a hydraulic fracture initiated in the control well, 48/13-2. Less steep sets of semi-open fissures can be identified in the lower half of the interval, and a bounding surface is visible at 9003′ AHBDF. The lithology beneath this surface is noticeably harder, which is apparant both from the amplitude profile and from the reduction in width of the drilling groove on the southern or lowside of the hole.
The CBIL was run on drill pipe, and logged through Oil Based Mud.

* C.B.I.L., a registered service mark of Western Atlas International Inc.

Figure 3

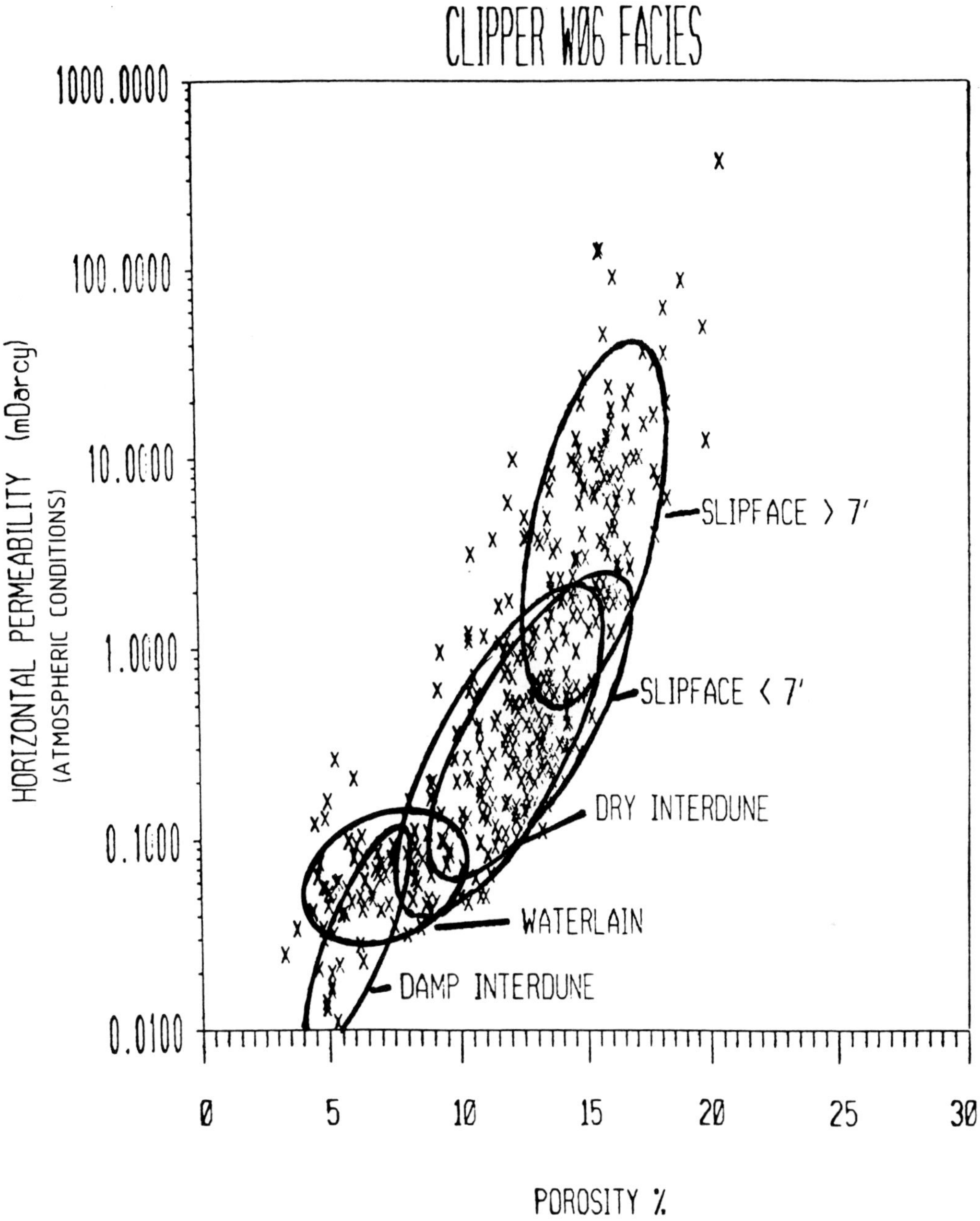

Poro-perm plot of core data from Clipper Well W06, showing distribution by lithofaces.

Porosity points, in the 14-16% range have a remarkable spread of core derived horizontal permeabilities, emphasising the difficulty in predicting well productivity after perforating and completing conventionally.

Figure 4

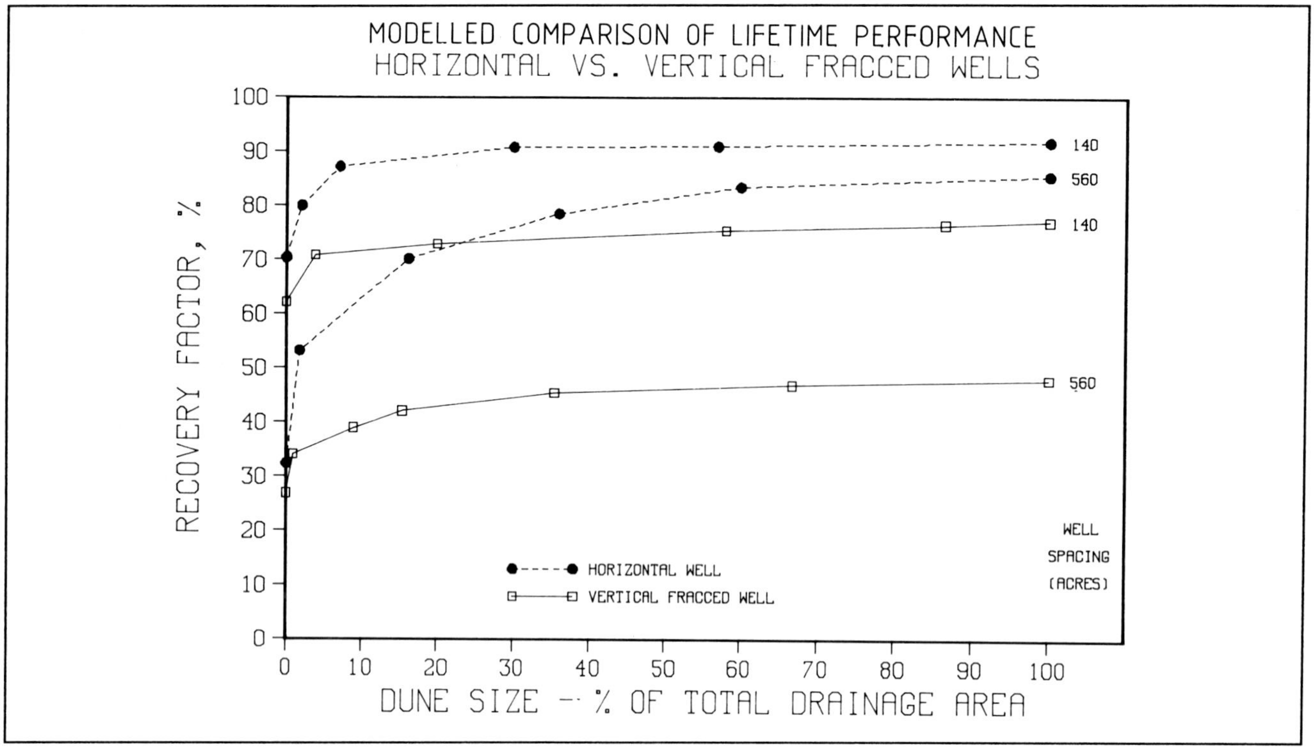

Figure 5. Sector model simulations for horizontal and vertical fracced wells showing the recovery factor dependence on dune sizes in connection with the wells and well spacing. Horizontal wells achieve a higher recovery with less wells.

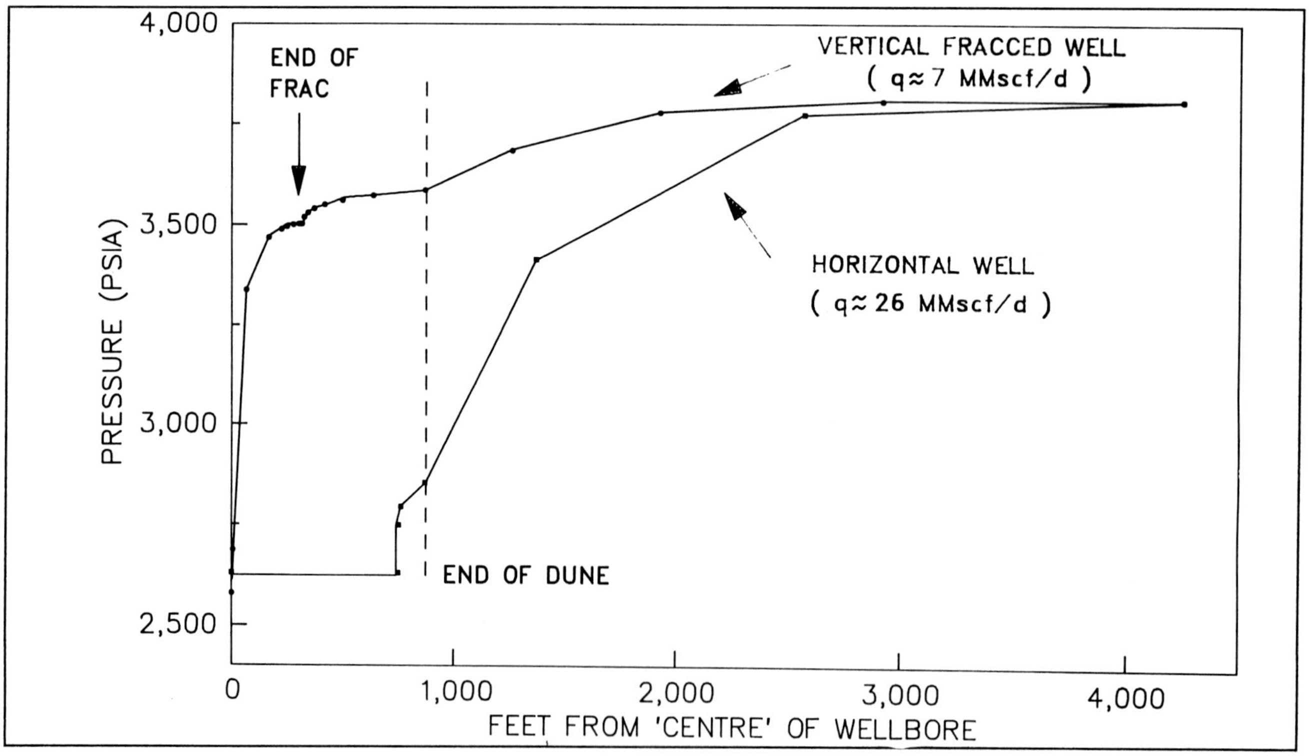

Figure 6. Pressure distribution from centre of the well bore for a horizontal and vertical fracced well. For the horizontal well the major part of the pressure drop takes place over the dune matrix interface while in the case of the fracced well the inflow performance of the well is the bottle neck.

Barque azimuth display (near top reservoir) from 3D seismic data revealing the late reactivation of a NNE-SSW (long arrows) and ENE-WSW fault system (short arrows), superimposed on the older NW-SE and NNW-SSE regional trends. The arrowed faults are thought to represent strike-slip movements of re-activated buried faults.

Figure 7

A Multidisciplinary Effort to Optimize the Ultimate Recovery of the St. Joseph field—NW Borneo

Peter Haalebos

The St. Joseph field is a large, structurally and stratigraphically complex oil field situated offshore West-Sabah.

The field was discovered in 1975 and since then 8 appraisal wells were drilled.

This has led to the field's current active development with twenty-one production wells having been drilled from four jackets.

Production commenced in 1981 and to date about 18% of the fields ultimate recovery of 186 MMstb have been developed.

In 1985, 3D seismic was acquired.

Detailed interpretation of the 3D seismic data set, in-depth sedimentological studies followed by 3D fullfield reservoir simulation have resulted in an optimization plan of the total field development.

Two additional jackets and a pressure maintenance scheme are planned to be implemented to enhance oil recovery.

This paper illustrates how a dedicated multidisciplinary approach in the use of modern evaluation techniques resulted in a significant increase in oil reserves and an optimised plan to develop these reserves.

Role of 3D Seismic Interpretation in Reservoir Identification and Characterization, Mississippi Canyon Block 109 Field, Offshore Gulf of Mexico

D. A. Herron, W. W. Wilson, and M. T. Currie

The Mississippi Canyon Block 109 field, which is located approximately 18 miles offshore from the Mississippi Delta in water depth of 950-1100 feet, was discovered in 1984 and declared commercial by BP Exploration in 1989. The field consists of middle and upper Pliocene reservoir sands, with oil accumulations in both structural-stratigraphic and entirely stratigraphic traps. A 3D seismic survey, acquired and processed in late 1984-1985 and reprocessed in 1987, has been and continues to be instrumental in the development of this field.

The 3D survey was acquired after drilling of the MC 109 #1 discovery well in 1984 and was first used to resolve the complex structure of the field. Integration of the results of initial drilling with this structural interpretation strongly suggested that there was a significant stratigraphic component to entrapment of hydrocarbons in the field. In conjunction with a revised depositional model for the reservoir sands, a new interpretation of the reprocessed 3D survey was used to identify these reservoirs and describe them in terms of reflection character and amplitude. The new interpretation was confirmed by the successful drilling of the MC 109 #3 and #3 ST wells in 1988, after which Block 109 was transferred to production for additional detailed work and assessment of development potential. The MC 109 #4 and #4 ST wells were subsequently drilled to confirm hydrocarbon pays in an untested fault block, and the MC 109 #5 well was drilled to obtain information on fluid properties and additional geological data needed for field development plans.

All of the wells in Block 109 were tied to the 3D survey, using both sonic and density logs to extract seismic wavelets with associated phase, time shift and spatial positioning information for optimal well ties to seismic. These established an excellent correlation of the seismic wavelet to hydrocarbon pay, thereby enabling detailed calculation of hydrocarbon volumes by time-amplitude analysis of the 3D seismic data. Reserves calculated by this technique were judged to be sufficient for commerciality, and the Block 109 field was sanctioned for development in late 1989.

Cognac Revisited: Mississippi Canyon Block 194 Field Redevelopment Planning-

M. Kohli and C. B. Speice

Mississippi Canyon Block 194 Field was discovered in 1975 and the initial development completed during 1981. Production began in 1979 during Produce While Drill operations, but did not reach full capacity until 1982. Total production to date exceeds 150 MMBE. During the last few years, sufficient production data have been available to aid in geologic and reservoir interpretations; however, the most powerful tool to aid in understanding the field was acquired in 1987—the 3-D seismic survey.

A redevelopment program was begun in 1989 with a planned 13-well program which has expanded to a 20-well program developing nearly 33 MMBE and identifying an additional 25 MMBE to be developed at a later date when slots become available. The report will discuss how the total data package—production, reservoir, geology and geophysics—were used to identify and evaluate drilling opportunities. Included in the opportunities are wells for bypassed oil due to poor sweep and/or small faults, new fault blocks and deeper sands. Based on the work at Cognac, a good 3-D survey can help monitor production, detect faults of 50± feet of throw and reduce dry hole/marginal well risks.

Modern Technology in an Old Area—Bay Marchand Field Revisited

W. L Abriel
P. S. Neale
J. S. Tissue
R. M. Wright

INTRODUCTION

Bay Marchand Field, a giant Gulf of Mexico oil field discovered in 1949, is undergoing renewed drilling activity as the result of a recently acquired 3-D seismic survey. The field, situated over a large salt diapir, is characterized by complex fault systems and typical Gulf Coast regressive sedimentation. As of 1989, over 700 wells had produced 518 MMBO and 379 BCFG from the field. The 3-D survey, which took nearly two years to plan, shoot. and process, covers over 60 square miles (Fig. 1). It was shot for the following objectives:

(1) delineate new pools;

(2) review mature, drilled areas for additional opportunities;

(3) assist reservoir management primarily in the form of secondary recovery.

3-D SEISMIC SURVEY ACQUISITION

Geophysically, the 3-D survey was designed to cover all CDP bins in a uniform manner, provide for maximum horizontal and vertical resolution, have consistent offset and azimuth distribution, and consist of the best geometry through careful straight shooting. The difficulty of obtaining such full coverage was heightened by numerous surface facilities dotting the field (Fig. 2), but the data was successfully acquired through state-of-the-art techniques.

GEOLOGICAL BENEFITS

To date numerous benefits have been obtained by close cooperation of the geological, geophysical, and engineering disciplines. Some of the benefits are:

Structurally

(1) A better definition of the salt sediment interface has exhibited very striking and geologically unusual salt shapes (Fig. 3). This has led to new places to drill.

(2) A drastically different fault pattern was mapped which has led to new reservoir geometries that better match engineering measurements and calculations. This in turn suggests new well locations in some cases and cancellation of unnecessary locations in others.

Stratigraphically

(1) We have a better understanding of the paleoenvironments. The relative amplitude nature of the seismic data, along with the good horizontal resolution, allows us to recognize depositional features with different sand characteristics (Fig. 4).

(2) We received a new concept and better understanding of E-log sand correlations. The ability to exactly tie true vertical sections between wells leads to changes in log correlations and recognition of unconformities.

(3) We obtained a better understanding of sand distribution and thickness through the use of seismic amplitudes.

RESERVOIR MANAGEMENT

Putting it all together in the form of better reservoir description, the engineers now have a better model on which to do reservoir simulation studies and locate water injection wells for secondary recovery.

CONCLUSIONS

The overall benefits of the 3-D seismic data has been a better understanding of the field, better cooperation between the earth scientists and engineers, but most importantly an increase in reserves as well as daily and ultimate production.

REFERENCES

Brown, A. R., Wright, R. M., Burkhart, K D., and Abriel, W. L, Interactive Seismic Mapping of Net Producible Gas Sand in the Gulf of Mexico; Geophysics. Vol. 49 No 6 (June 1984).

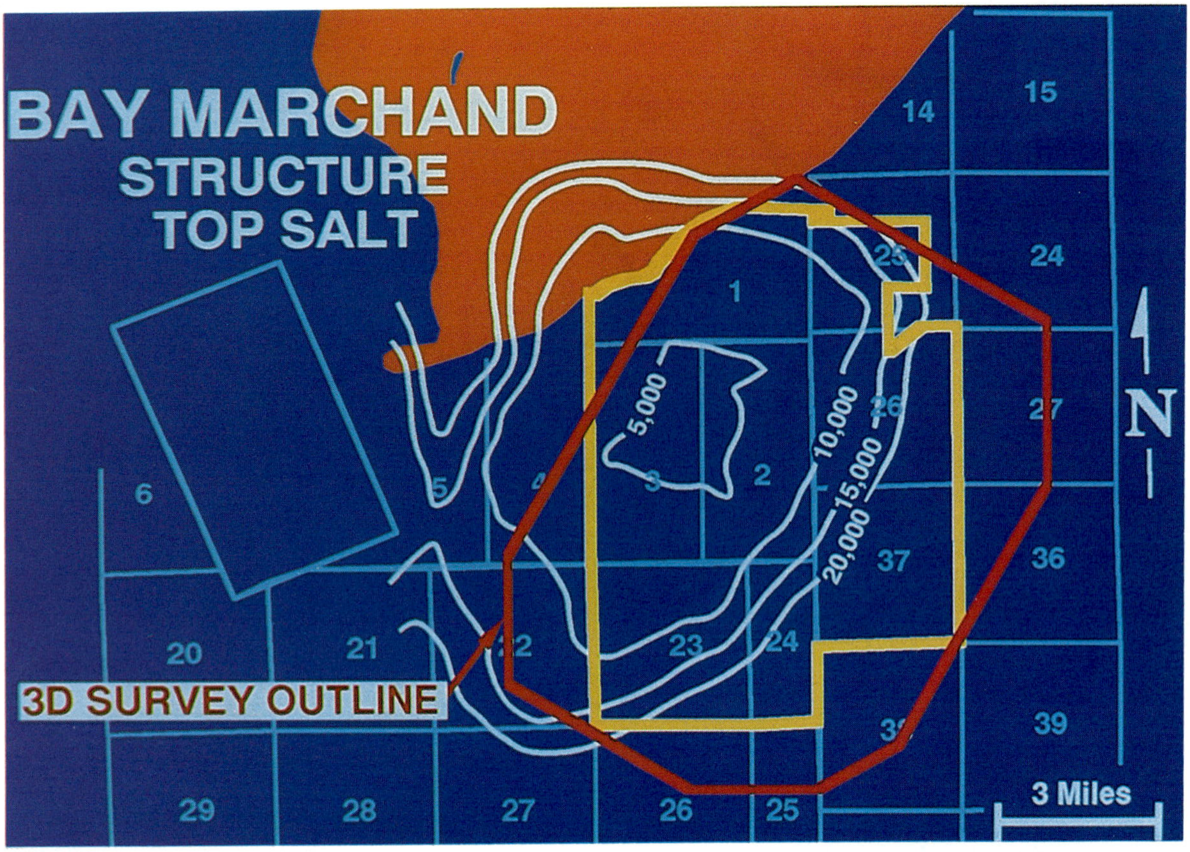

Figure 1

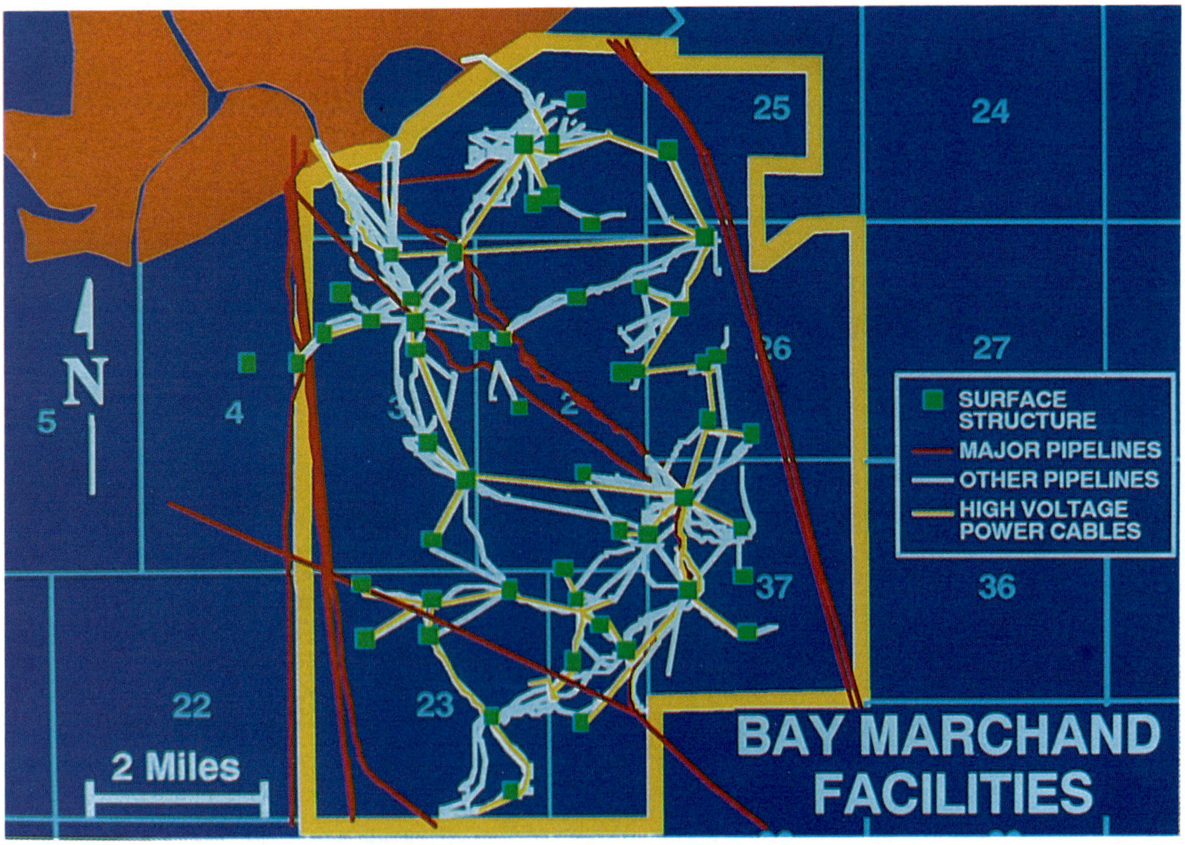

Figure 2

Figure 3

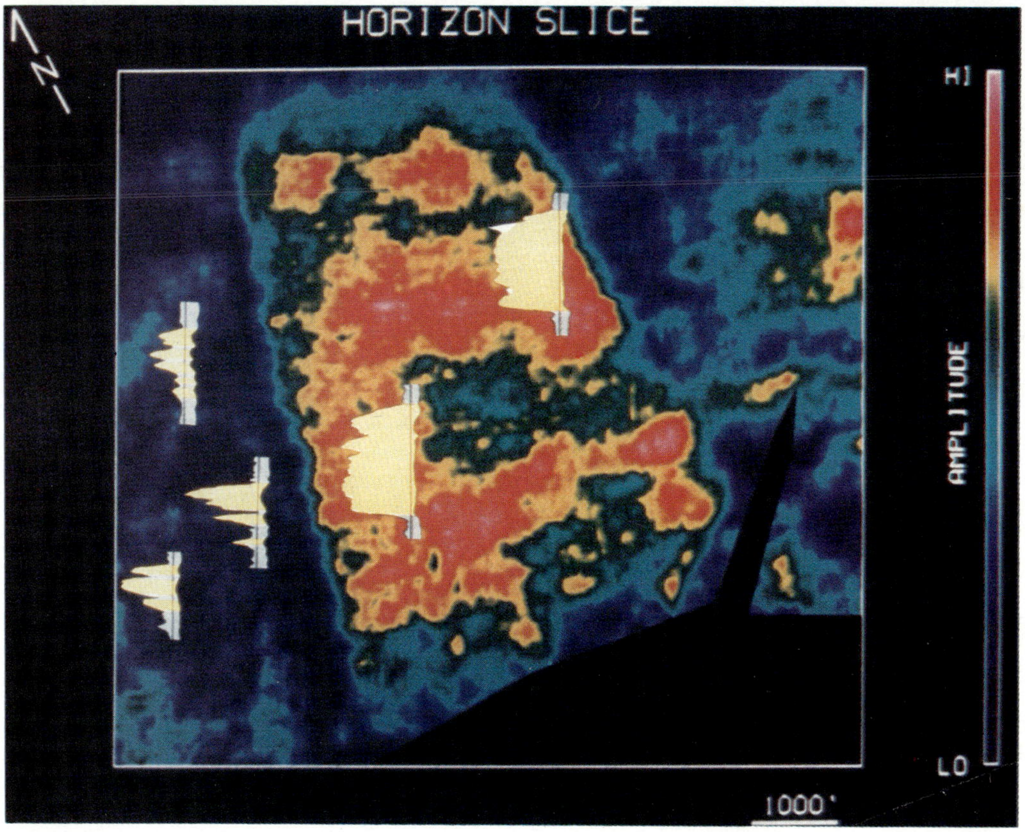

Figure 4